NONLINEAR CONTROL ENGINEERING

To my parents and grandparents

NONLINEAR CONTROL ENGINEERING

by

Professor D. P. ATHERTON, B.Eng., Ph.D.

University of New Brunswick, Fredericton, New Brunswick, Canada

VNR

VAN NOSTRAND REINHOLD COMPANY LONDON

New York Cincinnati Toronto Melbourne

VAN NOSTRAND REINHOLD COMPANY LIMITED
Molly Millar's Lane, Wokingham, Berks.

INTERNATIONAL OFFICES
New York Cincinnati Toronto Melbourne

Library of Congress Catalog Card No. 74-32159

ISBN 0 442 30017 4

First published 1975

16 15 14 13 12 11 10 9 8 7 6 5 4 3 2 1

Library of Congress Cataloging in Publication Data

Atherton, D P
Nonlinear control engineering.

Includes bibliographical references and index.
1. Automatic control. 2. Control theory.
3. Nonlinear theories. I. Title.
TJ213.A82 629.8'312 74-32159
ISBN 0-442-30017-4

Printed in Great Britain by
J. W. Arrowsmith Ltd., Winterstoke Road, Bristol BS3 2NT

Preface

The theory of automatic control has advanced significantly in recent years particularly with respect to the use of time domain methods in stability analysis and in optimal control. Although these techniques provide particularly elegant mathematical solutions to many problems they do have shortcomings from a practical viewpoint. The differential equations used to describe a particular process are only approximations to the true behaviour, and nonlinearities which are often neglected may become significant and the practical implementation of an optimal scheme may prove uneconomic. Further, the approach to the solution is such that in many cases the designer does not obtain a practical feeling for the system behaviour. On the other hand, without such theories space flights would not have been successfully accomplished nor would the computer control of industrial processes have advanced so rapidly. In many practical problems the optimum solution lies in a region where the performance criterion is relatively flat, rather than at a sharp minimum, so that the increased benefits in moving from a near optimum to optimum solution may be quite small. Nonlinearities occur in all physical systems and an understanding of the effects that these nonlinearities have on various types of signals is important to engineers in many fields from the design of car suspension systems to communication systems. Although this book is concerned with the analysis and design of nonlinear feedback systems a considerable amount of material is presented on nonlinear signal processing.

The primary aim of the book is to apply describing function methods to the analysis of nonlinear systems. However, three chapters are included on other topics, namely chapter 2 on Phase Plane Analysis, chapter 6 on Exact Analytical Methods for Relay Systems and chapter 7 on Random Signal Theory. The approach adopted in the development of the describing function theory has been to start with single input describing functions, and then systematically develop the procedure to the general multiple input case. In this way the book can be used by those seeking an introduction to describing functions as well as by the specialist wishing to pursue the topic in depth. Since the describing function approach is an approximate procedure it was felt desirable to include some methods of exact analysis with which the describing function solutions could be compared. The phase plane analysis of second order nonlinear systems is not only useful in its own right but is an ideal method for introducing various aspects of nonlinear behaviour and providing a basic understanding of nonlinear systems. In addition it tends to complement the describing function in that describing function methods are normally more accurate for higher order systems. The exact methods of analysis presented for relay systems are both significant in their own right and provide an ideal vehicle for comparing the accuracy of describing function solutions for high order systems. The wide use made of probabilistic methods throughout the text, both for the description of actual random signals and in analyses for deterministic signals, indicates that a relatively good background in random signal analysis is required. Chapter 7 on this topic was therefore thought desirable.

In chapter 3 the basic single sinusoidal input describing function is expounded in depth and use is made of the probabilistic representation to introduce the calculation of the describing function from the nonlinearity and the probability density of the input sinusoid. Chapter 4 is concerned with the applications of the sinusoidal input describing function in the analysis and design of feedback systems. Derivation of the closed loop frequency response using a universal chart which eliminates a large amount of the computation associated with other methods is given. Sinusoidal plus bias inputs to a nonlinearity, which lead directly to the concept of a modified nonlinearity, are considered in chapter 5. Consideration of sinusoidal plus other inputs to a nonlinearity follows and the situations where such two input describing functions can be used to predict certain types of nonlinear behaviour in systems are discussed. Chapter 8 studies the transmission of random signals through nonlinear characteristics and the analysis of feedback systems with random and other inputs. The general approach to the determination of the linear gain model of a nonlinear characteristic is presented in chapter 10 together with extensions to various particular situations. This and the remaining chapters 9, 11, 12, 13 and 14 are of a more specialist nature. Chapter 9 deals with the application of describing functions in system transient analysis, and chapters 11 and 12 with applications using the general multivariable nonlinearity model derived in chapter 10 in single and multivariable feedback systems. Chapter 13 considers the behaviour of double valued nonlinearities with random inputs and finally chapter 14 is concerned with the application of describing function methods in the analysis of a.c. control systems. The material of the last six chapters is taken from relatively recent research work and is presented in a unified form for the first time. With the current trend back to the use of frequency response methods in the design of multivariable systems, chapter 12 should become increasingly significant. Appendix A contains detailed tables of the various describing functions and appendix C a brief review of mathematical stability theory with particular emphasis on the work of Popov, the results of which may be easily compared with those of the describing function.

The majority of the material presented has been used in courses either at the undergraduate or graduate level. The book is arranged so that it can be followed without difficulty by a person who has taken a first course on conventional control theory and has the usual mathematical background in transform calculus and complex variables. A term course at the undergraduate or graduate level following the basic control course would normally consist of chapters 1, 2, 3, 4, 7 and parts of chapter 8. This content may be reduced by omitting the probabilistic aspects of chapter 3 and some parts of chapter 7 especially if the basic aspects are covered elsewhere, say a communications course. A second course would then normally begin with chapter 5 followed by the remaining chapters, or selected topics therefrom, which are written so that they may be taken in any order. No computer programs are listed within the text although many of the procedures for evaluating describing functions and solving the nonlinear algebraic equations from the describing function or Tsypkin's method have been implemented in such a manner.

Even in a book of this size, the reader will no doubt find some omission of relevant work and the time required to produce such a volume ensures that it will never be quite up to date in this dynamic age. Where possible I have therefore tried to add the most recent publications to a chapter bibliography. In the process of giving courses based on this material to several classes of students, many interesting discussions have led to changes in the presentations and the correcting of some errors. This work owes a lot to the efforts of many former students both from their presence in my courses or their research work. I have drawn heavily on the work of many former research students especially Drs. Balasubramanian, Fieguth and Mr. Ramani. Dr. Balasubramanian has kindly read the whole manuscript, spared many hours for discussion and supplied several of the problems, indeed without his encouragement and support this task might have proved too onerous. I am also indebted to Drs. V. P. Singh and S. S. Lamba for helpful discussions and submissions on the universal chart and jump resonance, respectively. The support of my research, from which many results are taken, by both the National Research Council of Canada and Defence Research Board of Canada is gratefully acknowledged as are a sabbatical leave from the University of New Brunswick and the facilities extended to me by the University of Sussex. Unfortunately, errors, for which I take complete responsibility, will no doubt remain and I would therefore be grateful to receive any information on such matters from the reader.

Thanks are due to several typists, but mainly to Mrs. V. Gesner and Mrs. S. E. Ward, especially for their valiant efforts in typing the Greek symbols and complicated tables.

Finally, this work would have proved impossible without the support and understanding of my wife and family through a gestation period with a duration probably unheard of in the biological world.

Derek P. Atherton
Fredericton
November, 1974

Contents

Chapter 8 RDFs and Nonlinear System Analysis

Chapter 9 Describing Function Analysis of Transient Behaviour

Acknowledgements

Tables 2.1–2.3 are published by permission of Addison Wesley Publishing Co. from 'Control and Dynamic Systems' by Takahashi, Rabins & Auslander.

Figure 2.11 is published by permission of Ronald Press Co. from 'Analysis and Control of Nonlinear Systems' by Y. H. Ku.

Figure 2.12 is published by permission of McGraw-Hill from 'Introduction to Nonlinear Network Theory' by Leon O. Chua.

Figure 2.20 is published by permission of The English University Press from 'Feedback Theory and its Applications' by P. H. Hammond.

Figures 5.27, 5.28, 5.30, 5.31, are reproduced from Y. P. Singh, Ph.D. thesis, Elec. Eng., 1968 I. I. T. Kharagpur.

Figure 5.37 is published by permission of McGraw-Hill from 'Multiple-input Describing Functions and Nonlinear System Design' by A. Gelb & W. E. Vander Velde.

Figure 6.4 is published by permission of McGraw-Hill from 'Feedback control systems' by J. C. Gille, M. J. Pelegrin and P. Decaulne.

Table 7.1 is reproduced from Table E.2-1 of 'Analytical design of linear feedback controls' by G. C. Newton, L. A. Gould, and J. F. Kaiser. (Wiley, New York, 1957.)

Chapter 1

Fundamentals of Nonlinear Systems

1.1 Introduction

The science of automatic control viewed in a wide sense is concerned with the identification, analysis and design of dynamic systems. Although it is normally true that the design criterion will include some aspect relating to the control of the system and that a satisfactory design may not be accomplished without using feedback, the basic fact remains that the control engineer spends a considerable time in the analysis, as distinct from the control, of systems. The mathematical 'tool' of the control engineer is the theory of dynamic systems, whether his interests are in, say, the control of chemical processes or in biological control systems. The majority of the available theory is concerned with linear time invariant, continuous or discrete, systems. A linear system is one whose performance obeys the principle of superposition. This principle states that if an input $r_1(t)$ produces the output $c_1(t)$ and an input $r_2(t)$ the output $c_2(t)$, then for all a, b, $r_1(t)$ and $r_2(t)$ the response to $ar_1(t) + br_2(t)$ will be $ac_1(t) + bc_2(t)$. Further the system is defined as time invariant if the input $r_1(t + \tau)$ produces the output $c_1(t + \tau)$ for all values of τ and input functions $r_1(t)$. Thus for this class of systems the time origin is of no consequence and, in principle, the response to any input can be found from the response to some other known input with non zero energy at all frequencies. The elementary input most commonly used as the basis function is the unit impulse, the response to which is known as the impulse response or weighting function.

Unfortunately no physical system belongs to the class of linear time invariant systems. This is true if for no other reasons than that there are always some limits to the range of physical variables and that components deteriorate with time. This statement does not of course mean that the theory of linear time invariant systems is of no use, but simply that it has limitations and in some situations it is essential to have theories which consider the effects of nonlinearity and/or time dependence. The material presented in this book is concerned with nonlinear systems, or more specifically the application of describing function methods in the analysis and design of nonlinear systems.

A nonlinear system may be defined as one to which the principle of superposition does not apply. Thus there is no means by which the response to a particular input can be found if the response to some other input is known. As a result of this fundamental and important difficulty with nonlinear systems no uniform approach to their analysis is possible. Available analytical methods are concerned with solutions for a specific type of input. It is because describing function theory can provide answers to so many of these problems, particularly for systems with more than one input, that the approach is important. Admittedly the method is an approximate one but it does provide answers to facets of nonlinear behaviour for which no other theoretical solutions are known. In this respect approximate methods are better than no theoretical approaches at all.

1.2 Models of physical systems

The implementation of a control scheme involves several systematic steps which might be categorised as follows:

1. Preliminary study and performance specification
2. Modelling
3. Analysis
4. Simulation
5. Design and optimisation
6. Physical realisation

The above steps will not necessarily be undertaken in exactly the given order, and certainly the trial and error procedures which may be used in the latter categories can involve repeated jumping from one step to another. The modelling stage, that is the mathematical formulation of the relationships between the various variables of the system, is extremely important since the results produced at the later stages, except those from the physical realisation, are obtained using the model equations. There is thus no point in applying sophisticated optimisation techniques to the design of a poorly modelled system. The models for the elements of a system may be obtained from the basic physical laws governing their behaviour or by experimental measurement, a process also referred to as identification. The form of the model determined for a particular device may depend upon how it was obtained and may also be chosen according to how it is going to be used in a particular analysis. The designer must always remember the assumptions made during the determination of a model and thus be aware over what range of variations of the variables it is sufficiently accurate and to what degree of accuracy. Many analytical techniques become at least extremely tedious with complex models so that the ability to represent the significant behaviour of a device in as simple a manner as possible is a desirable skill.

It is not the purpose of this book to consider device modelling, indeed the model representation in terms of a block diagram, frequency response data, or differential equation is the starting point for the theories presented here. Since describing function theory involves approximations, to obtain an appreciation of the errors that may result it is necessary to compare describing function and other solutions for the mathematical model. Thus, when the designer applies an approximate method of analysis to an actual system he may be more readily aware of whether the errors are attributable to his model or his analysis. Indeed if the expected accuracy for the models of some components in a system is no better than twenty per cent an approximate and easy to apply theoretical method of around five per cent accuracy is a truly valuable tool.

In most situations models of devices may be obtained in terms of one or more separable linear and nonlinear parts with the nonlinear elements static functions of one or more input variables. A power amplifier may be modelled by a saturation nonlinearity followed by a linear transfer function representing the dynamic behaviour. A field controlled motor with constant armature current and nonlinear friction forces on the output shaft could be modelled by the relationships

$$v = Rc + L\dot{c},$$

$$T = f_1(c),$$

$$T = J\ddot{\theta} + f_2(\dot{\theta}) + T_L,$$

for which the corresponding block diagram is shown in figure 1.1; v is the input

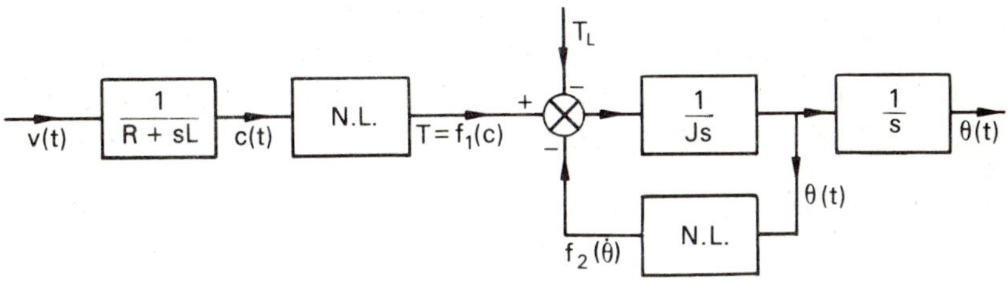

Figure 1.1 Block diagram of field controlled motor

voltage, c the field current, R and L the resistance and inductance of the field winding, T the motor torque, T_L the load torque, J the polar moment of inertia, θ the output shaft position and the nonlinearities $f_1(c)$ and $f_2(\dot{\theta})$ are explicit functions of c and $\dot{\theta}$ respectively. The nonlinear differential equation

$$(\ddot{x})^{1/3} + \dot{x}|\dot{x}| + x = r,$$

for which the relationship between x and r is one of an implicit dynamic nonlinearity can be represented by the closed loop formulation of figure 1.2. In the well known Van der Pol equation

$$\ddot{x} - \mu(1 - x^2)\dot{x} + x = 0$$

the nonlinear term $(1 - x^2)\dot{x}$ is an explicit function of the two related variables x and $\dot{x}$ or simply an explicit dynamic nonlinearity.

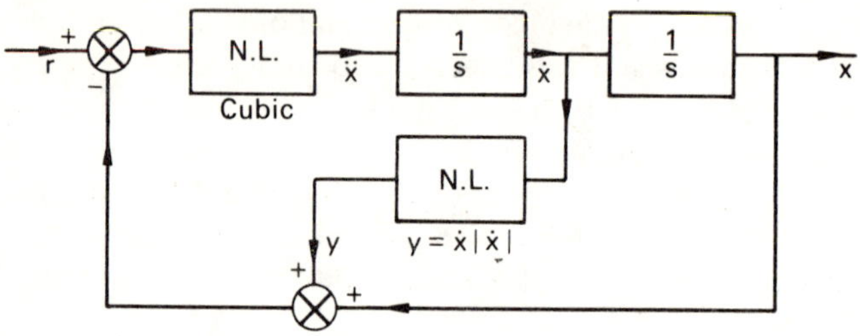

Figure 1.2 Block diagram of implicit dynamic nonlinearity

1.3 Nonlinear characteristics

The static nonlinear characteristic which is an explicit function of one input variable is by far the most common form of nonlinear model of a part of or a complete control system component. System nonlinearities are often classified as inherent or intentional nonlinearities. Inherent nonlinearities are those which exist in the 'presumably linear' components selected to perform a function other than a nonlinear one within the system. Often the designer would be happier if they did not exist. Typically such characteristics are saturation in amplifiers and motors, dead zones in valves, nonlinear friction, backlash in gears, and nonlinear transducers such as synchros. On the other hand intentional nonlinearities are those which are deliberately introduced into the system to perform a specific nonlinear function. They have been chosen to be used during the design because of their economic, weight, reliability, performance or other advantages. Some common forms of such devices are relays, thyristor circuits and limiters to restrict the range of a variable.

Static nonlinear characteristics describing many of the above effects are single valued functions of the input variable, as shown in figure 1.3. Such a characteristic will be referred to as a single valued nonlinearity, SVNL, and for analysis purposes must normally be described by some mathematical function. One function, as for example a power law, may be used to define the characteristic over the complete range or a set of simply defined functions, such as straight lines, may be used over different regions of the range of input values. Two common linear segmented characteristics used to approximate dead zone and saturation, and usually referred to as ideal dead zone and saturation characteristics, are shown in figure 1.4. If the SVNL is an odd function, that is $n(x) = -n(-x)$, it is referred to as a symmetrical odd characteristic and a system containing such nonlinearities only as a symmetrical odd system.

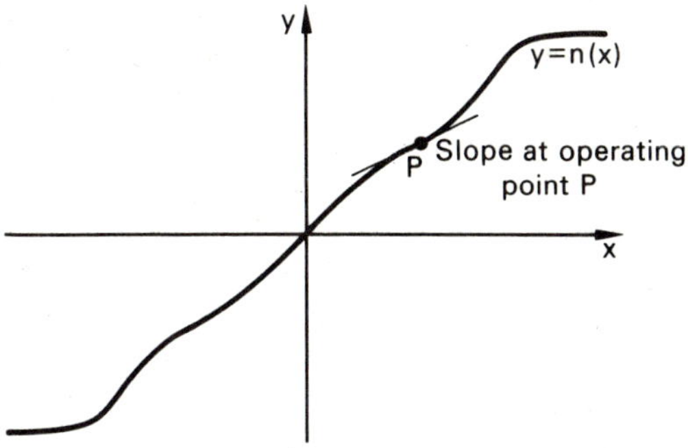

Figure 1.3 Single valued nonlinearity

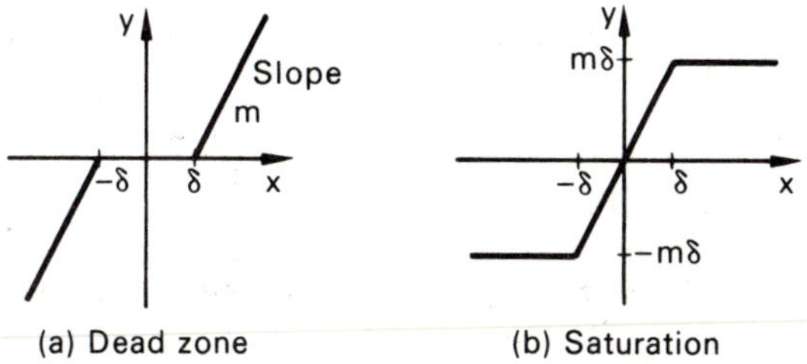

Figure 1.4 Ideal dead zone and saturation characteristics

Nonlinear friction forces are particularly common in control systems and a typical relationship between the friction force and speed, which is assumed to have odd symmetry, is shown in figure 1.5(a). The precise shape of the curve is influenced by a considerable number of factors such as the materials involved, the temperature and the pressure. An ideal approximation to the friction characteristic which is shown in figure 1.5(b), consists of the three component parts stiction, Coulomb friction, and viscous friction shown in figure 1.5(c). The

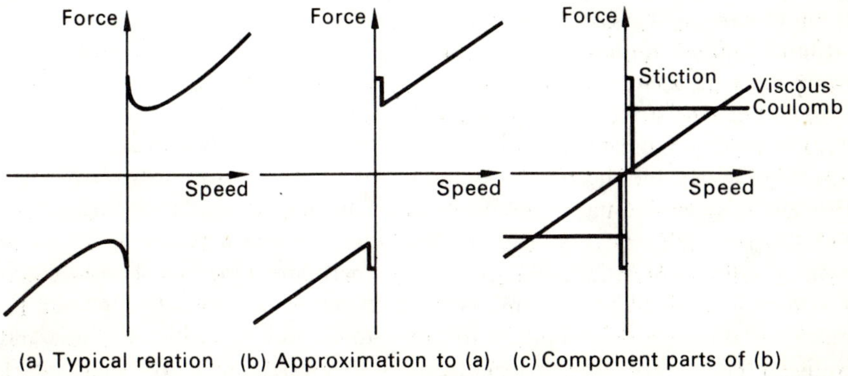

Figure 1.5 Friction characteristics

stiction force is assumed to exist at zero speed only, indeed the name is an abbreviation for static friction, Coulomb friction is a constant force independent of the speed and the viscous friction force is linearly proportional to speed.

Electromechanical relays are frequently used in control systems where the control strategy requires control signals with only two or three levels or if a cheap means of power amplification is required. When the control current through the relay field winding exceeds a certain value the magnetic force produced moves an armature from the open position, where it is restrained by a mechanical force, to the closed position. Due to magnetic hysteresis the relay will not return to the open position until the current drops below the value required to close the relay. The static characteristic of the relay thus appears as shown in figure 1.6(a). For input currents between the pull-in and drop-out

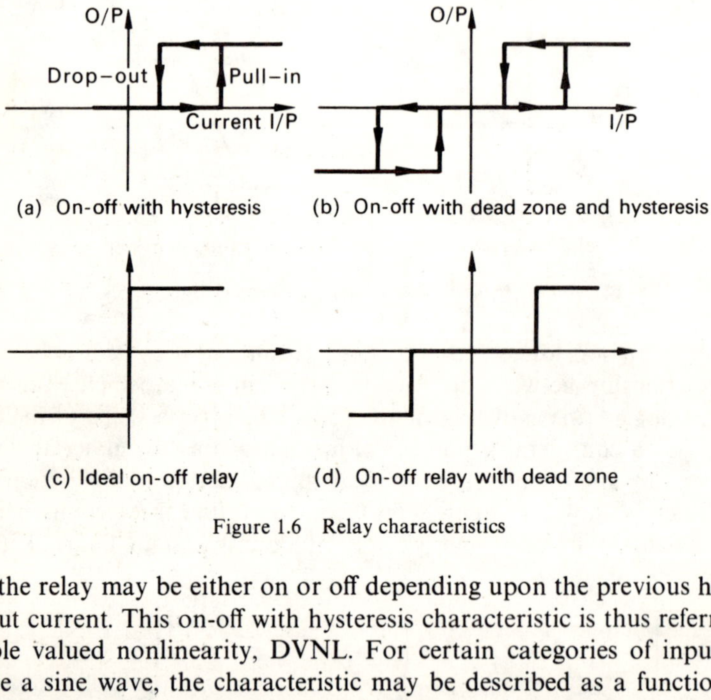

Figure 1.6 Relay characteristics

values the relay may be either on or off depending upon the previous history of the input current. This on-off with hysteresis characteristic is thus referred to as a double valued nonlinearity, DVNL. For certain categories of input, as for example a sine wave, the characteristic may be described as a function of the input and its derivative. Relays can also be obtained with the bidirectional static characteristic of figure 1.6(b), which is rather inappropriately known as an on-off with dead zone and hysteresis characteristic. Since a soft iron armature is attracted by an electromagnet independent of the magnet's polarity, the armature of this type of relay has its dead zone position centred between the other two positions with a polarising field applied by a permanent magnet. The weakening or strengthening of the total magnetic field by the current through the field winding coil causes the armature to move to the other positions. These static nonlinear characteristics are usually used to model the relay behaviour for any input, however since they neglect dynamic hysteresis effects, their accuracy deteriorates with increasing input frequency. In addition, if the field winding voltage rather than current is considered as the input, the time constant of the field winding may not be negligible. In certain applications the effect of the hysteresis in the characteristic may be unimportant so that the single valued nonlinearity models shown in figure 1.6(c) and (d) may be used.

An extremely complex form of nonlinear behaviour to model is the backlash which exists in gears. Figure 1.7 shows the mechanism of backlash where play in the gears allows relative movement to take place between the contact surfaces A and B. Where extremely accurate position control is required, for example

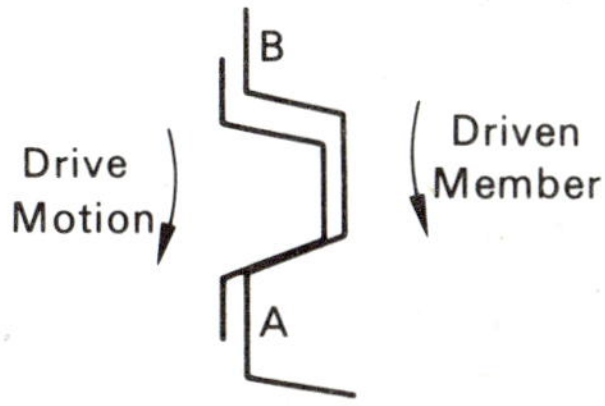

Figure 1.7 Illustration of backlash

with a radio telescope, to avoid the inaccuracies caused by backlash two entirely independent drive systems, one for each direction, have been employed. The two simple models commonly used for backlash are the friction controlled and inertia controlled characteristics. For the friction controlled backlash characteristic it is assumed that the driven member has a high friction to inertia ratio so that when the drive member starts to decelerate the surfaces will remain in contact at A. If on the other hand the friction to inertia ratio of the driven member is low then we have inertia controlled backlash where as soon as the drive member starts to decelerate, the driven member holds the maximum velocity until it reaches the other side of the backlash at B. Assuming no bouncing takes place the surfaces remain in contact at B until the drive member reaches a maximum speed in the reverse direction or starts to accelerate in the forward direction. Obviously in the general case for a random drive motion all kinds of impacting and changes in contact between the surfaces at A and B can take place.

Hysteresis in a magnetic material is a further nonlinear phenomenon which is exceedingly difficult to model for a general input. As is well known the actual path of an hysteresis loop even for a sinusoidal input depends upon the past history and the frequency of the sine wave. Nonlinear characteristics for which a large number of possible output values may exist for an input value selected at any time instant, are known as multivalued nonlinearities, MVNL.

1.4 Behaviour of nonlinear systems

Since a nonlinear system does not obey the law of superposition, analyses and measurements have to be carried out for many types and magnitudes of inputs in order to discover the various facets of its behaviour. The control engineer must therefore consider all the possible variations in parameters and inputs to which the system may be subjected in situ. Qualitative ideas as to how a nonlinear system will behave for one type of input, given its performance with respect to another, do exist, although any confidence in such comparisons can only be obtained from considerable experience of the various aspects of nonlinear behaviour which are associated with the different types of input.

Some of the distinctive properties of nonlinear systems are mentioned here, beginning with those of a system with no external inputs. Such a system is usually referred to as an autonomous system and mathematically it may be described by one, or a set of, homogeneous nonlinear differential equations. Firstly, the response of the system will be dependent on the initial conditions and will not be a linear combination of exponentials or time weighted exponentials as is the case for a linear system. The response shape can be markedly different, in some cases it may converge to an equilibrium point in other cases it may become unbounded, that is the system will be unstable. Multiple equilibrium points may exist so that when a steady state is attained it may not be the same for all initial conditions. In a position control system Coulomb friction introduces a 'band' of equilibrium points so that for a step input the steady state error depends upon the input magnitude. This loss of steady state accuracy is very undesirable. A further possibility is that the response may converge to a periodic motion (oscillation) which is known as a limit cycle. Unlike the linear second order system with no damping, where the oscillation amplitude depends upon the initial conditions, in a nonlinear system the same limit cycle can result from many different initial conditions. The Van der Pol equation, which was mentioned in section 1.2, was derived to describe the oscillations in an electronic oscillator and is thus a well known example of a nonlinear differential equation exhibiting a limit cycle.

When a periodic, usually sinusoidal, input is applied to a nonlinear system it may exhibit additional unique characteristics. The most obvious is that for a sinusoidal input the output, unlike the case of a linear system, will not be sinusoidal. The nonlinearity produces distortion so that new frequencies, which are harmonics of the input frequency, will be produced. If a good low pass filter exists after the nonlinearity and before the output the distortion in the actual output signal may be quite small; thus the amount of harmonic content in the signals at the various points within the closed loop system will vary quite considerably. Indeed we shall see that describing function theory is based on the assumption of a low distortion in the signal at the input to the nonlinearity. In some cases for a given amplitude and frequency of an input sinusoid a nonlinear system may have more than one stable periodic state; it is thus known as a multimodal system. A slight change in the input amplitude or frequency can cause the system to change states, a phenomenon referred to as jump resonance. In other instances a sinusoidal input may start a limit cycle in a system at either a related or unrelated frequency. When the limit cycle which is excited is a submultiple of the input frequency the behaviour is known as subharmonic resonance. Should the system possess a limit cycle with no input the sinusoidal input may quench the limit cycle or synchronise it to a frequency related to that of the input. These properties can be undesirable in control systems and the control engineer may wish to avoid their occurrence. On the other hand they may sometimes be used to advantage in devices such as detectors, amplifiers and oscillators.

Many practical control systems are subject to inputs or disturbances which are random rather than deterministic. To study the performance of these systems statistical methods must be used for typical sample input functions. Nonlinear systems when subjected to random inputs in general show similar patterns of behaviour, apart from those specifically related to periodicity, as they do when subjected to sinusoidal inputs. The system may thus have more than one solution for a set of given conditions, the random input may excite or quench a limit cycle, or it may make the system unstable. Similar situations also apply when a system has more than one input.

1.5 Methods of studying nonlinear systems

The performance of nonlinear systems can be studied by hardware, simulation, and/or analysis. Strictly speaking the latter two cases consider the performance of the system model. Measurement and testing of the actual system, which certainly will provide the most reliable information on how it will behave, is normally the final stage of design. Building a system and testing it without analysis or simulation can be both dangerous and expensive. In addition if it did not perform as expected no tested concepts would be available for its compensation. Suitable hardware techniques can sometimes be applied in the early stages of a design if scaled models can be used, as for example with ships and aircraft.

Modern analog, digital and hybrid computers are such that exhaustive simulations of complex systems can be undertaken. However, to simulate 'blindly' without having first done some preliminary analyses can be both expensive and unrewarding. Each computer run simply provides a solution for the given parameters and inputs. Theoretical approaches are necessary to indicate, at least qualitatively, the input magnitudes which will reveal all the possible facets of the system behaviour, the effects of disturbances, different parameter values and system configuration changes. Since this analysis and design of a nonlinear system will almost certainly involve approximations, computer simulations are essential to a complete study. However, the choice of the computer and how it is used should be strongly influenced by the analysis. Digital simulation languages such as the IBM Continuous Systems Modelling Program are often very suitable and economic for a short number of runs with initial condition problems for stable systems, since they avoid the patching and scaling associated with analog computers, however they can be far less efficient for determining periodic modes unless the initial conditions are well chosen.

Before discussing the analysis of nonlinear systems by describing function methods, it is appropriate to consider briefly some of the other available methods.

Exact mathematical methods

These may be divided into those which attempt to solve nonlinear differential equations[1] and those which provide answers for the important question of stability.[2] Closed form solutions can be found for some nonlinear differential equations but it is rare that a control system problem fits one of these situations. Indeed the majority of the known solutions are restricted to homogeneous equations of first or second order. A large number of methods exist which express the solution of differential equations in some form of series expansion.[3] The problems of choosing the series form and obtaining satisfactory convergence are such that they have found little use in other than slightly nonlinear systems where the solution to the linear system provides the first term in the series. Thus, although these mathematical methods may occasionally find use in analysis they are of no assistance whatsoever in design.

A central question in the design of any nonlinear system is that of stability. Will the system be stable for all the forms of input to which it may be subjected? The work of Lyapunov[4] and more recently that of Popov[5] can provide exact answers to this question without solving the nonlinear differential equation. A brief review of these methods is given in appendix C. The Popov method avoids the major difficulty of that of Lyapunov, the choice of a positive definite scalar function of the state variables with certain required properties, and is therefore much easier to apply. In addition the Popov criterion is formulated in the frequency domain which makes for easier comparison with describing function methods and also allows use of measured frequency response data. The major disadvantage of these exact methods is that since they provide sufficient and not necessary conditions for stability they often provide overly conservative estimates for the system stability domain. Also their application is restricted almost entirely to systems with only one nonlinear characteristic and external input. On the other hand both methods may be of some assistance in design.

Phase plane

The phase plane method[6,7] is a graphical approach which can be used to obtain solutions for first and second order nonlinear differential equations only, and is described in detail in chapter 2. A further disadvantage of the phase plane method is that its use is restricted almost entirely to homogeneous equations. It can, however, be of assistance in design and for studies of systems with more than one nonlinear element.

Linearisation

In many control systems, particularly those which may be referred to as regulators, except during infrequent transient conditions such as start up or changes in set point, the system variables undergo only relatively small changes in value. For these conditions the nonlinear differential equations describing the system may be linearised by taking Taylor series approximations about the operating point up to first order derivative terms only. Since the operating point is an equilibrium point this procedure is analogous to that described in section 2.5 for determining the approximate motion of a nonlinear second order system about a singular point. Linearisation of a nonlinear characteristic which is a function of a single input variable thus corresponds to replacing it by a constant gain which is the slope of the nonlinearity at the operating point as shown in figure 1.3.

Consider, for example, the determination of the linearised model of the field controlled motor of figure 1.1 when it is used in a speed control system about a nominal speed of $\dot{\theta}_s$. The operating point on $f_2(\dot{\theta})$ is found with $\dot{\theta} = \dot{\theta}_s$ and on $f_1(c)$ when $f_1(c) = T_L + f_2(\dot{\theta}_s) = T_L + K_1\dot{\theta}_s$ as illustrated in figure 1.8(a) for two assumed nonlinear characteristics $f_1(c)$ and $f_2(\dot{\theta})$. The linearised block diagram of the motor for small speed variations about the nominal value then becomes as shown in figure 1.8(b). Obviously no nonlinear behaviour can be found by this procedure since we are reducing the nonlinear model to an approximate linear one valid for small variations only.

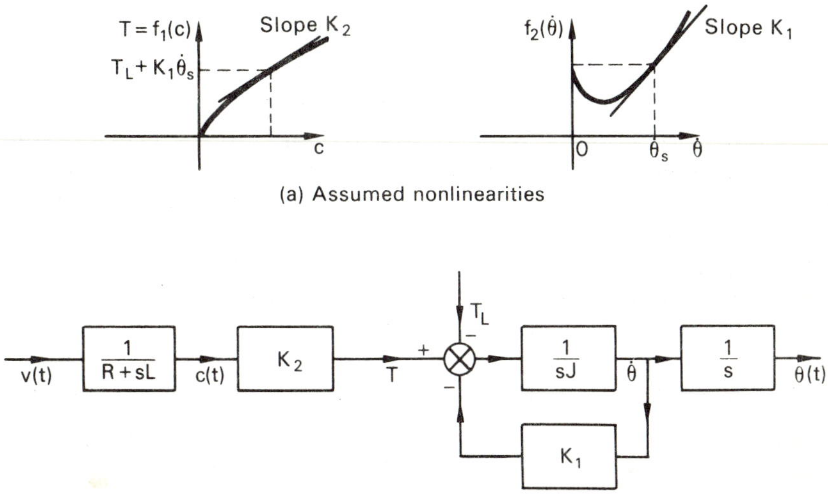

(a) Assumed nonlinearities

(b) Linearised block diagram

Figure 1.8 Linearised equivalent of figure 1.1

Quasilinearisation

The information contained in figure 1.3 about the nonlinear characteristic $y = n(x)$ may be presented in the alternate form of figure 1.9, which shows the gain $n(x)/x$ to x for constant values of x. In other words if x is a bias signal, γ, then the gain to the bias signal $K(\gamma)$ is a function of γ and equal to $n(\gamma)/\gamma$. Quasilinearisation is an extension of this concept to any input signal, that is we choose to represent the nonlinearity by a gain which is optimum in some sense for each 'magnitude' of the nonlinearity input.[8] The block diagram of figure 1.10 illustrates the procedure where the value of K used will depend not only on the nonlinear characteristic and the form of the input signal but also on the criterion of

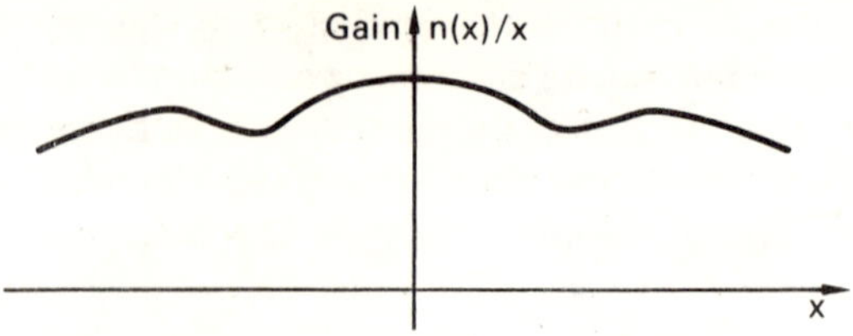

Figure 1.9 Gain $n(x)/x$ for SVNL of figure 1.3

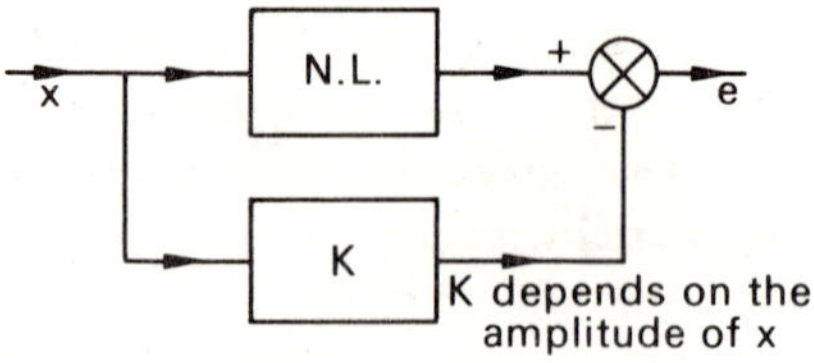

Figure 1.10 Illustration of quasilinearisation

optimality, that is what aspect of the error e we minimise. It will be appreciated that the error can only be made zero for a bias input or if the characteristic is linear. Since this representation of a nonlinearity maintains the dependence of the output on the input magnitude, its judicious use in nonlinear systems may be expected to predict true aspects of nonlinear behaviour. The notion can be extended for more than one input to a nonlinearity with each input represented by a separate equivalent gain.[9] This concept of quasilinearisation is the basis of the describing function method.[10]

1.6 The describing function approach

To obtain a quasilinear model for a nonlinear characteristic, the form of its input signal must be known. Unfortunately in a nonlinear feedback system the input to the nonlinearity is rarely known exactly since it is a function of the inputs to the system and the characteristics of the nonlinear feedback loop. Thus a specific input form to the nonlinearity must be assumed. Most control systems contain low pass plant transfer functions and since low pass filtering of any periodic signal tends to make it sinusoidal, periodic signals within the system might be expected to be approximately sinusoidal at the nonlinearity input. In addition it is also possible to show that the result of low pass filtering any random signal is to make it more Gaussian. These and other considerations then lead us to assume three basic forms of nonlinearity input signals namely, a bias or d.c. signal, a sinusoid, and a Gaussian random signal. For many situations we will also need to consider various combinations of these input signals. The relationships which give the quasi or equivalent linear gains for the nonlinearity are known as describing functions. In order to simplify references to describing functions for different types of input, the describing function abbreviated to DF is preceded by a letter or letters denoting the input, the letter S is used for a sinusoid, B for a bias signal and R, unless stated to the contrary, for a Gaussian random signal. Because of its familiar usage we make the one exception of denoting the sinusoidal input describing function by DF rather than SDF. Thus for example, RSDF, denotes the describing function for a random and sinusoidal input signal to a nonlinearity. The word, and, is used on purpose since in some of the later work we are concerned with obtaining describing functions for nonlinear characteristics which are functions of two or more signals not necessarily functions of the sum of two signals.

Obviously the assumption of a particular input signal to a nonlinearity in order to predict a certain form of behaviour in a nonlinear system is a particularly important aspect of the describing function approach and one which develops from experience. When the describing function model is used for the nonlinearity and the relevant system equations solved more than one solution may exist. It is therefore necessary to be able to investigate the stability of any solutions obtained by the describing function method. Investigation of the stability requires that all possible forms of perturbation of the solution should be considered. Even when a stable describing function solution exists for a particular set of system parameters, it is not necessary that it should always exist for those specific values. The solution can depend upon the initial conditions or equivalently how the system parameters were moved to their current values. A further difficulty of the describing function method is the lack of precise methods for estimating the accuracy of the solutions; again experience is a powerful ally. The great advantage of the approach however is that it is basically simple and gives the designer a relatively quick appreciation for the system behaviour. It indicates methods by which the system may be modified where necessary to meet performance criteria and may also be used for the design of compensators. More accurate solutions can be obtained by simulation and with the support of the knowledge gained by the describing function method considerable savings in time and expense can result.

1.7 A review of the contents

The aim of the book is to provide an in depth coverage of the role of describing functions in the analysis and design of nonlinear systems. Since describing function methods are approximate a knowledge of some other basic engineering approaches, which may then be used for comparative purposes, is extremely desirable. Chapter 2 of this book thus deals in some detail with the use of phase plane methods for the analysis of nonlinear second order systems. It contains a brief review of the classical theory for an autonomous nonlinear vector differential equation in two variables followed by methods for the plotting of phase plane trajectories. The procedure is then used to analyse in detail the effect of various forms of nonlinearity on the performance of a position control system.

The describing function approach has been developed in a systematic manner starting with the single sinusoidal input case and proceeding from there to more general and more sophisticated cases and applications. Chapter 3 is devoted entirely to methods for the evaluation of the single sinusoidal input describing function and the harmonic distortion products at the nonlinearity output. The various properties which the DF satisfies are emphasised and considerable emphasis is placed on the determination of the DF from the nonlinear characteristic and the probability density of the input sinusoid. This approach has many advantages over the more familiar technique of performing a Fourier analysis on the nonlinearity output waveform. Applications of the DF to the analysis of symmetrical odd nonlinear systems are considered in chapter 4. Also included are sections on systems with multiple nonlinearities, compensator design, estimation of the accuracy of DF solutions and a detailed presentation of methods for evaluating the closed loop frequency response.

Chapter 5 begins by considering evaluation of the SBDF and introduces the concept of a modified nonlinearity. Applications of the SBDF in the analysis of limit cycling and asymmetrical nonlinear systems are then considered. The use of the modified nonlinearity in determining describing functions for more than one input is then illustrated by considering a cubic characteristic with two sinusoidal inputs. Various applications of the incremental describing function, the IDF, and the SSDF in the analysis of periodic phenomena in nonlinear systems are then presented. Exact methods for the determination of periodic modes in relay systems are considered in chapter 6. Attention is focused on frequency domain methods for ease of comparison with the describing function approach. Asymmetrical as well as symmetrical oscillations are considered and closed form expressions are given for the various types of series which must be summed to obtain the required frequency response loci. Procedures for applying the method to systems with more than one relay characteristic are also presented. Chapter 7 which gives the basic theory for the analysis of linear systems with random inputs is included to provide the necessary background for understanding the procedures associated with the evaluation and application of describing functions for random inputs. Some required sections such as those on optimisation with constraints and expansions for joint density functions are not often found in similar introductory material. Chapter 8 is concerned with the evaluation of the RDF and the RBDF and the application of describing function methods to nonlinear systems with random inputs. Material is also included on the distortion produced by nonlinear elements, the accuracy of describing function methods and the design of optimum nonlinear systems.

Describing functions are normally used for the evaluation of steady state system behaviour, although their application in some transient problems is possible. Chapter 9 considers the use of describing function methods for the evaluation of step responses and the dynamics of limit cycles. Chapter 10 is concerned with the evaluation of the describing functions for multivariable nonlinearities where the nonlinearity output is simply any function of several variables. It is shown that when the inputs are unrelated, describing functions can be calculated in a relatively straightforward manner, particularly when the modified nonlinearity approach is used. Unfortunately for the important case where the inputs are related sinusoids the evaluation of describing functions usually becomes much more difficult. Chapter 11 considers the analysis of systems containing multivariable nonlinearities and chapter 12 examines the general problem of nonlinear multivariable systems. Also included is a brief introduction to the application of frequency response methods in linear multivariable systems. In previous chapters where random input signals have been treated the nonlinearities have been assumed to be single valued as the determination of describing functions for double and multivalued nonlinear characteristics with random inputs is an exceedingly complex problem. Chapter 13 presents some methods for evaluating RDFs for DVNLs. In many control systems the signals are transmitted in other than direct form, one example of which is the large number of small position control systems which have amplitude modulated signals. Chapter 14 discusses the effect of nonlinearity in these a.c. control systems and it provides a good demonstration of analysis with multivariable nonlinearities.

REFERENCES

1. Coddington, E. A. and Levinson, N.: *Theory of ordinary differential equations* (McGraw-Hill, New York, 1955, pp. 375–387).
2. Willems, J. L.: *Stability theory of dynamical systems* (Nelson, London, 1970).
3. Davis, H. T.: *Introduction to nonlinear differential and integral equations* (Dover Publications Inc., New York, 1962).
4. LaSalle, J. P. and Lefschetz, S.: *Stability by Liapunov's direct method* (Academic Press, New York, 1961).
5. Aizerman, M. A. and Gantmacher, F. R.: *Absolute stability of regulator systems* (Holden-Day, San Francisco, 1964).
6. Minorsky, N.: *Nonlinear oscillations* (Van Nostrand, Princeton, 1962).
7. Hayashi, C.: *Nonlinear oscillations in physical systems* (McGraw-Hill, New York, 1964).
8. Booton, R. C., Jr.: 'Nonlinear control systems with random inputs', *IRE Trans.*, 1954, CT-1, pp. 9–17.
9. Somerville, M. J. and Atherton, D. P.: 'Multi-gain representation for a single-valued nonlinearity with several inputs and the evaluation of their equivalent gains by a cursor method', *Proc. I.E.E.*, 1958, 105C, pp. 537–549.
10. Gelb, A. and Vander Velde, W. E.: *Multiple-input describing functions and nonlinear system design* (McGraw-Hill, New York, 1968).

Chapter 2

Phase Plane Analysis

2.1 Introduction

The behaviour of a linear system may be described by a set of first order differential equations known as the state equations, or simply the state equation when matrix notation is used. When the system is described by n first order equations, the state vector has n components and an n space is required to represent the motion. For a system with $n = 2$ the components of the state vector, which normally vary with time, can be shown on a state plane. If the system state variables are chosen equal to the phase variables, namely the system output and its first derivative, the state plane is known as a phase plane. Taking different initial conditions a family of phase trajectories, known as a phase portrait, which show the system response with time a parameter on the trajectory, can be obtained.

In this chapter we will be concerned with the applications of phase plane techniques to the analysis of transient phenomena, which can be defined either directly or indirectly by a set of initial conditions, of second order nonlinear differential equations. The main advantage of the approach is the availability of various methods which enable the trajectories to be sketched quite accurately without recourse to analytical solution of the differential equation, which in fact may not be possible. It is also particularly suited to the analysis of systems with linear segmented nonlinear characteristics as the phase plane can be divided into regions corresponding to operation on a particular linear segment of the nonlinearity. The result is that trajectories in the various regions of the phase plane are described by different linear differential equations. The main disadvantage of the method is that its application is essentially limited to second order systems. Some third order systems have been studied, using three-dimensional diagrams, but except for a few special cases, the method is not very satisfactory.

The DF method to be presented in the next chapter is on the other hand normally more suitable for the study of higher order systems. Since the DF method is an approximate technique, it is also useful to compare its accuracy for second order systems with the phase plane solutions. In many ways, therefore, the phase plane and DF approaches may be regarded as complementary, in that each is available for the study of systems which are most likely to be beyond the scope of the other.

2.2 The general second order linear system

Any linear time invariant system may be represented by the state equations[1]

$$\dot{\mathbf{x}} = \mathbf{A}\mathbf{x} + \mathbf{B}\mathbf{u}, \tag{2.1}$$

$$\mathbf{y} = \mathbf{C}\mathbf{x} + \mathbf{D}\mathbf{u},$$

where $\mathbf{u}$, $\mathbf{y}$ and $\mathbf{x}$ are functions of time and respectively the input, output and state vectors, $\mathbf{A}$, $\mathbf{B}$, $\mathbf{C}$ and $\mathbf{D}$ are constant coefficient matrices and $\dot{\mathbf{x}}$ denotes the derivative of $\mathbf{x}$ with respect to time. Equations (2.1) are general and applicable to multivariable systems in which case if $\mathbf{x}$ is an n vector, $\mathbf{u}$ an r vector and $\mathbf{y}$ an m vector, the matrices $\mathbf{A}$, $\mathbf{B}$, $\mathbf{C}$ and $\mathbf{D}$ would be of order $n \times n$, $n \times r$, $m \times n$ and $m \times r$ respectively. In this chapter we will be concerned with single input single output systems only so that y and u become scalars and the equations become

$$\dot{\mathbf{x}} = \mathbf{A}\mathbf{x} + \mathbf{B}u, \qquad y = \mathbf{C}\mathbf{x} + Du, \tag{2.2}$$

where the matrices **B**, **C** and D become a vector, the transpose of a vector and a scalar respectively.

For the second order system with $\mathbf{x} = (x_1, x_2)^T$, where T denotes transpose,

$$\mathbf{A} = \begin{pmatrix} a & b \\ c & d \end{pmatrix}, \qquad \mathbf{B} = \begin{pmatrix} b_0 \\ b_1 \end{pmatrix}, \qquad \mathbf{C} = (c_0, c_1)$$

and $D = d_0$

$$\dot{x}_1 = ax_1 + bx_2 + b_0u, \tag{2.3}$$

$$\dot{x}_2 = cx_1 + dx_2 + b_1u, \tag{2.4}$$

$$y = c_0x_1 + c_1x_2 + d_0u, \tag{2.5}$$

where equations (2.3) and (2.4) are the state equations and equation (2.5) is the output equation. A useful concept for analysis of the autonomous system, that is when $u = 0$, is the velocity vector interpretation of the state equations[(2)] (2.3) and (2.4), which are in matrix notation

$$\dot{\mathbf{x}} = \mathbf{A}\mathbf{x}. \tag{2.6}$$

A 'representative point' P which traces out a trajectory in a state plane has a velocity vector with components $\dot{x}_1$ and $\dot{x}_2$ which are determined uniquely in terms of x_1 and x_2 from equation (2.6). The magnitude of the vector is $(\dot{x}_1^2 + \dot{x}_2^2)^{1/2}$ and its direction $\tan\theta = (\dot{x}_2/\dot{x}_1) = \mathrm{d}x_2/\mathrm{d}x_1$. Hence by means of the slope equation

$$\frac{\dot{x}_2}{\dot{x}_1} = \frac{\mathrm{d}x_2}{\mathrm{d}x_1} = \frac{cx_1 + dx_2}{ax_1 + bx_2} \tag{2.7}$$

line segments, or line elements, in the direction of the velocity vector can be constructed in the state plane, each such segment being tangent to the trajectory through its mid-point. If sufficient line segments are sketched then the shape of the trajectory can be guessed reasonably well. Points at which the velocity vector is zero, where the system can be in equilibrium, are known as singular points or singularities since the slope of a trajectory is 0/0 and thus undefined. For the linear system of equation (2.6) the only singular point is at the origin, provided $\det \mathbf{A} \neq 0$. Singular points are classified according to the patterns of trajectories in their immediate neighbourhood. In the case of equation (2.6) the behaviour is determined by the eigenvalues of the **A** matrix which are given by $\det(\mathbf{A} - \lambda\mathbf{I}) = 0$. Thus

$$\begin{vmatrix} a - \lambda & b \\ c & d - \lambda \end{vmatrix} = 0$$

that is

$$\lambda^2 - (a + d)\lambda + ad - bc = 0. \tag{2.8}$$

Figure 2.1 shows the dependence of the various singular points on the coefficients, $-(a + d)$ and $(ad - bc)$ of equation (2.8). It will be seen that (i) the parabola $(a + d)^2 = 4(ad - bc)$ separates the regions of real and complex eigenvalues (ii) on the parabola the roots are real and equal (iii) for $(ad - bc) < 0$ one eigenvalue is positive and the other negative (iv) for $(ad - bc) > 0$ and $(a + d) = 0$ the eigenvalues are imaginary.

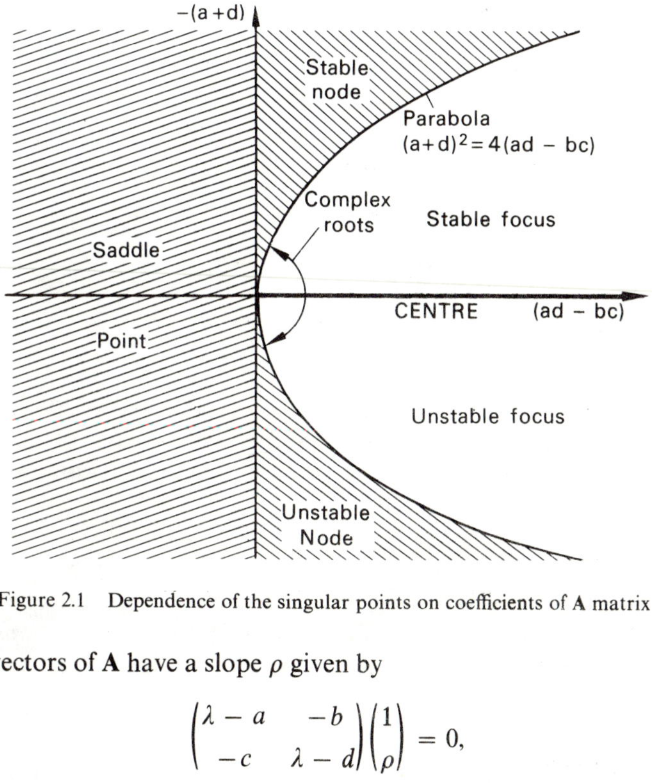

Figure 2.1 Dependence of the singular points on coefficients of **A** matrix

The eigenvectors of **A** have a slope ρ given by

$$\begin{pmatrix} \lambda - a & -b \\ -c & \lambda - d \end{pmatrix} \begin{pmatrix} 1 \\ \rho \end{pmatrix} = 0,$$

giving $\lambda - a - b\rho = 0$. This may be written

$$\lambda - \{(a + d)/2\} = b\rho + \{(a - d)/2\},$$

which on squaring and substituting for λ from equation (2.8) gives

$$bc - ad + \{(a + d)^2/4\} = b^2\rho^2 + b\rho(a - d) + \{(a - d)^2/4\}$$

$$b\rho^2 + \rho(a - d) - c = 0. \tag{2.9}$$

If the slope of a trajectory in the state plane is r, then from equation (2.7)

$$(ax_1 + bx_2)r = cx_1 + dx_2,$$

which is a line, known as an isocline since all trajectories cut it with the same slope r, of slope

$$\frac{x_2}{x_1} = \frac{ar - c}{d - br}.$$

If the slope of the isocline is r the same as that of the trajectory so that the trajectories travel down, rather than cut the isocline

$$r = (ar - c)/(d - br),$$

that is

$$br^2 + r(a - d) - c = 0. \tag{2.10}$$

Comparing with equation (2.9) we see that r and ρ are identical, so that an isocline which has trajectories along it is an eigenvector. Obviously, since r must be real, the situation only occurs for real eigenvalues.

For the linear second order differential equation

$$\ddot{x} + a_1\dot{x} + a_0 x = 0 \tag{2.11}$$

with the phase variables $x_1 = x$, $x_2 = \dot{x}$ chosen as the state variables the $\mathbf{A}$ matrix is given by

$$\mathbf{A} = \begin{pmatrix} 0 & 1 \\ -a_0 & -a_1 \end{pmatrix} \tag{2.12}$$

so that $a = 0$, $b = 1$ and equation (2.8) becomes identical with equations (2.9) and (2.10). Also the axes $-(a + d)$ and $(ad - bc)$ in figure 2.1 are a_1 and a_0 respectively. The trajectories for the various eigenvalue situations are discussed in more detail below.

2.2.1 *Real eigenvalues*

In this case $(a + d)^2 > 4(ad - bc)$ and there are two distinct real eigenvectors which are possible trajectories. Performing a nonsingular linear transformation from the state vector $\mathbf{x}$ to a new state vector $\mathbf{x}^*$ according to

$$\mathbf{x} = \mathbf{T}\mathbf{x}^*, \qquad \mathbf{x}^* = \mathbf{T}^{-1}\mathbf{x}, \tag{2.13}$$

equation (2.6) becomes

$$\dot{\mathbf{x}}^* = \mathbf{T}^{-1}\mathbf{A}\mathbf{T}\mathbf{x}^*.$$

If we choose the matrix $\mathbf{T} = (\mathbf{v}_1, \mathbf{v}_2)$ where the vectors $\mathbf{v}_1, \mathbf{v}_2$ are the eigenvectors of $\mathbf{A}$ it can be shown that[(3)]

$$\mathbf{T}^{-1}\mathbf{A}\mathbf{T} = \mathbf{\Lambda} = \begin{pmatrix} \lambda_1 & 0 \\ 0 & \lambda_2 \end{pmatrix}$$

where λ_1 and λ_2 are the eigenvalues of $\mathbf{A}$. $\mathbf{T}$ is known as the modal matrix of $\mathbf{A}$ since it decouples the two modes giving

$$\dot{\mathbf{x}}^* = \mathbf{\Lambda}\mathbf{x}^*, \tag{2.14}$$

which is the canonical form of equation (2.6). Since the expression for $\mathbf{\Lambda}$ involves the inverse of $\mathbf{T}$ diagonalisation is only possible when the inverse exists, that is when the eigenvectors $\mathbf{v}_1, \mathbf{v}_2$ are linearly independent. A set of vectors $\mathbf{v}_1, \mathbf{v}_2 \ldots \mathbf{v}_n$ is linearly dependent if there exists a set of scalars c_i (not all zero) such that $\sum_i c_i\mathbf{v}_i = 0$. It is easily shown that equation (2.14) has a solution of the form

$$(x_1^*)^{\lambda_2} = C(x_2^*)^{\lambda_1} \tag{2.15}$$

where C is a constant.

When λ_1 and λ_2 are both positive the singular point is an unstable node, when they are both negative it is a stable node and when they have alternate signs a saddle point. For the stable situation with λ_1 and λ_2 both negative, the motion tends tangentially to the slowest eigenvector, that is the eigenvector corresponding to the smallest eigenvalue magnitude, as the node is approached. State portraits for various conditions are shown in table 2.1.

2.2.2 *Equal real eigenvalues*

In this case $(a + d)^2 = 4(ad - bc)$, $\lambda_1 = \lambda_2$ and it is impossible to find two linearly independent eigenvectors and so uncouple the equations. It is however always possible to transform equation (2.6) to the Jordon canonical form, where one equation is independent of the other, that is

$$\dot{\mathbf{x}}^* = \mathbf{\Lambda}_J\mathbf{x}^* \tag{2.16}$$

where

$$\mathbf{\Lambda}_J = \begin{pmatrix} \lambda_1 & 1 \\ 0 & \lambda_1 \end{pmatrix}.$$

Here $\mathbf{T} = (\mathbf{v}_1, \mathbf{v}_2)$, where $\mathbf{v}_1$ is the eigenvector of $\mathbf{A}$ obtained from $(\mathbf{A} - \lambda_1\mathbf{I})\mathbf{v}_1 = 0$ and $\mathbf{v}_2$ (not an eigenvector) is obtained from $(\mathbf{A} - \lambda_1\mathbf{I})\mathbf{v}_2 = \mathbf{v}_1$.

Again the solution of equation (2.16) is easily shown to be of the form

$$x_2^* = Ce^{\lambda_1 x_1^*/x_2^*}. \tag{2.17}$$

Trajectories for various conditions are shown in table 2.2. The singular point is again a node.

2.2.3 *Complex eigenvalues*

Proceeding as in section 2.2.1, since the eigenvalues are complex conjugates $\lambda_1, \lambda_2 = -\sigma \pm j\omega$, the modal matrix $\mathbf{T} = (\mathbf{v}_1, \mathbf{v}_2)$ will have conjugate complex eigenvectors and the diagonal matrix $\mathbf{\Lambda}$ will be of the form

$$\mathbf{\Lambda} = \begin{pmatrix} -\sigma + j\omega & 0 \\ 0 & -\sigma - j\omega \end{pmatrix}.$$

In order to avoid complex quantities, the matrix $\mathbf{T}$ can be replaced by $\mathbf{T}_m = (\boldsymbol{\alpha}, \boldsymbol{\beta})$ where $\boldsymbol{\alpha}$ and $j\boldsymbol{\beta}$ are respectively the real and imaginary vector

TABLE 2.1

Motion of second order systems with distinct real eigenvalues (poles)

Both poles not zero	Poles	$\lambda_1=-1,\ \lambda_2=-4$	$\lambda_1=-2,\ \lambda_2=1$	$\lambda_1=1,\ \lambda_2=4$
	Arbitrary form	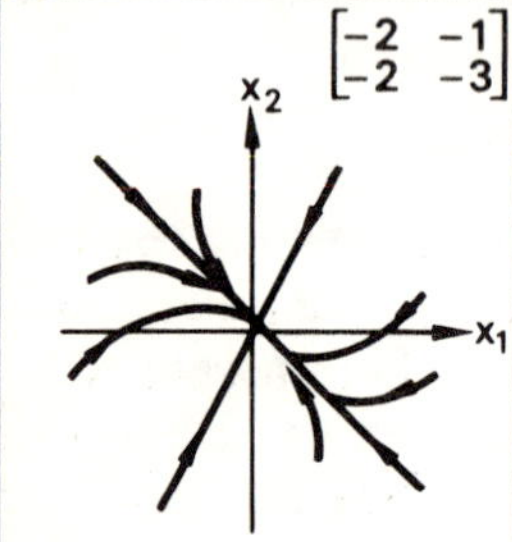 1(a) Stable node	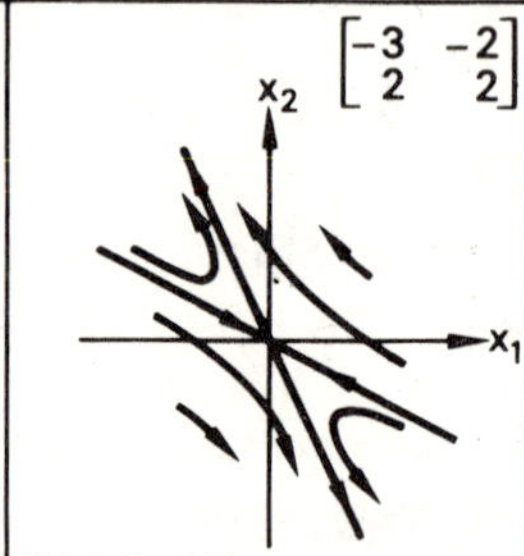 2(a) Saddle	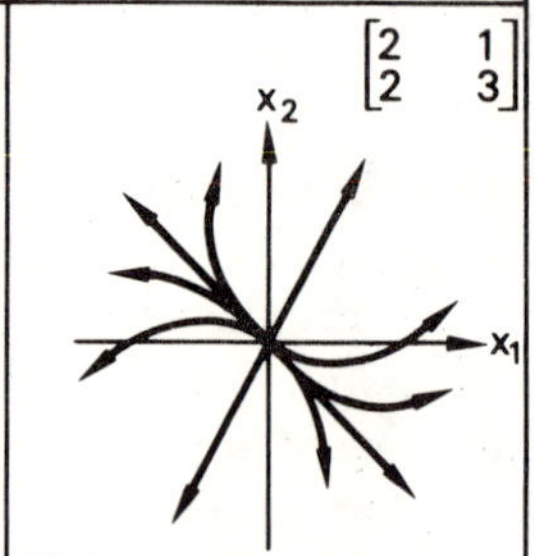 3(a) Unstable node
	Canonical form	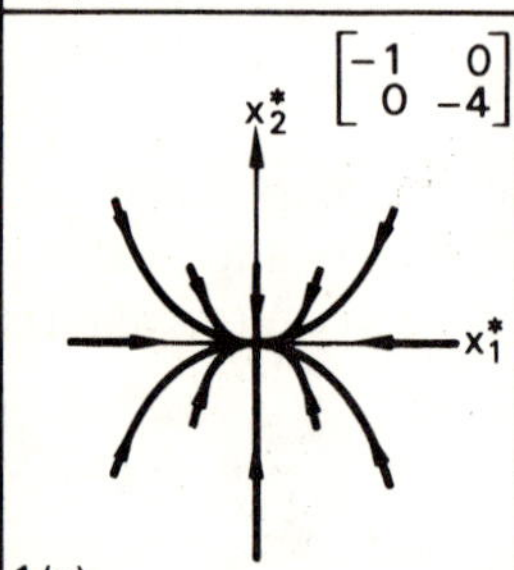 1(c)	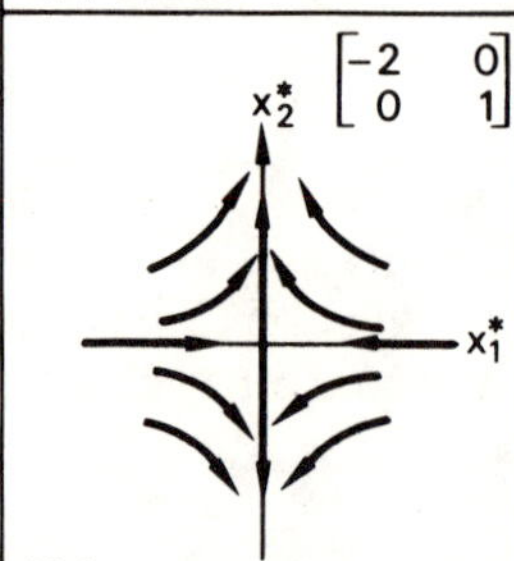 2(c)	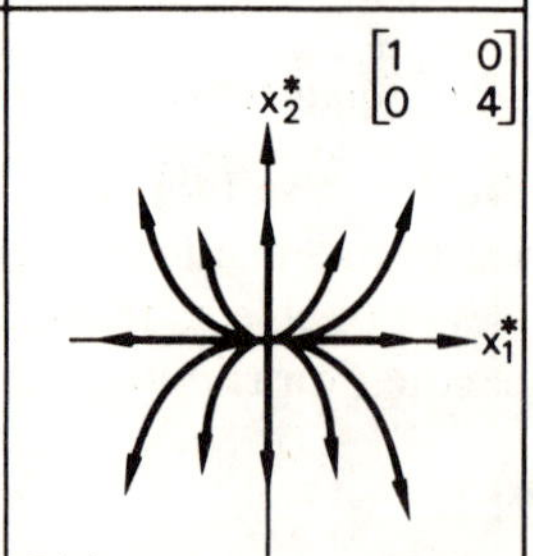 3(c)
One pole zero	Poles	$\lambda_1=-5,\ \lambda_2=0$		$\lambda_1=5,\ \lambda_2=0$
	Arbitrary form	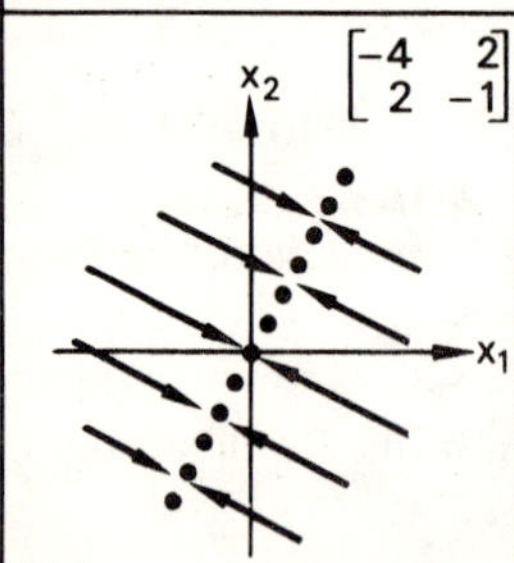 4(a)	—	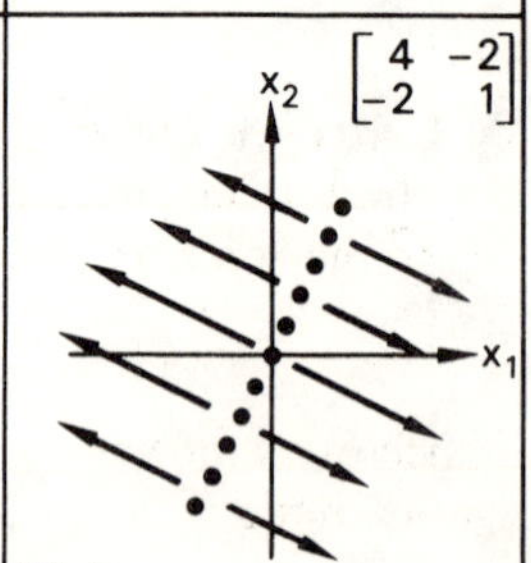 5(a)
	Canonical form	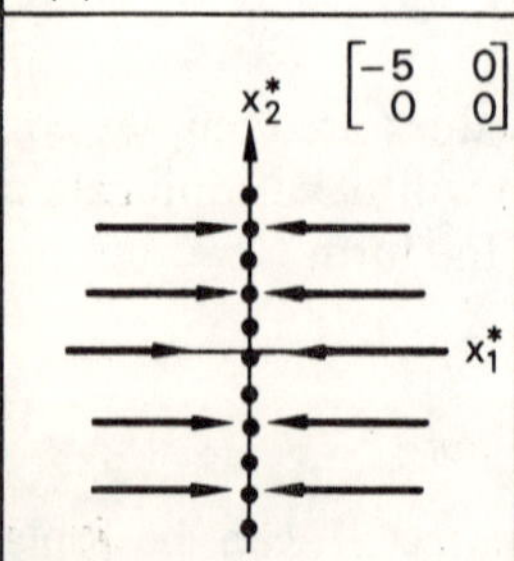 4(c)	—	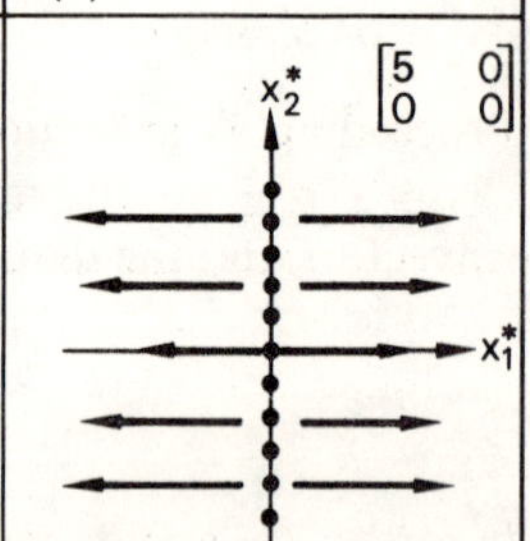 5(c)

TABLE 2.2

Motion of second order systems with equal eigenvalues (double poles)

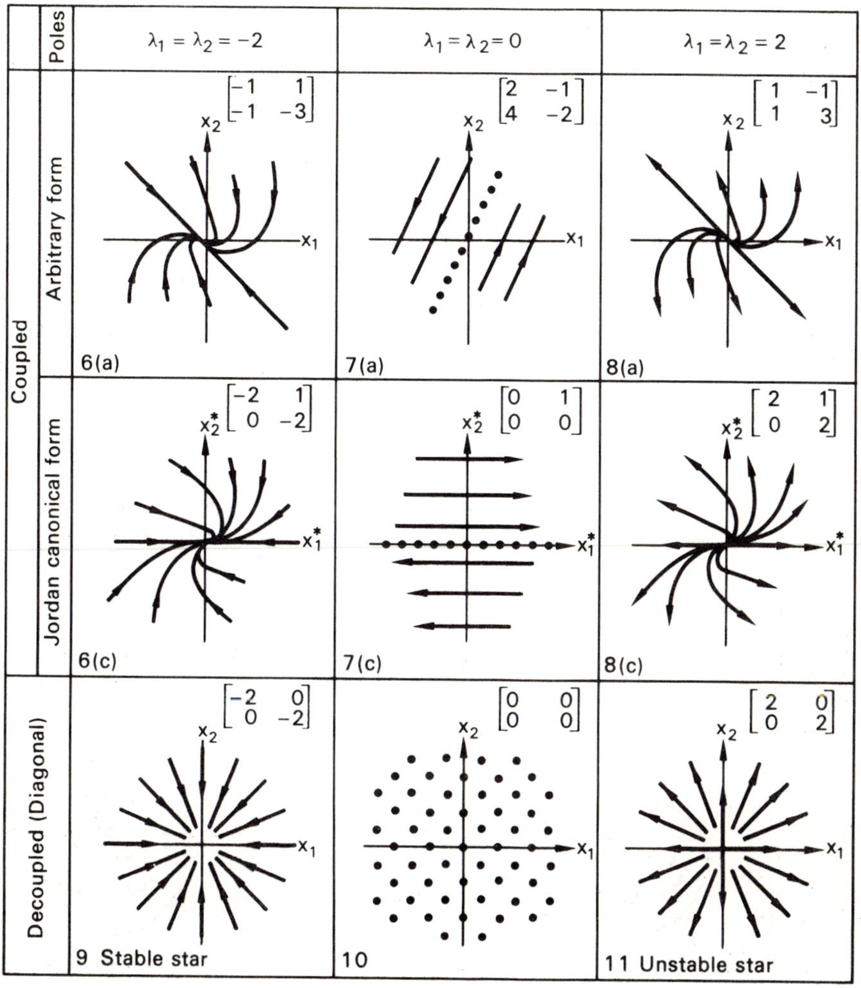

components of $\mathbf{v}_1$ and $\mathbf{v}_2$, that is

$$\mathbf{v}_1 = \boldsymbol{\alpha} + j\boldsymbol{\beta}, \qquad \mathbf{v}_2 = \boldsymbol{\alpha} - j\boldsymbol{\beta}.$$

This results in

$$\boldsymbol{\Lambda}_m = \begin{pmatrix} -\sigma & \omega \\ -\omega & -\sigma \end{pmatrix}$$

and

$$\dot{\mathbf{x}}^* = \boldsymbol{\Lambda}_m \mathbf{x}^*, \tag{2.18}$$

which is known as the modified canonical form.

Equation (2.18) has the solution

$$r = r_0 e^{(\theta - \theta_0)\sigma/\omega} \tag{2.19}$$

in polar coordinates, which is a logarithmic spiral. The singular point is a focus. For the special case of $\sigma = 0$ equation (2.19) reduces to the circle, $r = r_0$, that is

$$(x_1^*)^2 + (x_2^*)^2 = (x_{10}^*)^2 + (x_{20}^*)^2$$

about the singular point which is a centre or vortex. Table 2.3 shows the shape of the various trajectories.

TABLE 2.3

Motion of second order systems with complex eigenvalues

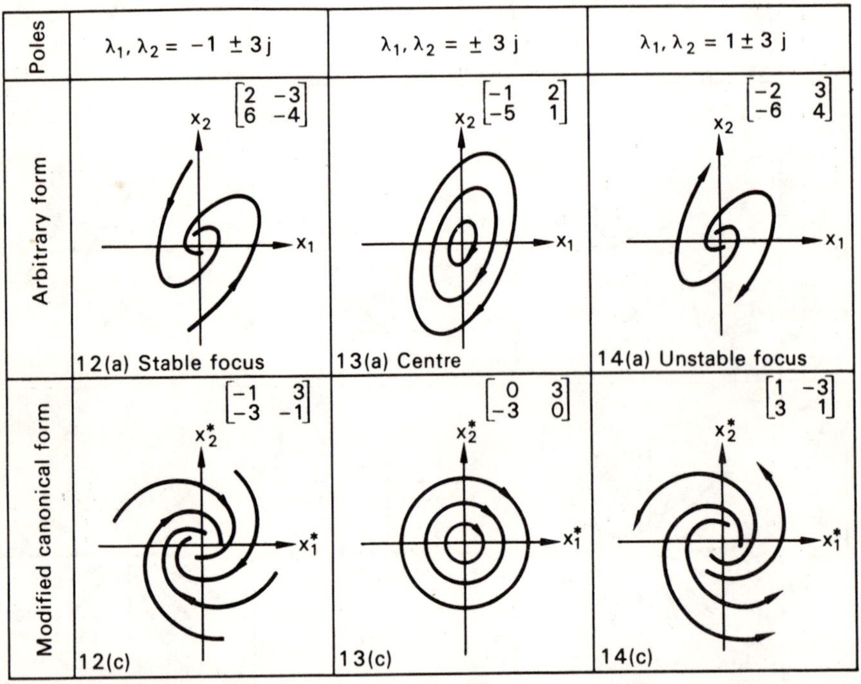

2.3 The linear second order differential equation

The general second order linear differential equation is of the form

$$(1/\omega_0^2)\ddot{x} + (2\zeta/\omega_0)\dot{x} + x = u(t), \tag{2.20}$$

where ω_0 is the undamped natural frequency, ζ the damping ratio, $u(t)$ a step input forcing function, and $x(t)$ the system response or output.

We consider below phase plane portraits for this equation, firstly when all the terms are present corresponding to a type 0 system, then when there is no term in x corresponding to a type 1 system and finally when there are no terms in $\dot{x}$ and x corresponding to a type 2 system.

2.3.1 *Type 0 system*

Initially the homogeneous form of equation (2.20), that is when the input $u(t) = 0$, is considered. Then

$$\ddot{x} + 2\zeta\omega_0\dot{x} + \omega_0^2 x = 0, \tag{2.21}$$

which may be written

$$\begin{aligned} \dot{x}_1 &= x_2, \\ \dot{x}_2 &= -\omega_0^2 x_1 - 2\zeta\omega_0 x_2, \end{aligned} \tag{2.22}$$

where $x_1 = x$.

Thus the coefficients of the **A** matrix are $a = 0$, $b = 1$, $c = -\omega_0^2$ and $d = -2\zeta\omega_0$ and the slope of trajectories in the phase plane is

$$\frac{\dot{x}_2}{\dot{x}_1} = \frac{dx_2}{dx_1} = \frac{-2\zeta\omega_0 x_2 - \omega_0^2 x_1}{x_2}. \tag{2.23}$$

In particular all trajectories cut the x_1 axis, i.e. $x_2 = 0$, at right angles (slope $= \infty$), the x_2 axis with a slope of $-2\zeta\omega_0$ and the straight line $2\zeta x_2 + \omega_0 x_1 = 0$ with zero slope. Trajectories have a slope r when crossing the isocline

$$-2\zeta\omega_0 x_2 - \omega_0^2 x_1 = x_2 r \tag{2.24}$$

and the slope of a trajectory and isocline are equal when

$$r = -\omega_0^2/(r + 2\zeta\omega_0),$$

that is

$$r = \{-\zeta \pm (\zeta^2 - 1)^{1/2}\}\omega_0. \tag{2.25}$$

Since the system **A** matrix has $a = 0$ and $b = 1$, equation (2.25) gives the eigenvalues as well as the slope of the eigenvectors. Analytical solutions for the time response can be obtained from the Laplace transform of equation (2.21), namely

$$(s^2 + 2\zeta s\omega_0 + \omega_0^2)X(s) = sx_{10} + x_{20} + 2\zeta\omega_0 x_{10}, \tag{2.26}$$

where $x_1 = x$, $x_2 = \dot{x}$ and the initial conditions are $x_1 = x_{10}$ and $x_2 = x_{20}$. The equation $(s^2 + 2\zeta s\omega_0 + \omega_0^2) = 0$ is known as the characteristic equation of the system. As the solutions depend upon ζ phase portraits are considered in more detail for various positive values of ζ, from which those for corresponding negative ζ are easily deduced using equation (2.22).

Zero damping $\zeta = 0$

In this case the eigenvalues or roots of the characteristic equation for ω_0^2 positive are purely imaginary, that is $s = \pm j\omega_0$ and

$$\dot{x}_2/\dot{x}_1 = dx_2/dx_1 = -\omega_0^2 x_1/x_2, \tag{2.27}$$

which on integration gives the ellipse

$$(x_2^2/\omega_0^2) + x_1^2 = (x_{20}^2/\omega_0^2) + x_{10}^2 = \text{constant} \tag{2.28}$$

about a centre at the origin, since $a + d = 0$ and $ad - bc = \omega_0^2$. Using equation (2.26)

$$X(s) = (sx_{10} + x_{20})/(s^2 + \omega_0^2)$$

which on inversion gives

$$x_1 = x = x_{10} \cos \omega_0 t + (x_{20}/\omega_0) \sin \omega_0 t, \tag{2.29}$$

$$x_2 = \dot{x} = -x_{10}\omega_0 \sin \omega_0 t + x_{20} \cos \omega_0 t, \tag{2.30}$$

from which time can be eliminated to produce equation (2.28).

Light damping $0 < \zeta < 1$

The eigenvalues are now complex with $s = \sigma \pm j\omega$ where $\sigma = -\zeta\omega_0$ and $\omega = (1 - \zeta^2)^{1/2}\omega_0$, the origin is a stable focus and equation (2.26) has the solutions

$$\begin{aligned} x_1 = \frac{\mathrm{e}^{-\zeta\omega_0 t}}{(1-\zeta^2)^{1/2}}[x_{10} \sin\{(1-\zeta^2)^{1/2}\omega_0 t + \phi\} \\ + (x_{20}/\omega_0) \sin\{(1-\zeta^2)^{1/2}\omega_0 t\}], \end{aligned} \tag{2.31}$$

$$\begin{aligned} x_2 = \frac{-\mathrm{e}^{-\zeta\omega_0 t}}{(1-\zeta^2)^{1/2}}[\omega_0 x_{10} \sin\{(1-\zeta^2)^{1/2}\omega_0 t\} \\ + x_{20} \sin\{(1-\zeta^2)^{1/2}\omega_0 t - \phi\}], \end{aligned} \tag{2.32}$$

where $\phi = \cos^{-1}\zeta$. When $x_{20} = 0$, it is seen that x_1 possesses alternate minima and maxima at times t_n given by

$$\sin\{(1-\zeta^2)^{1/2}\omega_0 t_n\} = 0,$$

that is

$$t_n = n\pi/\omega_0(1-\zeta^2)^{1/2} = n\pi/\omega, \tag{2.33}$$

with

$$x_1(t_n) = (-1)^n x_{10}\mathrm{e}^{-\zeta n\pi/(1-\zeta^2)^{1/2}}$$

Also since $e^{-(n+1)a} = (e^{-na})(e^{-a})$, the magnitudes of the overshoots and undershoots are simply related by

$$|x_1(t_{n+1})| = \Delta|x_1(t_n)|,$$

where

$$\begin{aligned} \Delta &= \mathrm{e}^{-\zeta\pi/(1-\zeta^2)^{1/2}}, \\ &= \mathrm{e}^{\sigma\pi/\omega}. \end{aligned} \tag{2.34}$$

Critical damping $\zeta = 1$

The eigenvalues are now real and both equal to $-\omega_0$, the origin is a node and the phase portrait has the single eigenvector $x_2 + \omega_0 x_1 = 0$.

Equation (2.26) has the solutions

$$x_1 = e^{-\omega_0 t}\{x_{10}(1 + \omega_0 t) + x_{20}t\}, \tag{2.35}$$

$$x_2 = e^{-\omega_0 t}\{-x_{10}\omega_0^2 t + x_{20}(1 - \omega_0 t)\}. \tag{2.36}$$

Eliminating t gives the phase plane equation

$$\frac{x_2 + \omega_0 x_1}{x_{20} + \omega_0 x_{10}} = \exp\left\{\frac{-\omega_0(x_1 x_{20} - x_2 x_{10})}{(x_1\omega_0 + x_2)(x_{10}\omega_0 + x_{20})}\right\}. \tag{2.37}$$

Provided $x_{20} + \omega_0 x_{10} \leq 0$ for $x_{10} < 0$, the response will have no overshoot and will be the fastest possible.*

Heavy damping $\zeta > 1{\cdot}0$

There are now two eigenvectors with slopes equal in value to the real eigenvalues $\{-\zeta \pm (\zeta^2 - 1)^{1/2}\}\omega_0$ and the origin is a node. Equation (2.26) has the solutions

$$x_1 = \frac{e^{-\zeta\omega_0 t}}{(\zeta^2 - 1)^{1/2}}[x_{10}\sinh\{(\zeta^2 - 1)^{1/2}\omega_0 t + \phi\}$$
$$+ (x_{20}/\omega_0)\sinh\{(\zeta^2 - 1)^{1/2}\omega_0 t\}], \tag{2.38}$$

$$x_2 = \frac{e^{-\zeta\omega_0 t}}{(\zeta^2 - 1)^{1/2}}[\omega_0 x_{10}\sinh\{(\zeta^2 - 1)^{1/2}\omega_0 t\}$$
$$+ x_{20}\sinh\{(\zeta^2 - 1)^{1/2}\omega_0 t - \phi\}], \tag{2.39}$$

where $\phi = \cosh^{-1}\zeta$. Table 2.4 summarises the above results showing for $\zeta = 0$, 0·4, 1 and 2 the responses $x_1(t)$ and phase plane trajectories for $x_{10} = -1$ and $x_{20} = 0$. The normalised parameter $x_{2N} = x_2/\omega_0$ is used as the ordinate of the phase plane.

When $u(t) = U$ a constant equation (2.18) is

$$\ddot{x} + 2\zeta\omega_0\dot{x} + \omega_0^2 x = \omega_0^2 U,$$

which can be written in the form

$$\frac{d^2(x - U)}{dt^2} + 2\zeta\omega_0\frac{d(x - U)}{dt} + \omega_0^2(x - U) = 0. \tag{2.40}$$

The phase plane portraits are thus identical to those for equation (2.21) but about the singular point $(U, 0)$ rather than the origin.

For an input of the form $u(t) = U + Vt$ equation (2.20) becomes

$$\ddot{x} + 2\zeta\omega_0\dot{x} + \omega_0^2 x = \omega_0^2(U + Vt),$$

* Theoretically the origin is only reached after infinite time so that the response time must be defined to within a certain steady state error, usually 2 per cent.

TABLE 2.4

Step responses of linear second order systems for various values of damping

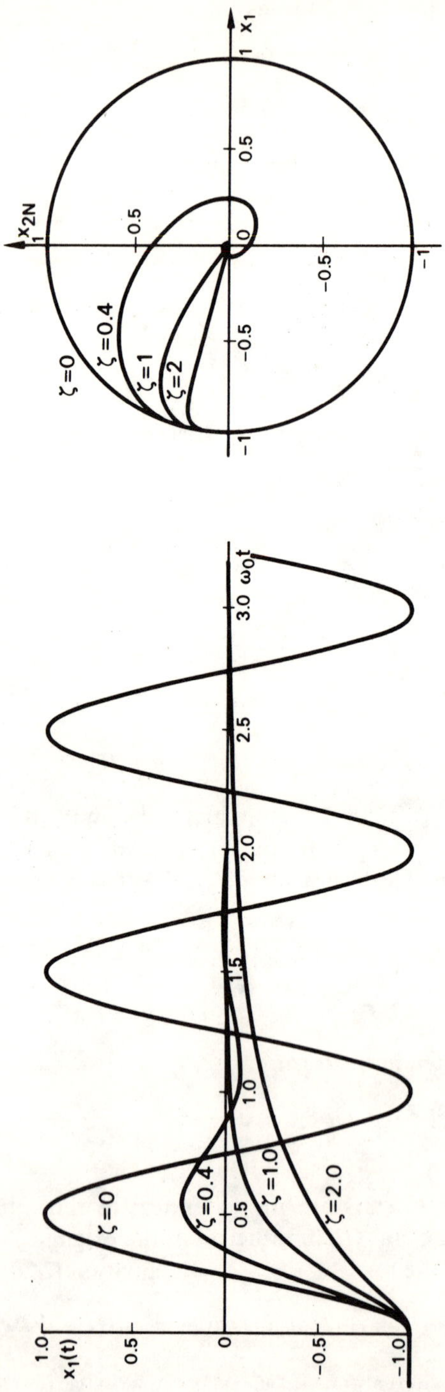

substituting $e(t) = x(t) - u(t)$ so that

$$x = e + U + Vt,$$
$$\dot{x} = \dot{e} + V,$$
$$\ddot{x} = \ddot{e},$$

gives

$$\ddot{e} + 2\zeta\omega_0\dot{e} + \omega_0^2 e = -2\zeta\omega_0 V. \tag{2.41}$$

Phase portraits can thus be drawn in the $e - \dot{e}$ plane where the behaviour will be as for equation (2.21) but about the singular point $(-2\zeta V/\omega_0, 0)$.

2.3.2 *Type* 1 *system*

In this case the differential equation is

$$(1/\omega_0^2)\ddot{x} + (2\zeta/\omega_0)\dot{x} = u(t), \tag{2.42}$$

which, when $u(t) = U$ a constant, may be written

$$\dot{x}_1 = x_2,$$
$$\dot{x}_2 = -\alpha x_2 + \omega_0^2 U,$$

so that

$$dx_2/dx_1 = (-\alpha x_2 + \omega_0^2 U)/x_2, \tag{2.43}$$

where $\alpha = 2\zeta\omega_0$ is assumed positive. Equation (2.43) has no singular points unless $U = 0$ in which case all points on the x_1 axis are singular points. Isoclines cut by trajectories of slope r are given by

$$x_2 = \omega_0^2 U/(r + \alpha),$$

which are lines parallel to the x_1 axis. In particular all trajectories cut the x_1 axis at 90° and tend asymptotically with zero slope to the line $x_2 = \omega_0^2 U/\alpha$.

Integrating equation (2.42) gives

$$x_1 = x_{10} + (x_{20}/\alpha)(1 - e^{-\alpha t}) + (\omega_0^2 U/\alpha^2)(\alpha t - 1 + e^{-\alpha t}), \tag{2.44}$$

$$x_2 = x_{20}e^{-\alpha t} + (\omega_0^2 U/\alpha)(1 - e^{-\alpha t}), \tag{2.45}$$

and eliminating t or from direct integration of equation (2.43) the phase plane equation is

$$x_1 - x_{10} = \{-(x_2 - x_{20})/\alpha\} + (\omega_0^2 U/\alpha^2)\log_e\{(\alpha x_{20} - \omega_0^2 U)/(\alpha x_2 - \omega_0^2 U)\}. \tag{2.46}$$

Figure 2.2 shows trajectories sketched for $U = 0$ and U finite. In the former case they are parallel straight lines; in the second case identical in shape and simply shifted along the x_1 axis.

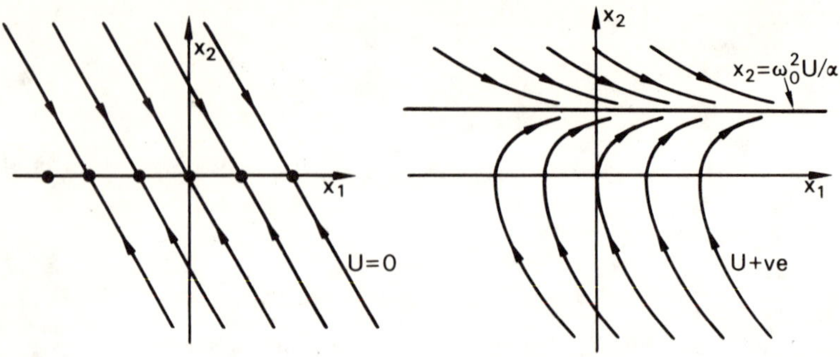

Figure 2.2 Phase trajectories for type 1 system

2.3.3 *Type 2 system*

Here we consider the differential equation

$$(1/\omega_0^2)\ddot{x} = u(t), \tag{2.47}$$

so that with $u(t) = U$

$$\dot{x}_1 = x_2,$$

$$\ddot{x}_2 = \omega_0^2 U,$$

$$\mathrm{d}x_2/\mathrm{d}x_1 = \omega_0^2 U/x_2. \tag{2.48}$$

Again equation (2.48) has no singular points unless $U = 0$, in which case all points on the x_1 axis are singular. Isoclines cut by trajectories of slope r are the lines $x_2 = \omega_0^2 U/r$ parallel to the x_1 axis. Trajectories cut the x_1 axis at 90°, have zero slope when $x_2 \to \infty$ and are mirrored about the x_1 axis.

Integrating equation (2.47) gives

$$x_1 = x_{10} + x_{20}t + \omega_0^2 U t^2/2, \tag{2.49}$$

$$x_2 = x_{20} + \omega_0^2 U t, \tag{2.50}$$

and eliminating t or integrating equation (2.48) directly gives the phase plane equation

$$x_1 - x_{10} = (1/2\omega_0^2 U)(x_2^2 - x_{20}^2), \tag{2.51}$$

which is the equation of a parabola if $U \neq 0$. Figure 2.3 shows sketches of the trajectories which are again parallel to each other. When $U = 0$, $x_2 = x_{20}$ from equation (2.50) and the trajectories are straight lines parallel to the x_1 axis moving to the right for $x_2 > 0$ and to the left for $x_2 < 0$.

2.4 Sketching of phase portraits for linear second order systems

The phase plane equations for the type 1 and 2 systems are not too difficult mathematically and often tractable analytical solutions can be obtained to problems. Techniques for the rapid sketching of phase plane portraits are particu-

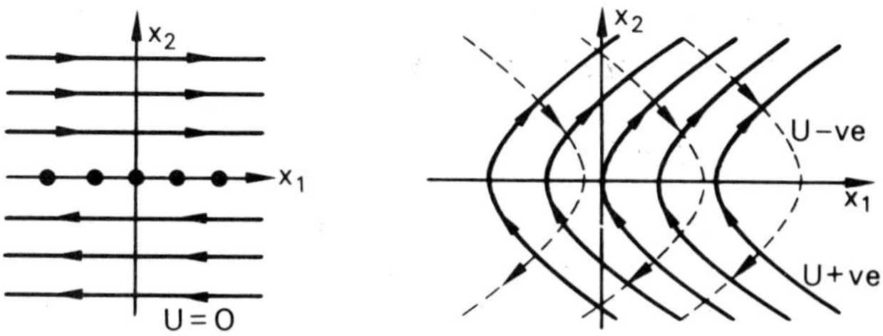

Figure 2.3 Phase trajectories for type 2 system

larly useful, however, expecially for the type 0 system and several procedures are outlined below. The isocline method given can be used in all cases.

2.4.1 *The method of isoclines*

Equation (2.24) gives

$$-2\zeta\omega_0 x_2 - \omega_0^2 x_1 = x_2 r \tag{2.24}$$

as the straight line which all trajectories cut with slope r. If values of r are selected the straight lines, called isoclines, given by the above equation can be drawn and the slope of a trajectory crossing an isocline marked with an arrow. When sufficient isoclines are drawn it is a relatively easy procedure to sketch a trajectory. Alternatively, the procedure shown in figure 2.4 can be used,[4] where the point P_1 at which a trajectory crosses the isocline OI_1 is found from the previous point P_0 on the trajectory by drawing from P_0 lines with slope corresponding to the slope of a trajectory through the isoclines OI_0 and OI_1. The midpoint of the

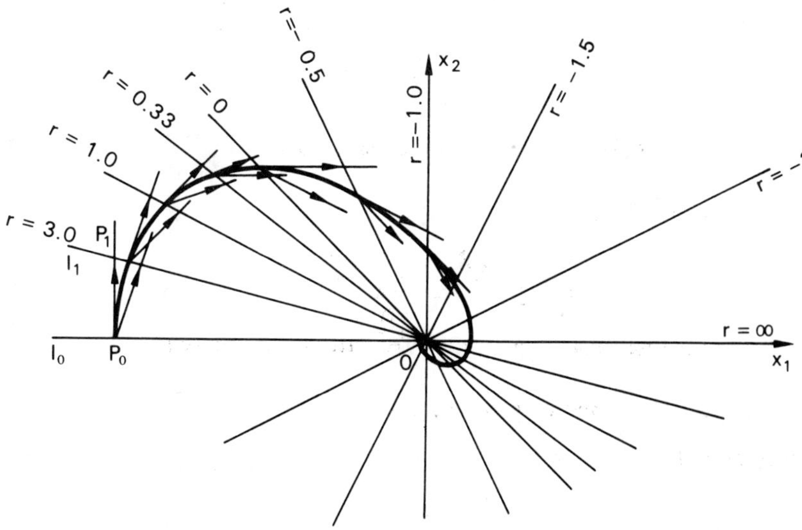

Figure 2.4 Sketch of a phase plane trajectory for $\ddot{x} + \dot{x} + x = 0$

intersection of the lines with OI_1 may then be taken as the next point P_1 on the trajectory. The procedure is then repeated.

2.4.2 *Liénard's construction*[5]

An alternative method of constructing the phase plane trajectories for the type 0 system is shown in figure 2.5. The vertical axis of the phase plane is first normalised with respect to ω_0 by taking $x_{2N} = x_2/\omega_0$ so that equation (2.23) becomes

$$dx_{2N}/dx_1 = -(2\zeta x_{2N} + x_1)/x_{2N}. \tag{2.52}$$

To determine the slope of a trajectory at any point $P(x_1', x_{2N}')$, figure 2.5, the following construction is performed.

(a) Draw the straight line $x_1 + 2\zeta x_{2N} = 0$. This is the isocline in the $x_1 - x_{2N}$ plane which trajectories cut with zero slope.
(b) A horizontal line is drawn from P to Q a point on $x_1 + 2\zeta x_{2N} = 0$.
(c) A vertical line is drawn from Q to meet the x_1 axis at R.
(d) A straight line is drawn from R to P.

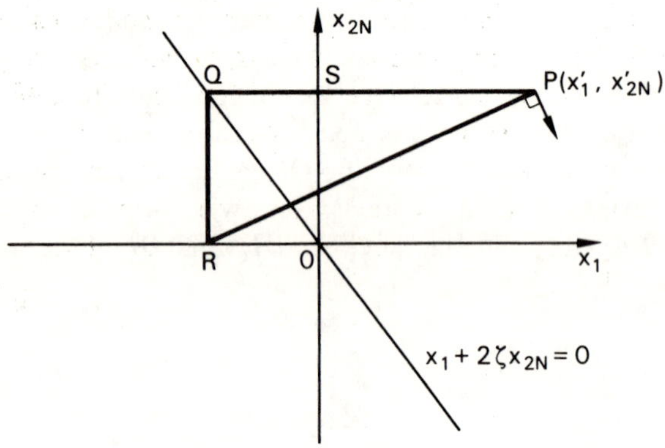

Figure 2.5 Illustration of Liénard's construction

Since the slope of RP is

$$\frac{QR}{PQ} = \frac{QR}{PS + SQ} = \frac{x_{2N}'}{x_1' + 2\zeta x_{2N}'},$$

it is normal to the line with slope given by equation (2.52) as required.

2.4.3 *Time scaling of a phase plane trajectory*

Since by definition $\dot{x}_1 = x_2$,

$$t_2 - t_1 = \int_{x_1(t_1)}^{x_1(t_2)} (1/x_2)\, dx_1.$$

Thus if a graph of $1/x_2$ is plotted against x_1, the time taken for the motion to go from $x_1(t_1)$ to $x_1(t_2)$ can be found by evaluating the area under the curve. This method is obviously not too accurate near the x_1 axis where $x_2 \to 0$. Alternatively a time scale can be put on a trajectory using the relatively simple construction[6] shown in figure 2.6. For a small time interval Δt corresponding to movement from P_0 $\{x_1(0), x_2(0)\}$ to P_1

$$\Delta t \approx \Delta x_1/\bar{x}_2,$$

where $\bar{x}_2$ the mean value of x_2 at P_0 and P_1 is given by

$$\bar{x}_2 = x_2(0) + \Delta x_2/2.$$

Thus eliminating $\bar{x}_2$ gives

$$2/\Delta t = \{2x_2(0) + \Delta x_2\}/\Delta x_1.$$

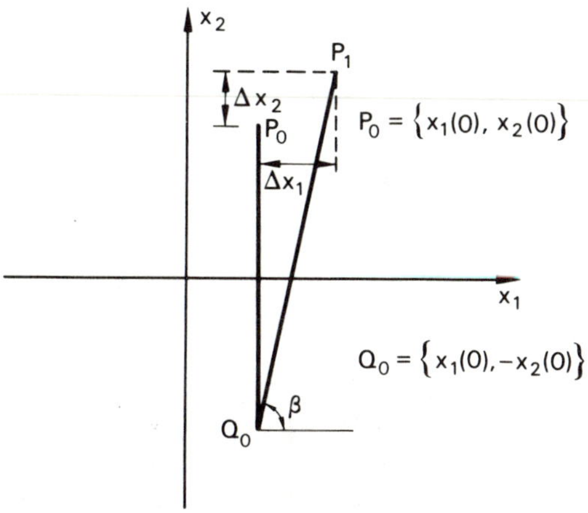

Figure 2.6 Time scaling of a trajectory

Referring to figure 2.6 it is seen that the slope of the line Q_0P_1 is

$$\tan\beta = \{2x_2(0) + \Delta x_2\}/\Delta x_1,$$

where $Q_0\{x_1(0), -x_2(0)\}$ is the mirror image of P_0. Hence by selecting a value of Δt corresponding to a value of β given by

$$\beta = \tan^{-1}(2/\Delta t)$$

of say 80° to 85° the trajectory can be time scaled by repeated use of the above construction.

2.5 The nonlinear second order differential equation

The general nonlinear form of the state equation (2.1) in two space is

$$\begin{aligned} \dot{x}_1 &= P(x_1, x_2), \\ \dot{x}_2 &= Q(x_1, x_2), \end{aligned} \tag{2.53}$$

so that singular points exist where

$$P(x_1, x_2) = Q(x_1, x_2) = 0,$$

which in general will have more than one solution. Lyapunov[7] has shown that the stability of and the behaviour of the trajectories in the neighbourhood of a singular point can be found from a linearised version of equations (2.53) about the singular point. Thus if a singular point of equations (2.53) is (x_{1s}, x_{2s}), expanding the equations in a Taylor series about the singular point the equations of the first approximation are

$$\begin{aligned} \dot{x}_1 &= \left.\frac{\partial P}{\partial x_1}\right|_{\substack{x_1 = x_{1s} \\ x_2 = x_{2s}}} (x_1 - x_{1s}) + \left.\frac{\partial P}{\partial x_2}\right|_{\substack{x_1 = x_{1s} \\ x_2 = x_{2s}}} (x_2 - x_{2s}), \\ \dot{x}_2 &= \left.\frac{\partial Q}{\partial x_1}\right|_{\substack{x_1 = x_{1s} \\ x_2 = x_{2s}}} (x_1 - x_{1s}) + \left.\frac{\partial Q}{\partial x_2}\right|_{\substack{x_1 = x_{1s} \\ x_2 = x_{2s}}} (x_2 - x_{2s}), \end{aligned} \tag{2.54}$$

which may be written

$$\begin{aligned} \dot{x}_1 &= ax_1 + bx_2 + c_1, \\ \dot{x}_2 &= cx_1 + dx_2 + c_2, \end{aligned}$$

where c_1 and c_2 are constants and a, b, c and d are the values of the partial derivatives at the singular point. The form of the singular point, which is unaffected by c_1 and c_2, can then be found using figure 2.1. There is however one exception[7] in that when the linearised equation indicates a centre, that is when $a + d = 0$ and $ad - bc$ is positive, the singular point of the nonlinear equation may be either a centre, or stable or unstable focus.

By choosing the phase variables as state variables the general nonlinear homogeneous second order differential equation

$$\ddot{x} + f(x, \dot{x}) = 0 \tag{2.55}$$

can with $x_1 = x$ be written

$$\begin{aligned} \dot{x}_1 &= x_2, \\ \dot{x}_2 &= -f(x_1, x_2). \end{aligned} \tag{2.56}$$

All the singular points thus lie on the x_1 axis, where

$$f(x_1, 0) = 0.$$

Linearising the equation about a singular point $(x_{1s}, 0)$ gives

$$\dot{x}_2/\dot{x}_1 = \{-a_0(x_1 - x_{1s}) - a_1 x_2\}/x_2 \tag{2.57}$$

where

$$a_0 = (\partial f/\partial x_1)\Big|_{\substack{x_1 = x_{1s} \\ x_2 = 0}} \quad \text{and} \quad a_1 = (\partial f/\partial x_2)\Big|_{\substack{x_1 = x_{1s} \\ x_2 = 0}}.$$

The form of the singular point can again be found directly from figure 2.1, which for this case has axes $-(a + d) = a_1$ and $(ad - bc) = a_0$. Away from the singular point the trajectories of a nonlinear system will obviously be quite different from those of the linearised system. Furthermore since a nonlinear system may have several singular points the phase portrait can be markedly dissimilar to that of a linear system with completely different response forms for different initial conditions. The phase trajectories corresponding to various initial conditions which give portraits in different regions of the phase plane will be divided by an asymptote called a separatrix.

2.5.1 *Limit cycles*

A limit cycle (8,9) is any closed trajectory in the phase plane which is approached asymptotically both from within and without by other trajectories either as $t \to \infty$ or $-\infty$. It is a periodic motion peculiar to nonlinear systems. Figure 2.7 illustrates the four possible forms of limit cycles, C. In case (a) the trajectories both from within and without approach C as $t \to \infty$, that is in the direction of the arrows which denote increasing time on the phase plane portrait. Any small perturbation of the motion from C in any manner will result in a return to C and the limit cycle is therefore stable. Correspondingly in case (b) since the trajectories are directed away from C both from within and without, or

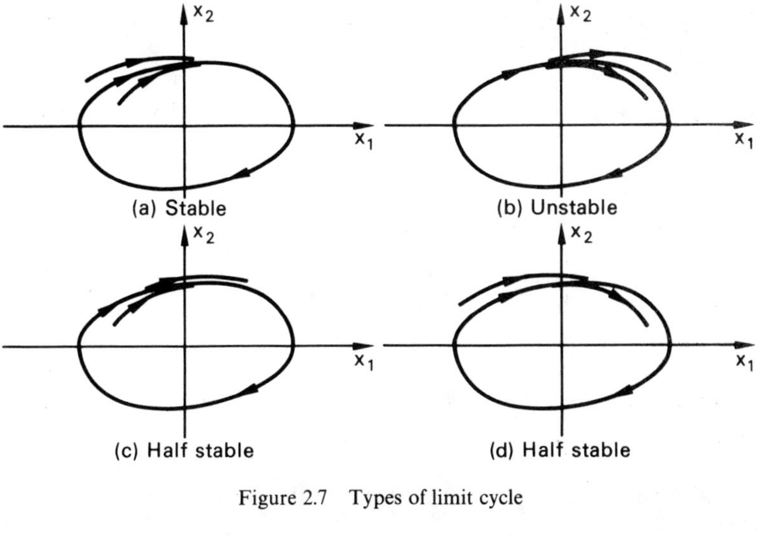

Figure 2.7 Types of limit cycle

alternatively C is only approached as $t \to -\infty$, the limit cycle is unstable. Cases (c) and (d) show the other possibilities where the limit cycle is approached from one direction as $t \to \infty$ and from the other as $t \to -\infty$. Here the limit cycles are designated as half stable.

Obviously it is important to know if a particular nonlinear system possesses a limit cycle. A precise answer to the question unfortunately does not exist; however, two theorems due to Poincaré[10,11] and the other to Bendixson[12] are of some assistance and easy to apply. Poincaré's theorem states that if the singularities of the nonlinear system are given the indices of $+1$ for a focus, node or centre and -1 for a saddle point then the singularities enclosed by a limit cycle must have an index sum of $+1$. The sign of the index will be seen to correspond with the sign of $(ad - bc)$ in figure 2.1. With reference to the system of equation (2.53) Bendixson's theorem states that no limit cycle can exist in a region of the phase plane for which $\partial P/\partial x_1 + \partial Q/\partial x_2$ has an invariant sign and is not identically zero.

2.6 Phase plane trajectories of second order nonlinear systems

Various methods are available for sketching phase plane trajectories of second order nonlinear systems. In general the approach used will depend upon the form of the function $f(x_1, x_2)$ of equation (2.56) although the method of isoclines and the delta method can be applied in all cases. The technique of Liénard has been extended by Pell[13] and Bickart[14] to cover a large number of possible functions $f(x_1, x_2)$.

2.6.1 *The method of isoclines*

This method for plotting phase plane trajectories, which was described in section 2.4.1 for the linear system, is also applicable to nonlinear systems. The isoclines, however, are no longer straight lines but loci given by the equation

$$rx_2 + f(x_1, x_2) = 0.$$

The plotting of these loci can be quite laborious, so that the method loses much of its earlier appeal.

2.6.2 *Liénard's construction*

This construction described earlier in section 2.4.2 can be applied to the restrictive class of nonlinear systems for which

$$f(x_1, x_2) = \omega_0^2\{x_1 + f_2(x_2)\},$$

where ω_0 is a constant. Again, replacing x_2 by $\omega_0 x_{2N}$ gives

$$\dot{x}_{2N}/\dot{x}_1 = -\{x_1 + f_2(\omega_0 x_{2N})\}/x_{2N}.$$

The procedure is exactly as before except that the isocline for zero slope which is drawn to start the construction is

$$x_1 + f_2(\omega_0 x_{2N}) = 0,$$

and not the straight line $x_1 + 2\zeta x_{2N} = 0$.

2.6.3 *Pell's method*

This method[13] can be used when the functions of x_1 and x_2 in the differential equation (2.1) are separable. Writing

$$f(x_1, x_2) = f_1(x_1) + f_2(x_2)$$

gives

$$\dot{x}_2/\dot{x}_1 = -\{f_1(x_1) + f_2(x_2)\}/x_2.$$

The construction to determine the slope of a trajectory through any point, $P(x_1', x_2')$ is shown in figure 2.8 and is as follows.

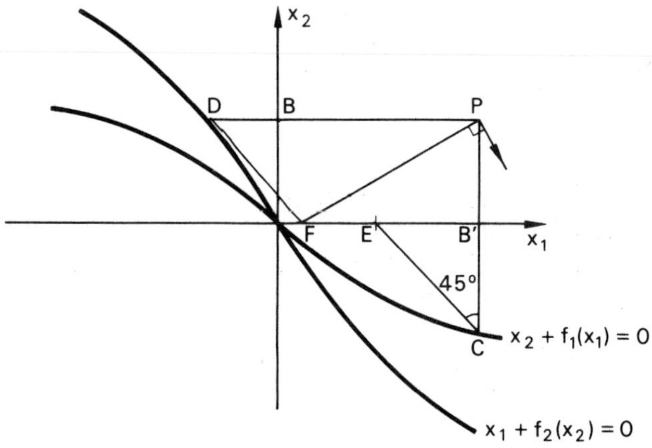

Figure 2.8 Construction for Pell's method

(a) The loci $x_2 + f_1(x_1) = 0$ and $x_1 + f_2(x_2) = 0$ are drawn in the phase plane.
(b) A vertical line is drawn through P meeting $x_2 + f_1(x_1) = 0$ at C and the x_1 axis at B'. The point E, where $B'E = B'C$, is marked on the x_1 axis.
(c) A horizontal line is drawn through P meeting $x_1 + f_2(x_2) = 0$ at D and the x_2 axis at B. The point F, where $FE = DB$, is marked on the x_1 axis.
(d) The direction of a trajectory through P is at right angles to FP since the slope of FP is

$$\frac{PB'}{FE + EB'} = \frac{x_2'}{f_2(x_2') + f_1(x_1')}.$$

2.6.4 *The delta method*

A trajectory is approximated in this method [15,16] by small arcs of a circle which are obtained by writing

$$f(x_1, x_2) = x_1 + \delta(x_1, x_2)$$

and then assuming $\delta(x_1, x_2)$ remains constant and equal to δ for small changes in x_1 and x_2. Thus

$$\dot{x}_2/\dot{x}_1 = -(x_1 + \delta)/x_2,$$

which after integration may be written in the form $(x_1 + \delta)^2 + x_2^2 = r^2$ a circle of radius r with its centre at the point $(-\delta, 0)$. To obtain the trajectory beyond P, the point (x_1', x_2') in figure 2.9, $\delta = \delta(x_1', x_2')$ is evaluated, the point $A(-\delta, 0)$ marked on the x_1 axis and a small arc with centre A drawn from P to P_1. The time required to travel from P to P_1 is

$$\int_{x_1'}^{x_1' + \Delta x_1} (1/x_2)\,\mathrm{d}x_1 = \sin^{-1}\left\{\frac{x_1' + \Delta x_1 + \delta}{r}\right\} - \sin^{-1}\left\{\frac{x_1' + \delta}{r}\right\},$$

so that a trajectory can also easily be time scaled.

When $f(x_1, x_2)$ is a ratio of terms in x_1, x_2 an extension of the above method, usually known as the double delta method, can be employed. Here one expresses the trajectory slope in the form

$$\frac{\dot{x}_2}{\dot{x}_1} = -\left\{\frac{x_1 + \delta_1(x_1, x_2)}{x_2 + \delta_2(x_1, x_2)}\right\} = -\left\{\frac{x_1 + \delta_1}{x_2 + \delta_2}\right\},$$

which on integration gives

$$(x_1 + \delta_1)^2 + (x_2 + \delta_2)^2 = r^2,$$

so that the centre of the small circular arc through P becomes the point $A(-\delta_1, -\delta_2)$.

2.6.5 *The Liénard plane*[5]

Consider the differential equation

$$\ddot{x} + f(x)\dot{x} + g(x) = 0. \tag{2.58}$$

Here

$$f(x_1, x_2) = f(x_1)x_2 + g(x_1)$$

and

$$\dot{x}_2/\dot{x}_1 = -\{x_2 f(x_1) + g(x_1)\}/x_2. \tag{2.59}$$

Introducing the new variable

$$y = x_2 + F(x_1), \tag{2.60}$$

where

$$F(x_1) = \int_0^{x_1} f(\alpha)\,\mathrm{d}\alpha,$$

so that

$$\dot{y}/\dot{x}_1 = (\dot{x}_2/\dot{x}_1) + f(x_1),$$

equation (2.59) becomes

$$\dot{y}/\dot{x}_1 = -g(x_1)/\{y - F(x_1)\}.$$

Trajectories can thus be sketched in the Liénard $x_1 - y$ plane using either the method of isoclines or Bickart's method and then transformed back using equation (2.60) to the original $x_1 - x_2$ phase plane. In particular, when $g(x_1) = x_1$, the simple construction illustrated in figure 2.10 can be used. The slope of a trajectory cutting the vertical line BC is obtained by drawing circular arcs from the centre A. The Liénard plane is particularly useful for the determination of limit cycles since a closed trajectory in the Liénard plane transforms to a closed

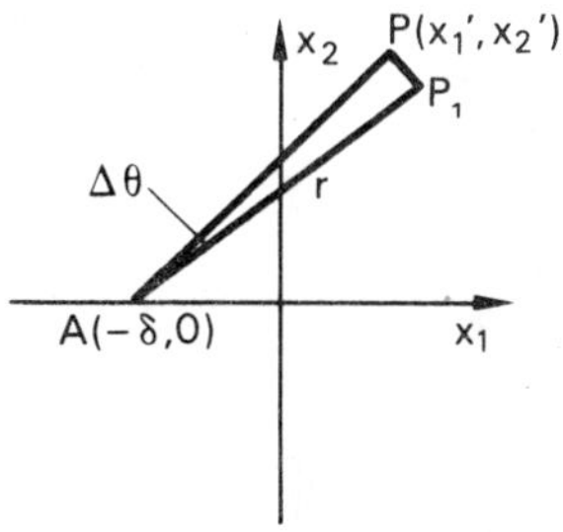

Figure 2.9 Construction of trajectories by the delta method

trajectory in the phase plane. Liénard showed that equation (2.58) will have a unique stable limit cycle if the following sufficient but not necessary conditions are met:

(i) $f(x)$ and $g(x)$ are analytic,
(ii) $g(x)$ has odd symmetry with $g(0) = 0$ and $xg(x) \geq 0$,
(iii) $f(x)$ has even symmetry and $f(0) < 0$, and
(iv) $F(x)$ $\{=\int_0^x f(x)\,dx\}$ has a unique root at $x = a$, $a > 0$ and is monotone increasing for $x > a$ tending to ∞ as $x \to \infty$.

2.7 Some particular nonlinear differential equations

To illustrate the techniques discussed and to provide some familiarity with phase plane portraits of nonlinear second order differential equations two examples are considered below.

2.7.1 *Nonlinear conservative system*

A conservative system is one in which the dissipation of energy is negligible and may be represented by the general differential equation

$$\ddot{x} + f(x) = 0,$$

that is

$$\dot{x}_1 = x_2$$

$$\dot{x}_2 = -f(x_1).$$

Hence, the slope of a phase plane trajectory is

$$dx_2/dx_1 = \dot{x}_2/\dot{x}_1 = -f(x_1)/x_2. \tag{2.61}$$

In particular, singular points occur on the x_1 axis where $f(x_1) = 0$, trajectories cross the x_2 axis with zero slope assuming $f(0) = 0$, and the x_1 axis orthogonally except at the possible singular points. The Taylor expansion of the numerator of equation (2.61) about the singular point at the origin gives $a_0 = \partial\{f(x)\}/\partial x)_{x_1=0}$ and $a_1 = 0$. The origin is thus either a centre ($a_0 > 0$) or saddle point ($a_0 < 0$).

Equation (2.61) can be integrated to give the phase plane constant energy equation

$$(x_2^2 - x_{20}^2)/2 + F(x_1) - F(x_{10}) = 0, \tag{2.62}$$

where $F(x_1) = \int_0^{x_1} f(x)\,dx$ and x_{10}, x_{20} are the initial values at $t = 0$ of x_1 and x_2. When equation (2.62) is rewritten in the form

$$x_2 = dx_1/dt = [x_{20}^2 + 2\{F(x_{10}) - F(x_1)\}]^{1/2}$$

further integration gives

$$t = \int_{x_{10}}^{x_1} \frac{dx_1}{[x_{20}^2 + 2\{F(x_{10}) - F(x_1)\}]^{1/2}}. \tag{2.63}$$

The period of oscillation, T, about the centre ($a_0 > 0$) at the origin if $f(x_1)$ has odd symmetry is equal to $4t$ where t can be obtained from equation (2.63) with $x_1 = x_{20} = 0$. t will be a function of x_{10} the amplitude of oscillation.

As a specific example the case $f(x_1) = x_1 - x_1^3/6$ is considered. Singularities exist where

$$x_1 - x_1^3/6 = 0, \qquad \text{i.e. } x_1 = 0 \text{ or } \pm\sqrt{6},$$

and the corresponding values of a_0 are 1, -2, and -2. Therefore the origin is a centre and saddle points exist at $(\pm\sqrt{6}, 0)$. The slope of trajectories in the phase plane is given by

$$\dot{x}_2/\dot{x}_1 = -\{x_1 - x_1^3/6\}/x_2, \tag{2.64}$$

which in particular is 0 when $x_1 = 0$, ∞ when $x_2 = 0$, and -1 along the curve $x_2 = x_1 - x_1^3/6$. This curve is shown dotted on the phase portrait of figure 2.11

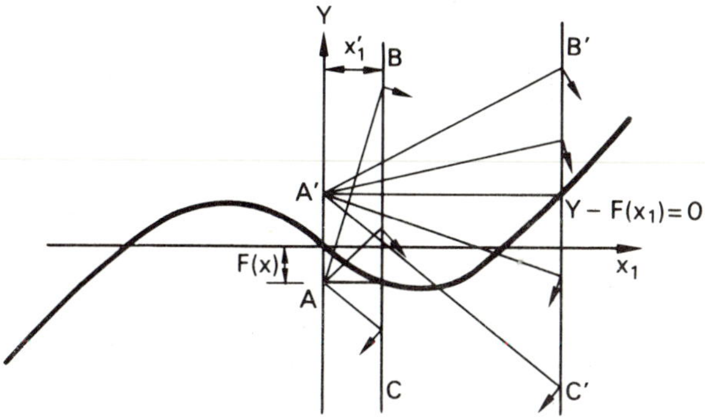

Figure 2.10 Construction of trajectories in the Liénard plane

which was sketched using Pell's method. Since

$$F(x_1) = \int_0^{x_1} (x - x^3/6)\,\mathrm{d}x = x_1^2/2 - x_1^4/24,$$

a closed trajectory crossing the x_1 axis at $(x_{10}, 0)$ will using equation (2.62) cut the x_2 axis at $(0, x_2)$ where

$$x_2^2/2 - x_{10}^2/2 + x_{10}^4/24 = 0,$$

that is

$$x_2 = x_{10}\{1 - x_{10}^2/12\}^{1/2}.$$

The slope of the trajectories along the separatrices which pass through the saddle points can be obtained by applying L'Hospital's rule to equation (2.64). Further differentiation of the right hand side gives

$$\dot{x}_2/\dot{x}_1 = \{-(1 - x_1^2/2)\dot{x}_1\}/\dot{x}_2,$$

that is

$$\dot{x}_2/\dot{x}_1 = \{x_1^2/2 - 1\}^{1/2}|_{x_1 = \pm\sqrt{6}} = \pm\sqrt{2}.$$

2.7.2 *System with nonlinear damping*

An example of a system with nonlinear damping is the well known Van der Pol equation[17]

$$\ddot{x} - \mu(1 - x^2)\dot{x} + x = 0,$$

where μ is a positive constant. This equation has the same form as equation (2.58) with $f(x) = -\mu(1 - x^2)$ and $g(x) = x$. Since $\int f(x)\,dx = -\mu(x - x^3/3)$ has a root at $x = \sqrt{3}$ the equation satisfies the sufficient conditions for possessing a limit cycle as given in section 2.6.5. Also applying Bendixson's criterion $\partial P/\partial x_1 + \partial Q/\partial x_2 = \mu(1 - x_1^2)$, which is positive for $|x_1| < 1$ and negative for $|x_1| > 1$ so that a limit cycle could exist. Now $\dot{x}_2/\dot{x}_1 = \{\mu(1 - x_1^2)x_2 - x_1\}/x_2$ so that the only singular point is at the origin where

$$a_0 = \frac{\partial}{\partial x_1}\{-\mu(1 - x_1^2)x_2 + x_1\}|_{x_1 = x_2 = 0} = 1$$

and

$$a_1 = \frac{\partial}{\partial x_2}\{-\mu(1 - x_1^2)x_2 + x_1\}|_{x_1 = x_2 = 0} = -\mu.$$

The form of the singular point thus depends upon μ, being an unstable focus for $\mu < 2$ and an unstable node for $\mu \geq 2$. The limit cycle of the equation obtained by the isocline method is shown in the phase plane, figure 2.12, for four different values of μ.

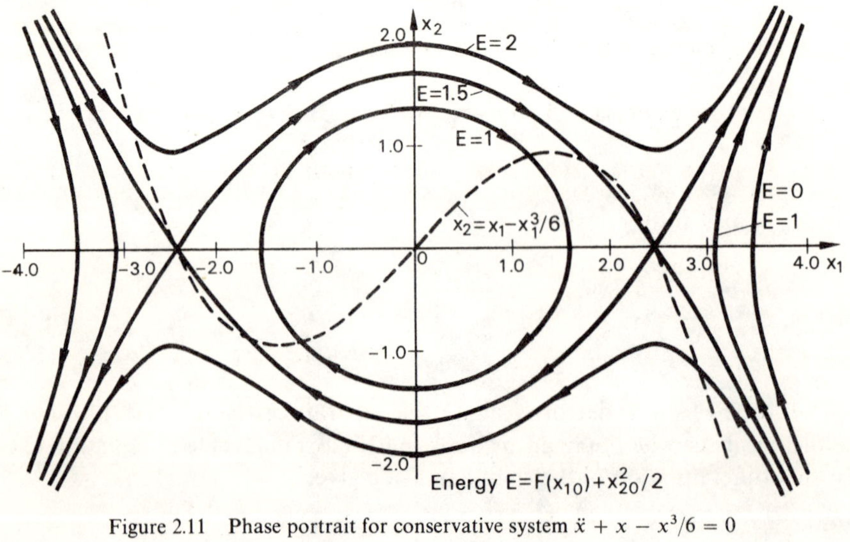

Figure 2.11 Phase portrait for conservative system $\ddot{x} + x - x^3/6 = 0$

2.8 The effect of nonlinearity on the step response of a position control system

A position control system can often be approximated by a second order differential equation with the result that its performance to step inputs can be studied using the phase plane. In addition, when nonlinear phenomena occurring in such a system can be represented by linear segmented characteristics the phase plane analysis may be considerably simplified. The procedure is to divide the phase plane up into various regions each of which corresponds to motion on a particular linear segment of the nonlinearity. The motion is then described by different linear differential equations in the various regions of the phase plane and solutions for particular initial values are obtained by matching the conditions at the region boundaries. Singular points for trajectories in a particular region may thus lie outside that region, in which case the singularity is known as a virtual singular point. The approach is used below to illustrate the effect of various nonlinear phenomena on the performance of a specific position control system.

2.8.1 *The basic system*

A schematic diagram of the basic system is shown in figure 2.13. The signal at the input to the power amplifier is

$$K(\theta_i - \theta_0) - k_T R\dot{\theta}_M/R_1 \text{ volts},$$

where K is the sensitivity in volts/rad. of the position transducers (assumed identical), k_T is the tachogenerator constant in volts/rad./sec. and R and R_1 are the resistances in the operational amplifier. The motor torque T_M, since $\theta_M = n\theta_0$ where n is the gear ratio, is given by

$$T_M = KK_T G\{(\theta_i - \theta_0) - nk\dot{\theta}_0/K\} \text{ newton metres,} \tag{2.65}$$

where G is the amplifier gain in mA/V, K_T is the motor torque constant in newton metres/mA, and $k = k_T R/R_1$. The equation of motion for the output shaft is

$$(J_L + n^2 J_M)\ddot{\theta}_0 = n(T_M - F_M\dot{\vartheta}_M) - F_L\dot{\theta}_0,$$

where J_L is the inertia of the load (output) shaft in newton metres sec.^{2}, J_M is the inertia of the motor shaft in newton metres sec.^{2}, F_L is the viscous friction coefficient in newton metres/rad./sec. for the load shaft, and F_M is the viscous friction coefficient in newton metres/rad./sec. for the motor shaft. This reduces to

$$J\ddot{\theta}_0 + F\dot{\theta}_0 = nT_M, \tag{2.66}$$

where

$$J = J_L + n^2 J_M \quad \text{and} \quad F = F_L + n^2 F_M.$$

Substituting for T_M from equation (2.65) gives

$$J\ddot{\theta}_0 + F\dot{\theta}_0 = nKK_T G(\theta_i - \theta_0) - n^2 kK_T G\dot{\theta}_0.$$

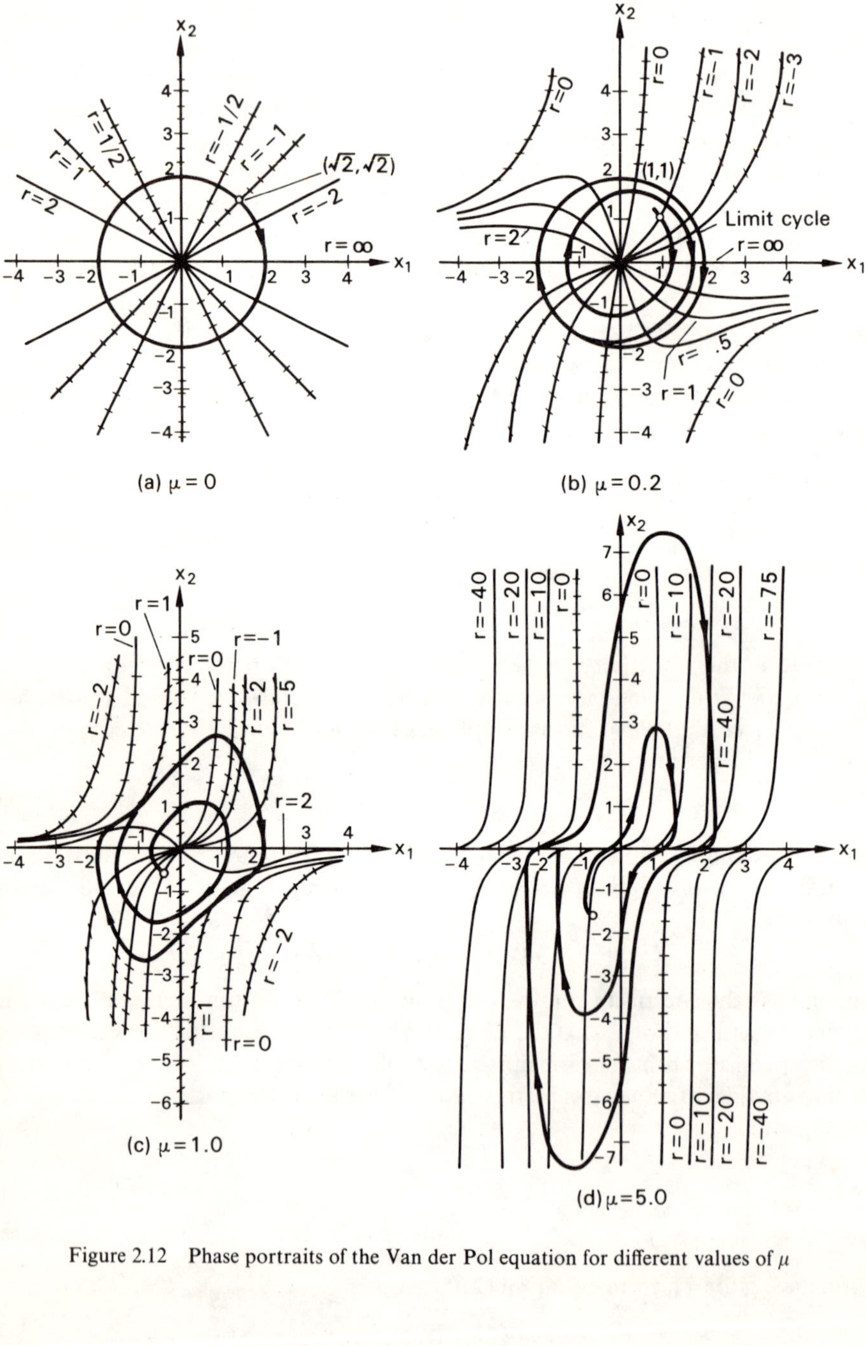

Figure 2.12 Phase portraits of the Van der Pol equation for different values of μ

Taking the Laplace transform and assuming all initial conditions zero gives

$$\left\{\frac{Js^2}{nKK_TG} + s\left(\frac{F + n^2kK_TG}{nKK_TG}\right) + 1\right\}\theta_0(s) = \theta_i(s). \tag{2.67}$$

Comparing the above with the common form for the second order differential equation given in equation (2.20) gives

$$\omega_0^2 = nKK_TG/J \quad \text{and} \quad 2\zeta/\omega_0 = (F + n^2kK_TG)/nKK_TG,$$

from which

$$\zeta = \left\{\frac{nKK_TG}{4J}\right\}^{1/2}\left(\frac{F}{nKK_TG} + \frac{nk}{K}\right) = \zeta_v + \zeta_T,$$

where $\zeta_v = F/2(nKK_TGJ)^{1/2}$ is the damping coefficient due to the viscous friction and $\zeta_T = (n^3k^2K_TG/4KJ)^{1/2}$ is the damping coefficient due to the velocity feedback from the tachogenerator. A block diagram of the system in terms of the parameters ω_0, ζ_v and ζ_T is shown in figure 2.14.

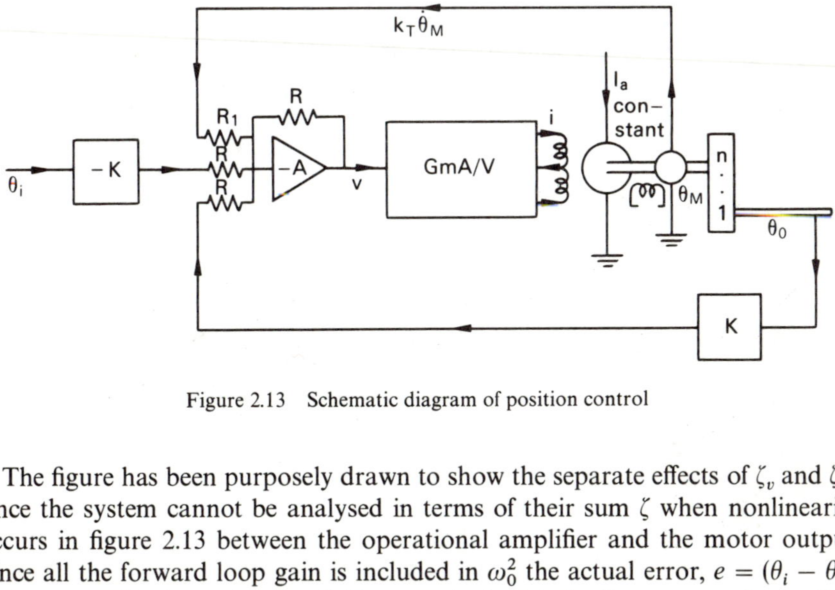

Figure 2.13 Schematic diagram of position control

The figure has been purposely drawn to show the separate effects of ζ_v and ζ_T, since the system cannot be analysed in terms of their sum ζ when nonlinearity occurs in figure 2.13 between the operational amplifier and the motor output. Since all the forward loop gain is included in ω_0^2 the actual error, $e = (\theta_i - \theta_0)$, and a normalised torque signal, T, occur in the block diagram, as shown. The actual torque, T_M, is a factor KK_TG greater than the normalised torque signal (see equation (2.65)) and the system error voltage a factor K greater than the error.

2.8.2 *The effect of Coulomb friction*

When Coulomb friction is present equation (2.66) becomes

$$J\ddot{\theta}_0 + F\dot{\theta}_0 = n\{T_M - T_c \operatorname{sgn} \dot{\theta}_0\},$$

where T_c is the motor torque required to overcome the Coulomb friction and

$$\operatorname{sgn}\dot{\theta}_0 = \begin{cases} +1 & \dot{\theta}_0 > 0 \\ -1 & \dot{\theta}_0 < 0 \end{cases}.$$

The resulting equation of motion is

$$(1/\omega_0^2)\ddot{\theta}_0 + (2\zeta/\omega_0)\dot{\theta}_0 + \theta_0 + (T_c/KK_TG)\operatorname{sgn}\dot{\theta}_0 = \theta_i, \tag{2.68}$$

which may be written

$$\ddot{\theta}_0 + 2\zeta\omega_0\dot{\theta}_0 + \omega_0^2\theta_0 + \omega_0^2 d\operatorname{sgn}\dot{\theta}_0 = \omega_0^2\theta_i,$$

where $d = T_c/KK_TG$ is the error in radians required for the motor to give a torque equal to T_c.

With $\theta_i = 0$, $x_1 = \theta_0$ and $x_2 = \dot{\theta}_0$ the equation can be written

$$\dot{x}_1 = x_2,$$

$$\dot{x}_2 = -\omega_0^2 x_1 - 2\zeta\omega_0 x_2 - d\omega_0^2\operatorname{sgn}x_2.$$

Motion in the phase plane is thus described by two linear differential equations one valid for x_2 positive and the other for x_2 negative. The line $x_2 = 0$ is therefore referred to as a switching line. In both cases the phase portraits are identical with those of equation (2.22) but about an origin shifted to the singular point 0_1 $(-d, 0)$ for $x_2 > 0$ and 0_2 $(d, 0)$ for $x_2 < 0$. Further, any point on the x_1 axis between these singular points is an equilibrium point at which trajectories terminate.

For the initial conditions $x_{10} = -u_0$ and $x_{20} = 0$ with $u_0 \gg d$ the phase plane trajectory will terminate at 0_1 when $\zeta \geq 1$ since the linear response has no overshoot. With $0 \leq \zeta < 1$ however the trajectory will terminate anywhere between 0_1 and 0_2, the exact point being dependent upon ζ and the magnitude of u_0. Consider a trajectory starting from the initial condition P_0 $(-u_0, 0)$, figure 2.15. Since x_2 is positive for the first portion of the response the motion is about the focus at 0_1 to the point P_1 $(u_1, 0)$ where

$$u_1 = (u_0 - d)\Delta - d$$

and

$$\Delta = \exp\{-\zeta\pi/(1 - \zeta^2)^{1/2}\}.$$

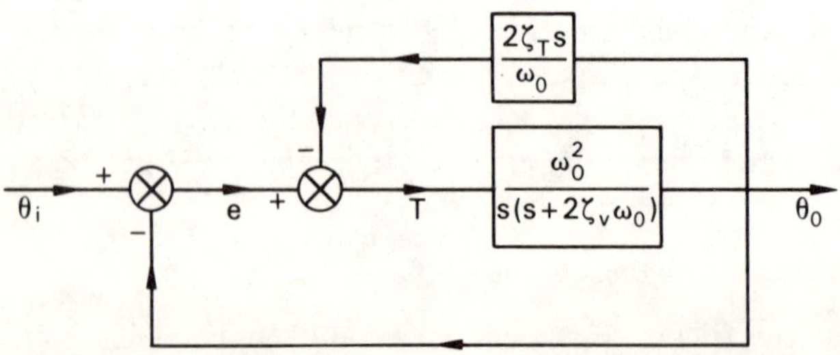

Figure 2.14 Block diagram for linear position control of figure 2.13

The next section of the response is about 0_2 as focus to $P_2\ (-u_2, 0)$ where

$$u_2 = (u_1 - d)\Delta - d.$$

This procedure is continued until the final equilibrium point $P_n\ (u_n, 0)$ is found with $|u_n| < d$.

The effect of stiction on the response is to extend the length of the equilibrium zone on the x_1 axis to $\pm d'$ from $\pm d$. Although Coulomb friction damps the response, which may easily be appreciated by evaluating a step response for $\zeta = 0$, its existence is highly undesirable due to its effect on the steady state accuracy of the system.

2.8.3 *The effect of error limiting*

When the error signal, e, passes through a nonlinear characteristic $f_e(e)$ before being added to the velocity feedback signal, the equation of motion becomes

$$\ddot{\theta}_0 + 2\zeta\omega_0\dot{\theta}_0 = \omega_0^2 f_e(\theta_i - \theta_0). \tag{2.69}$$

In particular when $f_e(e)$ is an ideal saturation characteristic given by

$$f_e(e) = \begin{cases} m_e e & |e| < \delta_e, \\ h_e & e > \delta_e, \\ -h_e & e < -\delta_e, \end{cases}$$

where $m_e = h_e/\delta_e$, the phase plane can be divided into three regions corresponding to operation on the three linear segments of the saturation characteristic as shown in figure 2.16. With $\theta_i = 0$, $x_1 = \theta_0$ and $x_2 = \dot{\theta}_0$ we have for

region I $\quad x_1 < -\delta_e \qquad \mathrm{d}x_2/\mathrm{d}x_1 = (-2\zeta\omega_0 x_2 + \omega_0^2 h_e)/x_2,$

region II $\quad |x_1| < \delta_e \qquad \mathrm{d}x_2/\mathrm{d}x_1 = (-2\zeta\omega_0 x_2 - m_e\omega_0^2 x_1)/x_2,$

region III $\quad x_1 > \delta_e \qquad \mathrm{d}x_2/\mathrm{d}x_1 = (-2\zeta\omega_0 x_2 - \omega_0^2 h_e)/x_2.$

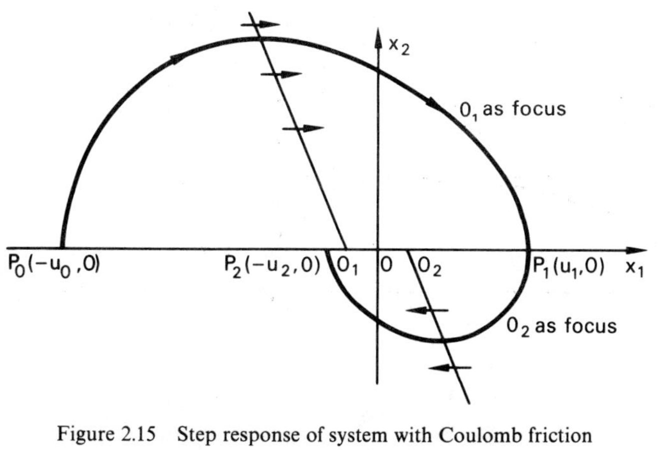

Figure 2.15 Step response of system with Coulomb friction

The equations in regions I and III are thus of the form of equation (2.43) for the type 1 system and in region II of the form of equation (2.23) for the type 0 system. In particular we note that for an initial condition $x_{10} = -u_0, x_{20} = 0$ with $u_0 \gg \delta_e$, which corresponds to a large step input, a maximum velocity of $x_2 = h_e\omega_0/2\zeta$ will be attained in region I. Also if $\zeta = 1$ and $m_e = 1$ no overshoot in the above response will occur since region II is entered with the initial conditions $(-h_e, h_e\omega_0/2)$, a point below the eigenvector. The technique of introducing error saturation to limit the system velocity is useful where damage may be caused by large speeds.

2.8.4 *The effect of torque saturation*[18,19]

When nonlinear operation takes place in the power amplifier or the motor the equation of motion is

$$\ddot{\theta}_0 + 2\zeta_v\omega_0\dot{\theta}_0 = \omega_0^2 f_T(\theta_i - \theta_0 - 2\zeta_T\dot{\theta}_0/\omega_0). \tag{2.70}$$

Again assuming $\theta_i = 0$ and the nonlinearity to be ideal saturation with limit levels $\pm h_T$ occurring at input levels of $\pm\delta_T$ the phase plane can be divided into three regions, in this case by the straight lines

$$-x_1 - 2\zeta_T x_2/\omega_0 = \pm\delta_T,$$

as shown in figure 2.17. Thus for

region I $\quad x_1 + 2\zeta_T x_2/\omega_0 < -\delta_T \qquad \mathrm{d}x_2/\mathrm{d}x_1 = (-2\zeta_v\omega_0 x_2 + \omega_0^2 h_T)/x_2,$

region II $\quad |x_1 + 2\zeta_T x_2/\omega_0| < \delta_T \qquad \mathrm{d}x_2/\mathrm{d}x_1 = \{-2(\zeta_v + m_T\zeta_T)\omega_0 x_2 - m_T\omega_0^2 x_1\}/x_2,$

region III $\quad x_1 + 2\zeta_T x_2/\omega_0 > \delta_T \qquad \mathrm{d}x_2/\mathrm{d}x_1 = (-2\zeta_v\omega_0 x_2 - \omega_0^2 h_T)/x_2,$

where

$$m_T = h_T/\delta_T.$$

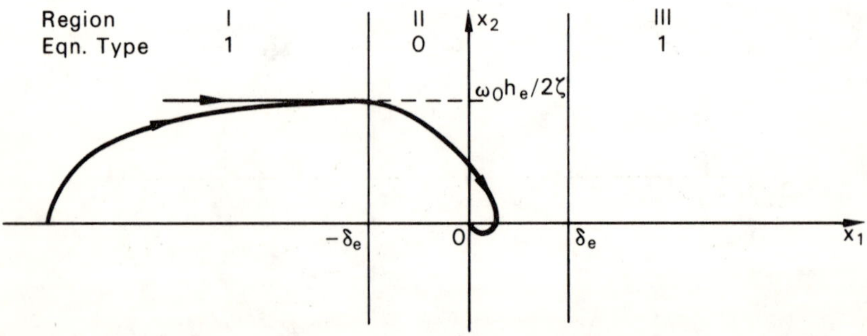

Figure 2.16 Step response of system with error limiting

The equations are again of type 1 form for regions I and III and of type 0 form for region II. In this case however we see that the switching lines are affected by δ_T and the damping in regions I and III depends upon ζ_v only. When δ_T is small approximate solutions can be obtained by neglecting region II and using the single boundary line

$$x_1 + 2\zeta_T x_2/\omega_0 = 0$$

to separate regions I and III. In cases where ζ_v may be neglected the equations describing the motion in regions I and III are of type 2 form and the trajectories become parabolae.

2.8.5 *The effect of error limiting, torque saturation and Coulomb friction*

The equation of motion for this case is seen to be

$$\ddot{\theta}_0 + 2\zeta_v\omega_0\dot{\theta}_0 + d\omega_0^2 \operatorname{sgn} \dot{\theta}_0 = \omega_0^2 f_T\{f_e(\theta_i - \theta_0) - 2\zeta_T\dot{\theta}_0/\omega_0\}. \tag{2.71}$$

Taking $\theta_i = 0$, f_T and f_e the ideal saturation characteristics described previously, the phase plane will be divided into various regions by the lines

$$x_2 = 0, \qquad \text{for Coulomb friction,}$$

$$x_1 = \pm\delta_e, \qquad \text{for error limiting,}$$

$$m_e x_1 - 2\zeta_T x_2/\omega_0 = \pm\delta_T, \qquad \text{for torque saturation,}$$

$$\pm h_e - 2\zeta_T x_2/\omega_0 = \pm\delta_T, \qquad \text{for error limiting and torque saturation.}$$

The various regions of the phase plane are shown in figure 2.18 assuming $d < \delta_T < \delta_e$. There are now twelve regions of different behaviour, six for $x_2 > 0$ and six for $x_2 < 0$, for which the equations of motion taking the top sign for $x_2 > 0$ and the bottom for $x_2 < 0$ are

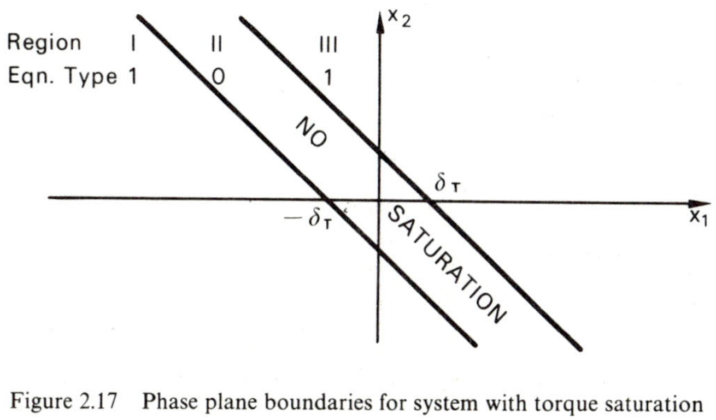

Figure 2.17 Phase plane boundaries for system with torque saturation

regions I and I′, error limiting and torque saturation,

$$dx_2/dx_1 = (-2\zeta_v\omega_0 x_2 \pm \omega_0^2 h_T \mp d\omega_0^2)/x_2,$$

regions II and II′, error limiting only,

$$dx_2/dx_1 = \{-2(\zeta_v + m_T\zeta_T)\omega_0 x_2 \mp m_T\omega_0^2 h_e \mp d\omega_0^2\}/x_2,$$

regions III and III′, torque saturation only,

$$dx_2/dx_1 = (-2\zeta_v\omega_0 x_2 \pm \omega_0^2 h_T \mp d\omega_0^2)/x_2,$$

regions IV and IV′, no error limiting or torque saturation,

$$dx_2/dx_1 = \{-2(\zeta_v + m_T\zeta_T)\omega_0 x_2 - m_e m_T\omega_0^2 x_1 \mp d\omega_0^2\}/x_2,$$

regions V and V′, torque saturation only,

$$dx_2/dx_1 = (-2\zeta_v\omega_0 x_2 \mp \omega_0^2 h_T \mp d\omega_0^2)/x_2,$$

regions VI and VI′, error limiting and torque saturation,

$$dx_2/dx_1 = (-2\zeta_v\omega_0 x_2 \mp \omega_0^2 h_T \mp d\omega_0^2)/x_2.$$

When the torque signal is saturated the equation of motion is independent of the magnitude of the error signal so that trajectories are described by the same equation at each side of the boundary $x_1 = \pm\delta_e$ where it divides region I from III and V from VI. The boundary lines between the various regions have therefore been shown chain dotted when they are not also switching lines.

2.8.6 *The effect of backlash*[20]

When backlash exists in the gears connecting the motor and load shafts of figure 2.13, the block diagram is as shown in figure 2.19. θ_0 is the actual ouput

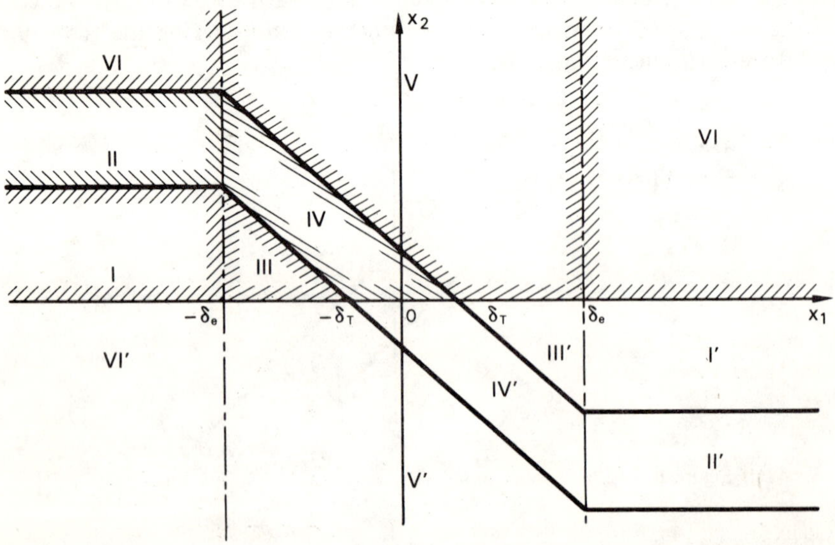

Figure 2.18 Phase plane boundaries for system with error limiting, torque saturation and Coulomb friction

position and $\theta'_0 = \theta_M/n$ the output which would exist with no backlash. The nonlinear relationship between θ_0 and θ'_0 assuming ideal friction controlled backlash is also shown in figure 2.19.

The equation of motion written in terms of θ'_0 is thus

$$\frac{\ddot{\theta}'_0}{\{f_1(\omega_0)\}^2} + 2\frac{f_2(\zeta)}{f_1(\omega_0)}\dot{\theta}'_0 = \theta_i - \theta_0 = \theta_i - f_b(\theta'_0), \tag{2.72}$$

where the nonlinear functions $f_1(\omega_0)$ and $f_2(\zeta)$ have two different values; one for the normal motion when both motor and output shafts are moving together and the other for motion of the motor shaft within the backlash region where the output shaft is at rest. Ideally, for friction controlled backlash the friction to inertia ratio for the output shaft should be large so that the change in $f_1(\omega_0)$ between the two regions is normally small.

Motion outside the backlash region

Taking $f_1(\omega_0) = \omega_0$, $f_2(\zeta) = \zeta_1$ in this region and with $\theta_i = 0$ equation (2.72) becomes

$$\ddot{\theta}' + 2\zeta_1\omega_0\dot{\theta}'_0 = -\omega_0^2\{\theta'_0 - (b/2)\,\mathrm{sgn}\,\dot{\theta}'_0\},$$

since from the backlash characteristic $\theta_0 = \theta'_0 - (b/2)\,\mathrm{sgn}\,\dot{\theta}'_0$ when the shafts are in contact. With $x'_1 = \theta'_0$ and $x'_2 = \dot{\theta}'_0$ the equation may be written

$$\dot{x}'_1 = x'_2,$$

$$\dot{x}'_2 = -2\zeta_1\omega_0 x'_2 - \omega_0^2\{x'_1 - (b/2)\,\mathrm{sgn}\,x'_2\},$$

giving

$$\frac{\dot{x}'_2}{\dot{x}'_1\omega_0} = \frac{-\{x'_1 - (b/2)\,\mathrm{sgn}\,x'_2\} - 2\zeta_1 x'_2/\omega_0}{x'_2/\omega_0}.$$

Substituting the normalised coordinates $X = x'_1/b$ and $Y = x'_2/b\omega_0$ gives

$$\frac{\dot{Y}}{\dot{X}} = \frac{-X + 0{\cdot}5\,\mathrm{sgn}\,Y - 2\zeta_1 Y}{Y} \tag{2.73}$$

as the equation for motion outside the backlash region in the $X - Y$ phase plane. The motion is thus of the form of equation (2.23) for the type 0 system, about the singular points $(0{\cdot}5, 0)$ when Y is positive and $(-0{\cdot}5, 0)$ when Y is negative rather than the origin for all Y.

Motion in the backlash region

In the ideal case the shafts lose contact when $\dot{\theta}'_0$ reverses and remain separated with the output shaft at rest until θ'_0 has moved through the backlash, a distance of b. If θ'_{0b} is the value of θ'_0 when $x'_2 = 0$ then during the motion of the motor shaft through the backlash the output position θ_0 remains at $\theta'_{0b} - (b/2)\,\mathrm{sgn}\,\dot{\theta}'_{0b-}$ where $\dot{\theta}'_{0b-}$ is the velocity of θ'_0 before reversing. Taking $f_1(\omega_0) = \mu\omega_0$, where

$\mu \geq 1$, and $f_2(\zeta) = \zeta_2$ in this region the equation of motion is

$$\ddot{\theta}'_0 + 2\zeta_2\mu\omega_0\dot{\theta}'_0 = -\mu^2\omega^2_0\{\theta'_{0b} - (b/2)\,\mathrm{sgn}\,\dot{\theta}'_{0b-}\}.$$

Using the same substitutions as above the equation becomes

$$\frac{\dot{Y}}{\dot{X}} = \frac{-\mu^2(X_0 - 0{\cdot}5\,\mathrm{sgn}\,Y_{0-}) - 2\zeta_2\mu Y}{Y} \tag{2.74}$$

in the $X - Y$ plane, where $X_0 = \theta'_{0b}/b$ and Y_{0-} is the normalised velocity of the output shaft just before coming to rest. The motion is thus of the form of equation (2.43) for a type 1 system.

Figure 2.20 shows two trajectories sketched for a system governed by the above equations with $\mu = 1$ and $\zeta_1 = \zeta_2 = 0{\cdot}125$. One trajectory spirals outwards and the other inwards indicating the existence of a limit cycle somewhere between.

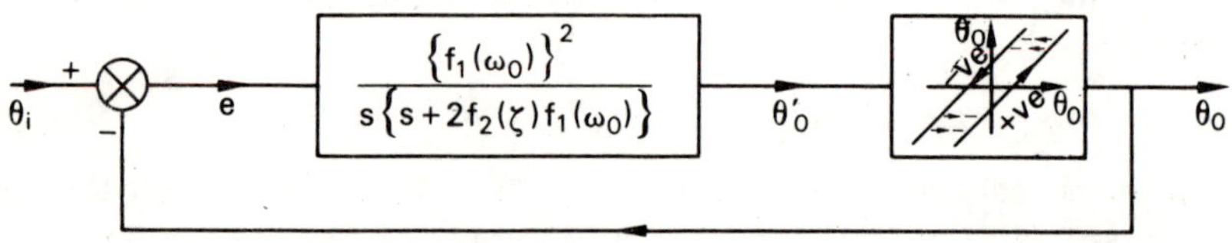

Figure 2.19 Block diagram of system with backlash

An exact analysis using equations (2.73) and (2.74) shows that the limit cycle cuts the X axis at $\pm 2{\cdot}29$. This corresponds to an oscillation amplitude of $2{\cdot}29b$ for θ'_0, which from the backlash characteristic gives an oscillation amplitude of $1{\cdot}79b$ for θ_0. The effect of the backlash is thus to make the system unstable in that it exhibits a stable limit cycle.

The analyses of systems with backlash using idealised representations such as the friction controlled characteristic as above or the inertia controlled characteristic are much simplified. For more detailed considerations the reader is referred to specialised work elsewhere.(21,22)

2.9 Some particular examples

To illustrate the application of phase plane techniques a few examples are now considered.

Example 1(23)

What is the maximum allowable magnitude of step input for which the position control system with torque saturation discussed above will have no overshoot if $m_T = 1$ and the damping, due to velocity feedback only, is critical?

The solution is obtained by proceeding backwards along the trajectory from the origin. Referring to figure 2.21, the final path is from $P_0(u_0, v_0)$ to 0 governed

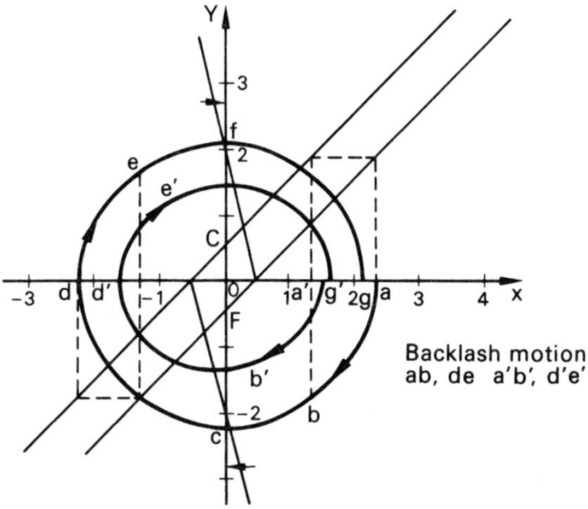

Figure 2.20 Phase plane for system of figure 2.19

by equations (2.35) and (2.36). That is

$$x_1 = [v_0 t + u_0(1 + \omega_0 t)]\mathrm{e}^{-\omega_0 t}, \tag{2.35}$$

$$x_2 = [v_0(1 - \omega_0 t) - u_0\omega_0^2 t]\mathrm{e}^{-\omega_0 t}, \tag{2.36}$$

where u_0 and v_0 are the initial values of x and $\dot{x}$. Putting $x_1 = x_2 = 0$ gives

$$t = \frac{-u_0}{v_0 + u_0\omega_0} = \frac{v_0}{\omega_0(v_0 + u_0\omega_0)}, \qquad \text{so that } v_0 = -u_0\omega_0.$$

Thus P_0 lies on the eigenvector of the linear motion, a fact which could be deduced from the phase plane topology.

The point P_0 lies on one of the boundaries

$$-x_1 - (2x_2/\omega_0) = \pm h_T,$$

which gives $-u_0 + 2u_0 = \pm h_T$; $u_0 = \pm h_T$.

Thus assuming a negative initial condition, P_0 is the point $(-h_T, h_T\omega_0)$. The path P_1P_0 for the type 2 system motion of equation (2.51) and the boundary condition for $P_1(u_1, v_1)$ give

$$\omega_0^2 h_T^2 - v_1^2 = -2\omega_0^2 h_T(-h_T - u_1),$$

and

$$-u - (2v_1/\omega_0) = -h_T,$$

from which $u_1 = -5h_T$ and $v_1 = 3h_T\omega_0$. The solution for $P_2(u_2, v_2)$ is obtained from the boundary condition and equations (2.35) and (2.36) for the linear motion, that is

$$-u_2 - (2v_2/\omega_0) = h_T, \tag{2.75}$$

$$-5h_T = [v_2 t + u_2(1 + \omega_0 t)]\mathrm{e}^{-\omega_0 t}, \tag{2.76}$$

and

$$3h_T\omega_0 = [v_2(1 - \omega_0 t) - u_2\omega_0^2 t]e^{-\omega_0 t}. \tag{2.77}$$

Multiplying equation (2.76) by ω_0 and adding equation (2.77) gives

$$-2h_T\omega_0 = (v_2 + u_2\omega_0)e^{-\omega_0 t},$$

i.e.

$$e^{\omega_0 t} = -(v_2 + u_2\omega_0)/2h_T\omega_0,$$

which on using equation (2.75) gives

$$e^{\omega_0 t} = (v_2 + h_T\omega_0)/2h_T\omega_0.$$

Substituting back for $\omega_0 t$ in equation (2.77) and using equation (2.75) gives

$$3h_T\omega_0 = \frac{2v_2h_T\omega_0}{v_2 + h_T\omega_0} + 2h_T\omega_0 \log_e\left(\frac{v_2 + h_T\omega_0}{2h_T\omega_0}\right),$$

$$1 + \frac{2h_T\omega_0}{v_2 + h_T\omega_0} = 2\log_e\left(\frac{v_2 + h_T\omega_0}{2h_T\omega_0}\right),$$

with the solution

$$(v_2 + h_T\omega_0)/2h_T\omega_0 \approx 2{\cdot}09,$$

that is $v_2 = 3{\cdot}18h_T\omega_0$ and $u_2 = -7{\cdot}36h_T$. The initial condition $P_3(u_3, 0)$ is finally given by

$$-(3{\cdot}18h_T\omega_0)^2 = 2\omega_0^2h_T(u_3 + 7{\cdot}36h_T),$$

i.e.

$$u_3 = -h_T\{(3{\cdot}18^2/2) + 7{\cdot}36\} = -12{\cdot}42h_T.$$

An approximate solution can be found extremely quickly if the width of the linear regime is neglected and the single switching boundary, shown dotted in figure 2.21, is used. Again proceeding backwards from the origin, the point $P_4(u_4, v_4)$ on the changeover boundary is given by

$$-u_4 - 2v_4/\omega_0 = 0,$$

and

$$v_4^2 = -2\omega_0^2h_Tu_4,$$

that is

$$v_4 = 4\omega_0h_T \quad \text{and} \quad u_4 = -8h_T.$$

Thus from the symmetry of the parabolae the initial input point P_5 is $(-16h_T, 0)$, a result which differs from the exact solution by approximately 25 per cent.

Example 2

A position control system has error limiting, torque saturation, Coulomb friction and negligible viscous friction. The system is critically damped by velocity feedback for linear operation. Determine the required relationship between h_T, h_e, δ_e and d if the system must have no overshoot in the step response. δ_T is small and may be assumed zero.

Since $\zeta_v = 0$ and $\zeta_T = 1$ the equation of motion is

$$\ddot{\theta}_0 + \omega_0^2 d \operatorname{sgn} \dot{\theta}_0 = \omega_0^2 f_T\{f_e(-\theta_0) - 2\dot{\theta}_0/\omega_0\},$$

with $\theta_i = 0$. As $\delta_T \approx 0$ there is a single switching line between the positive and negative levels of torque saturation given by

$$-m_e x_1 - 2x_2/\omega_0 = 0 \qquad \text{for } |x_1| < \delta_e,$$

and

$$\pm h_e - 2x_2/\omega_0 = 0 \qquad \text{for } |x_1| > \delta_e,$$

where

$$x_1 = \theta_0 \quad \text{and} \quad x_2 = \dot{\theta}_0.$$

The equation of motion is

$$\begin{aligned} \dot{x}_1 &= x_2, \\ \dot{x}_2 &= -\omega_0^2 d \operatorname{sgn} x_2 \pm \omega_0^2 h_T, \end{aligned} \tag{2.78}$$

where the positive sign is for motion to the left of the switching line and the negative sign for that to the right as shown in figure 2.22. Obviously for large

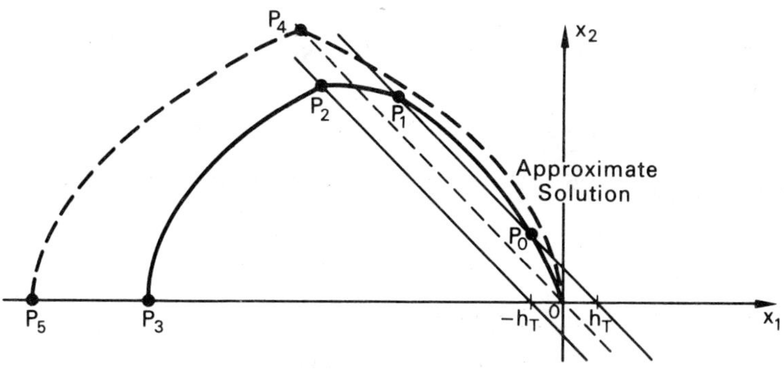

Figure 2.21 Phase plane for example 1, section 2.9

step inputs a constant velocity given by the equation of the switching line for $|x_1| > \delta_e$ will be obtained. The motion will be essentially along the switching line with rapid switching between positive and negative torque taking place. Thus for any large step input trajectories reach the line $x_1 = -\delta_e$ with velocity $x_2 = h_e\omega_0/2$.

The solution of equations (2.78) in the phase plane is

$$x_2^2 - x_{20}^2 = -2\omega_0^2(h_T + d)(x_1 - x_{10}), \qquad \text{(see equation 2.51)}$$

so that for no overshoot

$$(h_e\omega_0/2)^2 \leq 2\omega_0^2(h_T + d)\delta_e,$$

that is

$$h_e^2 \leq 8\delta_e(h_T + d).$$

Note that when $d = 0$ and $h_e = \delta_e$ the solution becomes $h_e < 8h_T$ a result which can be derived from the previous example.

Example 3

A position control system with torque saturation and backlash has $h_T = 1$, $\delta_T \approx 0$ and $b = 2$. The equation of motion is $\ddot{\theta}_0' + a\dot{\theta}_0' = Kf_T\{-\theta_0 - \dot{\theta}_0'/20\}$ where $\theta_0 = f_b(\theta_0')$ is the backlash characteristic. For motion outside the backlash region $K = 20$ and $a = 1$ and for motion within the backlash region $K = 24$ and $a = 0$. Determine the amplitude of the system limit cycle.

Let x_1 and x_2 denote the position and velocity of the output shaft, that is θ_0 and $\dot{\theta}_0$ respectively. In this case as a variation to the procedure of section 2.8.6 we will find the motion in the output shaft $x_1 - x_2$ phase plane.

Considering the behaviour from the instant the motor reverses from a negative to positive velocity the output shaft will remain at rest until the motor moves through $b = 2$.

The equation of motion in the backlash region is

$$\dot{x}_2'/\dot{x}_1' = 24/x_2',$$

where $x_1' = \theta_0'$ and $x_2 = \dot{\theta}_0'$. In moving through b from rest the velocity will reach x_2', where

$$(x_2')^2 = 2 \times 24 \times 2 = 96, \qquad x_2' = 9{\cdot}8.$$

The output shaft gains this velocity instantaneously and both shafts then move together with

$$x_1 = x_1' - \operatorname{sgn} x_2, \qquad x_2 = x_2',$$

and governed by the equation of motion

$$\dot{x}_2/\dot{x}_1 = -1 \pm 20/x_2, \tag{2.79}$$

since $K = 20$ and $a = 1$. Switching between the ± 20 takes place at the line

$$-x_1 - x_2/20 = 0, \qquad \text{since } \delta_T = 0,$$

as shown in figure 2.23.

From equation (2.46) the solution of equation (2.79) is

$$x_1 - x_{10} = -(x_2 - x_{20}) \pm 20 \log_e\{(x_{20} \mp 20)/(x_2 \mp 20)\}.$$

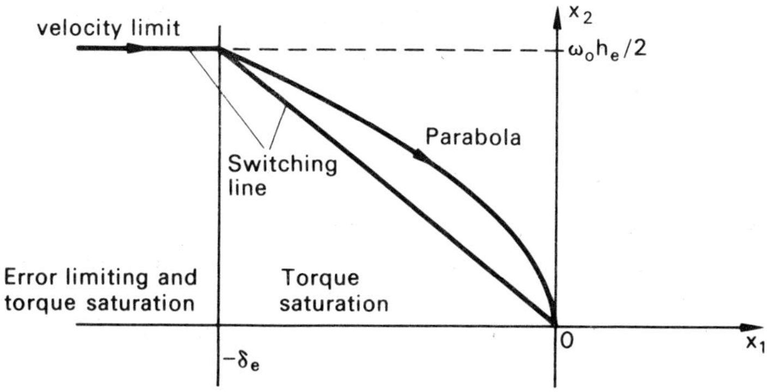

Figure 2.22 Phase plane for example 2, section 2.9

Assuming the points $B(-u_0, v_0)$, $C(-u_1, v_1)$ and $D(u_0, 0)$ lie on a limit cycle at the end of the backlash, at the switching line and at the start of the backlash respectively, as shown in figure 2.23, then the equation of motion from B to C gives

$$u_0 - u_1 = -v_1 + v_0 + 20\log_e\{(v_0 - 20)/(v_1 - 20)\} \tag{2.80}$$

and from C to D gives

$$u_0 + u_1 = v_1 - 20\log_e\{(v_1 + 20)/20\}. \tag{2.81}$$

Subtracting equation (2.80) from (2.81) gives

$$2u_1 = 2v_1 - v_0 + 20\log_e\{20(v_1 - 20)/(v_1 + 20)(v_0 - 20)\}. \tag{2.82}$$

Now C lies on the switching line, so that

$$u_1 - v_1/20 = 0,$$

and at B

$$v_0 = x_2' = 9{\cdot}8.$$

Substituting these values in equation (2.82) gives

$$v_1/10 = 2v_1 - 9{\cdot}8 + 20\log_e(20/10{\cdot}2) + 20\log_e\{(20 - v_1)/(20 + v_1)\},$$

which reduces to

$$20\log_e\{(20 + v_1)/(20 - v_1)\} = 1{\cdot}9v_1 + 3{\cdot}66,$$

with the solution $v_1 = 10{\cdot}8$. Using this value gives $u_1 = 0{\cdot}54$ and an amplitude of oscillation for x_1 of $u_0 = 1{\cdot}62$.

2.10 Plants with transfer function zeros

The transfer function corresponding to equation (2.20) is

$$\frac{X(s)}{U(s)} = \frac{\omega_0^2}{s^2 + 2\zeta\omega_0 s + \omega_0^2}, \tag{2.83}$$

which is seen to have no zeros. In many situations one is required to find the behaviour of second order systems having plants with a zero such that the transfer function has the form

$$\frac{X(s)}{U(s)} = \frac{\omega_0^2(b_0 + b_1 s)}{s^2 + 2\zeta\omega_0 s + \omega_0^2}. \tag{2.84}$$

In a physical system the frequency response gain falls off at high frequencies so that it is sufficient to consider a transfer function with one zero only.

If the differential equation corresponding to equation (2.84) is written as two first order equations in terms of the phase variables one obtains

$$\dot{x}_1 = x_2,$$
$$\dot{x}_2 = -\omega_0^2 x_1 - 2\zeta\omega_0 x_2 + b_0\omega_0^2 u + b_1\omega_0^2 \dot{u}.$$

The second equation contains a term involving the derivative of the input u, so that step changes in u will cause discontinuities in x_2. Such derivative terms do not occur in the normal form of the state equations (2.1) and the state variables cannot therefore change by a finite amount instantaneously unless the control input contains impulses. Since in general it will be advantageous to have the output as one state variable, choosing

$$\begin{aligned} x_1 &= x, \\ x_2 &= \dot{x} - b_1\omega_0^2 u, \end{aligned} \tag{2.85}$$

the state equations become

$$\begin{aligned} \dot{x}_1 &= x_2 + b_1\omega_0^2 u, \\ \dot{x}_2 &= -\omega_0^2 x_1 - 2\zeta\omega_0 x_2 - 2\zeta\omega_0^3 b_1 u + b_0\omega_0^2 u. \end{aligned} \tag{2.86}$$

The **A** matrix and the eigenvalues are seen to be the same as the case for no zero present in the transfer function. Taking the Laplace transform of the above equations for any input $u(t)$ and writing the result in matrix form gives

$$\mathbf{X}(s) = (s\mathbf{I} - \mathbf{A})^{-1}\mathbf{x}(0) + (s\mathbf{I} - \mathbf{A})^{-1}\mathbf{B}U(s), \tag{2.87}$$

where

$$\mathbf{X}(s) = \{X_1(s), X_2(s)\}^T, \qquad \mathbf{x}(0) = (x_{10}, x_{20})^T,$$

$$(s\mathbf{I} - \mathbf{A})^{-1} = \begin{pmatrix} \dfrac{s + 2\zeta\omega_0}{s^2 + 2\zeta s\omega_0 + \omega_0^2} & \dfrac{1}{s^2 + 2\zeta s\omega_0 + \omega_0^2} \\ \dfrac{-\omega_0^2}{s^2 + 2\zeta s\omega_0 + \omega_0^2} & \dfrac{s}{s^2 + 2\zeta s\omega_0 + \omega_0^2} \end{pmatrix},$$

and

$$\mathbf{B}U(s) = \begin{pmatrix} \omega_0^2 b_1 U(s) \\ (b_0 - 2\zeta\omega_0 b_1)\omega_0^2 U(s) \end{pmatrix}.$$

Solutions for $u(t)$ a step input are thus easily found and are summarised below for systems of type 1 and 2. Results for the type 0 system are not given since for the general case the expressions are rather unwieldy. When trajectories for the system are required in this state plane graphical rather than analytical techniques are often preferable.

Type 1 system

In this case equation (2.84) has the form

$$X(s)/U(s) = (b_0 + b_1 s)/s(s + \alpha), \tag{2.88}$$

for which

$$(s\mathbf{I} - \mathbf{A})^{-1} = \begin{pmatrix} 1/s & 1/s(s+\alpha) \\ 0 & 1/(s+\alpha) \end{pmatrix}$$

and

$$\mathbf{B}U(s) = \begin{pmatrix} b_1 U/s \\ (b_0 - b_1\alpha)U/s \end{pmatrix},$$

with the state variables $x_1 = x$, $x_2 = \dot{x} - b_1 u$ and $u(t)$ a step function of magnitude U.

The solution for the state variables is

$$x_1 = x_{10} + (x_{20}/\alpha)(1 - e^{-\alpha t}) + k_1 t + (k_2/\alpha^2)(\alpha t - 1 + e^{-\alpha t}), \tag{2.89}$$

$$x_2 = x_{20}e^{-\alpha t} + (k_2/\alpha)(1 - e^{-\alpha t}), \tag{2.90}$$

where

$$k_1 = b_1 U \quad \text{and} \quad k_2 = (b_0 - b_1\alpha)U.$$

The equations are of course identical with equations (2.44) and (2.45) respectively when $b_0 = \omega_0^2$ and $b_1 = 0$. The corresponding state plane equation is

$$x_1 - x_{10} = -\left(\frac{x_2 - x_{20}}{\alpha}\right) + \left(\frac{k_2 + k_1\alpha}{\alpha^2}\right)\log_e\left(\frac{\alpha x_{20} - k_2}{\alpha x_2 - k_2}\right). \tag{2.91}$$

Type 2 system

This corresponds to equation (2.88) with $\alpha = 0$. The solutions for x_1 and x_2 can thus be obtained by taking the limit as $\alpha \to 0$ of equations (2.89) and (2.90). This gives

$$x_1 = x_{10} + x_{20}t + k_1 t + k_2 t^2/2, \tag{2.92}$$

$$x_2 = x_{20} + k_2 t, \tag{2.93}$$

where

$$k_1 = b_1 U \quad \text{and} \quad k_2 = b_0 U.$$

Eliminating t gives the state plane equation

$$x_1 - x_{10} = (1/2k_2)(x_2^2 - x_{20}^2) + (k_1/k_2)(x_2 - x_{20}), \tag{2.94}$$

which is again a parabola although now it is not symmetrical about the x_1 axis but about the line $x_2 = -b_1 U$.

2.11 Relay systems

The relay is a unique nonlinear element in that it gives an output waveform which can have two or three distinct values only, independent of the form of the input signal which simply controls the time of switching between the levels. It is this property which enables periodic modes in relay systems of any order to be accurately determined by the methods presented in chapter 6. The phase plane analysis of second order relay systems is straightforward, since the characteristics are linear segmented and an extension to include the effects of a pure time delay[24,25] in the loop is also possible. By assuming $\delta_T = 0$ the torque saturation in two examples of section 2.9 has been approximated by an ideal relay.

Some useful results using graphical techniques similar to those used in the phase plane have been obtained for third order systems although the approach is usually tedious. Reference 26 is an excellent work for further reading on the time domain analysis and synthesis of relay systems. Several examples of second order relay systems are considered below based on the general block diagram of figure 2.24.

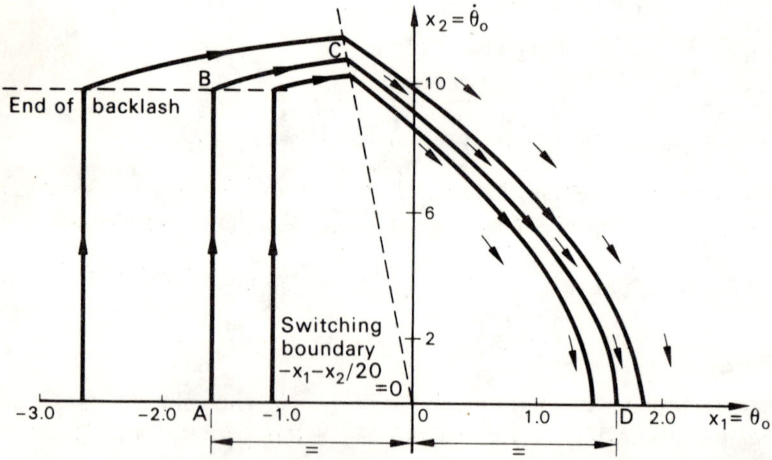

Figure 2.23 Upper half of phase plane for example 3, section 2.9

Example 1

Determine the amplitude and frequency of the limit cycle for the system of figure 2.24 if $G(s) = K/s^2$ and $H(s) = 1 + s\lambda$.

The relay switches when its input e_i has

(a) a positive rate of change and is equal to $-(\delta - \Delta)$ or $(\delta + \Delta)$,

(b) a negative rate of change and is equal to $-(\delta + \Delta)$ or $(\delta - \Delta)$. Choosing the phase variables $x_1 = x$ and $x_2 = \dot{x}$ gives

$$e_i = -x_1 - \lambda x_2.$$

Since $\dot{e}_i = -x_2 - \lambda \dot{x}_2$, and $\dot{x}_2$ is K times the relay output which is either $\pm h$ or 0, the maxima and minima of e_i for a zero relay output (i.e. for a given region in the phase plane) occurs at $x_2 = 0$. Thus the switching boundaries for zero relay output are

$$-x_1 - \lambda x_2 = \delta - \Delta, \qquad x_2 > 0,$$

$$-x_1 - \lambda x_2 = -(\delta - \Delta), \qquad x_2 < 0,$$

$$-x_1 - \lambda x_2 = \delta + \Delta, \qquad x_2 < 0,$$

$$-x_1 - \lambda x_2 = -(\delta + \Delta), \qquad x_2 > 0.$$

The response of the system, sketched for an initial value of x_1 in figure 2.25, shows that a limit cycle is reached when the trajectory cuts the switching boundaries at the points S_1, S_2, S_3, and S_4 the four corners of a rectangle.

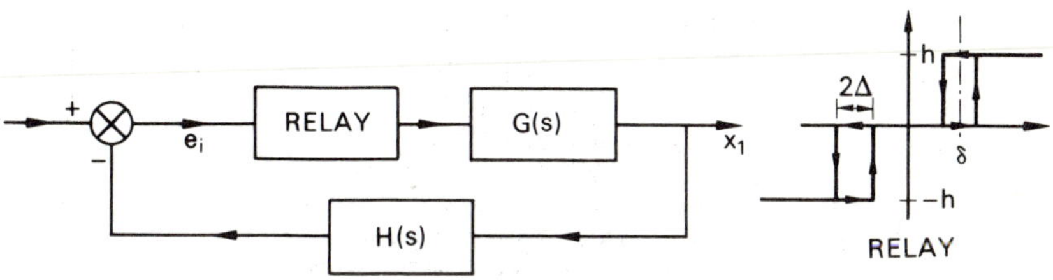

Figure 2.24 Block diagram for relay system

Obviously no oscillation would exist if $G(s)$ had a damping term. If S_1 is the point $(-u_1, v_1)$ then by symmetry

$$u_1 = \{(\delta + \Delta) + (\delta - \Delta)\}/2 = \delta,$$

and from the equation of the boundary through S_1,

$$\delta - \lambda v_1 = \delta - \Delta, \qquad v_1 = \Delta/\lambda.$$

The parabola passing through S_1 meets the x_1 axis at $P(-u_0, 0)$ where

$$\Delta^2/\lambda^2 = 2hK(u_0 - \delta), \qquad u_0 = \delta + \Delta^2/2hK\lambda^2,$$

which is the amplitude of the limit cycle. The time, t_1, taken to travel from P to S_1 is given by

$$t_1 = v_1/Kh = \Delta/Kh\lambda$$

and from S_1 to S_2, t_2, is given by

$$t_2 = S_1S_2/v_1 = 2\delta\lambda/\Delta.$$

The period of the limit cycle is

$$4t_1 + 2t_2 = 4\Delta/Kh\lambda + 4\delta\lambda/\Delta,$$

giving a frequency of $Kh\Delta\lambda/4(\Delta^2 + Kh\delta\lambda^2)$ cycles per second.

Example 2

What effect upon the limit cycle of the previous example has the introduction of a time delay of T_d seconds in the constant component of the feedback transfer function $H(s)$, that is $H(s) = e^{-sT_d} + s\lambda$.

In this case the input signal e_i to the relay is given by

$$e_i = -x_1' - \lambda x_2,$$

where

$$x_1'(t) = x_1(t - T_d).$$

Referring to figure 2.26 and considering a phase plane trajectory starting from P, it is seen that when P reaches P_1, x_1' will be at P_1', the position of x_1, T_d seconds

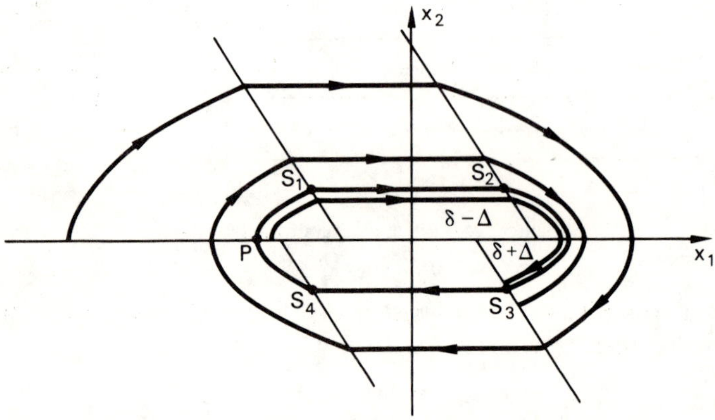

Figure 2.25 Phase plane for example 1, section 2.11

previously. Thus when switching takes place at the point $S_1(-u_1, v_1)$ on the trajectory of x_1', x_1 is at the point $T_1(-u_2, v_1)$ where

$$S_1T_1 = v_1T_d - hKT_d^2/2. \tag{2.95}$$

Hence for given values of Kh and T_d a switching boundary for x_1 can be plotted as shown in the figure if further points S_1' are taken on the x_1' switching boundary. Similarly when x_1' is at $S_2(u_3, v_1)$ the corresponding position of x_1 is $T_2(u_4, v_1)$ where since the previous motion is at constant velocity

$$S_2T_2 = v_1T_d. \tag{2.96}$$

Obviously from symmetry the system has a limit cycle when $0T_1 = 0T_2$, that is $u_2 = u_4$. Now

$$u_1 - \lambda v_1 = \delta - \Delta, \tag{2.97}$$

$$-u_3 - \lambda v_1 = -(\delta + \Delta), \tag{2.98}$$

and using equations (2.95) and (2.96) gives

$$u_2 = u_1 - v_1 T_d + hKT_d^2/2,$$

$$u_4 = u_3 + v_1 T_d,$$

so that $u_1 - u_3 = 2v_1 T_d - hKT_d^2/2$. Substituting for u_1 and u_3 from equations (2.97) and (2.98) gives

$$-2\Delta + 2\lambda v_1 = 2v_1 T_d - hKT_d^2/2,$$

that is

$$v_1 = (hKT_d^2 - 4\Delta)/4(T_d - \lambda),$$

and

$$u_2 = \delta - \Delta + hKT_d^2/2 - (hKT_d^2 - 4\Delta)/4 = \delta + hKT_d^2/4.$$

The limit cycle amplitude u_0 is given by

$$v_1^2 = 2hK(u_0 - u_2),$$

which on substituting for v_1 and u_2 gives

$$u_0 = \frac{1}{2hK}\left\{\frac{hKT_d^2 - 4\Delta}{4(T_d - \lambda)}\right\}^2 + \delta + hKT_d^2/4,$$

and the period of oscillation is

$$4\left\{\frac{v_1}{hK} + \frac{u_2}{v_1}\right\} = 4\left\{\frac{hKT_d^2 - 4\Delta}{4hK(T_d - \lambda)} + \frac{(T_d - \lambda)(KhT_d^2 + 4\delta)}{(hKT_d^2 - 4\Delta)}\right\}.$$

4

Example 3

Determine the response of the system of figure 2.24 to a large step input if $G(s) = 16/s(s + 2)$, $H(s) = 1 + s/8$, $\Delta = 0$, $\delta = 1/2$ and $h = 1$.

Choosing the phase variables $x_1 = x$ and $x_2 = \dot{x}$, a phase plane trajectory from equation (2.46) is given by

$$x_1 - x_{10} = -\left(\frac{x_2 - x_{20}}{2}\right) + 4U \log_e\left(\frac{2x_{20} - 16U}{2x_2 - 16U}\right),$$

where $U = \pm 1$ or 0 and the switching boundaries are

$$-x_1 - x_2/8 = \pm 1/2.$$

For a large step input x_2 reaches a maximum value of $16/2 = 8$ and the first switching occurs at $P_1(-u_1, 8)$ where

$$u_1 = 8/8 + 1/2 = 3/2.$$

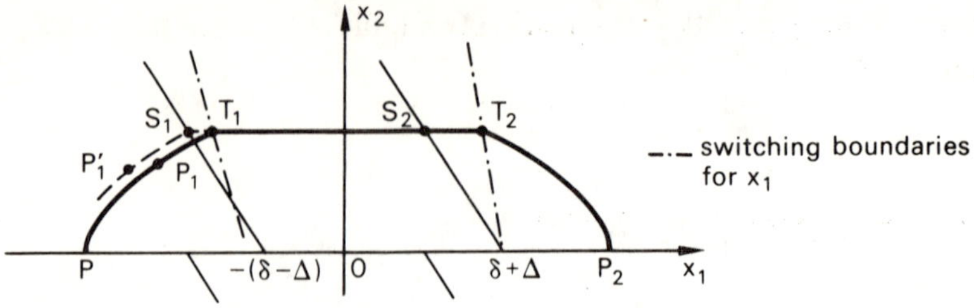

Figure 2.26 Phase plane for example 2, section 2.11

Between P_1 and $P_2(u_2, v_2)$, figure 2.27, the trajectory is a straight line of slope -2, so that

$$u_2 = 1/2 - v_2/8,$$

and

$$-2(u_2 + 3/2) = (v_2 - 8),$$

giving

$$v_2 = 16/3 \quad \text{and} \quad u_2 = -1/6.$$

Thus for the trajectory from P_2 to $P_3(u_3, v_3)$

$$u_3 + 1/6 = -v_3/2 + 8/3 - 4\log_e\{40/3(x_2 + 8)\},$$

and on the switching boundary

$$u_3 + v_3/8 = 1/2.$$

Solving the equations gives $u_3 = 0{\cdot}45$ and $v_3 = 0{\cdot}4$. On either side of the switching boundary below P_3 motion is towards the boundary, so that the resulting path from P_3 will be approximately along the switching boundary accomplished by rapid switching of the relay between -1 and 0. Equilibrium is reached at $P_4(1/2, 0)$.

Example 4

Determine the limit cycle in the system of figure 2.24 if $G(s) = (b_0 + b_1 s)/s^2$, $H(s) = 1$ and $\delta = 0$.

In this case the plant has a transfer function containing a zero and to show the motion of the output x we choose the state variables $x_1 = x$, $x_2 = \dot{x} - b_1 U$, where $U = \pm h$.

The state plane equation (2.94) is

$$x_1 - x_{10} = \pm(1/2b_0 h)(x_2^2 - x_{20}^2) + (b_1/b_0)(x_2 - x_{20}),$$

and the switching boundary $x_1 = \Delta$ for $x_2 > 0$ and $x_1 = -\Delta$ for $x_2 < 0$.

Referring to figure 2.28(a) for a limit cycle the trajectory starting at $(-\Delta, -v_1)$ must pass through (Δ, v_1) so that

$$\Delta - (-\Delta) = \{v_1^2 - (-v_1)^2\}/2b_0 h + b_1\{v_1 - (-v_1)\}/b_0,$$

that is

$$v_1 = b_0\Delta/b_1.$$

Also from the state plane equations (2.86)

$$dx_2/dx_1 = b_0U/(x_2 + b_1U),$$

so that $dx_2/dx_1 = \infty$ when $x_2 = -b_1U$, and the amplitude x_{1M} of the limit cycle is given by

$$\begin{aligned} x_{1M} &= \Delta - (b_1^2h^2 - b_0^2\Delta^2/b_1^2)/2b_0h + b_1(b_1h - b_0\Delta/b_1)/b_0, \\ &= \{2b_0b_1^2h\Delta - b_1^4h^2 + b_0^2\Delta^2 + 2b_1^4h^2 - 2b_1^2b_0h\Delta\}/2b_0b_1^2h, \\ &= (b_1^4h^2 + b_0^2\Delta^2)/2b_0b_1^2h. \end{aligned}$$

Using equation (2.93) the period of the limit cycle is

$$2\left(\frac{v_1 - (-v_1)}{b_0h}\right) = \left(\frac{4}{b_0h}\right)\left(\frac{b_0\Delta}{b_1}\right) = 4\Delta/b_1h.$$

If it is not required to show the output motion on a state plane the period for the limit cycle can also be obtained by considering a system with $G(s) = 1/s^2$, $H(s) = b_0 + b_1s$ and choosing the phase variables $x_1' = x'$ and $x_2' = \dot{x}'$ where x' is the output of $G(s)$. The situation is shown in figure 2.28(b). The period of

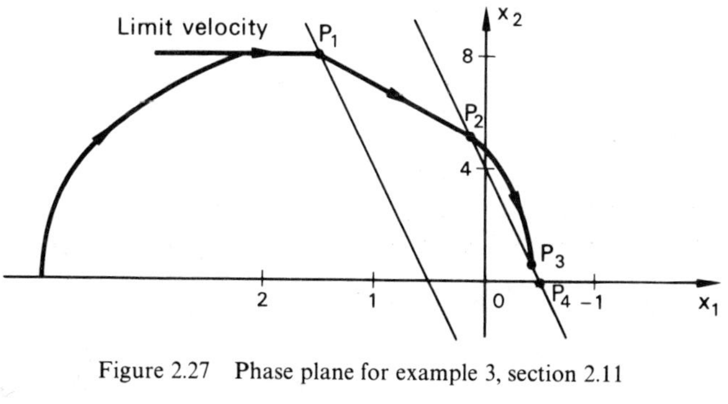

Figure 2.27 Phase plane for example 3, section 2.11

the limit cycle is $4(\Delta/b_1)/h = 4\Delta/b_1h$ as before and since $x_1 = b_0x' + b_1\dot{x}'$, x_1 has its maximum value of x_{1M} when $b_0x' + b_1\dot{x}_2' = 0$. Thus $x_1 = x_{1M}$, when $x_2' = b_1h/b_0$ and the corresponding value of x_1' from the parabola is given by

$$\begin{aligned} (b_1h/b_0)^2 &= -2h(x_1' - \Delta^2/2b_1^2h), \\ x_1' &= -(b_1^2h/2b_0^2) + (\Delta^2/2hb_1^2). \end{aligned}$$

Hence

$$\begin{aligned} x_{1M} &= b_0x_1' + b_1x_2', \\ &= -(b_1^2h/2b_0) + (b_0\Delta^2/2hb_1^2) + b_1^2h/b_0 = (b_1^4h^2 + b_0^2\Delta^2)/2b_0b_1^2h, \end{aligned}$$

as before.

2.12 An optimum second order system

Often a desirable characteristic for a position control system is that it should respond in a minimum time and without overshoot to a step input. To possess this optimum behaviour it is obvious that maximum acceleration is required from the motor over the initial portion of the response and then a maximum deceleration to bring the system to rest at the equilibrium point. The problem remains however of where and how to implement the change from maximum acceleration to maximum deceleration. Phase plane analysis shows that there is only one possible trajectory with maximum deceleration which passes through the origin so that the response should be such that for all input step magnitudes switching takes place to give subsequent motion along that trajectory.

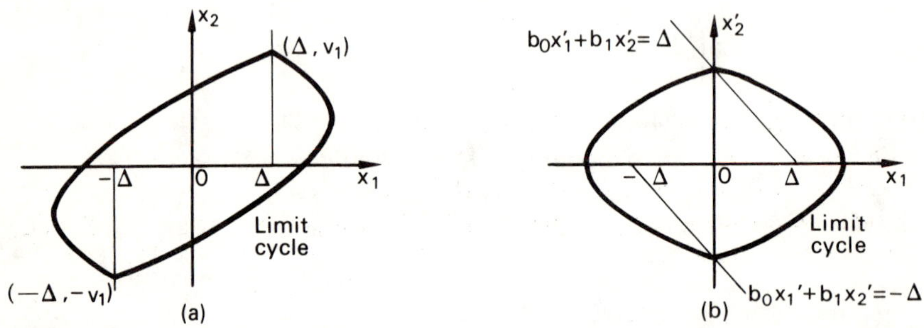

Figure 2.28 State planes for example 4, section 2.11

As a particular example the position control system of figure 2.29 with negligible friction is considered in more detail. The deceleration trajectory which passes through the origin, figure 2.30 may be described by the equation

$$x_2 + (2a|x_1|)^{1/2} \operatorname{sgn} x_1 = 0$$

or, alternatively, by

$$x_2|x_2| + 2ax_1 = 0.$$

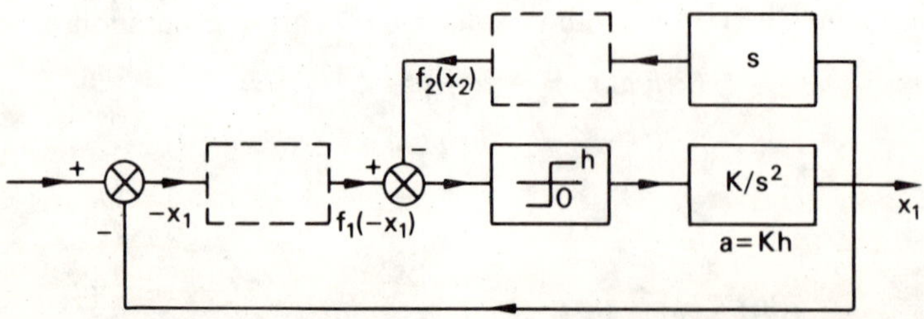

Figure 2.29 Block diagram of optimum second order system

As the ideal relay switches when its input e_i is zero, the deceleration parabola will be the switching boundary if

$$e_i = -x_2 - (2a|x_1|)^{1/2} \operatorname{sgn} x_1 \tag{2.99}$$

or

$$e_i = -x_2|x_2| - 2ax_1. \tag{2.100}$$

The value of e_i given by equation (2.99) can be obtained if the nonlinearity $f_1(x_1)$ is inserted in the error channel and the velocity feedback loop remains a linear unit gain as shown dotted in figure 2.29, where

$$f_1(x_1) = (2a|x_1|)^{1/2} \operatorname{sgn} x_1$$

is a soft spring characteristic which is the mirror image in the x_2 axis of the switching boundary. In this form the system is usually referred to as the SERME system,[27] the mnemonic SERME 'sign error root modulus error' defining the characteristic $f_1(x_1)$.

The alternative form for e_i, equation (2.100), can be obtained if x_2 is passed through the characteristic, $f_2(x_2)$ where

$$f_2(x_2) = (1/2a)x_2|x_2|$$

is a hard spring characteristic and the error channel remains a unit linear gain. Again this scheme is usually known by the mnemonic $V|V|$ (V = velocity).

Either method can, of course, be used for synthesis of the optimum system, the former approach is usually preferred since inevitable saturation of the characteristic $f_1(x_1)$ for large inputs results in the switching boundary shown dotted, and a step response slower than the optimum but with no overshoot. On the other hand saturation of $f_2(x_2)$ in the $V|V|$ system results in a step response which may have considerable overshoot.

The system has a minimum time response from any initial state (x_{10}, x_{20}) and as seen from the phase plane only one switching is required. This example is a particular case of the more general situation of an nth order plant subject to relay control where it has been shown[28] that provided the plant has real eigenvalues the time optimal transition requires only $(n - 1)$ switchings.

In a practical system to ensure no oscillation in the steady state the relay must have a small dead zone or linear range and the slope of $f_1(x_1)$ must be finite as x_1 tends to zero; or if the $V|V|$ system is used $f_2(x_2)$ must have a finite non zero slope as x_2 tends to zero. Since $f_1(x_1)$ and $f_2(x_2)$ are normally approximated by linear segmented characteristics this small amplitude behaviour is easily accomplished.

The characteristics $f_1(x_1)$ and $f_2(x_2)$ are seen to depend upon the value of a, which may not be known accurately in some cases. The effect of an incorrect estimate a' for a is shown in figure 2.31. For $a' > a$ the step response will possess an overshoot and for $a' < a$ the step response will have no overshoot but will be slower than the optimum with the path to the origin being along the switching boundary. This path is followed with the relay switching rapidly from one state

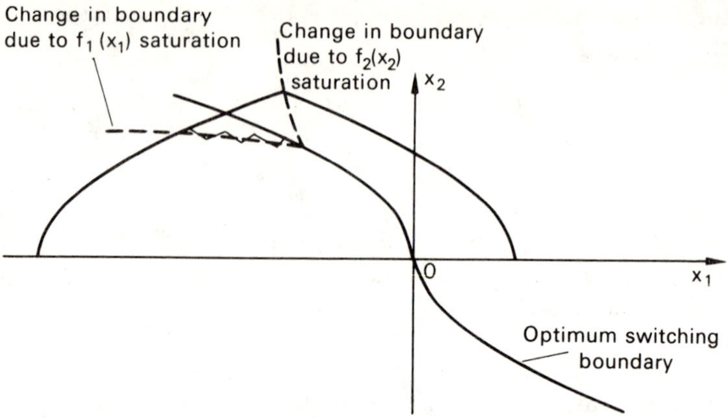

Figure 2.30 Switching boundary for optimum system

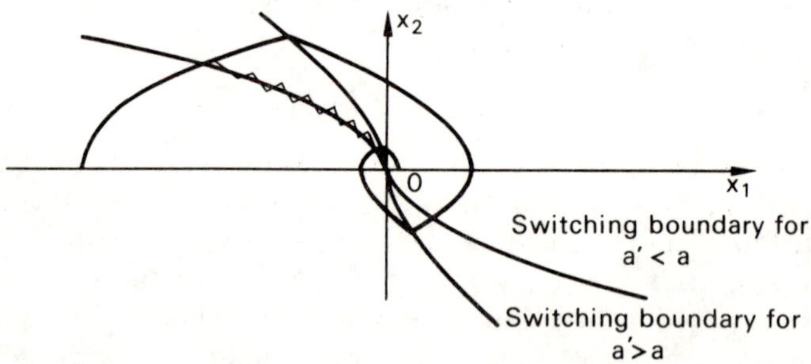

Figure 2.31 Phase portraits for incorrect values of *a*

to the other a phenomenon known as chatter. If the imperfections in the relay are vanishingly small the chatter amplitude tends to zero and the motion is often called sliding.[26] The time taken in this ideal case is calculated assuming the motion to be along the switching boundary. When chatter takes place and there is no average motion in the state plane it is referred to as noncontrolling chatter, but when an average motion exists, as for example in figure 2.31, it is called controlling chatter.

REFERENCES

1. Cooper, G. R. and McGillem, C. D.: *Methods of signal and system analysis* (Holt, Rinehart and Winston, New York, 1967, Chapter 8).
2. Seifert, W. W. and Steeg, C. W.: *Control systems engineering* (McGraw-Hill, New York, 1960, Chapter 13).
3. Takahashi, Y., Rabins, M. J. and Auslander, D. M.: *Control and dynamic systems* (Addison-Wesley, Reading, Mass., 1970, Chapter 3).
4. Andronov, A. A. and Chaikin, C. E.: *Theory of oscillations* (Princeton University Press, New Jersey, 1949).

5. Liénard, A.: 'Etude des oscillations entretenues', *Revue Générale de l'Electricité*, 23, 1928, pp. 901–946.
6. Cunningham, W. J.: *Introduction to nonlinear analysis* (McGraw-Hill, New York, 1958, pp. 36–39).
7. Coddington, E. A. and Levinson, N.: *Theory of ordinary differential equations* (McGraw-Hill, New York, 1955, pp. 375–387).
8. Minorsky, N.: *Nonlinear oscillations* (Van Nostrand, New York, 1962).
9. Struble, R. A.: *Nonlinear differential equations* (McGraw-Hill, New York, 1962, Chapter 6).
10. Poincaré, H.: *Les méthodes nouvelles de la mécanique céleste*, Gauthier-Villars, vols. I, II, Paris, 1892, 1893 (Reprinted by Dover Press, New York, 1957.)
11. Gibson, J. E.: *Nonlinear automatic control* (McGraw-Hill, New York, 1963, pp. 275–278).
12. Bendixson, I.: 'Sur les courbes definies par les équations differentielles', *Acta Mathematica*, 24, 1901, pp. 1–88.
13. Pell, W. H.: 'Graphical solution of single degree of freedom vibration problem with arbitrary damping and restoring forces', *Trans. A.S.M.E.*, 79, 1957, pp. 311–312.
14. Bickart, T. A.: 'An extension of Pell's method of phase trajectory construction', *I.E.E.E. Trans.*, AC-10, 1965, pp. 218–219.
15. Jacobsen, L. S.: 'On a general method of solving second order differential equations by phase plane displacements', *Trans. A.S.M.E.*, 74, 1952, pp. 543–553.
16. Buland, R. N.: 'Analysis of nonlinear servos by phase plane delta method', *Journal Franklin Institute*, 257, 1954, pp. 37–48.
17. Van der Pol, B.: 'Relaxation oscillations', *Phil. Mag.*, 2, 1926, pp. 978–992.
18. West, J. C., Douce, J. L. and Naylor, R.: 'The effects of some nonlinear elements on the transient performance of a simple r.p.c. system possessing torque limitation', *Proc. I.E.E.*, 101, 1954, pp. 156–165.
19. West, J. C. and Dalton, I. R.: 'The step function response of an r.p.c. servomechanism possessing torque limitation', *ibid.*, pp. 166–173.
20. Hammond, P. H.: *Feedback theory and its applications* (E.U.P., London, 1958, Chapter 10).
21. Freeman, E. A.: 'The effect of speed dependent friction and backlash on the stability of automatic control systems', *Trans. A.I.E.E.*, 77(II), 1958, pp. 680–691.
22. Freeman, E. A.: 'The stabilization of control systems with backlash using a high frequency on off loop', *Proc. I.E.E.*, 107 C, 1960, pp. 150–157.
23. West, J. C.: *Textbook of servomechanisms* (English Universities Press Ltd., London, 1953, pp. 212–217).
24. Thaler, G. J. and Pastel, M. P.: *Analysis and design of nonlinear and feedback control systems* (McGraw-Hill, New York, 1962, Chapter 6).
25. El-Geldawi, F.: 'On the phase plane analysis of relay control system with dead time', *Int. J. Control*, 11, 1970, pp. 123–131.
26. Flügge-Lotz, I.: *Discontinuous and optimal control* (McGraw-Hill, New York, 1968).
27. West, J. C.: *Analytical techniques for nonlinear control systems* (E.U.P., London, 1960, Chapter 6).
28. McCausland, I.: *Introduction to optimal control* (Wiley, New York, 1969, Chapter 6).

BIBLIOGRAPHY

Andronov, A. A., Vitt, A. A. and Khaikin, S.E .: *Theory of oscillations* (Pergamon, Oxford, 1966).

Bauer, W. L.: 'Dauerschwingungsformen in einem regelkreis mit mehrpunktregler berechnung mit hilfe einer grenzzyklenmatrix', *Regelungstechnik und Prozeß-Datenverarbeitung*, 20, 1972, pp. 428–435.

Blaquiere, A.: *Nonlinear systems analysis* (Academic Press, New York, 1966).

Chua, L. O.: *Introduction to nonlinear network theory* (McGraw-Hill, New York, 1969).

Coddington, E. A. and Levinson, N.: *Theory of ordinary differential equations* (McGraw-Hill, New York, 1955).

Cosgriff, R. L.: *Nonlinear control systems* (McGraw-Hill, New York, 1958).

Cunningham, W. J.: *Introduction to nonlinear analysis* (McGraw-Hill, New York, 1958).

Gibson, J. E.: *Nonlinear automatic control* (McGraw-Hill, New York, 1963, Chapter 7).

Graham, D. and McRuer, D.: *Analysis of nonlinear control systems* (Wiley, New York, 1961).

Hayashi, C.: *Nonlinear oscillations in physical systems* (McGraw-Hill, New York, 1964).

Higgins, T. J.: 'A resumé of the development and literature of nonlinear control system theory', *Trans. A.S.M.E.*, 1957, 79, pp. 445–449.
Kalman, R. E.: 'Phase plane analysis of automatic control systems with nonlinear gain elements', *Trans. A.I.E.E.*, 1954, 73(II), pp. 383–390.
Kalman, R. E.: 'Analysis and design principles of second and higher order saturating servomechanisms', *Trans. A.I.E.E.*, 1955, 74(II), pp. 294–308.
Minorsky, N.: *Nonlinear oscillations* (Van Nostrand, Princeton, 1962).
Naslin, P.: *The dynamics of linear and nonlinear systems* (Blackie, London, 1965, Chapter 9).
Stoker, J. J.: *Nonlinear vibrations* (Interscience Publishers, Inc., New York, 1950).
Thaler, G. J. and Pastel, M. P.: *Analysis and design of nonlinear feedback control systems* (McGraw-Hill, New York, 1962).
Van der Pol, B.: 'Nonlinear theory of electric oscillations', *Proc. I.R.E.*, 1934, 22, pp. 1051–1086.
West, J. C.: *Analytical techniques for nonlinear control systems* (E.U.P., London, 1960).

PROBLEMS

P.2.1 Use Liénard's construction to compare the response to step inputs of magnitude 1, 2 and 4 to systems described by the differential equations:
(a) $\ddot{x} + \dot{x} + x = u(t)$
(b) $\ddot{x} + \dot{x}|\dot{x}| + x = u(t)$.

P.2.2 Sketch phase plane trajectories for the differential equation

$$\ddot{x} + \dot{x} - (1/4)\dot{x}|\dot{x}| + x = 0$$

and estimate the amplitude of any possible limit cycles.

P.2.3 Sketch phase plane trajectories for the conservative system described by

$$\ddot{x} + \sin x = 0.$$

P.2.4 Evaluate the period of an oscillation about the centre at the origin for the conservative system

$$\ddot{x} + f(x) = 0$$

if
(a) $f(x) = x$
(b) $f(x) = x - x^3/6$
(c) $f(x) = h \operatorname{sgn} x$
(d) $f(x) = \sin x$.

P.2.5 For the differential equation

$$\ddot{x} - (0{\cdot}5 - 3\dot{x}^2)\dot{x} + x + x^2 = 0,$$

(a) determine the singular points
(b) sketch trajectories near to the singular points
(c) discuss the possibility of a limit cycle.

P.2.6 A second order system is described by the differential equation

$$(1/\omega^2)\ddot{\theta}_0 + (2\zeta/\omega_0)\dot{\theta}_0 + \theta_0 = 0.$$

Determine for $\zeta = 1$ the response for the initial conditions $\dot{\theta}_0 = v_0$ and $\theta_0 = -u_0$. What is the maximum allowable value of v_0 if
(a) the response must not overshoot the origin
(b) the overshoot must not be greater than u_0.

P.2.7 When $\zeta > 1$ the above differential equation has possible trajectory paths along isoclines in the phase plane. If $P_1(-u_1, v_1)$ is a point on one of these isoclines and $P_2(-u_1, v_2)$ a point on the other, calculate in terms of ζ the ratio of the time to travel from P_1 to the origin to the time to travel from P_2 to the origin if $v_2 > v_1$.

P.2.8 For the above differential equation with $\zeta < 1$ and initial conditions $\dot{\theta}_0 = 0$ and $\theta_0 = -u_0$, determine the times taken to reach the various maxima and minima of $\dot{\theta}_0$ in the response.

P.2.9 Show how (a) Coulomb friction and (b) viscous friction stabilise a simple control system (i.e. no other damping exists). Discuss the disadvantages in the use of these methods of stabilisation. Such a position control system has viscous friction equivalent to a quarter critical damping and Coulomb friction equivalent to the torque produced by a 4° error. Determine the magnitude of the first overshoot and the final steady state error if a 64° input step is applied to the system.

P.2.10 A d.c. position control system employs a pair of synchros as the error detecting means (i.e. the output voltage is proportional to K sine (error)). The system has a natural frequency of 6 hz and is stabilised by velocity feedback such that it is critically damped in the linear regime. What is the maximum rate of constant velocity input the system can follow?

P.2.11 A simple servomechanism (with $\zeta_v = 0$), stabilised by velocity feedback, includes a motor whose field current is controlled by an electronic amplifier, giving a final output acceleration of 0·12 radians/sec^2/ma, and an error detector of sensitivity 300 mV/degree. Give the specification of an amplifier and nonlinear element so that the servomechanism will have an undamped natural frequency of 8 hz, a maximum velocity of 60 r.p.m. and be critically damped in the linear regime.

P.2.12 A particular second order position control system with $\zeta_v = 0$ and $\zeta_T = 1$ includes an amplifier which saturates when the control signal exceeds a given amplitude h. Estimate the magnitude of input step θ, such that the first overshoot is of magnitude $\theta/2$ (i.e. neglect time spent in the linear regime).

P.2.13 A second order relay position control system has a feedforward transfer function of a relay in series with $9/s^2$ and a feedback transfer function of $1 + s/6$. Assuming the relay has the characteristic

$$\begin{aligned} e_0 &= h, & e_i &> h \\ &= 0, & |e_i| &< h \\ &= -h, & e_i &< -h \end{aligned}$$

determine

(a) the phase plane boundaries

(b) the equations of a phase plane trajectory when $|e_i|$ is (i) less than h and (ii) greater than h.

(c) for an input step of $4h$ (i) the maximum velocity and (ii) the magnitude of the first overshoot. (Assume zero initial output velocity.) Sketch the phase plane trajectory and time response for this input and calculate the time taken to the first overshoot.

P.2.14 Determine an equation for the boundary of the backlash region on a $\theta_0 - \dot{\theta}_0$ phase plane for the system considered in figures 2.19 and 2.20.

P.2.15 The block diagram of a digital servomechanism (neglecting the sampling) with an ideal relay is shown in figure P.2.1. If u is the maximum input step

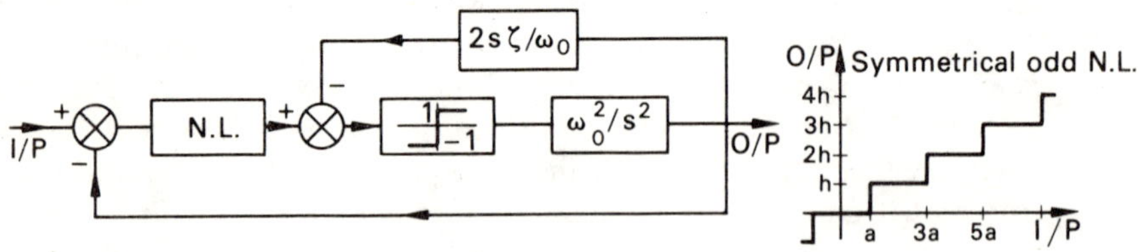

Figure P.2.1

magnitude for no overshoot determine the relationship between u, h and a. What is the value of u for (i) $a = 1, h = 2$ and (ii) $a = 1, h = 4$. Calculate the magnitude of successive over and undershoots for conditions (i) if an input equal to twice the calculated value of u is used. Suggest possible methods of modifying the system so that no overshoot will exist for any step input. Discuss the relative advantages and disadvantages of these methods.

P.2.16 A control system has a feedforward transfer consisting of an on-off relay with dead zone and an integrator of transfer function K/s. The relay deadband is $\pm\delta$ and it gives an output of $\pm h$ for input magnitudes greater than $\pm\delta$. The feedback transfer function is a time delay of T seconds. Determine the maximum possible value of K in terms of T, δ and h for stable operation of the system. What effect have variations in K and T on the frequency and amplitude of output oscillations when the system is unstable?

P.2.17 Determine the equation of the nonlinear characteristic $n(y)$ in figure P.2.2 if the system is to have the fastest response possible with no overshoot for all input step magnitudes.

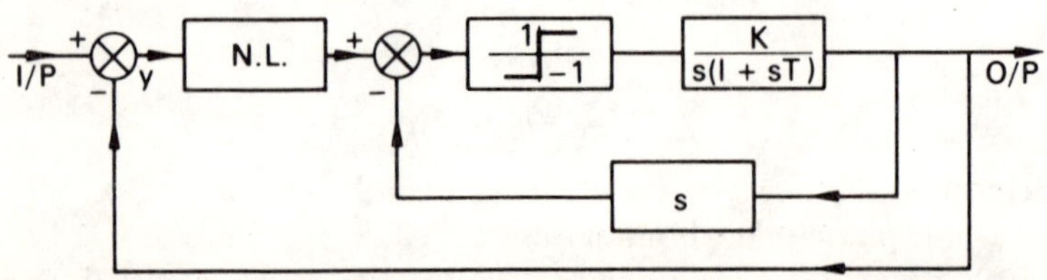

Figure P.2.2

P.2.18 The block diagram of a 'SERME' system is shown in figure P.2.3. To simplify construction the required nonlinear characteristic in the error channel is approximated by the linear segmented characteristic shown. Determine the required parameters of the characteristic if the area in the phase plane between the ideal switching boundary and the approximate one is to be a minimum and the system must have no overshoot

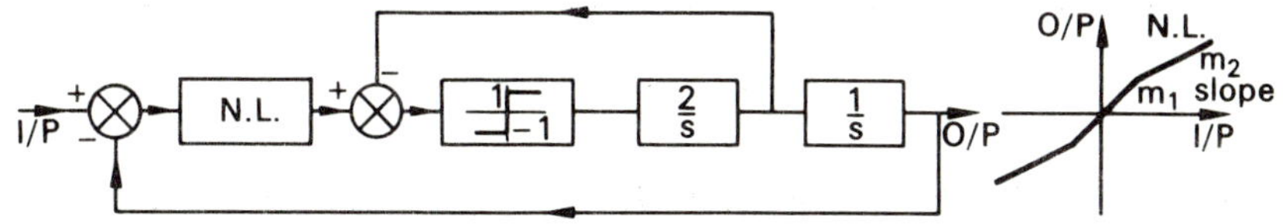

Figure P.2.3

for any input magnitude less than or equal to 32 units. Obtain the exact step responses on the phase plane for inputs of 32, 24 and 8 units respectively and compare the times of response with those of the ideal system.

P.2.19 The block diagram of a position control system is shown in figure P.2.4. Sketch step responses on the phase plane for inputs of 20 units and 10 units respectively. Specify the motion exactly for both cases as far as the first

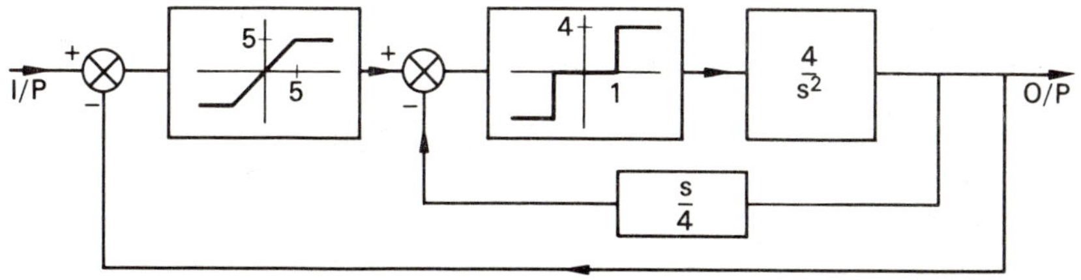

Figure P.2.4

undershoot, giving the times spent in the various modes and the magnitudes of the first overshoot and undershoot. Sketch the corresponding output position and velocity versus time curves for the 20 unit input.

P.2.20 The block diagram of a nonlinear control system is shown in figure P.2.6. Draw the switching boundaries on the phase plane and determine the step response on the phase plane to inputs of 5 and 9 units. Show also the velocity and position versus time waveforms and evaluate the response times. Is the system more or less damped for larger inputs? Give a physical explanation.

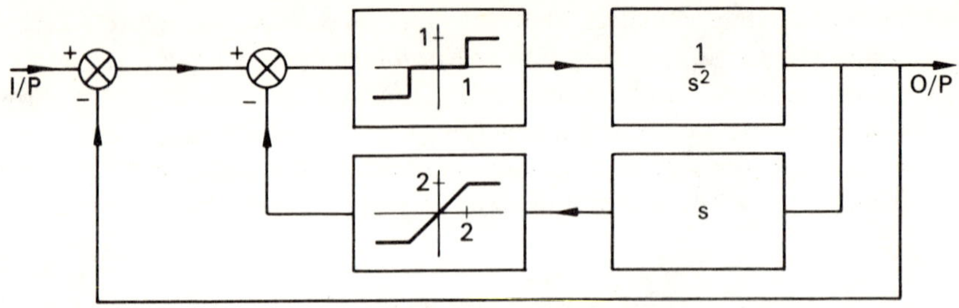

Figure P.2.5

P.2.21 Figure P.2.6 shows the block diagram of a relay control system with nonlinear velocity feedback. Draw the switching boundaries on the phase plane and discuss the step response behaviour of the system. Determine the amplitude and frequency of any limit cycles from the phase plane.

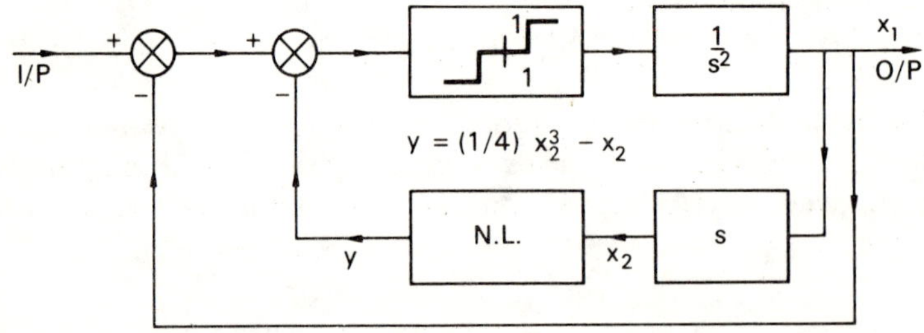

Figure P.2.6

Chapter 3

The Describing Function

3.1 Introduction

The original and most widely used describing function is that for a single sinusoidal input. It is therefore usually referred to simply as the describing function, abbreviated to DF. As its name implies, the DF is a quasilinear representation for a nonlinear element subjected to a sinusoidal input, and its use in the analysis of a nonlinear system is thus based on the assumption that the nonlinear element has a sinusoidal input.

The frequency response approach is a powerful tool for the analysis and design of linear systems, so that it does not seem surprising that methods were sought for extending its use to nonlinear systems. The outcome, the describing function method, was proposed in an essentially similar manner by P. J. Daniel and A. Tustin(1) in England, J. Dutilh(2) in France, W. Oppelt(3) in Germany, L. Goldfarb(4) in Russia and R. Kochenburger(5) in the United States.

The block diagram of figure 3.1 shows a simple feedback loop with a nonlinearity and linear elements having transfer functions $G_1(s)$, $G_2(s)$ and $H(s)$. The

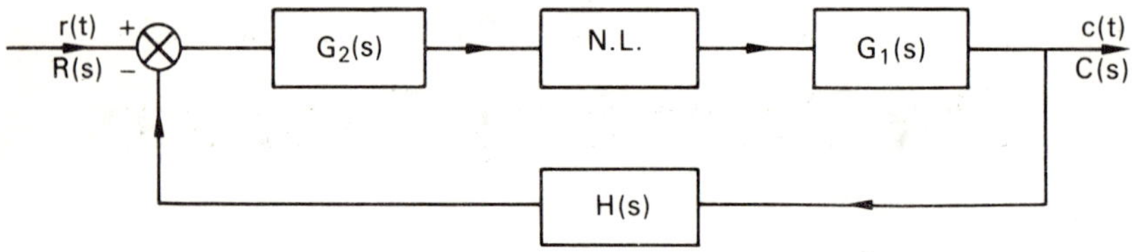

Figure 3.1 Block diagram of nonlinear system

input to the nonlinearity will be sinusoidal only if the system has a sinusoidal input and there is no feedback signal since this must of necessity contain harmonics of the input signal generated by the nonlinearity. The linear elements $G_1(s)$, $G_2(s)$ and $H(s)$, however, may be such as to reduce the harmonics fed back to the nonlinearity input to a small fraction of the total signal. For example, if the nonlinearity is an ideal relay then a sinusoidal input will produce a symmetrical square wave output which has a harmonic content, such that each odd harmonic of order n has a magnitude $(1/n)$th the fundamental magnitude. In addition, if $G_1(s)G_2(s)H(s) = K/s^2$, then the nth harmonic content fed back to the relay input will only have a magnitude of $1/n^3$ that of the fundamental. The harmonic content at the input to the relay is thus seen to be quite small and an analysis which assumes the input to be sinusoidal, thus neglecting harmonics other than the fundamental in the relay output, should produce quite accurate results.

To facilitate the analysis of a nonlinear system as described above, the DF of a nonlinear element is defined as the complex ratio of the fundamental component of the output to the sinusoidal input. The accuracy of an analysis based on the DF is seen to depend upon two factors namely the distortion produced by the nonlinearity assuming a sinusoidal input and the frequency characteristic of the linear elements $G_1(s)G_2(s)H(s)$. Evidently a low pass frequency characteristic is desirable to attenuate the harmonics more than the fundamental. With the above limitations it is clear that the DF can be used for investigating self-oscillations and the frequency response of nonlinear systems.

The DF of a dynamic nonlinear element, whose output depends upon both the input sinusoid and its derivatives, will be a function of both the amplitude and frequency of the input sinusoid. However, many nonlinear elements found in practical systems are static nonlinearities with output, y, for an input, x, defined by $y = n(x)$ and it is with such characteristics we are initially concerned.

In fact, in this chapter we further restrict ourselves to considering only those static characteristics which have odd symmetry that is $n(x) = -n(-x)$. Although derivation of the DF for asymmetrical characteristics is straightforward, it is generally necessary to obtain the DF in the presence of a bias signal, which due to the asymmetry, normally exists at all points in the feedback loop. This situation thus becomes one of a dual input to the nonlinearity, namely bias plus sinusoid, and its consideration is deferred to chapter 5.

3.2 DF evaluation

A symmetrical odd DVNL with a sinusoidal input x and corresponding output waveform $y(\theta)$, where $\theta = \omega t$, are shown in figure 3.2. The nonlinearity is defined by

$$y = n_1(x) \quad \text{for } \mathrm{d}x/\mathrm{d}\theta > 0,$$
$$y = n_2(x) \quad \text{for } \mathrm{d}x/\mathrm{d}\theta < 0. \tag{3.1}$$

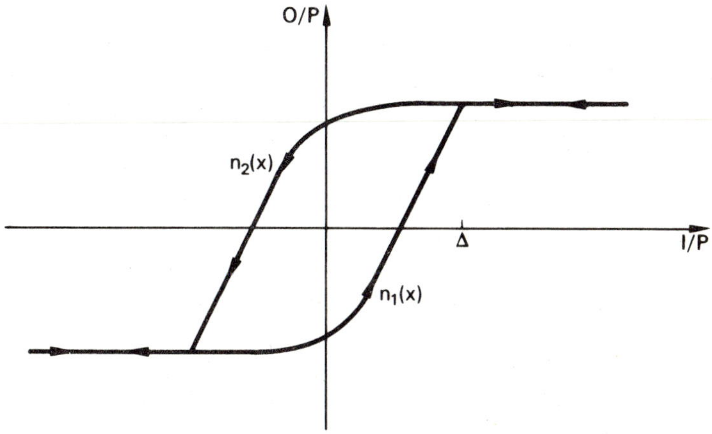

Figure 3.2 Odd symmetrical DVNL

In addition, since it has odd symmetry

$$n_1(x) = -n_2(-x). \tag{3.2}$$

The output harmonics and the DF can be evaluated either by Fourier analysis of the output waveform $y(\theta)$, which is the more common procedure, or directly from the nonlinearity $\{n_1(x), n_2(x)\}$ using orthogonal polynomials.[6,7] To facilitate the latter approach the input x has been chosen equal to $a \cos \theta$.

input = a cos θ

3.2.1 *Fourier analysis of the output waveform*

The output, $y(\theta)$, which is periodic may be expanded in a Fourier series, that is

$$y(\theta) = \sum_{s=0}^{\infty} a_s \cos s\theta + b_s \sin s\theta, \tag{3.3}$$

where

$$a_s = (\varepsilon_s/2\pi) \int_{-\pi}^{\pi} y(\theta) \cos s\theta \, d\theta, \tag{3.4}$$

$$b_s = (1/\pi) \int_{-\pi}^{\pi} y(\theta) \sin s\theta \, d\theta, \tag{3.5}$$

and the Neumann factor

$$\begin{aligned} \varepsilon_s &= 1 \quad \text{for } s = 0, \\ &= 2 \quad \text{for } s \geq 1. \end{aligned} \tag{3.6}$$

In particular it can be shown from equations (3.4) and (3.5) for the SVNL that since $y(\theta) = y(-\theta) = -y(-\pi + \theta)$, only the coefficients a_s for s odd will be nonzero, therefore

$$a_s = (4/\pi) \int_0^{\pi/2} y(\theta) \cos s\theta \, d\theta \quad \text{for } s \text{ odd}. \tag{3.7}$$

Thus, in this case, the DF, $N(a)$, is real and given by

$$N(a) = a_1/a = (4/a\pi) \int_0^{\pi/2} y(\theta) \cos \theta \, d\theta.$$

For the DVNL, since $y(\theta) = -y(-\pi + \theta)$, it can be shown from equations (3.4) and (3.5) that for s even $a_s = b_s = 0$ and

$$a_s = (2/\pi) \int_0^{\pi} y(\theta) \cos s\theta \, d\theta \quad \text{for } s \text{ odd}, \tag{3.8}$$

$$b_s = (2/\pi) \int_0^{\pi} y(\theta) \sin s\theta \, d\theta \quad \text{for } s \text{ odd}. \tag{3.9}$$

The fundamental output from the nonlinearity is $a_1 \cos \theta + b_1 \sin \theta$, so that the DF is complex and given by

$$N(a) = (a_1 - jb_1)/a. \tag{3.10}$$

If in rectangular and polar coordinates $N(a)$ is denoted by

$$N(a) = N_p(a) + jN_q(a) = M(a)e^{j\psi(a)}, \tag{3.11}$$

then

$$\begin{aligned} &N_p(a) = a_1/a, \qquad N_q(a) = -b_1/a, \\ &M(a) = (a_1^2 + b_1^2)^{1/2}/a \quad \text{and} \quad \psi(a) = -\tan^{-1}(b_1/a_1). \end{aligned} \tag{3.12}$$

The complex DF can if desired be evaluated from the single integral expression

$$N(a) = (2/a\pi) \int_0^{\pi} y(\theta) e^{-j\theta} \, d\theta. \tag{3.13}$$

3.2.2 *Orthogonal polynomials*

Substituting for θ in terms of x, that is $\theta = \cos^{-1}(x/a)$ in equation (3.4) and remembering that $y(\theta) = n_1(x)|_{x=a\cos\theta}$ for $-\pi \le \theta \le 0$ and $y(\theta) = n_2(x)|_{x=a\cos\theta}$ for $0 \le \theta \le \pi$ gives

$$a_s = (1/\pi)\left[\int_{-a}^{a} n_1(x)\cos\{s\cos^{-1}(x/a)\}\{+(a^2 - x^2)^{-1/2}\}\,dx + \int_{a}^{-a} n_2(x)\cos\{s\cos^{-1}(x/a)\}\{-(a^2 - x^2)^{-1/2}\}\,dx\right],$$

that is

$$a_s = (1/\pi)\int_{-a}^{a}\{n_1(x) + n_2(x)\}\cos\{s\cos^{-1}(x/a)\}(a^2 - x^2)^{-1/2}\,dx.$$

Denoting $n_1(x) + n_2(x)$ by $2n_p(x)$ and noting that $\cos\{s\cos^{-1}(x/a)\}$ is the Chebyshev polynomial $T_s(x/a)$ and $\{\pi(a^2 - x^2)^{1/2}\}^{-1}$ is the amplitude probability density function $r(x)$ for a sinusoid (see section 7.9.1) a_s is given by

$$a_s = 2\int_{-a}^{a} n_p(x)T_s(x/a)r(x)\,dx. \tag{3.14}$$

Similarly equation (3.5) gives

$$b_s = (1/\pi)\left[\int_{-a}^{a} n_1(x)\sin\{s\cos^{-1}(x/a)\}(a^2 - x^2)^{-1/2}\,dx + \int_{a}^{a} n_2(x)\sin\{s\cos^{-1}(x/a)\}\{-(a^2 - x^2)^{-1/2}\}\,dx\right].$$

We note however that over the range of the first integral $\cos^{-1}(x/a)$ lies between $-\pi$ and 0 and over the range of the second between 0 and π. Thus for example if we wish to evaluate b_1, $\sin\{\cos^{-1}(x/a)\} = -(a^2 - x^2)^{1/2}/a$ in the first integral and $+(a^2 - x^2)^{1/2}/a$ in the second. We therefore define the function $D_s(x/a)$ equal to $\sin\{s\cos^{-1}(x/a)\}$ with $\cos^{-1}(x/a)$ in the principal range from 0 to π. $D_s(x/a)$ is related to the Chebyshev polynomial of the second kind $U_s(x/a)$ by

$$D_s(x/a) = \{1 - (x^2/a^2)\}^{1/2}U_{s-1}(x/a).$$

Making this substitution for $\sin\{s\cos^{-1}(x/a)\}$ and letting $2n_q(x) = n_2(x) - n_1(x)$, which is the resulting nonlinear function obtained on combining the integrals, we have

$$b_s = 2\int_{-a}^{a} n_q(x)D_s(x/a)r(x)\,dx. \tag{3.15}$$

The nonlinear characteristics $n_p(x)$ and $n_q(x)$ will be called the equivalent in phase and quadrature nonlinearities respectively. It must be emphasised that for the DVNL care should be used in evaluating the DF from equations (3.14) and (3.15). For amplitudes $a < \Delta$ in figure 3.2 the input signal will follow a different characteristic so that $n_p(x)$ and $n_q(x)$ are not constant for all input amplitudes.

Now for symmetrical odd nonlinearities we note that $n_p(x)$ has odd symmetry and $n_q(x)$ even symmetry. In addition we see from table 3.1, which summarises some of the properties of the Chebyshev polynomials $T_s(x)$, $U_s(x)$ and the related

TABLE 3.1
Chebyshev polynomials and the related function $D_s(x)$

The recursion relationships for $T_s(x)$ and $U_s(x)$ are identical, that is

$$T_{s+1}(x) = 2xT_s(x) - T_{s-1}(x)$$

First few polynomials

s	$T_s(x)$	$U_s(x)$
0	1	1
1	x	$2x$
2	$2x^2 - 1$	$4x^2 - 1$
3	$4x^3 - 3x$	$8x^3 - 4x$
4	$8x^4 - 8x^2 + 1$	$16x^4 - 12x^2 + 1$
5	$16x^5 - 20x^3 + 5x$	$32x^5 - 32x^3 + 6x$

Relationship between $D_s(x)$ and $U_s(x)$

$$D_s(x) = (1 - x^2)^{1/2}U_{s-1}(x)$$

Integral relationships

$$\int T_s(x/a)r(x)\,dx = -(1/s\pi)D_s(x/a)$$

$$\int D_s(x/a)r(x)\,dx = (1/s\pi)T_s(x/a)$$

$$\int U_s(x/a)\,dx = (a/s)T_s(x/a)$$

Orthogonality relations

$$\int_{-a}^{a} T_j(x/a)T_k(x/a)r(x)\,dx = \delta_{jk}/(\varepsilon_j\varepsilon_k)^{1/2}$$

$$\int_{-a}^{a} D_j(x/a)D_k(x/a)r(x)\,dx = \delta_{jk}/2$$

$$\int_{-a}^{a} U_j(x/a)U_k(x/a)(a^2 - x^2)^{1/2}\,dx = \pi\delta_{jk}/2$$

function $D_s(x)$, that $T_s(x)$ and $U_s(x)$ are even for s even and odd for s odd. Therefore, as $r(x)$ is an even function, we may write

$$a_s = 4\int_0^a n_p(x)T_s(x/a)r(x)\,dx, \tag{3.16}$$

$$b_s = 4\int_0^a n_q(x)D_s(x/a)r(x)\,dx, \tag{3.17}$$

or

$$b_s = (4/a\pi)\int_0^a n_q(x)U_{s-1}(x/a)\,\mathrm{d}x. \tag{3.18}$$

Alternatively on resubstituting $x = a\cos\theta$ we have

$$a_s = (4/\pi)\int_0^{\pi/2} n_p(a\cos\theta)\cos s\theta\,\mathrm{d}\theta, \tag{3.19}$$

$$b_s = (4/\pi)\int_0^{\pi/2} n_q(a\cos\theta)\sin s\theta\,\mathrm{d}\theta. \tag{3.20}$$

Obviously for a SVNL $n_p(x) = n(x)$, $n_p(a\cos\theta) = y(\theta)$ and $n_q(x) = 0$. For a DVNL $n_p(x)$ is that SVNL which yields the same harmonic output terms as the a_s or in phase terms from the DVNL and in this case $n_p(a\cos\theta) \neq y(\theta)$. $n_q(x)$ however is not that SVNL which yields the b_s or quadrature terms when it has an input $a\sin\theta$. Such a nonlinearity $n_{qa}(x)$ may be found and shown to be given by

$$n_{qa}(x) = n_q(y)|_{y=(a^2-x^2)^{1/2}} \quad \text{for } y > 0.$$

In general $n_{qa}(x)$ is not very convenient to use as the characteristic depends upon the input amplitude a. The magnitude of the quadrature terms is seen to depend upon the area enclosed by the DVNL of which $n_q(x)$ is a measure.

In particular the fundamental output coefficients required for finding the DF are

$$a_1 = (4/a)\int_0^a x n_p(x) r(x)\,\mathrm{d}x, \tag{3.21}$$

and

$$b_1 = (4/a\pi)\int_0^a n_q(x)\,\mathrm{d}x. \tag{3.22}$$

Substituting for $r(x)$ and integrating equation (3.21) by parts, gives the alternate expression

$$a_1 = (4/\pi)n_p(0^+) + (4/a\pi)\int_0^a n_p'(x)(a^2 - x^2)^{1/2}\,\mathrm{d}x, \tag{3.23}$$

where

$$n_p'(x) = \mathrm{d}\{n_p(x)\}/\mathrm{d}x \quad \text{and} \quad n_p(0^+) = \lim_{\varepsilon\to 0} n_p(\varepsilon), \quad \varepsilon \text{ positive.}$$

Equation (3.22) is also seen to have the physical interpretation

$$b_1 = (1/a\pi) \qquad \text{(Area of DVNL loop).}$$

It is often useful to be able to compare the relative values of the DFs for various nonlinearities. Equations (3.21) and (3.22) being in terms of the known characteristics $n_p(x)$ and $n_q(x)$ are therefore particularly useful in this respect. In particular we see that

(a) for two SVNLs $n_1(x)$ and $n_2(x)$ with $n_1(x) < n_2(x)$ for all $x < a$ then $N_1(a) < N_2(a)$ for any a, and

(b) for a SVNL $n(x)$ with $k_1 x < n(x) < k_2 x$ for all $x < a$ then $k_1 < N_1(a) < k_2$ for any a.

For a SVNL the output coefficients a_s can also be found from the integral

$$a_s = (1/\pi)\int_{-\infty}^{\infty} N(j\omega) j^s J_s(a\omega)\,\mathrm{d}\omega, \tag{3.24}$$

where $N(j\omega)$ is the Fourier transform of $n(x)$, that is $N(j\omega) = \int_{-\infty}^{\infty} n(x)\mathrm{e}^{-j\omega x}\,\mathrm{d}x$, and J_s the Bessel function of order s. Equation (3.24) is a result of the transform method, which is normally used for multiple inputs, and is studied in chapter 10. The equation can be obtained directly by taking a double Fourier transform of equation (3.14), since $j^s J_s(a\omega)$ and $T_s(x/a)r(x)$ are Fourier transform pairs.

3.3 Alternative definitions for the DF

It is shown in the following sections that the DF for a SVNL may also be found from two alternative definitions. These alternatives namely the correlation and mean squared error definitions are more general in that the terminology is applicable to either deterministic or random inputs. The equivalence of these two definitions for any input is proved in section 8.2.

3.3.1 *Minimisation of mean squared error definition*

The DF of a SVNL is defined as that value of a variable gain K, having the same input as the nonlinearity, which minimises the mean squared value of the error between the output from the nonlinearity and that from the variable gain. The situation is shown diagrammatically in figure 3.3.

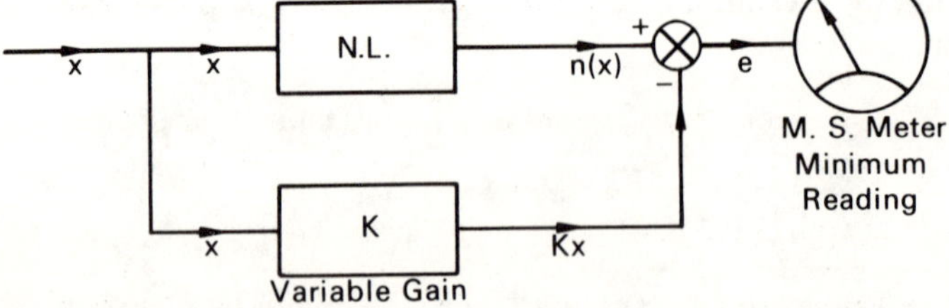

Figure 3.3 Illustration of mean squared error DF definition

The error or difference e between the output from the nonlinearity and that from the linear gain for the input $x = a\cos\theta$ is given by

$$e = n(a\cos\theta) - Ka\cos\theta = y(\theta) - Ka\cos\theta.$$

As the signal is periodic its mean squared value (or variance) is

$$\overline{e^2} = (1/2\pi)\int_0^{2\pi} \{y(\theta) - Ka\cos\theta\}^2\, d\theta,$$

where the bar denotes taking a time average.

Differentiating and setting $(d\overline{e^2}/dK) = 0$ gives

$$\int_0^{2\pi} 2\{y(\theta) - Ka\cos\theta\}\{-a\cos\theta\}\, d\theta = 0,$$

which gives for the optimum value of K,

$$K = (1/a\pi)\int_0^{2\pi} y(\theta)\cos\theta\, d\theta = a_1/a = N(a). \tag{3.25}$$

Further differentiation of $(d\overline{e^2}/dK)$ shows that this choice of K does in fact produce a minimum.

This definition may be extended for the case of the DVNL where instead of minimising with respect to a single gain K we minimise with respect to two gains K_1 and K_2. The input to K_1 is taken as the nonlinearity input $x = a\cos\theta$ and that to K_2 as $a\sin\theta = -(dx/dt)/\omega$. The optimum values of K_1 and K_2 to minimise $\overline{e^2}$ then become $K_1 = a_1/a$ and $K_2 = b_1/a$ respectively.

The above minimisation property is well known in the theory of Fourier series.

3.3.2 *Correlation definition*

The DF of a SVNL is defined as the covariance of the given input signal and the total output divided by the variance of the input.

The covariance of two signals is the mean value of their product, that is the value of their crosscorrelation function for zero delay. For two periodic signals $x(t)$ and $y(t)$ of period T the crosscorrelation function $R_{xy}(\tau)$ is defined by

$$\begin{aligned} R_{xy}(\tau) &= (1/T)\int_0^T x(t)y(t+\tau)\, dt, \\ &= (1/2\pi)\int_0^{2\pi} x(\theta)y(\theta+\omega\tau)\, d\theta, \end{aligned} \tag{3.26}$$

where τ is a time delay, $\omega = 2\pi/T$ and $\theta = \omega t$.

Similarly the variance is the zero delay value of the autocorrelation function, where the autocorrelation function $R_{xx}(\tau)$ is given by equation (3.26) with $y(\theta)$ equal to $x(\theta)$. Hence this definition of $N(a)$ becomes

$$N(a) = R_{xy}(0)/R_{xx}(0), \tag{3.27}$$

where

$$x(\theta) = a\cos\theta \quad \text{and} \quad y(\theta) = \sum_{s=1(2)}^{\infty} a_s \cos s\theta.$$

Thus

$$R_{xy}(\tau) = (1/2\pi)\int_0^{2\pi} a\cos\theta\left\{\sum_{s=1}^{\infty} a_s \cos s(\theta+\omega\tau)\right\} d\theta,$$

$$= (aa_1 \cos\omega\tau)/2,$$

as

$$\int_0^{2\pi} \cos n\theta \cos m\theta \, d\theta = \begin{cases} \pi \text{ if } m = n \\ 0 \text{ if } m \neq n \end{cases}.$$

Similarly we can show that

$$R_{yy}(\tau) = \sum_{s=1(2)}^{\infty} (a_s^2 \cos s\omega\tau)/2,$$

$$R_{xx}(\tau) = (a^2 \cos\omega\tau)/2,$$

$$\frac{R_{xy}(0)}{R_{xx}(0)} = \frac{a_1}{a} = N(a) \quad \text{as required.}$$

For this case we see there exists the more general relationship

$$R_{xy}(\tau)/R_{xx}(\tau) = R_{xy}(0)/R_{xx}(0),$$

a result which is not valid for any signal x. It is true for the sinusoid since it belongs to the class of separable processes (see section 7.9.1). In addition it is easy to show that $R_{xe}(\tau) = 0$ where $e(\theta) = y(\theta) - x(\theta)$ is the error in figure 3.3.

For a DVNL, we can show that

$$R_{xy}(0)/R_{xx}(0) = (a_1/a) = N_p(a).$$

The above definition may not therefore be applied to a DVNL as the result is the real part of the DF not the DF itself.

3.4 DFs for certain nonlinearities

To illustrate the application of the previously derived equations, DFs are calculated below for several analytically defined nonlinearities. A detailed list of DFs and harmonic coefficients for many odd symmetrical nonlinearities is given in appendix A.1.

3.4.1 *Ideal saturation*

The nonlinearity defined by

$$\begin{aligned} n(x) &= mx \qquad -\delta < x < \delta, \\ &\quad m\delta \qquad x \geq \delta, \\ &\quad -m\delta \qquad x \leq -\delta, \end{aligned} \tag{3.28}$$

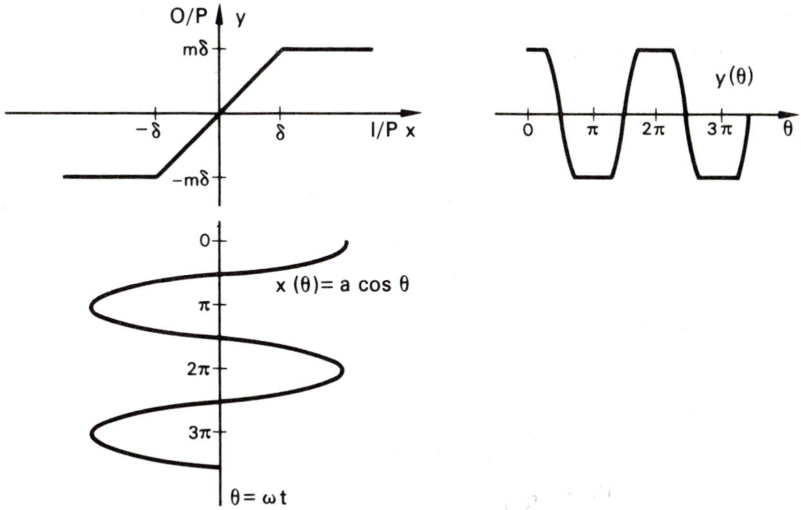

Figure 3.4 Saturation waveforms

and its output $y(\theta)$ for an input $x = a\cos\theta$ are shown in figure 3.4. We see that

$$\begin{aligned} y(\theta) &= m\delta \qquad && \text{for } 0 \le \theta < \beta, \\ & ma\cos\theta && \text{for } \beta < \theta \le \pi/2, \end{aligned}$$

where $\beta = \cos^{-1}(\delta/a)$, so that equation (3.7) gives for $a > \delta$

$$a_s = (4m/\pi)\left\{\int_0^\beta \delta\cos s\theta\,\mathrm{d}\theta + \int_\beta^{\pi/2} a\cos\theta\cos s\theta\,\mathrm{d}\theta\right\},$$

which on integrating becomes

$$\begin{aligned} a_s &= (4m/\pi)\left\{\frac{\delta\sin s\beta}{s} - \frac{a\sin(s+1)\beta}{2(s+1)} - \frac{a\sin(s-1)\beta}{2(s-1)}\right\}, \\ &= (2am/s\pi)\left\{\frac{\sin(s+1)\beta}{s+1} - \frac{\sin(s-1)\beta}{s-1}\right\}, \end{aligned}$$

or in terms of α, where $\alpha = \sin^{-1}(\delta/a) = (\pi/2) - \beta$,

$$a_s = (-1)^{(s-1)/2}(2am/s\pi)\left\{\frac{\sin(s+1)\alpha}{s+1} + \frac{\sin(s-1)\alpha}{s-1}\right\}. \tag{3.29}$$

In particular,

$$a_1 = \lim_{s\to 1} a_s = (am/\pi)(2\alpha + \sin 2\alpha).$$

Obviously for $a < \delta$, $a_1 = ma$ and $a_s = 0$ for $s > 1$. The DF for saturation is therefore

$$\begin{aligned} N(a) &= m && a < \delta, \\ & (m/\pi)(2\alpha + \sin 2\alpha) && a > \delta, \end{aligned}$$

with $\alpha = \sin^{-1}(\delta/a)$. As it is possible to obtain results for several other nonlinearities from those for saturation we let

$$a_s = amf_s(\rho) \quad \text{for } s > 1 \text{ and } \rho = \delta/a,$$

so that for $s > 1$,

$$f_s(\rho) = (-1)^{(s-1)/2}(2/s\pi)\left[\frac{\sin\{(s+1)\sin^{-1}\rho\}}{s+1} + \frac{\sin\{(s-1)\sin^{-1}\rho\}}{s-1}\right] \quad \text{for } |\rho| < 1,$$

$$= 0, \quad \text{for } |\rho| > 1,$$

and

$$\begin{aligned} f_1(\rho) &= 1 && \text{for } \rho > 1, \\ &(2/\pi)\{\sin^{-1}\rho + \rho(1-\rho^2)^{1/2}\} && \text{for } |\rho| < 1, \\ &-1 && \text{for } \rho < -1. \end{aligned}$$

The function $f_1(\rho)$ for $\rho > 0$ is shown in figure 3.5. The DF for saturation can also easily be found from equation (3.23), that is

$$N(a) = a_1/a = (2m/\pi)\int_{-\delta}^{\delta} (a^2 - x^2)^{1/2}\,dx,$$

$$= (2m/\pi)\{\sin^{-1}(\delta/a) + \delta(a^2 - \delta^2)^{1/2}/a^2\} \text{ as before.}$$

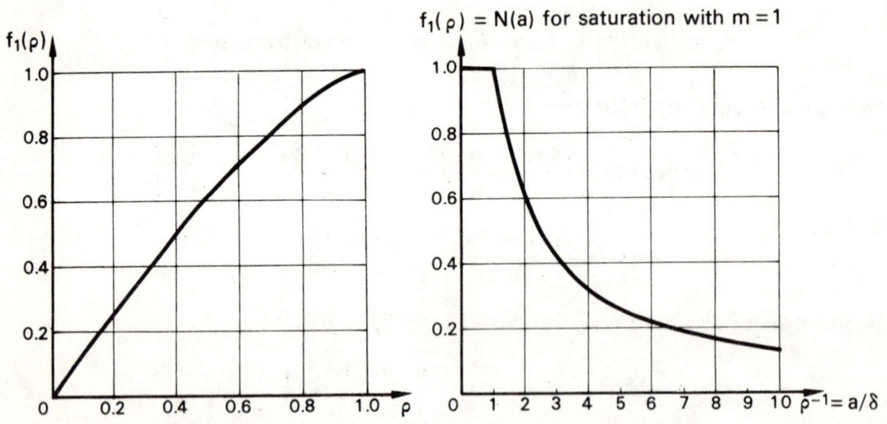

Figure 3.5 Saturation function $f_1(\rho)$

3.4.2 *Odd power law*

Any odd power of x can be expressed as a series in Chebyshev polynomials so that the orthogonal property of the polynomials can be used to obtain the coefficients a_s without evaluating any integrals.

Consider

$$n(x) = \sum_{k=1(2)}^{\infty} c_k T_k(x/a);$$

then

$$a_s = 2\int_{-a}^{a} \sum_{k=1(2)}^{\infty} c_k T_k(x/a) T_s(x/a) r(x)\,\mathrm{d}x.$$

Using the orthogonal relationship (see table 3.1)

$$\int_{-a}^{a} T_k(x/a) T_s(x/a) r(x)\,\mathrm{d}x = \delta_{sk}/(\varepsilon_s \varepsilon_k)^{1/2},$$

we obtain $a_s = c_s$. Thus if for example $n(x) = x^5$, we take

$$x^5 = a_1 T_1(x/a) + a_3 T_3(x/a) + a_5 T_5(x/a),$$

substituting for T_1, T_3 and T_5 and comparing the coefficients of x, x^3 and x^5 gives

$$16a_5/a^5 = 1,$$

$$-20a_5/a^3 + 4a_3/a^3 = 0,$$

$$5a_5/a - 3a_3/a + a_1/a = 0.$$

Solving one obtains

$$a_1 = 5a^5/8,\ a_3 = 5a^5/16 \quad \text{and} \quad a_5 = a^5/16$$

so that $N(a) = 5a^4/8$.

3.4.3 *On-off relay with hysteresis*

The characteristic, the output waveform $y(\theta)$ and the equivalent in phase and quadrature nonlinearities $n_p(x)$ and $n_q(x)$ for $a > \Delta$ are shown in figure 3.6. As $n_p(x)$ has an on-off relay with dead zone characteristic the in phase harmonic components produced by the on-off with hysteresis and on-off with dead zone relays are equal.

Using equation (3.16) we obtain

$$a_s = 4\int_{\Delta}^{a} hT_s(x/a) r(x)\,\mathrm{d}x,$$

$$= (4h/s\pi) D_s(\Delta/a),$$

and in particular $a_1 = (4h/\pi) D_1(\Delta/a) = 4h(a^2 - \Delta^2)^{1/2}/a\pi$. Using equation (3.17) we obtain

$$b_s = 4\int_{0}^{\Delta} hD_s(x/a) r(x)\,\mathrm{d}x,$$

$$= (4h/s\pi) T_s(\Delta/a),$$

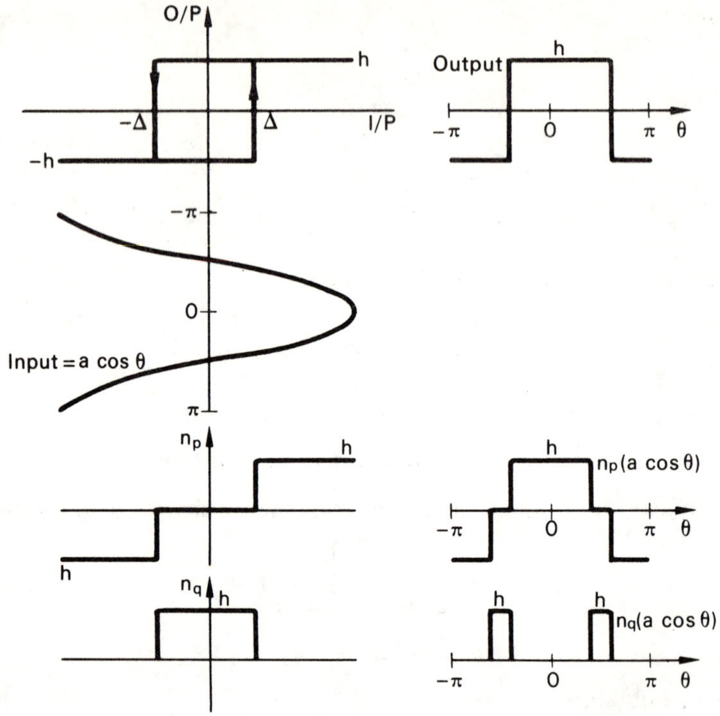

Figure 3.6 On-off relay with hysteresis waveforms

and in particular $b_1 = (4h/\pi)T_1(\Delta/a) = 4h\Delta/a\pi$. Thus $N(a) = (4h/a^2\pi)\{(a^2 - \Delta^2)^{1/2} - j\Delta\}$.

3.5 DFs for graphically defined characteristics

Unfortunately most inherent nonlinearities which occur in practice cannot be defined by mathematical equations. The existence of nonlinearity can be detected by harmonic or d.c. signal tests. In the former case experimental results can be used to obtain the DF provided direct access to the nonlinearity input and output signals is possible, otherwise allowance for possible dynamic effects of other system elements involved in the measurement may not be straightforward. When a graph of $n(x)$ is available, several approximate methods may be used to find the harmonic behaviour of the nonlinearity. These may be placed into two categories:

(a) Curve fitting—approximating $n(x)$ by analytical, including linear segmented expressions which then allow exact integration of equations (3.16) and (3.17).

(b) Approximate integration—the integrals of equations (3.16) and (3.17) are obtained numerically using the graph of $n(x)$. Some of the possible methods are given below, which to use in a particular case will be influenced by:

(i) The required accuracy of the result; here it should be remembered that the DF approach is an approximate method.
(ii) The simplicity of the approximate solution.
(iii) The rapidity with which the approximate solution can be obtained.

Since for symmetrical odd nonlinearities $n_p(x)$ has odd symmetry and $n_q(x)$ even symmetry, results can be obtained by just considering the behaviour for $x > 0$. In the following we therefore assume the above properties and use defining equations valid for $x > 0$ only, although the solutions, which exist for s odd only, are given for the complete nonlinearity.

3.5.1 *Polynomial expansion*

With $n_p(x)$ expanded in a power series

$$n_p(x) = \sum_{m=0}^{\infty} A_m x^m. \tag{3.31}$$

$a_s = 0$ for $s + m$ odd and, using equation (3.19), for $s + m$ even gives

$$a_s = (4/\pi) \int_0^{\pi/2} \sum_{m=0}^{\infty} A_m a^m \cos^m \theta \cos s\theta \, d\theta.$$

Integrating, we obtain

$$a_s = \sum_{m=0}^{\infty} \{A_m a^m \Gamma(m + 1)\}/[2^{m-1}\Gamma\{(2 + s + m)/2\}\Gamma\{(2 + m - s)/2\}], \tag{3.32}$$

where Γ denotes the gamma function.

This result is also valid for fractional values of m. Since $\Gamma(n) = \infty$ for any integer $n \leq 0$, $a_s = 0$ for $s > m$. Thus, the mth harmonic is the highest harmonic output from x^m, when m is integral.

In particular for a SVNL defined by equation (3.31)

$$N(a) = \sum_{m=0}^{\infty} \{A_m a^{m-1}\Gamma(m + 1)\}/[2^{m-1}\Gamma\{(3 + m)/2\}\Gamma\{(1 + m)/2\}],$$

$$= \sum_{m=0}^{\infty} 2A_m a^{m-1} \pi^{-1/2} \Gamma\{(m + 2)/2\}/\Gamma\{(m + 3)/2\}, \tag{3.33}$$

where we have made use of the duplication formula for the gamma function

$$\Gamma(n) = 2^{n-1}\pi^{-1/2}\Gamma(n/2)\Gamma\{(n + 1)/2\}.$$

Similarly, with $n_q(x)$ expanded in the power series

$$n_q(x) = \sum_{m=0}^{\infty} B_m x^m,$$

$b_s = 0$ for $s + m$ even and equation (3.18) for $s + m$ odd gives

$$b_s = (4/a\pi) \int_0^{a} \sum_{m=0}^{\infty} B_m x^m U_{s-1}(x/a) \, dx.$$

Replacing the Chebyshev polynomial by its power series expansion

$$U_{s-1}(x) = \sum_{k=0}^{s-1} u_k x^k$$

gives

$$b_s = (4/\pi) \sum_{m=0}^{\infty} B_m a^m \sum_{k=0}^{s-1} a^k u_k/(m + k + 1), \qquad (3.34)$$

and

$$N_q(a) = -(4/a\pi) \sum_{m=0}^{\infty} B_m a^m/(m + 1). \qquad (3.35)$$

3.5.2 *Fourier series expansion*

Let $n_p(x)$ for $x > 0$ be approximated by a Fourier sine series of period T. When $n_p(x)$ is not a periodic function, which is the usual situation, T will be chosen such that $T/2 > a_m$, where a_m is the maximum input amplitude for which the DF is required.

Thus for

$$n_p(x) = \sum_{m=1}^{\infty} A_m \sin mcx \qquad (3.36)$$

where $c = 2\pi/T$, equation (3.19) gives

$$a_s = (2/\pi) \int_0^{\pi} \sum_{m=1}^{\infty} A_m \sin(mca \cos\theta) \cos s\theta \, d\theta.$$

Substituting

$$\sin(z \cos\theta) = -2 \sum_{p=1}^{\infty} (-1)^p J_{2p-1}(z) \cos\{(2p - 1)\theta\},$$

we find that all the terms in the series have zero value when multiplied by $\cos s\theta$ and integrated apart from the term for which $s = 2p - 1$.

Therefore

$$a_s = 2 \sum_{m=1}^{\infty} A_m \sin(s\pi/2) J_s(mca), \qquad (3.37)$$

and

$$N_p(a) = (2/a) \sum_{m=1}^{\infty} A_m J_1(mca).$$

Similarly a Fourier sine or cosine series can be used to represent $n_q(x)$ but unfortunately whichever form is chosen all the terms in the series expansion for $\sin(mca \cos\theta)$ or $\cos(mca \cos\theta)$ contribute to the integral. The simplicity of the previous case is thus lost and the method loses much of its appeal for a DVNL.

3.5.3 *Quantised approximation*

A quantised approximation to a nonlinear characteristic for $x > 0$ is shown in figure 3.7. This approach[8] has the advantage that the approximation can be used for $n_p(x)$ and $n_q(x)$, is simple and straightforward, whilst good accuracy can be obtained using relatively few quantised levels. For many characteristics five steps within the input amplitude produces an accuracy of better than 3 per cent for the magnitudes of the fundamental components. In addition it is a relatively simple matter to write a computer program for use with any nonlinearity.

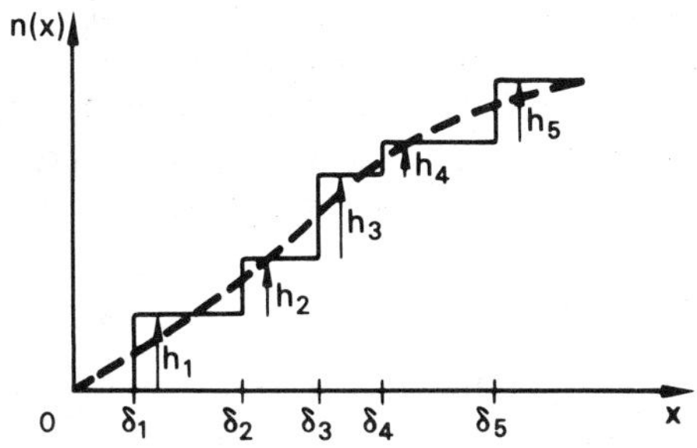

Figure 3.7 A quantised approximation

Improved accuracy can be obtained by increasing the number of quantised levels at the expense of providing more input data and an increased computation time.

Assuming the quantisation characteristic shown represents $n_p(x)$, then from equation (3.16)

$$a_s = 4\left\{\int_{\delta_1}^{\delta_2} h_1 T_s(x/a)r(x)\,\mathrm{d}x + \int_{\delta_2}^{\delta_3} (h_1 + h_2)T_s(x/a)r(x)\,\mathrm{d}x\right.$$

$$\left. + \cdots \int_{\delta_m}^{a} (h_1 + h_2 + \cdots h_m)T_s(x/a)r(x)\,\mathrm{d}x\right\},$$

which on integration gives

$$a_s = (4/s\pi) \sum_{m=1}^{M} h_m D_s(\delta_m/a), \tag{3.39}$$

and in particular

$$N_p(a) = (4/a^2\pi) \sum_{m=1}^{M} h_m(a^2 - \delta_m^2)^{1/2}. \tag{3.40}$$

Similarly, if the quantisation characteristic of figure 3.7 represents $n_q(x)$, equation (3.17) gives

$$b_s = 4\left\{\int_{\delta_1}^{\delta_2} h_1 D_s(x/a)r(x)\,\mathrm{d}x + \int_{\delta_2}^{\delta_3} (h_1 + h_2)D_s(x/a)r(x)\,\mathrm{d}x\right.$$

$$\left. + \cdots \int_{\delta_m}^{a} (h_1 + h_2 + \cdots h_m)D_s(x/a)r(x)\,\mathrm{d}x\right\},$$

that is

$$b_s = (4/s\pi) \sum_{m=1}^{M} h_m\{1 - T_s(\delta_m/a)\}, \tag{3.41}$$

and in particular

$$N_q(a) = -(4/a^2\pi) \sum_{m=1}^{M} h_m(a - \delta_m). \tag{3.42}$$

As results for the above equations are required for various values of a, the calculations can be performed either by keeping constant values of δ_m/a as different values of a are used, that is altering the quantisation approximation for each value of a, or alternatively using the same approximation for all values of a. Obviously in this latter case the accuracy will decrease as a is decreased because fewer steps are involved in the approximation.

3.5.4 *Graphical integration*

Graphical integration[9,10,11] using cursors is based on the relationship

$$I_s = \int_{x=0}^{\theta} n(x)\,\mathrm{d}\{A_s(x)\} \simeq \sum_{m=1}^{M} n(x_m)A_\theta/M, \tag{3.43}$$

where A_θ is the total change in $A_s(x)$ between 0 and θ, M is the number of equal intervals into which A_θ is divided, and x_m the values of x at the mid points of these intervals.

Equation (3.16) can be written in the form

$$a_s = 4\int_0^a n_p(x)\,\mathrm{d}\{A_{sa}(x)\}, \tag{3.44}$$

where

$$\begin{aligned} A_{sa}(x) &= -(1/s\pi)D_s(x/a) \qquad \text{for } s \geq 1, \\ &= -(1/\pi)\cos^{-1}(x/a) \quad \text{for } s = 0, \end{aligned} \tag{3.45}$$

and equation (3.17) in the form

$$b_s = 4\int_0^a n_q(x)\,\mathrm{d}\{A_{sb}(x)\}, \tag{3.46}$$

where

$$A_{sb}(x) = (1/s\pi)T_s(x/a) \quad \text{for } s \geq 1. \tag{3.47}$$

Graphs of the functions $T_s(x/a)$ and $D_s(x/a)$ up to $s = 5$ over the range $0 \leq x/a \leq 1$ are shown in figure 3.8. Even values of s are included here as they will be made use of later. All the curves are symmetrical about $x = 0$; $T_s(x/a)$ has odd symmetry for s odd and even symmetry for s even, whilst $D_s(x/a)$ has even symmetry for s odd and odd symmetry for s even. The total change A_θ in the

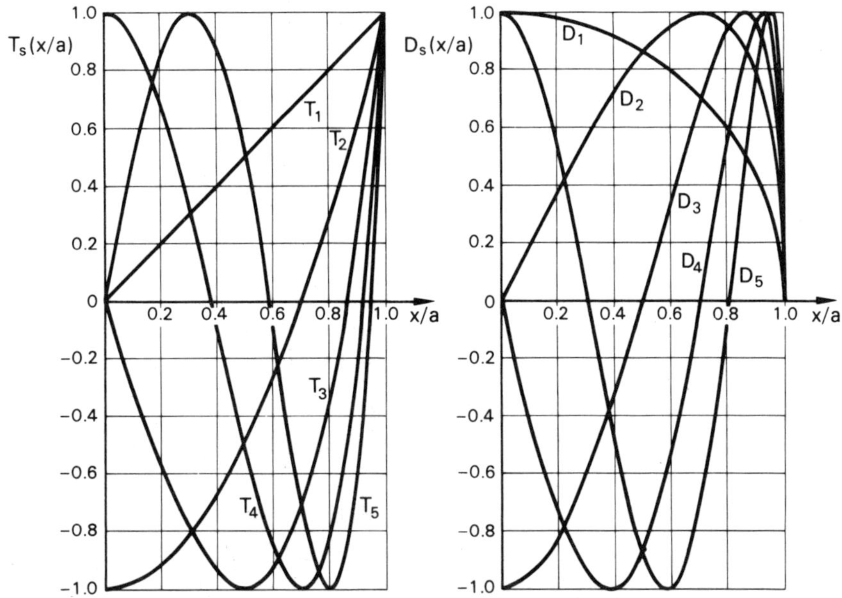

Figure 3.8 Graphs of the functions $T_s(x/a)$ and $D_s(x/a)$

functions $A_{sa}(x)$ and $A_{sb}(x)$ as (x/a) goes from 0 to 1 is $1/\pi$, for $s \geq 1$ and 1/2 for $s = 0$, so that equation (3.43) becomes

$$I_s = (1/\pi) \sum_{m=1}^{M} n(x_m)/M \qquad s \geq 1.$$

The required values of x/a to give equal increments of 0·1 or 0·2 in $T_s(x/a)$ and $D_s(x/a)$ for $0 < x/a < 1$ are given in table 3.2. A cursor for $D_1(x/a)$ which can be used for any input amplitude, a, is shown in figure 3.9. The input amplitude is marked on the vertical axis and the cursor lines intersect the amplitude levels at distances of a times the values given in table 3.2. As an example the value of $n(x_3)$ for an input amplitude of 2·0 is shown to be 0·85 in figure 3.9. In using the cursors for graphical integration, it should be noted that the sign of $n(x_m)$ is positive when the nonlinearity and the slope of $A_s(x)$ have the same sign.

To illustrate the method, the DF for $a = 4$ of the DVNL shown in figure 3.10 is evaluated. As the equivalent in phase and quadrature nonlinearities $n_p(x)$ and $n_q(x)$, which are also shown, are linear segmented it is not necessary to move the cursors over them; the values of $n_p(x_m)$ and $n_q(x_m)$ for the required values of x_m are easily written down. Using the approximation, we have

$$a_1 = (4/\pi) \sum_{m=1}^{M} n_p(x_m)/M,$$

TABLE 3.2

Values of (x/a) for equal increments in $T_s(x/a)$ and $D_s(x/a)$

(a) *Increments of* 0·1

	Gradient of $T_s(x/a)$, of $D_s(x/a)$	Values of x/a									
$T_1(x/a)$	+	0·050	0·150	0·250	0·350	0·450	0·550	0·650	0·750	0·850	0·950
$T_2(x/a)$	+	0·158	0·274	0·354	0·418	0·474	0·524	0·570	0·612	0·652	0·689
	+	0·725	0·753	0·791	0·822	0·851	0·880	0·908	0·935	0·962	0·987
$T_3(x/a)$	−	0·017	0·050	0·084	0·119	0·155	0·173	0·234	0·278	0·332	0·406
	+	0·538	0·651	0·692	0·725	0·753	0·775	0·800	0·821	0·840	0·853
	+	0·874	0·890	0·905	0·917	0·933	0·946	0·959	0·971	0·983	0·994
$\cos^{-1}(x/a)$†	−	0·079	0·233	0·383	0·523	0·649	0·760	0·853	0·924	0·972	0·997
$D_1(x/a)$	−	0·312	0·527	0·661	0·760	0·835	0·893	0·937	0·968	0·989	0·999
$D_2(x/a)$	+	0·023	0·075	0·126	0·178	0·231	0·287	0·346	0·411	0·486	0·586
	−	0·810	0·874	0·911	0·938	0·958	0·973	0·984	0·992	0·997	1·000
$D_3(x/a)$	+	0·105	0·184	0·239	0·284	0·324	0·360	0·394	0·425	0·456	0·487
	+	0·513	0·543	0·571	0·599	0·628	0·658	0·689	0·722	0·759	0·808
	−	0·914	0·943	0·960	0·972	0·981	0·988	0·993	0·996	0·999	1·000

† $\cos^{-1}(x/a) = \lim_{s\to 0} s^{-1} D_s(x/a)$.

TABLE 3.2 (*continued*)

(b) *Increments of* 0·2

	Gradient of $T_s(x/a)$, of $D_s(x/a)$	Values of x/a				
$T_1(x/a)$	+	0·100	0·300	0·500	0·700	0·900
$T_2(x/a)$	+	0·224	0·387	0·500	0·592	0·671
	+	0·742	0·806	0·866	0·922	0·975
$T_3(x/a)$	−	0·033	0·101	0·174	0·256	0·365
	+	0·624	0·709	0·766	0·811	0·849
	+	0·882	0·912	0·940	0·965	0·989
$T_4(x/a)$	−	0·113	0·198	0·259	0·311	0·359
	−	0·406	0·452	0·500	0·553	0·623
	+	0·782	0·833	0·866	0·892	0·914
	+	0·933	0·950	0·966	0·980	0·994
$T_5(x/a)$	+	0·020	0·061	0·105	0·154	0·222
	−	0·393	0·456	0·500	0·537	0·571
	−	0·604	0·636	0·669	0·706	0·753
	+	0·859	0·892	0·914	0·930	0·945
	+	0·957	0·968	0·978	0·987	0·996
$\cos^{-1}(x/a)$	−	0·156	0·454	0·707	0·891	0·988
$D_1(x/a)$	−	0·436	0·714	0·866	0·954	0·995
$D_2(x/a)$	+	0·050	0·152	0·259	0·378	0·531
	−	0·847	0·926	0·966	0·988	0·999
$D_3(x/a)$	+	0·150	0·262	0·342	0·410	0·471
	+	0·529	0·585	0·643	0·705	0·781
	−	0·931	0·967	0·988	0·995	0·999
$D_4(x/a)$	−	0·025	0·076	0·131	0·193	0·276
	+	0·484	0·558	0·609	0·651	0·689
	+	0·725	0·759	0·793	0·830	0·875
	−	0·961	0·981	0·991	0·997	1·000
$D_5(x/a)$	−	0·090	0·159	0·208	0·251	0·290
	−	0·328	0·366	0·407	0·452	0·512
	+	0·658	0·709	0·743	0·772	0·797
	+	0·821	0·843	0·866	0·890	0·919
	−	0·975	0·988	0·995	0·998	1·000

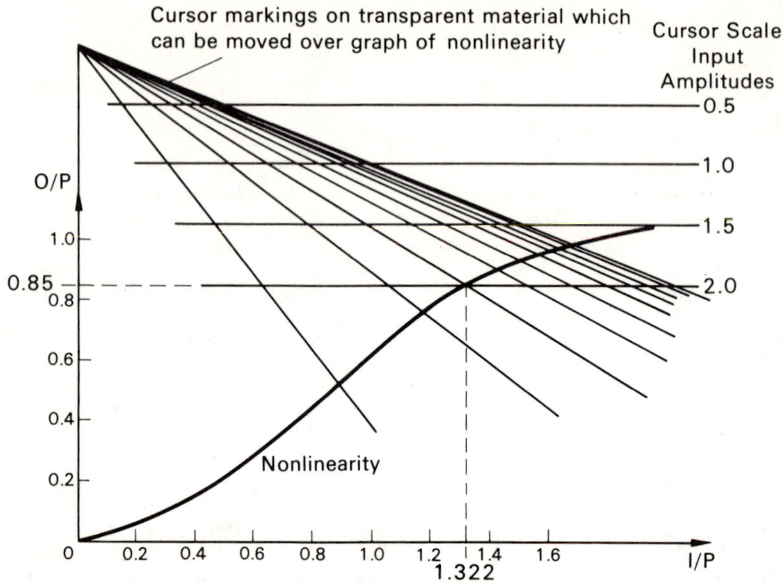

Figure 3.9 Illustration of cursor use

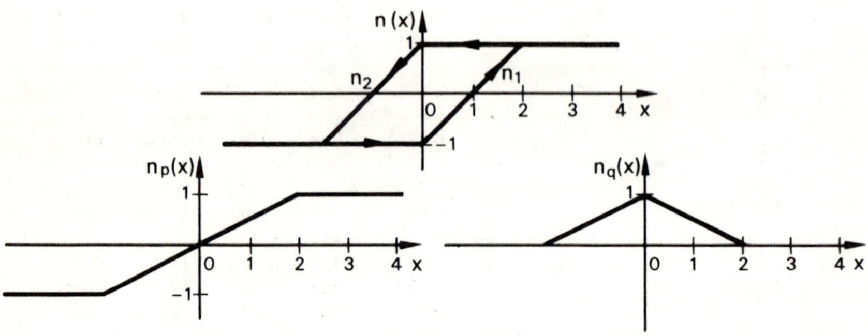

Figure 3.10 DVNL with its in phase and quadrature nonlinearities

where the points x_m for $a = 4$ and $M = 10$ are 4 times the values tabulated for $D_1(x/a)$ in 0·1 increments. That is

x_m	1·248	2·108	2·644	3·040	3·340	3·572	3·748	3·872	3·956	3·996
$n_p(x_m)$	0·624	1·0	1·0	1·0	1·0	1·0	1·0	1·0	1·0	1·0

Therefore $a_1 = (4/10\pi)(9{\cdot}624) = 3{\cdot}85/\pi$.

Similarly, using the tabulated values for $T_1(x/a)$ we have

$$b_1 = (4/10\pi)(0{\cdot}9 + 0{\cdot}7 + 0{\cdot}5 + 0{\cdot}3 + 0{\cdot}1 + 5 \times 0),$$

$$= 1{\cdot}00/\pi.$$

The exact solutions for comparison are

$$a_1 = 3{\cdot}83/\pi \quad \text{and} \quad b_1 = 1{\cdot}00/\pi.$$

3.5.5 *Approximate analytical DF solution*

Substituting $u = x/a$ in equation (3.21) we obtain

$$N_p(a) = (4/a\pi)\int_0^1 \frac{un_p(au)}{(1-u^2)^{1/2}}\,\mathrm{d}u.$$

Using the Steklov[12] approximation

$$\int_{-1}^{1} \frac{F(u)}{(1-u^2)^{1/2}}\,\mathrm{d}u \simeq (\pi/6)\{F(1) + F(-1) + 2F(1/2) + 2F(-1/2)\}, \tag{3.48}$$

with $F(u) = un_p(au)$ an even function of u,

$$N_p(a) \simeq (2/3a)\{n_p(a) + n_p(a/2)\}. \tag{3.49}$$

The approximation is exact for a cubic characteristic and gives errors of less than 5 per cent for saturation. An even better approximation

$$N_p(a) \simeq (1/3a)\{n_p(a) + n_p(a/2) + \sqrt{3}n_p(\sqrt{3}a/2)\} \tag{3.50}$$

has been given by Silberberg.[13]

A similar approximate solution for $N_q(a)$ can be obtained by applying Simpson's rule to equation (3.22) giving[14]

$$N_q(a) = -(2/3\pi)\{n_q(0) + 4n_q(a/2) + n_q(a)\}. \tag{3.51}$$

The above approximations are particularly useful for obtaining quick estimates of DFs, however the results can be considerably in error for characteristics with rapid slope changes.

3.6 Multiple nonlinearities

In many systems nonlinearities occur at more than one position in the system. If the combination of nonlinear and linear elements is such that the input to all the nonlinearities will be approximately sinusoidal for the situation under investigation each nonlinearity can be replaced by its DF (see section 4.7). Occasionally nonlinear characteristics occur directly in parallel or cascade, in some cases possibly separated by a linear element whose dynamics may be neglected.

3.6.1 *Nonlinearities in series*

The usual procedure for analysing a series combination of nonlinearities is to obtain the overall nonlinearity, as illustrated in figure 3.11 and then evaluate the DF of the new nonlinearity. This is a straightforward procedure for single and double valued characteristics.

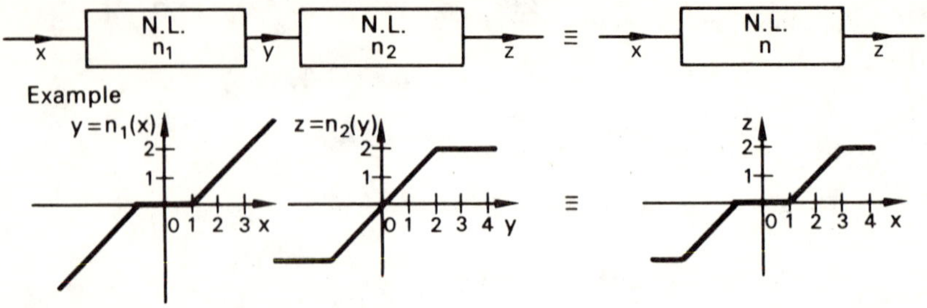

Figure 3.11 Equivalent of a series nonlinear combination

The DF of a series combination is not equal to the product of the DFs for the separate nonlinearities. This is true whether we define the DF of the second nonlinearity ignoring the distortion in its input or taking it into account by using the more general DF definition of equation (3.27). Since the former case is self evident we consider the latter situation for SVNLs in more detail here.

Referring to the block diagram of figure 3.11

$$N_1(a) = R_{xy}(0)/R_{xx}(0) = \overline{x(t)y(t)}/\overline{x^2(t)}.$$

Also we may write

$$y(t) = N_1(a)x(t) + \delta_1(t),$$

where $\delta_1(t)$ is uncorrelated with $x(t)$ for all τ.

Using the same definition for the second nonlinearity with the nonsinusoidal input $y(t)$,

$$N_2(y) = \overline{y(t)z(t)}/\overline{y^2(t)}$$

and

$$z(t) = N_2(y)y(t) + \delta_2(t),$$

where in this case since $y(t)$ is nonsinusoidal $\delta_2(t)$ is uncorrelated with $y(t)$ for zero τ only.

Similarly for the cascade structure

$$N(a) = \overline{x(t)z(t)}/\overline{x^2(t)},$$

which on substituting for $z(t)$ and $y(t)$ and making use of the fact that $\overline{\delta_1(t)x(t)} = 0$, gives

$$N(a) = N_1(a)N_2(y) + \overline{\delta_2(t)x(t)}/\overline{x^2(t)}. \tag{3.52}$$

To provide further insight into the situation, particularly with regard to the significance of the DF for nonsinusoidal inputs, we consider the example of two cubic nonlinearities in series. In this case

$$y(t) = x^3(t),$$

$$z(t) = y^3(t) = x^9(t).$$

Thus from equation (3.33)

$$N(a) = 2a^8\pi^{-1/2}\Gamma(11/2)/\Gamma(6) = 63a^8/128.$$

Now

$$y(t) = (a\cos\theta)^3 = (3a^3\cos\theta + a^3\cos 3\theta)/4,$$

where $\theta = \omega t$, so that

$$N_1(a) = 3a^2/4 \quad \text{and} \quad \delta_1(t) = (a^3\cos 3\theta)/4.$$

Also,

$$\begin{aligned} z(t) &= \{(3a^3\cos\theta + a^3\cos 3\theta)/4\}^3, \\ &= a^9(63\cos\theta + 42\cos 3\theta + 18\cos 5\theta + 4{\cdot}5\cos 7\theta, + 0{\cdot}5\cos 9\theta)/128, \end{aligned}$$

so that

$$\begin{aligned} N_2(y) &= \overline{y(t)z(t)}/\overline{y^2(t)}, \\ &= \{a^{12}(63 \times 3 + 42 \times 1)/8 \times 128\}/\{a^6(3^2 + 1^2)/32\}, \\ &= 231a^6/320, \end{aligned}$$

and

$$\delta_2(t) = a^9(-6{\cdot}3\cos\theta + 18{\cdot}9\cos 3\theta + 18\cos 5\theta + 4{\cdot}5\cos 7\theta + 0{\cdot}5\cos 9\theta)/128.$$

Hence

$$\begin{aligned} \overline{y(t)\delta_2(t)} &= a^{12}(-6{\cdot}3 \times 3 + 18{\cdot}9)/8 \times 128 = 0 \text{ as expected}, \\ \overline{x(t)\delta_2(t)} &= -6{\cdot}3a^{10}/256, \end{aligned}$$

and

$$N_1(a)N_2(y) = 69{\cdot}3a^8/128,$$

which is 10 per cent greater than $N(a)$. Using equation (3.52), we confirm that

$$N(a) = 69{\cdot}3a^8/128 - 6{\cdot}3a^8/128 = 63a^8/128.$$

If we neglect the distortion at the input to the second nonlinearity the overall DF is

$$N_1(a)N_2\{aN_1(a)\} = (3a^2/4)(3/4)(3a^3/4)^2 = 40{\cdot}5a^8/128,$$

which is less than $N(a)$ by approximately 30 per cent.

The above procedure becomes extremely unwieldy for the general situation where $y(t)$ contains an infinite number of harmonics.

3.6.2 *Nonlinearities in parallel*

A parallel combination of two nonlinearities is shown in figure 3.12. Obviously since both nonlinearities have the same input they are equivalent to the single

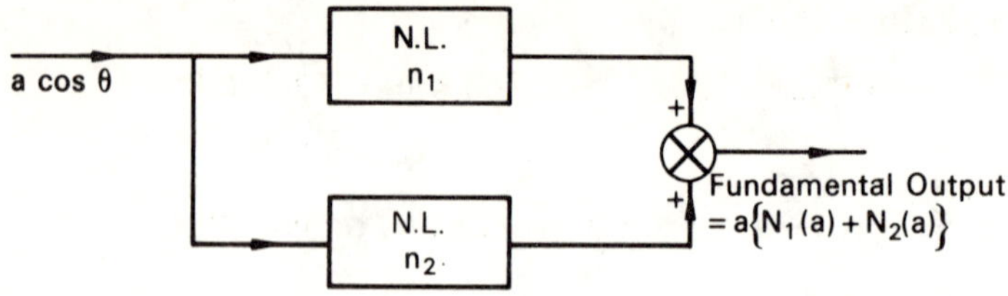

Figure 3.12 Nonlinearities in parallel

nonlinearity, $n(x)$, where

$$n(x) = n_1(x) + n_2(x). \tag{3.53}$$

Thus the harmonic content of the output waveform can be obtained from a summation of the harmonic outputs of the individual nonlinearities, that is

$$a_s = a_{1s} + a_{2s},$$

$$b_s = b_{1s} + b_{2s},$$

and

$$N(a) = N_1(a) + N_2(a).$$

Although the above configuration of nonlinearities occurs infrequently in practical systems the concepts are extremely useful in the synthesis of nonlinearities for simulation purposes and in evaluating DFs. The same technique can be extended to cover the case of any number of nonlinearities in parallel. To illustrate the procedure for obtaining DFs three examples are considered.

Dead space

We see from figure 3.13 that in this case, the characteristic can be obtained with

$$n_1(x) = \text{linear gain of unit magnitude}$$

$$-n_2(x) = \text{saturation of unit gain and limit level } \delta.$$

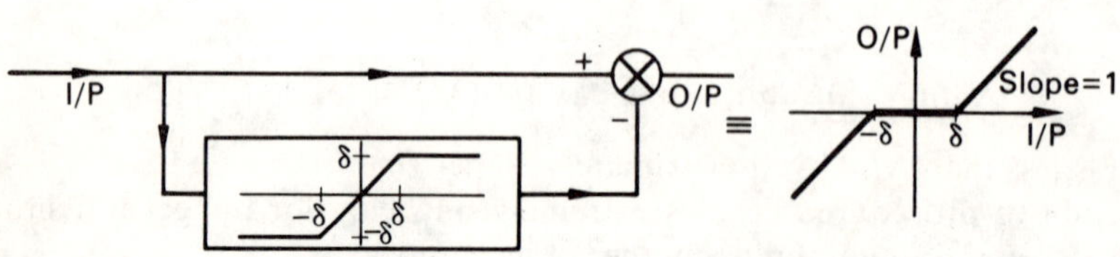

Figure 3.13 Dead space synthesis

Therefore

$$N(a) = 1 - f_1(\delta/a).$$

Limited field of view

We see from figure 3.14 that for this example

$$n_1(x) = \text{saturation of gain 2 and break point } \delta,$$
$$-n_2(x) = \text{saturation of gain 1 and break point } 2\delta.$$

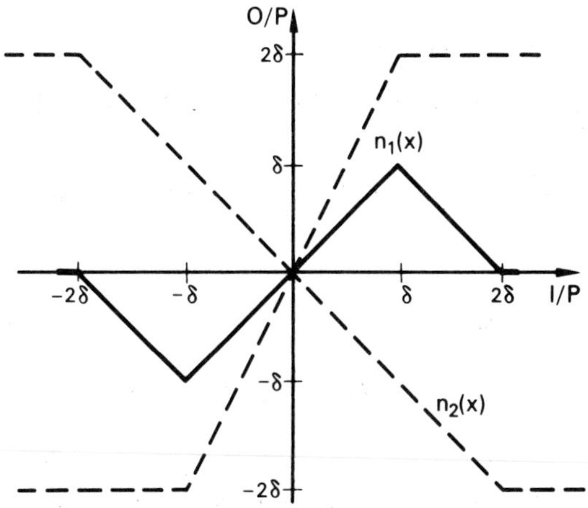

Figure 3.14 Limited field of view synthesis

Therefore

$$N(a) = 2f_1(\delta/a) - f_1(2\delta/a).$$

Backlash

The friction controlled backlash characteristic and its equivalent in phase and quadrature nonlinearities are shown in figure 3.15 for an input amplitude

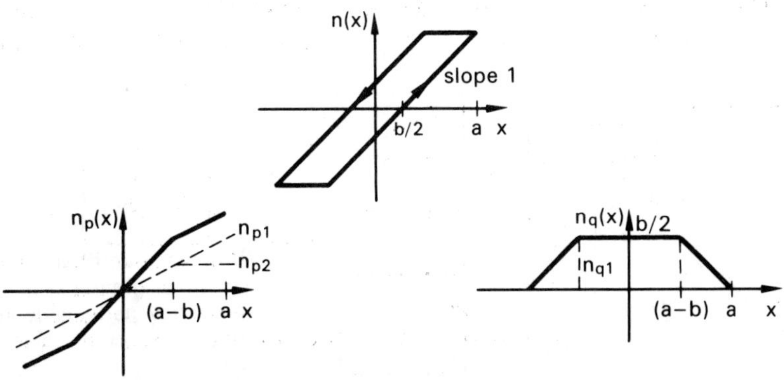

Figure 3.15 Backlash synthesis

$a > b$. Here we see that $n_p(x)$ can be synthesised from

$n_{p1}(x)$ = linear gain of 1/2,

$n_{p2}(x)$ = saturation of gain 1/2 and limit level $(a - b)$.

Therefore

$$N_p(a) = (1/2)[1 + f_1\{(a - b)/a\}].$$

Also $n_q(x)$ can be synthesised from

$n_{q1}(x)$ = quadrature nonlinearity for on-off relay with hysteresis; $h = b/2$, $\Delta = (a - b)$,

$n_{q2}(x)$ = nonlinearity for additional contribution from $(a - b)$ to a.

Therefore

$$N_q(a) = -\{2b(a - b)/a^2\pi\} - (4/a^2\pi)\int_{a-b}^{a} \{(a - x)/2\}\, dx,$$

$$= -\{2b(a - b)/a^2\pi\} - b^2/a^2\pi = b(b - 2a)/a^2\pi.$$

Thus

$$N(a) = (1/2)[1 + f_1\{(a - b)/a\}] + jb(b - 2a)/a^2\pi \quad \text{for } a > b.$$

3.7 Experimental measurement of the DF

For characteristics which are known to be single valued the output harmonics and DF are easily obtained using a wave analyser. However, for double valued characteristics, the phase of the output fundamental relative to the input sinusoid must also be measured. The principles of two useful measuring techniques are outlined below.

3.7.1 *A.c. potentiometer*

A schematic diagram of this method is given in figure 3.16. The oscillator gives two outputs $p = A \sin \omega t$ the 'in phase' component and $q = A \cos \omega t$ the

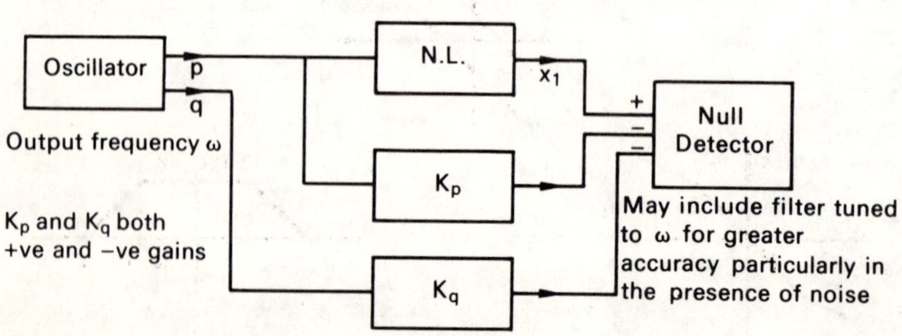

Figure 3.16. A.C. potentiometer block diagram

'quadrature' component. The former signal is used as the reference input to the nonlinearity. The fundamental output from the nonlinearity $x = X\sin(\omega t + \phi)$ can then be balanced out by suitable adjustments of the gains K_p and K_q giving

$$X = A(K_p^2 + K_q^2)^{1/2}$$

and

$$\phi = \tan^{-1}(K_q/K_p).$$

If a two phase, or variable phase, oscillator is not available a quadrature signal for a particular frequency, ω, can be generated by feeding the reference to a network giving 90° phase shift at the frequency, ω. A particularly suitable network is the all pass network of transfer function $(1 - sT)/(1 + sT)$, with $T = 1/\omega$, as the output is not attenuated. This method can also be applied to measure the phase shift to a particular frequency produced by a nonlinearity in a closed loop situation.

In addition the above method can be extended for measurement of the phase of the higher harmonics.[(15)] Here we generate harmonic references using a SVNL fed with the reference input. Then, employing the above phase shift network tuned to the harmonic k of interest by selecting $T = 1/k\omega$, we can use the in phase and quadrature harmonics to null the kth harmonic output of the nonlinearity.

3.7.2 *Thermocouple wattmeter*

The thermocouple wattmeter measures thc average value of the product of two signals, namely a reference signal and an unknown x. If in figure 3.16 the a.c. potentiometer is replaced by the thermocouple wattmeter, first the signal p and then q may be used as reference signals for the unknown x. If the corresponding wattmeter readings are W_p and W_q respectively, then

$$W_p = XA\cos\phi$$

and

$$W_q = XA\sin\phi,$$

so that

$$X = (W_p^2 + W_q^2)^{1/2}/A \quad \text{and} \quad \tan\phi = W_q/W_p.$$

The wattmeter is usually calibrated before results are taken by connecting the known signal p to both input channels. The sensitivity of the instrument is adjusted so that a full scale deflection is obtained with some nominal value p_1 of p on the unknown channel. To avoid damage to the thermocouples these instruments normally have protective drives so that nonlinear operation will occur and incorrect results will be obtained when the peak value of the signal applied to the unknown channel exceeds that of p_1 used in the sensitivity adjustment. Thus for the measurement of unknown signals with high distortion or imbedded in large amounts of noise the sensitivity has to be reduced to avoid nonlinear operation, resulting in readings which are low on the scale and relatively inaccurate.

3.8 Identification of a nonlinearity by a harmonic analysis

The problem here is to determine the nonlinearity given information on its harmonic output. This situation can arise in practice when harmonic measurements have been made and the characteristic is needed for further analytical work, or when nonlinear compensation techniques are used so that the nonlinearity corresponding to a particular DF is required. We consider two situations below, namely the case where we know all the harmonic output terms for a specific input amplitude and secondly where we know the DF.[16,17]

3.8.1 *From harmonic response*

Let

$$n_p(x) = \sum_{j=0}^{\infty} \alpha_j T_j(x/a),$$

$$n_q(x) = \sum_{j=0}^{\infty} \beta_j D_j(x/a).$$

Substituting these expressions in equations (3.14) and (3.15) and using the orthogonal relationships for $T_s(x/a)$ and $D_s(x/a)$ given in table 3.1 we obtain

$$a_s = \alpha_s \quad \text{and} \quad b_s = \beta_s.$$

Thence

$$n_p(x) = \sum_{s=0}^{\infty} a_s T_s(x/a),$$

$$n_q(x) = \sum_{s=1}^{\infty} b_s D_s(x/a).$$

It can further be shown that if $n_p(x)$ and $n_q(x)$ are approximated by $n_{pa}(x)$ and $n_{qa}(x)$ which contain a finite number of terms, that is

$$n_{pa}(x) = \sum_{s=0}^{M} a'_s T_s(x/a),$$

$$n_{qa}(x) = \sum_{s=1}^{M} b'_s D_s(x/a),$$

then the best approximation in the mean squared error sense for any M, where the error is the difference in output from the nonlinearity and its approximation, is obtained by taking $a'_s = a_s$ and $b'_s = b_s$. Therefore the approximations for

$n_p(x)$ and $n_q(x)$ which are best in the above sense are

$$n_p(x) = \sum_{s=0}^{M} a_s T_s(x/a), \tag{3.55}$$

$$n_q(x) = \sum_{s=1}^{M} b_s D_s(x/a). \tag{3.56}$$

Thus from a measurement of the output harmonics up to order M, we are able to approximately synthesise the nonlinearity. The approximations obtained for different inputs a for nonlinearities with characteristics independent of the input amplitude obviously only become identical over their common range as M tends to infinity, whilst for nonlinearities with characteristics dependent upon the input amplitude a the identification is only valid for that value of a. Usually acceptable identifications are obtained for values of M of order ten.

3.8.2 *From DF*

Equations (3.21) and (3.22) for the symmetrical odd nonlinearity are

$$N_p(a) = (4/a^2) \int_0^a x n_p(x) r(x)\, \mathrm{d}x,$$

$$N_q(a) = -(4/a^2 \pi) \int_0^a n_q(x)\, \mathrm{d}x.$$

It would appear therefore that if we can solve these integral equations for $n_p(x)$ and $n_q(x)$ given $N_p(a)$ and $N_q(a)$ the nonlinearity will be identified. This procedure is proposed in reference 16, however the important point that $n_p(x)$ and $n_q(x)$ depend upon the amplitude a for the DVNL is ignored. The above equations should of course strictly be written

$$N_p(a) = (4/a^2) \int_0^a x n_p(x, a) r(x)\, \mathrm{d}x,$$

$$N_q(a) = -(4/a^2 \pi) \int_0^a n_q(x, a)\, \mathrm{d}x,$$

which unfortunately cannot be solved.

Thus restricting our consideration to SVNLs,

$$N(a) = (4/a^2) \int_0^a x n(x) r(x)\, \mathrm{d}x,$$

which is a special case of Volterra's integral equation of the first kind and has the solution

$$n(x) = (1/2x) \frac{\mathrm{d}}{\mathrm{d}x} \left\{ \int_0^x z^3 N(z)(x^2 - z^2)^{-1/2}\, \mathrm{d}z \right\},$$

$$= (\pi/2x) \frac{\mathrm{d}}{\mathrm{d}x} \left\{ \int_0^x z^3 N(z) r(z)\, \mathrm{d}z \right\}. \tag{3.57}$$

Since $\int_0^\infty r(z)\,dz = 1/2$ if $N(z)$ satisfies the above equation so does $\{N(z) + K/z^3\}$ where K is a constant; thus different DFs may give the same nonlinearity $n(x)$.

In many practical situations where we wish to solve this inverse problem the given DF is available in graphical rather than analytical form. Several approximate procedures are then available to determine $n(x)$, two of which are given below.

Polynomial expansion

From equation (3.33) it is seen that the DF for a polynomial characteristic is also a polynomial. Thus, assuming a polynomial curve fit to $N(a)$ has been obtained, we can find $n(x)$ by comparing the coefficients of the DF for a polynomial nonlinearity with those obtained in the curve fit. This process is slightly easier than solving equation (3.57).

Assuming

$$N(a) = \sum_{m=0}^{\infty} c_m a^m,$$

and comparing coefficients with the DF solution (equation (3.33)) for

$$n(x) = \sum_{m=0}^{\infty} A_m x^m,$$

we obtain

$$c_{m-1} = 2A_m \pi^{-1/2} \Gamma\{(m+2)/2\}/\Gamma\{(m+3)/2\},$$

that is,

$$A_m = \pi^{1/2} c_{m-1} \Gamma\{(m+3)/2\}/2\Gamma\{(m+2)/2\}.$$

Quantised nonlinearity

Here we determine the quantised characteristic $n(x)$ whose DF values for certain discrete amplitudes lie on a given DF curve. To obtain $n(x)$ for $x \le a$ we divide a into M equal intervals and assume steps in the characteristic $n(x)$ of height h_m at the points $x = (2m-1)a/2M$ where m takes values from 1 to M. Then from equation (3.40)

$$N(ka/M) = \frac{4M}{ka\pi} \sum_{m=1}^{k} h_m \left\{1 - \left(\frac{2m-1}{2k}\right)^2\right\}^{1/2},$$

and taking values of k from 1 to M we have

$$N(a/M) = (4M/a\pi)(\sqrt{3}h_1/2),$$
$$N(2a/M) = (4M/2a\pi)\{(\sqrt{15}h_1/4) + (\sqrt{7}h_2/4)\}.$$

By successively solving the equations

$$h_1 = a\pi N(a/M)/2\sqrt{3}\mathrm{M},$$
$$h_2 = (2a\pi/\sqrt{7}M)\{N_p(2a/M) - (3\sqrt{5}/4)N_p(a/M)\} \quad \text{etc.}$$

This approach is particularly amenable to computer solution, where by choosing a large enough M, sufficient accuracy for any specific problem can be obtained.

A procedure similar to the above for fitting the DF information to a linear segmented characteristic may be used as also may repeated application of the approximate DF formula of equation (3.49).[18]

3.9 Other DFs

DFs different to the one introduced have been defined for SVNLs and used in the analysis of nonlinear feedback systems. Probably the two which have attracted most attention are the r.m.s. DF[19] and its truncated version the corrected conventional DF.[20]

For odd SVNLs these DFs are defined by

$$
\begin{aligned}
N_{\text{r.m.s.}}(a) &= (\text{R.m.s. output})/(\text{R.m.s. of sine input}),\\
&= \left[(1/2\pi)\int_0^{2\pi}\{y(\theta)\}^2\,\mathrm{d}\theta\right]^{1/2}\Big/(a/\sqrt{2}),\\
&= \left[\int_{-a}^{a}\{n(x)\}^2 r(x)\,\mathrm{d}x\right]^{1/2}\Big/(a/\sqrt{2}),\\
&= \left[\sum_{k=1(2)}^{\infty} a_k^2\right]^{1/2}\Big/a,
\end{aligned}
\tag{3.58}
$$

and

$$N_{\text{ccDF}}(a) = (a_1^2 + a_3^2)^{1/2}/a.$$

Further suggestions have included using the definition of section 3.3.1 with error criteria other than the one of mean squared error; such as average or absolute error. A more complicated DF is presented in reference 21, where the limit cycle waveform is obtained more accurately after the amplitude and frequency of the oscillation have been found approximately.

For most nonlinearities these other possible DFs are often more difficult to calculate and the resulting expressions usually more complex. The major consideration as to their use however, apart from simplicity of their derivation, is the accuracy with which they can be expected to predict solutions, a subject which is taken up in detail in section 4.8.

3.10 Frequency dependent DFs

In some control situations components exist whose response to sinusoidal inputs depends upon both the frequency and the amplitude of the signal. Mathematically this means that the behaviour of the device is described by a nonlinear differential equation, rather than a linear differential equation as is the case for a linear dynamic element, or a nonlinear function, as is the case for the static nonlinearity considered earlier. The component thus has a different DF for each input frequency, that is the DF is frequency dependent and may be denoted by $N(a, \omega)$.

It is not necessary at this time to study the properties of, and techniques for, the evaluation of DFs for frequency dependent nonlinearities. The problem is discussed in detail in chapter 10 where it is treated as a specific case of a DF for a nonlinear function of two variables. The reader should realise however that a frequency dependent DF, can be used to approximate a dynamic nonlinearity, that it can be calculated using equations (3.4), (3.5) and (3.10), if the output waveform $y(\theta)$ is known for all the required values of a and ω, and that $N(a, \omega)$ can be measured experimentally as described in section 3.7. Approximate solutions for some nonlinear problems can also often be obtained by using a frequency dependent DF to represent a cascade combination of linear and nonlinear elements, for example where a lead network exists between two static nonlinear characteristics.

REFERENCES

1. Tustin, A.: 'The effects of backlash and of speed dependent friction on the stability of closed cycle control systems', *J.I.E.E.*, 1947, pt. II, 94, pp. 143–151.
2. Dutilh, J.: 'Theorie des servomechanisms a relais', *Onde Elec.*, 1950, pp. 438–445.
3. Oppelt, W.: 'Locus curve method for regulators with friction', *Z. Deut. Ingr.*, *Berlin*, 1948, 90, pp. 179–183, Translated in Report 1691, National Bureau of Standards, Washington, 1952.
4. Goldfarb, L. C.: 'On some nonlinear phenomena in regulatory system', *Automatika i Telemekhanika*, 1947, pp. 349–383, Translation: U. Oldenburger (ed.), 'Frequency Response', The Macmillan Co., New York, 1956.
5. Kochenburger, R. J.: 'A frequency response method for analysing and synthesizing contactor servomechanisms', *Trans A.I.E.E.*, 1950, 69, pp. 270–283.
6. Atherton, D. P.: 'Determination of nonlinearity from its harmonic response', *Electronic Letters*, 1966, 2, p. 152.
7. Atherton, D. P. and Jain, N. K.: 'Identification of double-valued nonlinearities from their harmonic response', *Electronic Letters*, 1967, 3, pp. 485–486.
8. Atherton, D. P.: 'Rapid evaluation of the auto-correlation function of the output of single-valued nonlinearities in response to sinusoidal and Gaussian signals', *Proc. I.E.E.*, 1962, 109C, pp. 656–664.
9. Douce, J. L.: 'A note on the evaluation and response of a nonlinear element to sinusoidal and random signals', *Proc. I.E.E.*, 1957, 105C, pp. 88–92.
10. Somerville, M. J. and Atherton, D. P.: 'Multi-gain representation for a single-valued nonlinearity with several inputs and the evaluation of their equivalent gains by a cursor method', *Proc. I.E.E.*, 1958, 105C, pp. 537–549.
11. Atherton, D. P.: 'The evaluation of the response of single-valued nonlinearities to several inputs', *Proc. I.E.E.*, 1962, 109C, pp. 146–157.
12. Macmillan, R. H.: *Nonlinear control systems analysis* (Pergamon, Oxford, 1962, p. 75).
13. Silberberg, M. Y.: 'A note on the describing function of an element with coulomb, static and viscous friction', *Trans. A.I.E.E., Appl. Ind.*, 1957, pt. II, pp. 423–425.
14. McBride, L. E., Jr.: 'Approximate describing function for double-valued nonlinearities', *Electronic Letters*, 1965, 1, pp. 72–73.
15. Jain, J. K. and Atherton, D. P.: 'Development and application of a generalized harmonic analyser', *International Electronics Conference, Toronto*, 1967, paper no. 67072.
16. Gibson, J. E.: *Nonlinear automatic control* (McGraw-Hill, New York, 1963, pp. 565–571).
17. Gibson, J. E. and di Tada, E. S.: 'On the inverse describing function problem', *Proc. 2nd I.F.A.C. Conference Basle*, 1963, published in *Automatic and Remote Control* (Butterworth, London, pp. 29–35).
18. Gelb, A. and Vander Velde, W. E.: *Multiple-input describing functions and nonlinear system design* (McGraw-Hill, New York, p. 97).
19. Gibson, J. E. and Prasanna-Kumar, K. S.: 'A new r.m.s. describing function for single-valued nonlinearities', *Proc. I.R.E.*, 1961, 49, p. 1321.

20. Rankine, R. R. Jr. and d'Azzo, J. J.: 'A corrected-conventional describing function for intermediate-order nonlinear control systems', *Proc. Northeast Electronics Res. Eng. Meeting*, 1964, pp. 38–39.
21. Jopling, A. D. and Johnson, R. A.: 'The elliptic describing function', *Proc. Joint Automatic Control Conference, Stanford, California*, 1964, pp. 175–184.

PROBLEMS

P.3.1 Determine the DF for all input amplitudes to the nonlinearity shown in figure P.3.1 using the result $f_1(\rho)$ for a saturation characteristic.

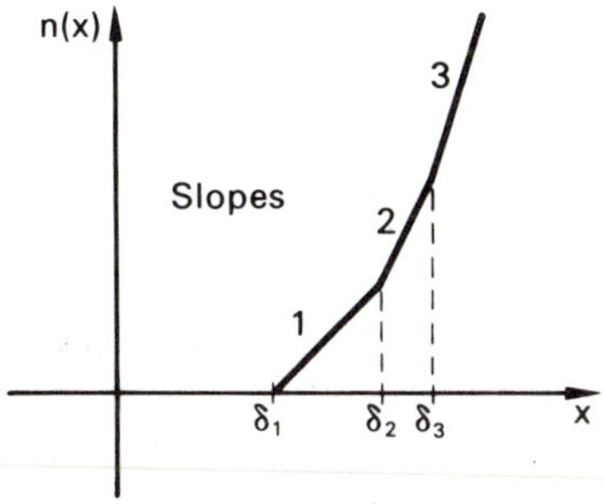

Figure P.3.1

P.3.2 Use equations (3.8) and (3.9) to determine the DF of the on-off relay with dead zone and hysteresis. Obtain the in phase and quadrature characteristics n_p and n_q and check the expression for the DF using the results of section 3.5.3. Calculate the numerical value for $h = 2$, $\delta = 3/4$, $\Delta = 1/4$, $a = 3$ and compare with an approximate result obtained by the graphical approach of section 3.5.4.

P.3.3 Obtain an expression of the form $n(x, \dot{x})$ which describes the output of the nonlinearity of problem P.3.2 for a sinusoidal input x.

P.3.4 Determine the equation of a nonlinear element which gives an output of $1{\cdot}2 \sin \omega t + 0{\cdot}3 \sin 3\omega t$ for an input of $2 \sin \omega t$.

P.3.5 Determine the equation of a nonlinear element which gives an output of $2 \sin(\omega t - 30°) + 0{\cdot}5 \sin(3\omega t - 60°) + 0{\cdot}1 \sin(5\omega t - 90°)$ for an input of (a) $4 \sin \omega t$ and (b) $4 \cos \omega t$.

P.3.6 Using the definition of section 3.3.2 determine the DF of an ideal on-off relay to a periodic triangular input waveform. Check your result by evaluating the covariance from the individual terms of the Fourier series expansions for the input and output waveforms. Is the definition applicable for all values of τ?

P.3.7 Check the result of the previous problem by evaluating the DF from the general expression

$$K = (1/\sigma^2) \int_{-\infty}^{\infty} x n(x) p(x)\, dx,$$

where the input x has a mean square value of σ^2 and an amplitude probability density of $p(x)$.

P.3.8 Determine the DF for the nonlinear block diagram of figure P.3.2.

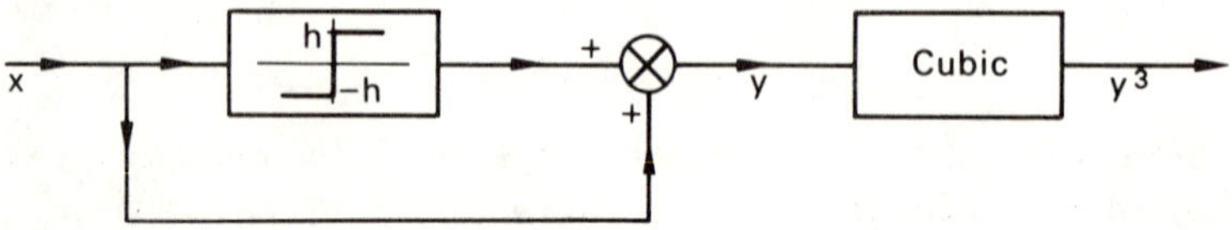

Figure P.3.2

P.3.9 The voltage supplied to an a.c. machine is varied using controlled rectifiers. With an input voltage $x = a\cos\theta$ the output voltage $y(\theta)$ is given by

$$y(\theta) = 0 \qquad -\pi/2 \le \theta < -\alpha$$

$$a\cos\theta \qquad -\alpha < \theta \le \pi/2$$

and $y(\theta + \pi) = -y(\theta)$. Determine using Fourier analysis the fundamental of $y(\theta)$. Show that $y(\theta)$ can be represented as the product of x and a pulse function. Determine the nonlinear characteristic which will produce this pulse function from x for $0 < \alpha < \pi/2$ and thus draw a nonlinear block diagram which gives $y(\theta)$ from x. Express the output of the nonlinear block as a function of x and $\dot{x}$.

P.3.10 Use equations (3.4) and (3.5) to obtain the DF for the asymmetrical nonlinearity of figure P.3.3 and show that it has no phase shift. Determine n_p and n_q and synthesise a SVNL which will give the same fundamental output. Does this nonlinearity give any other output harmonics which are identical to those of the original nonlinearity?

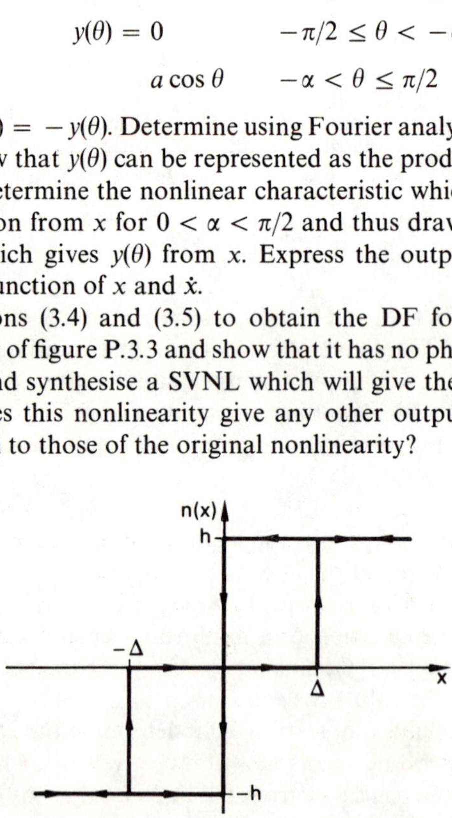

Figure P.3.3

P.3.11 Use equation (3.23) to find the DF for a symmetrical odd linear segmented nonlinearity having breakpoints for positive inputs at $\alpha_1, \alpha_2, \ldots, \alpha_n$ and slopes k_1 from the origin to α_1 and $k_2, \ldots, k_n$ between the breakpoints and k_{n+1} beyond α_n. Give the result for $a > \alpha_n$.

P.3.12 Use equation (3.32) to find a_s for an ideal relay.

P.3.13 Show that the DF of the asymmetrical DVNL for which both $n_p(x)$ and $n_q(x)$ have odd symmetry is the same as the DF of the SVNL $n(x) = n_p(x)$.

P.3.14 Find the DF of the DVNL given by

$$
\begin{aligned}
n_1(x) &= -1 && x < 0 \\
&\quad -\cos x && 0 < x < \pi \\
&\quad 1 && x > \pi \\
n_2(x) &= 1 && x > 0 \\
&\quad \cos x && -\pi < x < 0 \\
&\quad -1 && x < -\pi
\end{aligned}
$$

P.3.15 Find for the on-off relay with dead zone

(a) The r.m.s. DF, $N_{\text{r.m.s.}}(a)$.

(b) The corrected conventional DF, $N_{\text{ccDF}}(a)$.

(c) The gain which using the DF definition of section 3.3.1 minimises the mean absolute error.

(d) As (c) but minimises the average error.

P.3.16 Show that the quadrature component of the DF for a DVNL is proportional to $1/a^2$ for those amplitudes, a, beyond the double valued region.

P.3.17 Synthesise the general linear piecewise characteristic given by

$$
\begin{aligned}
y &= m_1 x \qquad 0 \le x \le \delta_1 \\
&\quad m_{M+1}x + \delta_1(m_1 - m_2) + \delta_2(m_2 - m_3) + \cdots \delta_M(m_M - m_{M+1}) \\
&\qquad\qquad\qquad\qquad \delta_M \le x \le \delta_{M+1}
\end{aligned}
$$

from saturation characteristics. Hence, or otherwise, determine $N_p(a)$ and a_s.

P.3.18 Determine a_1 and b_1 for an input amplitude of 5 units to the series combination of a backlash characteristic and a uniform quantiser. Take $b = 4$ for the backlash, $h = 2$ and $\delta = 1$ for the quantiser.

P.3.19 When $n_p(x)$ and $n_q(x)$ are linear segmented characteristics integrals of the form $\int xT_s(x/a)r(x)\,dx$ and $\int xD_s(x/a)r(x)\,dx$ must be evaluated. Show how this can be done using the relationships given in table 3.1 rather than by making the substitution $x = a\cos\theta$. In particular show that for saturation

$$
a_s = \frac{4m}{\pi}\left\{\frac{\delta D_s(\delta/a)}{s} - \frac{aD_{s+1}(\delta/a)}{2(s+1)} - \frac{aD_{s-1}(\delta/a)}{2(s-1)}\right\}.
$$

P.3.20 Obtain the in phase and quadrature nonlinearities for the nonlinearity shown in figure P.3.4. Show that the in phase nonlinearity may be represented by the parallel combination of two common nonlinear characteristics. Hence, making use of appendix A1, write down the describing function for the nonlinearity for input amplitudes greater than 2 units.

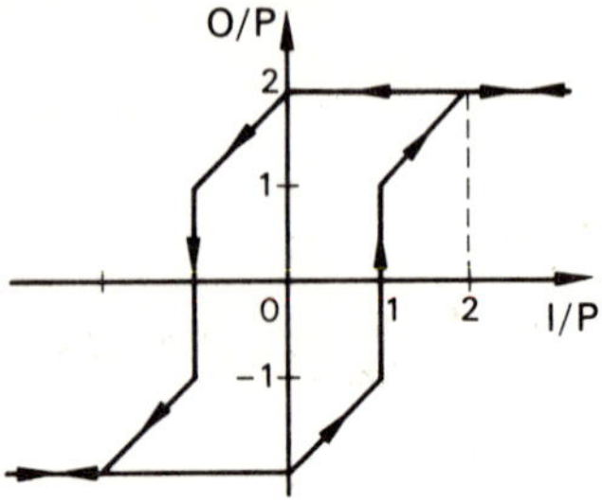

Figure P.3.4

P.3.21 Evaluate the DF for the nonlinearity shown in figure P.3.5.

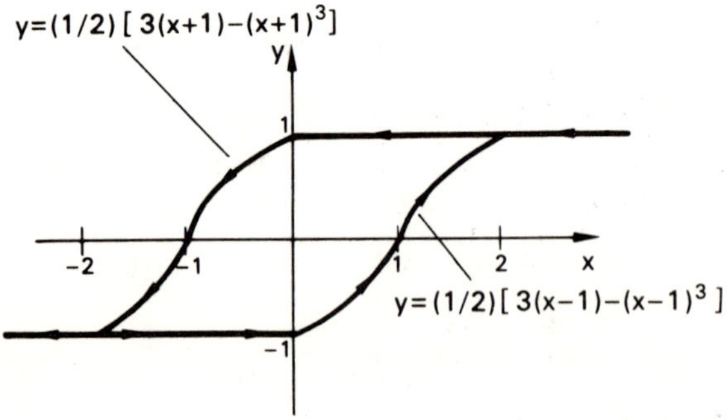

Figure P.3.5

P.3.22 Evaluate the DF for the nonlinear network shown in figure P.3.6.

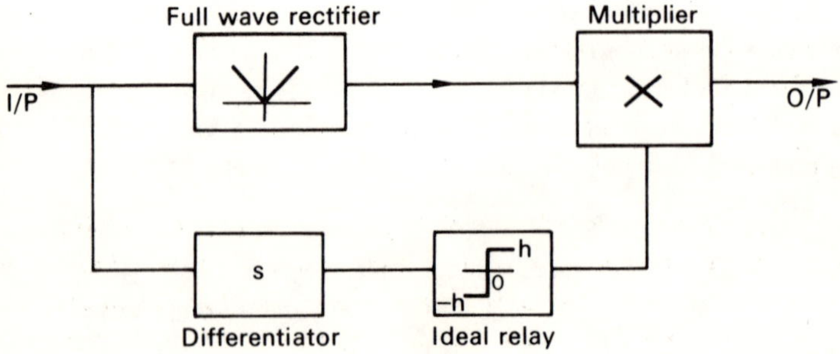

Figure P.3.6

Chapter 4

Applications of the DF Method

4.1 Introduction

In view of the fact that the frequency response approach is such a powerful tool for the designer of linear systems it was inevitable that an effort would be made to extend the approach to nonlinear systems. This did in fact take place in several countries simultaneously as mentioned in section 3.1, resulting in the first harmonic or DF method. It was the forerunner of what may be called quasilinear techniques, where a nonlinear element is replaced by the best linear gain approximation. The value of this linear gain is dependent upon the criterion chosen for the definition of the best approximation, the magnitude and type (i.e. sinusoidal, random, etc.) of input signal and the shape of the nonlinear characteristic. The usual criterion taken is one which minimises the mean squared error between the output from the equivalent linear gain and that of the nonlinear element. It has been shown in section 3.3.1 that for a sinusoidal input this is identical to neglecting all the output harmonics except the fundamental from the nonlinear characteristic.

In this chapter the stability and frequency response of nonlinear systems is investigated using the first harmonic method. In studying the stability of a system one is investigating whether a symmetrical oscillation, assumed sinusoidal at the nonlinearity input, can exist in the closed loop. The effect of auxiliary inputs on the system stability cannot be considered at this stage as in general their presence will violate the assumption that the input to the nonlinearity is a single sinusoid. Situations requiring the assumption of the sinusoidal oscillation together with other signals at the nonlinearity input are considered in later chapters. Also only systems with odd symmetrical characteristics can be considered, as asymmetrical characteristics will normally cause bias signals at other points in the loop, including the nonlinearity input. Similar considerations restrict the closed loop frequency response determination to nonlinear systems with a symmetrical nonlinearity, which continues to operate symmetrically with the input applied, and with no external inputs other than the forcing sine wave.

In addition however the basic assumption of a single sinusoidal input to the nonlinearity will be invalidated if the system is unstable in the autonomous state, for example, possessing a limit cycle when no sinusoidal input is applied to the system or if the presence of the sinusoidal input induces a limit cycle at some frequency different from that of the input. These latter situations which require the assumption of two sinusoidal signals at the input to the nonlinearity, are considered in section 5.11.2.

4.2 The condition for stability

A closed loop system containing a simple nonlinear characteristic is shown in the block diagram of figure 4.1. If the loop is opened at the nonlinear characteristic

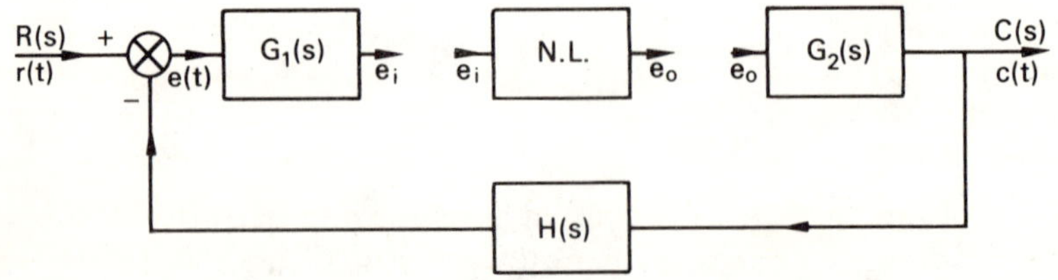

Figure 4.1 General block diagram

and a sinusoidal signal $E_0 \sin \omega t$ injected at the input to $G_2(s)$, then the sinusoidal signal e_i fed back to the nonlinearity input, with $r(t) = 0$, is given by

$$e_i = -g(\omega)E_0 \sin\{\omega t + \theta(\omega)\},$$

where

$$g(\omega) = |G(j\omega)| = |G_1(j\omega)G_2(j\omega)H(j\omega)|,$$

and

$$\theta(\omega) = \angle G(j\omega) = \angle G_1(j\omega) + \angle G_2(j\omega) + \angle H(j\omega).$$

Now in passing through the nonlinearity the sinusoid e_i assumed of amplitude a has a gain of

$$N(a) = M(a)e^{j\psi(a)},$$

so that the output from the nonlinearity is

$$-M(a)g(\omega)E_0 \sin\{\omega t + \theta(\omega) + \psi(a)\},$$
$$= M(a)g(\omega)E_0 \sin\{\omega t + \pi + \theta(\omega) + \psi(a)\}.$$

The system will oscillate if the above signal is equal to $E_0 \sin \omega t$, that is if

$$M(a)g(\omega) = 1,$$

and (4.1)

$$\theta(\omega) + \psi(a) = (2n + 1)\pi,$$

where n is an integer, or in one equation if

$$G(j\omega) = -1/N(a). \quad (4.2)$$

Oscillation thus occurs if the product of the open loop frequency response and the DF is equal to minus one. As in general $G(j\omega)$ is a complex function of ω and $N(a)$ a complex function of a, the solution of equation (4.2) gives both the frequency and amplitude of the oscillation.

Thus, the stability of the nonlinear system can be investigated using any of the conventional linear techniques with the nonlinearity replaced by an amplitude dependent gain, its DF $N(a)$. These methods[1] include use of the Routh criterion, Nyquist diagrams, Nichols chart, the parameter plane and several others. Which approach is used in a specific situation will depend both upon the problem under consideration and personal preference. It should be noted that since the DF is a description of the behaviour of the nonlinearity to a sinusoidal input, s plane techniques such as the root locus and the parameter plane which are based on the characteristic equation $1 + N(a)G(s) = 0$ are not valid for $s \neq j\omega$. For small excursions of s from the imaginary axis however the equation usually yields suitable qualitative, if not quantitative, information on the system.

When the frequency response locus $G(j\omega)$ and the negative reciprocal of the DF, $-1/N(a)$, are plotted on a Nyquist diagram any intersection will be a solution of equation (4.2). The amplitude of oscillation a_0 and its frequency ω_0 can then be read off the $-1/N(a)$ and $G(j\omega)$ loci as shown in figure 4.2(a). The oscillation may be stable or unstable as noted for limit cycles of second order systems in section 2.5.1, where in this context half stable cases are classified as unstable. The stability of the oscillation is discussed further in the next section.

If on the other hand no intersection of the loci $G(j\omega)$ and $-1/N(a)$ exists the stability of the system is assessed using the normal Nyquist criterion with respect to any point on the $-1/N(a)$ locus rather than the point $(-1, 0)$. It should however be remembered that the DF method is an approximate technique and a near miss of the $-1/N(a)$ and $G(j\omega)$ loci may in fact be a hit as far as the actual system is concerned.

Frequency response approaches, such as the Nyquist diagram, are also suitable for synthesising linear, and occasionally nonlinear, compensating networks to improve the relative stability of the system. For example, a series compensating network $G_c(s)$ can be added so that the new open loop frequency response locus $G(j\omega)G_c(j\omega)$ does not intersect or is more distant from the $-1/N(a)$ locus. For a stable system such as that shown in figure 4.2(b) the further apart the

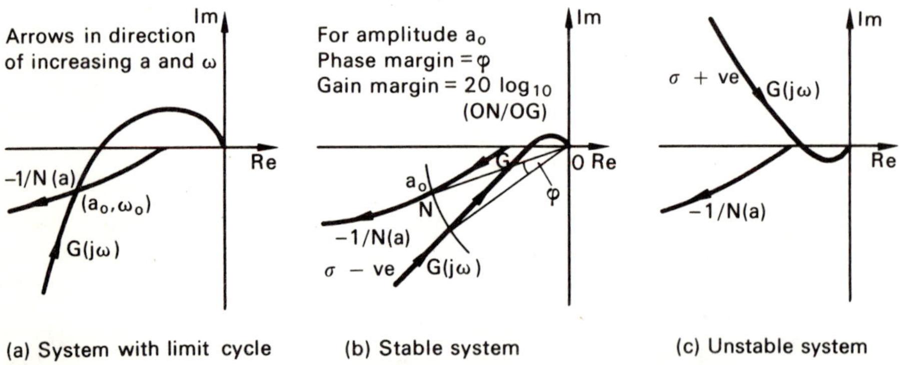

Figure 4.2 Illustration of stability, gain and phase margins

$-1/N(a)$ and $G(j\omega)$ loci the more stable the system, in that for a given magnitude of sinusoidal input the peak resonance will be lower, and the response less oscillatory for a given step input. Quantitative results regarding the transient behaviour cannot however be deduced from frequency response diagrams.

An indication of the relative stability, as for a linear system, can be obtained from the gain and phase margins. However a gain and phase margin exists for each value of a on the $-1/N(a)$ locus as indicated in figure 4.2(b). The minimum values of these quantities, which normally occur for different values of a, may be used in a system specification.

Bode diagrams are particularly useful for systems with a SVNL since $\psi(a)$ is zero and the effect of the gain change $M(a)$ of the nonlinearity can be assessed by vertical movement of the 0 dB level on the gain frequency characteristic.

When the DF is frequency dependent, possible solutions to the equation

$$G(j\omega) = -1/N(a, \omega), \tag{4.3}$$

which gives the frequency and amplitude of a limit cycle can be investigated in a similar manner to those for equation (4.2).

4.3 The stability of a limit cycle

When a solution to equation (4.2) or (4.3) is obtained does it correspond to a stable or unstable oscillatory condition? If following a small perturbation in either amplitude or frequency the limit cycle returns to the original solution state, the limit cycle is said to be stable, whereas if the amplitude and/or frequency continue to move away from this equilibrium state, the limit cycle solution is unstable. This problem is one of limit cycle dynamics, which is discussed in more depth in section 9.6, however a satisfactory solution[2] can be obtained by ignoring the effects of the rate of change of frequency and damping.

Equation (4.2) or (4.3) may be written in the form

$$X(a, \omega) + jY(a, \omega) = 0 \tag{4.4}$$

and assumed to have a limit cycle solution $a = a_0 \exp(j\omega_0 t)$. Considering a small perturbation in a_0 to $a_0 + \Delta a$ and in ω_0 to $\omega_0 + \Delta\omega + j\Delta\sigma$, a becomes $(a_0 + \Delta a)\exp\{j(\omega_0 + \Delta\omega + j\Delta\sigma)t\}$. If $\Delta\sigma$ is positive the real part of the exponential is negative and a is positively damped (i.e. decaying). Now

$$X(a_0 + \Delta a, \omega_0 + \Delta\omega + j\Delta\sigma) + jY(a_0 + \Delta a, \omega_0 + \Delta\omega + j\Delta\sigma) = 0$$

so expanding in a Taylor series to first order terms about the equilibrium state (a_0, ω_0) and removing the limit cycle solution equation $X(a_0, \omega_0) + jY(a_0, \omega_0) = 0$ gives

$$\left.\frac{\partial X}{\partial a}\right|_s \Delta a + \left.\frac{\partial X}{\partial \omega}\right|_s \Delta\omega + j\left.\frac{\partial X}{\partial \omega}\right|_s \Delta\sigma + j\left.\frac{\partial Y}{\partial a}\right|_s \Delta a + j\left.\frac{\partial Y}{\partial \omega}\right|_s \Delta\omega - \left.\frac{\partial Y}{\partial \omega}\right|_s \Delta\sigma = 0,$$

where $|_s$ denotes that the partial derivative is evaluated at the solution (a_0, ω_0). Separating the equation into its real and imaginary parts gives

$$\left.\frac{\partial X}{\partial a}\right|_s \Delta a + \left.\frac{\partial X}{\partial \omega}\right|_s \Delta\omega - \left.\frac{\partial Y}{\partial \omega}\right|_s \Delta\sigma = 0, \tag{4.5}$$

and

$$\left.\frac{\partial Y}{\partial a}\right|_s \Delta a + \left.\frac{\partial Y}{\partial \omega}\right|_s \Delta\omega + \left.\frac{\partial X}{\partial \omega}\right|_s \Delta\sigma = 0. \tag{4.6}$$

Eliminating $\Delta\omega$ from equations (4.5) and (4.6) one has

$$\left.\left(\frac{\partial X}{\partial a}\frac{\partial Y}{\partial \omega} - \frac{\partial X}{\partial \omega}\frac{\partial Y}{\partial a}\right)\right|_s \Delta a = \left\{\left(\left.\frac{\partial X}{\partial \omega}\right|_s\right)^2 + \left(\left.\frac{\partial Y}{\partial \omega}\right|_s\right)^2\right\}\Delta\sigma.$$

For the limit cycle to be stable a positive Δa must lead to a positive $\Delta\sigma$ and vice versa, so that a necessary condition for the limit cycle to be stable is

$$\left.\left(\frac{\partial X}{\partial a}\frac{\partial Y}{\partial \omega} - \frac{\partial X}{\partial \omega}\frac{\partial Y}{\partial a}\right)\right|_s > 0. \tag{4.7}$$

Although this is a suitable general result it is often easier to apply if it is extended to allow a graphical interpretation. Assuming a frequency independent DF let

$$G(j\omega) = U(\omega) + jV(\omega),$$

and

$$C(a) = -1/N(a) = P(a) + jQ(a),$$

then

$$G(j\omega) + \{1/N(a)\} = U(\omega) + jV(\omega) - P(a) - jQ(a),$$

so that

$$X(a, \omega) = U(\omega) - P(a) \quad \text{and} \quad Y(a, \omega) = V(\omega) - Q(a).$$

Substituting in equation (4.7), the Loeb condition for a stable limit cycle becomes

$$\left(\frac{\partial U}{\partial \omega}\frac{\partial Q}{\partial a} - \frac{\partial V}{\partial \omega}\frac{\partial P}{\partial a}\right)\bigg|_s > 0, \tag{4.8}$$

which, if $N(a)$ is real, reduces further to

$$\left(\frac{\partial V}{\partial \omega}\frac{\partial P}{\partial a}\right)\bigg|_s < 0. \tag{4.9}$$

Examination of the left hand side of equation (4.8) shows that it is equal to the vector product $\partial \mathbf{G}/\partial \omega \times \partial \mathbf{C}/\partial a$, which may be denoted by $\mathbf{g} \times \mathbf{c}$. The necessary condition for a stable limit cycle thus requires the vectors $(\mathbf{g}, \mathbf{c}, \mathbf{v})$, where $\mathbf{v} = \mathbf{g} \times \mathbf{c}$, to form a right hand triple at the solution point.

A similar graphical criterion* may also be arrived at by logical argument under the same conditions that apply to the analytical solution. For a stable limit cycle, if a small increase in amplitude along the $C(a)$ locus from the solution is considered, then the movement should be into the region of a stable system configuration and for a small decrease in amplitude along the $C(a)$ locus, the movement should be into the region of an unstable system configuration. In both cases the perturbation moves the system into a state which will counteract the perturbation so as to restore the system to the limit cycle solution. The locations of the loci $G(\Delta\sigma + j\omega)$ and $G(-\Delta\sigma + j\omega)$ with respect to the $G(j\omega)$ locus, where $\Delta\sigma$ is a small positive quantity, are easily found from the rules of the s to $G(s)$ plane mapping.

Consider for example the situation portrayed in figure 4.3 where equation (4.2) has three solutions at the points A, B and C respectively. For small increments in a along $C(a)$ from these solutions, movement is into regions of $\Delta\sigma$, negative, positive, negative respectively. The limit cycle solutions at A and C are therefore stable and at B unstable. The autonomous system will attain the limit cycle at A without any initial disturbance, since it is unstable for small values of a. A

* This criterion, see reference 3, can give incorrect results for high order systems with multiple limit cycle solutions.

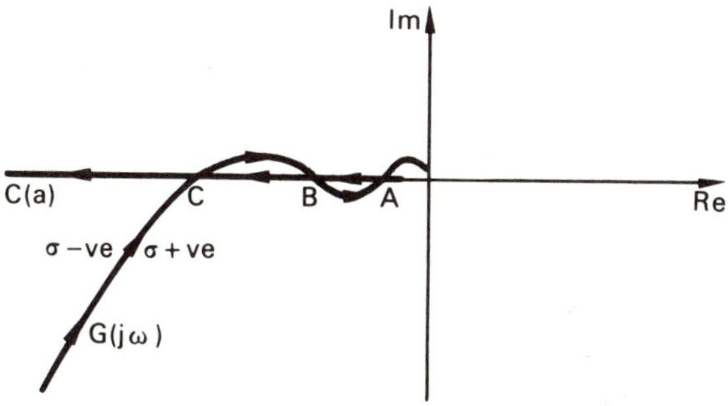

Figure 4.3 Illustration of limit cycle stability

disturbance large enough to move a on $C(a)$ beyond B will cause the solution at C to be reached and held when the disturbance is removed.

An alternative method for investigating the stability of a limit cycle which uses the incremental describing function (IDF) is described in section 5.10.

4.4 Applications of the DF to stability analysis

To illustrate applications of the material presented in the previous sections several examples are considered below using various techniques of linear control theory.

4.4.1 *Single loop system with a SVNL*

General results regarding the stability of the simple system of figure 4.1 can be found when the nonlinearity is any single valued odd characteristic. When the nonlinear characteristic is replaced by a linear gain K, that is $y = n(x) = Kx$, the range of K for which the linear system is stable, say $k_1 < K < k_2$, can easily be found. The system is then said to be stable for values of K inside the open interval (k_1, k_2) and the straight lines $y = k_1x$, $y = k_2x$ define the bounds of the region of stable linear gains, which is known as the Hurwitz sector and is shown in figure 4.4.

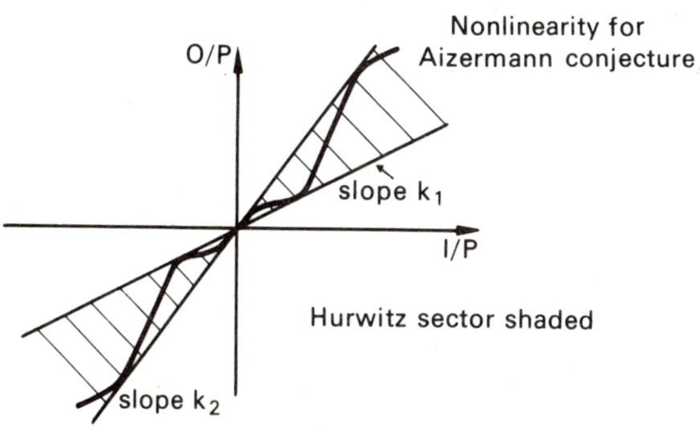

Figure 4.4 Hurwitz sector

By DF theory the corresponding nonlinear system will be stable if $k_1 < N(a) < k_2$ for all a. This inequality will be satisfied (see section 3.2.2) if $k_1 x < n(x) < k_2 x$, that is if the nonlinearity $n(x)$ is contained within the Hurwitz sector for all x. In particular if $k_1 = 0$ and $n(x)$ is monotonic with a maximum slope at the origin, that is $n'(0) > n'(x)$ for all $x \neq 0$, the system will be stable if $0 < n'(0) < k_2$.

Aizerman[4] in an early paper put forward the conjecture that a simple nonlinear system as discussed above would be stable provided the nonlinearity was confined to the Hurwitz sector. This implies that the result obtained above by the approximate DF theory is exact. Although it has not been found possible using exact methods of stability theory to disprove Aizerman's conjecture in general, several specific examples have been cited[5,6] to which it does not apply and thus for which the DF solution is incorrect. The accuracy of the DF approach is discussed in a later section where a distortion criterion is suggested as a suitable indicator of the validity of the results. Due to the form of the linear transfer functions used in the various counterexamples of Aizerman's conjecture cited, the distortion criterion rejects the DF results as inaccurate.

As a particular example assume the open loop transfer function $G(s) = 40/\{s(s+2)(s+8)\}$. Taking a linear gain K in series with $G(s)$ gives the characteristic equation

$$s(s+2)(s+8) + 40K = 0,$$

and applying Routh's criterion the sufficient condition for stability is $0 < K < 4$. If the nonlinearity is a saturation characteristic with unit gain in the linear region the system will be stable since $N(a) \leq 1$ for all a, but if it is a cubic characteristic, $y = x^3$, for which $N(a) = 3a^2/4$ then an oscillation is possible with amplitude $a = (4 \times 4/3)^{1/2} = 4/\sqrt{3}$. Figure 4.5 shows a sketch of the $G(j\omega)$ locus and the describing functions for the cubic and saturation characteristics. The frequency of the possible oscillation at A for the cubic nonlinearity is 4 rads/sec, the value of ω when $\text{Im}\{G(j\omega)\} = 0$.

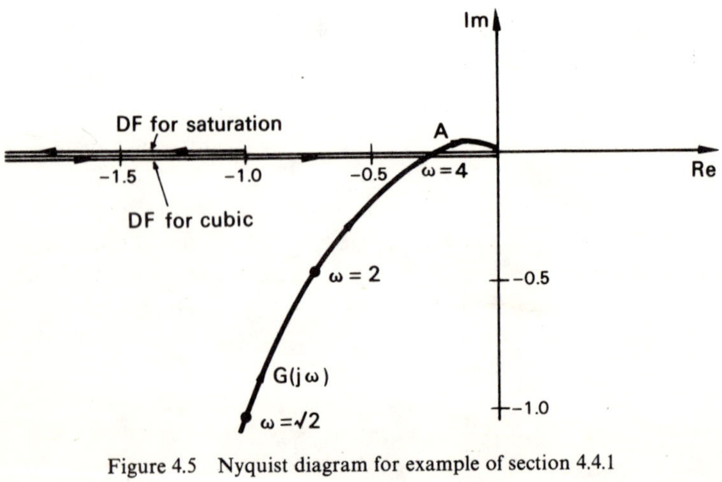

Figure 4.5 Nyquist diagram for example of section 4.4.1

Two points should be noted about the limit cycle for the cubic. First, the oscillation is not self-starting, as the slope $n'(0)$ of $n(x)$ at $x = 0$ is less than 4, and if the oscillation amplitude is produced by some external disturbance the system will not return to equilibrium if the disturbance is immediately removed. Since the loop can be made unstable by inputs and disturbances above a certain magnitude, which, as will be shown later, may be random, its properties as a control system may be unsatisfactory. Also, it is seen in figure 4.5 that the describing function locus intersects the frequency response locus at A with the increasing amplitude direction moving into the positive $\Delta\sigma$ region, the oscillation corresponding to this intersection is therefore unstable. Since no other intersection of the describing function and frequency locus exists the oscillation will build up from A to an amplitude determined by other nonlinear phenomena which will eventually occur in the system.

4.4.2 *System containing an on-off relay with hysteresis*

The block diagram of the system considered is shown in figure 4.6 with $G(s) = 10/(1 + s)(1 + s/4)(1 + s/8)$. If the reference input is not zero, since the linear transfer function $G(s)$ has a finite gain of 10 at zero frequency, a d.c. or bias signal will exist at the nonlinearity input. The analysis then requires evaluation of the DF in the presence of a d.c. signal, a topic which is covered in the next chapter.

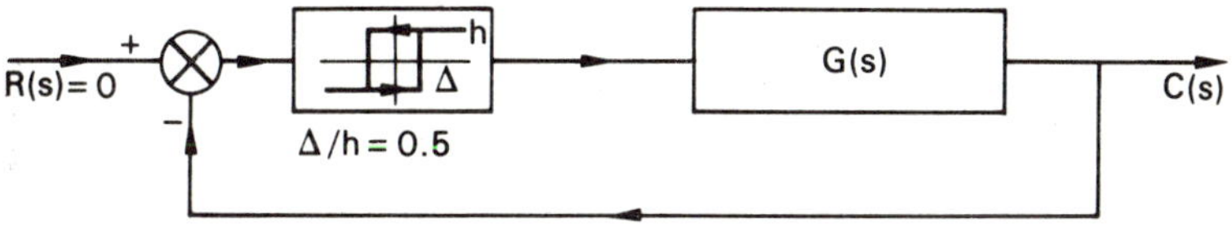

Figure 4.6 Block diagram for example of section 4.4.2

The DF, $N(a)$, for the nonlinearity with zero bias is

$$N(a) = (4h/a\pi)e^{-j\theta} \quad \text{for } a > \Delta,$$

where $\theta = \sin^{-1}(\Delta/a)$. This gives

$$C(a) = \{-1/N(a)\} = (-a\pi/4h)e^{j\theta} = (-a\pi/4h)(\cos\theta + j\sin\theta)$$
$$= (-\pi/4h)(a^2 - \Delta^2)^{1/2} - j(\pi\Delta/4h). \qquad (4.10)$$

The imaginary part of $C(a)$ is independent of the input amplitude, a, so that the locus is a straight line parallel to the real axis of the Nyquist diagram. The assumed polar locus $G(j\omega)$ and the $C(a)$ locus are sketched in figure 4.7 and intersect at a single point. The solution is a stable oscillation as the $G(j\omega)$ locus is crossed by the describing function for increasing amplitude in the direction of σ negative.

In this particular case, since the $C(a)$ locus is a straight line, an exact solution for the values of a and ω at the point of intersection is more easily obtained analytically than graphically.

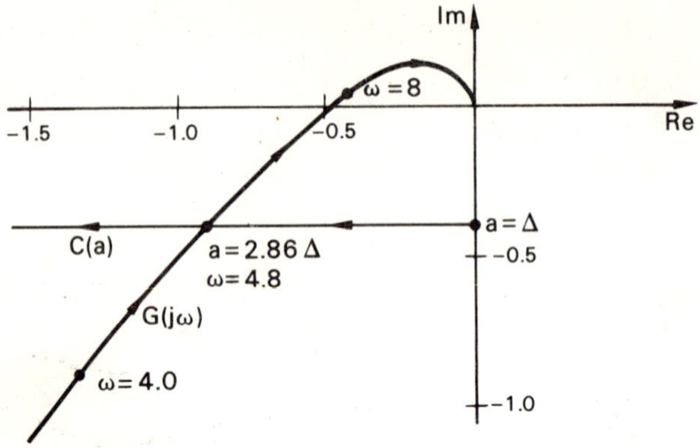

Figure 4.7 Nyquist diagram for example of figure 4.6

Now

$$G(s) = \frac{10}{(1 + s)(1 + s/4)(1 + s/8)} = \frac{320}{(s + 1)(s + 4)(s + 8)},$$

giving

$$G(j\omega) = \frac{320\{32 - 13\omega^2 - j\omega(44 - \omega^2)\}}{(1 + \omega^2)(16 + \omega^2)(64 + \omega^2)}.$$

The loci intersect where $\text{Im}\{G(j\omega)\} = -\pi\Delta/4h$. Taking $\Delta/h = 0{\cdot}5$ gives

$$\frac{320\omega(44 - \omega^2)}{(1 + \omega^2)(16 + \omega^2)(64 + \omega^2)} = \frac{\pi}{8}, \qquad \text{i.e. } \omega = 4{\cdot}8 \text{ rads/sec.}$$

The amplitude, a, of the oscillation at the input to the nonlinearity is then given by

$$\frac{\pi(a^2 - \Delta^2)^{1/2}}{4h} = \frac{320(13\omega^2 - 32)}{(1 + \omega^2)(16 + \omega^2)(64 + \omega^2)}, \qquad \text{with } \omega = 4{\cdot}8,$$

so that

$$a = 1{\cdot}43h \text{ or } 2{\cdot}86\Delta.$$

4.4.3 *A second order system*

To provide a comparison with the phase plane approach example 1 of section 2.11 is solved using the DF. The block diagram of the system is shown in figure 2.25 and the linear transfer functions are $G(s) = K/s^2$ and $H(s) = 1 + s\lambda$ respectively. As a closed form analytical solution in terms of all the parameters is extremely involved the following values are assumed, $h = 4$, $\Delta = 1/2$ and $\delta = 1$ for the relay and $K = 1$.

It has been stated earlier that the accuracy of the DF method may be expected to improve when the linear elements provide improved low pass filtering, thus solutions to the problem are considered as λ is varied. For the relay,

$$N(a) = (8/a^2\pi)[\{a^2 - (3/2)^2\}^{1/2} + \{a^2 - (1/2)^2\}^{1/2} - j],$$

and for the linear elements,

$$G(j\omega)H(j\omega) = -(1 + j\omega\lambda)/\omega^2.$$

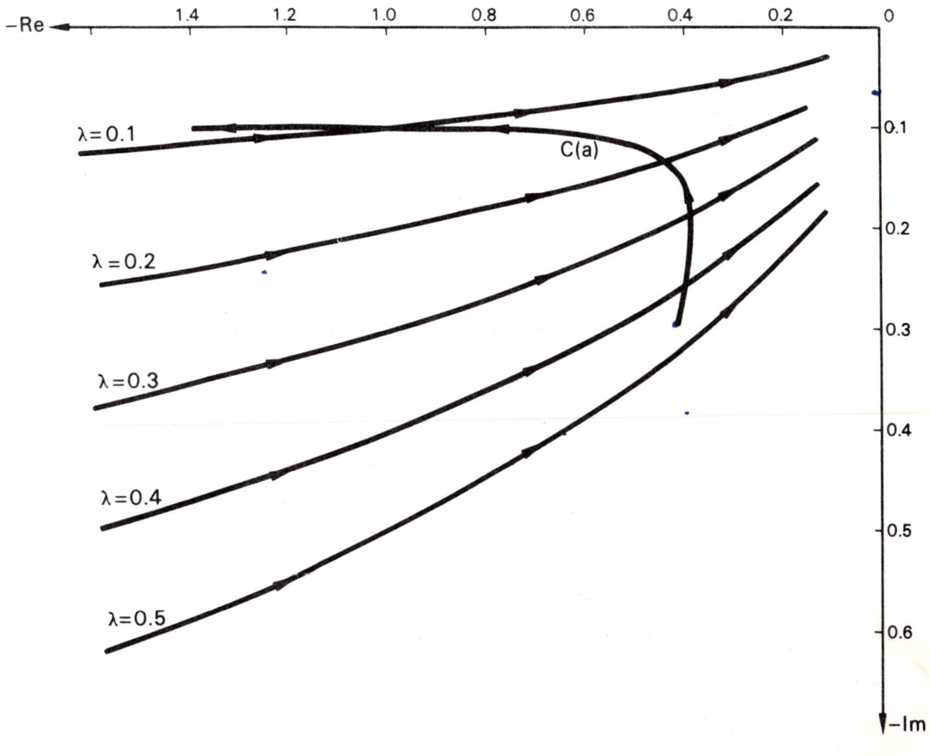

Figure 4.8 DF and frequency response loci

Figure 4.8 shows graphs of the $C(a)$ locus and $G(j\omega)H(j\omega)$ for various values of λ. Numerical solution of the above equations gives the results shown in figure 4.9 for a and ω as functions of λ. The exact solutions for the fundamental input amplitude to the nonlinearity and ω, obtained from section 2.11, are also given. As expected the approximate DF solutions for both a and ω become more accurate as λ decreases and, in fact, no DF solution exists for $\lambda > 0{\cdot}45$.

4.4.4 *A multiloop system*

The block diagram in normalised form, with unit gain through the limiter in the linear regime, of figure 4.10 represents a position control system with torque saturation. The control signal contains terms proportional to the error signal,

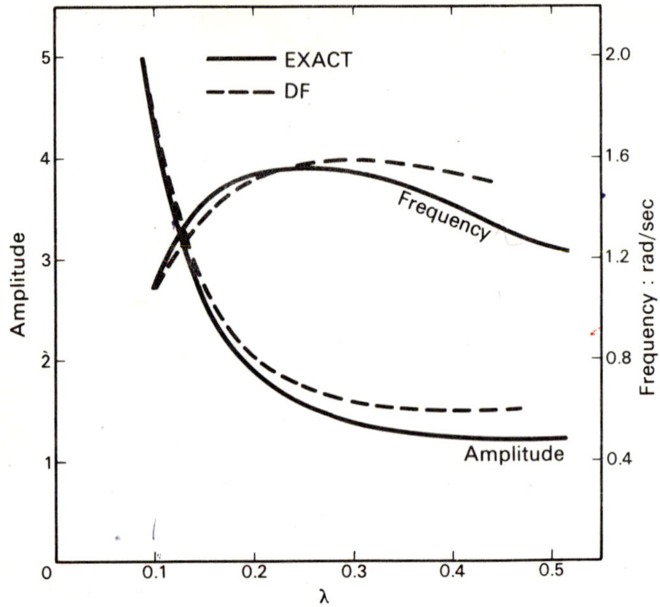

Figure 4.9 Comparison of exact and DF solutions

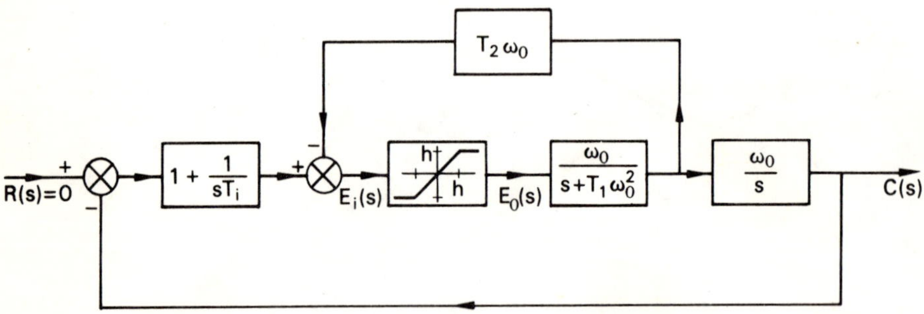

Figure 4.10 Block diagram of system considered in section 4.4.3

the integral of the error and the output velocity. The system is of second order when no integral of error term is present and has a damping factor for small signals of $(T_1 + T_2)\omega_0$ where ω_0 is the undamped natural frequency. The integral control time constant, $T_i = 8$ secs, is included in the system to eliminate any steady state error when a constant disturbance torque exists on the output shaft. The problem is to determine the minimum value of the ratio T_1/T_2 for which the system will always be stable if $\omega_0 = 2$ rads/sec and the system is critically damped, that is $(T_1 + T_2)\omega_0 = 2$, in the linear regime with no integral of error present. To obtain the required open loop transfer function $G(s)$ the block diagram must be manipulated using the conventional techniques into the single loop form of figure 4.1, or alternatively if the loop is assumed broken at the control signal, a signal $E_0(s)$ injected and the return signal $E_i(s)$ calculated then $G(s) = -E_i(s)/E_0(s)$. Using this second procedure one obtains the equations

$$C(s) = \frac{\omega_0^2 E_0(s)}{s(s + T_1\omega_0^2)},$$

and

$$E_i(s) = -C(s)\left(\frac{8s + 1}{8s}\right) - T_2 sC(s).$$

Eliminating $C(s)$ gives

$$G(s) = \frac{-E_i(s)}{E_0(s)} = \frac{\omega_0^2(8s^2 T_2 + 8s + 1)}{8s^2(s + T_1\omega_0^2)},$$

and substituting for ω_0

$$G(s) = \frac{8s^2 T_2 + 8s + 1}{2s^2(s + 4T_1)}.$$

The locus $G(j\omega)$ thus tends to ∞ along $-180°$ when ω tends to 0 and tends to 0 along $-90°$ when ω tends to ∞. Hence the locus can only cut the negative real axis from above, that is from positive to negative imaginary values as ω increases, and the system will be unstable in the linear regime if the axis is crossed within the range -1 to 0. The $C(a)$ function for the saturation characteristic lies on the negative real axis from -1 to $-\infty$, so that oscillations are possible with saturation present if $G(j\omega)$ crosses the negative real axis in this range.

To determine the condition for stability

$$G(j\omega) = \frac{1 + 8j\omega - 8\omega^2 T_2}{-2\omega^2(4T_1 + j\omega)},$$

giving

$$\text{Im}\{G(j\omega)\} = \frac{j\omega(32T_1 - 1 + 8\omega^2 T_2)}{-2\omega^2(\omega^2 + 16T_1^2)}.$$

Therefore

$$\text{Im}\{G(j\omega)\} = 0 \qquad \text{when } \omega = \infty \quad \text{or} \quad \omega^2 = (1 - 32T_1)/8T_2.$$

The last solution is real only if $T_1 < 1/32$, so that no intersection with the real axis will occur if $T_1 > 1/32$. But $(T_1 + T_2)\omega_0 = 2$ for critical damping, giving $T_2 = 31/32$ when $T_1 = 1/32$. Thus to ensure stability it is required that $T_1/T_2 > 1/31$. Alternatively, applying Rouths' criterion to the characteristic equation

$$1 + KG(s) = 0$$

which becomes $2s^3 + 8s^2(T_1 + KT_2) + 8sK + K = 0$, with $K = N(a)$, and gives $T_1 + KT_2 > 1/32$ for stability. Substituting $T_1 = 1 - T_2$ gives $T_2(1 - K) < 31/32$ so that for stability $T_2 < 31/32$, since $N(a)$ tends to its minimum value of zero as a tends to infinity. Hence, as before, $T_1/T_2 > 1/31$ for stability.

4.4.5 *System containing a relay with dead zone*

Here the stability of a feedback loop containing an on-off relay with dead zone and a linear transfer function $G(s) = K/s(s + 2)(s + \alpha)$, as shown in figure 4.11, is studied using the parameter plane. The parameter plane is not considered in many introductory control systems texts so that a brief review of the approach is given first.

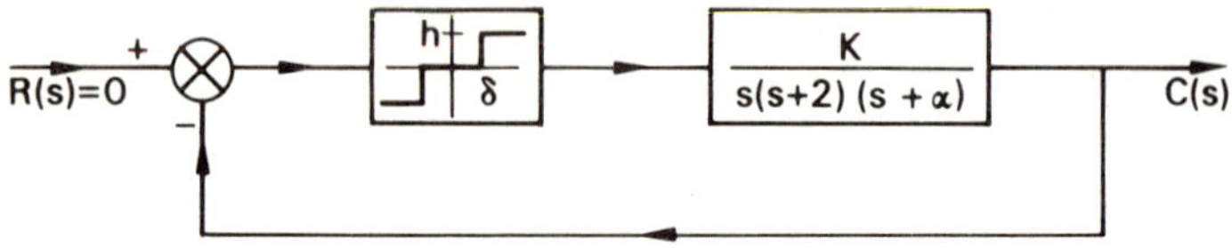

Figure 4.11 Block diagram of example considered in section 4.4.4

The procedure is based on the system characteristic equation $1 + KG(s) = 0$. If the equation has two unknown parameters[7] α and β then it may be written in the form

$$f_1(s) + \alpha f_2(s) + \beta f_3(s) = 0, \tag{4.11}$$

where the functions f_1, f_2 and f_3 are known. Often the coefficients α and β are associated with a particular power of s in the equation, for example where $f_2(s) = s$ and $f_3(s) = s^0 = 1$. The roots of the equation must be either real, that is of the form $s = \gamma$, or complex pairs with $s = \sigma \pm j\omega = -\zeta\omega_0 \pm j\omega_0(1 - \zeta^2)^{1/2}$. Substitution of these values of s in equation (4.11) gives a set of loci in the $\alpha - \beta$ parameter plane, in the first case for various values of γ and in the second for various values of ζ with ω_0 a parameter on the loci. Thus for chosen values of α and β defining a point P in the parameter plane the roots of the characteristic equation are found from the loci passing through P. For stability studies the location of the locus for $\zeta = 0$ is of sole importance and the technique may be regarded as a graphical approach to the Routh criterion which is helpful to design if the parameters α and β are variables. When using the method for investigating stability of a nonlinear system the parameters α and β can be the real and imaginary components, $N_p(a)$ and $N_q(a)$, of the DF.

The side of the $\zeta = 0$ locus which corresponds to stable operation in the parameter plane may be determined by drawing another locus for ζ small, using the Routh criterion, or alternatively by the following mapping theorem.[7] If both the $j\omega$ axis in the s plane and the $\zeta = 0$ locus in the parameter plane are traced for ω increasing then corresponding regions are on the same sides of the two curves if J is negative and on opposite sides if J is positive. The Jacobian J is given by

$$J = \frac{\partial R}{\partial \alpha}\frac{\partial I}{\partial \beta} - \frac{\partial R}{\partial \beta}\frac{\partial I}{\partial \alpha}, \tag{4.12}$$

where $R(\alpha, \beta, \omega) = 0$ and $I(\alpha, \beta, \omega) = 0$ are respectively the real and imaginary equations obtained from equation (4.11) when $s = j\omega$ is substituted.

Returning to the example, $N(a) = (2h/\pi\delta)\sin 2\theta$ where $\theta = \cos^{-1}\delta/a$, and the characteristic equation is

$$s(s+2)(s+\alpha) + KN(a) = 0.$$

Substituting $s = j\omega$ and β for $KN(a)$ gives

$$R(\alpha, \beta, \omega) = -\omega^2(2+\alpha) + \beta = 0,$$

$$I(\alpha, \beta, \omega) = -\omega^3 + 2\alpha\omega = 0.$$

Thus $\beta = \omega^2(2+\alpha)$, $\alpha = \omega^2/2$ and using equation (4.12) $J = -2\omega$ which is negative for ω positive. In this simple case ω can be eliminated easily to give

$$\beta = 4\alpha + 2\alpha^2$$

as the equation of the $\zeta = 0$ locus in the parameter plane, which is shown in figure 4.12. The stable side of the locus, determined by the above theorem, is shown shaded.

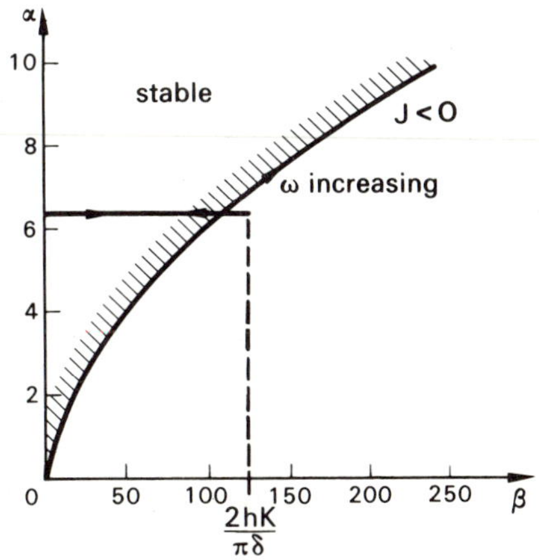

Figure 4.12 Parameter plane diagram for system of figure 4.11

For a given value of α the line showing the variation of β with amplitude a, which has a maximum value of $(2hK/\pi\delta)$ when $a = \sqrt{2}\delta$, can be drawn as shown. Obviously the system will be stable provided

$$(2hK/\pi\delta) < 4\alpha + 2\alpha^2,$$

that is

$$hK < \pi\delta\alpha(2+\alpha). \tag{4.13}$$

4.5 System compensation using the DF

Because of all the variables involved, including economic as well as performance criteria, system design is a complex task often requiring considerable experience and know how. Compensation is usually the final part of the design;

the system components are known, some possibly with alternatives having slightly different characteristics, and adjustments are required to meet the performance specifications. For linear systems the usual procedures are to introduce compensating networks into the low power signal channels or to provide additional feedback paths which often require auxiliary transducers to provide the feedback signal. As seen in chapter 2 nonlinear elements may also be intentionally introduced to provide specific characteristics.

In this section we are concerned with the design of compensators to meet satisfactory frequency response behaviour. This may include stabilising a system, possibly one initially possessing a limit cycle, providing a satisfactory gain and phase margin, or adjusting a limit cycle to a certain amplitude and/or frequency. Although there is no relationship between transient and frequency response for a nonlinear system it is usually possible to assess qualitatively the effect of a frequency domain action on the corresponding transient response. A further check that the transient behaviour is satisfactory however would normally have to be made.

Most compensation techniques for nonlinear systems, since designs are accomplished using say Bode or Nyquist diagrams, follow in a fairly straightforward manner from the concepts for linear systems. Linear compensation(8) may be achieved using phase lead networks, lag networks or additional feedback paths. If it is desired to introduce such features for a given range of amplitudes only, it can be done by introducing nonlinear feedback paths, cascade compensators which are switched to different transfer function values according to a signal amplitude or by using nonlinear feedback networks around a constant gain(9) (see problem P4.13). When switched compensators are used the system is usually referred to as a variable structure system.(10)

Several ingenious nonlinear compensating networks such as the Clegg integrator,(11) the Span filter (12) and the driven limiter(13) are described in the literature. Since these fit into the category of frequency dependent nonlinearities they are discussed in more detail in section 11.6. They all belong to a class of nonlinear networks(14) for which the DF, $N(a, \omega)$, is actually independent of a. The networks have also been used for compensating linear systems primarily because their gain and phase characteristics can be 'independently' adjusted.

Another obvious approach for nonlinear system compensation is to introduce an additional nonlinearity either in series or parallel with the existing nonlinearity to produce approximately linear or a preferable nonlinear behaviour. The physical properties of an actual system, however, often put practical limitations on many of the ideas of this type which may be conceived from purely theoretical considerations. A particularly interesting form of parallel nonlinear compensation known as nonlinear model feedback(15) is illustrated in figure 4.13. By implementing the compensator as shown in (c) the difficulty and expense of producing a high power signal at the output of N_2 is avoided. If G_3 is chosen equal to G_2 then the characteristic equation becomes

$$1 + \{N_1(a) + N_2(a)\}G_1(s)G_2(s) = 0,$$

which reduces to the linear form $1 + G_1(s)G_2(s) = 0$ if $N_2(a) = 1 - N_1(a)$.

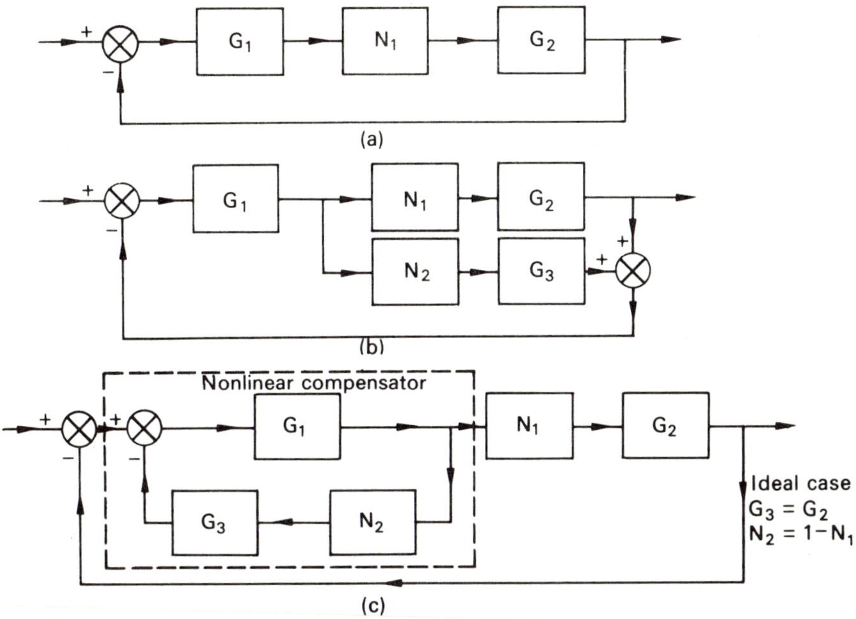

Figure 4.13 Illustration of nonlinear model feedback

In many situations the accuracy of a DF analysis of nonlinear systems with additional nonlinear compensators is open to question since the required assumption of a sinusoidal input to both nonlinearities may not be justified. Because both nonlinearities have the same input the nonlinear model feedback technique does not suffer from this disadvantage. To illustrate the various ideas discussed above, several examples are considered below.

4.5.1 *Compensation of a system containing a relay with dead zone*

Again the block diagram of figure 4.11 is considered. The system is required to have a phase margin of 30° with K, h/δ and α chosen to meet the steady state accuracy requirements, at values of 160, 2 and 8 respectively.

Since the DF of the nonlinearity has a maximum value of $2h/\pi\delta$, the maximum frequency independent loop gain is $hK/\pi\delta\alpha = 40/\pi = 22{\cdot}1$ dB, and the system will be stable if the gain margin of $1/s(1 + s/2)(1 + s/8)$ is greater than 22·1 dB. The Bode diagram of $1/j\omega(1 + j\omega/2)(1 + j\omega/8)$ is shown in figure 4.14 from which it can be seen that the phase lag is 180° at $\omega = 4$ rads/sec corresponding to a gain margin of only 20 dB (18 dB on the Bode approximation) and the system is unstable. This result agrees with equation (4.13), since with the correct system parameters substituted it gives as the condition for stability

$$2 \times 160 < \pi \times 8 \times 10, \qquad \text{i.e. } 4 < \pi,$$

which is not true.

Examination of the Bode diagram suggests the system can be stabilised using a phase advance network $G_c(s) = (1 + sT)/(1 + \alpha sT)$ in cascade with $G(s)$. The

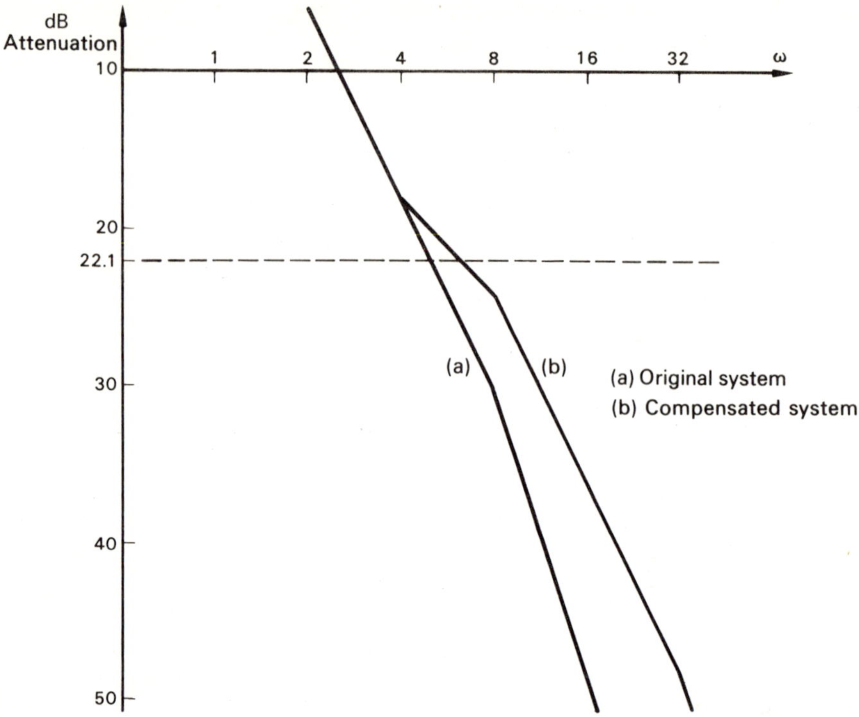

Figure 4.14 Bode diagram for example of section 4.5.1

network $G_c(s)$ provides a maximum angle of lead of $\tan^{-1}\{(1-\alpha)/2\sqrt{\alpha}\}$ at an angular frequency of $\omega_c = 1/\sqrt{\alpha}T$. A reasonable choice for α is $\alpha = 1/8$ and it is then necessary to select a possible value for T. For a suitable choice of T the phase lag of the compensated system $G(s)G_c(s)$ will increase continuously with increasing frequency and the minimum phase margin for all a will be obtained when $a = \delta\sqrt{2}$. Addition of the phase advance network reduces the slope of the gain frequency characteristic around the 180° phase lag frequency so that the 22·1 dB down point will occur at a somewhat higher frequency on the compensated frequency response characteristic $G_c(j\omega)G(j\omega)$ than on the uncompensated characteristic $G(j\omega)$. Thus to produce an approximately maximum phase margin, choice of ω_c in the range 8 to 16 rads/sec, corresponding to T between $1/2\sqrt{2}$ and $1/4\sqrt{2}$, would seem appropriate.

The compensated gain frequency characteristic for $T = 1/4$ sec is shown in figure 4.14. The phase margin is 28° and the gain is 34·8 dB down when the phase shift is 180°, which is a gain margin of (34·8 − 22·1) = 12·7 dB. Comparative figures for other possible values of T are 27° and 9·3 dB for $T = 1/2\sqrt{2}$ sec and 24° and 15·0 dB for $T = 1/4\sqrt{2}$ sec. Thus T within the range $1/2\sqrt{2} > T > 1/4$ would produce a satisfactory system design.

Due to the dead zone δ the above system has steady state errors in response to step inputs. A method of eliminating this error is to shunt the relay with a linear amplifier of relatively low gain which is operative for small signals up to an amplitude of approximately δ. This technique is known as dual mode control.[16]

4.5.2 *Nonlinear compensation*

As an illustration of nonlinear model feedback the position control system shown in figure 4.15, which was discussed in reference 17, is considered.

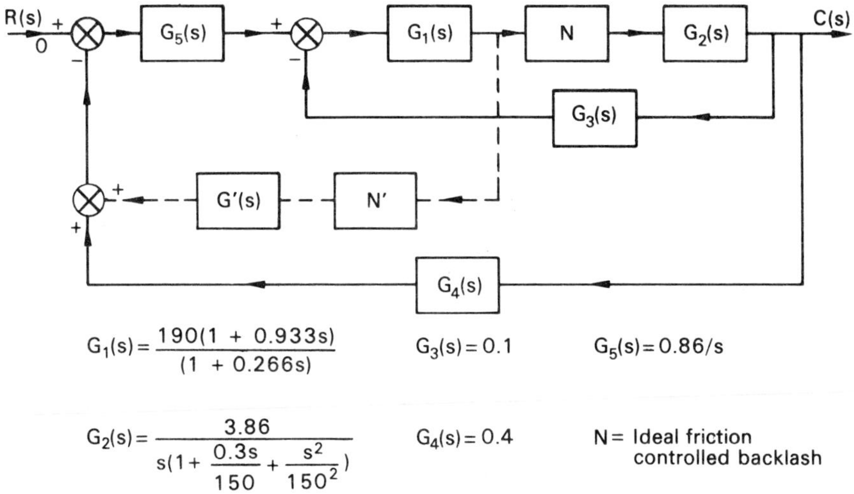

Figure 4.15 Block diagram for example of nonlinear stabilisation

The compensating loop, added to stop the low frequency oscillation caused by the backlash nonlinearity N, consists of the nonlinearity N' and the linear network $G'(s)$. The closed loop relationship between $C(s)$ and $R(s)$ is

$$\left\{\left(R - G_4C - \frac{G'N'C}{G_2N}\right)G_5 - G_3C\right\}G_1G_2N = C,$$

which gives

$$\frac{C}{R} = \frac{G_1G_2G_5N}{1 + NG_1G_2(G_3 + G_4G_5) + G_1G_5G'N'},$$

where the G's are functions of s and the N's of the amplitude a. Thus the condition for a limit cycle is

$$1 + NG_1G_2(G_3 + G_4G_5) + G_1G_5G'N' = 0, \tag{4.14}$$

and if

$$G' = G_2(G_3 + G_4G_5)/G_5, \tag{4.15}$$

the condition becomes

$$-1/(N + N') = G_1G_2(G_3 + G_4G_5) = G, \tag{4.16}$$

with N and N' effectively in parallel, and the DF approach for a system with a single nonlinearity is applicable.

In practice it may be difficult to satisfy equation (4.15) exactly. Provided, however, the agreement is close near to the frequencies of interest, in this case near to the fairly low frequency of the backlash oscillation, and, in addition, the compensating loop does not dominate the high frequency loop gain, use of equation (4.16) is justified. When doubt exists it is not too difficult to deal with equation (4.14) as discussed in section 4.6 for systems with multiple nonlinearities.

Figure 4.16 shows the locus $G(j\omega)$ on the Nyquist diagram for the transfer function values given in figure 4.15. To eliminate the limit cycle it is seen that

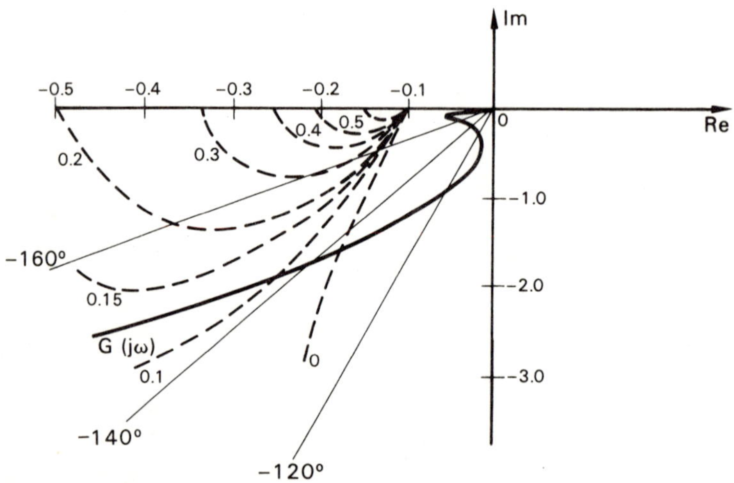

Figure 4.16 Nyquist diagram for system of figure 4.13

the DF, $\{N(a) + N'(a)\}$, of the combined nonlinearity must produce a smaller phase lag than that of $N(a)$ alone so that the DF of $N'(a)$ must have a leading phase angle. A nonlinearity having a DF $x\{1 - N(a)\}$, where x may be regarded as the per unit compensation, since the system becomes linear for $x = 1{\cdot}0$, will provide a suitable phase lead. A circuit for obtaining the DF, $\{1 - N(a)\}$, is shown in figure 4.17. The relationship between the sinusoidal input and the voltage across the capacitor can be made identical over a given frequency range

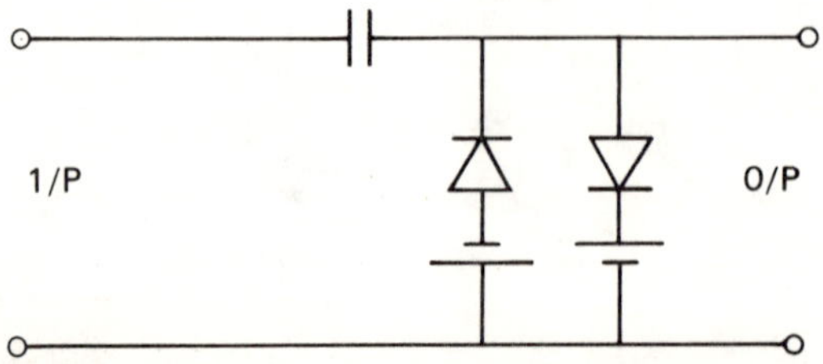

Figure 4.17 Network giving DF of $\{1 - N(a)\}$ for example of section 4.5.2

to the ideal friction controlled backlash characteristic of N. To realise $(1 - N)$ the output is taken from across the diodes. The loci $C(a, x) = -1/\{N(a) + N'(a)\} = -1/\{x + (1 - x)N(a)\}$ are shown in figure 4.16 for various values of x. It is seen that the system will be stable for values of x greater than 0·15.

In the actual system stabilisation was achieved using a value of 0·20 for x and with $G'(s)$ equal to $0{\cdot}45(s + 3{\cdot}44)/(s + 0{\cdot}075)$ instead of the value $0{\cdot}45(s + 3{\cdot}44)/s(1 + 0{\cdot}30s/150 + s^2/150^2)$ required by equation (4.15). The behaviour of both these transfer functions in the region of the oscillation frequency of 12 rads/sec of the unstabilised system is approximately $0{\cdot}45(s + 3{\cdot}44)/s$. Since $G(j\omega)$ has a high gain and $-180°$ phase shift at low frequencies, linear cascade stabilisation of the system is only possible using a pure derivative transfer function, which results in a finite velocity constant and unsatisfactory steady state performance.

4.5.3 *The use of a nonlinear integrator*

The system shown in the block diagram of figure 4.10 and discussed in section 4.4.4 is again considered. Here it is required to stabilise the system when the viscous friction is negligible, that is $T_1 = 0$, and to maintain the steady state accuracy against load disturbances provided by the integral control. The open loop transfer function is

$$G(s) = (1 + sT_2 + 1/sT_i)(\omega_0^2/s^2), \tag{4.17}$$

so that at low frequencies $G(j\omega)$ has a phase lag of 270° and its locus must cut the negative real axis. Thus due to the presence of the saturation the system can always be driven unstable. The $G(j\omega)$ locus is shown in figure 4.18 for values of $T_2\omega_0 = 2$

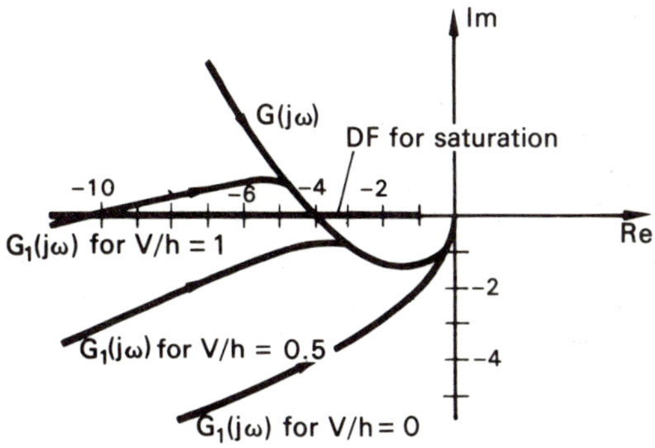

Figure 4.18 Nyquist diagram for example of section 4.5.3

and $T_i = T_2$. It cuts the real axis at -4 where the angular frequency $\omega = \omega_0/2$. An interesting technique for stabilising this system is given in reference 18, where it is proposed that the integration be converted to a limited integration using nonlinear feedback. The form of the limited integrator together with its steady state output waveform in response to a sinusoidal input is shown in figure 4.19.

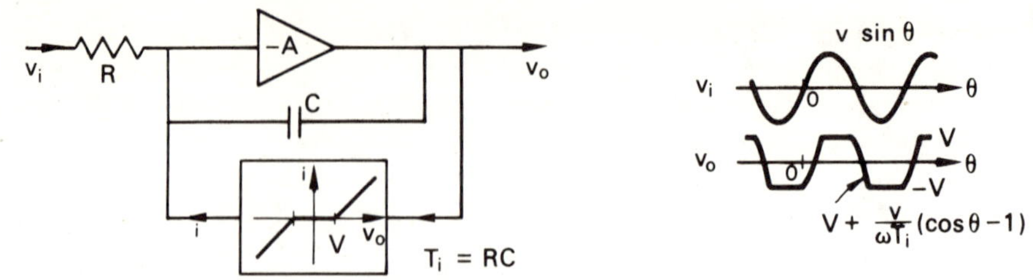

Figure 4.19 Limited integrator circuit and waveforms

At very low frequencies the output waveform becomes essentially a square wave with fundamental in phase with the input sinusoid.

The transfer function

$$G_1(s) = (\omega_0^2/s^2)\{1 + sT_2 + \lim(1/sT_i)\} \tag{4.18}$$

thus has a phase lag of only 180° at low frequencies. It should be noted that clipping of either the input or output waveforms of an integrator does not produce the desired limited integration action of the circuit of figure 4.19.

Fourier analysis of the output waveform of the limited integrator gives for the fundamental

$$(2/\pi)\{(2V - B)\cos A + (B/4)(3 + \cos 2A)\}\sin\theta - (2/\pi)[(2V - B)\sin A + (B/2)\{A + (\sin 2A)/2\}]\cos\theta, \tag{4.19}$$

where

$$B = v/\omega T_i \quad \text{and} \quad A = \cos^{-1}\{1 - (2V/B)\}.$$

The fundamental output thus depends upon both the magnitude and frequency of the input signal. However, as the original system is unstable only for large input amplitudes a, to the nonlinearity, it may be assumed for stability considerations that the fundamental output from the torque limiting characteristic has a peak value of $4h/\pi$. This is in error by less than 3 per cent for an amplitude $>2{\cdot}4$ times the saturation level, h. The corresponding peak magnitude of the fundamental v, at the input to the limited integrator is thus $4h\omega_0^2/\pi\omega^2$. Substituting this value of v in equation (4.19) the effective fundamental transfer function $\lim(1/j\omega T_i)$ for the limited integrator becomes

$$\frac{\omega^2}{2\omega_0^2}\left[\left(\frac{2V}{h} - \frac{4\omega_0^2}{\pi\omega^3 T_i}\right)\cos A + \frac{\omega_0^2}{\pi\omega^3 T_i}(3 + \cos 2A) - j\left\{\left(\frac{2V}{h} - \frac{4\omega_0^2}{\pi\omega^3 T_i}\right)\sin A + \frac{2\omega_0^2}{\pi\omega^3 T_i}\left(A + \frac{\sin 2A}{2}\right)\right\}\right], \tag{4.20}$$

where

$$A = \cos^{-1}(1 - V\pi\omega^3 T_i/2\omega_0^2 h).$$

The new frequency response locus $G_1(j\omega)$ is shown in figure 4.18 for V/h equal to 1 and 0·5, from which it is seen that the latter value results in a stable system as there is no intersection with the $C(a)$ locus. With this choice of V the integrator can hold a steady state output up to $0{\cdot}5h$ which would appear adequate for normal load disturbance correction.

It is also of interest to note from equation (4.19) that if the input amplitude v to the limited integrator increases linearly with frequency, so that without limits on the integrator the output increases in frequency but has a constant magnitude, the waveform appearing at the limited integrator output increases in frequency but has a fixed shape. Hence the series combination of a differentiator sT_i and the limited integrator $\lim(1/sT_i)$ is a frequency independent nonlinear characteristic. Therefore writing $v = a\omega T_i$, that is $B = a$, in equation (4.19) and dividing by a, the DF, $N_i(a)$, of the series combination is obtained as a function of V/a. The limited integrator may thus be considered as the series combination of an integrator $1/sT_i$ and the DF, $N_i(a)$, which gives a phase lead. The locus $C_i(a) = -1/N_i(a)$ is shown in figure 4.20.

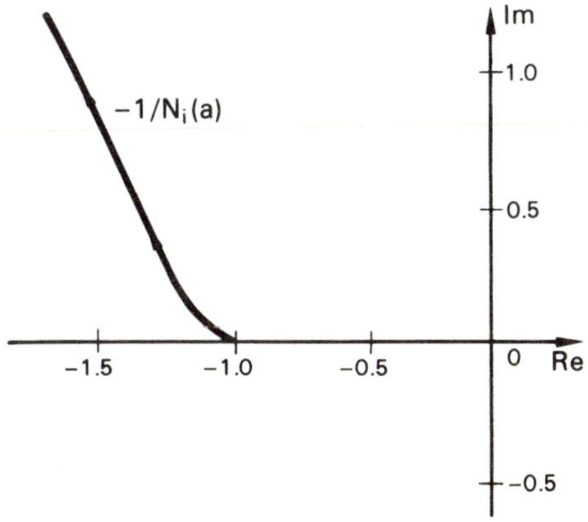

Figure 4.20 $C(a)$ locus for series combination of differentiator and limited integrator

4.6 Stability of systems with multiple nonlinearities

Systems containing direct series and parallel combinations of nonlinear elements can be analysed using the overall DF as explained in section 3.6. In particular cases, analysis of systems with two nonlinearities can be done using suitable approximations for the range of amplitudes and frequencies of interest. This was done for example in the previous section where the saturation nonlinearity output was taken as a square wave.

Obviously, no matter what the system configuration is or how many nonlinear elements exist, if an assumption that the inputs to the nonlinearities are sinusoidal is in good agreement with the actual situation under investigation, each nonlinearity may be replaced by its DF and a solution sought for the set of nonlinear algebraic equations obtained. Unfortunately the above assumption is probably valid in only a relatively small number of situations where multiple nonlinearities occur in practice. For example, it is not unusual to find a nonlinear process controlled by a nonlinear function of the error signal which passes through a non-low pass transfer function network before reaching the process. In cases such as this and in various other situations analyses may be performed using a frequency dependent DF for a combination of linear and nonlinear elements or a nonlinear subsystem.

Equations for limit cycles in systems which contain either a series or parallel combination of linear and nonlinear elements and satisfy the DF sinusoidal input assumption are given below.

4.6.1 *Series combinations of linear and nonlinear elements*

The block diagram of the general system is shown in figure 4.21. When all the nonlinear elements are single valued, application of the gain and phase relationships, as given in equation (4.1) for a single nonlinearity, show that a limit cycle

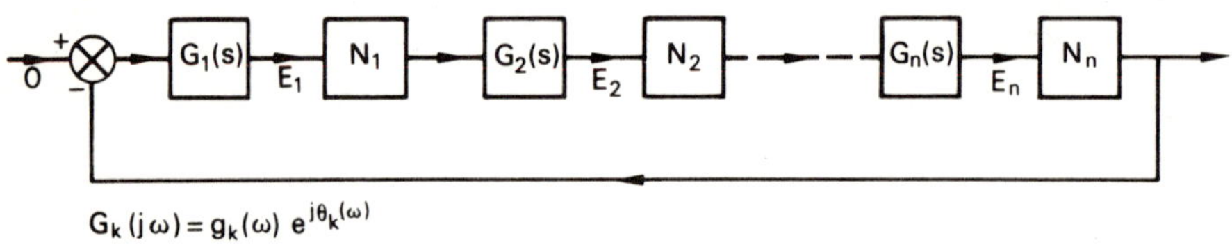

Figure 4.21 Series combination of linear and nonlinear elements

will exist if

$$g_1(\omega)M_1(E_1)g_2(\omega)M_2(E_2)\cdots g_n(\omega)M_n(E_n) = 1, \tag{4.21}$$

and

$$\theta_1(\omega) + \theta_2(\omega) + \cdots \theta_n(\omega) = (2k + 1)\pi, \tag{4.22}$$

where k is an integer. In addition, the sinusoidal amplitudes E_k are related by

$$E_k = E_{k-1}M_{k-1}(E_{k-1})g_k(\omega). \tag{4.23}$$

Since equation (4.22) is independent of the nonlinear elements, it yields directly all the possible solutions $\omega = \omega_0$ for the angular frequency of a limit cycle. These values of ω_0 are easily obtained from the Bode plots of the phase relationships for the linear elements. The magnitude of any limit cycle can then be obtained from a graph of

$$X(\omega_0, E_1) = g_1(\omega_0)M_1(E_1)g_2(\omega_0)M_2\{E_1M_1(E_1)g_2(\omega_0)\}g_3(\omega_0)M_3\{\cdots\} \tag{4.24}$$

versus E_1; the value of E_1 for which $X(\omega_0, E_1)$ is unity being the required solution. If no solution exists the stability of the system can be investigated by applying the Nyquist criterion to the corresponding linear system with all the nonlinearities replaced by a nominal gain within their Hurwitz sector. Usually if

$$X(\omega_0, E_1) < 1$$

for all E_1 and all ω_0, the system will be stable, and if

$$X(\omega_0, E_1) > 1$$

for all E_1 and all ω_0, the system will be unstable.

Theoretically the above approach can be used for double valued nonlinearities, but in practice the solution of equations (4.21) to (4.23) is not as easy when more than one nonlinearity has a phase shift, as equation (4.22) is no longer a function of ω alone. If, for example, N_1 gives a phase lead of $\psi_1(E_1)$ as well as a gain $M_1(E_1)$ then equation (4.22) becomes

$$\theta_1(\omega) + \theta_2(\omega) + \cdots \theta_n(\omega) + \psi_1(E_1) = (2k + 1)\pi, \tag{4.25}$$

and equation (4.21) remains unchanged. Thus $X(\omega_0, E_1)$ given by equation (4.24) must now be plotted using values of ω_0 and E_1 which satisfy equation (4.25); so that ω_0 in this case is a variable parameter on the $X(\omega_0, E_1)$ locus.

For the particular case of a symmetrical system with $N_i(E_i) = N(E_i) = M(E_i)e^{j\psi(E_i)}$ and $G_i(s) = G(s)$ for all $i = 1$ to n, it can be argued that due to the symmetry the values of E_i at all the nonlinearity inputs must be equal to E if a limit cycle exists. Thus, for this special case, the amplitude and phase equations become

$$g(\omega)M(E) = 1,$$

$$\theta(\omega) + \psi(E) = (2m + 1)\pi/n.$$

4.6.2 *Parallel combinations of linear and nonlinear elements*[19]

A block diagram of the general system is shown in figure 4.22. When all the nonlinear elements are single valued, by breaking the loop at A, the condition for a limit cycle is seen to be

$$\sum_{i=1}^{n} G_i(j\omega)F_i(j\omega)N_i(E_i) = -1 + j0. \tag{4.26}$$

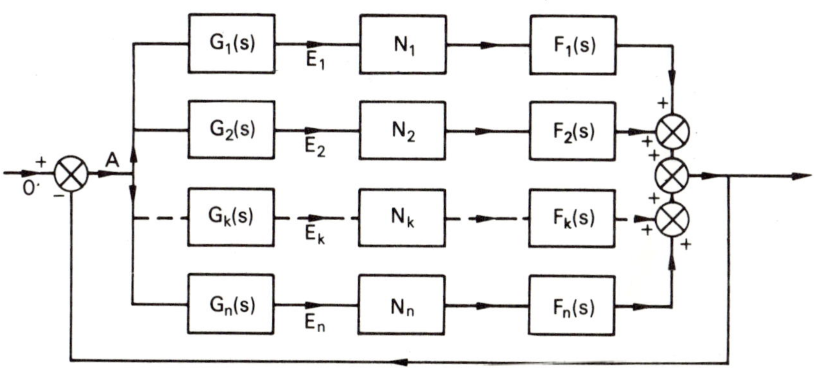

Figure 4.22 Parallel combination of linear and nonlinear elements

Separating the real and imaginary parts equation (4.26) can be replaced by the two equations

$$\sum_{i=1}^{n} g_i(\omega) f_i(\omega) M_i(E_i) \cos\{\theta_i(\omega) + \lambda_i(\omega)\} = -1 \tag{4.27}$$

and

$$\sum_{i=1}^{n} g_i(\omega) f_i(\omega) M_i(E_i) \sin\{\theta_i(\omega) + \lambda_i(\omega)\} = 0, \tag{4.28}$$

where

$$G_i(j\omega) = g_i(\omega) e^{j\theta_i(\omega)},$$

and

$$F_i(j\omega) = f_i(\omega) e^{j\lambda_i(\omega)}.$$

In addition, the E_i's can be expressed as a function of any particular E_i, say E_k, by

$$E_i = \{g_i(\omega)/g_k(\omega)\} E_k. \tag{4.29}$$

Substitution of equation (4.29) in equations (4.27) and (4.28) yields two equations in the variables E_k and ω. Any solution $E_k = E_0$ and $\omega = \omega_0$ of these equations gives the amplitude E_0 at the input to N_k and angular frequency ω_0 of a limit cycle. The most systematic way to test for a solution of the equations is to select a value of ω, find the corresponding values of E_k and plot the two relationships in the $E_k - \omega$ plane to find any intersection. When no intersection exists the system will be stable if for all E_k and ω satisfying equation (4.28) with (4.29) substituted, the left hand side of equation (4.27) with (4.29) substituted is less than unity.†

Again the results may be simplified if the system has some symmetry. It is also of interest to note that limit cycles in multivariable systems can be evaluated from similar equations[20,21,22] and that the block diagram of a nonlinear autonomous multivariable system may be rearranged into the form of figure 4.22. Throughout the above analysis it has been assumed that any limit cycle which may exist has a single frequency, a fact which may not always be true in practice. This point as well as oscillations in multivariable systems are covered in greater detail in chapter 11.

As a specific example the evaluation of the limit cycle in the system of figure 4.23 is considered. Without the derivative feedback paths the system is unstable as the d.c. loop gain K is greater than 3, whilst with them the system is stable in the linear mode for all values of K. Assuming the limit cycle has an angular frequency ω_0 and amplitude A at the system output, then from equation (4.26)

$$\frac{60}{j\omega_0(1 + j\omega_0)(2 + j\omega_0)} + \frac{60 j\omega_0 N(\omega_0^2 A)}{(1 + j\omega_0)(2 + j\omega_0)} + \frac{60 N(\omega_0 A)}{(1 + j\omega_0)(2 + j\omega_0)} = -1 + j0, \tag{4.30}$$

† The usual case; dependent upon the linear elements.

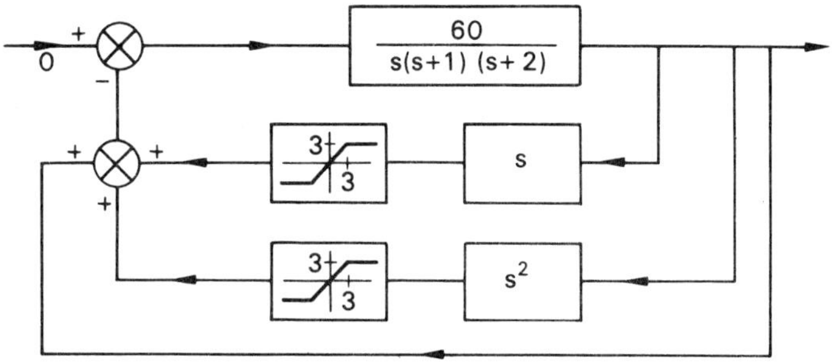

Figure 4.23 Example of system with two nonlinear elements considered in section 4.6.2

where $N(a)$ is the DF for saturation which for large a may be approximated by

$$N(a) = 4h/a\pi, \quad \text{that is } N(a) = 12/a\pi \text{ for } a \gg 3. \tag{4.31}$$

Substituting this value in equation (4.30) and rearranging gives

$$\frac{60}{(1+\omega_0^2)(4+\omega_0^2)}\{-3 + j(\omega_0^2 - 2)/\omega_0\}\{(1 - 12/\pi A) + j12/\pi A\} = -1 + j0.$$

The imaginary part gives

$$\pi A/12 = (3\omega_0 + \omega_0^2 - 2)/(\omega_0^2 - 2), \tag{4.32}$$

and substituting this value in the real part gives

$$\omega_0^3 + 3\omega_0^2 = 60 + 2\omega_0,$$

which has the solution $\omega_0 = 3{\cdot}26$. Substituting this value of ω_0 back in equation (4.32) gives

$$A = 12\{1 + 3\omega_0/(\omega_0^2 - 2)\}/\pi = 8{\cdot}15.$$

Thus $\omega_0 A \gg 3$ and the error in using equation (4.31) for the saturation DF is negligible, when compared with the neglecting of harmonics other than the fundamental for the transfer function $s/(s+1)(s+2)$ around the second derivative feedback loop, which is not a good low pass filter. An accuracy of around 10 per cent in the values for ω_0 and A is thus the best that can be expected.

The above solution can also be obtained directly from the characteristic equation of the system with the DFs and $s = j\omega_0$ substituted, that is

$$(j\omega_0)^3 + \{3 + 60(12/\pi\omega_0^2 A)\}(j\omega_0)^2 + \{2 + 60(12/\pi\omega_0 A)\}j\omega_0 + 60 = 0,$$

which gives

$$3\omega_0^2 + 60(12/\pi A) = 60,$$

and

$$\omega_0^3 - 2\omega_0 - 60(12/\pi A) = 0,$$

with the solution $\omega_0 = 3{\cdot}26$, $A = 8{\cdot}15$.

4.7 Closed loop frequency response

When a stable autonomous nonlinear closed loop system is subjected to a sinusoidal forcing function, the presence of which does not cause instability, then the frequency response can be calculated approximately using the DF method. From the block diagram of figure 4.1 the closed loop frequency response is

$$\frac{C(j\omega)}{R(j\omega)} = \frac{N(e_i, \omega)G_1(j\omega)G_2(j\omega)}{1 + N(e_i, \omega)G_1(j\omega)G_2(j\omega)H(j\omega)}. \tag{4.33}$$

However, before this expression can be evaluated it is necessary to obtain the magnitude and phase of e_i, the assumed sinusoidal input to the nonlinearity, corresponding to the given magnitude and frequency of the sinusoidal input r. Here we restrict our attention to frequency independent nonlinearities although extension of the techniques presented to frequency dependent nonlinearities is straightforward.(23) Several methods(24,25,26,27) are available for the solution of the problem; the only difference between them being the form in which the two nonlinear algebraic equations for the system are expressed and the graphical technique used for their solution. It is possible for a certain choice of linear element transfer functions for the nonlinear equations to have multiple(28) rather than a single solution over a range of input frequencies and amplitudes. This results in the now well known phenomenon of jump resonance which is discussed later in more detail. Other methods(29,30) have been proposed for the frequency response evaluation of specific nonlinear systems; their non-generality precludes them from further discussion here.

4.7.1. *Graphical method for a SVNL*

Here we follow the approach of reference 25. Again, consider the closed loop of figure 4.1 opened at the nonlinearity and a signal $e_0 = E_0 \sin \omega t$ injected at the input to $G_2(s)$. Taking the input signal r equal to $R \sin(\omega t + \beta)$ the signal, e_i, fed back to the nonlinearity input is given by

$$e_i = Rg_1 \sin(\omega t + \theta_1 + \beta) - g_1 g_2 h E_0 \sin(\omega t + \theta_1 + \theta_2 + \theta_3),$$

where g_1 and θ_1, etc. denote $|G_1(j\omega)|$ and $\angle G_1(j\omega)$ respectively.

Assuming the nonlinearity is single valued e_i can be taken equal to $E_i \sin \omega t$ giving

$$E_i \sin \omega t = Rg_1 \sin(\omega t + \theta_1 + \beta) - g_1 g_2 h E_0 \sin(\omega t + \theta_1 + \theta_2 + \theta_3).$$

Hence

$$E_i = g_1 R \cos(\theta_1 + \beta) - g_1 g_2 h E_0 \cos(\theta_1 + \theta_2 + \theta_3), \tag{4.34}$$

and

$$0 = g_1 R \sin(\theta_1 + \beta) - g_1 g_2 h E_0 \sin(\theta_1 + \theta_2 + \theta_3). \tag{4.35}$$

Eliminating β between these equations gives

$$(E_i/R) = g_1[1 - \{g_2h(E_0/R)\sin(\theta_1 + \theta_2 + \theta_3)\}^2]^{1/2} - g_1g_2h(E_0/R)\cos(\theta_1 + \theta_2 + \theta_3). \tag{4.36}$$

Thus for any particular angular frequency ω_c a locus showing the possible relationships between E_0/R and E_i/R can be drawn using equation (4.36). To determine the particular operating point or points on the locus the relationship

$$E_0/R = E_iN(E_i)/R \tag{4.37}$$

for the nonlinear characteristic with the given system input amplitude R is also plotted. The fundamental output signal from the system is then $E_{0c}g_{2c}\sin(\omega_c t + \theta_{2c})$, where E_{0c} is obtained from the value of E_0/R at the intersection(s) of equations (4.36) and (4.37) and g_{2c} and θ_{2c} are the gain and phase shift respectively of $G_2(s)$ at the frequency ω_c. The magnitude of the closed loop gain for a frequency ω_c is then $E_{0c}g_{2c}/R$ and the phase shift $\theta_{2c} - \beta_c$ with β_c evaluated from equation (4.35).

4.7.2 *Jump resonance*

Figure 4.24 shows typical frequency loci obtained from equation (4.36) for the lightly damped second order system with $G_1(s) = k(1 + sT)$, $G_2(s) = \omega_0^2/s^2$ and $H(s) = 1$, considered in reference 25. Loci, corresponding to equation (4.37),

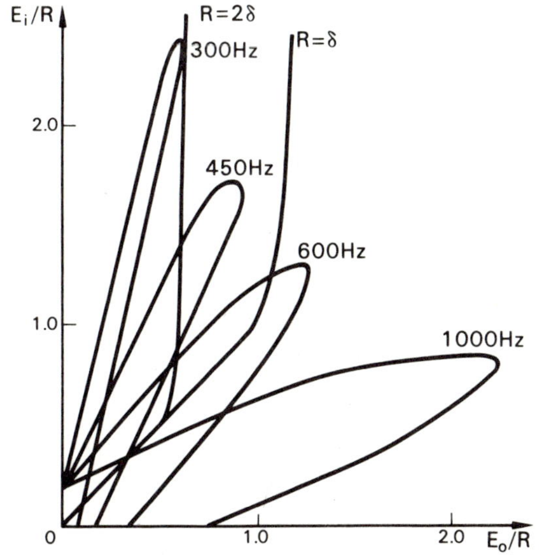

Figure 4.24 Typical curves for a lightly damped second order system

for an ideal saturation characteristic with saturation levels $\pm\delta$ and $m = 1$ are also drawn for two input amplitudes R relative to δ. To explain the jump phenomena consider the similar loci drawn in figure 4.25. For a low frequency

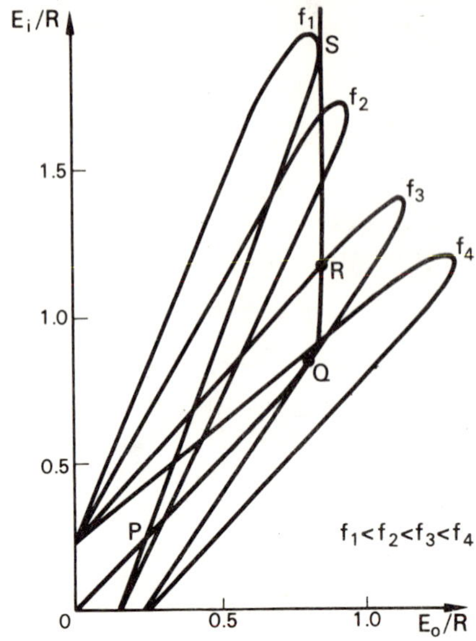

Figure 4.25 Illustration for the explanation of the jump phenomena

input the system error and the input to the nonlinearity are small so that as the frequency is increased the operating point moves from 0 on the linear portion of the nonlinearity locus through P at the frequency f_1 to the point Q at the frequency f_3. However at the point Q the nonlinearity locus and frequency locus are tangential so that further increase in frequency causes the operating point to jump from the lower side of the frequency locus at Q to the upper side at R. This jump from Q to R on the saturated region of the nonlinearity locus corresponds to a relatively small change in E_0/R and thus only a small jump in the closed loop gain frequency characteristic. The operating point remains on the upper side of the frequency loci for further increase in frequency to f_4 and above. When the frequency is decreased from f_4 the operating point remains on the upper side of the frequency loci until the frequency f_1 is reached at the tangential point S. A further decrease in frequency causes the operating point to jump to linear operation at P; a much larger change in E_0/R occurring than at the increasing frequency jump. A sketch of the resulting closed loop gain frequency characteristic for a given input to the system of figure 4.24 is shown in figure 4.26. The full curve shows the theoretical solution which gives three possible gains for frequencies f in the range $f_1 < f < f_3$. The dotted lines show the jumps which take place at the frequencies f_1 and f_3 in a practical frequency response test since the third solution is unstable (see section 4.7.4). As an increase in magnitude of the signal at the input to the saturation characteristic reduces the magnitude of the DF, the effect of increasing the input R is to reduce the system bandwidth and to lower the frequencies f_1 and f_3 at which jumps take place.

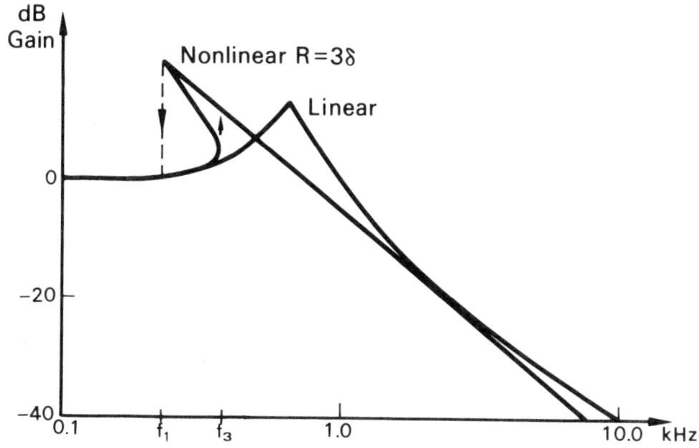

Figure 4.26 Closed loop frequency response

If the saturation nonlinearity is replaced by a hard spring nonlinear characteristic, that is one for which the DF increases with increasing input amplitude, the peak obtained in the calculated frequency response 'bends' in the opposite direction.[31] This is illustrated in figure 4.27 which also shows that the larger jump now occurs when the frequency is increasing.

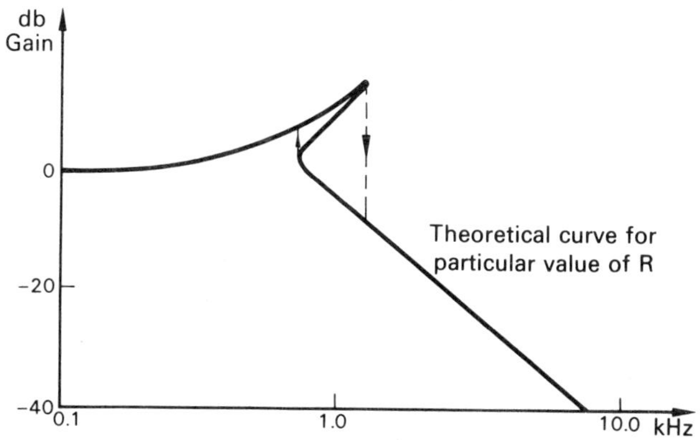

Figure 4.27 Typical frequency response for system with hard spring nonlinearity

4.7.3 *A method for a DVNL*

For a nonlinear characteristic having a complex DF or one which is frequency dependent a different approach is required to calculate the closed loop frequency response.[32,33] From figure 4.1

$$C(j\omega) = E_i(j\omega)N(e_i, \omega)G_2(j\omega), \tag{4.38}$$

using this value in equation (4.33) and rearranging gives

$$R(j\omega) = E_i(j\omega)[\{G_1(j\omega)\}^{-1} + N(e_i, \omega)G_2(j\omega)H(j\omega)]. \tag{4.39}$$

Taking $e_i = E_i \sin \omega_c t$, for any particular input frequency ω_c, the right hand side of equation (4.39) can be evaluated for a complex $N(e_i, \omega)$ as the input magnitude E_i to the nonlinearity varies. For a constant amplitude of the input signal $r = R\sin(\omega t + \beta)$ the vector locus is a circle of radius R as β, which is the phase difference with respect to e_i, takes the values 0 to 2π. The points of intersection of this circle of radius R and the ω_c locus give the possible operating conditions for the given input magnitude and frequency. The fundamental output from the system and the closed loop frequency response can then be calculated in a similar manner to that described in the previous section. Figure 4.28 shows the procedure. For illustration purposes the curves have been sketched for the SVNL system of figure 4.25 using a normalised input and with the corresponding points marked accordingly. If a hard spring characteristic is present the loci bend in the other direction as shown dotted for the frequency f_4.

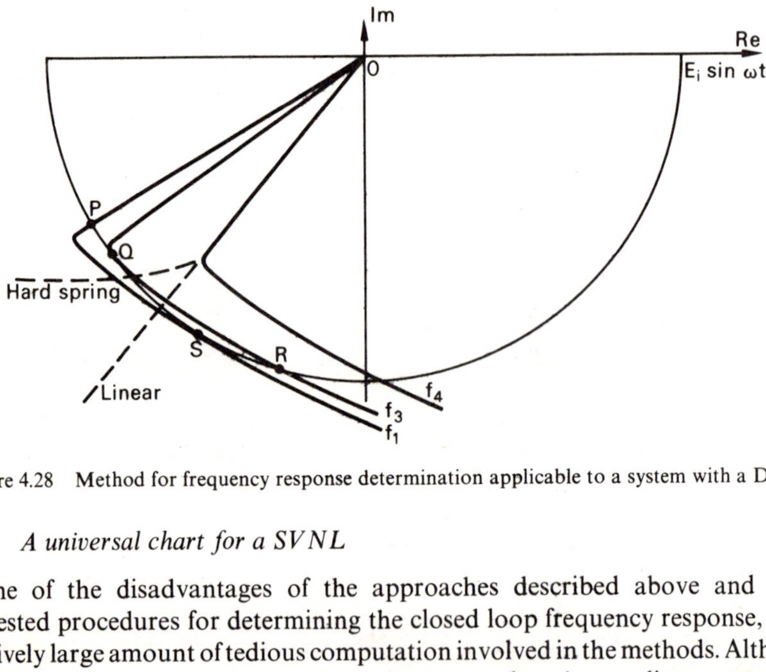

Figure 4.28 Method for frequency response determination applicable to a system with a DVNL

4.7.4 *A universal chart for a SVNL*

One of the disadvantages of the approaches described above and other suggested procedures for determining the closed loop frequency response, is the relatively large amount of tedious computation involved in the methods. Although digital computer programs can be written to solve the nonlinear equations they do not give the designer a clear understanding of the system behaviour and there are inherent difficulties in such routines when multiple solutions exist. A technique is described in reference 34 which enables the frequency response to be found from two simple loci, one for the linear elements and the other for the nonlinearity plotted on the same graph sheet; and a universal chart valid for any system. Since intersections between lines on the graph and chart have to be located one of them should be available on transparent paper.

Assuming a SVNL, and ignoring distortion terms, the phasor diagram for the block diagram of figure 4.1 is as shown in figure 4.29. In the diagram the

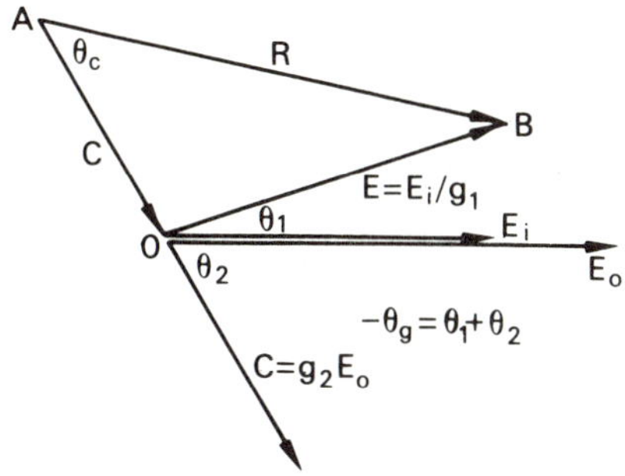

Figure 4.29 Phasor diagram for figure 4.1

phase angles θ_1 and θ_2 of $G_1(j\omega)$ and $G_2(j\omega)$ are shown as lagging angles and their combined phase lag of $-(\theta_1 + \theta_2)$ is denoted by θ_g. Thus for an input sinusoid of known magnitude R and frequency ω, if the line AB is fixed the point O must move on a circle of which AB is a chord since θ_g is constant for a fixed ω.

From the triangle OAB

$$\frac{E}{R} = \frac{\sin\theta_c}{\sin\theta_g} \quad \text{and} \quad \frac{C}{R} = \frac{\sin(\theta_g - \theta_c)}{\sin\theta_g}.$$

Hence

$$\begin{aligned}\log(E/R) &= \log E - \log R = \log\sin\theta_c - \log\sin\theta_g,\\ \log(C/R) &= \log C - \log R = \log\sin(\theta_g - \theta_c) - \log\sin\theta_g.\end{aligned} \tag{4.40}$$

Thus graphs of $\log(C/R)$ versus $\log(E/R)$ can be drawn for fixed values of θ_g with θ_c a parameter. These loci are those of the universal chart and are shown in figure 4.30. If the origin is shifted to the point $O'(-\log R, -\log R)$ on the chart it then represents the variation of $\log C$ with $\log E$ for a given loop phase shift θ_g.

For the linear elements the relationship $\log g_2$ versus $\log g_1$ with frequency a parameter and for the nonlinearity $\log E_0$ versus $\log E_i$ can be plotted as shown in figure 4.31. Now from figure 4.1, $E = E_i/g_1$ and $C = E_0 g_2$ so that

$$\begin{aligned}\log E &= \log E_i - \log g_1,\\ \log C &= \log E_0 + \log g_2.\end{aligned} \tag{4.41}$$

Therefore for a given frequency ω_1 at the point O_1 on the linear relationship G in figure 4.31 the nonlinearity locus N represents the relationship $\log C$ versus $\log E$ as defined by equation (4.41). Aligning the origin O' of the universal chart of figure 4.30 and the origin O_1 of figure 4.31 and maintaining the axes parallel, the intersection(s) of N and the θ_g curve for the given frequency ω_1 yield the solutions for C and E of equations (4.40) and (4.41). The closed loop response can thus be obtained for any frequency and amplitude by aligning different origins on the locus G for frequency and on the chart for amplitude.

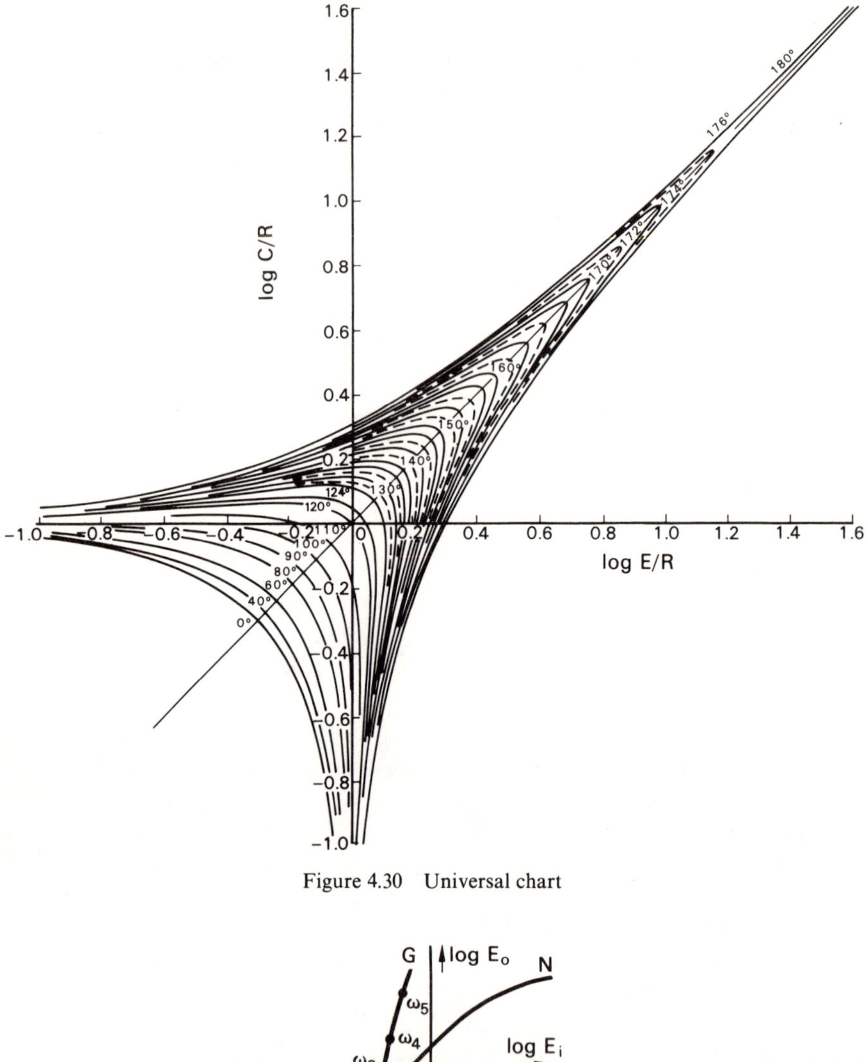

Figure 4.30 Universal chart

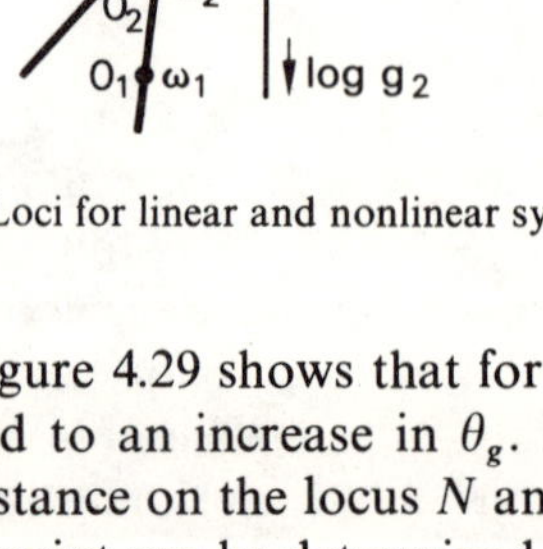

Figure 4.31 Loci for linear and nonlinear system elements

The phasor diagram of figure 4.29 shows that for a given R an increase in E and therefore C should lead to an increase in θ_g. By perturbing the solution point(s) obtained a small distance on the locus N and testing for this condition, the stability of the solution point can be determined.

Example[34]

Consider a system having an ideal saturation characteristic with $m = 1$ and $\delta = 1{\cdot}5$, $G_1(s) = 0{\cdot}631$ and $G_2(s) = 40/s(s+1)$. Figure 4.32 shows the linear locus G, which is parallel to the vertical axis as $g_1(\omega) = 0{\cdot}631$ a constant, and

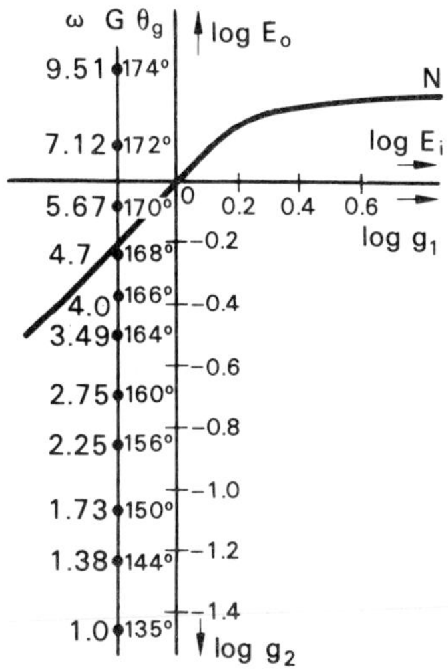

Figure 4.32 System characteristics for example of section 4.7.4

the nonlinear locus N. For an input of amplitude 1·25 and frequency 4 rads/sec the required origins are at $O'(-0{\cdot}097, -0{\cdot}097)$ on the universal chart and at O_1 on G, where the closed loop phase shift is 166°. Figure 4.33 shows the origins

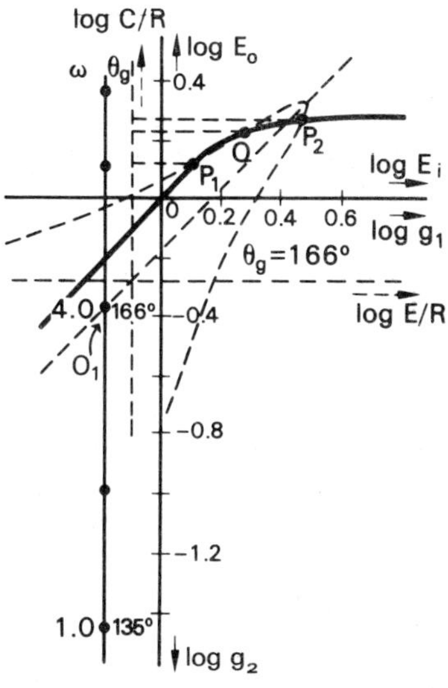

Figure 4.33 Superposition of universal chart and figure 4.32

O' and O_1 aligned for these conditions and the $\theta_g = 166°$ locus is seen to intersect N three times at the points P_1, Q and P_2. The closed loop gain for these solutions can be read directly off the graph as shown. The point Q corresponds to an unstable solution since the small movement from Q along N in the direction of C and E increasing is into a region of smaller θ_g on the chart.

4.7.5 *Region of jump resonance*

The multimodal region of a lightly damped feedback system containing a soft-spring nonlinearity can be represented on the $R - \omega$ plane as illustrated in figure 4.34. This system will exhibit a multivalued response for sinusoidal signals represented by all the points within the region bounded by the two jump loci.

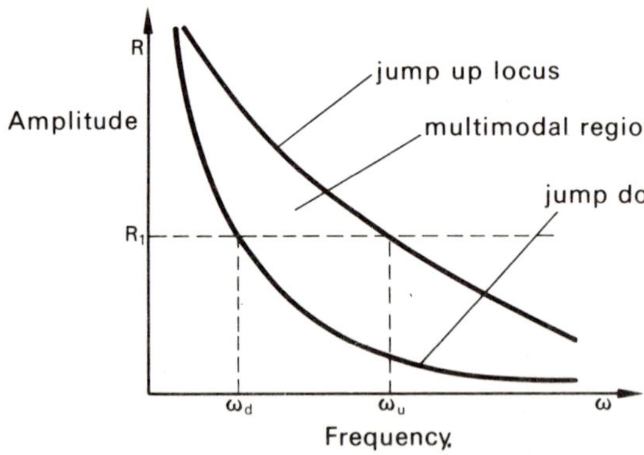

Figure 4.34 Multimodal region in R-ω plane

Although jump resonance is normally undesirable in a control system, several situations where jump resonance has been used to advantage are described in the literature.[35,36] Use of the above methods to determine whether or not a system possesses a jump resonance is not too easy, whilst the design of a compensator to prohibit jump resonance or to make it occur at a particular amplitude or frequency is almost impossible. Experimental investigations undertaken with some of the earlier theoretical work discussed above, suggested that jump resonance only occurs in lightly damped systems and is thus probably dependent on the location of the open loop frequency response locus in the complex plane.

Two satisfactory solutions to this problem have been found which in fact turn out to be identical although the initial approach is quite different. Here we describe the method of Hatanaka[37] which has been extended to cover the cases of double valued and frequency dependent nonlinearities.[38,39] The second approach which uses the incremental describing function, IDF, is discussed in section 5.11.2.

Consider for convenience the system of figure 4.1 with $H(s) = G_1(s) = 1$ and $G_2(s) = G(s)$. A system for which $G_1(s)$ and $H(s)$ are not unity can easily be put into the above form by taking $G(s) = G_1(s)G_2(s)H(s)$ and passing the input signal $r(t)$ through $G_1(s)$ for the correct system input. From the triangle OAB of the phasor diagram of figure 4.29, if ω is constant so is θ_g and thus for a stable solution, E should increase if R increases. Jump resonance will therefore occur in the region where

$$\left(\frac{\partial R}{\partial E}\right)_{\omega\ \mathrm{const.}} < 0, \tag{4.42}$$

as illustrated in figure 4.35. The graph shows a typical plot of the E versus R solution for a system possessing jump resonance. At the jump resonance point

$$\left(\frac{\partial R}{\partial E}\right)_{\omega\ \mathrm{const.}} = 0. \tag{4.43}$$

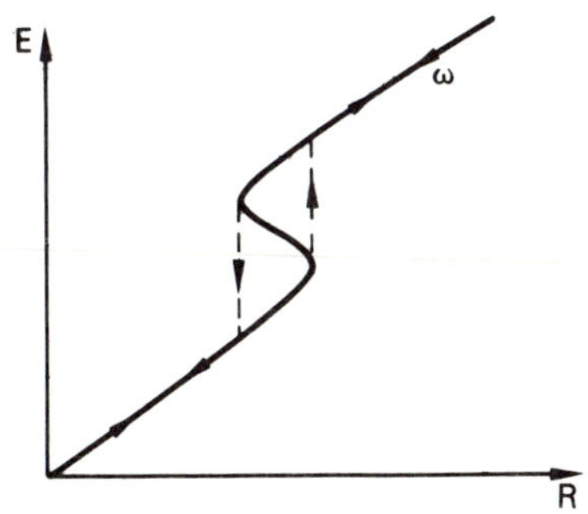

Figure 4.35 E versus R locus for system with jump resonance

Now

$$\frac{E(j\omega)}{R(j\omega)} = \frac{1}{1 + N(E)G(j\omega)},$$

and substituting $G(j\omega) = U(\omega) + jV(\omega)$ gives

$$R = EN(E)\left[\left\{U(\omega) + \frac{1}{N(E)}\right\}^2 + \{V(\omega)\}^2\right]^{1/2}.$$

Taking the partial derivative with respect to E and substituting in equation (4.43) gives

$$U(\omega) + \frac{1}{N(E)} \quad U(\omega) + \frac{1}{\mathrm{d}\{EN(E)\}/\mathrm{d}E} \quad + \{V(\omega)\}^2 = 0.$$

For a given E this equation describes a locus on the Nyquist diagram ($U - V$ plane). Denoting $U(\omega)$ by x, $V(\omega)$ by y, $N(E)$ by K_1 and $\mathrm{d}\{EN(E)\}/\mathrm{d}E = N(E) + EN'(E)$ by K_2 the equation can be written

$$\left(x + \frac{K_1 + K_2}{2K_1K_2}\right)^2 + y^2 = \left(\frac{K_1 - K_2}{2K_1K_2}\right)^2, \tag{4.44}$$

which for a given amplitude E is a circle with centre on the negative real axis at the point

$$\left\{-\left(\frac{K_1 + K_2}{2K_1K_2}\right), 0\right\} \quad \text{and} \quad \text{radius } (|K_1 - K_2|)/2K_1K_2.$$

If for a particular value of E the circle in the complex plane given by equation (4.44) is cut by the $G(j\omega)$ locus at frequencies ω_1 and ω_2, then jump resonance can occur in the system for this value of E at these frequencies. To ensure that no jump resonance takes place the $G(j\omega)$ locus must not enter the region occupied by these circles for all values of E. It can be shown that the envelope of this family of circular contours $f(x, y, E) = 0$, since E is a parameter on the curve, must also satisfy the relationship

$$\frac{\partial f(x, y, E)}{\partial E} = 0. \tag{4.45}$$

Writing equation (4.44) in the form $f(x, y, E) = 0$, differentiating partially with respect to E to obtain equation (4.45) and then solving the two equations, the envelope of the circles is obtained in the parametric form

$$\left.\begin{aligned} x &= \frac{-(K_1'K_2 + K_1K_2')}{K_1'K_2^2 + K_2'K_1^2}, \\ y &= \pm\frac{(K_1'K_2')^{1/2}(K_1 - K_2)}{K_1'K_2^2 + K_2'K_1^2}, \end{aligned}\right\} \tag{4.46}$$

where $'$ denotes the derivative with respect to the amplitude E.

As a specific example consider the quintic nonlinearity x^5. Denoting the input amplitude by a rather than E, we have

$$K_1 = N(a) = 5a^4/8 \quad \text{and} \quad K_2 = \frac{\mathrm{d}}{\mathrm{d}a}\{5a^5/8\} = 25a^4/8$$

so that $K_1' = 5a^3/2$ and $K_2' = 25a^3/2$. Substituting in equation (4.46) gives

$$x = \frac{-a^7(\frac{25}{8} \times \frac{5}{2} + \frac{5}{8} \times \frac{25}{2})}{a^{11}\{\frac{5}{2} \times (\frac{25}{8})^2 + \frac{25}{2} \times (\frac{5}{8})^2\}} = \frac{-8}{15a^4},$$

$$y = \frac{\pm a^7(5\sqrt{5}/2)(-5/2)}{a^{11}\{\frac{5}{2} \times (\frac{25}{8})^2 + \frac{25}{2} \times (\frac{5}{8})^2\}} = \pm\frac{16\sqrt{5}}{75a^4}.$$

Eliminating a gives

$$y = \pm(2/\sqrt{5})x.$$

Since for any a the circular locus has a radius smaller than the distance from the origin of its centre, the above equation defining the envelope corresponds to two lines at $\tan^{-1}(2/\sqrt{5}) \sim 42°$ to the negative real axis. Thus if the $G(j\omega)$ locus for a system containing the quintic nonlinearity does not enter the region bounded by the radius vectors at 138° and 222° in the complex plane then jump resonance will not occur.

4.8 Accuracy of the DF method

In the previous sections we have shown how the DF may be used in the analysis and design of nonlinear systems. The methods are relatively easy to apply since they are only basic extensions of the frequency domain approach used for linear systems. Unfortunately the method is an approximate one so there is always some doubt as to the accuracy of the results obtained. The engineer is continually making approximations based on experience and rules of thumb, and it is with these aspects in mind that we examine DF accuracy.

First it should always be remembered that to illustrate theoretical methods, block diagrams with precisely defined transfer characteristics are analysed, whereas in practice these descriptions are approximations to the actual behaviour. Thus any analysis of a physical system, due to the inaccuracies involved in finding theoretical models of devices, is an approximate one and the ability of the engineer to assess the relative and interactive effects of the various approximations is extremely important.

What do we mean by DF accuracy and how do we define it? To shed some light on this problem let us consider the determination of limit cycles in symmetrical nonlinear systems. The actual limit cycle at the input to the nonlinearity, and of course at other points in the system, will be a complex periodic waveform. How do we compare the exact solution for the nonlinearity input, which we will assume is known, and the sinusoid obtained by the DF approach? Obviously as regards frequency there is no problem; the frequency of the sinusoid can be compared with the fundamental frequency of the periodic waveform. What about the wave shape, do we compare simple properties like its fundamental, peak, mean square value with those of the sine wave, or do we take a more precise look and use an error criterion involving the error between the two waveforms over say a period of the exact solution?

In reference 40 two properties of the DF approach are proved, first that the DF solution gives an approximation to the fundamental of the actual waveform and secondly that it requires $|G(jn\omega_0)| \ll |G(j\omega_0)|$, $n = 3, 5, 7, \ldots$, the so called filter hypothesis, to obtain an accurate solution. An accuracy criterion should therefore involve a comparison of the amplitude of the DF solution for the nonlinearity input with the fundamental amplitude of the exact solution. Presumably, in addition, some measure of the distortion in the exact solution waveform should also be included. Similar comments apply to the evaluation of the closed loop frequency response using the DF, where the calculated output should be regarded as an approximation to the fundamental output. At present there is no rigorous method available which enables such a measure of the accuracy of the DF solution to be found without actually defeating the purpose and finding the exact solution.

Several other possible DFs were suggested in section 3.9 where it was pointed out that although their definitions are straightforward they usually led to more complicated expressions. There is also no strong evidence to show that their use will in general produce more accurate results. Unlike the normal DF it is not known either under what conditions their accuracy is improved or what aspect of the system waveform, such as the peak or the mean square value, the solutions approximate. Although it might be possible to find a specific application where some other DF yields slightly better results, it only proves its superiority for that particular case and gives no real confidence for other applications. In addition when making comparisons it is important to relate similar quantities from the various methods. For example, if the mean square value of the limit cycle at the output of the nonlinearity is required the solution using say the r.m.s. DF should be compared with the nonlinearity r.m.s. output value calculated using the DF solution for the input amplitude.

A few theoretical papers(41,42,43) have investigated the validity and accuracy of the DF method for limit cycle determination but unfortunately they have not yielded any results of true practical significance. Johnson (42,44) adopted an approach similar to the perturbation theory of Poincaré(45) and sought solutions for the oscillation $x(t)$ at the nonlinearity input and its fundamental frequency ω in terms of the series

$$x(t) = x_0(t) + \mu x_1(t) + (\mu^2/2!)x_2(t) + (\mu^3/3!)x_3(t) + \cdots$$

$$\omega = \omega_0(1 + \mu\lambda_1 + (\mu^2/2!)\lambda_2 + (\mu^3/3!)\lambda_3 + \cdots),$$

with μ a parameter to account for the nonlinearity $n(x)$ and defined by the equation

$$\mu f_d(x) = n(x) - N(a)x,$$

where $f_d(x)$ are the distortion terms. For $\mu = 1$ the exact solution is obtained and for $\mu = 0$ the linearised system solution. The solutions for $x_0(t)$ and ω_0 are thus those obtained by the DF method. Johnson's most important result was to show that the first frequency correction term λ_1 is zero. He was also able to obtain equations for the second frequency correction term λ_2 and the first amplitude correction term, B, where B is the amplitude of the component of x_1 at frequency ω_0 (as $\lambda_1 = 0$). Due to the complicated nature of the equations and the resulting difficulty of obtaining solutions for λ_2 and B, the method appears to have little practical value when compared with direct analog or digital computation solutions.

4.8.1 *A distortion criterion*

The only approach remaining, to try and obtain an estimate of the accuracy of the DF method is therefore the standard engineering procedure of checking the accuracy of earlier assumptions when additional information has been obtained. For the DF method this means opening the loop at the nonlinearity input, applying to the nonlinearity the sinusoidal DF solution signal and comparing with it the actual signal fed back to the nonlinearity, which of course includes the solution term plus higher harmonics. An easily calculated 'figure of merit' is the per unit distortion d defined as the ratio of the r.m.s. value of the harmonics to the r.m.s. value of the fundamental in the fedback signal. In order to investigate what indication of the DF accuracy this simply derived figure of merit might give, the only approach is to determine its value and the DF solution for systems with known exact solutions.

Two situations are discussed below: a conservative second order system including several different nonlinear elements and a relay system with various plant transfer functions. Exact solutions for relay systems are discussed in chapter 6 and at that time additional comparisons are made.

Second order conservative system

Consider a system[46] described by the differential equation

$$(\ddot{x}/\omega_0^2) + n(x) = 0, \tag{4.47}$$

where $n(x)$ is an odd SVNL. Using equation (2.63) the period of oscillation for the closed cycle is, T, where

$$T = 4\int_{x_{10}}^{0} \frac{dx_1}{\{2[F(x_{10}) - F(x_1)]\}^{1/2}},$$

where $x = x_1$, $\dot{x} = x_2$, $x_1 = x_{10}$ when $x_2 = 0$ and

$$F(x_1) = \int_0^{x_1} \omega_0^2 n(x)\,dx, \quad \text{an even function of } x.$$

Writing $x = x_1$, $x_{10} = -a_m$ and $m(x) = \int_0^x n(x)\,dx$ the expression becomes

$$T = \frac{4}{\omega_0\sqrt{2}}\int_0^{a_m} \frac{dx}{\{m(a_m) - m(x)\}^{1/2}}. \tag{4.48}$$

Alternatively if the nonlinearity is replaced by its DF, $N(a)$, equation (4.47) becomes

$$(\ddot{x}/\omega_0^2) + N(a)x = 0,$$

with the solution $x = a\cos\omega t$ and period of oscillation $T_{\text{d.f.}}$ given by

$$T_{\text{d.f.}} = 2\pi/\omega = 2\pi/\omega_0\{N(a)\}^{1/2}. \tag{4.49}$$

The amplitude and hence the period of oscillation will depend upon the initial conditions as a closed cycle can exist for any value of a. This is indicated by the DF method as the frequency and $C(a)$ loci are coincident along part of the negative real axis and so intersect at an infinite number of points. In particular for

(a) On-off characteristic

$$n(x) = h\,\text{sgn}\,x, \quad \text{giving } m(x) = hx\,\text{sgn}\,x.$$

Therefore equation (4.48) gives

$$T = \frac{4}{\omega_0\sqrt{2}}\int_0^{a_m} \frac{dx}{(ha_m - hx)^{1/2}} = (5{\cdot}657/\omega_0)(a_m/h)^{1/2},$$

and equation (4.49)

$$T_{\text{d.f.}} = (2\pi/\omega_0)(a\pi/4h)^{1/2} = (5{\cdot}567/\omega_0)(a/h)^{1/2}.$$

For simplicity we compare T and $T_{\text{d.f.}}$ assuming the fundamental amplitude a and the peak a_m of the oscillation to be equivalent. The period of oscillation predicted by the DF is thus in error by about 0·016 per unit. Also

$$d = \{(1/3^3)^2 + (1/5^3)^2 + (1/7^3)^2 + \cdots\}^{1/2} = 0{\cdot}038.$$

(b) Cubic characteristic

$$n(x) = Ax^3, \text{ so that } m(x) = Ax^4/4.$$

Equation (4.48) gives

$$T = \frac{8}{\omega_0\sqrt{2A}} \int_0^{a_m} \frac{\mathrm{d}x}{(a_m^4 - x^4)^{1/2}},$$

and substituting $x = a_m \cos\theta$,

$$T = \frac{8}{\omega_0\sqrt{2A}} \int_0^{\pi/2} \frac{a_m \sin\theta \, \mathrm{d}\theta}{a_m \sin\theta\{a_m^2 + a_m^2 \cos^2\theta\}^{1/2}}$$

$$= \frac{4}{a_m\omega_0\sqrt{A}} \int_0^{\pi/2} \{1 - (1/2)\sin^2\theta\}^{-1/2} \, \mathrm{d}\theta.$$

By definition the complete elliptic integral

$$K(k) = \int_0^{\pi/2} (1 - k^2 \sin^2\theta)^{-1/2} \, \mathrm{d}\theta,$$

so that

$$T = 4K(1/\sqrt{2})/a_m\omega_0\sqrt{A} = (4/a_m\omega_0\sqrt{A})(1{\cdot}854).$$

Using equation (4.49)

$$T_{\text{d.f.}} = (2\pi/\omega_0)(4/3a^2A)^{1/2} = (4/a\omega_0\sqrt{A})(1{\cdot}814).$$

The error in this case is thus 0·022 per unit. Also, since the only harmonic produced by a cubic with a sinusoidal input is the third $d = 1/3^3 = 0{\cdot}037$.

(c) Sinusoidal characteristic

Here $n(x) = \sin x$, giving $m(x) = -\cos x + 1$. Substituting in equation (4.48)

$$T = \frac{4}{\omega_0\sqrt{2}} \int_0^{a_m} \frac{\mathrm{d}x}{(\cos x - \cos a_m)^{1/2}}$$

$$= \frac{4}{\omega_0\sqrt{2}} \int_0^{a_m} \frac{\mathrm{d}x}{\{\sin^2(a_m/2) - \sin^2(x/2)\}^{1/2}},$$

and putting $\sin\theta = \sin(x/2)/\sin(a_m/2)$ gives

$$T = \frac{4}{\omega_0\sqrt{2}} \int_0^{\pi/2} \frac{2\cos\theta \sin(a_m/2) \, \mathrm{d}\theta}{\cos(x/2)\sin(a_m/2)\cos\theta}$$

$$= (4/\omega_0) \int_0^{\pi/2} \{1 - \sin^2(a_m/2)\sin^2\theta\}^{-1/2} \, \mathrm{d}\theta,$$

that is

$$T = (4/\omega_0)K\{\sin(a_m/2)\}.$$

Using equation (4.49),

$$T_{\text{d.f.}} = (2\pi/\omega_0)\{a/2J_1(a)\}^{1/2} = (4/\omega_0)\{a\pi^2/8J_1(a)\}^{1/2}.$$

The distortion d is given by

$$d = \left\{\left(\frac{J_3(a)}{9J_1(a)}\right)^2 + \left(\frac{J_5(a)}{25J_1(a)}\right)^2 + \cdots\right\}^{1/2}.$$

The values of T, $T_{\text{d.f.}}$ and d initially increase monotonically with a, and become infinite at values of a of π, 3·83 and 3·83 respectively, where the latter value is the first zero of $J_1(a)$. Up to an amplitude a of 130° where $d = 0{\cdot}036$ the error between T and $T_{\text{d.f.}}$ is less than 0·02 per unit. Beyond this value of a however both the error and d increase rapidly.

A relay system

Consider the relay system of figure 4.6 with a linear transfer function $G(s)$. For the relay

$$C(a) = -(\pi/4h)\{(a^2 - \Delta^2)^{1/2} + j\Delta\}, \tag{4.10}$$

which is shown plotted in figure 4.7,

(a) First order plant

$$G(s) = K/(1 + sT).$$

The loci $G(j\omega)$ and $C(a)$ do not intersect yet an exact analysis* indicates a limit cycle of frequency

$$\omega = \pi/\{2T\tanh^{-1}(\Delta/hK)\}.$$

For this case

$$d = \left\{\frac{1 + \omega^2T^2}{3^2(1 + 3^2\omega^2T^2)} + \frac{1 + \omega^2T^2}{5^2(1 + 5^2\omega^2T^2)} + \cdots\right\}^{1/2},$$

so that

$$d \geq \{(1/3)^4 + (1/5)^4 + (1/7)^4 + \cdots\}^{1/2} \sim 0{\cdot}12.$$

For a system with $hK \gg \Delta$, the exact solution gives $\omega T < 1$, which yields a much higher value of d.

(b) Second order system

$$G(s) = K/s(s + \alpha).$$

Equating the imaginary parts of $G(j\omega)$ and $C(a)$ gives

$$K\alpha/\{\omega(\omega^2 + \alpha^2)\} = \pi\Delta/4h,$$

* See chapter 6.

which reduces to

$$(1 + \lambda^2) = (4/\pi x), \tag{4.50}$$

where $\lambda = \omega/\alpha$ and $x = \alpha^2\Delta/hK$. The exact solution is given by

$$x = (\pi/2\lambda) - \tanh(\pi/2\lambda). \tag{4.51}$$

The solutions of equations (4.50) and (4.51) are shown in figure 4.36. The per unit error, ε, in the fundamental amplitude at the nonlinearity input is given by

$$\varepsilon = 1 - \frac{\lambda_E(1 + \lambda_E^2)^{1/2}}{\lambda_D(1 + \lambda_D^2)^{1/2}},$$

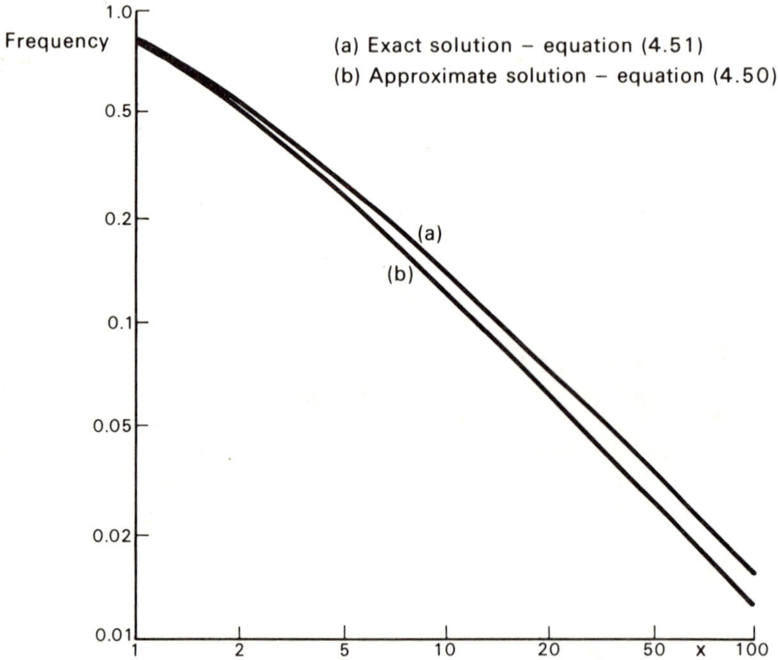

Figure 4.36 Comparison of DF and exact solutions for frequency

and the distortion, d, by

$$d = \left\{ \frac{1 + \lambda_D^2}{9^2(1 + 9\lambda_D^2)} + \frac{1 + \lambda_D^2}{25^2(1 + 25\lambda_D^2)} + \cdots \right\}^{1/2},$$

where $\alpha\lambda_D$ and $\alpha\lambda_E$ are the DF and exact solutions for the oscillation frequency for a given value of x. Figure 4.37 shows graphs of ε, d and the per unit frequency error $(\omega_D - \omega_E)/\omega_E$ again as a function of x.

From the above examples it is apparent that the distortion criterion can give a good indication of the validity and accuracy of the DF method. As a general

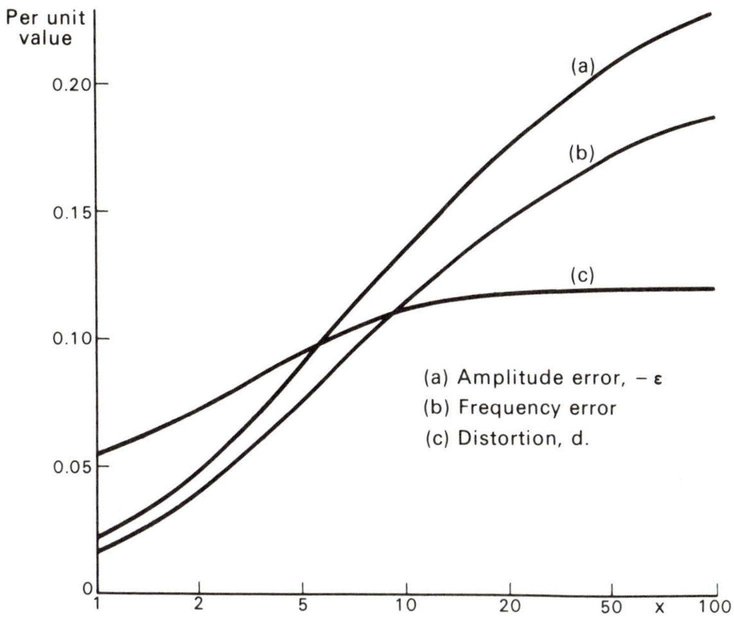

Figure 4.37 Graphs of distortion, frequency and amplitude errors

conclusion DF results for values of d much greater than 0·1 are probably unreliable whilst for values of d less than 0·1 they should be adequate for engineering purposes with crrors probably no greater than 10 per cent. The only known situations where the distortion criterion may tend to give false indications are if it is used for systems with a $G(s)$ having two or more resonances,[6,47] a situation where the gain frequency characteristic does not obey the filter hypothesis and suggests the possibility of peculiar behaviour.

4.8.2 *Extensions of the DF method*

Extensions of the DF method requiring considerably more labour can be used to determine limit cycles in nonlinear systems more accurately and also in principle to evaluate closed loop frequency responses. In the refined DF method[48] the exact waveform fedback to the nonlinearity from which the distortion d is calculated, is assumed as a new input to the nonlinearity and the new fundamental output component evaluated. Since the new input signal consists of higher harmonics in addition to the fundamental, the fundamental term in the output will not normally be in phase with that of the input even for a SVNL (see section 5.8). The refined DF is then defined as fundamental output/fundamental input and used in the normal way to calculate a limit cycle. Often use of the exact waveform as the new input to the nonlinearity causes computational difficulties so that as an alternative it may be approximated by ignoring all terms bar the fundamental and lowest harmonic.

An alternative approach is to assume a nonlinearity input waveform consisting of computing the Fourier series of the nonlinearity output, or equivalently balance around the loop for each of the frequencies. Again the practical difficulties of computing the Fourier series of the nonlinearity output, or equivalently multiple input DFs for harmonically related inputs makes the method, though so simple in concept, extremely difficult and laborious. The approach is considered further in section 5.13 as a use of the two sinusoidal input DFs.

4.9 The DF and transient response

DFs for sinusoidal and random inputs are descriptions of the nonlinearity based on averages and are normally used for predicting steady state system behaviour. Attempts have been made to extend DF methods(49,50,51,52,53) to the determination of system step responses although it is obviously not as clear in such situations what waveform should be assumed for the nonlinearity input. The error criterion and time over which it is evaluated are also of importance. The various methods are discussed in more detail in chapter 9.

It is appropriate, however, to make the following points here:

(i) To compute a transient response with reasonable accuracy using a DF approach may require a significant amount of computation.

(ii) It is doubtful with so many alternative methods such as the phase plane, piecewise linear analysis, step by step procedures,(54) numerical methods(55) and various forms of simulation whether the approach is practically advantageous.

(iii) If as is often the case a quick estimate of the behaviour is required, qualitative results using the normal DF are usually acceptable. Two reasons account for this, first the fact that the shape of the DF input amplitude relationship is primarily influenced by the nonlinearity rather than the input waveform, for example for saturation all DFs give a decreased gain as the amplitude increases, and second the relative insensitivity of closed loop characteristics to changes in the nominally high open loop gain.

Consider for example the position control system of figure 4.38. With the nonlinearity replaced by its DF, $N(a)$, the closed loop transfer function is

$$\left\{\frac{s^2}{N(a)K_1K_2} + \frac{sT}{K_1} + 1\right\} C(s) = R(s),$$

that is

$$\frac{C(s)}{R(s)} = \frac{\omega_0^2}{s^2 + 2\zeta\omega_0 + \omega_0^2},$$

where $\omega_0 = K_1K_2N(a)$, $\zeta = TK_2N(a)/2K_1$ and are both proportional to $N(a)$. Thus if the nonlinearity is a saturation characteristic $N(a)$ will decrease with increased magnitude of input step so that the response will be slower and more oscillatory. For a hard spring characteristic the reverse will apply, $N(a)$ will

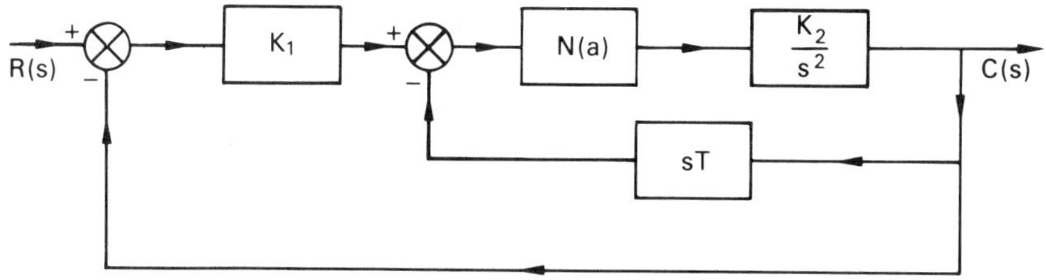

Figure 4.38 Position control with torque saturation

increase with increased step input and the response will be faster and more damped.

REFERENCES

1. Dorf, R. C.: *Modern control systems* (Addison-Wesley, Reading, Mass., 1967).
2. Loeb, J. M.: 'Recent advances in nonlinear servo theory', in R. Oldenburger (ed.), *Frequency response* (The Macmillan Press, New York, 1956, pp. 260–268).
3. Choudhury, S. K. and Atherton, D. P.: *Limit cycles in high order nonlinear systems, Proc. I.E.E.*, 1974, 121, pp. 717–724.
4. Aizerman, M. A.: 'On the effect of nonlinear functions of several variables on the stability of automatic control systems (in Russian)', *Automatika i Telemekhanika*, 1947, 8, pp. 64–72.
5. Dewey, A. G. and Jury, E. I.: 'A note on Aizerman's conjecture', *I.E.E.E. Trans.*, 1965, AC-10, pp. 482–483.
6. Fitts, R. E.: 'Two counterexamples to Aizerman's conjecture', *I.E.E.E. Trans.*, 1966, AC-11, pp. 553–556.
7. Siljak, D. D.: *Nonlinear systems* (Wiley, New York, 1969, Chapter 1).
8. Atkinson, P.: *Feedback control theory for engineers* (Heinemann, London, 1968, Chapters 11 and 12).
9. Gelb, A. and Vander Velde, W. E.: *Multiple input describing functions and nonlinear system design* (McGraw-Hill, New York, 1968, pp. 151–154).
10. Smetanin, N. A.: 'Harmonic linearization of a component with variable structure', *Automation and Remote Control*, 1965, 26, pp. 998–1001.
11. Clegg, J. C.: 'A nonlinear integrator for servomechanisms', *Trans. A.I.E.E.*, 1958, pt. II, 77, pp. 41–42.
12. Foster, W. C., Gieseking, D. L. and Waymeyer, W. K.: 'A nonlinear filter for independent gain and phase (with application)', *Trans. A.S.M.E.*, 1966, pp. 475–482.
13. Bailey, A. R.: 'Stabilisation of control systems by the use of driven limiters', *Proc. I.E.E.*, 1966, 113, pp. 169–174.
14. Balasubramanian, R. and Atherton, D. P.: 'Performance of feedback systems with nonlinear phase-advance compensators to Gaussian inputs', *Electronics Letters*, 1967, 3, pp. 563–565.
15. Mishkin, E. and Braun, L. Jr.: *Adaptive control systems* (McGraw Hill, New York, 1961, pp. 204–216).
16. Gibson, J. E. and McVey, E. S.: 'Stability analysis of dual-mode servomechanisms', *Trans. A.I.E.E.*, 1960, 79, pt. II, pp. 173–179.
17. Shulkind, D.: 'Accuracy requirements for nonlinear compensation of backlash', *I.R.E. Trans.*, 1960, AC-5, pp. 79–84.
18. West, J. C. and Somerville, M. J.: 'Integral control with torque limitation', *Proc. I.E.E.*, 1956, 103, p. 407.
19. Jud, H. G.: 'Limit cycle determination for parallel linear and nonlinear elements', *I.E.E.E. Trans.*, 1964, AC-9, pp. 183–184.
20. Atherton, D. P.: 'Conditions for periodicity in control systems containing several relays', *Proc. I.F.A.C.*, 1966, London, Paper 28E.
21. Mees, A.: 'Describing functions, circle criteria and multiloop feedback systems', *Proc. I.E.E.*, 1973, 120, pp. 126–130.

22. Hopkins, W. G.: 'On the prediction of limit cycles in systems containing multiple nonlinearities', *I.E.E.E. Southeast Conference*, 1973, Louisville, Kentucky, U.S.A., Paper M-2.
23. Stein, W. A. and Thaler, G. J.: 'Obtaining the frequency response characteristics of a nonlinear servomechanism from an amplitude and frequency sensitive describing function', *Trans. A.I.E.E.*, 1958, 77, pt. II, pp. 91–96.
24. Levinson, E.: 'Some saturation phenomena in servomechanisms with emphasis on the tachometer stabilized system', *Trans. A.I.E.E.*, 1953, pt. II, pp. 1–9.
25. West, J. C. and Douce, J. L.: 'The frequency response of a certain class of nonlinear feedback systems', *British Journal Applied Physics*, 1954, 5, pp. 204–210.
26. Deekshatulu, et al.: 'The frequency response and jump resonance criteria of nonlinear control systems', *Int. J. of Control*, 1968, 7, pp. 265–270.
27. McAllister, A. S.: 'A graphical method for finding the frequency response of nonlinear closed loop systems', *Trans. A.I.E.E.*, 1961, pt. II, pp. 268–277.
28. Jayawant, B. V. and Rea, D. P.: 'Multimodal and transitional behaviour in second order nonlinear systems', *Proc. I.E.E.*, 1968, 115, pp. 863–876.
29. Prince, L. T.: 'A generalised method for determining the closed loop frequency response of nonlinear systems, *Trans. A.I.E.E.*, 1954, 73, pt. II, pp. 217–224.
30. Ogata, K.: 'An analytic method for finding the closed loop frequency response of nonlinear feedback control systems', *Trans. A.I.E.E.*, 1957, 76, pt. II, pp. 277–285.
31. West, J. C. and Nikiforuk, P.: 'The behaviour of a remote position control servomechanism with hard spring nonlinear characteristics', *Proc. I.E.E.*, 1954, 101, pt. II, p. 481.
32. Hammond, P. H.: *Feedback theory and its applications* (E.U.P., London, 1958, Chapter 11, pp. 258–263).
33. Siljak, D. D.: *Nonlinear systems* (Wiley, New York, 1969, Chapter 7).
34. Singh, Y. P.: 'Graphical method for finding the closed loop frequency response of nonlinear feedback control systems', *Proc. I.E.E.*, 1965, 112, pp. 2167–2170.
35. West, J. C., Jayawant, B. V. and Rea, D. P.: 'Transition characteristics of the jump phenomenon in nonlinear resonant circuits', *Proc. I.E.E.*, 1967, 114, pp. 381–392.
36. Lamba, S. S. and Kavanagh, R. J.: 'The phenomenon of isolated jump resonance and its applications', *Proc. I.E.E.*, 1971, 118, pp. 1047–1050.
37. Hatanaka, H.: 'The frequency responses and jump resonance phenomena of nonlinear feedback control systems', *Trans. A.S.M.E., J. Basic Eng.*, 1963, D-85, pp. 236–242.
38. Fukuma, A. and Matsubara, M.: 'Jump resonance criteria of nonlinear systems', *I.E.E.E. Trans.*, 1966, AC-11, pp. 699–706.
39. Lamba, S. S. and Kavanagh, R. J.: 'Jump resonance criteria for systems containing double-valued and frequency dependent nonlinearities', *Proc. I.E.E.*, 1969, 116, pp. 1225–1228.
40. Popov, E. P. and Pal'tov, N. P.: 'Priblizhennye metody issledovania nelineinych avtomaticheskikh sistem', *Gosudarstvennoe Izdatelstvo Phizikio, Mathematicheskoi Literatur, Moscow*, 1960. English translation: Approximate methods for analyzing nonlinear automatic systems, Wright Patterson Air Force Base, Ohio, Translation Services Branch, Foreign Technology Division, 1963.
41. Bogoliubov, N. N. and Mitropolsky, Y. A.: *Asymptotic methods in the theory of nonlinear oscillations* (translated from Russian) (Gordon and Breach, New York, 1961).
42. Johnson, E. C.: 'Sinusoidal analysis of feedback control systems containing nonlinear elements', *Trans. A.I.E.E.*, 1952, 71, pt. II, pp. 169–181.
43. Bass, R. W.: 'Mathematical legitimacy of equivalent linearization by describing functions', *Proc. I.F.A.C.*, Moscow, 1960 (Butterworth, London, pp. 2074–2083).
44. Gibson, J. E.: *Nonlinear automatic control* (McGraw-Hill, New York, 1962).
45. Poincaré, H.: *Les méthodes nouvelles de la mécanique céleste* (Gauthier-Villars, Paris, 1892, Vol. 1; reprinted by Dover, New York, 1957).
46. West, J. C.: 'The use of frequency response analysis in nonlinear control systems', in *Regelungstechnik, moderne theorien und ihre verwendbarkeit*, 1957.
47. Jones, N. B.: 'Simultaneous oscillations at unrelated frequencies in a feedback loop containing only one nonlinearity', *Int. J. Control*, 1967, 5, pp. 331–343.
48. Gelb, A. and Vander Velde, W. E.: *Multiple input describing functions and nonlinear systems design* (McGraw-Hill, New York, 1968, pp. 174–179).
49. Grensted, P. E. W.: 'Analysis of the transient response of nonlinear control systems), *Trans. A.S.M.E.*, 1958, 80, pp. 427–432.
50. Bickart, T. A.: 'The exponential describing function in the analysis of nonlinear systems', *I.E.E.E. Trans.*, 1966, AC-11, pp. 491–497.

51. Chen, K.: 'Quasi-linearization techniques for transient study of nonlinear feedback control systems', *Trans. A.I.E.E.*, 1956, 75, pt. II, pp. 354–365.
52. Freeman, E. A. and Cox, C. S.: 'Dynamic analysis of nonlinear control systems', *Proc. I.E.E.*, 114, 1967, pp. 129–138.
53. Freeman, E. A. and Cox, C. A.: 'Evaluation of transient gain functions for nonlinear elements and their use in step function analysis', *Ibid.*, pp. 1772–1780.
54. Stout, T. M.: 'A step by step method for transient analysis of feedback systems with one nonlinear element', *Trans. A.I.E.E.*, 1956, 75, pt. II, pp. 278–289.
55. Gibson, J. E.: *Nonlinear automatic control* (McGraw-Hill, New York, 1962, Chapter 4).

BIBLIOGRAPHY

Aizerman, M. A.: *Theory of automatic control* (Pergamon Press, New York, 1963, 5).

Andrievskii, V. R.: 'The estimation of self-oscillation parameters in nonlinear automatic control systems', *Automation and Remote Control*, 1961, 22, pp. 146–150.

Atherton, D. P. and Jain, A. K.: 'Identification of double-valued nonlinearities from their harmonic response', *Electron. Lett.*, 1967, 3, pp. 485–486.

Bass, R. W.: 'Equivalent linearization, nonlinear circuit synthesis and the stabilization and optimization of control systems', *Proc. Symp. Nonlinear Circuit Analysis, M.R.I. Series, Polytechnic Institute of Brooklyn, New York*, 1956, pp. 163–198.

Bass, R. W.: 'Mathematical legitimacy of equivalent linearization by describing functions', *Proc. I.F.A.C.*, Moscow, 1960 (Butterworth Scientific Publications, London, pp. 2074–2083).

Bogoliubov, N. N. and Mitropolsky, Y. A.: *Asymptotic methods in the theory of nonlinear oscillations* (translated from Russian) (Gordon and Breach, Science Publishers, Inc., New York, 1961).

Bonenn, Z.: 'Stability of forced oscillations in nonlinear feedback systems', *I.R.E. Trans.*, 1958, AC-6, pp. 109–111.

Bonenn, Z.: 'Connection between various methods of investigating absolute stability', *I.R.E. Trans.*, 1960, AC-5, pp. 330–331.

Casserly, G. and Truxal, J. G.: 'Measurement and stabilization of nonlinear feedback systems', *I.R.E. Conv. Record*, 1956, 4, 52–60.

Chestnut, H.: 'Approximate frequency-response methods for representing saturation and dead band', in R. Oldenburger (ed.), *Frequency Response* (The Macmillan Press, New York, 1956, pp. 269–287).

Choksy, N. H.: 'A new analytic stability criterion for use with describing functions', *J.A.C.C. Proc.*, 1963, 6, pp. 481–484.

Davison, E. J.: 'Application of the describing function technique in a single-loop feedback system with two nonlinearities', *I.E.E.E. Trans.*, 1968, AC-13, pp. 168–170.

Davison, E. J. and Constantinescu, D.: 'A describing function technique for multiple nonlinearities in a single loop feedback system', *I.E.E.E. Trans.*, 1971, AC-16, pp. 56–60.

Deekshatulu, B. L. and Krishna, G.: 'Evaluation of harmonic response of nonlinear components', *I.E.E.E. Summer General Meeting and Nuclear Radiation Effects Conference*, Toronto, June 1963, CP63-904.

Denn, M. M. and Black, J. R.: 'Estimation of nonlinear oscillations by an energy method', *Chemical Engineering Science*, 1973, 28, pp. 1515–1518.

Drosdziok, S.: 'Nichtlineare schwingungen in nickel', *Acustica*, 1973, 28, pp. 267–271.

Embler, J. N. and Weaver, C. H.: 'A new method for compensating contactor servomechanisms', *Trans. A.I.E.E.*, 1957, 76, pt. II, pp. 165–170.

Engels, D. A.: 'The describing function for a constrained range integration process with bang-bang input and dead zone', *I.E.E.E. Trans.*, 1967, AC-12, pp. 582–585.

Freeman, E. A.: 'Stability analysis of control systems having two nonlinear elements', *Proc. I.E.E.*, 1962, 109C, pp. 665–675.

Freeman, E. A. and Barney, G. C.: 'Divergent oscillations and their excitation in control systems with two saturation-type non-linear elements', *Proc. I.E.E.*, 1963, 110, pp. 1096–1106.

Gelb, A.: 'Limit cycles in symmetric multiple nonlinearity systems', *I.R.E. Trans.*, 1963, AC-8, pp. 177–178.

Glatenok, I. V.: 'Evaluation of the region of the true periodic solution found approximately by the harmonic balance method', *Automation and Remote Control*, 1958, 18, pp. 1175–1178.

Graham, D. and McRuer, D.: *Analysis of nonlinear control systems* (John Wiley & Sons, Inc., New York, 1961).

Gran, R. and Rimer, M.: 'Stability analysis of systems with multiple nonlinearities', *I.E.E.E. Trans.*, 1965, AC-10, pp. 94–97.

Gray, J. O. and Taylor, P. M.: 'Computer aided design using describing function methods', *I.E.E. Conf. Publication* No. 111, 1974, pp. 92–98.

Grayson, P.: 'Stability and oscillations in systems containing one or more single-valued symmetric nonlinearities', *J. Franklin Inst.*, 1965, 280, pp. 52–64.

Grensted, P. E. W.: 'Frequency response methods applied to nonlinear systems', pp. 103–141, *Progress in Control Engineering*, Vol. I, Academic Press, New York, 1962.

Grensted, P. E. W.: 'The frequency-response analysis of non-linear systems', *Proc. I.E.E.*, 1955, 102C, pp. 244–253.

Grumondz, V. T.: 'G. V. Kamenkov's method as a means for justifying the method of harmonic linearization', *Automation and Remote Control*, 1973, 34, pp. 983–985.

Haas, V. B., Jr.: 'Coulomb friction in feedback control systems', *Trans. A.I.E.E.*, 1953, 72, pt. II, pp. 119–123.

Halstenberg, R. V.: 'Combined hysteresis and nonlinear gains in complex control systems', *I.R.E. Trans.*, 1958, AC-3, pp. 51–58.

Hill, J. C.: 'Closed-loop response of nonlinear systems by a functional transformation approach', *I.R.E. Trans.*, 1962, AC-7, pp. 39–44.

Holtzman, J. M.: 'Contraction maps and equivalent linearization', *Bell System Tech. J.*, 1967, 46, pp. 2405–2435.

Hopkin, A. M. and Ogata, K.: 'An analytic frequency-response solution for a higher-order servomechanism with a nonlinear control element', *Trans. A.S.M.E., J. Basic Eng.*, 1959, D-81, pp. 41–45.

Hsu, J. C. and Meyer, A. U.: *Modern control principles and applications* (McGraw-Hill, New York, 1968, Chapter 6).

Jayawant, B. V. and Rea, D. P.: 'Multimodal and transitional behaviour of second order nonlinear systems', *Proc. I.E.E.*, 1968, 115, pp. 863–876.

Jonande, R. K. and Weygendt, C. N.: 'Limit cycles of higher order nonlinear autonomous systems', *J. of Franklin Inst.*, 1971, 291, pp. 195–210.

Kislyakov, V. S.: 'Application of the Krylov–Bogoliubov method of constructing asymptotic approximations to the investigation of systems with lag', *Automation and Remote Control*, 1960, 21, pp. 307–315.

Kislyakov, V. S.: 'Foundations for the application of the harmonic linearization method to an investigation of periodic oscillations in systems with lag', *Automation and Remote Control*, 1961, 21, pp. 1051–1056.

Kochenburger, R. J.: 'Limiting in feedback control systems', *Trans. A.I.E.E.*, 1953, 72, pt. II, pp. 180–194.

Kuruppu, U. S.: 'A describing function for backlash', *Int. J. Control*, 1967, 5, pp. 33–45.

Lamba, S. S.: 'Partial avoidance of jump resonance in feedback systems', *Int. J. of Control*, 1968, 8, pp. 537–542.

Leonhard, A.: 'Extension of the describing function method to the investigation of parametric oscillations', *Proc. I.F.A.C.*, Basel, Switzerland, 1963, pp. 187/1–187/7.

Levinson, E.: 'Gain-phase relations of non-linear circuits', *I.R.E. Conv. Record*, 1958, pt. IV, pp. 141–159.

Levinson, E.: 'Non-linear feedback control systems', *Parts 3–5, Electro-Technology*, 1962, Sept. pp. 136–146, Oct. pp. 131–140, Nov. pp. 71–82.

Lhote, F.: 'Sur les auto-oscillations de certains systèmes asservis non lineaires', *C.R. Acad. Sc.* Paris, 1967, 264, pp. 337–339.

Magnus, K.: 'On a method of investigating nonlinear systems of oscillations and of servomechanisms', *N.A.S.A.* translation of original article in German, 1955, OTS 59-16776, 1958.

Manabe, S.: 'Analyze nonlinear controls with saturation functions', *Control Eng.*, 1961, 8, pp. 83–86.

Marstrander, J. H. and Lueng, R. E.: 'Limit cycle determination for instrument servomechanism with backlash', *I.E.E.E. Trans.*, 1969, AC-14, pp. 756–757.

Marzawa, S.: 'Perfect Fourier series solution for feedback control system with a piecewise-linear element subjected to a sinusoidal input', *Bulletin of J.S.M.E.*, 1961, 4, pp. 499–507.

Marzawa, S.: 'Perfect Fourier series solution for feedback control system with a piecewise-linear element subjected to sinusoidal input', *Bulletin of J.S.M.E.*, 1964, 7, pp. 52–62.

Mees, A. I.: 'Periodic waveforms in forced non-linear systems', *Int. J. of Control*, 1973, 18, pp. 1169–1188.

Mees, A. I.: 'The describing function matrix', *J. Inst. Maths. Applics.*, 1972, 10, pp. 49–67.

Mikhail, S. L. and Fett, G. H.: 'Frequency response of nonlinear closed-loop feedback control systems', *Trans. A.I.E.E.*, 1958, 77, pt. II, pp. 436–438.
Mishkin, E. and Truxal, J. G.: 'Nonlinear compensation networks for feedback systems', *I.R.E. Conv. Record*, 1957, pt. 4. pp. 3–7.
Moruzzi, R. L. and Tuteur, F. B.: 'Nonlinear servomechanism of limited dynamic range', *Trans. A.I.E.E.*, 1960, 79, pt. II, pp. 314–320.
Mufti, I. H.: 'A note on the application of harmonic linearization', *I.E.E.E. Trans.*, 1963, AC-8, pp. 175–177.
Naslin, P.: *The dynamics of linear and non-linear systems* (Blackie, London, 1965, Chapter 8).
Nichols, N. B.: 'Backlash in a velocity lag servomechanism', *Trans. A.I.E.E.*, 1954, 73, pt. II, pp. 462–467.
Numakura, T., Mivra, T. and Kamoi, A.: 'Analyses of servomechanisms by describing function method on a plane with parameters as coordinates', *A.I.E.E. Paper* CP-57-1035, 1957.
Paquet et al.: 'Synthesis of nonlinear servosystems with a piecewise linear component', *Inst. Engrs. Australia, Electrical Eng. Trans.*, 1969, 5, pp. 219–221.
Petelin, D. P.: 'Approximate determination of the auto-oscillations in the automatic control system for a synchronous motor', *Automation and Remote Control*, 1959, 20, pp. 13–19.
Rangacharyulu, M. A. V. and Srinivasan, P.: 'A note on the period of oscillation of non-linear systems', *Journal of Sound and Vibration*, 1973, 29, pp. 207–213.
Riemenschneider, A. L. and Mastascusa, E. J.: 'A correlation between Fukuma–Matsubara and circle criteria for determination of jump resonance', *I.E.E.E. Trans.*, 1970, AC-15, pp. 507–508. Comments on the above paper by Lamba, S. S. et al., *I.E.E.E. Trans.*, 1972, AC-17, pp. 176–178.
Sandberg, I. W.: 'On the response of nonlinear control systems to periodic input signals', *Bell System Tech. J.*, 1964, 43, pp. 911–926.
Seltzer, S. M.: 'Analysis of a control system containing two non-linearities', *Int. J. of Control*, 1971, 13, pp. 1057–1064.
Sen, A. K.: 'The use of an R.M.S. describing function for the study of low order systems incorporating backlash', *Int. J. of Control*, 1967, 6, pp. 75–80.
Silberberg, M. Y. and Schultheiss, P. M.: 'Stability criteria for some nonlinear servomechanisms', *A.I.E.E. Paper* CP62-976, 1962.
Siljak, D. D.: 'Analysis and synthesis of feedback control systems in the parameter plane', *A.I.E.E. Trans.*, 1964, 83, pt. II, pp. 449–473.
Siljak, D. D.: 'Analysis of asymmetrical nonlinear oscillations in the parameter plane', *I.E.E.E. Trans.*, 1966, AC-11, pp. 239–247.
Slemon, G. R.: 'A method of approximate steady-state analysis for nonlinear networks', *J. I.E.E., London*, 1953, 100, pt. I, pp. 275–287.
Snelsire, R. and Clemson, S. C.: 'On the use of the describing function as a stability criterion', *I.E.E.E. Trans.*, 1967, AC-12, pp. 788–789.
Starkermann, R.: *Die harmonische linearisierung, Vols. 1 and 2* (Bibliographisches Institut, Mannheim, 1970).
Stein, W. A. and Thaler, G. J.: 'Obtaining the frequency response characteristics of a nonlinear servomechanism from an amplitude and frequency-sensitive describing function', *Trans. A.I.E.E.*, 1958, 77, pt. II, pp. 91–96.
Stoker, J. J.: *Non-linear vibrations in mechanical and electrical systems* (Interscience Publishers, Inc., New York, 1950).
Taft, V. A.: 'Stability of periodic conditions in automatic control systems found approximately on the basis of a filter hypothesis', *Automation and Remote Control*, 1959, 19, pp. 550–555.
Teodorescu, D.: 'Describing function series: a new means for nonlinear control system analysis', *Proc. I.E.E.*, 1970, 117, pp. 2175–2180.
Thaler, G. J. and Brown, R. G.: *Analysis and design of feedback control systems, 2nd edition* (McGraw-Hill, New York, 1960, Chapter 13).
Thaler, G. J. and Pastel, M. P.: *Analysis and design of nonlinear feedback control systems* (McGraw-Hill, New York, 1962, Chapter 4).
Thomas, C. H.: 'Stability characteristics of closed-loop systems with dead-band', in R. Oldenburger (ed.), *Frequency Response* (The Macmillan Press, New York, 1956, pp. 288–305).
Truxal, J. G.: *Control system synthesis* (McGraw-Hill, New York, 1955, Chapter 10).
Urabe, M.: 'Galerkin's procedure for nonlinear periodic systems', *Arch. Rational Mech. Anal.*, 1965, 20, pp. 120–152.
Wedlake, W. E. and Vacroux, A. G.: 'A new method for finding the frequency response of nonlinear control systems', *I.E.E.E. Trans.*, 1970, AC-15, pp. 512–513.
West, J. C.: 'The describing function technique applied to non-linear systems', *Process Control and Automation*, 1960, 7, pp. 421–426 and pp. 469–473.

PROBLEMS

P.4.1 Determine, using the describing function, the maximum value of K for stable operation of the system of question P.2.16. What are the amplitude and frequency of oscillation for values of K greater than that calculated above. Comment on the accuracy of the result.

P.4.2 Determine the amplitude and frequency of the limit cycle for example 2, section 2.11, by the DF method using the numerical values $h = 4$, $\Delta = 1/2$, $\delta = 1$, $K = 1$, $\lambda = 1/4$ and $T_d = 1/8$. Compare your result with the exact solution.

P.4.3 Discuss the effect of variations in T on possible limit cycles in the system of figure 4.1, if $G_1(s) = (s + 1)/s$, $G_2(s) = K/s^2$, $H(s) = 1 + sT$ and N.L. is an on-off relay with hysteresis. What is the amplitude of the output oscillation when $T = 0{\cdot}25$ secs, $K = 10$, $h = 4$ and $\Delta = 1$?

P.4.4 A position control system with negligible viscous friction and coulomb friction equivalent to an error of d radians is stabilised by velocity feedback. The velocity feedback signal is inadvertently connected in the wrong sense so that the equation of motion is governed by

$$(1/\omega_0^2)\ddot{\theta}_0 - (2\zeta/\omega_0)\dot{\theta}_0 + \theta_0 = \theta_i - d\,\mathrm{sgn}(\dot{\theta}_0).$$

Determine the amplitude and frequency of any possible limit cycles for $0 < \zeta < 1$, using
(a) the phase plane
(b) the describing function.
Is the limit cycle stable? Compare the results for $\zeta = 0{\cdot}25$.

P.4.5 Determine the maximum allowable value of T_d for the system of figure 4.1 to be stable if $G_1(s) = 1$, $G_2(s) = 10/s(s + 1)$, $H(s) = e^{-sT_d}$ and N.L. is an ideal saturation characteristic. If $T_d = 0{\cdot}5$ sec determine the frequency, amplitude at the input to the nonlinearity and stability of any possible oscillation. Check your result using the Routh criterion and the Padé approximation.

$$e^{-x} \approx \frac{1 - (3/4)x + (1/4)x^2 - (1/24)x^3}{1 + (1/4)x}.$$

P.4.6 A closed loop system contains a nonlinear element $n(x) = x^3$ and a linear transfer function $G(s) = 10/s(s + 1)^2$. Determine the limit cycle frequency using (a) the DF and (b) the refined DF.

P.4.7 A closed loop system of the form shown in figure 4.1 has $G_1(s) = H(s) = 1$, $G_2(s) = Ke^{-sT_d}/s$ and $n(x) = x^3$. Will the system possess jump resonance in its frequency response if (a) $T_d = 0$ and (b) T_d is finite. If $T_d = 1$ sec determine in terms of K the minimum input amplitude for which jump resonance will occur.

P.4.8 A system of the form shown in figure 4.1 has $G_1(s) = 1$, $G_2(s) = 1{\cdot}5/s(s + 1)$, $H(s) = e^{-sT_d}$ and $n(x)$ an on-off relay with dead zone having $\delta = h = 1$. Determine the maximum value, T_{dm}, of T_d for which the system is stable.

What is the amplitude and frequency of the limit cycle if $T_d = 1{\cdot}5T_{dm}$. Design a phase lead compensating network to stabilise the system for $T_d = T_{dm}$ with a phase margin of 30°.

P.4.9 A closed loop system contains an on-off relay with dead zone and a linear transfer function $G(s) = K\omega_0^2/s(s^2 + 2\zeta s\omega_0 + \omega_0^2)$. Show that the system will be stable if

$$0 < K < (\pi\zeta\omega_0\delta)/h.$$

If $K = 2\pi\zeta\omega_0\delta/h$ show that the system has a limit cycle with amplitude approximately $3{\cdot}9\delta$ and frequency ω_0.

P.4.10 Determine expressions which yield the amplitude and frequency of the limit cycle in a system containing a friction controlled backlash nonlinearity with width b and a linear transfer function $G(s) = 4\omega_0^2/s(4s + \omega_0)$.

P.4.11 The frictional damping in a position control was found to be proportional to the velocity squared (i.e. $kv|v|$). To establish the constant of proportionality, k, the velocity feedback signal was adjusted to a value which would give critical damping in the absence of friction, its sign reversed and the amplitude, a, at the output and frequency ω_0 of the resulting limit cycle measured. What is the value of k assuming the system to be of second order?

P.4.12 For the system of figure 4.1 sketch the open loop frequency response locus $G(j\omega)$ if $G_1(s) = H(s) = 1$ and $G_2(s) = G(s) = K/(s - 3)(s^2 + 2s + 10)$. If $K = 60$ and the nonlinearity is a saturation characteristic discuss the stability of the system using the Nyquist criterion for all magnitudes of input amplitude to the nonlinearity. Determine the frequency and amplitude at the intersection of $G(j\omega)$ and $C(a)$, and the stability of the limit cycle by the Loeb criterion (equation (4.9)). Do you expect this answer to be correct in the light of the Nyquist criterion results? Can you give an explanation for the incorrect result of the Loeb criterion?

P.4.13 A network has a feedforward gain of K and a feedback transfer function of $N/(1 + sT)$. Determine the closed loop transfer function. What nonlinear characteristics would you suggest for N if the network is to provide (a) phase lead at large signal levels only and (b) phase lead at low signal levels only?

P.4.14 Determine the amplitude and frequency using describing functions of any limit cycles existing in the system shown in figure P.2.6.

P.4.15 Using the describing function discuss the stability of the differential equation

$$\dddot{x} + 3\ddot{x} + \dot{x} - (\dot{x}^3/2) + x = 0.$$

P.4.16 Using the describing function discuss the stability of the differential difference equation

$$\ddot{x}(t) - \mu\{1 - \dot{x}^2(t)\}\dot{x}(t) + x(t - 1) = 0.$$

P.4.17 Show, using a DF representation for the nonlinearities, that any autonomous system consisting of a combination of linear and nonlinear blocks can be represented by a set of equations which in matrix form may be written

$$\{\mathbf{G}(j\omega)\mathbf{N}(\mathbf{X}) + I\}\mathbf{X} = \mathbf{O}.$$

$\mathbf{G}(j\omega)$ is a $n \times n$ matrix representing the linear dynamics and $\mathbf{N}(\mathbf{X})$ is a diagonal matrix of the n DFs and I is the identity matrix. The vector $\mathbf{X}$ is a complex representation for the sinusoidal inputs to the nonlinearities. What is the condition for a limit cycle?

P.4.18 Two nonlinear systems satisfying the conditions of question P.4.18 are shown in figures P.4.1 and P.4.2. Determine for each system the $\mathbf{G}(j\omega)$ matrix. If in figure P.4.2 both nonlinearities are ideal relays with ± 1 outputs $G_{11} = G_{22} = 10/(1+s)^3$ and $G_{12} = G_{21} = 5/(1+s)^3$ what will be the frequency of any possible limit cycles?

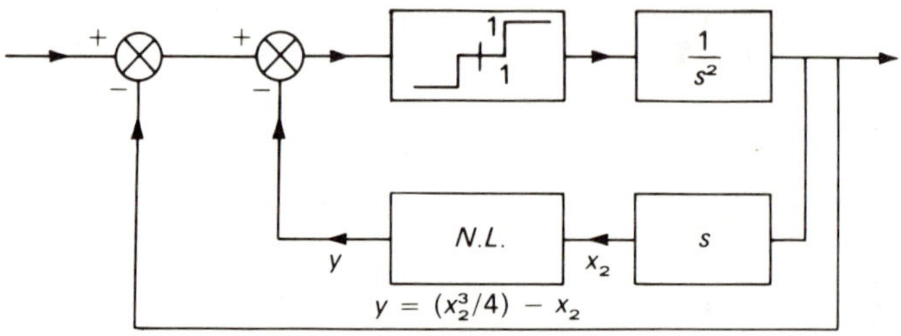

Figure P.4.1

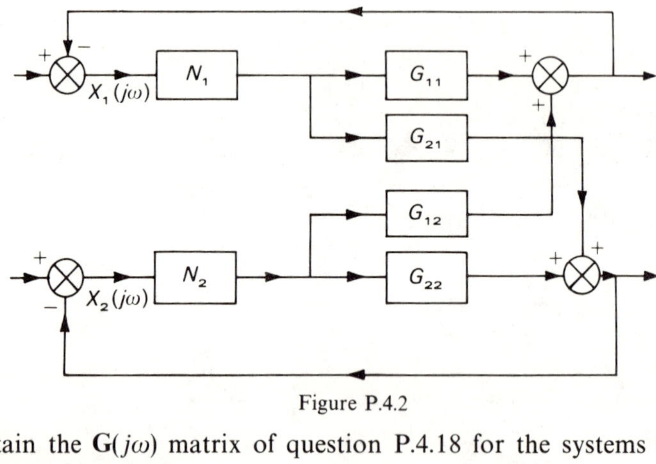

Figure P.4.2

P.4.19 Obtain the $\mathbf{G}(j\omega)$ matrix of question P.4.18 for the systems shown in figures 4.21 and 4.22.

P.4.20 A system has a feedforward path consisting of an on-off relay with dead zone, with $\delta = h = 1$, and a transfer function $G(s) = 10/s(s+2)^2$. The feedback path has a transfer function $H(s) = e^{-sT_d}$. Determine the maximum value, T_{dm}, of T_d for which the system is stable. What is the amplitude and frequency of the limit cycle when $T_d = T_{dm}$? Design a phase lead network to stabilise the system and give a minimum phase margin of about 30° when $T_d = T_{dm}/4$.

Chapter 5

Evaluation and Applications of SBDFs and SSDFs

5.1 Introduction

In this chapter the describing function for a sinusoid plus bias (steady or d.c.) signal, the SBDF, is introduced. In many systems it is necessary to model the nonlinearity input in this manner in order to analyse the behaviour. Systems containing asymmetrical nonlinearities or symmetrical nonlinear systems when subject to bias inputs, either as constant reference signals or disturbances, fall into this category. Several applications of the SBDF in the determination of steady state behaviour of free and forced systems are discussed. It is also indicated that DF methods applied to asymmetrical systems usually produce less accurate solutions due to their neglect of the harmonics, as in this case the lowest is normally the second rather than the third. The concept of modified nonlinearities is introduced during the derivation of the SBDF. Of particular importance is the '0' modified nonlinearity which may be used in many cases to replace the original characteristic when the bias becomes a slowly varying or other signal.

The two-sinusoidal input describing function, denoted by SSDF, is derived for the simple case of the cubic nonlinearity by direct expansion. The output terms at the same frequency as the two-sinusoidal inputs are found to depend upon whether the input frequencies are commensurate. For incommensurate frequencies the SSDF is seen to be the same as the DF of the modified nonlinearity, a result which is shown to be true in general in chapter 10. The method is used here, however, to evaluate SSDFs for several common nonlinearities. For commensurate frequencies, solutions for the SSDF are relatively complicated and a more detailed discussion of their determination is also deferred to the above-mentioned chapter. A special case of the SSDF, the incremental describing function denoted by IDF, which is the DF for one of the sinusoids as its amplitude tends to zero, is studied in detail both for commensurate and incommensurate frequencies. The latter case gives the same result as the DF for the bias in the SBDF as the bias tends to zero.

Applications of the SSDF and IDF, which include aspects such as the induction and quenching of limit cycles, jump resonance, signal stabilisation and subharmonic oscillations are discussed in depth.

5.2 The SBDF

In this case the input z to the nonlinearity is given by

$$z = x + \gamma,$$

where $x = a\cos\theta$ is the harmonic signal and γ the constant bias signal. The harmonic output coefficients a_s and b_s can still be evaluated from equations (3.4) and (3.5) provided the correct output waveform $y(\theta)$, which for the SVNL is $n(\gamma + a\cos\theta)$, is used. Alternatively in terms of the input signals and the in phase and quadrature nonlinearities equations (3.14) and (3.15) become

$$a_s = \varepsilon_s \int_{-a}^{a} n_p(x + \gamma) T_s(x/a) r(x)\, \mathrm{d}x, \tag{5.1}$$

and

$$b_s = 2\int_{-a}^{a} n_q(x+\gamma)D_s(x/a)r(x)\,\mathrm{d}x, \tag{5.2}$$

where, as in the single input case, $n_p(z)$ and $n_q(z)$ may be different characteristics for the various magnitudes of a and γ. Since a_s and b_s are functions of a and γ we will also use the notation

$$a_s = \varepsilon_s n_p(a, \gamma)_s, \tag{5.3}$$

$$b_s = 2n_q(a, \gamma)_s, \tag{5.4}$$

and refer to $n_p(a, \gamma)_s$ and $n_q(a, \gamma)_s$ as modified nonlinearities[1,2] of order s when plotted against γ for fixed values of a.

The SBDF consists of two components namely the gain to the sinusoid $N(a, \gamma)$ and the gain to the bias signal $N_\gamma(a, \gamma)$ where

$$N(a,\gamma) = N_p(a,\gamma) + jN_q(a,\gamma) = (a_1 - jb_1)/a, \tag{5.5}$$

and

$$N_\gamma(a,\gamma) = a_0/\gamma = n_p(a,\gamma)_0/\gamma. \tag{5.6}$$

It can easily be shown that for SVNLs these gains satisfy mean squared error and correlation definitions similar to those of section 3.3. Referring to figure 5.1 the

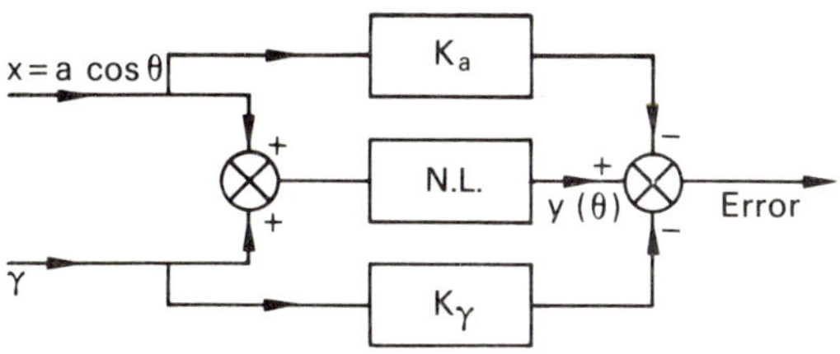

Figure 5.1 Linear gain model of nonlinearity

optimum values of K_a and K_γ to give a minimum mean squared error are $N(a, \gamma)$ and $N_\gamma(a, \gamma)$ respectively. In addition the adjustment of the two gains is orthogonal so that the optimum value of K_a is $N(a, \gamma)$, independent of the setting of K_γ and vice versa. Similarly with reference to the correlation definition it can be shown that

$$N(a,\gamma) = (\overline{y(\theta)a\cos\theta})/(a^2/2), \tag{5.7}$$

and

$$N_\gamma(a,\gamma) = (\overline{y(\theta)\gamma})/\gamma^2. \tag{5.8}$$

The modified nonlinearity of order zero, $n_p(a, \gamma)_0$, is usually the modified characteristic of most interest. Thus hereafter it will simply be referred to as the modified nonlinearity and denoted by $n_p(a, \gamma)$. In addition, its slope $\mathrm{d}\{n_p(a, \gamma)\}/\mathrm{d}\gamma$, denoted by $g(a, \gamma)$ is called the d.c. incremental gain. It is shown later how modified nonlinearities can be used to determine describing functions for multiple input situations, although, since additional complications are encountered for DVNLs they are discussed separately in chapter 13.

For the SVNL

$$n(a, \gamma) = \int_{-a}^{a} n(x + \gamma)r(x)\,\mathrm{d}x, \tag{5.9}$$

and

$$g(a, \gamma) = \frac{\mathrm{d}}{\mathrm{d}\gamma}\left\{\int_{-a}^{a} n(x + \gamma)r(x)\,\mathrm{d}x\right\}. \tag{5.10}$$

It is shown in section 10.7.2 that $g(a, \gamma)$ gives the gain to any small signal in the presence of x and γ, provided the signal is unrelated to x. In particular the gain to any small signal in the presence of the sinusoid x only, $g(a, 0)$, is the incremental describing function (IDF) denoted by $N_{i\gamma}(a)$. From equation (5.10) we see that

$$N_{i\gamma}(a) = g(a, 0) = \lim_{\gamma \to 0} N_{\gamma}(a, \gamma) = \lim_{\gamma \to 0} \frac{\mathrm{d}}{\mathrm{d}\gamma}\left\{\int_{-a}^{a} n(x + \gamma)r(x)\,\mathrm{d}x\right\}$$

$$= \int_{-a}^{a} n'(x)r(x)\,\mathrm{d}x, \tag{5.11}$$

where $n'(x) = \mathrm{d}\{n(x)\}/\mathrm{d}x$. Thus we may also interpret $g(a, 0)$ as the bias output from the nonlinearity $n'(x)$ with input x.

Integrating equation (3.14) for $s = 1$ by parts, the equivalent of equation (3.23) over the range $-a$ to a is

$$N(a) = (2/a^2)\int_{-a}^{a} (a^2 - x^2)n'(x)r(x)\,\mathrm{d}x, \tag{5.12}$$

so that

$$a^2 N(a) = (2/\pi)\int_{-a}^{a} n'(x)(a^2 - x^2)^{1/2}\,\mathrm{d}x.$$

Differentiating both sides with respect to a gives

$$2aN(a) + a^2\{\mathrm{d}N(a)/\mathrm{d}a\} = 2\int_{-a}^{a} an'(x)r(x)\,\mathrm{d}x. \tag{5.13}$$

Hence

$$N_{i\gamma}(a) = N(a) + (a/2)\{\mathrm{d}N(a)/\mathrm{d}a\}. \tag{5.14}$$

Thus the gain to any small unrelated input in the presence of the sinusoid x can be evaluated directly from the DF of x.

5.3 Some analytical solutions

Expressions are derived below for the modified nonlinearities corresponding to several common nonlinear characteristics. Here our attention of necessity is not restricted to characteristics with odd symmetry. Additional solutions for SBDFs are tabulated in appendix A2.

5.3.1 *Integer power law*

Let $n(x) = x^m$; then

$$n(a, \gamma)_s = \int_{-a}^{a} (x + \gamma)^m T_s(x/a) r(x)\,dx.$$

Expanding $(x + \gamma)^m$ and denoting the binomial coefficient $m!/(m - k)!k!$ by ${}_mC_k$

$$n(a, \gamma)_s = \sum_{k=0}^{m} \int_{-a}^{a} {}_mC_k x^k \gamma^{m-k} T_s(x/a) r(x)\,dx,$$

which on substituting $x = a\cos\theta$ and integrating gives

$$n(a, \gamma)_s = \sum_{k=\phi(2)}^{m} \frac{{}_mC_k \gamma^{m-k} (a/2)^k \Gamma(k + 1)}{\Gamma\{1 + (k + s)/2\}\Gamma\{1 + (k - s)/2\}}, \tag{5.15}$$

where

$$\phi = \begin{cases} 0 & \text{if } m + s \text{ is odd} \\ 1 & \text{if } m + s \text{ is even} \end{cases}.$$

Thus for a fixed value of a if

(i) m is odd, $n(a, \gamma)_s$ is an odd $\{$even$\}$ polynomial in γ if s is even $\{$odd$\}$ of highest power m $\{m - 1\}$;

(ii) m is even, $n(a, \gamma)_s$ is an even $\{$odd$\}$ polynomial in γ if s is even $\{$odd$\}$ of highest power m $\{m - 1\}$.

In particular

$$N_\gamma(a, \gamma) = \sum_{k=0}^{m} {}_mC_k \mu_k \gamma^{m-k-1},$$

$$g(a, \gamma) = \sum_{k=0}^{m} (m - k)\,{}_mC_k \mu_k \gamma^{m-k-1},$$

and

$$N(a, \gamma) = (2/a^2) \sum_{k=0}^{m} {}_mC_k \mu_{k+1} \gamma^{m-k},$$

where

$$\mu_k = \int_{-a}^{a} x^k r(x)\,dx = \frac{a^k \Gamma\{1 + (k + 1)/2\}}{\pi^{1/2}\Gamma\{1 + (k + 2)/2\}} \quad \text{if } k \text{ even},$$

is the kth moment of the amplitude probability density for a sinusoid (see section 7.9.1).

5.3.2 *Harmonic nonlinearity*

For $n(x) = \sin mx$,

$$n(a, \gamma)_s = \int_{-a}^{a} \sin m(x + \gamma) T_s(x/a) r(x)\,dx.$$

Expanding $\sin m(x + \gamma)$ and making the substitution $x = a\cos\theta$ gives

$$n(a, \gamma)_s = (1/\pi)\left[\int_0^{\pi} \cos m\gamma \sin(ma\cos\theta)\cos s\theta\,d\theta \right.$$
$$\left. + \int_0^{\pi} \sin m\gamma \cos(ma\cos\theta)\cos s\theta\,d\theta\right].$$

Substituting

$$\cos(z\cos\theta) = J_0(z) + 2\sum_{p=1}^{\infty} (-1)^p J_{2p}(z)\cos 2p\theta,$$

$$\sin(z\cos\theta) = -2\sum_{p=1}^{\infty} (-1)^p J_{2p-1}(z)\cos(2p - 1)\theta,$$

and integrating, we obtain

$$n(a, \gamma)_s = \begin{cases} (-1)^{s/2} J_s(ma)\sin m\gamma & \text{for } s \text{ even,} \\ (-1)^{(s-1)/2} J_s(ma)\cos m\gamma & \text{for } s \text{ odd,} \end{cases}$$

or, in one expression,

$$n(a, \gamma)_s = (-1)^s J_s(ma)\sin\{m\gamma - (s\pi/2)\}. \tag{5.16}$$

In particular

$$N_\gamma(a, \gamma) = \{J_0(ma)\sin m\gamma\}/\gamma,$$
$$g(a, \gamma) = mJ_0(ma)\cos m\gamma,$$

and

$$N(a, \gamma) = (2/a)J_1(ma)\cos m\gamma.$$

It is seen from equation (5.16) that the modified nonlinearity $n(a, \gamma)$ as a function of γ for a constant value of a differs only in amplitude from the original characteristic $n(x)$—a property which will be found to be very useful in later work.

In a similar manner we can show that if

$$n(x) = \cos mx,$$
$$n(a, \gamma)_s = (-1)^s J_s(ma)\cos\{m\gamma - (s\pi/2)\}.$$

5.3.3 *Ideal saturation*

For saturation, with $h = m\delta$

$$n(x+\gamma) = \begin{cases} m(x+\gamma) & \text{for } -(\delta+\gamma) < x < \delta-\gamma \\ h & \text{for } x \geq \delta-\gamma \\ -h & \text{for } x \leq -(\delta+\gamma). \end{cases}$$

Therefore

$$n(a,\gamma)_s = \int_{-a}^{\mu} -hT_s(x/a)r(x)\,\mathrm{d}x + \int_{\mu}^{\delta-\gamma} m(x+\gamma)T_s(x/a)r(x)\,\mathrm{d}x$$
$$+ \int_{\delta-\gamma}^{a} hT_s(x/a)r(x)\,\mathrm{d}x,$$

where

$$\mu = \begin{cases} -a & \text{for } a < \delta+\gamma \\ -(\delta+\gamma) & \text{for } a > \delta+\gamma. \end{cases}$$

Integrating

$$n(a,\gamma)_s = (h/s\pi)[D_s(x/a)]_{-a}^{\mu} - (h/s\pi)[D_s(x/a)]_{\delta-\gamma}^{a} - (m\gamma/s\pi)[D_s(x/a)]_{\mu}^{\delta-\gamma}$$
$$- (ma/2\pi)\left[\frac{D_{s-1}(x/a)}{s-1} + \frac{D_{s+1}(x/a)}{s+1}\right]_{\mu}^{\delta-\gamma}$$

for $s \geq 1$, which has the solution

$$n(a,\gamma)_s = \left(\frac{h+m\gamma}{s\pi}\right)D_s\{-(\delta+\gamma)/a\} + \left(\frac{h-m\gamma}{s\pi}\right)D_s\{(\delta-\gamma)/a\}$$
$$+ (ma/2\pi)\left[\frac{D_{s-1}\{-(\delta+\gamma)/a\}}{s-1} + \frac{D_{s+1}\{-(\delta+\gamma)/a\}}{s+1}\right.$$
$$\left. - \frac{D_{s-1}\{(\delta-\gamma)/a\}}{s-1} - \frac{D_{s+1}\{(\delta-\gamma)/a\}}{s+1}\right] \quad \text{for } s \geq 1,$$

where $D_s(x)$ is the function defined in section 3.2.2 for $|x| < 1$ and we note that $\{D_s(x)/s\}|_{s=0} = \cos^{-1} x$. So that the result is valid for all values of a and γ we define for $|x| > 1$, $D_s(x) = 0$ if $s \neq 0$ and for $s = 0$ take

$$\cos^{-1} x = \begin{cases} \pi & \text{for } x < -1 \\ 0 & \text{for } x > 1 \end{cases}.$$

Rearranging the terms in the above equation gives

$$n(a,\gamma)_s = \frac{am}{\pi}\left[\frac{(\delta+\gamma)}{as}D_s\{-(\delta+\gamma)/a\} + \frac{D_{s+1}\{-(\delta+\gamma)/a\}}{2(s+1)} + \frac{D_{s-1}\{-(\delta+\gamma)/a\}}{2(s-1)} + \frac{(\delta-\gamma)}{as}D_s\{(\delta-\gamma)/a\} - \frac{D_{s+1}\{(\delta-\gamma)/a\}}{2(s+1)} - \frac{D_{s-1}\{(\delta-\gamma)/a\}}{2(s-1)}\right],$$

$$= (am/4)[f_s\{(\delta-\gamma)/a\} - f_s\{-(\delta+\gamma)/a\}], \tag{5.17}$$

where $f_s(\rho)$ is the saturation function of section 3.4.1 and may also be evaluated from

$$f_s(\rho) = \frac{4}{\pi}\left\{\frac{\rho D_s(\rho)}{s} - \frac{D_{s+1}(\rho)}{2(s+1)} - \frac{D_{s-1}(\rho)}{2(s-1)}\right\}.$$

In particular, with $m = 1$

$$a_1 = 2n(a,\gamma)_1 = (a/2)[f_1\{(h-\gamma)/a\} + f_1\{(h+\gamma)/a\}]. \tag{5.18}$$

Similarly for $s = 0$, we obtain

$$n(a,\gamma) = \left(\frac{h+\gamma}{\pi}\right)\cos^{-1}\{-(h+\gamma)/a\} + \left(\frac{h-\gamma}{\pi}\right)\cos^{-1}\{(h-\gamma)/a\} - h + (a/\pi h)[D_1\{-(h+\gamma)/a\} - D_1\{(h-\gamma)/a\}],$$

where $\cos^{-1} x$ takes the values given above for $|x| > 1$, or

$$n(a,\gamma) = (a/2)[f_0\{(h+\gamma)/a\} - f_0\{(h-\gamma)/a\}], \tag{5.19}$$

where

$$f_0(\rho) = \begin{cases} (2/\pi)\{\rho\sin^{-1}\rho + (1-\rho^2)^{1/2}\} & \text{for } |\rho| \le 1, \\ |\rho| & \text{for } |\rho| > 1. \end{cases}$$

The d.c. incremental gain can be obtained using equation (5.10) with the result

$$g(a,\gamma) = (1/\pi)[\cos^{-1}\{-(h+\gamma)/a\} - \cos^{-1}\{(h-\gamma)/a\}].$$

5.3.4 *On-off relay with hysteresis*

Using the equivalent in phase and quadrature nonlinearities of figure 3.6, we have for $|\gamma| < a - \Delta$

$$n_p(a,\gamma)_s = \int_{-a}^{-\Delta-\gamma} -hT_s(x/a)r(x)\,dx + \int_{\Delta-\gamma}^{a} hT_s(x/a)r(x)\,dx,$$

$$= (h/s\pi)[D_s\{-(\Delta+\gamma)/a\} + D_s\{(\Delta-\gamma)/a\}] \quad \text{for } s \ge 1, \tag{5.20}$$

and

$$n(a,\gamma) = (h/\pi)[\cos^{-1}\{(\Delta-\gamma)/a\} + \cos^{-1}\{-(\Delta+\gamma)/a\} - \pi],$$
$$= (h/\pi)[\sin^{-1}\{(\Delta+\gamma)/a\} - \sin^{-1}\{(\Delta-\gamma)/a\}] \quad (5.21)$$

also

$$n_q(a,\gamma)_s = (h/s\pi)[T_s\{(\Delta-\gamma)/a\} - T_s\{-(\Delta+\gamma)/a\}]. \quad (5.22)$$

Obviously, for $|\gamma| > a - \Delta$, $n_p(a,\gamma)_s = n_q(a,\gamma)_s = 0$ and $n(a,\gamma) = h\,\text{sgn}\,\gamma$. For $a < \Delta$, $n(a,\gamma) = h\,\text{sgn}\,\gamma$ for $|\gamma| > \Delta - a$ and for $|\gamma| < \Delta - a$ is either $\pm h$ depending upon whether γ is increasing or decreasing; thus $n(a,\gamma)$ is an on-off relay with hysteresis of width $(\Delta - a)$. Modified nonlinearities $n(a,\gamma)$ are shown in figure 5.2 for various magnitudes of a/Δ for Δ finite and zero.

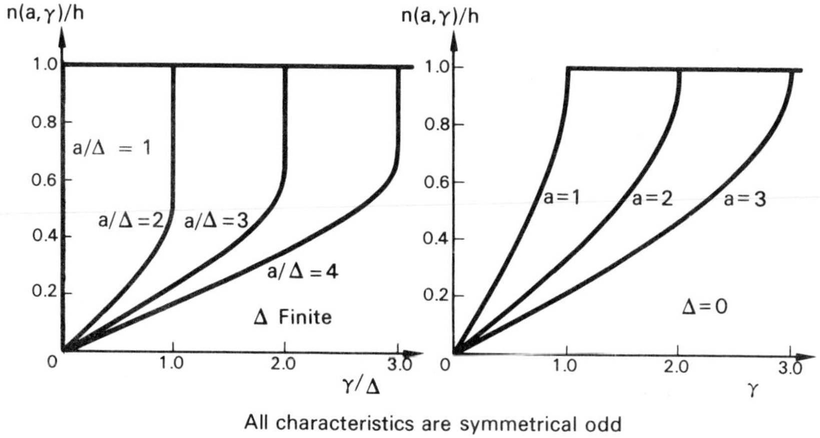

Figure 5.2 Modified nonlinearities for an on-off relay with and without hysteresis

Also we can show that

$$g(a,\gamma) = (h/\pi)[\{a^2 - (\Delta+\gamma)^2\}^{-1/2} + \{a^2 - (\Delta-\gamma)^2\}^{-1/2}].$$

5.4 Approximate solutions for sinusoidal modified nonlinearities

The results of the previous section show that analytical solutions can be obtained for modified nonlinearities $n(a,\gamma)_s$ of those asymmetrical SVNLs which can be approximated by a power or Fourier series. It is important to note that in the case of symmetrical SVNLs for a given value of a, $n(a,\gamma)_s$ as a function of γ is even (odd) if s is even (odd) for $n(x)$ odd, and $n(a,\gamma)_s$ is even (odd) if s is odd (even) for $n(x)$ even. For DVNLs, as with the single sinusoidal input case, simple solutions for $n_q(a,\gamma)_s$ can be found for $n_q(x)$ a power series but they are more involved when $n_q(x)$ is taken as a Fourier series. In some cases the origin of an asymmetrical characteristic can be shifted so that the results available for a symmetrical characteristic may be used as illustrated in figure 5.3. Although any asymmetrical

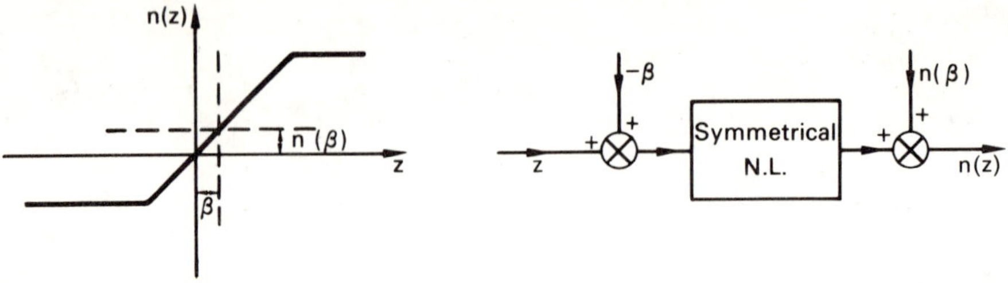

Figure 5.3 Illustration of origin shift to make use of symmetry

characteristic can be replaced by a parallel combination (i.e. sum) of odd and even nonlinearities, the approach is rarely useful as SBDF expressions for even symmetrical functions are often complicated. Some general techniques which may be used for graphically defined asymmetrical characteristics are given below.

Consider the general asymmetrical DVNL separated into its equivalent in phase and quadrature characteristics which are assumed valid for all the input amplitudes to be used. For the input $z = x + \gamma$ with $x = a\cos\theta$, we may consider $n_p(z)$ and $n_q(z)$ divided into two parts with respect to a new origin 0_γ, the point $(\gamma, 0)$ as shown in figure 5.4, so that

$$n_{p+}(x) = n_p(z) \quad \text{for } z > \gamma,$$
$$n_{p-}(x) = n_p(z) \quad \text{for } z < \gamma,$$

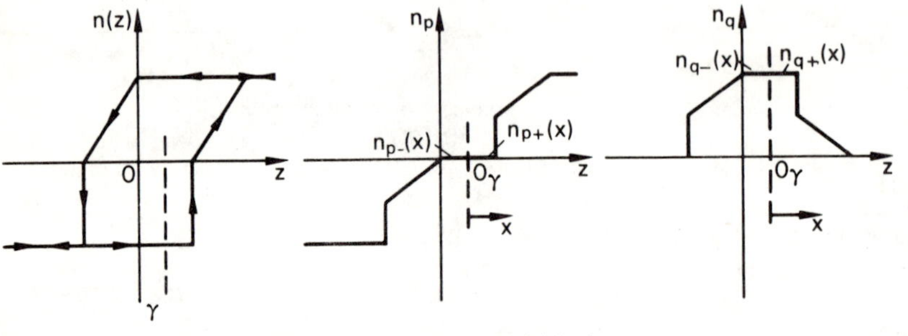

Figure 5.4 New origin with bias input

and similarly for $n_q(x)$. Then

$$\left.\begin{aligned} a_0 &= n(a,\gamma) = a_{+0} + a_{-0} \\ a_s &= 2n_p(a,\gamma)_s = a_{+s} + a_{-s} \\ b_s &= 2n_q(a,\gamma)_s = b_{+s} + b_{-s} \end{aligned}\right\}, \tag{5.23}$$

where a_{+s} is the sth harmonic output coefficient from $n_{p+}(x)$, etc. It should be noted that as

$$\int_{-a}^{a} T_s(x/a) r(x)\,\mathrm{d}x = 0, \quad \text{for } s \neq 0,$$

the coefficients a_s may also be evaluated with respect to the origin $\{\gamma, n_p(\gamma)\}$ on the nonlinear characteristic $n_p(x)$. If the bias term is evaluated with respect to this origin then the actual output bias is greater by $n_p(\gamma)$. It is not possible, however, to evaluate the coefficients b_s with respect to the origin $\{\gamma, n_q(\gamma)\}$. In addition a_{-s} and b_{-s} may be evaluated from similar nonlinear characteristics $m_{p+}(x)$ and $m_{q+}(x)$ in the first quadrant,[3] so that general results only need be obtained for single-sided first quadrant characteristics. If

$$m_{p+}(x) = \pm n_{p-}(-x),$$

then

$$a_{-s} = \begin{cases} \pm a_{+s} & \text{for } m_{p+}(x) \text{ if } s \text{ is even,} \\ \mp a_{+s} & \text{for } m_{p+}(x) \text{ if } s \text{ is odd,} \end{cases} \tag{5.24}$$

and similarly if

$$m_{q+}(x) = \pm n_{q-}(-x),$$

$$b_{-s} = \begin{cases} \mp b_{+s} & \text{for } m_{q+}(x) \text{ if } s \text{ is even,} \\ \pm b_{+s} & \text{for } m_{q+}(x) \text{ if } s \text{ is odd.} \end{cases} \tag{5.25}$$

For the nonlinear characteristics $n_{p+}(x)$ and $n_{q+}(x)$, a_{+s} and b_{+s} may be evaluated approximately by using (a) a quantised approximation, (b) graphical integration, or if the SBDF only is required, by (c) approximate analytical solution of the integrals.

(a) Quantised approximation

Proceeding as in section 3.5.3 we obtain the results

$$\left.\begin{aligned} a_{+0} &= (1/\pi) \sum_{m=1}^{M} h_m \cos^{-1}(\delta_m/a) \\ a_{+s} &= (2/s\pi) \sum_{m=1}^{M} h_m D_s(\delta_m/a) \\ b_{+s} &= (2/s\pi) \sum_{m=1}^{M} h_m\{1 - T_s(\delta_m/a)\} \end{aligned}\right\}. \tag{5.26}$$

It should be remembered here that h_m is positive along the positive direction of the ordinate axis.

(b) Graphical integration

Cursor values are given in table 3.2 for s both odd and even. The procedure follows that described in section 3.5.4.

(c) Approximate analytical SBDF solution

In this case we make use of the Steklov approximation to find a_{+0} as well as a_{+1}.

Now

$$a_{+0} = \int_0^a n_{p+}(x)r(x)\,\mathrm{d}x,$$

which on substituting for $r(x)$ and putting $u = x/a$, gives

$$a_{+0} = (1/\pi)\int_0^1 \frac{n_{p+}(au)}{(1-u^2)^{1/2}}\,\mathrm{d}u.$$

Thus

$$a_{+0} \approx (1/6)\{n_{p+}(a) + 2n_{p+}(a/2)\}.$$

Similarly, from section (3.5.5)

$$a_{+1} \approx (1/3)\{n_{p+}(a) + n_{p+}(a/2)\},$$
$$b_{+1} \approx +(1/3\pi)\{n_{q+}(0) + 4n_{q+}(a/2) + n_{q+}(a)\}.$$

The following two examples illustrate the various methods.

Example 1

Evaluate the SBDF for the asymmetrical saturation characteristic

$$n(x) = \begin{cases} 1 & x > 1 \\ x & -2 < x < 1. \\ -2 & x < -2 \end{cases}$$

Using the procedure illustrated in figure 5.3, $\beta = -1/2$ and the symmetrical saturation characteristic has limit levels of $\pm 3/2$. Thus from the results of section 5.3.3

$$\gamma N_\gamma(a,\gamma) = -1/2 + (a/2)\{f_0[(2+\gamma)/a] - f_0[(1-\gamma)/a]\},$$

and

$$N(a,\gamma) = (1/2)\{f_1[(2+\gamma)/a] + f_1[(1-\gamma)/a]\}.$$

Figure 5.5 shows $n(a,\gamma) = \gamma N_\gamma(a,\gamma)$ and $N(a,\gamma)$ plotted against γ for various values of a. $n(a,\gamma)$ has odd symmetry about the point $(-1/2, -1/2)$ and $N(a,\gamma)$ even symmetry about $(-1/2, 0)$. Also marked for $\gamma > -1/2$ are solutions obtained using the approximate analytical solutions.

Example 2

Determine a_s and b_s for values of s from 0 to 3 for an input of $1 + 3\cos\theta$ to the DVNL of figure 3.10.

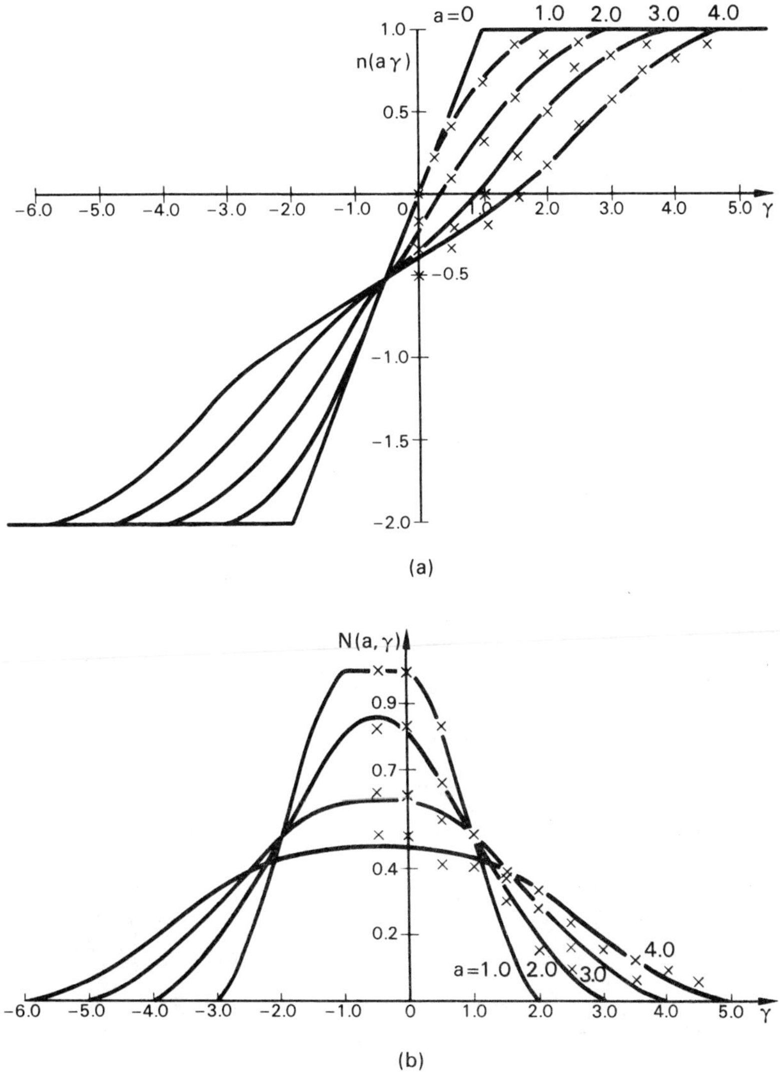

Figure 5.5 SBDF for an asymmetrical saturation

Table 5.1 shows the procedure and solutions obtained by graphical integration on $n_p(x)$ and $n_q(x)$ about the origin (1, 0). The theoretical solutions are given by

$$a_s = \varepsilon_s\left\{\int_{-3}^{1} \{(x+1)/2\}T_s(x/a)r(x)\,dx + \int_{1}^{3} T_s(x/a)r(x)\,dx\right\},$$

and

$$b_s = (2/a\pi)\left\{\int_{-3}^{-1} \{(x+3)/2\}U_{s-1}(x/a)\,dx + \int_{-1}^{1} \{(1-x)/2\}U_{s-1}(x/a)\,dx\right\}.$$

TABLE 5.1
Graphical integration solution for example 2, section 5.4

a_0—10 increment cursor

$\cos^{-1} x/3$	Slope	n_{p+}	Slope	n_{p-}
0·237	−	−0·618	−	−0·381
0·699	−	−0·850	−	−0·152
1·149	−	−1·000	−	0·074
1·569	−	−1·000	−	0·285
1·947	−	−1·000	−	0·473
2·180	−	−1·000	−	0·590
2·559	−	−1·000	−	0·780
2·772	−	−1·000	−	0·886
2·916	−	−1·000	−	0·958
2·991	−	−1·000	−	0·995
		−9·468		4·508

$$a_0 = \frac{-(-9{\cdot}468 + 4{\cdot}508)}{20} = 0{\cdot}248$$

a_1—10 increment cursor

$D_1(x/3)$	D_1 Slope	n_{p+}	D_1 Slope	n_{p-}
0·936	−	0·968	+	0·032
1·581	−	−1·000	+	−0·290
1·983	−	−1·000	+	−0·492
2·280	−	−1·000	+	−0·640
2·505	−	−1·000	+	−0·752
2·679	−	−1·000	+	−0·840
2·811	−	−1·000	+	−0·905
2·904	−	−1·000	+	−0·952
2·967	−	−1·000	+	−0·984
2·997	−	−1·000	+	−0·998
		−9·968		−6·855

$$a_1 = \frac{-4}{\pi}\left\{\frac{-9{\cdot}968 - 6{\cdot}855}{20}\right\} = 1{\cdot}069$$

b_1—10 increment cursor

$T_1(x/3)$	T_1 Slope	n_{q+}	T_1 Slope	n_{q-}
0·150	+	0·425	+	0·575
0·450	+	0·275	+	0·725
0·750	+	0·125	+	0·875
1·050	+	0	+	0·975
1·350	+	0	+	0·825
1·650	+	0	+	0·675
1·950	+	0	+	0·525
2·250	+	0	+	0·375
2·550	+	0	+	0·225
2·850	+	0	+	0·075
		0·825		5·850

$$b_1 = \frac{4}{\pi}\left\{\frac{0{\cdot}825 + 5{\cdot}850}{20}\right\} = 0{\cdot}425$$

a_2—5 increment cursor

$D_2(x/3)$	D_2 Slope	n_{p+}	D_2 Slope	n_{p-}
0·150	+	0·575	+	0·425
0·456	+	0·728	+	0·272
0·777	+	0·888	+	0·111
1.134	+	1·000	+	−0·077
1·593	+	1·000	+	−0·296
2·541	−	−1·000	−	0·770
2·778	−	−1·000	−	0·889
2·898	−	−1·000	−	0·949
2·964	−	−1·000	−	0·982
2·997	−	−1·000	−	0·999
		−0·809		5·024

$$a_2 = \frac{-4}{\pi}\left\{\frac{-0{\cdot}809 + 5{\cdot}024}{20}\right\} = -0{\cdot}269$$

b_2—5 increment cursor

$T_2(x/3)$	T_2 Slope	n_{q+}	T_2 Slope	n_{q-}
0·672	+	0·164	−	−0·836
1·161	+	0	−	−0·919
1·500	+	0	−	−0·750
1·776	+	0	−	−0·612
2·013	+	0	−	−0·444
2·226	+	0	−	−0·387
2·418	+	0	−	−0·291
2·598	+	0	−	−0·201
2·766	+	0	−	−0·117
2·925	+	0	−	−0·037
		0·164		−4·644

$$b_2 = \frac{4}{\pi}\left\{\frac{0{\cdot}164 - 4{\cdot}644}{20}\right\} = -0{\cdot}285$$

TABLE 5.1—*continued*

a_3—5 increment cursor					b_3—5 increment cursor				
	D_3		D_3			T_3		T_3	
$D_3(x/3)$	Slope	n_{p+}	Slope	n_{p-}	$T_3(x/3)$	Slope	n_{q+}	Slope	n_{q-}
0·450	+	0·725	−	−0·275	0·099	−	−0·450	−	−0·549
0·786	+	0·893	−	−0·107	0·303	−	−0·349	−	−0·652
1·026	+	1·000	−	0·013	0·522	−	−0·239	−	−0·761
1·230	+	1·000	−	0·115	0·768	−	−0·116	−	−0·884
1·413	+	1·000	−	0·206	1·095	−	0	−	−0·952
1·587	+	1·000	−	0·294	1·872	+	0	+	0·564
1·755	+	1·000	−	0·377	2·127	+	0	+	0·436
1·929	+	1·000	−	0·465	2·298	+	0	+	0·351
2·115	+	1·000	−	0·556	2·433	+	0	+	0·284
2·343	+	1·000	−	0·672	2·547	+	0	+	0·226
2·793	−	−1·000	+	−0·896	2·646	+	0	+	0·177
2·901	−	−1·000	+	−0·951	2·736	+	0	+	0·132
2·964	−	−1·000	+	−0·982	2·820	+	0	+	0·090
2·985	−	−1·000	+	−0·992	2·895	+	0	+	0·052
2·997	−	−1·000	+	−0·999	2·967	+	0	+	0·017
		4·618		−2·504			−1·154		−1·469

$$a_3 = \frac{-4}{\pi}\left\{\frac{4{\cdot}618 - 2{\cdot}504}{30}\right\} = 0{\cdot}090 \qquad b_3 = \frac{4}{\pi}\left\{\frac{-1{\cdot}154 - 1{\cdot}469}{30}\right\} = -0{\cdot}111$$

These expressions evaluated for the individual values of s give the solutions of table 5.2 which are compared with the cursor results. Also included in the table are the results obtained using the Steklov approximation for a_{+s}, which is

$$a_{+s} \approx (1/3)\{n_p(a)T_s(1) + 2n_p(a/2)T_s(1/2)\}, \qquad s \geq 1,$$

TABLE 5.2

Comparison of results example 2, section 5.4

	a_0	a_1	b_1	a_2	b_2	a_3	b_3
Theory	0·247	1·062	0·425	−0·268	−0·282	−0·089	−0·110
Graphical integration	0·248	1·069	0·425	−0·269	−0·285	−0·090	−0·111
Approximate analytical	0·250	1·083	0·425	−0·250	−0·318	−0·167	−0·106

and Simpson's rule for b_{+s}, which gives

$$b_{+s} \approx (1/3\pi)\{n_q(a)U_{s-1}(1) + 4n_q(a/2)U_{s-1}(1/2) + n_q(0)U_{s-1}(0)\}.$$

Due to the fact that the functions $n_p(x)T_s(x/a)$ and $n_q(x)U_{s-1}(x/a)$ may have derivatives which vary appreciably, the approximations are not recommended for values of $s > 1$. The results of table 5.2 tend to confirm the statement for this particular case.

5.5 Asymmetrical oscillations

Consider the general block diagram of figure 5.6 where n is a frequency independent asymmetrical nonlinearity and $u(t)$ a constant disturbance of magnitude U. Proceeding in a similar manner to section 4.2 there are now two loop conditions to be satisfied for a limit cycle to exist, one as previously for the fundamental of the oscillation, higher harmonics again being neglected, and one for the bias signals. As superposition applies for the linear elements the equations are

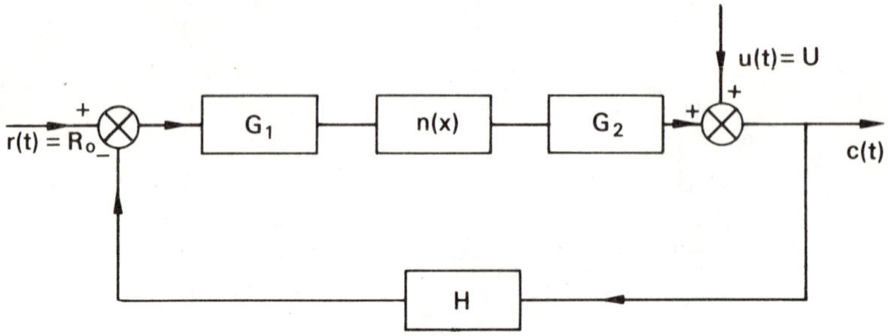

Figure 5.6 General asymmetrical system

$$1 + G(j\omega)\,N(a, \gamma) = 0, \tag{5.27}$$

and

$$\gamma\{1 + N_\gamma(a, \gamma)G(0)\} = R_0G_1(0) - UG_1(0)H(0), \tag{5.28}$$

where $G(j\omega) = G_1(j\omega)G_2(j\omega)H(j\omega)$ and $r(t) = R_0$ a constant. Equation (5.28) is often simplified if an integrator or differentiator exists in the loop since in the steady state the bias input to the former and output of the latter must be zero. For an assumed value of a equation (5.28) yields the corresponding value(s) of γ, say $\gamma = \gamma(a)$, so that $N(a, \gamma)$ for use in equation (5.27) is $N\{a, \gamma(a)\}$ evaluated for these allowable combinations of a and γ only. If $\gamma(a)$ is multivalued yielding more than one value of γ for a given a, then more than one $N\{a, \gamma(a)\}$ locus exists. With $N\{a, \gamma(a)\}$ known, stability of the system can be investigated using the standard procedures of chapter 4. For example, any intersections of the loci $G(j\omega)$ and $C\{a, \gamma(a)\} = -1/N\{a, \gamma(a)\}$ on the Nyquist diagram yield the values of the amplitude a, for which the corresponding value of γ is easily found, and the frequency ω of possible limit cycles. The stability of the limit cycle can be determined using the intersection criterion of section 4.3 or the IBDF,* $g(a, \gamma)$ of equation (5.10).

Due to the presence of a lower harmonic, the second, in asymmetrical systems, solutions obtained by DF techniques tend to be less accurate than those for symmetrical systems. Indications are however that the distortion criterion, which of course takes the second harmonic into account, again gives a good estimate

* The incremental describing function in the presence of sine and bias.

of the accuracy. Compensation methods similar to those outlined in section 4.5 can be used for asymmetrical systems. The procedure may become more tedious if the compensator has a d.c. gain other than one of unity since it can then, dependent upon the particular system configuration, alter the combination of values of a and γ satisfying equation (5.28). The compensated frequency response locus has then to be studied with respect to the new nonlinearity locus $C\{a, \gamma_c(a)\}$.

The above approach can also be used for frequency dependent nonlinearities where $N(a, \gamma)$ and $N_\gamma(a, \gamma)$ become $N(a, \gamma, \omega)$ and $N_\gamma(a, \gamma, \omega)$ respectively. Equation (5.27) is then studied for those values of a, γ and ω which satisfy equation (5.28).

In summary the additional variable γ involved in asymmetrical systems makes analysis and design more tedious. This can, of course, be avoided to a large extent if the various graphical relationships for the equations are produced, or even the set of nonlinear equations solved, by digital computer.

5.5.1 *System with a SVNL*

Again referring to figure 5.6 with $r(t) = R_0$ and assuming that additional bias signals may be introduced at any point within the system, the equilibrium or operating point may theoretically be moved to any position on the nonlinearity. The maximum and minimum possible gains through the nonlinearity are those values corresponding to the maximum and minimum slopes. The DF condition for stability is thus

$$K_1 < n'(x) < K_2 \quad \text{for all } x,$$

where K_1 and K_2 are the slopes of lines defining the Hurwitz sector. The above relationship is known as the Kalman conjecture,[4] a result which has again been shown by experiment to be invalid in the previously cited work of Fitts.[5]

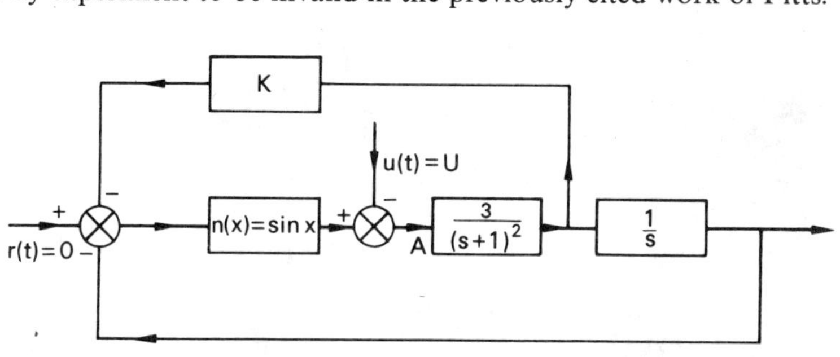

Figure 5.7 System considered in section 5.5.1

As a specific case we consider the system shown in figure 5.7 subject to the constraint that the disturbance U satisfies $|U| < 0{\cdot}5$. It is required to determine for what minimum value of K the system will always be stable and for smaller values of K, how the amplitude and frequency of the limit cycle vary with U and K.

Assuming a limit cycle with bias level γ and amplitude a to exist at the nonlinearity input, then from section 5.3.2

$$N_\gamma(a, \gamma) = J_0(a) \sin \gamma/\gamma,$$

and

$$N(a, \gamma) = 2J_1(a) \cos \gamma/a. \tag{5.29}$$

Since the system contains an integrator there can be no bias level at its input nor, therefore, at the point A in the steady state. The relationship equivalent to equation (5.28) is thus

$$J_0(a) \sin \gamma = U, \tag{5.30}$$

which on substituting in equation (5.29) gives

$$N(a, \gamma) = \{2J_1(a)/a\}[1 - \{U^2/J_0^2(a)\}]^{1/2}. \tag{5.31}$$

The system characteristic equation is

$$s^3 + 2s^2 + s\{1 + 3KN(a, \gamma)\} + 3N(a, \gamma) = 0,$$

which may be written

$$s^3 + 2s^2 + \alpha s + \beta = 0,$$

where $\alpha = 1 + 3KN(a, \gamma)$ and $\beta = 3N(a, \gamma)$, so that $\alpha = 1 + \beta K$. Figure 5.8 shows a parameter plane, which has the $\zeta = 0$ locus $\beta = 2\alpha$, alongside the $N(a, \gamma)$

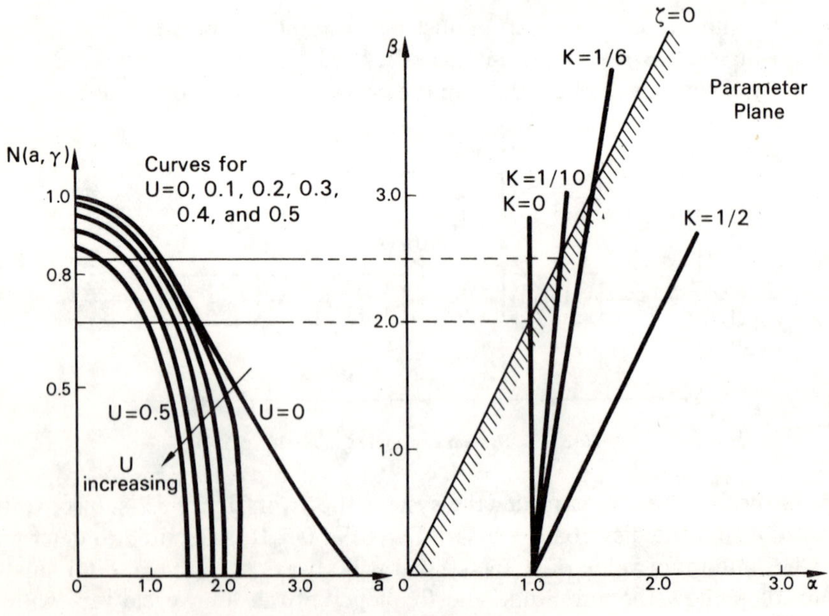

Figure 5.8 Parameter plane for system of figure 5.7

versus a relationship for various values of U in equation (5.31). The $N(a, \gamma)$ scale is chosen to be three times that of the β axis so that the value of β for the parameter plane can be obtained by projection from the $N(a, \gamma)$ axis. If the line $\alpha = 1 + \beta K$ is drawn on the parameter plane for a given value of K, then where it intersects the $\zeta = 0$ locus gives the value of β for instability and the frequency of the limit cycle. Projecting across to the $N(a, \gamma)$ loci gives the values of a and U which satisfy this condition. Figure 5.9 shows how the amplitude a and bias γ

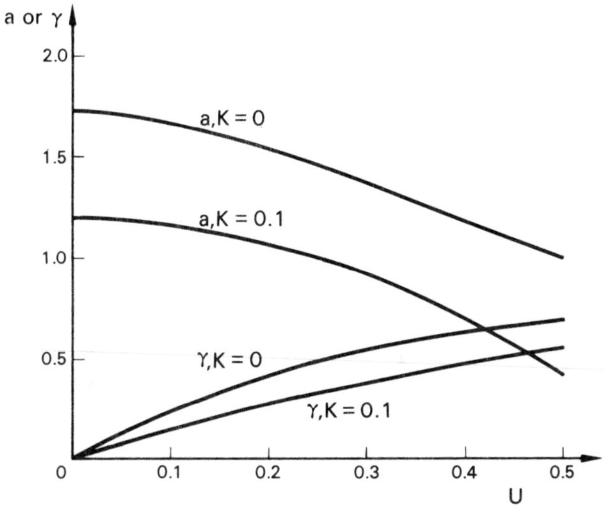

Figure 5.9 Limit cycle solutions for system of figure 5.7

vary with U for values of K equal to 0 and 0·1 with corresponding limit cycle frequencies of 1·0 and 1·12 rads/sec. The maximum value of $N(a, \gamma)$ is unity when $U = 0$ so that the system will be stable for $K > 1/6$. It can also be seen that if $K > 1/2$, the system will be stable for any nonlinearity $n(x)$ in the sector $[0, \infty]$, as the line $\alpha = 1 + \beta/2$ is parallel to the $\zeta = 0$ locus.

Perturbation along the $N(a, \gamma)$ locus by Δa from the limit cycle solution for a given U is seen in all cases to move the corresponding point in the parameter plane to the unstable region if Δa is negative and the stable region if Δa is positive. All the limit cycle solutions are therefore stable.

5.5.2 *A relay temperature control*

The block diagram of figure 5.10 shows the configuration of a relay temperature control where the controlled output oscillates about the desired level. Often such processes have a single dominant time constant;[6] however, a transfer function with a double time constant has been chosen so that the results obtained by DF analysis are more accurate. The SBDF for the relay of figure 5.10 can be obtained from the results of section 5.3.4 for the symmetrical on-off relay with hysteresis using the scheme of figure 5.3 with $\beta = 0$ and $n(\beta) = h$, followed by a

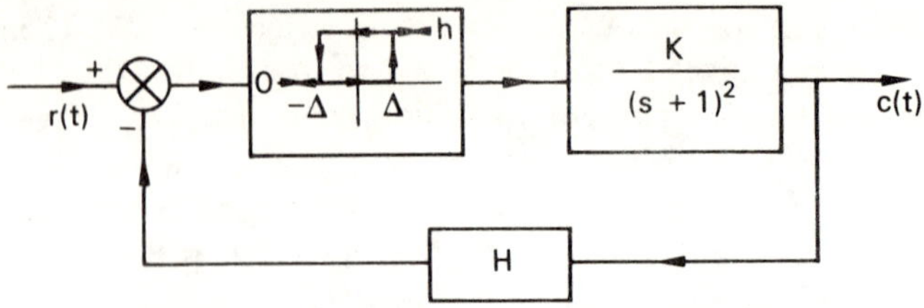

Figure 5.10 On-off temperature control system

gain of 1/2. Thus, for $|\gamma| < a - \Delta$

$$n(a, \gamma) = \frac{h}{2\pi}\left\{\sin^{-1}\left(\frac{\Delta + \gamma}{a}\right) - \sin^{-1}\left(\frac{\Delta - \gamma}{a}\right)\right\} + \frac{h}{2}, \qquad (5.32)$$

$$n_p(a, \gamma)_1 = (h/2\pi)[D_1\{-(\Delta + \gamma)/a\} + D_1\{(\Delta - \gamma)/a\}],$$

$$n_q(a, \gamma)_1 = h\Delta/a\pi.$$

Since we wish to examine the solutions for various set point values of R_0, rather than finding the values of a and γ which satisfy the bias equation

$$R_0 - KHn(a, \gamma) = \gamma$$

for each value of R_0, we investigate the behaviour for fixed values of $n(a, \gamma)$, the bias output from the relay. Assuming the relay operates with an on-off ratio of $p{:}1$ then

$$n(a, \gamma) = ph/(1 + p),$$

and from equation (5.32)

$$\sin^{-1}\{(\Delta + \gamma)/a\} - \sin^{-1}\{(\Delta - \gamma)/a\} = \pi(p - 1)/(p + 1),$$

that is

$$\theta_1 - \theta_2 = \pi(p - 1)/(p + 1) = 2\alpha,$$

where

$$\theta_1 = \sin^{-1}\{(\Delta + \gamma)/a\} \quad \text{and} \quad \theta_2 = \sin^{-1}\{(\Delta - \gamma)/a\}.$$

Thus

$$n_p(a, \gamma)_1 = h\{\cos\theta_1 + \cos\theta_2\}/2\pi,$$
$$= [h\cos\alpha\cos\{(\theta_1 + \theta_2)/2\}]/\pi. \qquad (5.33)$$

Now

$$\sin\theta_1 + \sin\theta_2 = 2\Delta/a = 2\sin\{(\theta_1 + \theta_2)/2\}\cos\{(\theta_1 - \theta_2)/2\},$$

giving

$$\sin\{(\theta_1 + \theta_2)/2\} = \Delta/a\cos\alpha,$$

and

$$\sin\theta_1 - \sin\theta_2 = 2\gamma/a = 2\cos\{(\theta_1 + \theta_2)/2\}\sin\{(\theta_1 - \theta_2)/2\},$$

giving

$$\cos\{(\theta_1 + \theta_2)/2\} = \gamma/a \sin\alpha.$$

Using these expressions in equation (5.33) gives

$$n_p(a, \gamma)_1 = \frac{h\cos\alpha}{\pi}\left\{1 - \frac{\Delta^2}{a^2\cos^2\alpha}\right\}^{1/2} = (h\gamma/a\pi)\cot\alpha.$$

Therefore

$$N(a, \gamma) = (2h/a^2\pi)\{(a^2\cos^2\alpha - \Delta^2)^{1/2} - j\Delta\},$$

and

$$C(a, \gamma) = -1/N(a, \gamma) = (-\pi/2h\cos^2\alpha)\{(a^2\cos^2\alpha - \Delta^2)^{1/2} + j\Delta\}.$$

The imaginary part of $C(a, \gamma)$ is thus independent of a and the $C(a, \gamma)$ loci for different values of on-off ratio p are straight lines parallel to the real axis. For γ positive the minimum value of a, a_m, is given by

$$a_m = \Delta + \gamma = \Delta + a_m\{1 - (\Delta^2/a_m^2\cos^2\alpha)\}^{1/2}\sin\alpha,$$

which reduces to $a_m^2\cos^2\alpha - 2a_m\Delta + \Delta^2\sec^2\alpha = 0$ with the solution

$$a_m = \Delta/\cos^2\alpha.$$

The $C(a, \gamma)$ loci thus start at the points

$$\left\{\frac{-\pi\Delta\sin\alpha}{2h\cos^3\alpha}, \frac{-\pi\Delta}{2h\cos^2\alpha}\right\}.$$

Sketches of the $C(a, \gamma)$ loci and the $G(j\omega)$ locus are shown in figure 5.11, where

$$G(j\omega) = \frac{KH(1 - \omega^2 - 2j\omega)}{(1 + \omega^2)^2}.$$

Since $C(a, \gamma)$ is an even function of α the loci for $p = p_1$ and $p = 1/p_1$ are the same. The frequency of oscillation, ω_0, is given by the solution of

$$\frac{-2\omega_0 KH}{(1 + \omega_0^2)^2} = \frac{-\pi\Delta}{2h\cos^2\alpha},$$

and the amplitude a_0 and bias γ from

$$\frac{(1 - \omega_0^2)KH}{(1 + \omega_0^2)^2} = \frac{-\pi(a_0^2\cos^2\alpha - \Delta^2)^{1/2}}{2h\cos^2\alpha} = \frac{-\pi\gamma}{2h}\cot\alpha. \tag{5.34}$$

To produce an oscillation with a given on-off ratio p, the required value of R_0 is

$$R_0 = \frac{KHph}{p + 1} + a_0\sin\alpha\left\{1 - \frac{\Delta^2}{a_0^2\cos^2\alpha}\right\}^{1/2}. \tag{5.35}$$

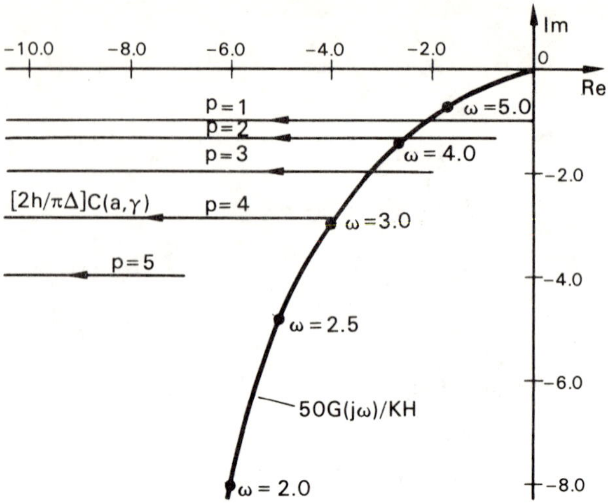

Figure 5.11 Polar loci for system of figure 5.10

It can be seen from figure 5.11 that for a given on-off ratio if KH or h/Δ are increased then ω_0 and a_0 are increased. Equations (5.34) and (5.35) also show that γ and R_0 increase.

5.5.3 *The modified nonlinearity in the presence of a limit cycle*

Consider the system of figure 5.12 which contains an ideal on-off relay. The bias equation is

$$n(a, \gamma) = U,$$

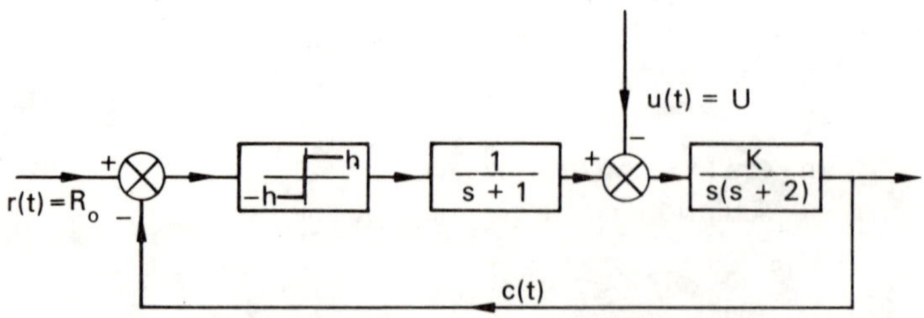

Figure 5.12 System considered in section 5.5.3

due to the presence of the integration term in the final transfer function. The value of U thus controls the on-off ratio of the oscillation and the value of R_0 the bias level of $c(t)$. The characteristic equation for the system is

$$s^3 + 3s^2 + 2s + KN(a, \gamma) = 0,$$

so that a stable limit cycle of frequency $\sqrt{2}$ rads/sec. exists when $N(a, \gamma) = 6/K$.

Now from the previous section with $\Delta = 0$,

$$n(a, \gamma) = (2h/\pi)\sin^{-1}(\gamma/a) \qquad |\gamma| < a, \tag{5.36}$$

and

$$N(a, \gamma) = (4h/a^2\pi)(a^2 - \gamma^2)^{1/2}.$$

$N(a, \gamma)$ has a maximum value when $a = \sqrt{2}\gamma$, and the condition for the limit cycle is

$$6/K = (4h/a^2\pi)(a^2 - \gamma^2)^{1/2}. \tag{5.37}$$

Substituting for γ from equation (5.36) and letting $y = n(a, \gamma)/h$ and $x = \gamma/h$ we obtain

$$6/K = (4/x\pi)\{1 - \sin^2(\pi y/2)\}^{1/2}\sin(\pi y/2),$$

that is

$$x = (K/3\pi)\sin \pi y. \tag{5.38}$$

When $y = 1/2$, $x = K/3\pi$, and if a higher value of y is demanded by U the system stops oscillating. The above x–y relationship is the d.c. input–d.c. output characteristic of the nonlinearity, or the modified nonlinearity, in the presence of the limit cycle. Using equation (5.38) in (5.36) gives

$$a/h = x/\sin\{(1/2)\sin^{-1}(3\pi x/K)\} \tag{5.39}$$

which gives the limit cycle amplitude in terms of the bias input to the nonlinearity. Figure 5.13 summarises the relationships of equations (5.38) and (5.39).

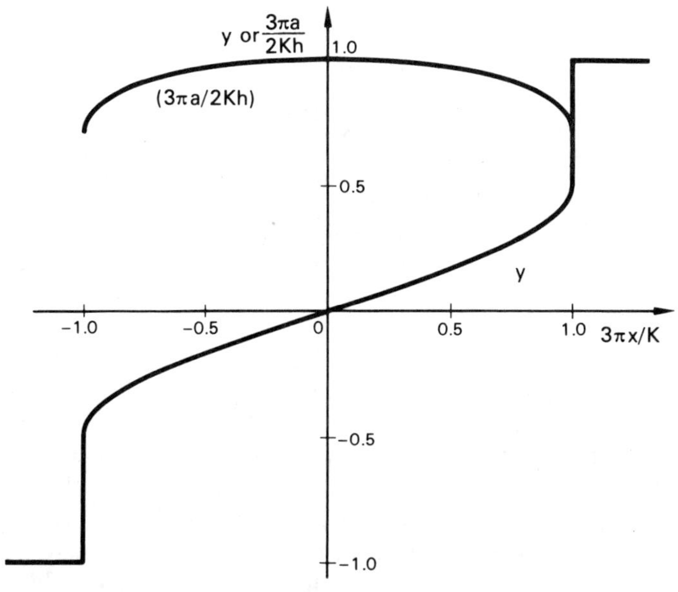

Figure 5.13 Modified nonlinearity and oscillation amplitude for system of figure 5.12

The modified nonlinearity of figure 5.13 shows the gain to a bias signal through the nonlinearity in the presence of the limit cycle. It can however be used[7,8,9] about the origin if $U = 0$, to describe the response of the nonlinearity to any slowly varying signal in the presence of the limit cycle. In particular, since the modified nonlinearity is approximately linear with a gain of $3/K$ for small inputs, the closed loop transfer function for a slowly varying input, $r(t)$, with a small component at the nonlinearity input is

$$\frac{C(s)}{R(s)} = \frac{(3/K)\{K/s(s+1)(s+2)\}}{1+(3/K)\{K/s(s+1)(s+2)\}},$$

that is

$$\frac{C(s)}{R(s)} = \frac{3}{s^3+3s^2+2s+3}, \tag{5.40}$$

which is independent of K. This property is known as limit cycle adaption,[10] for, if the gain K varies, the limit cycle amplitude changes to keep a constant (approximately so in practice) closed loop transfer function for $r(t)$. The technique has been used in the design of an autopilot for a high performance aircraft,[11] a system which is subject to large gain variations. The forced response of limit cycling systems is discussed in more detail in section 9.6.3.

5.6 The frequency response of asymmetrical systems

Consider again the system of figure 5.6. When $r(t) = R_0 + R \sin \omega t$, the closed loop frequency response, which is the relationship between the components of frequency ω in $c(t)$ and $r(t)$, is given by

$$\frac{C(j\omega)}{R(j\omega)} = \frac{G_1(j\omega)G_2(j\omega)N\{a,\gamma(a)\}}{1+N\{a,\gamma(a)\}G(j\omega)}, \tag{5.41}$$

where $\gamma = \gamma(a)$ is the relationship between γ and a which satisfies the bias equation (5.28). Thus for a given system configuration, $N\{a, \gamma(a)\}$ can be found as a function of a and used in place of $N(a)$ in any of the methods described in section 4.7 for the evaluation of the closed loop frequency response of a symmetrical system.

When determining the frequency response of a symmetrical nonlinear system it is not usually necessary to consider a bias input at the nonlinearity, as the system normally operates at the zero bias condition. This may not however be the case if the system nonlinearity, although symmetrical, has a negative slope for inputs above a certain magnitude. Nonlinear characteristics of this form can cause a d.c. jump[12] in the frequency response, due to the d.c. incremental gain of the nonlinearity in the presence of the forcing sinusoid changing sign.

A system which exhibits this behaviour is shown in figure 5.14, where the nonlinearity known as a limited field of view characteristic approximates the

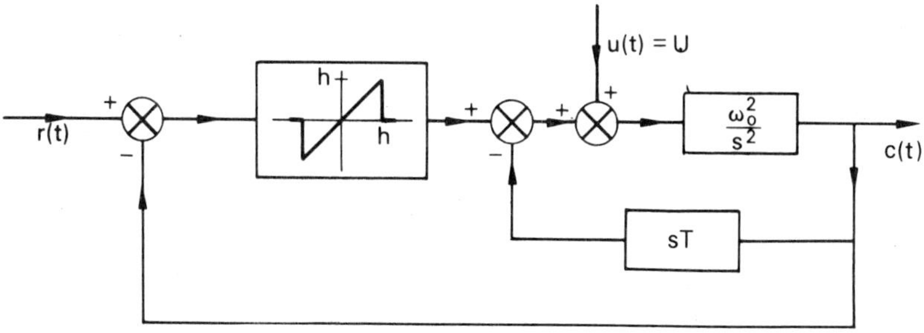

Figure 5.14 System which exhibits a d.c. jump in the frequency response

relationship between error signal and angular misalignment in an automatic-follow radar system. The nonlinearity can be seen to be equivalent to a parallel combination of ideal saturation and minus an on-off relay with dead zone. Figure 5.15 shows the relationships $n(a, \gamma)/h$ and $aN(a, \gamma)/h$ for various values of a/h versus γ/h for the nonlinearity. When U is zero the bias equation for the

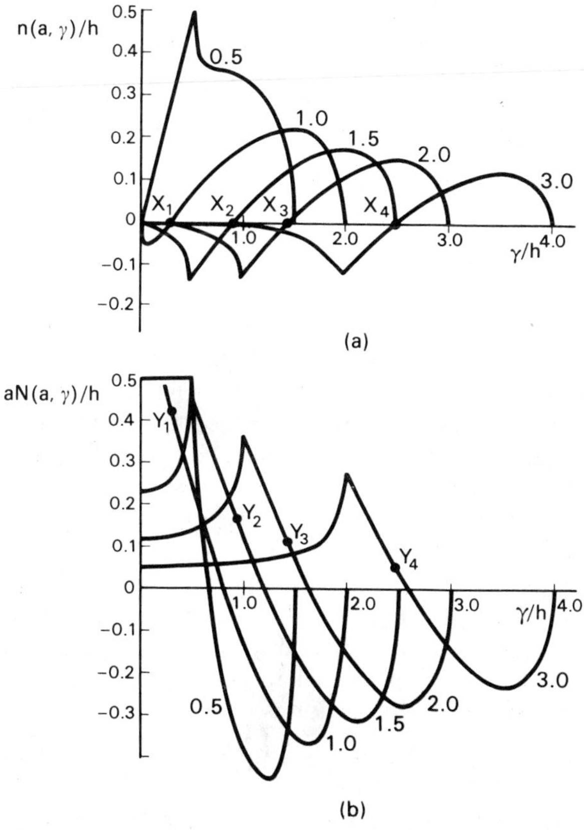

Figure 5.15 Sine plus bias characteristics for limited field of view nonlinearity

system is

$$n(a, \gamma) = 0,$$

which according to figure 5.15(a) is satisfied when $\gamma = 0$ for all a or when γ/h equals X_1 for $a/h = 1{\cdot}0$, X_2 for $a/h = 1{\cdot}5$ and so on. These finite values of γ/h which give a zero bias output from the nonlinearity in the presence of the sinusoid of magnitude a/h exist because of the specific shape of the nonlinearity. They do not occur with nonlinearities such as relays, saturation and dead space. The d.c. incremental gain, the slope of the $n(a, \gamma)/h$ characteristics, for zero bias output, is negative at the origin for $a/h > 1{\cdot}0$ and positive at the points X_1, X_2. For stable operation of the closed loop the d.c. incremental gain through the nonlinearity must be positive so that for a bias perturbation the loop feedback is negative. Operation at the points X_1, X_2, etc. is therefore stable. Figure 5.16

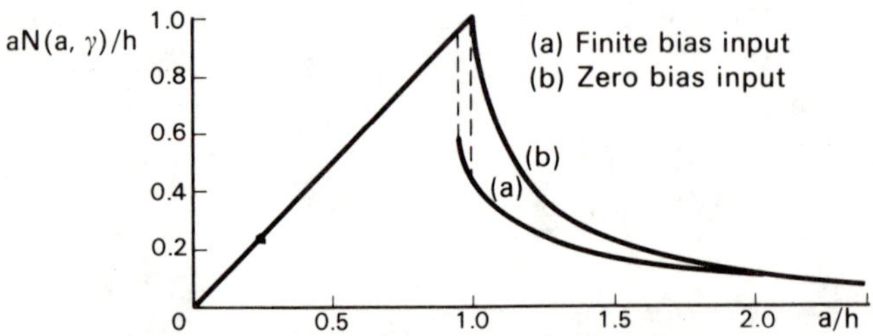

Figure 5.16 Possible fundamental output curves for zero bias output

shows $aN(a)/h$ the fundamental output with no input bias and $aN\{a, \gamma(a)\}/h$ the fundamental output when γ/h has the values X_1, X_2, etc., to give a zero bias nonlinearity output, versus a/h. The $aN\{a, \gamma(a)\}/h$ characteristic is seen to possess a small hysteresis in that for a/h increasing it does not exist until $a/h > 1{\cdot}0$ but for a/h decreasing it exists until a/h reaches approximately 0·95.

Therefore if a small sinusoidal input is applied to the system and its magnitude increased, when it produces a fundamental component a at the nonlinearity input so that $a/h > 1{\cdot}0$, the system jumps into an asymmetrical state with a bias $\gamma/h = X_1$ at the nonlinearity input. As the system input magnitude is further increased the bias γ/h continues to increase.

Figure 5.17 compares theoretical and experimental frequency responses for two different input signal magnitudes. Considering that the nonlinearity is particularly severe, producing a large amount of distortion, the agreement is quite good. The d.c. jump can occur with other similar nonlinearities. For example, for the harmonic characteristic sin mx of section 5.3.2

$$g(a, \gamma) = mJ_0(ma) \cos m\gamma,$$

giving $g(a, 0) = mJ_0(ma)$, which becomes negative when $ma > 2{\cdot}405$, the first zero of J_0. The $\gamma = 0$ condition is thus unstable for $a > 2{\cdot}405/m$.

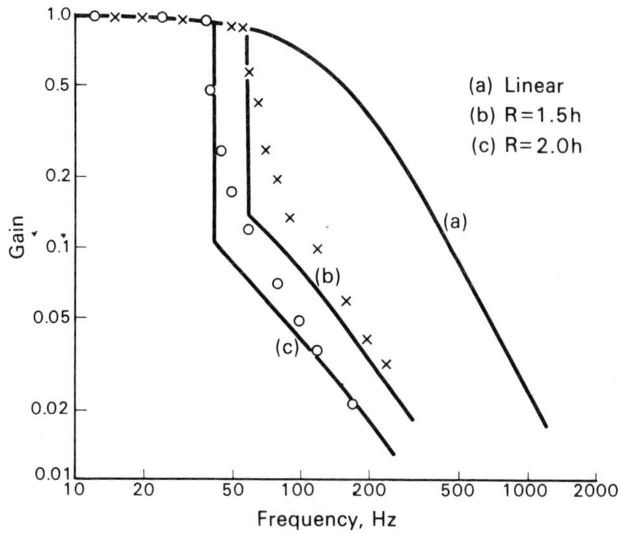

Figure 5.17 Frequency response for system of figure 5.14

5.7 Cubic nonlinearity with two-sinusoidal inputs

The response of a SVNL to two-sinusoidal inputs and the evaluation of the two-sinusoidal input describing function, denoted by SSDF, is introduced here by considering a cubic nonlinearity. A general approach to the problem is given in chapter 10. The choice of the cubic enables the physical aspects of the situation to be understood without the complication caused by involved mathematics, since for this case the output is easily evaluated by direct expansion.

Consider the input $x + y$, where $x = a\cos\theta_a$, $y = b\cos(\theta_b + \phi)$, $\theta_a = \omega_a t$ and $\theta_b = \omega_b t$. Then the output is

$$\begin{aligned} n(t) &= [a\cos\theta_a + b\cos(\theta_b + \phi)]^3, \\ &= a^3\cos^3\theta_a + 3a^2\cos^2\theta_a\{b\cos(\theta_b + \phi)\} + 3b^2\cos^2(\theta_b + \phi)\{a\cos\theta_a\} \\ &\quad + b^3\cos^3(\theta_b + \phi), \end{aligned}$$

which on expanding the powers of $\cos\theta$, may be written

$$\begin{aligned} n(t) = {} & \left(\frac{3a^3}{4} + \frac{3ab^2}{2}\right)\cos\theta_a + \left(\frac{3b^3}{4} + \frac{3ba^2}{2}\right)\cos(\theta_b + \phi) \\ & + \frac{3a^2b}{2}\cos 2\theta_a\cos(\theta_b + \phi) + \frac{3b^2a}{2}\cos 2(\theta_b + \phi)\cos\theta_a \\ & + \frac{a^3}{4}\cos 3\theta_a + \frac{b^3}{4}\cos 3(\theta_b + \phi), \end{aligned} \tag{5.42}$$

or on separating the cosine products

$$n(t) = (3a/4)(a^2 + 2b^2)\cos\theta_a + (3b/4)(b^2 + 2a^2)\cos(\theta_b + \phi)$$
$$+ (3a^2b/4)\cos(2\theta_a + \theta_b + \phi) + (3a^2b/4)\cos(2\theta_a - \theta_b - \phi)$$
$$+ (3b^2a/4)\cos(2\theta_b + 2\phi + \theta_a) + (3b^2a/4)\cos(2\theta_b + 2\phi - \theta_a)$$
$$+ (a^3/4)\cos 3\theta_a + (b^3/4)\cos 3(\theta_b + \phi). \tag{5.43}$$

It is shown in section 10.4.3 that the general form of this expression for any nonlinearity is

$$n(t) = \sum_{s=0}^{\infty}\sum_{k=0}^{\infty} \varepsilon_s\varepsilon_k\alpha_{sk}\cos s\theta_a \cos k(\theta_b + \phi), \tag{5.44}$$

and the output thus contains components of frequency $(s\omega_a \pm k\omega_b)$.

For the cubic only a finite number of the coefficients α_{sk} are non-zero, namely

$$\alpha_{10} = (3a/8)(a^2 + 2b^2), \qquad \alpha_{01} = (3b/8)(b^2 + 2a^2),$$
$$\alpha_{21} = 3a^2b/8, \qquad \alpha_{12} = 3b^2a/8,$$
$$\alpha_{30} = a^3/8, \qquad \alpha_{03} = b^3/8.$$

We also note that α_{mn} can be obtained from α_{nm} by replacing a by b and b by a.

To determine the SSDF the terms having the same frequencies ω_a and ω_b as those at the input are required. However, it can be seen from equation (5.43) that the magnitude of these terms will depend upon whether the frequencies ω_a and ω_b are related.

5.7.1 *SSDF for incommensurate frequencies*

Incommensurate frequencies are defined as those for which the ratio ω_a/ω_b is not a ratio of integers. In this case the terms in the output at frequencies ω_a and ω_b are

$$2\alpha_{10}\cos\theta_a = (3a/4)(a^2 + 2b^2)\cos\theta_a,$$

and

$$2\alpha_{01}\cos(\theta_b + \phi) = (3b/4)(b^2 + 2a^2)\cos(\theta_b + \phi).$$

The components $N_a(a, b)$ and $N_b(a, b)$ of the SSDF are therefore given by

$$N_a(a, b) = \frac{\text{output magnitude at frequency } \omega_a}{a} = \frac{2\alpha_{10}}{a} = \frac{3(a^2 + 2b^2)}{4} \tag{5.45}$$

and

$$N_b(a, b) = \frac{\text{output magnitude at frequency } \omega_b}{b} = \frac{2\alpha_{01}}{b} = \frac{3(b^2 + 2a^2)}{4}. \tag{5.46}$$

In particular we note that since the output terms are in phase with their corresponding inputs both $N_a(a, b)$ and $N_b(a, b)$ are real. Also

$$N_b(a, b) = [N_a(a, b]_{\substack{a \to b, \\ b \to a}}$$

and

$$\lim_{b \to 0} N_b(a, b) = N_{i\gamma}(a) = 3a^2/2. \tag{5.47}$$

5.7.2 *SSDF for commensurate frequencies*

Two sinusoids are commensurate if the ratio of their frequencies is a ratio of integers. It can be seen from equation (5.44) that for the general situation the term with coefficient α_{sk} will yield an output signal at the frequency ω_a if we choose k so that $k = (s \pm 1)(m/n)$ where $\omega_a/\omega_b = m/n$ the ratio of two integers. Similarly the term with coefficient α_{sk} will yield an output at frequency ω_b if we choose $k = (s \pm 1)(n/m)$.

Due to the limited number of finite coefficients α_{sk} for the cubic there exist only a few ratios m/n for which outputs, additional to the contributions of α_{10} and α_{01} at ω_a and ω_b respectively, are possible. From examination of equation (5.43) or using the above expressions it can be seen that these additional components are produced when

(i) $m/n = 1$ by α_{21} at ω_a and α_{12} at ω_b
(ii) $m/n = 1/3$ by α_{21} at ω_a and α_{30} at ω_b
(iii) $m/n = 3$ by α_{03} at ω_a and α_{12} at ω_b.

These three cases are therefore considered separately.

(i) $m/n = 1$

In this case $\omega_a = \omega_b$ and the output at this frequency is

$$\begin{aligned}(3/4)\{a(a^2 + 2b^2)\cos\theta_a + b(b^2 + 2a^2)\cos(\theta_a + \phi) + a^2 b\cos(\theta_a - \phi) \\ + b^2 a\cos(\theta_a + 2\phi)\} \\ = (3/4)[\{a(a^2 + 2b^2) + b(b^2 + 3a^2)\cos\phi + b^2 a\cos 2\phi)\}\cos\theta_a \\ - \{b(b^2 + a^2)\sin\phi + b^2 a\sin 2\phi\}\sin\theta_a], \\ = (3/4)[A^2(a + b\cos\phi)\cos\theta_a - A^2 b\sin\phi\sin\theta_a],\end{aligned} \tag{5.48}$$

where $A = (a^2 + b^2 + 2ab\cos\phi)^{1/2}$, the amplitude of the vector sum of the input signals.

The output can be specified by the single describing function

$$N_a(a|b) = M_a(a|b)e^{j\psi_a(a/b)},$$

where

$$M_a(a|b) = (3A^2/4a)\{(a + b\cos\phi)^2 + b^2\sin^2\phi\}^{1/2} = 3A^3/4a$$

and

$$\psi_a(a|b) = \tan^{-1}\left(\frac{b \sin \phi}{a + b \cos \phi}\right),$$

results which can also be obtained by combining the inputs and using the DF for a cubic. The notation $(a|b)$ is used in the above to denote that a and b have related frequencies.

(ii) $m/n = 1/3$

Here $\omega_a = \omega_b/3$ and the output at frequency ω_a is

$$(3/4)\{a(a^2 + 2b^2)\cos\theta_a + a^2 b\cos(-\theta_a - \phi)\}$$
$$= (3a/4)\{(a^2 + 2b^2 + ab\cos\phi)\cos\theta_a - ab\sin\phi\sin\theta_a\} \qquad (5.49)$$

and at frequency ω_b is

$$(1/4)\{3b(b^2 + 2a^2)\cos(\theta_b + \phi) + a^3\cos\theta_b\}$$
$$= (1/4)\{(a^3\cos\phi + 3b^3 + 6a^2 b)\cos(\theta_b + \phi) + a^3\sin\phi\sin(\theta_b + \phi)\}. \qquad (5.50)$$

Thus the magnitudes and phases of the SSDF are

$$M_a(a|b) = (3/4)\{(a^2 + b^2)(a^2 + 4b^2) + 2(a^2 + 2b^2)ab\cos\phi\}^{1/2},$$
$$\psi_a(a|b) = \tan^{-1}\{(ab\sin\phi)/(a^2 + 2b^2 + ab\cos\phi)\},$$

and

$$M_b(a|b) = (1/4b)\{(a^3\cos\phi + 3b^3 + 6a^2 b)^2 + a^6\sin^2\phi\}^{1/2},$$
$$\psi_b(a|b) = \tan^{-1}\{(-a^3\sin\phi)/(a^3\cos\phi + 3b^3 + 6a^2 b)\}.$$

(iii) $m/n = 3$

The results for this case, with $\omega_a = 3\omega_b$, are obviously similar to those obtained above for $m/n = 1/3$. If we change b to a and a to b in the above expressions we obtain the SSDF for an input $a\cos(\theta_a + \phi) + b\cos\theta_b = a\cos(\theta_a + \phi) + b\cos(\theta_a/3)$, so that for the input used it is also necessary to alter ϕ to -3ϕ in the solutions* for $m/n = 1/3$. For completeness, therefore, the results in this case are

$$M_a(a|b) = (1/4b)\{(b^3\cos 3\phi + 3a^3 + 6ab^2)^2 + b^6\sin^2 3\phi\}^{1/2},$$
$$\psi_a(a|b) = \tan^{-1}\{(b^3\sin 3\phi)/(b^3\cos 3\phi + 3a^3 + 6b^2 a)\},$$

and

$$M_b(a|b) = (3/4)\{(a^2 + b^2)(b^2 + 4a^2) + 2(b^2 + 2a^2)ab\cos 3\phi\}^{1/2},$$
$$\psi_b(a|b) = \tan^{-1}\{(-ab\sin 3\phi)/(b^2 + 2a^2 + ab\cos 3\phi)\}.$$

---

* This applies to the describing functions, it does *not* mean that to obtain the output at frequency ω_b we can interchange a and b and put $\phi = -3\phi$ in equation (5.49).

5.7.3 *The IDF*

For a small unrelated input, γ, we have using equation (5.14)

$$N_{i\gamma}(a) = (3a^2/4) + (a/2)\mathrm{d}(3a^2/4)/\mathrm{d}a = 3a^2/2,$$

which agrees with equation (5.47).

As special cases we extend our examination of the IDF to those situations where the small signal concerned may be related to the existing input signal. The cases for $m/n = 1$, 1/3 and 3 must therefore be considered.

$m/n = 1$

Using equation (5.48) we see that the change in the output at the frequency ω_a due to the small signal b, is

$$(3/4)[\{a(a^2 + 2b^2) + b(b^2 + 3a^2)\cos\phi + b^2 a\cos 2\phi\}\cos\theta_b$$
$$-\{b(b^2 + a^2)\sin\phi + b^2 a\sin 2\phi\}\sin\theta_b] - (3a^3/4)\cos\theta_b.$$

Therefore as $b \to 0$, the incremental output obtained by neglecting all terms of order greater than one in b, is

$$(3b/4)\{3a^2\cos\phi\cos\theta_b - a^2\sin\phi\sin\theta_b\}$$
$$= (3a^2b/4)\{(2 + \cos 2\phi)\cos(\theta_b + \phi) + \sin 2\phi\sin(\theta_b + \phi)\}.$$

Thus the IDF, $N_{ib}(a)$, is given by

$$N_{ib}(a) = (3a^2/4)\{(2 + \cos 2\phi) - j\sin 2\phi\},$$
$$= (3a^2/4)(2 + \mathrm{e}^{-j2\phi}).$$

$N_{ib}(a)$ is therefore complex and its locus in the complex plane is seen to be a circle of radius $3a^2/4$ with centre at the point $(3a^2/2, 0)$.

$m/n = 1/3$

Again examining the output at frequency ω_b, given by equation (5.50), it is seen that the change in output due to the addition of the small signal b is

$$(3b/4)\{(b^2 + 2a^2)\cos(\theta_b + \phi)\}.$$

Therefore

$$N_{ib}(a) = \lim_{b\to 0}\{3b(b^2 + 2a^2)/4\} = 3a^2/2.$$

$m/n = 3$

Interchanging a and b in equation (5.49) the output at frequency ω_b due to the addition of the small signal b is

$$(3b/4)\{(b^2 + 2a^2 + ab\cos\phi)\cos\theta_b - ab\sin\phi\sin\theta_b\}.$$

Remembering that this is the output for an input $a\cos(\theta_a + \phi) + b\cos\theta_b$,

$$N_{ib}(a) = \lim_{b\to 0} [(3/4)\{b^2 + 2a^2 + ab\cos\phi + jab\sin\phi\}],$$

$$= 3a^2/2.$$

In conclusion therefore for the cubic characteristic the IDF is real and equal to $N_{i\gamma}(a) = 3a^2/2$ for all small inputs except one with the same frequency ω_a as a.

5.8 Modified nonlinearities and the SSDF

Consideration of the simple cubic nonlinearity in the previous section has illustrated that the SSDF depends upon whether the input sinusoids are commensurate or not. When the signals are incommensurate the output from a general nonlinear characteristic consists of signal terms at the fundamental and harmonic frequencies of both the inputs, together with a theoretically infinite number of cross modulation frequencies, where the frequencies of all the terms are different. For commensurate inputs, however, all the above signal terms are not at unique frequencies so that at a particular frequency, say the fundamental of one input, there may be many output signal terms. Due to the frequency locking which takes place in nonlinear feedback loops the SSDF for related inputs is of extreme importance. Unfortunately the calculation of this SSDF for the general nonlinearity, since unlike the cubic it produces a theoretically infinite number of output signal terms, is more difficult and the general problem is deferred to chapter 10. To illustrate the use of the SSDF for commensurate inputs and the basic concepts involved, various applications are studied in section 5.13 with the cubic as the nonlinear characteristic.

Although the IDF for commensurate signals is a special case of the SSDF, general results for any nonlinearity are not too difficult to obtain, a fact which is possibly suggested by the relative simplicity of the results for the cubic. A general approach for the determination of IDFs is presented in section 5.9 and various applications of the IDF in feedback system analysis are considered in the following section, 5.10.

Here, however, we return to further investigation of the SSDF for incommensurate frequencies and in particular its evaluation making use of modified nonlinearities. Consider the cubic nonlinearity modified with respect to the input $x = a\cos\theta_a$, then

$$n(a, \gamma) = \int_{-a}^{a} (x + \gamma)^3 r(x)\,dx = \gamma^3 + 3a^2\gamma/2.$$

Assume that the input $y = b\cos(\theta_b + \phi)$ is now applied to this nonlinearity, the describing function for y is then

$$N_m(b) = (2/b^2)\int_{-b}^{b} (y^3 + 3a^2y/2)yr(y)\,dy,$$

$$= (2/b^2)\{(3b^4/8) + (3a^2b^2/4)\},$$

$$= (3/4)(b^2 + 2a^2) = N_b(a, b) \quad \text{from equation (5.46).}$$

A proof that the above procedure for obtaining the SSDF can be used for any nonlinearity is given in section 10.3. We therefore have the important result that the components of the SSDF are the DFs for one input of the nonlinearity modified with respect to the other input.

In fact, the more general result applies that the output signal $n(t)$ given by

$$n(t) = \sum_{s=0}^{\infty} \sum_{k=0}^{\infty} \varepsilon_s \varepsilon_k \alpha_{sk} \cos s\theta_a \cos k(\theta_b + \phi) \tag{5.44}$$

has

$$\alpha_{sk} = \int_{-b}^{b} \int_{-a}^{a} n(x + y) T_s(x/a) T_k(y/b) r(x) r(y) \, dx \, dy, \tag{5.51}$$

with $\alpha_{sk} = 0$ for all $s + k$ even for odd nonlinearities and for all $s + k$ odd for even nonlinearities.

The evaluation of the double integral for α_{sk} can be thought of as the two stage process

$$n(a, \gamma)_s = \int_{-a}^{a} n(x + \gamma) T_s(x/a) r(x) \, dx, \tag{5.52}$$

and

$$\alpha_{sk} = \int_{-b}^{b} n(a, y)_s T_k(y/b) r(y) \, dy, \tag{5.53}$$

or alternatively

$$n(b, \gamma)_k = \int_{-b}^{b} n(y + \gamma) T_k(y/b) r(y) \, dy, \tag{5.54}$$

and

$$\alpha_{sk} = \int_{-a}^{a} n(b, x)_k T_s(x/a) r(x) \, dx. \tag{5.55}$$

The SSDF is then given by

$$N_a(a, b) = 2\alpha_{10}/a \quad \text{and} \quad N_b(a, b) = 2\alpha_{01}/b.$$

Also from the symmetry of the above expressions it can be seen that

$$\alpha_{ks} = [\alpha_{sk}]_{\substack{a \to b \\ b \to a}} \quad \text{and} \quad N_b(a, b) = [N_a(a, b)]_{\substack{a \to b \\ b \to a}}.$$

Thus to find both components of the SSDF it is sufficient to determine one, say $N_a(a, b)$, for all amplitude ratios a/b, which can be done as outlined above by evaluating the double integral in either manner. It is shown in section 10.2, that the SSDF can also be evaluated from

$$N_a(a, b) = (j/a\pi) \int_{-\infty}^{\infty} N(j\omega) J_1(a\omega) J_0(b\omega) \, d\omega,$$

where $N(j\omega)$ is the Fourier transform of $n(x)$. This equation reduces to equation (3.24) with $b = 0$ as expected since $N(a) = \lim_{b \to 0} N_a(a, b)$. Although some analytical solutions of this integral are possible and it may also be used to find series expansions for the SSDF,[9] as done by a different approach in section 10.8, it is not considered further at this time. The modified nonlinearity approach is conceptually much simpler and approximate results can be found for any nonlinearity using the techniques described previously for the determination of modified nonlinearities and describing functions. Some analytical solutions are given below and additional results are listed in appendix A3.

5.8.1 *Integer power law*

For $n(x) = x^m$, from equation (5.15)

$$n(b, \gamma) = \sum_{k=0}^{m} {}_mC_k \mu_k(b) \gamma^{m-k},$$

where $\mu_k(b)$ denotes the kth moment for the sinusoidal distribution of amplitude b.

Now from appendix A1, the DF of x^n is $2\mu_{n+1}/a^2$, therefore

$$N_a(a, b) = (2/a^2) \sum_{k=0}^{m} {}_mC_k \mu_k(b) \mu_{m-k+1}(a).$$

In particular for $m = 3$,

$$\begin{aligned} N_a(a, b) &= (2/a^2)\{{}_3C_0 \mu_0(b) \mu_4(a) + {}_3C_2 \mu_2(b) \mu_2(a)\}, \\ &= (2/a^2)\{(3a^4/8) + (3a^2b^2/4)\} = (3/4)\{a^2 + 2b^2\}, \end{aligned}$$

as obtained in equation (5.45).

5.8.2 *Harmonic nonlinearity*

With $n(x) = \sin mx$

$$n(b, \gamma) = J_0(mb) \sin m\gamma$$

from equation (5.16).

Also from appendix A1 the DF of $\sin mx$ is $(2/a)J_1(ma)$, thus

$$N_a(a, b) = 2\alpha_{10}/a = (2/a)J_0(mb)J_1(ma).$$

Alternatively, from equation (5.16)

$$n(a, \gamma)_1 = J_1(ma) \cos m\gamma,$$

and from appendix A2 the d.c. output from $\cos mx$ with $\gamma = 0$ and an input sinusoid of amplitude b is $n(b, 0) = J_0(mb)$. Thus

$$\alpha_{10} = J_1(ma)J_0(mb),$$

and

$$N_a(a, b) = (2/a)J_0(mb)J_1(ma).$$

5.8.3 *Ideal relay*

From equations (5.20) and (5.21) with $\Delta = 0$, we have

$$n(a, \gamma) = \begin{cases} (2h/\pi)\sin^{-1}\gamma/a & |\gamma| < a \\ h \operatorname{sgn}\gamma & |\gamma| > a \end{cases}$$

and

$$n(a, \gamma)_1 = \begin{cases} (2h/\pi)\{1 - (\gamma^2/a^2)\}^{1/2} & |\gamma| < a \\ 0 & |\gamma| > a. \end{cases}$$

Assuming that $a > b$, then to avoid integration over two regions for the second integral the procedure of equations (5.52) and (5.53) is used, that is, the first integration is performed with respect to a, therefore

$$N_b(a, b) = (2/b^2\pi)\int_{-b}^{b} \frac{xn(a, x)}{(b^2 - x^2)^{1/2}}\,dx,$$

$$= (2/b^2\pi^2)\int_{-b}^{b} \frac{2hx\sin^{-1}(x/a)}{(b^2 - x^2)^{1/2}}\,dx.$$

Integrating by parts and recognising that $(b^2 - x^2)^{1/2}$ is zero at both limits gives

$$N_b(a, b) = (8h/b^2\pi^2)\int_0^b \{(b^2 - x^2)/(a^2 - x^2)\}^{1/2}\,dx,$$

$$= (8h/b^2\pi^2)\left[\int_0^b \frac{(a^2 - x^2)\,dx}{\{(a^2 - x^2)(b^2 - x^2)\}^{1/2}} - (a^2 - b^2)\int_0^b \frac{dx}{\{(a^2 - x^2)(b^2 - x^2)\}^{1/2}}\right].$$

Using the complete elliptic integrals

$$E(k) = \int_0^{\pi/2} (1 - k^2\sin^2\theta)^{1/2}\,d\theta \quad \text{and} \quad K(k) = \int_0^{\pi/2} (1 - k^2\sin^2\theta)^{-1/2}\,d\theta$$

we obtain the solution

$$N_b(a, b) = (8h/\pi^2k^2a)\{E(k) - (1 - k^2)K(k)\} = (8h/a\pi^2)B(k),$$

where $k = b/a < 1$.

Similarly, again integrating first with respect to a,

$$N_a(a, b) = (4h/a\pi^2)\int_{-b}^{b} \frac{\{1 - (x^2/a^2)\}^{1/2}}{(b^2 - x^2)^{1/2}}\,dx,$$

$$= (8h/a\pi^2)E(k).$$

5.9 General theory of the IDF

Consider the nonlinearity $n(x)$. For an input $x(t) + \delta(t)$ where $\delta(t)$ is small, the output may be approximated by the first two terms of a Taylor series, that is

$$n\{x(t) + \delta(t)\} \approx n\{x(t)\} + \delta(t)n'\{x(t)\},$$

where $n' = \mathrm{d}n(x)/\mathrm{d}x$.

The additional output due to the increment $\delta(t)$ is thus

$$\delta(t)n'\{x(t)\}.$$

Converting to an ensemble average the gain to the output correlated with $\delta(t)$, when $x(t) = a\cos\theta_a$, is

$$N_{i\delta}(a) = \frac{\int\int \delta^2 n'(x)p(x,\delta)\,\mathrm{d}x\,\mathrm{d}\delta}{\int \delta^2 p(\delta)\,\mathrm{d}\delta}.$$

When x and δ are independent $p(x, \delta) = r(x)p(\delta)$, and the above equation reduces to

$$N_{i\delta}(a) = \int_{-a}^{a} n'(x)r(x)\,\mathrm{d}x,$$

in agreement with equation (5.11). On the other hand when x and δ are related $p(x, \delta) \neq r(x)p(\delta)$ so that the above approach becomes more complex and is not considered further until section 10.4.

We can however proceed[13] in the time domain recognising that the above expression for the additional output may be interpreted as

$$\delta(t)\{\text{Output of nonlinear element } \mathrm{d}n(x)/\mathrm{d}x \text{ with input } x(t)\}.$$

For an input $x = a\cos\theta_a$ the output of the nonlinearity $n'(x) = \{\mathrm{d}n(x)/\mathrm{d}x\}$ is periodic and can be represented for the general case of a double valued asymmetrical nonlinearity by the Fourier series

$$n'\{x(t)\} = \sum_{s=0}^{\infty} c'_s \cos(s\theta_a + \phi'_s), \tag{5.56}$$

with $\phi'_0 = 0$.

Thus if $\delta(t) = b\cos(\theta_b + \phi)$,

$$\delta(t)n'\{x(t)\} = b\sum_{s=0}^{\infty} c'_s \cos(s\theta_a + \phi'_s)\cos(\theta_b + \phi),$$

$$= b\sum_{s=0}^{\infty} (c'_s/2)\{\cos(s\theta_a + \theta_b + \phi'_s + \phi) + \cos(s\theta_a - \theta_b + \phi'_s - \phi)\}.$$

If $x(t)$ and $\delta(t)$ are unrelated the only output at frequency ω_b, where $\theta_b = \omega_b t$, is $c'_0 b\cos(\theta_b + \phi)$ for $s = 0$. However if $\omega_a/\omega_b = m/n$ then the output at frequency

ω_b, provided an integer s equal to $2n/m$ exists is

$$c'_0 b\cos(\theta_b+\phi)+(c'_s b/2)\cos(\theta_b+\phi'_s-\phi)|_{s=2n/m}$$
$$= c'_0 b\cos(\theta_b+\phi)$$
$$+\,[(c'_s b/2)\{\cos(\phi'_s-2\phi)\cos(\theta_b+\phi)-\sin(\phi'_s-2\phi)\sin(\theta_b+\phi)\}]_{s=2n/m}.$$

Hence

$$N_{ib}(a)=\{c'_0+(c'_s/2)e^{j(\phi'_s-2\phi)}\}_{s=2n/m}. \tag{5.57}$$

The IDF can thus be complex for the SVNL for those frequencies of the incremental input which are an integral multiple of half the frequency of the existing signal. The incremental locus as ϕ varies is a circle of radius $(|c'_s|/2)$ with centre at the point c'_0 as shown in figure 5.18.

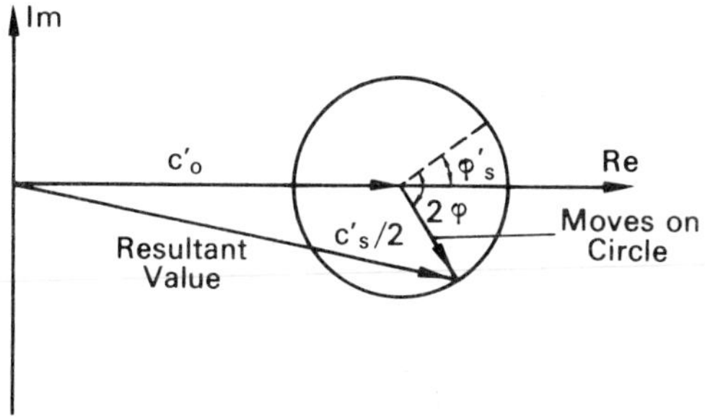

Figure 5.18 Polar locus of IDF

Since the output coefficients c'_s of the nonlinearity for $s=0$ and $2n/m$ are easily found, evaluation of the IDF is straightforward. Appendix A4 contains a listing for various nonlinearities of the coefficients c'_0, c'_s and ϕ'_s from which the IDF can be found for any ratio m/n.

5.9.1 *The special case $m/n = 1$*

It will be seen later that a particularly important application of the IDF involves the situation where the incremental input is at the same frequency as the existing input. This case is shown in the vector diagram of figure 5.19 for a DVNL. For the input $x(t)=a\cos\theta_a$ the output y has components y_p and y_q in phase and quadrature respectively with $x(t)$. When the perturbation $\delta(t)=b\cos(\theta_a+\phi)$ is added to $x(t)$ their vector sum, the new input, is a' which produces an output y' with in phase and quadrature components with respect to a of y'_p and y'_q. For the limiting condition of $b\to 0$ then $\cos\alpha\to 1$, $\sin\alpha\to(b/a)\sin\phi$ and $a'\to a+b\cos\phi$. The output signals y_{pn} and y_{qn} in phase and quadrature with a' are to a first approximation

$$y_{pn}=y_p+(\mathrm{d}y_p/\mathrm{d}a)b\cos\phi,$$
$$y_{qn}=y_q+(\mathrm{d}y_q/\mathrm{d}a)b\cos\phi,$$

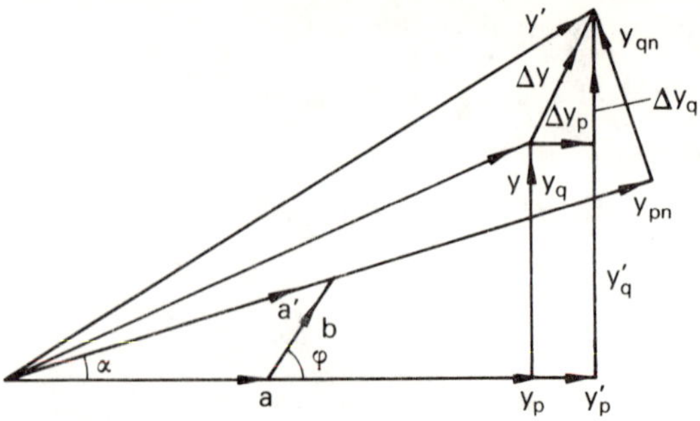

Figure 5.19 Vector diagram for IDF when $m/n = 1$

and the components Δy_p and Δy_q of Δy, the increment in output due to the addition of b, are given by

$$\Delta y_p = y_{pn} \cos \alpha - y_{qn} \sin \alpha - y_p \approx (\mathrm{d}y_p/\mathrm{d}a)b \cos \phi - (y_q/a)b \sin \phi,$$

$$\Delta y_q = y_{pn} \sin \alpha + y_{qn} \cos \alpha - y_q \approx (y_p/a)b \sin \phi + (\mathrm{d}y_q/\mathrm{d}a)b \cos \phi.$$

Now in terms of the DF

$$y_p = aN_p(a) \quad \text{and} \quad y_q = aN_q(a),$$

so that

$$\Delta y_p = b\{N_p(a) \cos \phi + a(\mathrm{d}N_p(a)/\mathrm{d}a) \cos \phi - N_q(a) \sin \phi\},$$

and

$$\Delta y_q = b\{N_p(a) \sin \phi + a(\mathrm{d}N_q(a)/\mathrm{d}a) \cos \phi + N_q(a) \cos \phi\}.$$

The IDF, using the direction of b as reference, is thus,

$$\begin{aligned} N_{ib}(a) &= \mathrm{e}^{-j\phi} \lim_{b \to 0} \{(\Delta y_p + j\Delta y_q)/b\}, \\ &= N_p(a) + (a/2)(\mathrm{d}N_p(a)/\mathrm{d}a) + j\{N_q(a) + (a/2)(\mathrm{d}N_q(a)/\mathrm{d}a)\} \\ &\quad + (a/2)\{(\mathrm{d}N_p(a)/\mathrm{d}a) + j(\mathrm{d}N_q(a)/\mathrm{d}a)\}\mathrm{e}^{-j2\phi}, \end{aligned} \tag{5.58}$$

and in particular for a SVNL this becomes

$$N_{ib}(a) = N(a) + (a/2)(\mathrm{d}N(a)/\mathrm{d}a)\{1 + \mathrm{e}^{-j2\phi}\}. \tag{5.59}$$

Equation (5.58) is valid for multivalued nonlinearities including those whose characteristic may change shape with the addition of the incremental input. If however we restrict ourselves to DVNLs then it can be shown that[13]

$$N_p(a) + (a/2)(\mathrm{d}N_p(a)/\mathrm{d}a) = c'_0 = a'_0,$$

$$N_q(a) + (a/2)(\mathrm{d}N_q(a)/\mathrm{d}a) = 0,$$

$$(a/2)(\mathrm{d}N_p(a)/\mathrm{d}a) = a'_2/2 = (c'_2/2) \cos \phi'_2,$$

$$(a/2)(\mathrm{d}N_q(a)/\mathrm{d}a) = -b'_2/2 = (c'_2/2) \sin \phi'_2,$$

and the above equation reduces to

$$N_{ib}(a) = c_0' + (c_2'/2)e^{j(\phi_2' - 2\phi)},$$

which is identical to equation (5.57) with $s = 2$.

5.9.2 *Ideal saturation*

An an example we consider the IDF for ideal saturation. In this case

$$n'(x) = dn(x)/dx = \begin{matrix} m & |x| < \delta, \\ 0 & |x| > \delta, \end{matrix}$$

and since the nonlinearity is single valued $c_s' = a_s'$ and $\phi_s = 0$. Using equation (5.26) the general expression for a quantised nonlinearity

$$a_0' = (2m/\pi)\{\cos^{-1} 0 - \cos^{-1}(\delta/a)\} = 2m\alpha/\pi,$$

where

$$\alpha = \sin^{-1}(\delta/a)$$

and

$$a_s' = (4m/s\pi)\{D_s(0) - D_s(\delta/a)\}.$$

In particular for the frequency ratio $n/m = 1$ and $s = 2$

$$a_2' = -(2m/\pi)\sin 2\alpha$$

and

$$N_{ib}(a) = (m/\pi)\{2\alpha - e^{-j2\phi}\sin 2\alpha\},$$

which for fixed a, and thus α, is a circle of radius $(m/\pi)\sin 2\alpha$ and centre at the point $2m\alpha/\pi$.

5.10 Applications of the IDF

A major difficulty in the analysis of nonlinear systems is that no uniform approach to their solution is possible and alternative methods have to be used to investigate different nonlinear phenomena. An understanding of the physical mechanisms of the different types of nonlinear behaviour is extremely important, in order that one can appreciate what the system might do and therefore use the appropriate tools to test the hypothesis. In addition it is not sufficient to obtain an analytical solution to a nonlinear problem; the solution, if it is to exist in practice must also be stable. If more than one stable solution exists determination of the conditions that lead to the various solutions is then necessary.

To assess the stability of a periodic solution the effect of the addition of a small sinusoidal perturbation $y = b\cos\omega_b t$ is considered. The loop gain to the perturbation in the presence of the sinusoid of amplitude, a, at the nonlinearity input is $N_{ib}(a)G(j\omega)$ and the characteristic equation of the perturbed system is

$$1 + N_{ib}(a)G(s) = 0. \tag{5.60}$$

The stability of the periodic solution can thus be evaluated by plotting the loop gain $N_{ib}(a)G(j\omega)$ on the Nyquist diagram. Two applications of the IDF are therefore apparent, namely determination of the stability of an autonomous or forced oscillation. For autonomous oscillations it is generally sufficient to consider an unrelated incremental input* so that the condition for stability becomes the locus $N_{i\gamma}(a)G(j\omega)|_{a=a_0}$ should not enclose the $(-1, 0)$ point. For the forced oscillation complete stability is only assured if $N_{ib}(a)G(j\omega)|_{a=a_f,\omega=\omega_f}$ does not enclose the $(-1, 0)$ point for all possible values of ω_b, the frequency of the perturbation b. From the theory of the IDF special cases have been seen to exist when $\omega_b = n\omega_f/2$ where n is an integer. In most practical control situations since $|G(j\omega)|$ and c'_s for the nonlinearity usually decrease as ω and s increase, the frequency $\omega_b = \omega_f$ is generally the one of concern. An exception is asymmetrical systems where c'_1 is not negligible and $\omega_b = \omega_f/2$ must be considered. When instability is predicted for this case it implies a second subharmonic oscillation, a topic which is discussed in more detail in section 5.13.

When $\omega_b = \omega_f$ and $G(j\omega) = ge^{j\theta}$ to ensure stability for a SVNL the IDF criterion can be interpreted as the vector $a'_2g/2$ from the point $a'_0ge^{j\theta}$ must not encircle the point $(-1, 0)$, thus

$$|a'_0ge^{j\theta} + 1| > a'_2g/2,$$

that is

$$1 + (a'_0g)^2 + 2a'_0g\cos\theta > (a'_2g)^2/4,$$

which on substituting for a'_0 and a'_2 gives

$$1 + g^2N^2 + g^2Na(\mathrm{d}N/\mathrm{d}a) + 2g\{N + (a/2)(\mathrm{d}N/\mathrm{d}a)\}\cos\theta > 0. \tag{5.61}$$

For the simple feedback configuration of figure 4.1 with a sinusoidal input of magnitude R, $G_1 = H = 1$ and $G_2 = G$

$$R^2 = a^2(1 + N^2g^2 + 2Ng\cos\theta).$$

Differentiating with respect to a

$$2R(\mathrm{d}R/\mathrm{d}a) = 2a(1 + N^2g^2 + 2Ng\cos\theta) + a^2\{2N(\mathrm{d}N/\mathrm{d}a)g^2 + 2(\mathrm{d}N/\mathrm{d}a)g\cos\theta\},$$

so that

$$(R/a)(\mathrm{d}R/\mathrm{d}a) = 1 + g^2N^2 + g^2aN(\mathrm{d}N/\mathrm{d}a) + 2g\{N + (a/2)(\mathrm{d}N/\mathrm{d}a)\}\cos\theta,$$

which is equal to the left hand side of equation (5.61). Thus since R and a are positive, the criterion for stability with b the same frequency as the forcing signal becomes

$$\mathrm{d}R/\mathrm{d}a > 0. \tag{5.62}$$

This result, which can be deduced from the IDF for more general nonlinear system configurations, is seen to be identical to that used by Hatanaka (see section 4.7.5) for investigating jump resonance.

* Possible exceptions to this rule are systems that have more than one limit cycle solution.

5.10.1 *Limit cycle stability*

Consider an on-off relay with dead zone in a closed loop system with a linear transfer function which may have either the form $G_1(j\omega)$ or $G_2(j\omega)$ illustrated in figure 5.20. For both transfer functions there are two possible limit cycles, one with an amplitude greater than and the other less than $\delta\sqrt{2}$.

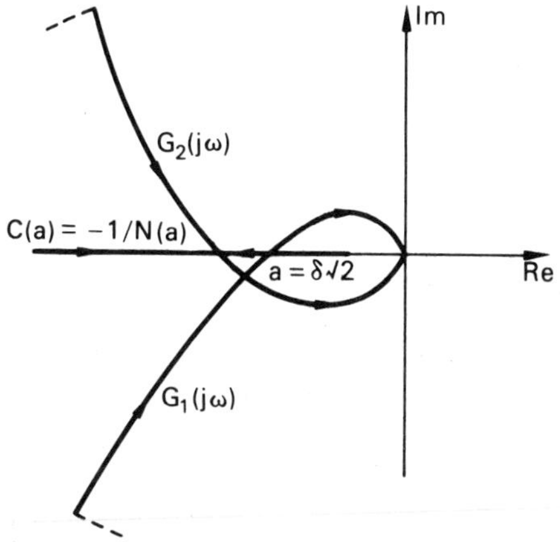

Figure 5.20 Illustration for limit cycle stability

Using the IDF stability criterion the oscillation of amplitude a with the transfer function $G_1(j\omega)$ will be stable if

$$N_{i\gamma}(a)|G_1(j\omega_0)| < 1,$$

which on substituting for $N_{i\gamma}(a)$ gives

$$(a/2)(\mathrm{d}N/\mathrm{d}a)|G_1(j\omega_0)| < 0 \quad \text{since } N(a)|G_1(j\omega_0)| = 1,$$

that is $\mathrm{d}N/\mathrm{d}a < 0$. The oscillation of amplitude less than $\delta\sqrt{2}$ is therefore unstable and that of amplitude greater than $\delta\sqrt{2}$ stable.

Similarly for $G_2(j\omega)$ the stability criterion requires

$$N_{i\gamma}(a)|G_2(j\omega)| > 1,$$

which reduces to

$$\mathrm{d}N/\mathrm{d}a > 0.$$

Thus in this case the oscillation of amplitude less than $\delta\sqrt{2}$ is stable and that of amplitude greater than $\delta\sqrt{2}$ unstable. Results which agree with those obtained from the Loeb criterion.

5.10.2 *Jump phenomena*

We have seen in section 4.7 that over a certain range of amplitude and frequency of an input harmonic signal to certain nonlinear systems three analytical solutions are possible, only two of which are stable. The IDF, as is obvious from its equivalence shown by equations (4.42) and (5.62) to the method of Hatanaka, can be used to find conditions for the existence of the unstable third solution. For b of frequency ω_f

$$N_{ib}(a) = N + (a/2)(\mathrm{d}N/\mathrm{d}a)(1 + \mathrm{e}^{-j2\phi}),$$

and for a SVNL the maximum phase shift of $N_{ib}(a)$ for a given a is α, where

$$\sin\alpha = \frac{|(a/2)(\mathrm{d}N/\mathrm{d}a)|}{N + (a/2)(\mathrm{d}N/\mathrm{d}a)}. \tag{5.63}$$

If α_m is the maximum possible value of α for all a then no unstable forced solution can occur if the open loop $G(j\omega)$ locus does not enter the region of the Nyquist diagram bounded by the radial lines $(180 \pm \alpha_m)$. For a polynomial nonlinearity of the form $n(x) = a|x|^{\mu} \operatorname{sgn} x$, it can be shown that

$$\alpha = \tan^{-1}\{(1 - \mu)/2\mu^{1/2}\},$$

which is independent of a, so that $\alpha_m = \alpha$ for any a. In particular for a cubic nonlinearity $\mu = 3$ and $\alpha = 30°$.

When the $G(j\omega)$ locus enters the region bounded by the lines $(180 \pm \alpha_m)$, jump resonance may occur in the frequency response. Figure 5.21 shows results obtained in reference 14 for the region of jump phenomena in a system containing a cubic nonlinearity and the linear locus $G(j\omega)$ illustrated. It is seen that once the frequency response enters the sector bounded by the radial lines at 150° and 210° the equation $G(j\omega) = -1/N_{ib}(a)$ for a given ω is satisfied for two values of a. The two values of a give the bounds on the region of jump phenomena shown in figure 5.21. It is not difficult to show from equation (5.57) that for any nonlinearity the IDF loci $-1/N_{ib}(a)$ are circles, given of course for a SVNL by equation (4.44).

5.11 Applications of the SSDF for unrelated inputs

A common use of SSDFs for unrelated signals is for the investigation of limit cycles in systems with a sinusoidal input. Occasionally, also, situations arise where a system has an input which is the sum of two unrelated sinusoids or two inputs, one usually being a disturbance, both of which are sinusoidal. Feedback loops can exist which support simultaneously two limit cycles of unrelated frequencies. Their performance can be studied using the SSDF.[15,16]

5.11.1 *Multiple limit cycles*

All that can be concluded from the multiple limit cycle situation such as that shown in figure 4.3, since the Loeb criterion is a necessary but not sufficient condition for a stable limit cycle, is that the system will have no oscillation at

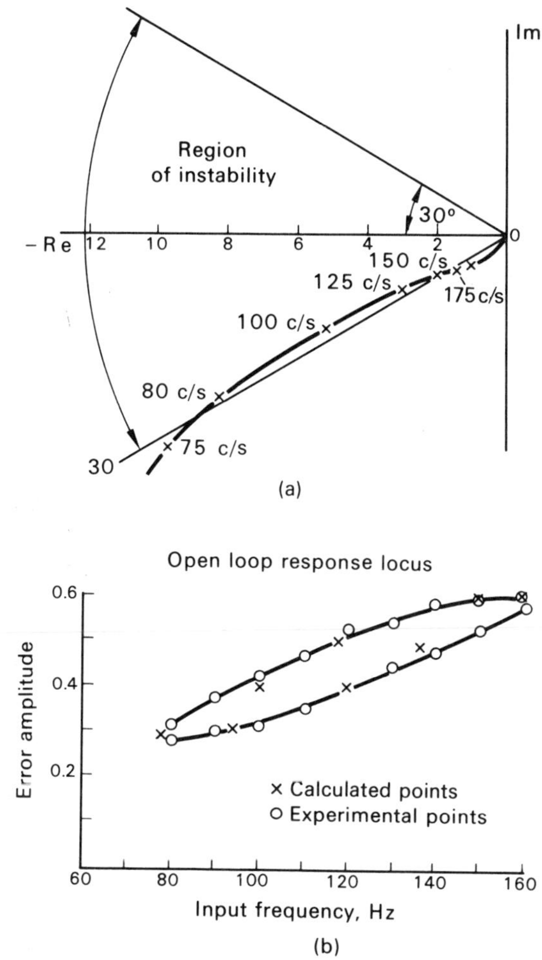

Figure 5.21 Region of jump phenomena

the single frequency, ω_b, given by the intersection of the $G(j\omega)$ and $C(a)$ loci at B. It is clear therefore that the system may behave in one of several ways namely (i) has a single frequency oscillation at both ω_a or ω_c depending on the initial conditions, (ii) has a single frequency oscillation at one of ω_a or ω_c, if only one is stable, (iii) has a combined mode at any two frequencies, (iv) has a combined mode at all three frequencies or (v) has no limit cycle.

To investigate this situation further requires the use of other multiple input DFs.[16] Also, as one might expect, the actual behaviour depends strongly on the form of the system transfer function $G(s)$ and the specified nonlinearity. We therefore consider a system with

$$G(s) = \frac{K(s^2 + 2{\cdot}8\zeta s + 1{\cdot}96)}{s(s + 1)^2(s^2 + 0{\cdot}384s + 2{\cdot}56)}$$

and an ideal relay nonlinearity. The Nyquist and root loci for this transfer function are shown in figure 5.22 for $K = 5{\cdot}0$ and $\zeta = 0{\cdot}7$. Considering first the possibility of single frequency limit cycles, these are given by the solution of

$$1 + N(a_n)G(j\omega_n) = 0, \quad \text{for } n = 1, 2 \text{ or } 3. \tag{5.64}$$

According to the IDF the solutions will be stable if

$$N_{i\gamma}(a_n)|G(j\omega_1)| < 1. \tag{5.65}$$

Now for the ideal relay

$$N(a_n) = 2N_{i\gamma}(a_n)$$

so that the condition for a stable limit cycle becomes

$$2g(\omega_n) > g(\omega_1) \quad \text{for } n = 1, 2 \text{ or } 3.$$

This condition is always satisfied for ω_1 and for ω_3 if $\alpha < 2$ where $\alpha = g(\omega_1)/g(\omega_3)$. For the parameters used in figure 5.22, $\alpha = 1{\cdot}86$ and the system has two stable single frequency limit cycle solutions at ω_1 and ω_3. The initial conditions, of course, determine which limit cycle is reached in a particular case. The location of the roots of the IDF characteristic equation, as shown in figure 5.22(b), clearly indicate why ω_3 is stable. When $\alpha > 2$ the IDF root on the portion of the locus which does not pass through ω_3 is in the right half of the s plane and the ω_3 oscillation is unstable. These results were confirmed experimentally by varying the parameters K and ζ of $G(s)$.(16) Since it can be shown that consideration of a synchronous perturbation (see section 9.5.2) normally gives a solution for stability in agreement with the Loeb criterion(16) a single frequency limit cycle is stable if it is predicted as stable by both the Loeb and IDF criteria.

For combined modes to exist the equations

$$\begin{aligned} 1 + N_{a_k}(a_k, a_n)G(j\omega_k) &= 0, \\ 1 + N_{a_n}(a_k, a_n)G(j\omega_n) &= 0 \end{aligned} \tag{5.66}$$

for $k \neq n$ and $n, k = 1, 2$ or 3 must have a solution. The stability of any solution can be determined from the ISSDF* characteristic equation

$$1 + N_{i\gamma}(a_k, a_n)G(s) = 0, \tag{5.67}$$

investigated for a bias perturbation and also one at the frequencies ω_k and ω_n of the solution. Although solutions to equations (5.66) exist in the above example for $k = 1$ and $n = 3$ the solutions may be shown to be unstable using equation (5.67)(16) for a bias perturbation.

When the nonlinearity is changed to a cubic, since it has

$$N(a) = N_{i\gamma}(a)/2,$$

the condition for a stable single oscillation from equations (5.64) and (5.65)

---

* The incremental DF in the presence of two sinusoids, $N_{i\gamma}(a_k, a_n)$ (see next section).

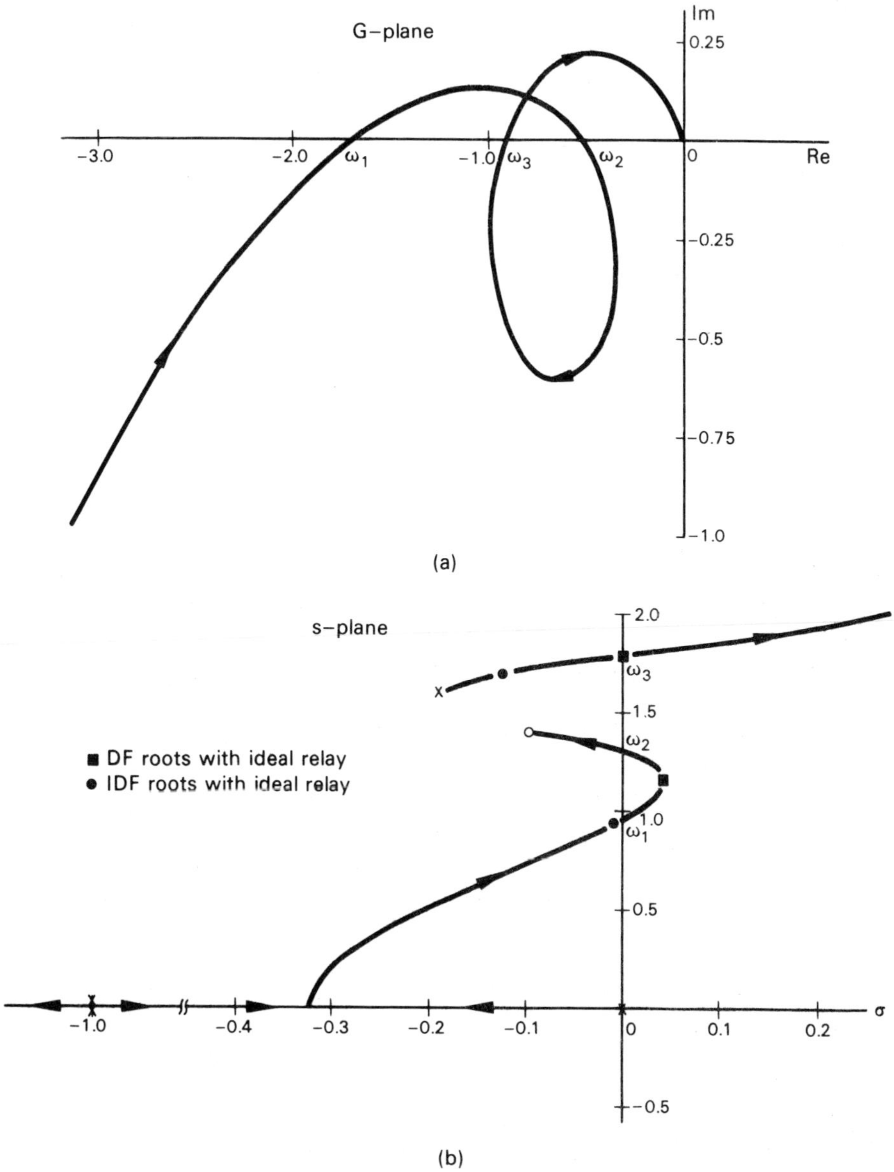

Figure 5.22 Polar frequency response and root locus diagrams for example $G(s)$ of section 5.11.1

becomes

$$g(\omega_n) > 2g(\omega_1), \quad \text{for } n = 1, 2 \text{ or } 3,$$

which cannot be satisfied for any n. Thus no single frequency limit cycle exists with a cubic nonlinearity although the solution at ω_2 satisfies the Loeb criterion. It can further be shown that although solutions exist to equations (5.66) for all the possible combined modes, and the triple mode from three equations similar to equations (5.66), the IDF equation shows them all to be unstable.[17] For a large enough disturbance the system goes unstable as $g(\omega_1)$ becomes greater than $|C(a)|$.

The example discussed points out the computations required to investigate the oscillations which may exist in a system with multiple limit cycle solutions. As the Loeb criterion is so easy to apply it is important to remember that all it does is to show which single frequency limit cycle modes are unstable. It does not prove, however, that such a mode may not exist as one component of a combined mode.[16]

5.11.2 *Harmonic response, limit cycle induction and quenching*

The forced sinusoidal response of nonlinear systems is particularly interesting since a variety of modes of operation may exist and in some cases may be obtained simply by varying the input amplitude. Apart from the normal forced response, which may possess unstable solutions as discussed in section 4.7, the possibility also exists that the input will excite an oscillation in the loop either at a frequency unrelated or related to that of the input. Not surprisingly investigations of the various situations proceed rather differently and the latter case of harmonic synchronised behaviour is deferred to section 5.13.2.

The variations of gain to a signal through a nonlinearity, which may be produced by an unrelated second signal, are readily appreciated from the modified nonlinearity concept so that physical situations where a sinusoidal input may induce an oscillation, or quench an existing one, are usually easily foreseen. Consider a system with a SVNL and a sinusoidal input to have a limit cycle at an unrelated frequency. This situation is usually referred to as a combined or almost periodic oscillation. If the forcing signal produces an amplitude a at the nonlinearity input and a limit cycle of amplitude b then the condition for the limit cycle is

$$1 + N_b(a, b)G(j\omega) = 0,$$

and since $N_b(a, b)$ has no phase shift the limit cycle frequency ω_0 is given by

$$\theta(\omega_0) = \angle G(j\omega_0) = (2n + 1)\pi. \tag{5.68}$$

Assuming only one such value of ω_0 exists, then for the limit cycle to be maintained

$$N_b(a, b) = 1/g(\omega_0), \tag{5.69}$$

a constant value.

For an input sinusoid of frequency ω_f

$$\frac{a}{R} = \frac{1}{|\{1 + N_a(a, b)G(j\omega_f)\}|}, \tag{5.70}$$

from which R can be obtained for those values of a and b which satisfy equation (5.69).

The universal chart approach presented earlier for evaluating the harmonic response can also be used when a combined oscillation exists. The nonlinear locus which is plotted must contain information on both the DF $N(a)$ and the SSDF $N_a(a, b)$ for those values of a and b which satisfy equation (5.69).

To illustrate the procedure consider a nonlinearity $n(x) = x - (x^3/6)$ in a system with linear elements such that equation (5.69) yields $\omega_0 = 1$ and $g(\omega_0) = 2$. For this nonlinearity

$$N(a) = 1 - (a^2/8),$$

$$N_a(a, b) = 1 - \{(a^2 + 2b^2)/8\},$$

$$N_b(a, b) = 1 - \{(b^2 + 2a^2)/8\}.$$

Substituting $N_b(a, b) = 1/g(\omega_0) = 1/2$ gives $b^2 = 4 - 2a^2$ and using this value of b in $N_a(a, b)$,

$$N_a(a, b) = 3a^2/8.$$

Solutions at points on the $N(a)$ locus will be stable, in the sense that no limit cycle can build up, provided $N_{i\gamma}(a) < 1/2$, that is

$$N_{i\gamma}(a) = N(a) + (a/2)\{dN(a)/da\} = 1 - (a^2/8) - (a^2/8) = 1 - (a^2/4) < 1/2,$$

giving $a > \sqrt{2}$. Similarly solutions on the $N_a(a, b)$ locus will be stable in the same sense if the ISSDF, $N_{i\gamma}(a, b) < 1/2$. It can be shown that[13,18]

$$N_{i\gamma}(a, b) = N_a(a, b) + \frac{a}{2}\frac{dN_a(a, b)}{da} = N_b(a, b) + \frac{b}{2}\frac{dN_b(a, b)}{db},$$

so that

$$N_{i\gamma}(a, b) = 1 - \{(a^2 + 2b^2)/8\} - (a^2/8) = 1 - \{(a^2 + b^2)/4\}.$$

But at a combined oscillation solution $b^2 = 4 - 2a^2$ so that $N_{i\gamma}(a, b) = a^2/4$ and a solution will be stable if $(a^2/4) < 1/2$, that is $a < \sqrt{2}$. The DF for determining the forced response is that shown in figure 5.23. For $a < \sqrt{2}$ the system exhibits combined oscillations as $N_a(a, b)$ is valid and the limit cycle amplitude adjusts to maintain $N_b(a, b) = 1/2$, and for $a > \sqrt{2}$ a synchronised harmonic response exists and the normal DF $N(a)$, is applicable. The unstable solution portions of the DF locus are shown dotted in the figure.

As with the single input DF an appreciation of the accuracy of a combined oscillation solution can be obtained by opening the loop at the nonlinearity and calculating the distortion in the fedback signal when the nonlinearity input is the determined two frequency solution. When the basic assumption of two independent sinusoidal inputs can be questioned the accuracy of the method is obviously in doubt. Dubious situations include when the two frequencies are almost small integral multiples of each other, or when their frequency ratio is within the range 0·33 to 3·0, since here significant distortion terms may be produced at a frequency lower than even both of the assumed input frequencies. The method, however, often gives a satisfactory qualitative picture of the system

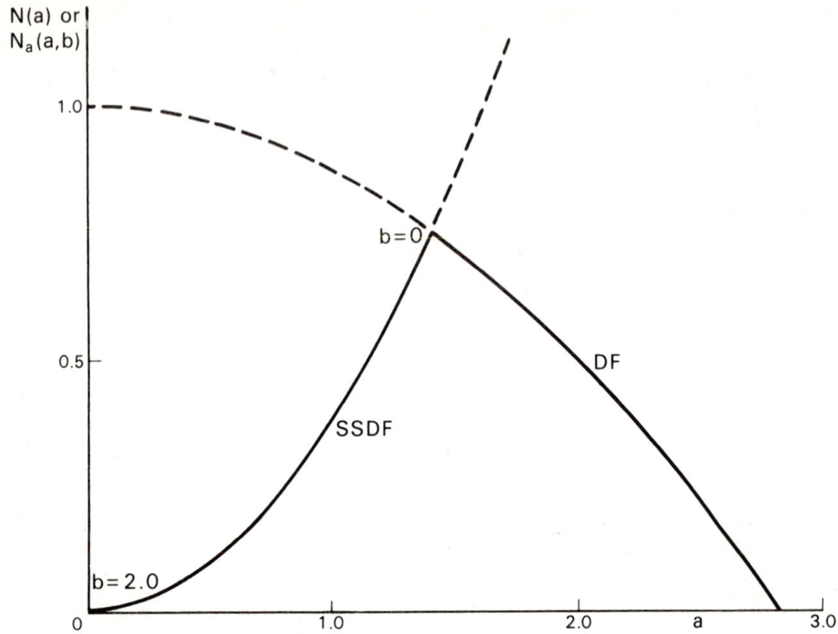

Figure 5.23 DF for $x - x^3/6$ in system with $\omega_0 = 1$, $g(\omega_0) = 2$

behaviour in those cases where the quantitative results are open to question. For example the existence of a range of system input amplitudes for which combined or forced oscillations may exist could be correctly indicated, although the input amplitudes at the boundaries may be markedly in error. Certainly a large number of control problems fall into the category which may be analysed sufficiently accurately for engineering purposes by the above procedure.

To illustrate a variety of situations which can occur the system of figure 5.24 is considered with $G(s) = 2/s(s + 1)^2$ and different nonlinearities N.

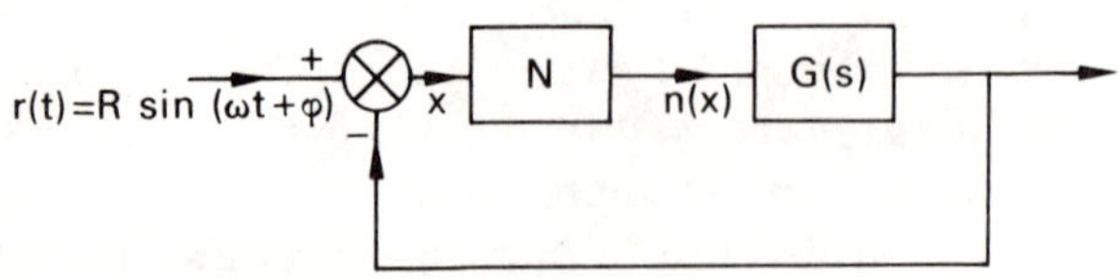

Figure 5.24 Nonlinear system

Example 1

Here N is taken as an ideal relay with $h = 1$. With $R = 0$ the system has a limit cycle of frequency $\omega_0 = 1$ rad/sec and $g(\omega_0) = 1$. Thus for the limit cycle to be maintained with an external input $N_b(a, b) = 1$. Computation of $N_a(a, b)$ as a function of a subject to $N_b(a, b) = 1$, can be done either analytically or graphically. From section 5.8.3

$$N_b(a, b) = \begin{cases} (8h/\pi^2 k^2 a)\{E(k) - (1 - k^2)K(k)\}, & k < 1, \\ (8h/\pi^2 ka)E(1/k), & k > 1, \end{cases}$$

where $k = b/a$. Figure 5.25 shows a graph of $\{aN_b(a, b)\}/h$ versus k. $N_b(a, b)$ is

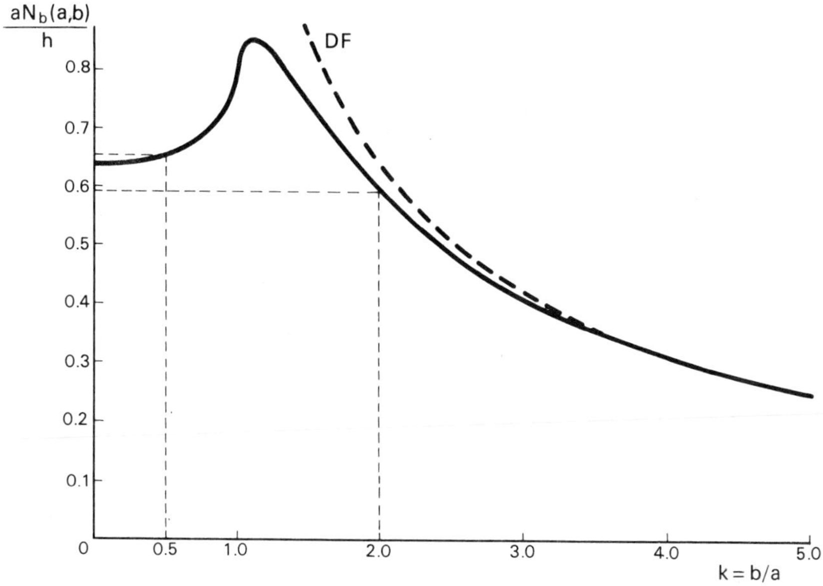

Figure 5.25 SSDF for ideal relay

seen to have a maximum value of $0{\cdot}855h/a$ when $k = 1{\cdot}11$, a result which can be shown by differentiating the above expression for $N_b(a, b)$ with $k > 1$. The graph may also, of course, be used to evaluate $N_a(a, b)$ by interchanging a and b. For example, with $h = 1$, $a = 1$ and $b = 2$, then $k = 2$, $N_b(a, b) = (0{\cdot}593/a) = 0{\cdot}593$ and $N_a(a, b) = 0{\cdot}656/b = 0{\cdot}328$. Thus to compute the required $N_a(a, b)$ the procedure is to

(a) Select a value of k and read the corresponding value of $aN_b(a, b)/h$ from the graph. Since $N_b(a, b)$ and h are known, in this case both unity, the value of a is found.

(b) Determine the ordinate of the graph corresponding to an abscissa of $1/k$. Its value is $\{bN_a(a, b)\}/h = \{kaN_a(a, b)\}/h$ from which $N_a(a, b)$ can be determined.

Table 5.3 summarises the results and figure 5.26 shows the complete relay fundamental output versus the input magnitude at the forcing frequency for the two modes of operation. The regions shown dotted correspond to unstable situations, the horizontal portion to the left of C_1 because it is the region of the DF for which the IDF is greater than unity, and the curved portion above C_2 because it is the region of the SSDF for which the ISSDF is greater than unity.

TABLE 5.3

Computation of relay DF for example 1

k	a	b	$1/k$	$kaN_a(a, b)$	$aN_a(a, b)$
∞	0	1·273	0	0·636	0
8·0	0·159	1·272	0·125	0·640	0·080
4·0	0·314	1·256	0·250	0·643	0·161
3·0	0·412	1·236	0·333	0·646	0·215
2·5	0·488	1·220	0·400	0·649	0·260
2·0	0·593	1·186	0·500	0·656	0·328
1·8	0·648	1·167	0·556	0·661	0·367
1·6	0·711	1·138	0·625	0·671	0·418
1·4	0·780	1·092	0·714	0·686	0·490
1·2	0·840	1·008	0·833	0·712	0·593
1·1	0·854	0·940	0·909	0·738	0·672
1·0	0·811	0·811	1·000	0·811	0·811
0·9	0·733	0·660	1·111	0·855	0·947
0·8	0·705	0·564	1·250	0·827	1·033
0·7	0·683	0·478	1·430	0·769	1·100
0·6	0·667	0·400	1·667	0·688	1·146
0·5	0·656	0·328	2·000	0·593	1·186
0·4	0·649	0·260	2·500	0·488	1·220
0·2	0·641	0·128	5·000	0·252	1·260
0	0·637	0	∞	0	1·273

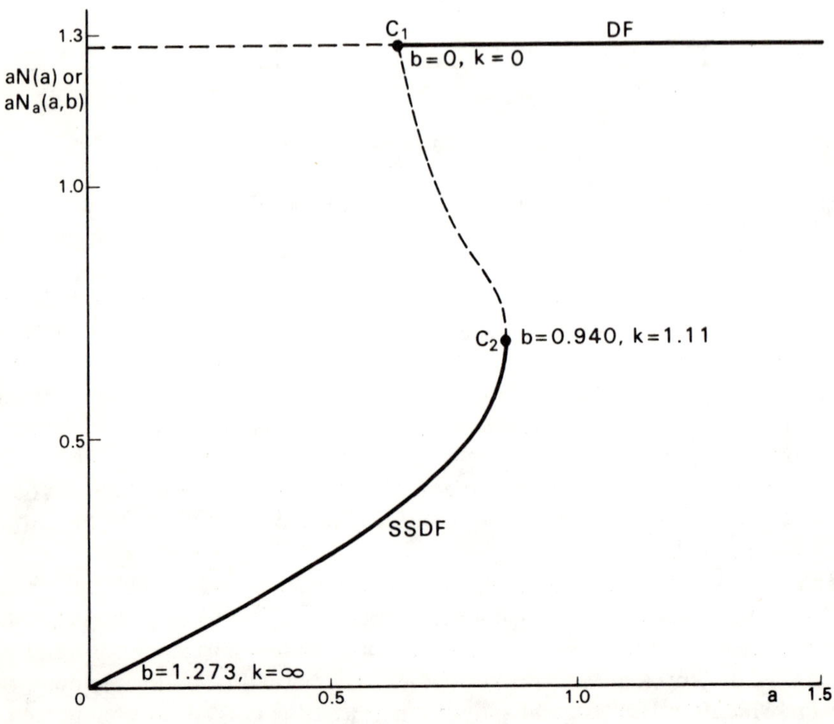

Figure 5.26 Relay output-input curve for example 1

With an input frequency of $\omega_f = 1{\cdot}43$ rads/sec, $\angle G(j\omega_f) = 200°$ and the value of θ_g for use on the chart is $360 - 200 = 160°$. Figure 5.27 shows the nonlinear characteristic positioned for five values of input amplitude R. For curve (i) corresponding to a small value of R the only stable solution is a combined

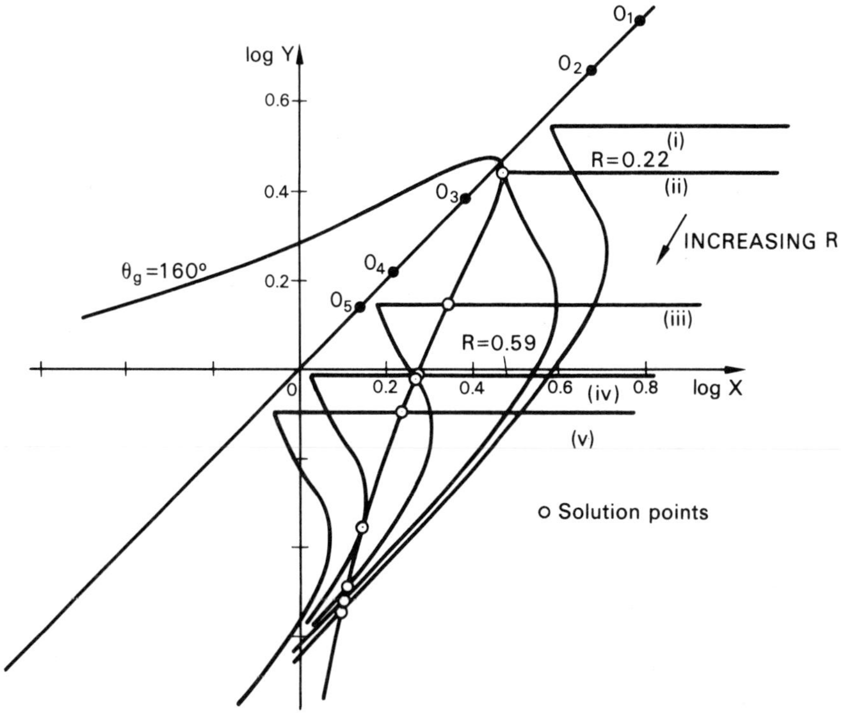

Figure 5.27 Loci for example 1, section 5.11.2, with $\omega_f = 1{\cdot}43$

oscillation, for curve (ii) a forced oscillation solution is also just possible, for curve (iii) two combined oscillation solutions, one of which is stable, and a forced oscillation solution exist, for curve (iv) a combined oscillation solution just exists and for curve (v) only the forced oscillation solution exists. This means in the actual system that if R is increased from zero the combined oscillation continues until $R = 0{\cdot}59$, curve (iv), and for larger values of R a synchronised forced oscillation exists. If R is decreased the synchronised oscillation will continue however until $R = 0{\cdot}22$, curve (ii), at which point the combined oscillation reappears.

Figure 5.28 shows the situation for various amplitudes R with $\omega_f = 0{\cdot}76$ for which $\theta_g = 165°$. With curve (i) a combined oscillation solution only exists, with curve (ii) no stable solution exists, a forced oscillation solution just exists with curve (iii) and finally with (iv) two forced oscillation solutions exist, one of which is stable. Tests on a simulated system showed combined oscillations on increasing R from zero to a value of 0·45, approximately the value of R for

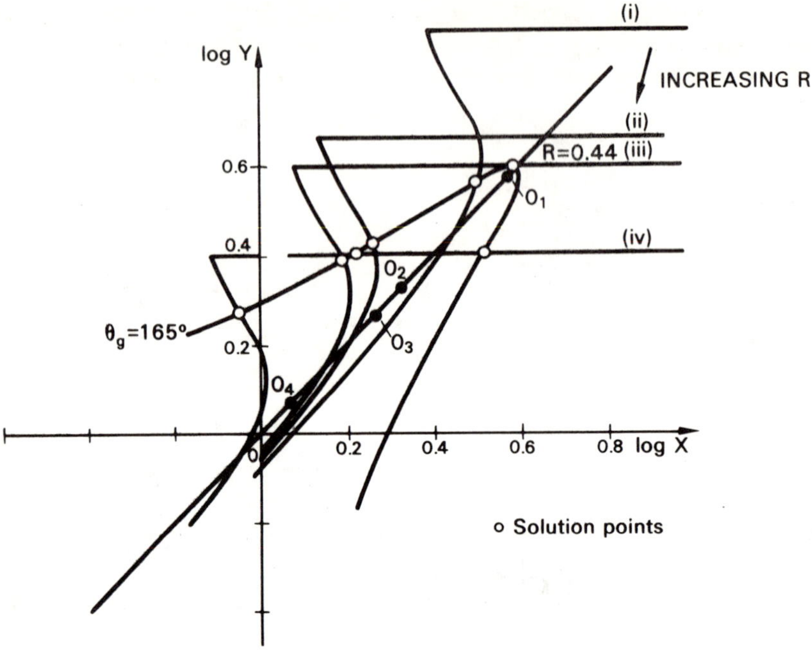

Figure 5.28 Loci for example 1 of section 5.11.2 with $\omega_f = 0{\cdot}76$

curve (iii), and for larger values of R a synchronised forced oscillation. On decreasing R desynchronisation again took place at $R = 0{\cdot}45$. The actual system thus exhibited a combined oscillation where no theoretical solution was available. This example is considered further in section 6.8.3 using exact methods of analysis applicable to relay systems.

Example 2

Consider N an on-off relay with dead zone having $\delta = h = 1$. The maximum value of the DF for the characteristic is $(2/\pi) = 0{\cdot}637$ so that in this case the system is stable in the autonomous state. The modified nonlinearity for the characteristic indicates, however, that the addition of another signal will increase the gain to a small signal, thus a limit cycle may be induced when an input is applied. For the characteristic

$$N(a) = 4(a^2 - 1)^{1/2}/a^2\pi \qquad a > 1$$

$$N_{i\gamma}(a) = 2/\pi(a^2 - 1)^{1/2} \qquad a > 1.$$

Thus $N_{i\gamma}(a)$ exceeds unity for $1 < a < 1{\cdot}19$ and solutions on $N(a)$ within these values of a are therefore unstable. If $N_a(a, b)$ is found for those values of a and b for which $N_b(a, b)$ equals unity, which as seen from the graphs in appendix A3 is only possible for a in the range $1 < a < 1{\cdot}3$, the complete DF can be plotted and the response for any input amplitude and frequency found as previously.

Again it is seen that for a range of input amplitudes combined or synchronised oscillations may exist, since a limit cycle can only be induced with a in the range $1 < a < 1{\cdot}19$, but if one exists quenching will not take place until a exceeds 1·3.

Example 3

In this case N is taken to be the cubic nonlinearity $n(x) = 2x^3/3$. For a sufficiently large input it is known that the system will become unstable; the object here is to investigate the magnitude of a harmonic input which will cause instability. For this characteristic we obtain

$$N(a) = a^2/2,$$

$$N_{i\gamma}(a) = a^2,$$

$$N_b(a, b) = (b^2 + 2a^2)/2 \quad \text{so that } b^2 = 2(1 - a^2) \quad \text{for } N_b(a, b) = 1,$$

$$N_a(a, b) = (a^2 + 2b^2)/2 = (4 - 3a^2)/2 \quad \text{with } b^2 = 2(1 - a^2),$$

and

$$N_{i\gamma}(a, b) = a^2 + b^2 = 2 - a^2 \quad \text{with } b^2 = 2(1 - a^2).$$

From these values it is seen that the stable region of $N(a)$ is for $a < 1$ and for $N_a(a, b)$ for $a > 1$. Values of $N_a(a, b)$ for $a > 1$ do not exist however for real b, so that stable solutions can only occur on $N(a)$ for $a < 1$ as shown in figure 5.29.

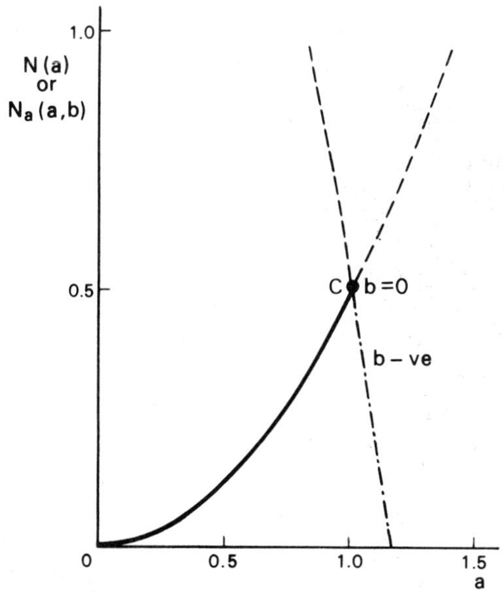

Figure 5.29 DF for cubic nonlinearity of example 3

Figures 5.30 and 5.31 show the linear and nonlinear system characteristics with dotted universal chart curves for various values of R for frequencies ω_f of 0·8 and 2·0 respectively. From the curves for $\omega_f = 0{\cdot}8$ it is seen that as R is increased a jump resonance takes place when $R = 0{\cdot}44$, corresponding to $X = a = 0{\cdot}71$, to a value of a greater than unity. Instability therefore occurs when R reaches 0·44. For $\omega_f = 2{\cdot}0$ as R is increased the curves show that a reaches unity for $R = 0{\cdot}92$. and the system will then go unstable without any previous jump behaviour.

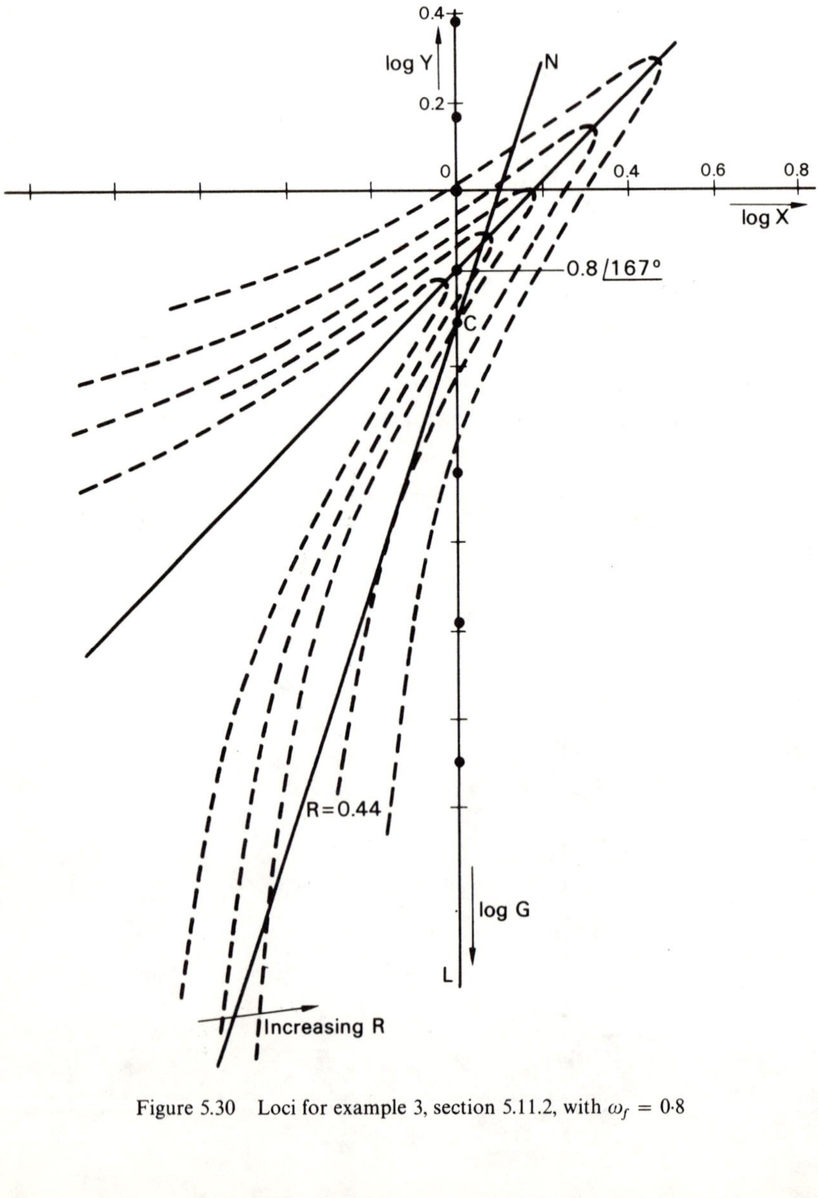

Figure 5.30 Loci for example 3, section 5.11.2, with $\omega_f = 0{\cdot}8$

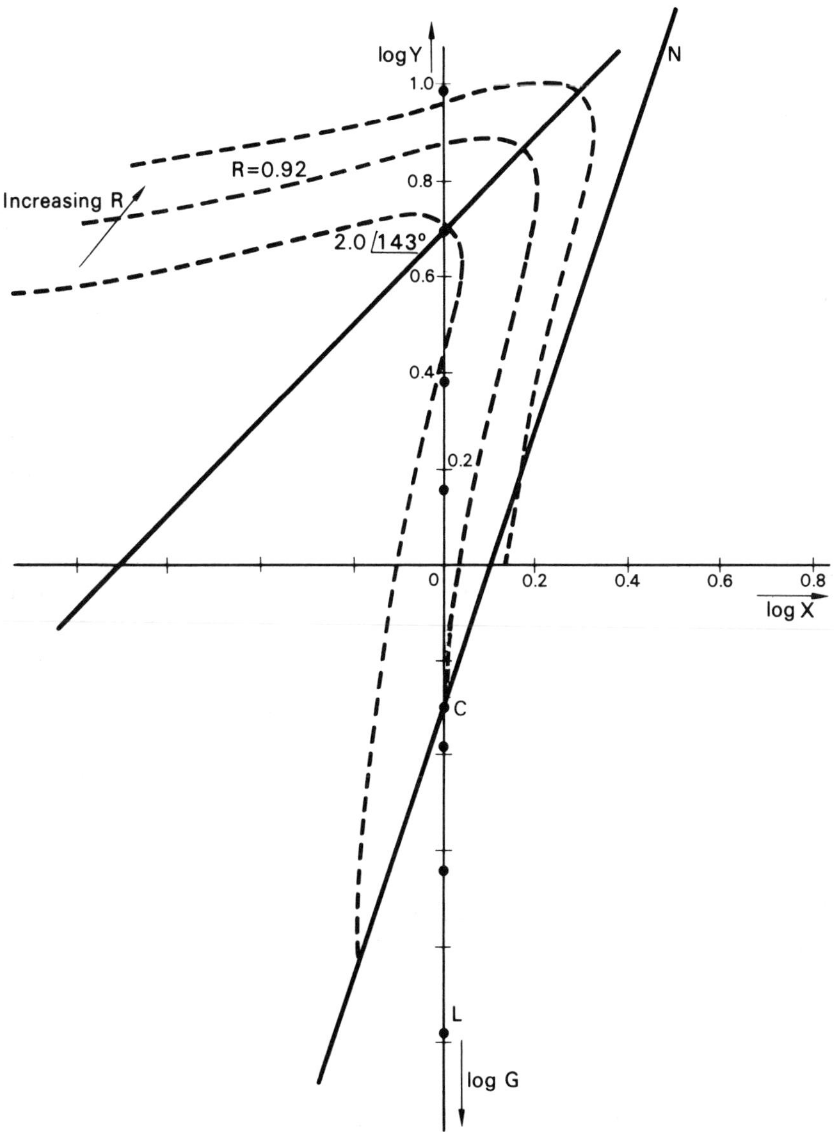

Figure 5.31 Loci for example 3, section 5.11.2, with $\omega_f = 2{\cdot}0$

5.12 Dither and signal stabilisation

In the theory of the SSDF and in some of the examples considered in the previous section we have seen how the DF for a particular signal is modified by the presence of another signal and the marked effect this can have on the system behaviour. The introduction of external signals to modify nonlinear system performance is therefore a logical consequence of the above observation. Such a signal is usually referred to as artificial dither or simply dither, although this latter term is also used to refer to unintentional noise which may be picked up by a system. High frequency dither signals have been used for several years to compensate for the effects of coulomb friction, dead zones in hydraulic valves and hysteresis effects.

The dither signal, x, is normally introduced immediately prior to the nonlinear element whose response to the system signal is to be varied and is chosen so that a negligible component of it will be fed back to the nonlinearity. The nonlinearity modified with respect to x is $n(a, \gamma)$, where

$$n(a, \gamma) = \int_{-a}^{a} n(x + \gamma)p(x)\, \mathrm{d}x, \tag{5.71}$$

and $p(x)$ is the amplitude probability density of the periodic dither signal x of amplitude a. The above expression is a direct extension of equation (5.9) to any input, x. It is also applicable when x is a random input as discussed in section 8.2. The major disadvantage of the introduction of dither is the excessive wear it may cause to system components by keeping them in rapid motion. The dither is required at a high frequency for two reasons. First, to ensure that the amount of perturbation it causes to the desired system outputs, and thus also feeds back to the nonlinearity input, is small. Secondly to justify substitution of the modified nonlinearity for the original characteristic in transient calculations, the dither frequency should be several times higher than the signal component frequencies. On the other hand, since the DF for a sinusoid describes an average behaviour, the modified nonlinearity of equation (5.71), as shown earlier, may be used to evaluate the DF for a sinusoid of any frequency. This is not the case however for a DVNL, where as shown in chapter 13, the frequency requirement of the second condition above must be satisfied.

Sinusoidal modified nonlinearities have been discussed in detail earlier, so the effect of other periodic waveforms is considered here. For a square wave

$$p(x) = (1/2)\{\delta(x - a) + \delta(x + a)\},$$

where δ is the impulse function, so that

$$\begin{aligned} n_{sq}(a, \gamma) &= \int_{-a}^{a} n(x + \gamma)(1/2)\{\delta(x - a) + \delta(x + a)\}\, \mathrm{d}x, \\ &= (1/2)\{n(a + \gamma) + n(-a + \gamma)\}. \end{aligned} \tag{5.72}$$

For a triangular or sawtooth waveform $p(x) = 1/2a$ so that

$$n_{tr}(a, \gamma) = (1/2a) \int_{-a}^{a} n(x + \gamma)\, \mathrm{d}x. \tag{5.73}$$

Table 5.4 gives results for $n_{sq}(a, \gamma)$ and $n_{tr}(a, \gamma)$ for several nonlinear characteristics. It is also of interest to consider the effect of both these waveforms on the arbitrary saturation characteristic of figure 5.32. Using equations (5.72) and (5.73) it is

TABLE 5.4

Square and triangle modified nonlinear characteristics

Nonlinearity: Name	No.—Appendix A1	$n_{sq}(a, \gamma)$—γ positive	$n_{tr}(a, \gamma)$—γ positive
Relay with dead zone	3	(a) $a < \delta$ 0 for $\gamma < \delta - a$ $h/2$ $\delta - a < \gamma < a + \delta$ h $\gamma > a + \delta$ (b) $a > \delta$ 0 for $\gamma > a - \delta$ $h/2$ $a - \delta < \gamma < a + \delta$ h $\gamma > a + \delta$	(a) $a < \delta$ 0 for $\gamma < \delta - a$ $(h/2a)(a - \delta + \gamma)$ $\delta - a < \gamma < a + \delta$ h $\gamma > a + \delta$ (b) $a > \delta$ $h\gamma/a$ for $\gamma < a - \delta$ $(h/2a)(a - \delta + \gamma)$ $a - \delta < \gamma < a + \delta$ h $\gamma > a + \delta$
Ideal relay	4	0 for $\gamma < a$ h $\gamma > a$	$h\gamma/a$ for $\gamma < a$ h $\gamma > a$
Preload	5	$m\gamma$ for $\gamma < a$ $m\gamma + h$ $\gamma > a$	$(h\gamma/a) + m\gamma$ for $\gamma < a$ $h + m\gamma$ $\gamma > a$
Ideal saturation	7	(a) $a < \delta$ $m\gamma$ for $\gamma < \delta - a$ $(m/2)(\gamma + \delta - a)$ $\delta - a < \gamma < a + \delta$ $m\delta$ $\gamma > a + \delta$ (b) $a > \delta$ 0 for $\gamma < a - \delta$ $(m/2)(\gamma + \delta - a)$ $a - \delta < \gamma < a + \delta$ $m\delta$ $\gamma > a + \delta$	(a) $a < \delta$ $m\gamma$ for $\gamma < \delta - a$ $(m/4a)\{2(a\gamma + a\delta + \delta\gamma) - a^2 - \delta^2 - \gamma^2\}$ $\delta - a < \gamma < a + \delta$ $m\delta$ $\gamma > a + \delta$ (b) $a > \delta$ $m\delta\gamma/a$ for $\gamma < a - \delta$ $(m/4a)\{2(a\gamma + a\delta + \delta\gamma) - a^2 - \delta^2 - \gamma^2\}$ $a - \delta < \gamma < a + \delta$ $m\delta$ $\gamma > a + \delta$
Polynomial	18 and 19	$(1/2)\{(a + \gamma)^n + (-a + \gamma)^n\}$	$\sum_{k=0}^{n} {}_nC_k \gamma^{n-k}\{a^{k+1} - (-a)^{k+1}\}/2a(k + 1)$
Harmonic	20	$A_m \cos ma \sin m\gamma$	$(A_m/am) \sin ma \sin m\gamma$
On-off relay with hysteresis	24	(a) $a < \Delta$ $\pm h$ for $\gamma < \Delta - a$ h $\gamma > \Delta - a$ (b) $a > \Delta$ 0 for $\gamma < a - \Delta$ h $\gamma > a - \Delta$	(a) $a < \Delta$ $\pm h$ for $\gamma < \Delta - a$ h $\gamma > \Delta - a$ (b) $a > \Delta$ $h\gamma/a$ for $\gamma < a - \Delta$ h $\gamma > a - \Delta$

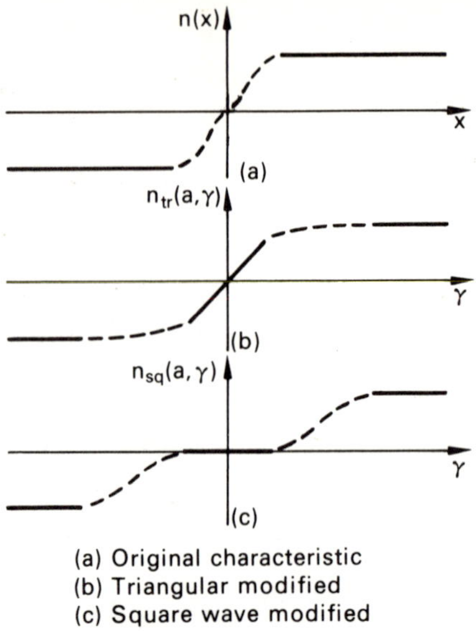

Figure 5.32 Modified saturation characteristics

easy to show that $n_{sq}(a, \gamma)$ and $n_{tr}(a, \gamma)$ have the forms shown in the figure. It is seen that for γ small and a sufficiently large, $n_{sq}(a, \gamma)$ is zero and $n_{tr}(a, \gamma)$ linear, independent of the shape of the saturation characteristic between the limit levels.

Modified nonlinearities can be displayed on an oscilloscope, when the modifying signal is of a sufficiently high frequency relative to the input signal, by connecting the input signal to the X plates and the low pass filtered output to the Y plates.[19] Ideally the low pass filter removes all signals in the output other than the fundamental and harmonics of the input signal.

5.12.1 *Some applications of dither*

Several applications of dither or signal stabilisation can be found in the literature.[20,21] For illustrative purposes two examples are considered below.

Example 1

Figure 5.33 shows an ideal relay system which has a limit cycle of frequency 1 rad/sec in the steady state. We are required to find what amplitude of dither signal $d(t)$ applied at the nonlinearity input will stop the limit cycle if $d(t)$ is (a) a triangular wave, (b) a sine wave and (c) a square wave. The lowest frequency component in $d(t)$ may be assumed to be much greater than unity. What are the disadvantages of using a square wave?

The equivalent nonlinear system with the modified nonlinearity replacing the relay is also shown in figure 5.33. At the limit cycle frequency of 1 rad/sec

$$|G(j\omega)| = K/2,$$

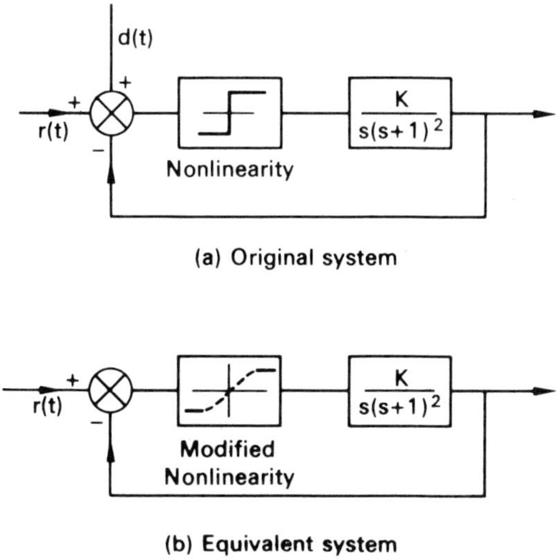

Figure 5.33 System with dither stabilisation

so that to maintain the limit cycle the modified nonlinearity must provide a gain of $-1/|G(j\omega)| = 2/K$.

(a) For $d(t)$ a triangular waveform of peak a

$$n_{tr}(a, \gamma) = \begin{cases} \gamma h/a & |\gamma| < a \\ h & \gamma > a \\ -h & \gamma < -a \end{cases},$$

an ideal saturation characteristic with limit level a. Thus the limit cycle will disappear when

$$(h/a) < (2/K), \qquad a > 0{\cdot}5Kh.$$

(b) For $d(t)$ a sinusoid of amplitude a, the SSDF $N_b(a, b)$ shown in figure 5.25 has a maximum value of $0{\cdot}855h/a$ when $b/a = 1{\cdot}11$. Thus in this case the limit cycle will cease when

$$0{\cdot}855h/a < 2/K, \qquad a > 0{\cdot}427Kh.$$

(c) For a square wave of amplitude a,

$$n_{sq}(a, \gamma) = \begin{cases} 0 & |\gamma| < a \\ h & \gamma > a \\ -h & \gamma < -a \end{cases},$$

an on-off relay with dead zone. The DF for a sine wave of amplitude b to $n_{sq}(a, \gamma)$ is thus $(4h/b^2\pi)\{b^2 - a^2\}^{1/2}$, which has a maximum value of $2\sqrt{2}h/\pi b$ when $b = \sqrt{2}a$. The limit cycle will therefore stop when

$$(2\sqrt{2}h/\pi\sqrt{2}a) < 2/K, \qquad a > 0{\cdot}318Kh.$$

Since the square wave dither signal produces a modified nonlinearity with a dead zone the stabilised system will possess steady state errors.

Example 2

Show how the introduction of a high frequency sinusoidal dither signal at the input of a cubic nonlinearity reduces the size of the region in the complex plane which $G(j\omega)$ should not enter to avoid jump resonance?

For the cubic

$$n(a, \gamma) = \gamma^3 + (3a^2\gamma/2),$$

so that

$$N_b(a, b) = (3/4)\{b^2 + 2a^2\}.$$

Using equation (5.59) with a signal DF of $N_b(a, b)$ the IDF for a small signal of the same frequency as b is thus

$$N_b(a, b) + (b/2)(\mathrm{d}N_b(a, b)/\mathrm{d}b)\{1 + \mathrm{e}^{-j2\phi}\}$$
$$= (3/4)(b^2 + 2a^2) + (b/2)(3b/2)(1 + \mathrm{e}^{-j2\phi}),$$

which has a maximum phase shift, α, of

$$\sin^{-1}\left\{\frac{3b^2/4}{(3/4)(b^2 + 2a^2) + (3b^2/4)}\right\},$$

that is $\alpha = \sin^{-1}\{b^2/2(b^2 + a^2)\}$. Hence as the amplitude of the dither signal a increases, α decreases from its value of 30° when $a = 0$. Correspondingly the region bounded by the radius vectors $(180 \pm \alpha)$ on the Nyquist diagram which $G(j\omega)$ must not enter to avoid jump resonance also reduces. As the addition of a sinusoidal signal to any nonlinearity tends in general to make $n(a, \gamma)$ smoother than the original characteristic, the above reduction in the region of nonlinear behaviour is expected.

5.13 Applications of the SSDF for related inputs

Although the generation of commensurate frequencies by separate individual sources is impossible, it is necessary to consider nonlinearities with commensurate inputs in feedback systems. First to handle periodic solutions approximated by a sine wave plus one or more harmonics and secondly because of a synchronised system oscillation which may be caused by an external input.

Synchronised phenomena exhibit themselves for periodic inputs in two different ways. The external signal may cause an autonomous stable loop to oscillate at a related frequency or it may cause a limit cycle which exists in the loop in the autonomous state to lock in at a frequency related to that of the input. The former situation is known as subharmonic or superharmonic oscillations and the latter as frequency entrainment or synchronisation.[22]

The major difficulty in the investigation of the above phenomena is caused by the complexity of the SSDF for related inputs. As seen in section 5.7.2 for the cubic characteristic the SSDF is a function of four input variables, namely the amplitudes a and b of the two sine waves, their phase difference ϕ and the frequency ratio m/n. When analytical results are not available graphical relationships showing the SSDF for a fixed frequency ratio are complicated because of the three other variables involved. Figure 5.34 shows some results[14] obtained for $m/n = 1/3$ for an ideal saturation characteristic with unit gain and $\delta = 1$. They were obtained by digital computation using the procedure of selecting a, b and ϕ, evaluating the input waveform $a \cos m\theta + b \cos(n\theta + \phi)$ at a large number of equidistant spot points in the period of a, the lower frequency, then evaluating the output waveform at the corresponding points and performing a Fourier analysis of it to obtain the magnitude and phase of the terms at frequencies ω_a and ω_b. Sufficient values of ϕ were taken to allow plotting of one of the loci shown. Then b or a was changed and the process repeated several times to obtain a reasonable number of loci for a fixed a or b. Although the speed of digital computers has increased by a thousand or more since the above work was done the method is still time consuming and expensive. It should also be pointed out that there is some uncertainty regarding the accuracy of the results obtained in the above reference, as the IDF loci presented at the same time were not all circular contours.

To illustrate some phenomena described above, two examples albeit with relatively simple nonlinearities to avoid calculation difficulties, are given below.

5.13.1 *Limit cycle determination*

Consider the feedback system of figure 5.24 with $r(t) = 0$, $n(x) = x - x^3/6$ and $G(s) = 2(1 - s)/s(1 + s)$. The DF for the nonlinearity $N(a) = 1 - a^2/8$, so that the characteristic equation is

$$s^2 + s\{(a^2/4) - 1\} + 2 - a^2/4 = 0,$$

giving $\omega = 1$ and $a = 2$ as the frequency and amplitude of the oscillation.

To determine the limit cycle more accurately a nonlinearity input x of the form

$$x = a \cos \theta_a + b \cos(3\theta_a + \phi)$$

may be assumed. Substituting in $n(x)$, expanding and collecting terms gives for the output at frequency ω_a

$$\{a - (a/8)(a^2 + 2b^2 + ab \cos \phi)\} \cos \theta_a + (1/8)a^2 b \sin \phi \sin \theta_a,$$

and at frequency $3\omega_a$

$$\{b - (b/8)(b^2 + 2a^2) - (a^3/24) \cos \phi\} \cos(3\theta_a + \phi) - (a^3/24) \sin \phi \sin(3\theta_a + \phi).$$

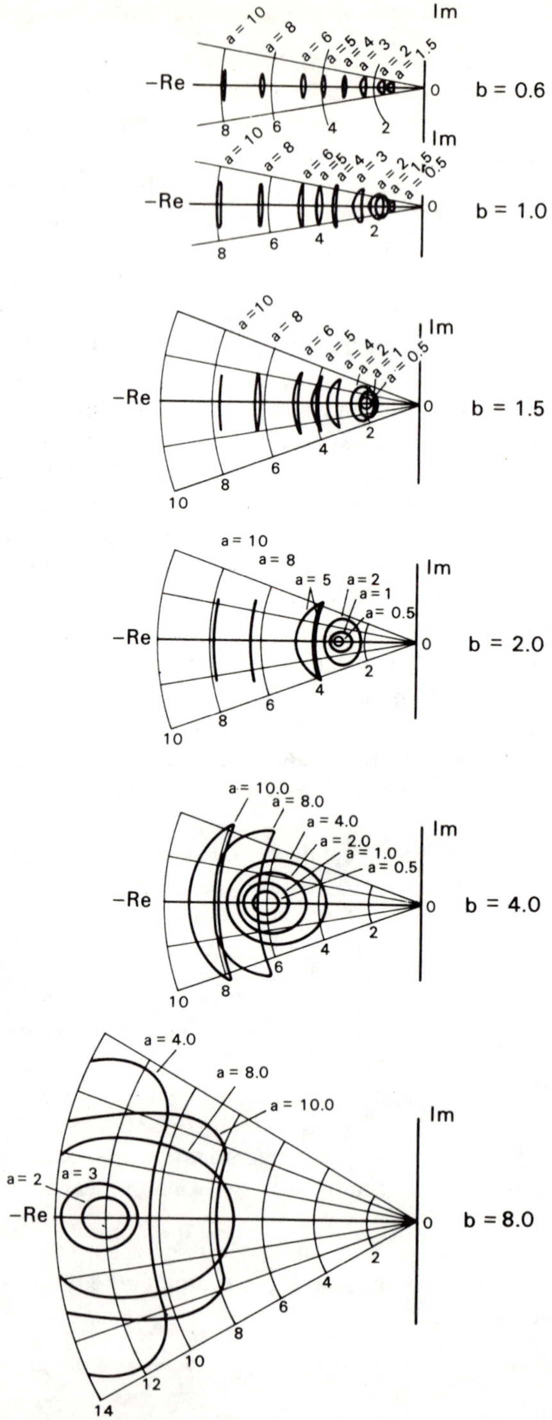

Figure 5.34 Negative inverse saturation SSDF for ω_2 with $m/n = \frac{1}{3}$

Therefore

$$M_a(a|b) = [\{1 - (1/8)(a^2 + 2b^2 + ab\cos\phi)\}^2 + \{(1/8)ab\sin\phi\}^2]^{1/2}, \tag{5.74}$$

$$M_b(a|b) = (1/b)[\{b - (b/8)(b^2 + 2a^2) + (a^3/24)\cos\phi\}^2 + \{(a^3/24)\sin\phi\}^2]^{1/2}, \tag{5.75}$$

$$\psi_a(a|b) = \tan^{-1}\{-ab\sin\phi/(8 - a^2 - 2b^2 - ab\cos\phi)\}, \tag{5.76}$$

$$\psi_b(a|b) = \tan^{-1}\{a^3\sin\phi/(24b - 3b^3 - 6a^2b - a^3\cos\phi)\}. \tag{5.77}$$

The loop equations for the limit cycle in this case are

$$M_a(a|b)g(\omega) = 1, \tag{5.78}$$

$$M_b(a|b)g(3\omega) = 1, \tag{5.79}$$

$$\psi_a(a|b) + \theta(\omega) = (2j + 1)\,180°, \tag{5.80}$$

$$\psi_b(a|b) + \theta(3\omega) = (2k + 1)\,180°, \tag{5.81}$$

where j and k are integers.

Using the expressions in equations (5.74) to (5.77) the above equations (5.78) to (5.81) can in principle be solved for a, b, ϕ and ω, although a trial and error hand calculated solution is very tedious.

A rough estimate for b and ϕ can be obtained using equations (5.79) and (5.81) only, with the DF solutions of $a = 2$, $\omega = 1$ and equations (5.75) and (5.77). The results are $\phi = -53°$ or $127°$ and $b = 0{\cdot}222$ in both cases. A digital computer solution of the nonlinear equations gives two possible answers namely $a = 2{\cdot}19$, $b = 0{\cdot}298$, $\phi = 138°$, $\omega = 0{\cdot}882$ and $a = 1{\cdot}88$, $b = 0{\cdot}166$, $\phi = -52°$, $\omega = 1{\cdot}06$. The former solution is the one for the stable limit cycle. The waveform $x(t)$ at the nonlinearity input computed from the above values of a, b, ϕ and ω is shown dotted in figure 5.35 together with that obtained by analog simulation. The agreement is seen to be much better than the DF solution. The nonlinearity output waveform, which is seen to contain a considerable amount of distortion, is also shown in the figure.

5.13.2 *Subharmonic oscillations*

It has been shown earlier how the application of an external signal to a stable autonomous nonlinear system can cause a limit cycle to occur. Basically the

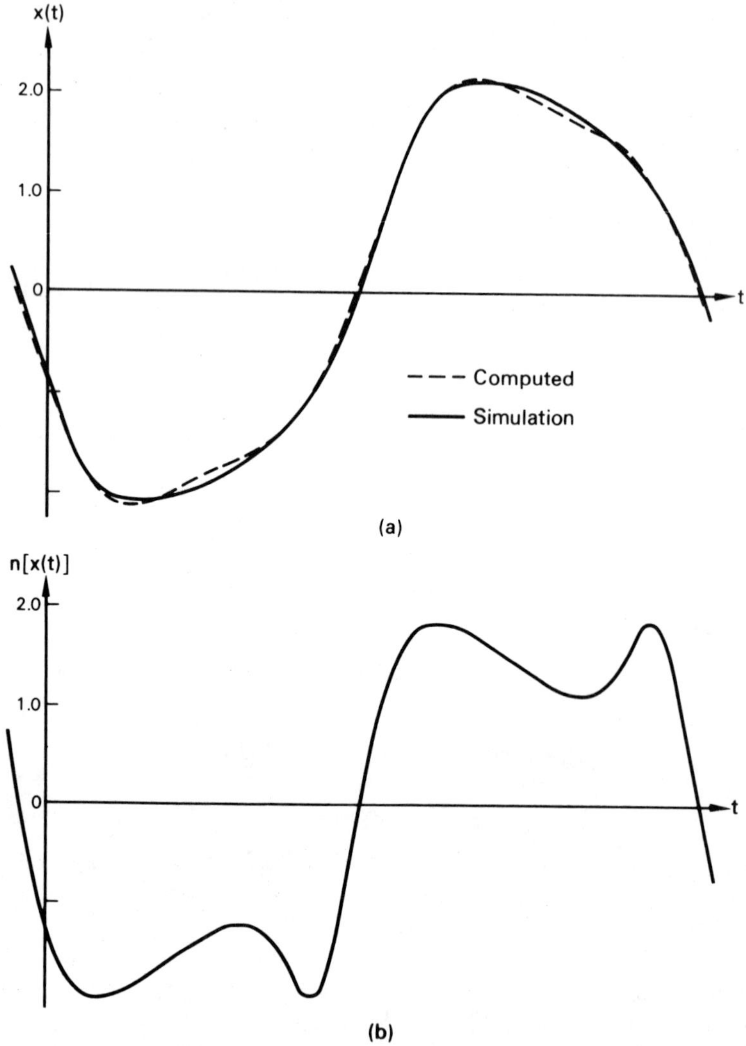

Figure 5.35 Oscillation in system with $x - x^3/6$ and $G(s) = \{2(1 - s)\}/\{s(1 + s)\}$

input signal alters the loop gain and the oscillation occurs at a frequency, determined by the loop linear elements, which may be unrelated to that of the input. If the oscillation were, however, to occur at a frequency related to that of the input, the oscillation signal could undergo a phase change in passing through the nonlinearity. This is the mechanism of subharmonic oscillations, where in effect the presence of the external signal enables an oscillation to occur at a subharmonic frequency by virtue of the support it gives to the oscillation phase shift through the nonlinearity. Subharmonic oscillations have been observed down to submultiples of over thirty of the input frequency,[23] although since the maximum phase shift of the SSDF normally decreases as the integer

n/m increases the high subharmonic number oscillations require for their existence an open loop $G(j\omega)$ which makes the closed loop linear system almost unstable. It can also be appreciated from the above physical explanation that the frequency of a subharmonic oscillation will be near to the system resonant frequency, thus its magnitude at the system output normally dominates that of the input component.

Consider a cubic nonlinearity with an input

$$x = a\cos\theta_a + b\cos(\theta_b + \phi) \text{ with } \theta_a/\theta_b = m/n = 3;$$

then from section 5.7.2 the output at frequency ω_b is as shown in the phasor diagram of figure 5.36. The phase shift α which is given by

$$\psi_b(a|b) = \alpha = \tan^{-1}\{-ab\sin 3\phi/(b^2 + 2a^2 + ab\cos 3\phi)\}$$
$$= \tan^{-1}\{-k\sin 3\phi/(k^2 + 2 + k\cos 3\phi)\}, \quad \text{where } k = b/a,$$

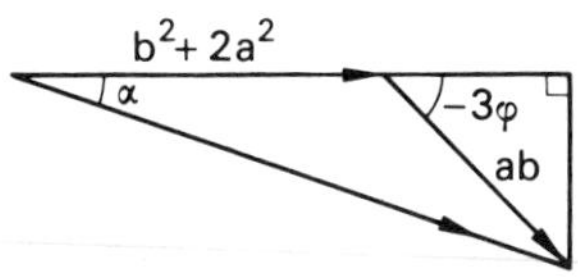

Figure 5.36 Output from cubic with fundamental plus third harmonic input

is seen to have a maximum value, as ϕ varies, of

$$\alpha' = \sin^{-1}\{k/(k^2 + 2)\},$$

when $3\phi = \pi/2 + \alpha'$. Thus, as k varies the absolute maximum value of α is $\alpha'_{\max}$ where $\alpha'_{\max} = \sin^{-1}(1/2\sqrt{2}) \sim 21°$ when $k = \sqrt{2}$. Thus if for a closed loop system of the form shown in figure 5.24, with $n(x) = x^3$, the open loop locus $G(j\omega)$ enters the region bounded by the radial lines at $(180° \pm 21°)$ then subharmonic oscillations are possible given by a solution of the equations

$$\theta(\omega_a/3) + \psi_b(a|b) = (2n + 1)\,180°, \tag{5.82}$$

$$M_b(a|b)g(\omega_a/3) = 1, \tag{5.83}$$

and

$$\frac{a}{R} = \frac{1}{|1 + N_a(a|b)G(j\omega_a)|}. \tag{5.84}$$

For a fixed input frequency, ω_a, values of k and ϕ which satisfy equation (5.82) can be found, and since

$$M_b(a|b) = (3/4a^2)\{(1 + k^2)(4 + k^2) + 2k(k^2 + 2)\cos 3\phi\}^{1/2},$$

the corresponding value of a for those values of k and ϕ can be obtained from equation (5.83). Finally the value of R corresponding to a particular combination of a, k and ϕ which satisfy equations (5.82) and (5.83) can be calculated from equation (5.84).

The region where subharmonic resonance occurs is shown in the $R - \omega$ plane of figure 5.37 for a system[24] of the above form with

$$G(s) = 0{\cdot}8(1 + sT)^2/s(1 + 5sT)^2 \quad \text{and} \quad T = 0{\cdot}0625.$$

The third subharmonic discussed above is not self-starting since the phase of the IDF $N_{ib}(a)$ for $m/n = 3$ is zero. From equation (5.57) it is seen that the phase of the IDF for any nonlinearity is finite when $2n/m$ is an integer, a condition which is satisfied for the second harmonic, $m/n = 2$, only. We then have the important result that no subharmonic apart from the second is self-starting.

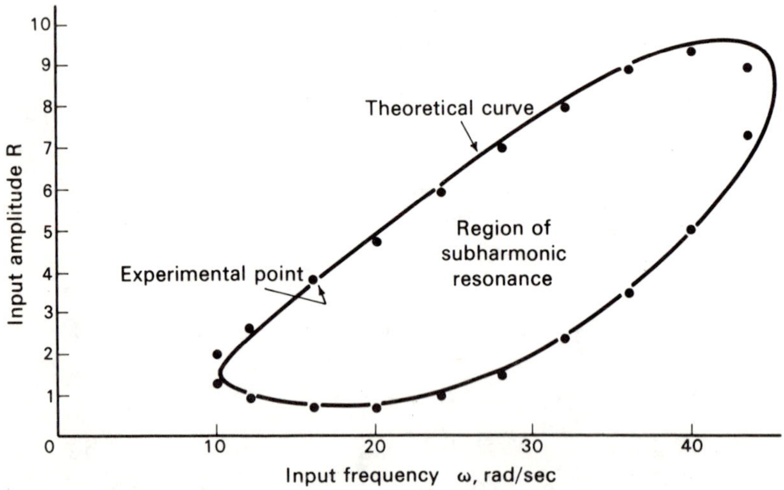

Figure 5.37 Region of subharmonic resonance for example system

Contrary to the often stated case even order subharmonics can exist in a system with no external bias signals and a symmetrical odd nonlinearity.[25] The regions of existence are however normally smaller than those for the odd subharmonics. If such a system maintains an even order subharmonic it is easy to prove that it must operate asymmetrically with a bias signal at the nonlinearity input. Thus initial conditions or disturbances applied to a harmonically forced symmetrical odd nonlinear system may excite either even or odd subharmonic oscillations. A theoretical determination of the conditions for an even subharmonic can be done assuming the nonlinearity input x of the form

$$x = \gamma + a \cos \theta_a + b \cos(\theta_b + \phi),$$

and proceeding as above. The additional variable γ, which also of course leads to an additional equation for the bias levels, usually makes the solution even more laborious.

As described earlier fairly detailed results have been computed for the SSDF of an ideal saturation characteristic. A value of 28° was found[14] for α_{max} with $m/n = 3$.

The stability of a subharmonic oscillation can be found by applying the normal DF arguments to the loop oscillation equations (5.82), (5.83) and the forced response equation (5.84). Superharmonic oscillations, in the sense that the higher harmonics in any nonlinear feedback system are always balanced around the loop, may be said to exist in all cases. Normally, and indeed for use of the DF, the conditions in most controlled processes are such that their magnitudes are small and they can be neglected.

REFERENCES

1. Somerville, M. J. and Atherton, D. P.: 'Multi-gain representation for a single-valued nonlinearity with several inputs, and the evaluation of their equivalent gains by a cursor method', *Proc. I.E.E.*, 1958, 105C, pp. 537–549.
2. Atherton, D. P.: 'The evaluation of the response of single-valued nonlinearities to several inputs', *Proc. I.E.E.*, 1962, 109C, pp. 146–157.
3. Atherton, D. P.: 'Rapid evaluation of the auto-correlation function of the output of single-valued nonlinearities in response to sinusoidal and Gaussian signals', *Proc. I.E.E.*, 1962, 109C, pp. 656–664.
4. Kalman, R. E.: 'Physical and mathematical mechanisms of instability in nonlinear automatic control systems', *Trans. A.S.M.E., J. Basic Eng.*, 1957, 79, pp. 553–556.
5. Fitts, R. E.: 'Two counter examples to Aizerman's conjecture', *I.E.E.E. Trans.*, 1966, AC-11, pp. 553–556.
6. Roots, W. K.: *Fundamentals of temperature control* (Academic Press, New York, 1969).
7. Gelb, A.: 'The dynamic input-output analysis of limit cycling control systems', *Proc. J.A.C.C.*, New York University, New York, 1962, pp. 9.3-1–9.3-11.
8. Gelb, A. and Vander Velde, W. E.: 'On limit cycling control systems', *I.E.E.E. Trans.*, 1963, AC-8, pp. 142–157.
9. Gelb, A. and Vander Velde, W. E.: *Multiple-input describing functions and nonlinear system design* (McGraw-Hill, New York, 1968, Chapters 5 and 6).
10. Gelb, A.: 'The foundation for limit cycling adaptive design', *Proc. Northeast Electronics Res. Eng. Meeting*, 1961, pp. 76–77.
11. Shuck, O. H.: 'Honeywell's history and philosophy in the adaptive control field', *Wright Air Develop. Center, Tech. Rept.*, 1959, Wright Patterson Air Force Base, Ohio.
12. Atherton, D. P.: 'Stability of a feedback system containing a limited field of view error detector', *Proc. I.E.E.*, 1961, 108C, pp. 265–270.
13. Singh, Y. P.: 'Some aspects of forced oscillations of nonlinear systems', *Ph.D. Thesis, Elect. Eng.*, 1968, I.I.T. Kharagpur.
14. West, J. C., Douce, J. L. and Livesley, R. K.: 'The dual input describing function and its use in the analysis of nonlinear feedback systems', *Proc. I.E.E.*, London, 1956, B-103, pp. 463–474.
15. Jones, N. B.: 'Simultaneous oscillations at unrelated frequencies in a feedback loop containing only one nonlinearity', *Int. J. of Control*, 1967, 4, pp. 331–343.
16. Choudhury, S. K. and Atherton, D. P.: 'Limit cycles in high order nonlinear systems', *Proc. I.E.E.*, 1974, 121, pp. 717–724.
17. Choudhury, S. K.: 'Stability of oscillations in nonlinear feedback systems', *M.Sc. Thesis, University of New Brunswick*, 1974.
18. Rajagopalan, R. K. and Singh, Y. P.: 'Analysis of harmonic and almost periodic oscillations in forced self-oscillating systems', *Paper 41-6, IVth I.F.A.C. Congress*, Warsaw, 1969.
19. Atherton, D. P.: 'Non-linear systems with multiple inputs', *Ph.D. Thesis, University of Manchester*, 1962.
20. Naslin, P.: 'A simplified theory of feedback control systems', *Part 12, Proc. Control and Automation*, 1959, 6, pp. 273–277.
21. Oldenberger, R. and Boyer, R. C.: 'Effects of extra sinusoidal inputs to nonlinear systems', *Trans. A.S.M.E., J. Basic Eng.*, 1962, D-84, pp. 559–570.
22. Hayashi, C.: *Nonlinear oscillations in physical systems* (McGraw-Hill, New York, 1964).
23. Douce, J. L. and King, R. E.: 'Instability of a nonlinear conditionally stable system subjected to a sinusoidal input', *Trans. A.I.E.E.*, 1959, 78, pp. 665–670.

24. Gelb, A. and Vander Velde, W. E.: *Multiple-input describing functions and nonlinear system design* (McGraw-Hill, New York, 1968, pp. 268–271).
25. Mufti, I. H.: 'A method for the exact determination of periodic motions in relay control systems', *Proc. J.A.C.C.*, Rensselaer Polytechnic Institute, Troy, New York, 1965, pp. 8–14.

BIBLIOGRAPHY

Amsler, B. E. and Gorozdos, R. E.: 'On the analysis of bi-stable control systems', *I.R.E. Trans.*, 1959, AC-4, December, pp. 46–58.

Atherton, D. P., Turnbull, G. F., Gelb, A. and Vander Velde, W. E.: 'Discussion of the double input describing function (D.I.D.F.) for unrelated sinusoidal signals', *I.E.E.E. Trans.*, 1964, AC-9, pp. 197–198.

Balasubramanian, R. and Atherton, D. P.: 'Prediction of jump resonance in systems containing certain multidimensional nonlinearities', *Proc. I.E.E.*, 1968, 115, pp. 1369–1372.

Bennett, W. R.: 'New results in the calculation of modulation products', *Bell System Tech. J.*, 1933, 12, pp. 228–243.

Bonenn, Z.: 'Stability of forced oscillations in nonlinear feedback systems', *I.R.E. Trans.*, 1958, AC-3, pp. 109–111.

Bonenn, Z.: 'Relative stability of oscillations in non-linear control systems', *Proc. I.F.A.C.*, Basel, Switzerland, 1963, pp. 214/1–4.

Douce, J. L. and King, R. E.: 'Instability of a nonlinear conditionally stable system subjected to a sinusoidal input', *Trans. A.I.E.E.*, 1959, 78, pt. II, pp. 665–670.

Gelb, A. and Blaschke, T. C.: 'Design for adaptive roll control of a missile', *J. Aeron. Astronautics*, 1962, 9, pp. 99–105.

Gelb, A. and Vander Velde, W. E.: *Multiple-input describing functions and nonlinear system design* (McGraw-Hill, New York, 1968).

Gibson, J. E.: *Nonlinear automatic control* (McGraw-Hill Book Company, New York, 1963).

Gibson, J. E. and Sridhar, R.: 'A new dual-input describing function and an application to the stability of forced oscillations', *Trans. A.I.E.E.*, 1963, 82, pt. II, pp. 65–70.

Gray, J. O. and Savvides, L. S.: 'Computer aided design using the two sinusoidal input describing function', *I.E.E. Conf.*, Publication No. 96, 1973, pp. 119–127.

Hayashi, C.: *Nonlinear oscillations in physical systems* (McGraw-Hill Book Company, New York, 1964).

Hsu, J. C. and Meyer, A. U.: *Modern control principles and applications* (McGraw-Hill, New York, 1968, Chapter 7).

Huey, R. M., Pawloff, O. and Glucharoff, T.: 'Extension of the dual-input describing-function technique to systems containing reactive nonlinearity', *Proc. I.E.E.*, 1960, 107C, pp. 334–341.

Kalb, R. M. and Bennett, W. R.: 'Ferromagnetic distortion of a two-frequency wave', *Bell System Tech. J.*, 1935, 14, pp. 322–359.

Li, Y. T. and Vander Velde, W. E.: 'Philosophy of nonlinear adaptive systems', *Proc. First I.F.A.C. Congr.*, Moscow, U.S.S.R., 1960 (Butterworth Scientific Publications, London).

Ludeke, C. A.: 'The generation and extinction of subharmonics', *Proc. Symp. Nonlinear Circuit Analysis*, 1953, 2, Polytechnic Institute of Brooklyn, New York.

Mahalanabis, A. K. and Nath, A. K.: 'On the dual-input describing functions of a nonlinear element', *I.E.E.E. Trans.*, 1965, AC-10, pp. 203–204.

Manohar, G. T. and Sellappan, R.: 'Signal stabilization of hysteresis type relay systems', *I.E.E.E. Trans.*, 1971, AC-16, pp. 79–80.

Marzawa, S.: 'Steady, forced vibration of unsymmetrical piecewise-linear system', *Bulletin of J.S.M.E.*, 1961, 4, pp. 201–229.

Musa, S. A.: 'Asynchronous excitation in nonlinear systems', *I.E.E.E. Trans.*, 1967, AC-12, pp. 625–627.

Nath., N. G., Nath, A. K. and Choudhury, A. K.: 'Studies on dual-input signal transmission through dead-zone, saturation and absquare type of nonlinearity', *Indian J. Pure Appl. Phys.*, 1964, 2, pp. 317–325.

Oldenberger, R.: 'Signal stabilization of a control system', *Trans. A.S.M.E.*, 1957, 79, pp. 1869–1872.

Oldenburger, R. and Liu, C. C.: 'Signal stabilization of a control system', *Trans. A.I.E.E.*, 1959, 78, pp. 96–100.

Oldenburger, R. and Nakada, R.: 'Signal stabilization of self-oscillating systems', *I.R.E. Trans.*, 1961, AC-6, pp. 319–325.

Oldenburger, R. and Boyer, R. C.: 'Effects of extra sinusoidal inputs to nonlinear systems', *Trans. A.S.M.E., J. Basic Eng.*, 1962, pp. 559–570.
Oldenburger, R. and Nicholls, R.: 'Stability of subharmonic oscillations in nonlinear systems', *Proc. J.A.C.C.*, Minneapolis, Minn., 1963, pp. 675–680.
Rajagopalan, R. K. and Mishra, J.: 'Dual input describing function of double valued nonlinearities', *Proc. I.E.E.*, 1969, 116, pp. 1764–1768.
Sea, R. G. and Vacroux, A. G.: 'Steady state analysis of nonlinear systems and multiple input describing functions', *Automatica*, 1969, 5, pp. 763–772.
West, J. C.: *Analytical techniques for non-linear control systems* (E.U.P., London, 1960).
West, J. C. and Douce, J. L.: 'The mechanism of sub-harmonic generation in a feedback system', *Proc. I.E.E.*, 1955, 102, pp. 569–574.

PROBLEMS

P.5.1 Prove that $N(a, \gamma)$ and $N_\gamma(a, \gamma)$ satisfy the correlation definitions of equations (5.7) and (5.8).

P.5.2 Consider the example discussed in section 5.5.2 with $h/\Delta = \pi/8$ and $KH = 200$. Evaluate the modified nonlinearity in the presence of the oscillation and find the values of p and γ when oscillation ceases.

P.5.3 Show that for a nonlinearity $n(x)$ with $n'(x)$ monotonic decreasing

$$N_{i\gamma}(a) < n(a)/a < N(a)$$

and

$$N_{i\gamma}(a) > N(a)/2.$$

P.5.4 Show that $-1/N_{ib}(a)$ is a circle for a given a with the equation

$$[x + \{(K_1 + K_2)/2K_1K_2\}]^2 + y^2 = \{(K_1 - K_2)/2K_1K_2\}^2$$

where $K_1 = N(a)$ and $K_2 = N(a) + aN'(a)$.

P.5.5 Discuss the stability of the system of question P.4.12 using the IDF.

P.5.6 Evaluate $N_a(a, b)$ for x^m, where m is an integer, using the modified nonlinearity approach of first finding $n(a, \gamma)_1$.

P.5.7 Prove that

$$N_{i\gamma}(a, b) = N_a(a, b) + (a/2)(\mathrm{d}N_a(a, b)/\mathrm{d}a) = N_b(a, b) + (b/2)(\mathrm{d}N_b(a, b)/\mathrm{d}b).$$

P.5.8 Figure P.5.1 shows the $C(a)$ and $G(j\omega)$ loci for a system with saturation and a conditionally stable plant transfer function. The loci intersect three times at the points P_1, P_2 and P_3 where the corresponding values of

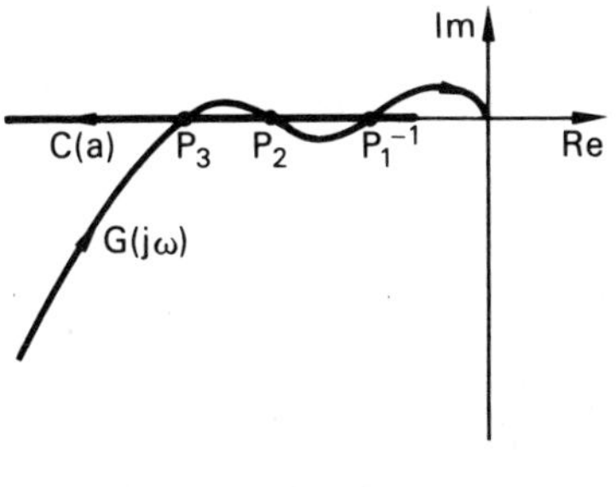

Figure P.5.1

amplitude and frequency are $a_1, \omega_1; a_2, \omega_2$ and a_3, ω_3. According to the Loeb criterion the solutions at the points P_1 and P_3 are stable. Use the IDF to show that the oscillation at P_1 is often unstable.

P.5.9 Prove that for a third order system with a transfer function $G(s) = K/(s^3 + a_2 s^2 + a_1 s + a_0)$ and a nonlinear characteristic $n(x)$, the condition for a limit cycle of amplitude a is $a_0 + KN(a) - a_2 a_1 = 0$. Show further that the limit cycle is predicted to be stable using (i) the IDF if $a_2 > 0$; $a_0 + KN_{i\gamma}(a) > 0$ and $a_1 a_2 - a_0 - KN_{i\gamma}(a) > 0$ and (ii) the Loeb criterion (equation (4.9)) if $a_1 a_2 - a_0 - KN_{i\gamma}(a) > 0$.

P.5.10 A nonlinear system consists of an ideal relay and a linear transfer function $Ke^{-sT_d}/s(s+1)$. Determine all the possible frequencies of oscillation and show using the IDF that only the one at the lowest frequency is stable.

P.5.11 Determine the effect of the high frequency triangular waveform perturbation, $d(t)$, on the oscillation frequency of the system shown in figure P.5.2.

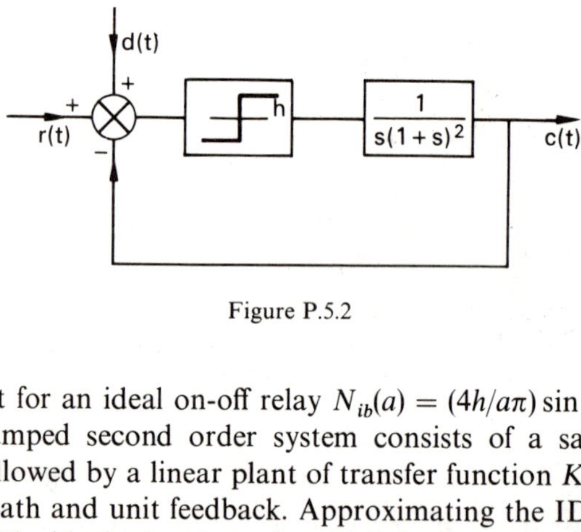

Figure P.5.2

P.5.12 Show that for an ideal on-off relay $N_{ib}(a) = (4h/a\pi)\sin\phi, \angle(\pi/2) - \phi$. A lightly damped second order system consists of a saturation characteristic followed by a linear plant of transfer function $K/s(1 + sT)$ in the forward path and unit feedback. Approximating the IDF for saturation by that of the ideal relay show that if jump phenomena takes place in the frequency response, the jump down will occur when the input and error phases are at 90°.

P.5.13 The block diagram of a nonlinear feedback system is shown in figure P.5.3. Using the parameter plane method find the amplitude and period of any oscillations. Are these oscillations stable?

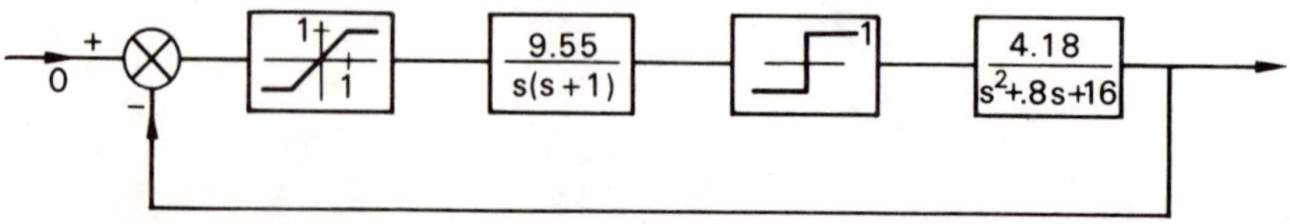

Figure P.5.3

P.5.14 Figure P.5.4 shows the block diagram of a nonlinear feedback system. Using the IDF find, if $d(t) = 0$,

(a) the minimum input frequency ω_m below which the system cannot exhibit jump behaviour, and

(b) the amplitudes of $e(t)$ at which jump resonance will occur when the input frequency is $2\omega_m$.

A high frequency triangular dither signal, $d(t)$, is added at the input as shown. What is the minimum amplitude, b, of the dither signal required to avoid jump behaviour when the input frequency is $2\omega_m$?

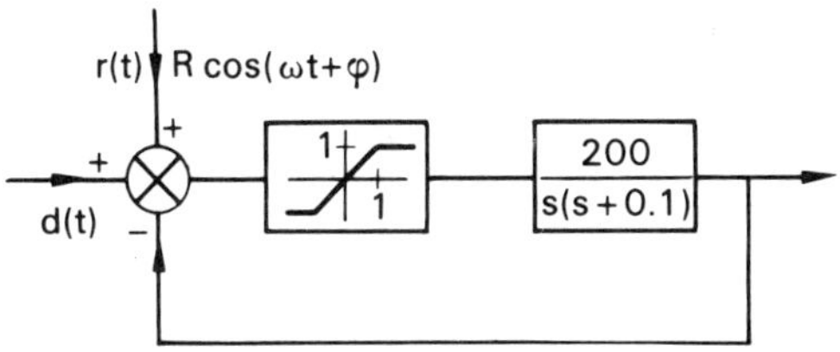

Figure P.5.4

P.5.15 Determine the gain versus frequency response of the asymmetrical nonlinear system with $\alpha = 0{\cdot}25$ shown in figure P.5.5 and compare the result with that of the linear system with $\alpha = 0$.

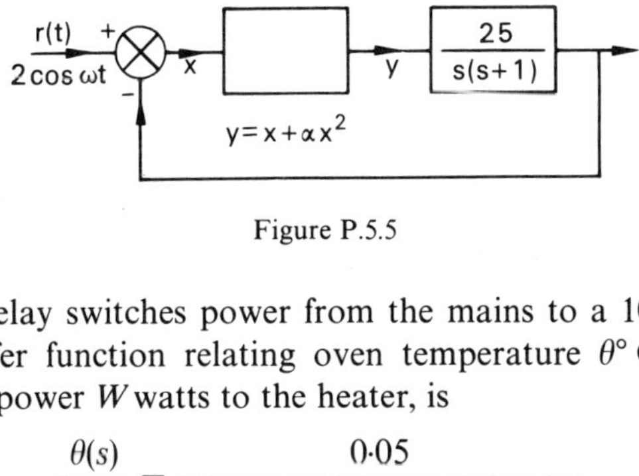

Figure P.5.5

P.5.16 An ideal relay switches power from the mains to a 10 kW oven heater. The transfer function relating oven temperature $\theta°$ C above ambient, and input power W watts to the heater, is

$$\frac{\theta(s)}{W(s)} = \frac{0{\cdot}05}{(I + sT_1)(I + sT_2)(I + sT_3)}$$

If $T_1 = 2T_2 = 4T_3 = 10$ secs. determine the peak amplitude and the frequency of the oscillations in temperature about a controlled level of 100° C above ambient.

Chapter 6

Periodic Modes in Relay Control Systems

6.1 Introduction

The importance of relay systems in control and their analysis using the phase plane and the DF method have already been discussed. Unlike other non-linearities relays are unique in that the waveform produced at their output has only three or two possible discrete values, depending upon whether the relay has or has not a dead zone. The relay input waveform thus only affects the times at which the relay switches between the various states of high, low and zero and it is this aspect of relay behaviour, which might be termed the 'limited control' of the relay output exercised by the input, that makes relay systems amenable to exact analysis. In this chapter we will be primarily concerned with the exact determination of free and forced oscillations in relay systems. Although approximate solutions can be obtained using DFs the availability of an exact method, albeit for a particular class of systems, provides an opportunity to assess DF accuracy and therefore to gain confidence in and appreciation for the use of DFs in other problems.

A simple relay control system is shown in figure 6.1. For a deterministic input the system response can be evaluated using a simple piecewise analysis. Consider for example a positive step input U to the system with an ideal relay. The relay output will become $+h$ and the response $c(t)$ of $G(s)$ to this input, assuming the

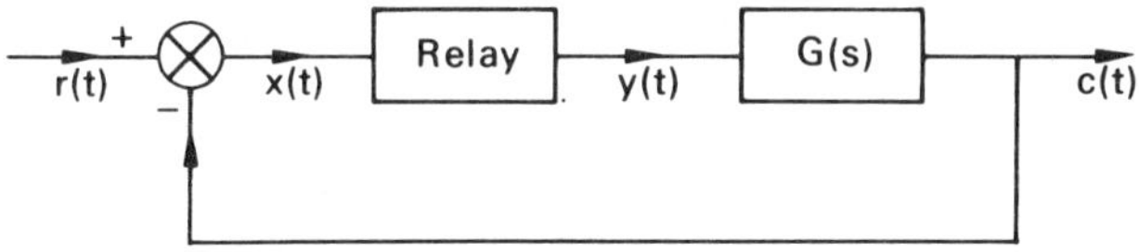

Figure 6.1 Basic relay control system

initial state of $G(s)$ is known, can be evaluated. The input to $G(s)$ will remain at $+h$ until the output $c(t)$ reaches U so that $x(t) = 0$. This will occur at say a time t equals t_1, which can be found from the expression for $c(t)$, and the relay output will then switch to $-h$. The procedure is now repeated $c(t)$ being evaluated for $G(s)$ having an input $-h$ and a known initial state at the time t_1. When $c(t)$ once more attains the value U say at time t equals t_2, the relay will again switch to $+h$. Continued application of the procedure yields the desired response. In principle this procedure can be used for any deterministic input and for more complex configurations possibly containing more than one relay type characteristic.

For harmonic inputs use of the above procedure can prove laborious since a large number of relay switchings may take place before the steady state is reached. Therefore, for such problems and also for the investigation of limit cycles, a simpler procedure is to assume a specific periodic output for the relay and to investigate conditions for its existence. A strong similarity with the DF procedure is thus apparent; a specific waveform, in this case at the nonlinearity output rather than the input, is assumed in order to commence the analysis. Needless to say the method has basically the same advantages and disadvantages as the DF procedure. It is relatively easy to apply and the effects of changing system parameters or adding linear networks is not too difficult to assess, however, like the DF one can only obtain information for those situations for which the initial assumption is valid, a fact of significance for forced oscillations in complex relay systems

Two basic approaches to the solution of the problem are possible, one of which may be called the time domain approach and the other the frequency domain approach. In the former the output $c(t)$ of $G(s)$ is determined for its assumed periodic input $y(t)$ as a closed form function of time and the necessary conditions it must satisfy, so that the relay output will be the assumed $y(t)$ investigated. This approach was first formulated by Hamel[(1)] and has also been given using alternative formulations by Bohn[(2)] and Chung.[(3)] For the frequency domain approach initially proposed by Tsypkin,[(4)] the only difference is that $y(t)$ is expressed as, and hence $c(t)$ determined as, a Fourier series; satisfaction of the necessary switching conditions again yielding a solution. Although as will be seen the solution procedures of Hamel and Tsypkin are almost identical, our main concern will be with frequency domain procedures because of their similarity to and ease of comparison with the DF method. It should however be pointed out that time domain methods do have advantages, possibly for example in relay systems involving other piecewise nonlinear characteristics[3,5] and certainly in the analysis of mode dependent systems, where the plant transfer function depends upon the relay output level.

6.2 Switching conditions

Figure 6.2 shows an on-off relay with dead zone and the basic periodic waveform which is assumed to occur at the relay output when periodic operation takes place in a symmetrical nonlinear system. The output waveform has four switches per cycle at the instants α/ω, $(\pi - \beta)/\omega$, $(\pi + \alpha)/\omega$ and $(2\pi - \beta)/\omega$

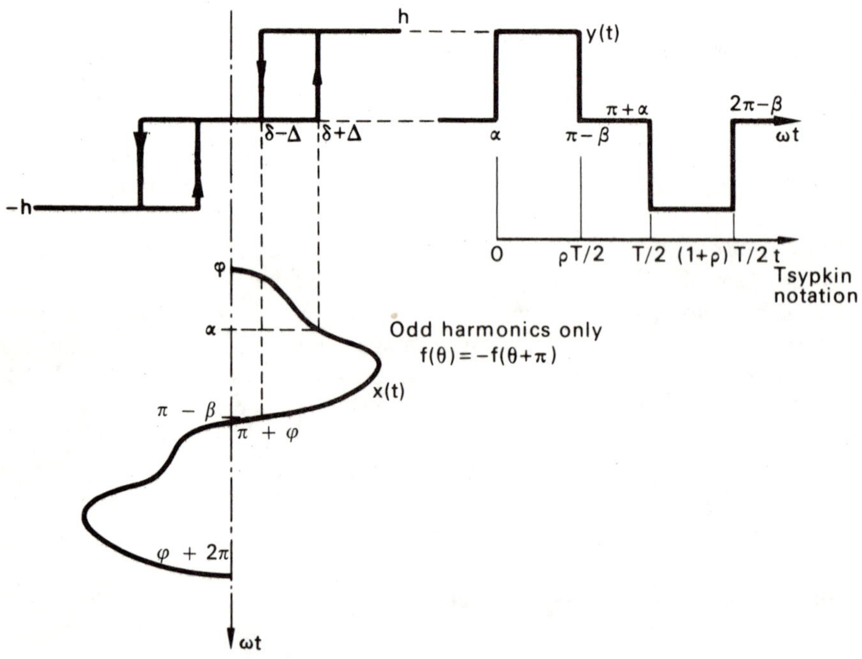

Figure 6.2 Typical relay input and output waveforms

where $\omega = 2\pi/T$ is the fundamental frequency of the waveform and T is the period. It will be appreciated that the assumed output waveform is that for the simplest form of periodic oscillation, although limit cycles with more complex switchings can exist for high order transfer functions $G(s)$. Also, asymmetrical limit cycles with four switchings per period are known, but occur very infrequently.[6] For multivariable relay systems and harmonically forced single loop systems complex periodic oscillations with more switchings per period are known to occur.[7] Although, in principle, the methods to be described can be extended to investigate such situations by assuming other periodic waveforms, larger numbers of nonlinear algebraic equations must be solved and the simplicity of the procedure is lost.

The relay output waveform shown can be expressed as a summation of step functions for a time domain analysis or by a Fourier series for a frequency domain approach. The series for the waveform of figure 6.2 is

$$y(t) = \sum_{n=1(2)}^{\infty} (4h/n\pi)\cos\{n(\alpha+\beta)/2\}\sin[n\omega t - \{n(\alpha-\beta)/2\}], \tag{6.1}$$

and the necessary switching conditions which must be satisfied are

$$\left.\begin{aligned} x(\alpha^-/\omega) &= \delta + \Delta \\ \dot{x}(\alpha^-/\omega) &> 0 \end{aligned}\right\}, \tag{6.2}$$

and

$$\left.\begin{aligned} x\{(\pi-\beta)^-/\omega\} &= \delta - \Delta \\ \dot{x}\{(\pi-\beta)^-/\omega\} &< 0 \end{aligned}\right\}, \tag{6.3}$$

where

$$\alpha^- = \lim_{\varepsilon\to 0}(\alpha - \varepsilon) \quad \text{with } \varepsilon > 0.$$

In addition $x(t)$ should further satisfy

$$x(t) < |\delta + \Delta| \quad \text{for } -\beta^-/\omega < t < \alpha^-/\omega,$$

$$x(t) > \delta - \Delta \quad \text{for } \alpha^-/\omega < t < (\pi-\beta)^-/\omega,$$

or $y(t)$ cannot have the assumed periodic form.

When the relay has a dead zone only, equations (6.2) and (6.3) still apply but with $\Delta = 0$.

For a relay with no dead zone, that is $\delta = 0$, switching is directly from the high to the low state so that $-\beta = \alpha$. The output Fourier series is then

$$y(t) = \sum_{n=1(2)}^{\infty} (4h/n\pi)\sin(n\omega t - n\alpha), \tag{6.4}$$

and equations (6.2) and (6.3) become identical to give the single switching condition

$$\left.\begin{aligned} x(\alpha^-/\omega) &= \Delta \\ \dot{x}(\alpha^-/\omega) &> 0 \end{aligned}\right\}, \tag{6.5}$$

with the condition that $x(t) > -\Delta$ for $\alpha^-/\omega < t < (\pi+\alpha)^-/\omega$.

The above switching equations have been written with an arbitrary time origin and therefore are particularly useful for the analysis of systems with multiple relays. In the original work by Tsypkin, where systems with only one relay were considered, a time scale with $\alpha = 0$ and the notation shown in figure

6.2 was used. Since $\beta = (1 - \rho)\pi$ this gives

$$y(t) = \sum_{n=1(2)}^{\infty} (4h/n\pi)\sin(n\rho\pi/2)\cos\{n\omega t - (n\rho\pi/2)\}, \tag{6.6}$$

and the switching conditions at $T/2$ and $\rho T/2$ are

$$\left.\begin{aligned} x(\pi^-/\omega) &= -(\Delta + \delta) \\ \dot{x}(\pi^-/\omega) &< 0, \end{aligned}\right\}, \tag{6.7}$$

$$\left.\begin{aligned} x(\rho\pi^-/\omega) &= \delta - \Delta \\ \dot{x}(\rho\pi^-/\omega) &< 0 \end{aligned}\right\}. \tag{6.8}$$

Again for a relay with no dead zone, $\delta = 0$, $\rho = 1$ and the single switching condition is

$$\left.\begin{aligned} x(\pi^-/\omega) &= -\Delta \\ \dot{x}(\pi^-/\omega) &< 0 \end{aligned}\right\}. \tag{6.9}$$

6.3 Time domain methods

Before proceeding with the analysis in the frequency domain it is appropriate to review briefly the time domain approaches of Hamel and Chung. This is done to show the relationship between the Hamel and Tsypkin loci and also to cover the time domain determination of closed form solutions for the steady state response of a network with a periodic input.

6.3.1 *Hamel's method*

For simplicity the relay is considered to have no dead zone although this is not a restriction on the basic method. It is further assumed that the transfer function $G(s)$ in figure 6.1 has distinct poles with none at the origin (i.e. no integration) so that it can be written

$$G(s) = \frac{P(s)}{Q(s)} = \sum_{i=1}^{n} \frac{A_i}{(s - p_i)},$$

where

$$A_i = P(p_i)/Q'(p_i),\ Q'(p_i) = \left.\frac{\mathrm{d}Q}{\mathrm{d}s}\right|_{s=p_i}.$$

Taking a time origin $t = 0$ at an instant at which the relay output $y(t)$, shown in figure 6.3, jumps from $-h$ to $+h$ and assuming that switching has been taking place for a long time prior to $t = 0$ so that the system has reached its steady state, then in the interval $0 < t < T/2$

$$y(t) = h\{u(t) + \sum_{k=1}^{\infty} (-1)^k[u\{t + (kT/2)\} - u\{t + [(k - 1)T/2]\}]\},$$

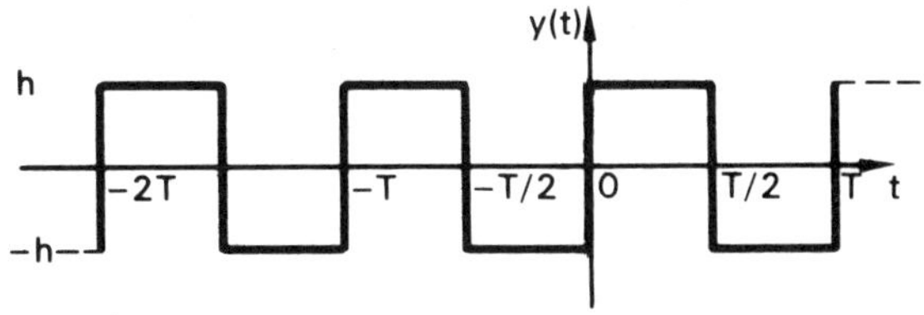

Figure 6.3 Relay output for Hamel's method

where $u(t)$ is the unit step function, and the response $c(t)$ is given by

$$c(t) = h\left\{q(t) + \sum_{k=1}^{\infty} (-1)^k[q\{t + (kT/2)\} - q\{t + [(k-1)T/2]\}]\right\},$$

where

$$q(t) = \mathscr{L}^{-1}\{G(s)/s\}, \text{ giving } q(t) = A_0 + \sum_{i=1}^{n} (A_i/p_i)e^{p_i t}.$$

Therefore

$$c(t) = h\left\{q(t) + \sum_{k=1}^{\infty} (-1)^k\left[\sum_{i=1}^{n} (A_i/p_i)e^{p_i\{t+(kT/2)\}}(1 - e^{-p_iT/2})\right]\right\}.$$

Reversing the order of the summations and summing the series in k gives

$$c(t) = h\left\{q(t) + \sum_{i=1}^{n} \frac{A_i e^{p_i t}(1 - e^{p_iT/2})}{p_i(1 + e^{p_iT/2})}\right\},$$

that is

$$c(t) = h\left\{q(t) - \sum_{i=1}^{n} (A_i/p_i)e^{p_i t}\tanh(p_iT/4)\right\}.$$

Thus at $t = 0^+$,

$$c(0^+) = h\left\{q(0^+) - \sum_{i=1}^{n} (A_i/p_i)\tanh(p_iT/4)\right\},$$

where $q(0^+) = \lim_{s\to\infty} G(s)$, the initial value of the step response of $G(s)$. Similarly

$$\dot{c}(0^+) = h\left\{g(0^+) - \sum_{i=1}^{n} A_i\tanh(p_iT/4)\right\},$$

where $g(0^+) = \lim_{s\to\infty} sG(s)$, the initial value of the impulse response of $G(s)$.

Now for a symmetrical periodic oscillation the values will be opposite at $t = 0^+$ and $T^+/2$, that is

$$c(T^+/2) = -c(0^+) \quad \text{and} \quad \dot{c}(T^+/2) = -\dot{c}(0^+).$$

Although switching takes place at $t = T^-/2$, we know that

$$c(T^+/2) - c(T^-/2) = -2h\lim_{s\to\infty} G(s) = -2hq(0^+),$$

and

$$\dot{c}(T^+/2) - \dot{c}(T^-/2) = -2h \lim_{s\to\infty} sG(s) = -2hg(0^+).$$

Therefore

$$x(T^-/2) = -c(T^-/2) = -2hq(0^+) + c(0^+),$$

and

$$\dot{x}(T^-/2) = -\dot{c}(T^-/2) = -2hg(0^+) + \dot{c}(0^+) \quad \text{for } r(t) = 0.$$

Hence

$$x(T^-/2) = h\left\{-q(0^+) - \sum_{i=1}^{n} (A_i/p_i)\tanh(p_iT/4)\right\}, \tag{6.10}$$

and

$$\dot{x}(T^-/2) = h\left\{-g(0^+) - \sum_{i=1}^{n} A_i \tanh(p_iT/4)\right\}. \tag{6.11}$$

Equations (6.10) and (6.11) with $h = 1$ define a Hamel locus graduated in T or ω, where $\omega = 2\pi/T$, in the $(x, \dot{x})$ plane. Thus using equation (6.9), since $\pi/\omega = T/2$, an intersection of the Hamel locus with the line $x = -\Delta$ corresponds to a possible periodic oscillation for the system if the intersection lies in the $\dot{x} < 0$ half plane.

If the transfer function $G(s)$ has pure integrations so that

$$G(s) = \frac{P(s)}{Q(s)} = \frac{p(s)}{s^nQ_n(s)},$$

where $Q_n(s)$ contains no zeros at the origin, the equations defining the Hamel locus with $h = 1$ become if order $P(s) <$ order $Q(s)$,

$$x(T/2) = -TP(0)/4Q_1(0) - \sum_{i=1}^{n-1} (A_i/p_i)\tanh(p_iT/4), \tag{6.12}$$

$$\dot{x}(T^-/2) = -g(0^+) - \sum_{i=1}^{n-1} A_i \tanh(p_iT/4), \tag{6.13}$$

where

$$A_i = P(p_i)/p_iQ'_1(p_i) \quad \text{for } n = 1,$$

and

$$x(T^-/2) = \frac{-T}{4}\lim_{s\to 0}\frac{\mathrm{d}}{\mathrm{d}s}\left\{\frac{P(s)}{Q_2(s)}\right\} - \sum_{i=1}^{n-2} (A_i/p_i)\tanh(p_iT/4), \tag{6.14}$$

$$\dot{x}(T^-/2) = -g(0^+) - \frac{TP(0)}{4Q_2(0)} - \sum_{i=1}^{n-2} A_i \tanh(p_iT/4), \tag{6.15}$$

where

$$A_i = P(p_i)/p_i^2Q'_2(p_i) \quad \text{for } n = 2.$$

When the transfer function $G(s)$ contains non-distinct poles the Hamel locus can be obtained in a similar manner, and if it contains a pure time delay either the input $y(t)$ to a plant without the delay may be expressed in a delayed form or the switching condition may be suitably modified. Also if a proportional plus derivative controller of transfer function $1 + \lambda s$ exists before the relay, the solution for the periodic oscillation can be found from the intersection of the Hamel locus and the line $x + \lambda\dot{x} = -\Delta$. The same solution is of course obtained from the intersection of the Hamel locus for $(1 + \lambda s)G(s)$ and the line $x = -\Delta$.

Hamel loci for various low order transfer functions, including some with complex poles, are plotted in reference 8. As a simple example consider

$$G(s) = \frac{1}{(s+1)(1+sa)} = \frac{1}{1-a}\left\{\frac{1}{s+1} - \frac{1}{s+(1/a)}\right\}.$$

From equations (6.10) and (6.11), with $h = 1$,

$$\begin{aligned} x(T^-/2) &= -\{1/(1-a)\}\{-\tanh(-T/4) + a\tanh(-T/4a)\} \\ &= -\{1/(1-a)\}\{\tanh(\pi/2\omega) - a\tanh(\pi/2a\omega)\}, \end{aligned} \tag{6.16}$$

and

$$\dot{x}(T^-/2) = \{1/(1-a)\}\{\tanh(\pi/2\omega) - \tanh(\pi/2a\omega)\}. \tag{6.17}$$

For the particular case of $a = 0$, a situation where $\dot{x}$ has a discontinuity at the switching instant, i.e. $\dot{x}(T^+/2) \neq \dot{x}(T^-/2)$

$$x(T^-/2) = -\tanh(\pi/2\omega), \tag{6.18}$$

$$\dot{x}(T^-/2) = -1 + \tanh(\pi/2\omega). \tag{6.19}$$

The loci are shown plotted for various values of a in figure 6.4.

6.3.2 *Chung's method*

In this approach the time domain expressions for the waveforms are obtained from the solution of the state equation. Thus if the state equation for the plant transfer function $G(s)$ is

$$\dot{\mathbf{z}}(t) = \mathbf{A}\mathbf{z}(t) + \mathbf{B}u(t),$$

where

$$\mathbf{z}(t) = \{z_1(t), z_2(t) \cdots z_n(t)\}^T,$$

then its solution is

$$\mathbf{z}(t) = \mathbf{\Phi}(t - t_0)\mathbf{z}(t_0) + \int_{t_0}^{t} \mathbf{\Phi}(t - \tau)\mathbf{B}u(\tau)\,d\tau, \tag{6.20}$$

where $\mathbf{\Phi}(t) = \exp \mathbf{A}t$ is the state transition matrix. As the required solution for $\mathbf{z}(t)$ is periodic, writing $\mathbf{z}(t) = \mathbf{z}(t + T)$ gives

$$\begin{aligned} &\mathbf{\Phi}(t - t_0)\mathbf{z}(t_0) + \int_{t_0}^{t} \mathbf{\Phi}(t - \tau)\mathbf{B}u(\tau)\,d\tau \\ &\qquad = \mathbf{\Phi}(t + T - t_0)\mathbf{z}(t_0) + \int_{t_0}^{t+T} \mathbf{\Phi}(t + T - \tau)\mathbf{B}u(\tau)\,d\tau, \end{aligned}$$

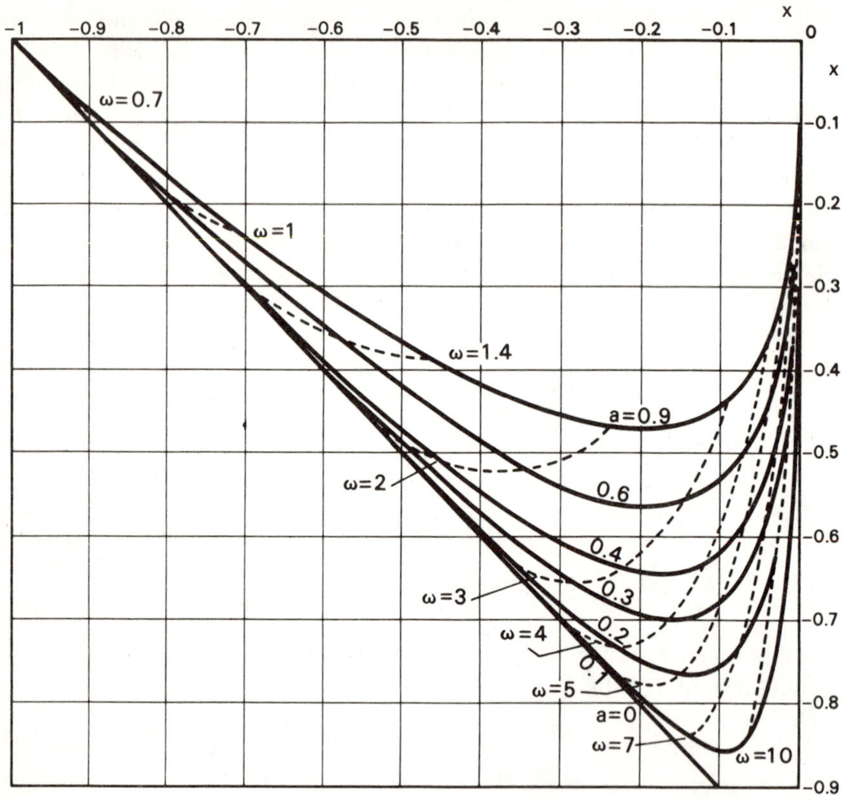

Figure 6.4 Hamel loci for $G(s) = 1/(s+1)(1+sa)$

which reduces to

$$\{\mathbf{\Phi}(-T) - \mathbf{I}\}\mathbf{z}(t_0) = \mathbf{\Phi}(t_0 - T)\left\{\int_{t_0}^{t+T} \mathbf{\Phi}(T-\tau)\mathbf{B}u(\tau)\,\mathrm{d}\tau - \int_{t_0}^{t} \mathbf{\Phi}(-\tau)\mathbf{B}u(\tau)\,\mathrm{d}\tau\right\}.$$

Changing the variable of integration in the second integral to $v = \tau + T$ and using $u(v) = u(v - T)$ gives

$$\{\mathbf{\Phi}(-T) - \mathbf{I}\}\mathbf{z}(t_0) = \mathbf{\Phi}(t_0 - T)\int_{t_0}^{t_0+T} \mathbf{\Phi}(T-\tau)\mathbf{B}u(\tau)\,\mathrm{d}\tau,$$

that is

$$\{\mathbf{\Phi}(-T) - \mathbf{I}\}\mathbf{z}(t_0) = \int_{t_0}^{t_0+T} \mathbf{\Phi}(t_0-\tau)\mathbf{B}u(\tau)\,\mathrm{d}\tau. \tag{6.21}$$

This equation gives the initial value of the state vector $\mathbf{z}(t_0)$ which can be used in equation (6.20) to find the vector $\mathbf{z}(t)$ at all instants of time. Unless multiple relays are involved in a system it is appropriate to choose $t_0 = 0$. When the relay input is expressed as a function of the state vector $\mathbf{z}(t)$, writing the necessary switching conditions gives a set of transcendental equations which can be solved for the required parameters. Solutions are usually most easily obtained by factoring the plant transfer function so that it consists of parallel combinations of simple transfer functions.

Again consider

$$G(s) = \frac{1}{(s+1)(as+1)} = \frac{1}{1-a}\left\{\frac{1}{s+1} - \frac{1}{s+(1/a)}\right\}.$$

Now for a transfer function $K/(s+\alpha)$, equation (6.21) with $t_0 = 0$ gives

$$(e^{\alpha T} - 1)z_{10} = \int_0^T e^{\alpha\tau} Ku(\tau)\,d\tau,$$

where $z_{10} = z_1(0)$, so that

$$z_{10} = \frac{K}{\alpha}\left\{\frac{1 + e^{\alpha T} - 2e^{\alpha t_j}}{1 - e^{\alpha T}}\right\}$$

for the square wave $u(t+T) = u(t)$,

$$u(t) = 1, \qquad 0 < t < t_j$$
$$\quad\;\; -1, \qquad t_j < t < T.$$

When $t_j = T/2$,

$$z_{10} = -z_1(T^-/2) = \frac{K(1 - e^{\alpha T/2})}{\alpha(1 + e^{\alpha T/2})} = -(K/\alpha)\tanh(\alpha T/4).$$

Using this value of z_{10} in the state equation solution gives

$$z_1(t) = z_{10}e^{-\alpha t} + (K/\alpha)(1 - e^{-\alpha t}), \qquad 0 < t < t_j,$$
$$z_1(t) = z_{10}e^{-\alpha t} - (K/\alpha)\{1 + e^{-\alpha t} - 2e^{\alpha(t_j - t)}\}, \qquad t_j < t < T.$$

Using the superposition of the initial values of z_{10} for the above $G(s)$ in the configuration of figure 6.1

$$x(T^-/2) = \{1/(1-a)\}\{-\tanh(\pi/2\omega) + a\tanh(\pi/2a\omega)\},$$

in agreement with equation (6.16). Equating $x(T^-/2)$ to $-\Delta$ is sufficient to give a solution, the value of $\dot{x}(T^-/2)$ which can be found by differentiation simply provides an incomplete validity test.

A short list of the time domain equations for some simple transfer functions is given in Table 6.1; additional results can be found in reference 3.

6.4 Tsypkin's method

Relays without dead zone

From equation (6.6) with $\rho = 1$,

$$y(t) = \sum_{n=1(2)}^{\infty} (4h/n\pi) \sin n\omega t. \tag{6.22}$$

If $\lim_{s\to\infty} G(s) \neq 0$, we can write

$$G(s) = G_1(s) + G(\infty), \tag{6.23}$$

where $\lim_{s\to\infty} G_1(s) = 0$. Then if $G_1(jn\omega) = g_{1n}e^{j\phi_{1n}}$,

$$c(t) = \sum_{n=1(2)}^{\infty} (4h/n\pi)g_{1n} \sin(n\omega t + \phi_{1n}) + G(\infty)y(t), \tag{6.24}$$

and

$$\dot{c}(t) = \sum_{n=1(2)}^{\infty} (4h\omega/\pi)g_{1n} \cos(n\omega t + \phi_{1n}) + G(\infty)\dot{y}(t). \tag{6.25}$$

At $t = \pi^-/\omega$, $y(t) = h$ and $\dot{y}(t) = 0$ since the impulse corresponding to the square wave derivative takes place at $t = \pi^+/\omega$. A discontinuity will occur in $\dot{c}(t)$ at the switching instants if $\lim_{s\to\infty} sG_1(s) \neq 0$. The magnitude of the discontinuity is

$$\dot{c}(\pi^+/\omega) - \dot{c}(\pi^-/\omega) = -2h \lim_{s\to\infty} sG_1(s).$$

Since a Fourier series of a piecewise continuous function converges to the midpoint of each finite discontinuity the value of $\dot{c}(t)$ at $t = \pi^-/\omega$ will be greater than that given by equation (6.25) by a factor $h \lim_{s\to\infty} sG_1(s)$.
Therefore

$$c(\pi^-/\omega) = \sum_{n=1(2)}^{\infty} (4h/n\pi)g_{1n} \sin(n\pi + \phi_{1n}) + hG(\infty),$$

and

$$\dot{c}(\pi^-/\omega) = \sum_{n=1(2)}^{\infty} (4h\omega/\pi)g_{1n} \cos(n\pi + \phi_{1n}) + h \lim_{s\to\infty} sG_1(s),$$

which may be written

$$c(\pi^-/\omega) = -(4h/\pi) \sum_{n=1(2)}^{\infty} (1/n) \operatorname{Im}\{G_1(j\omega n)\} + hG(\infty), \tag{6.26}$$

and

$$\dot{c}(\pi^-/\omega) = -(4h\omega/\pi) \sum_{n=1(2)}^{\infty} \operatorname{Re}\{G_1(j\omega n)\} + h \lim_{s\to\infty} sG_1(s), \tag{6.27}$$

where

$$\begin{aligned} G_1(jn\omega) &= g_{1n} \cos \phi_{1n} + jg_{1n} \sin \phi_{1n} = U_{G_1}(n\omega) + jV_{G_1}(n\omega) \\ &= \operatorname{Re}\{G_1(jn\omega)\} + j \operatorname{Im}\{G_1(jn\omega)\}. \end{aligned} \tag{6.28}$$

Tsypkin defined a locus $\Lambda(\omega)$ by

$$\Lambda(\omega) = (-1/\omega)\dot{c}(\pi^-/\omega) - jc(\pi^-/\omega), \tag{6.29}$$

so that

$$\text{Re}\{\Lambda(\omega)\} = (4h/\pi) \sum_{n=1(2)}^{\infty} U_{G_1}(n\omega) - (h/\omega) \lim_{s\to\infty} sG_1(s),$$

and

$$\text{Im}\{\Lambda(\omega)\} = (4h/\pi) \sum_{n=1(2)}^{\infty} (1/n)V_{G_1}(n\omega) - hG(\infty).$$

The coordinates of the Tsypkin locus by definition are $\{-\dot{c}(\pi^-/\omega)/\omega, -c(\pi^-/\omega)\}$ so that it is identical to the Hamel locus with axes interchanged, apart from the scale factor $1/\omega$ on the real axis of the Tsypkin locus.

Thus using the switching conditions of equation (6.9) the solution for the limit cycle is given by

$$\text{Re}\{\Lambda(\omega)\} < 0,$$

$$\text{Im}\{\Lambda(\omega)\} = -\Delta.$$

When $\lim_{s\to\infty} sG(s) = 0$, $\text{Re}\{\Lambda(\omega)\}$ and $\text{Im}\{\Lambda(\omega)\}$ are seen to be summed frequency loci. Defining the general summed frequency locus $A(\omega)$ by

$$\text{Re}\{A_G(\theta, \omega)\} = \sum_{n=1(2)}^{\infty} V_G(n\omega)\sin n\theta + U_G(n\omega)\cos n\theta, \tag{6.30}$$

and

$$\text{Im}\{A_G(\theta, \omega)\} = \sum_{n=1(2)}^{\infty} (1/n)\{V_G(n\omega)\cos n\theta - U_G(n\omega)\sin n\theta\}, \tag{6.31}$$

the solution for the limit cycle becomes

$$\text{Re}\{A_{G_1}(0, \omega)\} < (\pi/4\omega) \lim_{s\to\infty} sG_1(s), \tag{6.32}$$

$$\text{Im}\{A_{G_1}(0, \omega)\} = (\pi/4)\{G(\infty) - (\Delta/h)\}. \tag{6.33}$$

Equations (6.32) and (6.33) can be solved for a particular case by finding closed form expressions for the summations,

$$\text{Re}\{A_{G_1}(0, \omega)\} = \sum_{n=1(2)}^{\infty} U_{G_1}(n\omega),$$

and

$$\text{Im}\{A_{G_1}(0, \omega)\} = \sum_{n=1(2)}^{\infty} (1/n)V_{G_1}(n\omega),$$

or by summing a reasonable number of terms in the series for various magnitudes of ω to find the value which gives the solution. The latter procedure can be done numerically or by graphical construction from the $G_1(j\omega)$ locus as shown in figure 6.5. The similarity of this procedure to the describing function approach is evident especially when it is remembered that the negative reciprocal, $-1/N(a)$, of the relay describing function is a line distant $-(\pi\Delta/4h)$ from the negative real axis and parallel to it.

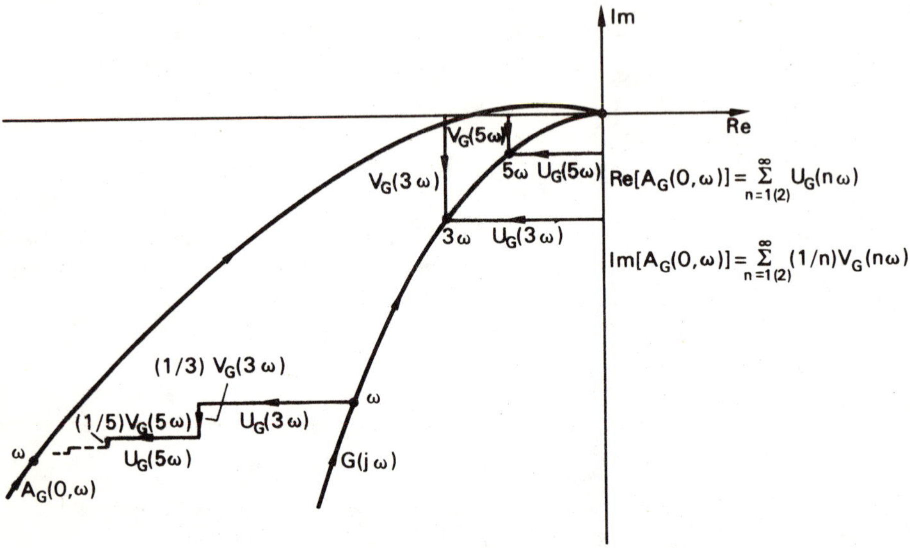

Figure 6.5 Construction of $A_G(0, \omega)$ from $G(j\omega)$

Example

Find the frequency of the limit cycle in the system of figure 6.1 if $G(s) = K/(s + \alpha)$ and the relay has no dead zone. In this case $G(\infty) = 0$ and

$$G(j\omega) = \frac{K}{\alpha + j\omega} = \frac{K(\alpha - j\omega)}{\alpha^2 + \omega^2}.$$

Therefore

$$U_G(n\omega) = \frac{(K/\alpha)}{1 + n^2\lambda^2} \quad \text{and} \quad \frac{V_G(n\omega)}{n} = \frac{-(K\lambda/\alpha)}{1 + n^2\lambda^2},$$

where $\lambda = \omega/\alpha$. Equations (6.30) to (6.33) give

$$(K/\alpha) \sum_{n=1(2)}^{\infty} \frac{1}{1 + n^2\lambda^2} < (\pi K/4\omega),$$

and

$$(K\lambda/\alpha) \sum_{n=1(2)}^{\infty} \frac{1}{1 + n^2\lambda^2} = (\pi\Delta/4h).$$

Now[9]

$$\sum_{n=1(2)}^{\infty} \frac{1}{1 + n^2\lambda^2} = (\pi/4\lambda)\tanh(\pi/2\lambda),$$

so that the above conditions become

$$\tanh(\pi\alpha/2\omega) < 1; \qquad \tanh(\pi\alpha/2\omega) = \alpha\Delta/Kh.$$

Thus the frequency of oscillation is given by

$$\omega = \pi\alpha/2\tanh^{-1}(\alpha\Delta/Kh),$$

and is valid provided $(Kh/\alpha\Delta) > 1$. If $(Kh/\alpha\Delta) < 1$, no oscillation takes place and the relay remains latched in one position.

The same solution is of course obtained by Hamel's method, equations (6.9), (6.10) and (6.11) yield directly the relevant closed form expression

$$x(T^-/2) = -(Kh/\alpha)\tan(\pi\alpha/2\omega) = -\Delta,$$

$$\dot{x}(T^-/2) = -Kh + Kh\tan(\pi\alpha/2\omega) < 0.$$

Relay with dead zone

In this case from equation (6.6)

$$y(t) = \sum_{n=1(2)}^{\infty} (4h/n\pi)\sin(n\rho\pi/2)\cos\{n\omega t - (n\rho\pi/2)\} \tag{6.6}$$

and the switching conditions are given by equations (6.7) and (6.8). Proceeding as in the previous case

$$c(t) = \sum_{n=1(2)}^{\infty} (4h/n\pi)g_{1n}\sin(n\rho\pi/2)\cos\{n\omega t - (n\rho\pi/2) + \phi_{1n}\} + G(\infty)y(t),$$

and

$$\dot{c}(t) = \sum_{n=1(2)}^{\infty} -(4h\omega/\pi)g_{1n}\sin(n\rho\pi/2)\sin\{n\omega t - (n\rho\pi/2) + \phi_{1n}\} + G(\infty)\dot{y}(t).$$

Evaluating $c(t)$ and $\dot{c}(t)$ at the switching instants and simplifying the equation gives

$$c(\pi^-/\omega) = -(2h/\pi)\sum_{n=1(2)}^{\infty}(1/n)\{\sin n\rho\pi U_{G_1}(n\omega) + (1-\cos n\rho\pi)V_{G_1}(n\omega)\},$$

$$\dot{c}(\pi^-/\omega) = -(2h\omega/\pi)\sum_{n=1(2)}^{\infty}\{(1-\cos n\rho\pi)U_{G_1}(n\omega) - \sin n\rho\pi V_{G_1}(n\omega)\}$$
$$+ (h/2)\lim_{s\to\infty} sG_1(s),$$

$$c(\rho\pi^-/\omega) = -(2h/\pi)\sum_{n=1(2)}^{\infty}(1/n)\{-\sin n\rho\pi U_{G_1}(n\omega) + (1-\cos n\rho\pi)V_{G_1}(n\omega)\}$$
$$+ hG(\infty),$$

$$\dot{c}(\rho\pi^-/\omega) = -(2h\omega/\pi)\sum_{n=1(2)}^{\infty}\{(1-\cos n\rho\pi)U_{G_1}(n\omega) + \sin n\rho\pi V_{G_1}(n\omega)\}$$
$$+ (h/2)\lim_{s\to\infty} sG_1(s).$$

Using the switching conditions of equations (6.7) and (6.8) we obtain

$$\mathrm{Im}\{A_{G_1}(0,\omega) - A_{G_1}(\rho\pi,\omega)\} = -(\delta + \Delta)\pi/2h,$$
$$\mathrm{Re}\{A_{G_1}(0,\omega) - A_{G_1}(\rho\pi,\omega)\} < (\pi/4\omega)\lim_{s\to\infty} sG_1(s),$$
$$\mathrm{Im}\{A_{G_1}(0,\omega) - A_{G_1}(-\rho\pi,\omega)\} = \{(\delta - \Delta)\pi/2h + (\pi/2)G(\infty),$$
$$\mathrm{Re}\{A_{G_1}(0,\omega) - A_{G_1}(-\rho\pi,\omega)\} < (\pi/4\omega)\lim_{s\to\infty} sG_1(s),$$

or in two equations the conditions for a limit cycle are that the loci

$$A_{G_1}(0,\omega) - A_{G_1}(\rho\pi,\omega) \text{ must have R.P.} < (\pi/4\omega)\lim_{s\to\infty} sG_1(s), \quad \text{I.P.} = -(\delta+\Delta)\pi/2h, \tag{6.34}$$

$$A_{G_1}(0,\omega) - A_{G_1}(-\rho\pi,\omega) \text{ must have R.P.} < (\pi/4\omega)\lim_{s\to\infty} sG_1(s), \quad \text{I.P.} = \{(\delta-\Delta)\pi/2h\} + (\pi/2)G(\infty), \tag{6.35}$$

where R.P. and I.P. denote the real part and imaginary part respectively. For $\lim_{s\to\infty} sG_1(s) = 0$, Tsypkin denoted the loci of equations (6.34) and (6.35) by $\pi\Lambda(\omega)/2h$ and $\pi\Lambda'(\omega)/2h$. To solve the equations graphically the loci $\Lambda(\omega)$ and $\Lambda'(\omega)$ can be plotted, with ω a parameter on the loci, for various values of ρ. Equations (6.34) and (6.35) then yield possible solution values for ρ and ω; which may be plotted in a $\rho - \omega$ plane to give the values of ρ and ω; which simultaneously satisfy both equations. Alternatively closed form expressions, to be discussed in the next section, may be found for $A_G(\theta, \omega)$ and the resulting nonlinear equations solved for ρ and ω.

6.5 The *A* loci

The loci which are functions of an angle θ and the frequency ω are defined by equations (6.30) and (6.31), that is

$$\mathrm{Re}\{A_G(\theta,\omega)\} = \sum_{n=1(2)}^{\infty} V_G(n\omega)\sin n\theta + U_G(n\omega)\cos n\theta, \tag{6.30}$$

and

$$\mathrm{Im}\{A_G(\theta,\omega)\} = \sum_{n=1(2)}^{\infty} (1/n)\{V_G(n\omega)\cos n\theta - U_G(n\omega)\sin n\theta\}, \tag{6.31}$$

where

$$A_G(\theta,\omega) = \mathrm{Re}\{A_G(\theta,\omega)\} + j\,\mathrm{Im}\{A_G(\theta,\omega)\}.$$

Properties

Using the above expressions the following properties are easily shown:

(a) the loci are periodic in θ with period 2π, that is

$$A_G(\theta + 2\pi, \omega) = A_G(\theta, \omega); \tag{6.36}$$

(b) the periodicity is odd, that is

$$A_G(\pi - \theta, \omega) = -A_G(-\theta, \omega); \tag{6.37}$$

(c) if $F(s) = G(s)e^{-sT_d}$

$$A_F(\theta, \omega) = A_G(\theta + \omega T_d, \omega). \tag{6.38}$$

For a transfer function with poles at $s = -\omega/\lambda$ the series for $\mathrm{Re}\, A_G(\theta, \omega)$ and $\mathrm{Im}\, A_G(\theta, \omega)$ involve sine and cosine series of the form

$$S_{j,k}(\theta, \lambda) = \sum_{n=1(2)}^{\infty} \frac{n^j \sin n\theta}{(1 + n^2\lambda^2)^k}, \tag{6.39}$$

$$C_{j,k}(\theta, \lambda) = \sum_{n=1(2)}^{\infty} \frac{n^j \cos n\theta}{(1 + n^2\lambda^2)^k}, \tag{6.40}$$

where j and k are integers, the latter being the multiplicity of the pole. Analytical expressions for the summations of these series for various values of j and k can be determined[10] and a list is given in table 6.2. Also included are the partial

TABLE 6.2

Summation of sine and cosine series odd terms only

(a) Sine Series

$$S_{j,k} = S_{j,k}(\theta, \lambda) = \sum_{n=1(2)}^{\infty} \frac{n^j \sin n\theta}{(1 + n^2\lambda^2)^k} \quad \text{for } 0 < \theta < \pi$$

General Relationships

$$S_{j,k}(-\theta, \lambda) = -S_{j,k}(\theta, \lambda)$$
$$S_{j,k} - S_{j,k+1} = \lambda^2 S_{j+2,k+1}$$
$$\frac{\partial}{\partial \theta} S_{j,k} = C_{j+1,k}$$
$$\frac{\partial}{\partial \lambda} S_{j,k} = 2k\lambda S_{j+2,k+1}$$

Series	Sum
$S_{-3,0}$	$(\pi^2\theta - \pi\theta^2)/8$
$S_{-1,0}$	$\pi/4$
$S_{1,1}$	$\{\pi \cosh[(\pi - 2\theta)/2\lambda]\}/4\lambda^2 \cosh[\pi/2\lambda]$
$S_{-1,1}$	$S_{-1,0} - \lambda^2 S_{1,1}$
$S_{1,2}$	$\{\pi\theta \sinh[(\pi - \theta)/2\lambda] + (\pi^2 \sinh[\theta/\lambda])/2 \cosh[\pi/2\lambda]\}/8\lambda^3 \cosh[\pi/2\lambda]$
$S_{-1,2}$	$S_{-1,1} - \lambda^2 S_{1,2}$
$S_{3,2}$	$\lambda^{-2}(S_{1,1} - S_{1,2})$
$S_{1,3}$	$\{\pi^2\lambda \sinh[\theta/\lambda] + 2\pi\theta\lambda \cosh[\pi/2\lambda] \sinh[(\pi - 2\theta)/2\lambda] - 2\pi^2\theta \cosh[\theta/\lambda] + 2\pi\theta^2 \cosh[\pi/2\lambda]$ $\times \cosh[(\pi - 2\theta)/2\lambda] + \pi^3 \sinh[\theta/\lambda] \tanh[\pi/2\lambda)\}/64\lambda^4 \cosh^2[\pi/2\lambda]$
$S_{-1,3}$	$S_{-1,2} - \lambda^2 S_{1,3}$
$S_{3,3}$	$\lambda^{-2}(S_{1,2} - S_{1,3})$

TABLE 6.2—*continued*

(b) Cosine Series

$$C_{j,k} = C_{j,k}(\theta, \lambda) = \sum_{n=1(2)}^{\infty} \frac{n^j \cos n\theta}{(1 + n^2\lambda^2)^k} \quad \text{for } 0 < \theta < \pi$$

General Relationships

$$C_{j,k}(-\theta, \lambda) = C_{j,k}(\theta, \lambda)$$
$$C_{j,k} - C_{j,k+1} = \lambda^2 C_{j+2,k+1}$$
$$\frac{\partial}{\partial\theta} C_{j,k} = -S_{j+1,k}$$
$$\frac{\partial}{\partial\lambda} C_{j,k} = -2k\lambda C_{j+2,k+1}$$

Series	Sum
$C_{-4,0}$	$(\pi^4 - 6\pi^2\theta^2 + 4\pi\theta^3)/96$
$C_{-2,0}$	$(\pi^2 - 2\pi\theta)/8$
$C_{0,1}$	$\{\pi \sinh[(\pi - 2\theta)/2\lambda]\}/4\lambda \cosh[\pi/2\lambda]$
$C_{-2,1}$	$C_{-2,0} - \lambda^2 C_{0,1}$
$C_{2,2}$	$\{\pi\lambda \sinh[(\pi - 2\theta)/2\lambda] - \pi\theta \cosh[(\pi - 2\theta)/2\lambda] + (\pi^2 \cosh[\theta/\lambda])/2 \cosh[\pi/2\lambda]\}/8\lambda^4 \cosh[\pi/2\lambda]$
$C_{0,2}$	$C_{0,1} - \lambda^2 C_{2,2}$
$C_{2,3}$	$\{2\pi\lambda^2 \sinh[(\pi - 2\theta)/2\lambda] \cosh[\pi/2\lambda] - \pi^2\lambda \cosh[\theta/\lambda] + 2\pi\theta\lambda \cosh[(\pi - 2\theta)/2\lambda] \cosh[\pi/2\lambda]$ $- 2\pi^2\vartheta \sinh[\theta/\lambda] - 2\pi\theta^2 \sinh[(\pi - 2\theta)/2\lambda] \cosh[\pi/2\lambda] + \pi^3 \cosh[\theta/\lambda] \tanh[\pi/2\lambda]\}$ $/64\lambda^5 \cosh^2[\pi/2\lambda]$
$C_{0,3}$	$C_{0,2} - \lambda^2 C_{2,3}$

derivatives of the series with respect to θ and λ which are required if, say, a Newton–Raphson[11] numerical solution is used to solve the resulting nonlinear algebraic equations. Table 6.3 shows how the A loci for various real pole transfer

TABLE 6.3

A loci for various transfer functions

Transfer function	Re(A)	lim $\lambda \to 0$	Im(A)	lim $\lambda \to 0$ $0 < \theta < \pi/2$
$1/sT$	$-\lambda^{-1}S_{-1,0}$	$-\infty$	$-\lambda^{-1}C_{-2,0}$	$-\infty$
$1/(1 + sT)$	$C_{0,1} - \lambda S_{1,1}$	0	$-S_{-1,1} - \lambda C_{0,1}$	$-\pi/4$
$1/sT(1 + sT)$	$-C_{0,1} - \lambda^{-1}S_{-1,1}$	$-\infty$	$S_{-1,1} - \lambda^{-1}C_{-2,1}$	$-\infty$
$1/s^2T^2$	$-\lambda^{-2}C_{-2,0}$	$-\infty$	$\lambda^{-2}S_{-3,0}$	$-\infty$
$1/(1 + sT)^2$	$-2\lambda S_{1,2} + C_{0,2} - \lambda^2 C_{2,2}$	0	$-2\lambda C_{0,2} - S_{-1,2} + \lambda^2 S_{1,2}$	$-\pi/4$
$sT/(1 + sT)^2$	$2\lambda^2 C_{2,2} + \lambda S_{1,2} - \lambda^3 S_{3,2}$	0	$\lambda C_{0,2} - \lambda^3 C_{2,2} - 2\lambda^2 S_{1,2}$	0
$1/s^3T^3$	$\lambda^{-3}S_{-3,0}$	$-\infty$	$\lambda^{-3}C_{-4,0}$	$-\infty$
$1/(1 + sT)^3$	$C_{0,3} - 3\lambda^2 C_{2,3} - 3\lambda S_{1,3} + \lambda^3 S_{3,3}$	0	$-3\lambda C_{0,3} + \lambda^3 C_{2,3} - S_{-1,3} + 3\lambda^2 S_{1,3}$	$-\pi/4$

where $\lambda = \omega T$

functions are obtained from $S_{j,k}$ and $C_{j,k}$. A loci for transfer functions with complex pairs of poles can be found by comparing time domain and A loci expressions for $c(\pi^-/\omega)$ and $\dot{c}(\pi^-/\omega)$ for a relay with dead zone.

Example

Determine the limit cycle in the system of figure 6.1 if the relay has no hysteresis, $\Delta = 0$, and $G(s) = (K/s)e^{-sT_d}$. (See problem P.2.14.)

From equations (6.34), (6.35) and (6.38) the solution for the limit cycle is given by

$$\{A_G(\omega T_d, \omega) - A_G(\rho\pi + \omega T_d, \omega)\} \text{ has R.P.} < (\pi K/4\omega),$$

$$\text{I.P.} = -\delta\pi/2h,$$

$$\{A_G(\omega T_d, \omega) - A_G(-\rho\pi + \omega T_d, \omega)\} \text{ has R.P.} < (\pi K/4\omega),$$

$$\text{I.P.} = \delta\pi/2h.$$

From table 6.3 for $G(s) = K/s$, $\lambda = \omega/K$ and

$$\text{Re}\{A_G(\theta, \omega)\} = -(K/\omega)S_{-1,0} = -(K\pi/4\omega), \quad \text{for } 0 < \theta < \pi,$$

and

$$\text{Im}\{A_G(\theta, \omega)\} = -(K/\omega)C_{-2,0} = -(K\pi/8\omega)(\pi - 2\theta), \quad \text{for } 0 < \theta < \pi.$$

Since the above expressions are only valid for $0 < \theta < \pi$ some assumptions must be made regarding the angles ωT_d, $\rho\pi + \omega T_d$ and $-\rho\pi + \omega T_d$. We assume therefore that

$$0 < \omega T_d < \pi,$$

$$\pi < \rho\pi + \omega T_d < 2\pi,$$

$$0 < \rho\pi - \omega T_d < \pi.$$

Substituting for $A_G(\theta, \omega)$ the conditions become

$$-(K\pi/2\omega) < (\pi K/4\omega),$$

$$-(K\pi/8\omega)\{\pi - 2\omega T_d + \pi - 2(\rho\pi + \omega T_d - \pi)\} = -\delta\pi/2h,$$

$$-(K\pi/2\omega) < (\pi K/4\omega),$$

$$-(K\pi/8\omega)\{\pi - 2\omega T_d + \pi - 2(\pi - \rho\pi + \omega T_d)\} = \delta\pi/2h.$$

The inequalities are true and the two other equations reduce to

$$\pi - \omega T_d - \rho\pi/2 = \delta\omega/Kh,$$

and

$$\omega T_d - \rho\pi/2 = \delta\omega/Kh,$$

with the solution $\omega = \pi/2T_d$, $\rho = (hKT_d - \delta)/hKT_d$. Returning to the original assumptions, the first is satisfied since $\omega T_d = \pi/2$, the second is satisfied if $1/2 < \rho < 3/2$ and the third if $1/2 < \rho < 1$. Since $\rho < 1$ from physical considerations the solution is valid if $\rho > 1/2$, which gives $K > 2\delta/hT_d$ as the condition for a limit cycle.

The output waveform being the integral of that shown in figure 6.2 delayed by T_d is a symmetrically clipped triangular wave of amplitude $(hKT_d - \delta)$.

6.6 Systems with multiple relays

It is a relatively simple procedure to extend the above method to find periodic modes in systems with more than one relay characteristic assuming that all the relay outputs have the same fundamental frequency and are expressible in the form of equation (6.1). Consider for example the output of relay k passing through a transfer function $-G_{jk}(s)$ to the input, $x_j(t)$, of relay j, then

$$y_k(t) = \sum_{n=1(2)}^{\infty} (4h_k/n\pi)\cos\{n(\alpha_k + \beta_k)/2\}\sin\{n\omega t - n(\alpha_k - \beta_k)/2\}, \tag{6.1}$$

and

$$x_j(t) = -\sum_{n=1(2)}^{\infty} (4h_k/n\pi)g_{jkn}\cos\{n(\alpha_k + \beta_k)/2\}\sin\{n\omega t - n(\alpha_k - \beta_k)/2 + \phi_{jkn}\}, \tag{6.41}$$

assuming for simplicity that $\lim_{s\to\infty} sG_{jk}(s) = 0$. If the relay j switches at $t = \alpha_j^-/\omega$ and $(\pi - \beta_j)^-/\omega$ then

$$x_j(\alpha_j^-/\omega) = \delta_j + \Delta_j,$$
$$\dot{x}_j(\alpha_j^-/\omega) > 0,$$
$$x_j\{(\pi - \beta)^-/\omega\} = \delta_j - \Delta_j,$$
$$\dot{x}_j\{(\pi - \beta)^-/\omega\} < 0.$$

Evaluating $x_j(t)$ and $\dot{x}_j(t)$ at these switching instants from equation (6.41) as done previously, and using the above equations we obtain the conditions

$$\{A_{G_{jk}}(\alpha_k - \alpha_j, \omega) + A_{G_{jk}}(-\beta_k - \alpha_j, \omega)\} \text{ has R.P.} < 0,$$
$$\text{I.P.} = -\pi(\delta_j + \Delta_j)/2h_k,$$

and

$$\{A_{G_{jk}}(\alpha_k + \beta_j, \omega) + A_{G_{jk}}(-\beta_k + \beta_j, \omega)\} \text{ has R.P.} < 0,$$
$$\text{I.P.} = \pi(\delta_j - \Delta_j)/2h_k.$$

Two possible configurations of relay and linear elements are shown in figure 6.6. To derive the conditions for a limit cycle it is only necessary to sum expressions of the above form for all the inputs to each relay. These equations for the systems

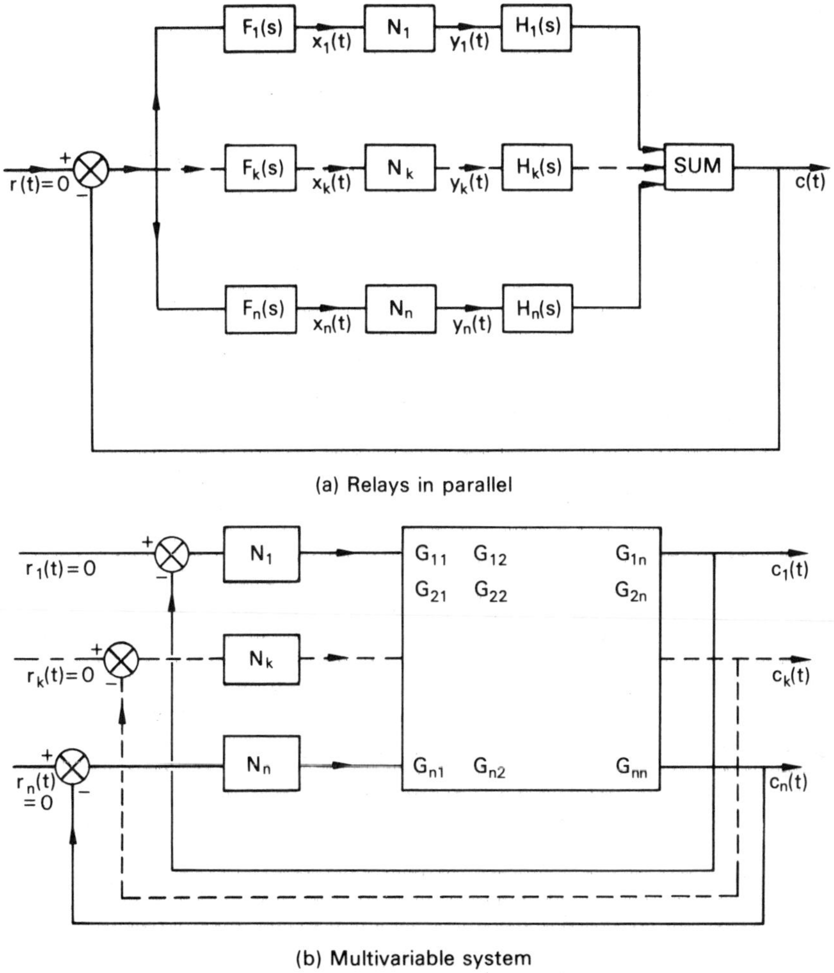

Figure 6.6 Systems with multiple relays

of figure 6.6 are identical if $G_{jk}(s) = F_j(s)H_k(s)$, and are for all relays $j = 1$ to n

$$\sum_{k=1}^{n} h_k\{A_{G_{jk}}(\alpha_k - \alpha_j, \omega) + A_{G_{jk}}(-\beta_k - \alpha_j, \omega)\}, \quad \text{R.P.} < 0,$$

$$\text{I.P.} = -\pi(\delta_j + \Delta_j)/2,$$

$$\sum_{k=1}^{n} h_k\{A_{G_{jk}}(\alpha_k + \beta_j, \omega) + A_{G_{jk}}(-\beta_k + \beta_j, \omega)\}, \quad \text{R.P.} < 0,$$

$$\text{I.P.} = \pi(\delta_j - \Delta_j)/2.$$

Assuming one relay switching instant as the time reference the number of equations is equal to the number of remaining unknown angles plus one, so that ω can be obtained.

Example

Determine the region of stability for the relay position control system with Coulomb friction shown in figure 6.7. Denoting the Coulomb friction as relay

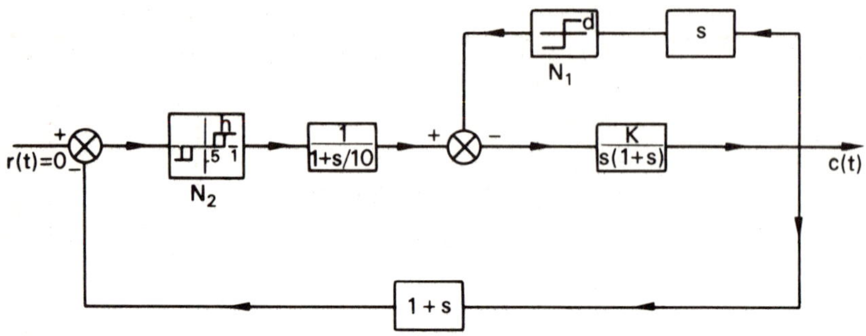

Figure 6.7 Block diagram of relay position control

characteristic 1, then using the above notation it is seen that

$$G_{11} = K/(1+s), \qquad G_{12} = -K/(1+s)\{1+(s/10)\},$$

$$G_{21} = -K/s, \qquad G_{22} = K/s\{1+(s/10)\}.$$

There are three equations which must be satisfied by any limit cycle, one for the Coulomb friction characteristic and two for the on-off relay with dead zone. The equations are

$$2dA_{G_{11}}(0,\omega) + h\{A_{G_{12}}(\alpha_2,\omega) + A_{G_{12}}(-\beta_2,\omega)\}, \quad \text{R.P.} < 0, \quad \text{I.P.} = 0,$$

$$2dA_{G_{21}}(-\alpha_2,\omega) + h\{A_{G_{22}}(0,\omega) + A_{G_{22}}(-\alpha_2-\beta_2,\omega)\}, \quad \text{R.P.} < 0, \quad \text{I.P.} = -\pi/2,$$

$$2dA_{G_{21}}(\beta_2,\omega) + h\{A_{G_{22}}(0,\omega) + A_{G_{22}}(\alpha_2+\beta_2,\omega)\}, \quad \text{R.P.} < 0, \quad \text{I.P.} = \pi/4,$$

since α_1 ($= -\beta_1$ for the Coulomb friction) has been chosen equal to zero. To find the stability boundary numerical solutions were sought for the above three nonlinear equations using a Newton–Raphson method on a digital computer. When no solutions exists to the above equations the system, assuming no other mode exists, is stable. Figure 6.8 shows the stability boundary obtained together with the results of an analog simulation.

6.7 Asymmetrical oscillations

Consider an asymmetrical oscillation to exist in the system shown in the block diagram of figure 6.1. The general form of the relay output waveform is shown in figure 6.9, where the notation for the angles α and β is different from

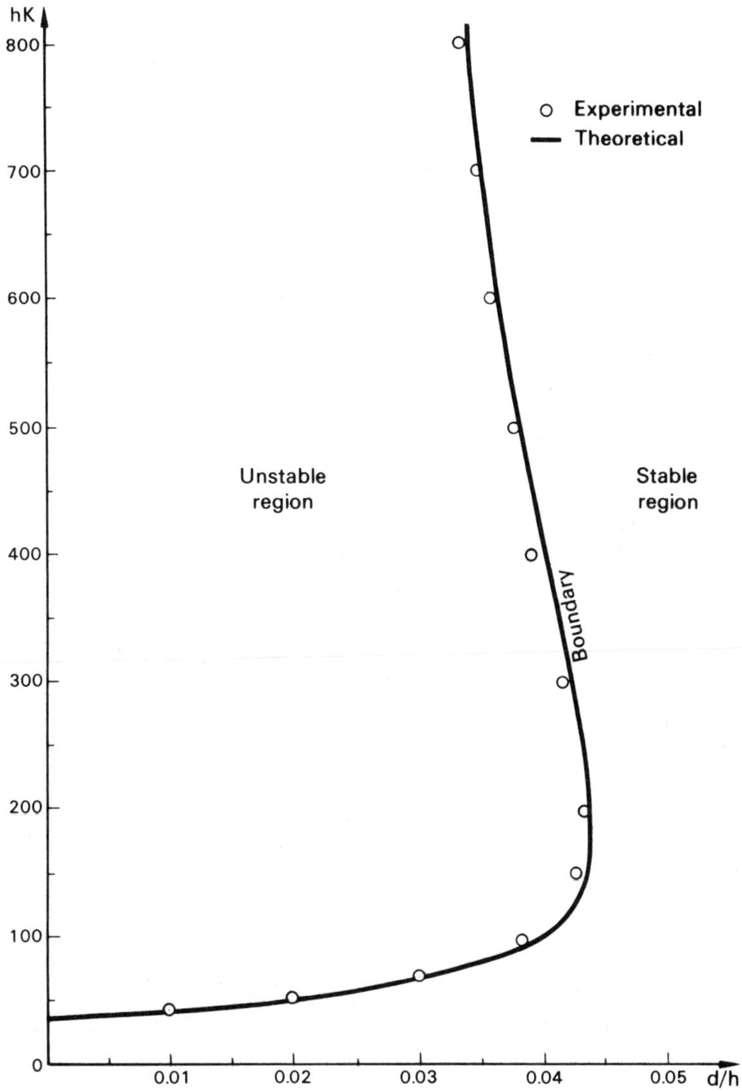

Figure 6.8 Stability boundary for system of figure 6.7

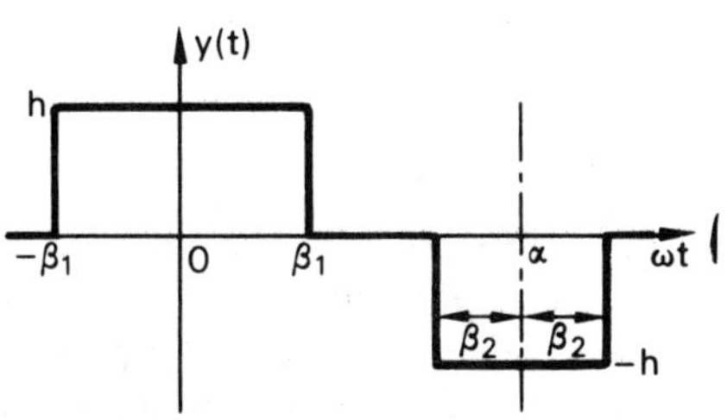

Figure 6.9 General asymmetrical relay output waveform

that used previously. The Fourier series for this signal is

$$y(t) = h(\beta_1 - \beta_2)/\pi + \sum_{n=1}^{\infty} a_n \cos n\omega t + b_n \sin n\omega t,$$

where

$$a_n = (2h/n\pi)(\sin n\beta_1 - \cos n\alpha \sin n\beta_2),$$

$$b_n = (2h/n\pi)(\sin n\alpha \sin n\beta_2),$$

and all terms are presented in the summation. The input to the relay assuming $\lim_{s\to\infty} sG(s) = 0$ is given by

$$x(t) = \gamma - \sum_{n=1}^{\infty} a_n g_n \cos(n\omega t + \phi_n) + b_n g_n \sin(n\omega t + \phi_n),$$

where

$$\gamma = r - \{hG(0)(\beta_1 - \beta_2)/\pi\}.$$

The values which $x(t)$ must satisfy at the switching instants, twice the number for the symmetrical case, yield the equations

$$\delta + \Delta = \gamma - \sum_{n=1}^{\infty} a_n g_n \cos(-n\beta_1 + \phi_n) + b_n g_n \sin(-n\beta_1 + \phi_n), \tag{6.42}$$

$$\delta - \Delta = \gamma - \sum_{n=1}^{\infty} a_n g_n \cos(n\beta_1 + \phi_n) + b_n g_n \sin(n\beta_1 + \phi_n), \tag{6.43}$$

$$-(\delta + \Delta) = \gamma - \sum_{n=1}^{\infty} a_n g_n \cos(n\alpha - n\beta_2 + \phi_n) + b_n g_n \sin(n\alpha - n\beta_2 + \phi_n), \tag{6.44}$$

$$-(\delta - \Delta) = \gamma - \sum_{n=1}^{\infty} a_n g_n \cos(n\alpha + n\beta_2 + \phi_n) + b_n g_n \sin(n\alpha + n\beta_2 + \phi_n). \tag{6.45}$$

If the values for a_n and b_n are substituted in the above expressions the summations can be expressed in terms of $S'_{j,k}$ and $C'_{j,k}$ where the angle θ can involve α, β_1 and β_2 and the prime denotes that the series contain terms for both odd and even values of n. Similarly, closed form expressions can be found for $\dot{x}(t)$ at the switching instants. Any limit cycles can thus be found, although somewhat laboriously.

For the relay with no dead zone $\delta = 0$, $\alpha = \pi$, $\beta_1 = \pi - \beta_2$ and only equations (6.42) and (6.43) are required to obtain a solution. Symmetrical oscillations can be found from the above formulation by setting $\alpha = \pi$, $\beta_1 = \beta_2$ and using equations (6.42) and (6.43) only, whilst if the relay has no dead zone $\alpha = \pi$, $\beta_1 = \beta_2 = \pi/2$ and equation (6.42) alone is required for a solution.

6.7.1 *Comparison with the DF method*

It has already been seen in section 6.4 for a symmetrical oscillation in a system with a relay having no dead zone that the solution, given by equations (6.32) and (6.33) when $\lim_{s\to\infty} sG_1(s) = 0$, corresponds to the intersection of $A_G(0, \omega)$ with the negative reciprocal of the DF, $C(a)$. Since $A_G(0, \omega)$ is a summed locus determined from $G(j\omega)$ the error in the DF solution is easily seen.

To show[12] that other solutions can be obtained by intersections with the relay $C(a)$ locus, consider those situations for which $\alpha = \pi$, $b_n = 0$ which covers all cases bar the asymmetrical oscillation of the three state relay. Adding and subtracting equations (6.42) and (6.43), which are the only ones required apart from the validity conditions on the derivatives, gives

$$\sum_{n=1}^{\infty} a_n U_G(n\omega) \cos n\beta_1 = \gamma - \delta, \tag{6.46}$$

$$\sum_{n=1}^{\infty} a_n V_G(n\omega) \sin n\beta_1 = -\Delta. \tag{6.47}$$

Now the same relay output can be produced by the bias plus fundamental input $\gamma + a \cos \omega t + b \sin \omega t$, in which case the switching conditions will be

$$\delta + \Delta = \gamma + a \cos \beta_1 - b \sin \beta_1,$$

$$\delta - \Delta = \gamma + a \cos \beta_1 + b \sin \beta_1,$$

giving

$$a = -(\gamma - \delta)/\cos \beta_1, \qquad b = -\Delta/\sin \beta_1.$$

Since $b_1 = 0$ for the conditions considered, from the definition of the DF, one obtains

$$C(a) = -(a/a_1) + j(b/a_1),$$

that is

$$C(a) = \{(\gamma - \delta)/a_1 \cos \beta_1\} - j\{\Delta/a_1 \sin \beta_1\}. \tag{6.48}$$

Thus from equations (6.46), (6.47) and (6.48) it can be seen that if the locus

$$\begin{aligned} G_m(\omega) &= \sum_{n=1}^{\infty} \frac{a_n \cos n\beta_1 U_G(n\omega)}{a_1 \cos \beta_1} + j \sum_{n=1}^{\infty} \frac{a_n \sin n\beta_1 V_G(n\omega)}{a_1 \sin \beta_1} \\ &= \mathrm{Re}\{G_m(\omega)\} + j\,\mathrm{Im}\{G_m(\omega)\}, \end{aligned} \tag{6.49}$$

is plotted on a polar plane together with $C(a)$ the stability of the system can be assessed. In particular any intersection of the two loci will give the amplitude and frequency of a possible limit cycle. It should be noted however that $G_m(\omega)$ is a function of ω and β_1 and $C(a)$ of a and β_1 so that the intersection must be for the same value of β_1 on the two loci.

Evaluating $\dot{x}(t)$ at $t = -\beta^-/\omega$ and β^-/ω in the usual manner gives the validity conditions

$$\sum_{n=1}^{\infty} V_G(n\omega) \sin 2n\beta_1 - U_G(n\omega)(1 - \cos 2n\beta_1) > 0, \tag{6.50}$$

$$\sum_{n=1}^{\infty} V_G(n\omega) \sin 2n\beta_1 + U_G(n\omega)(1 - \cos 2n\beta_1) < 0. \tag{6.51}$$

The method can be extended,[12] by assuming $\alpha \approx \pi$, to obtain approximate solutions for oscillations with relatively small asymmetry in a system with a three state relay. It may also be used to predict more accurately[12] oscillations in a system with hard saturation by assuming the saturation output waveform to be that obtained with a sinusoidal input and then taking all the harmonics into consideration.

To obtain closed form expressions for the left hand sides of equations (6.46) and (6.47) or for $G_m(\omega)$ again requires summation of series of the form

$$S'_{j,k}(\theta, \lambda) = \sum_{n=1}^{\infty} \frac{n^j \sin n\theta}{(1 + n^2\lambda^2)^k}, \tag{6.52}$$

and

$$C'_{j,k}(\theta, \lambda) = \sum_{n=1}^{\infty} \frac{n^j \cos n\theta}{(1 + n^2\lambda^2)^k}. \tag{6.53}$$

Table 6.4 provides a list of these summations.

TABLE 6.4
Summation of sine and cosine series all terms

(a) Sine series

$$S'_{j,k} = S'_{j,k}(\theta, \lambda) = \sum_{n=1}^{\infty} \frac{n^j \sin n\theta}{(1 + n^2\lambda^2)^k}, \quad 0 < \theta < 2\pi$$

General Relationships

$$\sum_{n=1}^{\infty} \frac{(-1)^n n^j \sin n\theta}{(1 + n^2\lambda^2)^k} = -\sum_{n=1}^{\infty} \frac{n^j \sin n(\pi - \theta)}{(1 + n^2\lambda^2)^k}, \quad -\pi < \theta < \pi$$

All relationships of table 6.2

Series	Sum
$S'_{-3,0}$	$\{(\theta - \pi)^3 - \pi^2\theta + \pi^3\}/12$
$S'_{-1,0}$	$(\pi - \theta)/2$
$S'_{1,1}$	$[\pi \sinh\{(\pi - \theta)/\lambda\}]/2\lambda^2 \sinh(\pi/\lambda)$
$S'_{-1,1}$	$S'_{-1,0} - \lambda^2 S'_{1,1}$
$S'_{1,2}$	$\pi[\theta \cosh\{(\pi - \theta)/\lambda\} \sinh(\pi/\lambda) - \pi \sinh(\theta/\lambda)]/4\lambda^3 \sinh^2(\pi/\lambda)$
$S'_{-1,2}$	$S'_{-1,1} - \lambda^2 S'_{1,2}$
$S'_{3,2}$	$\lambda^{-2}(S'_{1,1} - S'_{1,2})$

(b) Cosine series

$$C'_{j,k} = C'_{j,k}(\theta, \lambda) = \sum_{n=1}^{\infty} \frac{n^j \cos n\theta}{(1 + n^2\lambda^2)^k}, \quad 0 \le \theta < 2\pi$$

General Relationships

$$\sum_{n=1}^{\infty} \frac{(-1)^n n^j \cos n\theta}{(1 + n^2\lambda^2)^k} = \sum_{n=1}^{\infty} \frac{n^j \cos n(\pi - \theta)}{(1 + n^2\lambda^2)^k}, \quad -\pi \le \theta < \pi$$

All relationships of table 6.2

Series	Sum
$C'_{-4,0}$	$\{2\pi^2(\theta - \pi)^2 - (\theta - \pi)^4\}/48 - (7\pi^4/720)$
$C'_{-2,0}$	$(\theta - \pi)^2/4 - (\pi^2/12)$
$C'_{0,1}$	$\{\pi[\cosh\{(\pi - \theta)/\lambda\}]/2\lambda \sinh(\pi/\lambda)\} - (1/2)$
$C'_{-2,1}$	$C'_{-2,0} - \lambda^2 C'_{0,1}$
$C'_{2,2}$	$\pi[\sinh(\pi/\lambda)[\lambda \cosh\{(\pi - \theta)/\lambda\} - \theta \sinh\{(\pi - \theta)/\lambda\}] - \pi \cosh(\theta/\lambda)]/4\lambda^4 \sinh^2(\pi/\lambda)$
$C'_{0,2}$	$C_{0,1} - \lambda^2 C_{2,2}$

Example

The system discussed in section 5.5.3, the block diagram for which is given in figure 5.12, is reconsidered.

The bias output from the relay is

$$(1/2\pi)\{2\beta_1 h - h(2\pi - 2\beta_1)\} = h(2\beta_1 - \pi)/\pi = U,$$

giving

$$\beta_1 = (U + h)\pi/2h. \tag{6.54}$$

The open loop transfer function $G(s)$ is given by

$$G(s) = \frac{K}{s(s+1)(s+2)} = K\left\{\frac{1/2}{s} + \frac{1/2}{s+2} - \frac{1}{s+1}\right\},$$

so that

$$U_G(n\omega) = K\left\{\frac{1/4}{1 + n^2(\omega/2)^2} - \frac{1}{1 + n^2\omega^2}\right\},$$

and

$$V_G(n\omega) = K\left\{\frac{-1/2}{n\omega} - \frac{(1/8)n\omega}{1 + n^2(\omega/2)^2} + \frac{n\omega}{1 + n^2\omega^2}\right\}.$$

Also for the relay $\alpha = \pi$, $\beta_2 = \pi - \beta_1$, therefore

$$a_n = \frac{2h}{n\pi}\{\sin n\beta_1 - \cos n\pi \sin n(\pi - \beta_1)\} = \frac{4h}{n\pi}\sin n\beta_1.$$

Substituting for a_n and $V_G(n\omega)$ in equation (6.47) and writing β for β_1 gives

$$\frac{2hK}{\pi}\sum_{n=1}^{\infty}\left[\frac{1}{n^2\omega} + \frac{\omega}{4\{(1 + n^2(\omega/2)^2\}} - \frac{2\omega}{(1 + n^2\omega^2)}\right]\sin^2 n\beta = 0,$$

that is

$$(1/2\omega)\{C'_{-2,0}(0,\omega) - C'_{-2,0}(2\beta,\omega)\} + (\omega/8)\{C'_{0,1}(0,\omega/2) - C'_{0,1}(2\beta,\omega/2)\}$$
$$- \omega\{C'_{0,1}(0,\omega) - C'_{0,1}(2\beta,\omega)\} = 0.$$

Substituting for the series from table 6.4 yields

$$\frac{1}{2\omega}\left\{\frac{\pi^2}{4} - \frac{(2\beta - \pi)^2}{4}\right\} + \frac{\pi}{8}\left\{\frac{\cosh(2\pi/\omega) - \cosh\{(2\pi - 4\beta)/\omega\}}{\sinh(2\pi/\omega)}\right\}$$
$$- \frac{\pi}{2}\left\{\frac{\cosh(\pi/\omega) - \cosh\{(\pi - 2\beta)/\omega\}}{\sinh(\pi/\omega)}\right\} = 0. \tag{6.55}$$

Substituting for a_n and $U_G(n\omega)$ in equation (6.46) gives

$$\frac{4hK}{\pi}\sum_{n=1}^{\infty}\left[\frac{(1/8)\sin 2n\beta}{n\{1 + n^2(\omega/2)^2\}} - \frac{(1/2)\sin 2n\beta}{n(1 + n^2\omega^2)}\right] = \gamma,$$

that is

$$(hK/2\pi)\{S'_{-1,1}(2\beta, \omega/2) - 4S'_{-1,1}(2\beta, \omega)\} = \gamma,$$

which on substituting for the series reduces to

$$\frac{hK}{2\pi}\left[3(\beta - \pi/2) + \frac{2\pi \sinh\{(\pi - 2\beta)/\omega\}}{\sinh(\pi/\omega)} - \frac{\pi \sinh\{(2\pi - 4\beta)/\omega\}}{2\sinh(2\pi/\omega)}\right] = \gamma. \quad (6.56)$$

Thus for a given U/h equations (6.54), (6.55) and (6.56) can be solved for β_1, ω and γ/hK respectively. If solutions are obtained for various values of U/h the exact modified nonlinearity in the presence of the oscillation, a graph of U versus γ, can be plotted for various values of K.

For the particular case of $K = 30$, $U/h = 1/3$ the equations give $\beta = 2\pi/3$, $\omega = 1{\cdot}315$ and $\gamma = 2{\cdot}90$. The corresponding values obtained by the DF method are $\beta = 2\pi/3$, $\omega = 1{\cdot}414$ and $\gamma = 2{\cdot}75$, results which are in error by an amount less than the distortion factor d. If the solution values are substituted in the validity conditions of equations (6.50) and (6.51) they will be found to be satisfied.

6.8 Stability of a limit cycle

By applying a small perturbation to the limit cycle and considering the stability of the perturbed system using sampled data theory, Tsypkin[8] has shown that any symmetrical limit cycle predicted for the system of figure 6.1, having a relay with no dead zone, will be stable provided

$$\frac{\mathrm{d}\,\mathrm{Im}\{\Lambda(\omega_0)\}}{\mathrm{d}\omega_0} > 0,$$

where ω_0 is the predicted limit cycle frequency. Figure 6.10 illustrates the situation, where the limit cycles at A and C may be stable and that at B is unstable. Again the similarity with the DF approach is apparent.

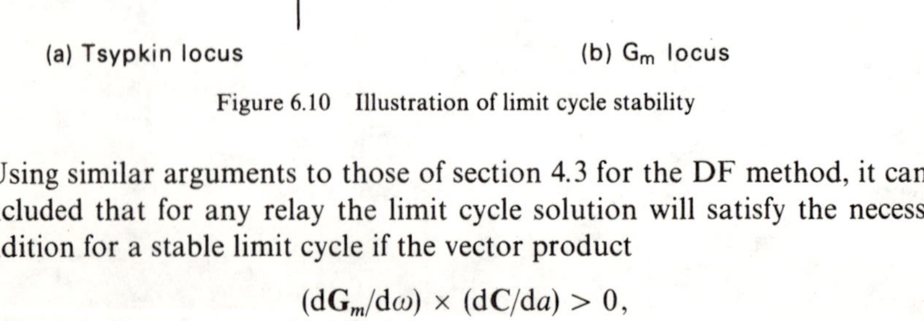

Figure 6.10 Illustration of limit cycle stability

Using similar arguments to those of section 4.3 for the DF method, it can be concluded that for any relay the limit cycle solution will satisfy the necessary condition for a stable limit cycle if the vector product

$$(\mathrm{d}\mathbf{G}_m/\mathrm{d}\omega) \times (\mathrm{d}\mathbf{C}/\mathrm{d}a) > 0,$$

where $\mathbf{C}(a)$ is the negative reciprocal of the DF and $\mathbf{G}_m(\omega)$ the modified frequency locus given by equation (6.49).

As the above condition can be interpreted as an exact equivalent of the Loeb criterion for a relay system there are situations where it will be satisfied, since it is only a necessary condition, and the predicted limit cycle will be unstable. This problem is discussed from a DF point of view in section 5.11.1 and again similar arguments apply. To check the stability of the limit cycle the exact IDF (see section 6.9.3) characteristic equation must also be examined.

6.9 Forced response of relay systems

Consider the system of figure 6.1 to have an input $r(t)$, not necessarily sinusoidal, of period $2\pi/\omega_f$. For a relay with no dead zone, when synchronisation takes place the relay output may be assumed to have the waveshape of figure 6.2 with period $2\pi/\omega_f$ and $\alpha = -\beta$. At the switching instant $t = \pi^-/\omega_f$,

$$x(\pi^-/\omega_f) = r(\pi^-/\omega_f) - c(\pi^-/\omega_f),$$

$$\dot{x}(\pi^-/\omega_f) = \dot{r}(\pi^-/\omega_f) - \dot{c}(\pi^-/\omega_f),$$

where $c(\pi^-/\omega_f)$ and $\dot{c}(\pi^-/\omega_f)$ are given by equations (6.26) and (6.27) respectively. Thus

$$\begin{aligned}(1/\omega_f)\dot{x}(\pi^-/\omega_f) + jx(\pi^-/\omega_f) &= (1/\omega_f)\dot{r}(\pi^-/\omega_f) + jr(\pi^-/\omega_f)\\ &\quad - (1/\omega_f)\dot{c}(\pi^-/\omega_f) - jc(\pi^-/\omega_f)\\ &= R(\pi + \phi) + \Lambda(\omega_f),\end{aligned}$$

where

$$R(\pi + \phi) = (1/\omega_f)\dot{r}(\pi^-/\omega_f) + jr(\pi^-/\omega_f),$$

and $\Lambda(\omega)$ is the Tsypkin locus defined by equation (6.29). $R(\pi + \phi)$ is a closed contour in the complex plane, for example, if $r(t) = R\sin(\omega_f t + \phi)$

$$jr(\pi^-/\omega_f) = jR\sin(\pi + \phi) = -jR\sin\phi,$$

$$(1/\omega_f)\dot{r}(\pi^-/\omega_f) = R\cos(\pi + \phi) = -R\cos\phi,$$

giving

$$R(\pi + \phi) = -R(\cos\phi + j\sin\phi) = -Re^{j\phi} \text{ a circle.}$$

Since $x(\pi^-/\omega_f)$ and $\dot{x}(\pi^-/\omega_f)$ must still satisfy the switching conditions of equation (6.9), the solution for a synchronised forced oscillation is given by the intersection of the line $-j\Delta$ and the locus $R(\pi + \phi) + \Lambda(\omega_f)$. A simple graphical procedure for finding the solution is shown in figure 6.11. It is seen from the figure that two solutions at points A and B corresponding to different phase angles ϕ are indicated. In section 6.9.2 it is shown that the solution at A is stable and that at B unstable, although both solutions satisfy the validity condition $\mathrm{Re}\{R(\pi + \phi) + \Lambda(\omega_f)\} < 0$. The minimum input amplitude $R_{\min}$ required to produce a synchronised forced oscillation at the assumed frequency

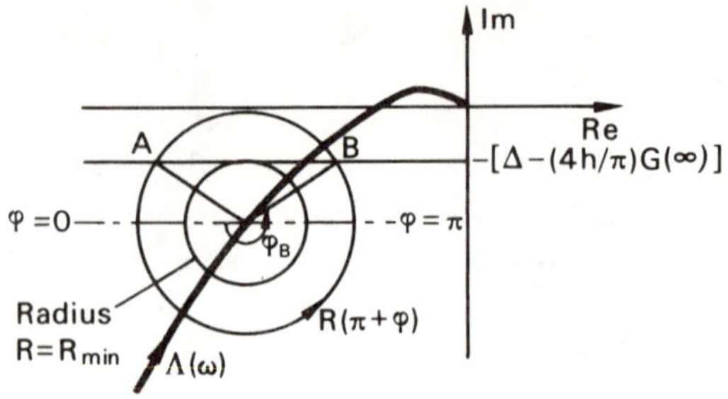

Figure 6.11 Forced oscillation solution for relay with no dead zone

is also shown in the figure. If the frequency locus $A_G(\theta, \omega)$ is used then the condition for a forced synchronised oscillation is given by the intersection of the locus

$$\{A_{G_1}(0, \omega_f) + (\pi/4h)R(\pi + \phi)\} \quad \text{with the line} \quad (j\pi/4)\{G(\infty) - (\Delta/h)\},$$

provided the real value of the solution point is less than $(\pi/4\omega) \lim_{s\to\infty} sG_1(s)$.

Using exactly the same procedure as above for a relay with dead zone, the two conditions which must be satisfied follow from equations (6.34) and (6.35) and are

$$\{A_{G_1}(0, \omega_f) - A_{G_1}(\rho\pi, \omega_f)\} + (\pi/2h)R(\pi + \phi) \text{ has}$$

$$\text{I.P.} = -(\delta + \Delta)\pi/2h, \quad \text{R.P.} < (\pi/4\omega) \lim_{s\to\infty} sG_1(s), \tag{6.57}$$

$$\{A_{G_1}(0, \omega_f) - A_{G_1}(-\rho\pi, \omega_f)\} + (\pi/2h)R(\rho\pi + \phi) \text{ has}$$

$$\text{I.P.} = \{(\delta - \Delta)/2h\} + (\pi/2)G(\infty), \quad \text{R.P.} < (\pi/4\omega) \lim_{s\to\infty} sG_1(s). \tag{6.58}$$

$R(\rho\pi + \phi)$ is a locus with the same shape as $R(\pi + \phi)$, the only difference being the location of the $\phi = 0$ point as indicated in figure 6.12 which illustrates a graphical solution procedure. For each value of ρ chosen, solutions to equations (6.57) and (6.58) separately may exist for two values of ϕ as shown in the figure. From graphs in the $\rho - \phi$ plane of those solutions the values of ρ and ϕ, normally two, which satisfy both equations simultaneously can be found.

An alternative graphical procedure which can be used to find synchronised forced oscillations in a system containing an on-off relay with dead zone is given in reference 13. Using Tsypkin's notation for the relay output waveform of figure 6.2 but with $(\rho T/2)$ denoted by t_1, the output $c(t)$ may be found as discussed in section 6.3. $c(t)$ will be a function of t, t_1 and T where for a given input frequency T is known. Showing the dependence on t_1 by denoting $c(t)$ by $c(t, t_1)$ the switching conditions at $t = 0$ and t_1 give

$$R \sin \phi - c(0, t_1) = \delta + \Delta, \tag{6.59}$$

$$R \sin(\omega t_1 + \phi) - c(t_1, t_1) = \delta - \Delta, \tag{6.60}$$

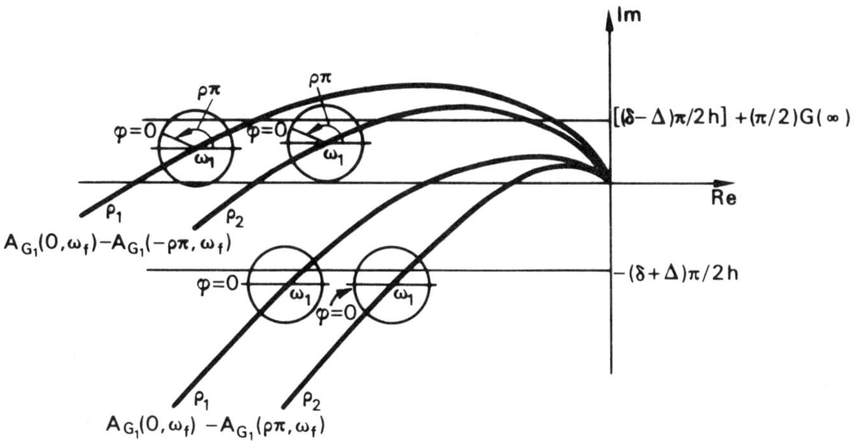

Figure 6.12 Forced oscillation solution for relay with dead zone

where $r(t) = R\sin(\omega t + \phi)$. Similar equations can of course be written in the frequency domain involving an angle rather than t_1. The parameter plane $X = R\cos\phi$, $Y = R\sin\phi$ is used by the authors(13) to solve these equations, which give

$$X = \frac{c(t_1, t_1) + (\delta - \Delta) + \{c(0, t_1) + \delta + \Delta\}\cos\omega t_1}{\sin\omega t_1} = X_G + X_R,$$

$$Y = \delta + \Delta + c(0, t_1) = Y_G + Y_R,$$

where

$$X_G = \frac{c(t_1, t_1) + c(0, t_1)\cos\omega t_1}{\sin\omega t_1}, \qquad X_R = \frac{(\delta - \Delta) + (\delta + \Delta)\cos\omega t_1}{\sin\omega t_1},$$

$$Y_G = c(0, t_1), \qquad Y_R = (\delta + \Delta).$$

The two loci, one for the linear transfer function, Γ_G with coordinates (X_G, Y_G) and the other for the relay Γ_R with coordinates (X_R, Y_R) can be plotted in the parameter plane. t_1, or if a frequency approach is used an angle, will be a parameter on the loci. The intersection of the sum $\Gamma = \Gamma_G + \Gamma_R$ with the circle of radius R gives the required values of t_1 and ϕ.

It is also possible to use the DF formulation of section 6.7.1 to study forced oscillations.(7) Equations (6.46) and (6.47) for this case become

$$-\delta = \sum_{n=1}^{\infty} a_n U_G(n\omega)\cos n\beta_1 - (1/2)\{R(\beta_1 + \phi) - R(-\beta_1 + \phi)\}, \qquad (6.61)$$

$$-\Delta = \sum_{n=1}^{\infty} a_n V_G(n\omega)\sin n\beta_1 + (1/2)\{R(\beta_1 + \phi) - R(-\beta_1 + \phi)\}, \qquad (6.62)$$

which after manipulation show that the solution for a synchronised forced oscillation is given by

$$C(a) = G_m(\omega) - R\{K_1(\beta_1)\sin\phi - jK_2(\beta_1)\cos\phi\}, \tag{6.63}$$

with $r(t) = R\sin(\omega t + \phi)$, and $K_1(\beta_1) = K_2(\beta_1) = \pi/4h\sin\beta_1$. Again equation (6.63) can be solved graphically.

Obviously the above procedures can be extended to cover asymmetrical forced oscillations in relay systems although the number of equations which have to be solved doubles. Relatively simple graphical solutions, similar to those described above, can however be used for a relay with no dead zone.

6.9.1 *Subharmonic oscillations*

To investigate the possibility of subharmonic oscillations the relay output is assumed to maintain the usual waveform of figure 6.2 with period $2\pi/\omega_1$, when the forcing frequency $\omega_f = m\omega_1$. Typical waveforms for the third subharmonic in the system of figure 6.1 containing a relay with no dead zone are shown in figure 6.13. The required loci to obtain a solution are easily shown to be:

$$\{A_{G_1}(0, \omega_1) + (\pi/4h)R'(m\pi + \phi)\}$$

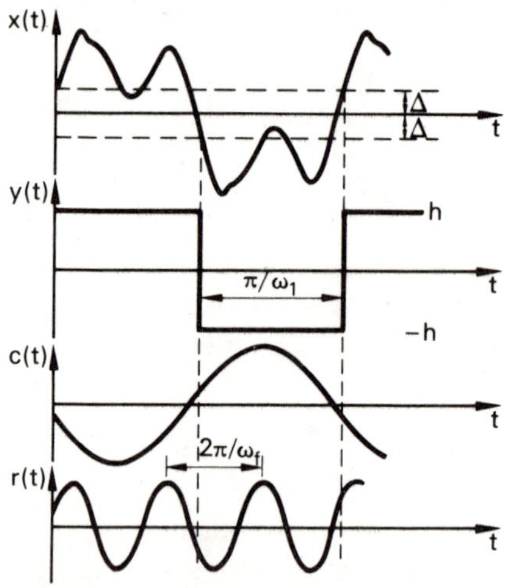

Figure 6.13 Typical third subharmonic waveforms with assumed $y(t)$

for the relay with no dead zone and

$$\{A_{G_1}(0, \omega_1) - A_{G_1}(\rho\pi, \omega_1) + (\pi/2h)R'(m\pi + \phi)\},$$

$$\{A_{G_1}(0, \omega_1) - A_{G_1}(-\rho\pi, \omega_1) + (\pi/2h)R'(\rho m\pi + \phi)\}$$

for the relay with dead zone where

$$R'(\rho m\pi + \phi) = (1/\omega_1)\dot{r}(\rho\pi^-/\omega_1) + jr(\rho\pi^-/\omega_1)$$

$$= R\{m\cos(\rho m\pi + \phi) + j\sin(\rho m\pi + \phi)\},$$

which is the equation of an ellipse in the complex plane. Solutions are obtained from the intersections of the above loci with the same lines as in the previous section. For the assumed symmetrical operation it can easily be shown from the switching conditions that, as expected, only odd subharmonics are possible. The graphical solution procedure for a relay without dead zone is shown in figure 6.14, where C and D are two possible solutions for the third subharmonic

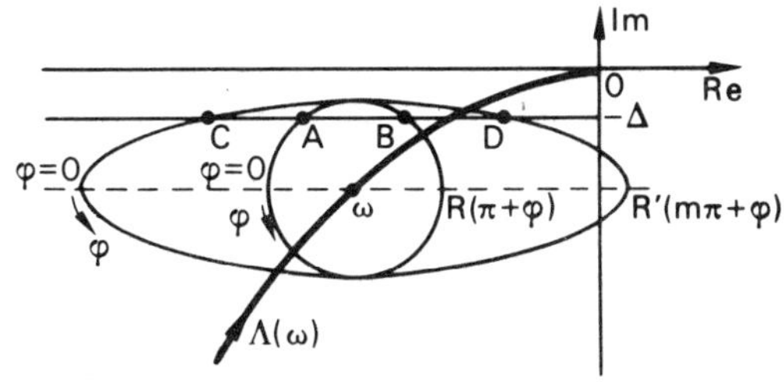

Figure 6.14 Solution procedure for subharmonic oscillation

with the assumed input amplitude R and frequency $3\omega_1$. Again in section 6.9.2 the solution at C will be shown to be stable and that at D unstable. Also shown in the figure are the solutions A and B for the synchronised oscillation, $\omega_f = \omega_1$ with the same input amplitude R. The phase angles ϕ_C and ϕ_D at C and D are equal to those at A and B, ϕ_A and ϕ_B, respectively. It can also be seen from the diagram that the minimum input amplitude $R'_{\min}$ at a frequency $m\omega_1$ to produce a subharmonic oscillation at frequency ω_1 is equal to the minimum amplitude $R_{\min}$ at frequency ω_1 which produces a synchronous oscillation. A diagram indicating the ranges of amplitude and frequency within which the various order subharmonics occur is sketched in figure 6.15 for the typical case of a relay with no dead zone and a plant transfer function of the form $1/s(1 + s)(1 + as)$. Obviously from the above discussion the boundaries are similar in shape, that about $m\omega_1$ being scaled horizontally by a factor m relative to that about ω_1. Regions are seen to be present where more than one subharmonic may exist, the order which exists in practice will depend upon the initial conditions.

There is an upper limit to the subharmonic regions,[14] which is shown dotted in figure 6.15, due to the assumed waveform for $y(t)$ in figure 6.13 becoming

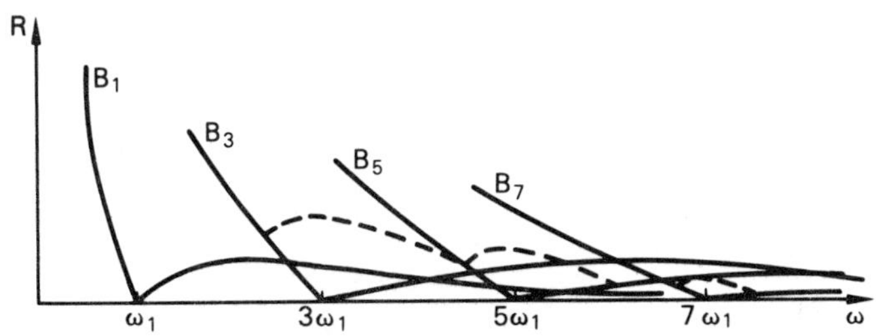

Figure 6.15 Region of subharmonic oscillations

invalid as $x(t)$ attains a value less than $-\Delta$ in the interval $0 < t < \pi/\omega_1$. The boundary can be found theoretically by calculating $x(t)$ for the assumed $y(t)$ and determining when $x(t)$ fails to satisfy the continuous time validity condition. On the other hand subharmonics can exist[7] for larger values of R, which break this validity condition, where the relay output assumes a more complex form with additional switchings per period. Typical waveforms observed in a system containing a relay with dead zone and no hysteresis are shown in figure 6.16. Waveforms (a) show the normal third order subharmonic and (b) the more complex form with a larger input amplitude.

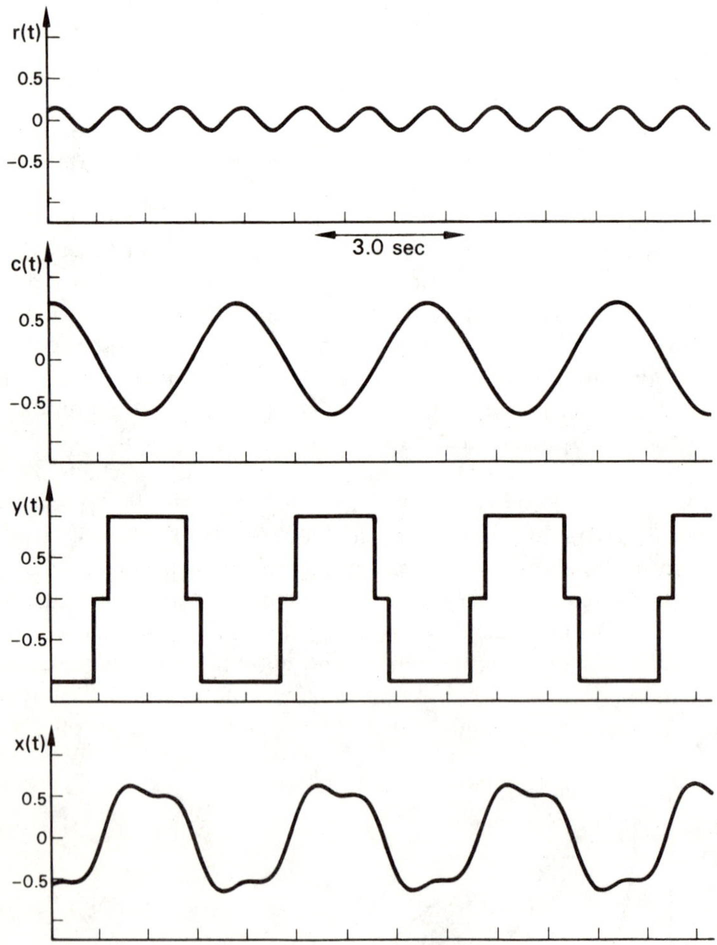

(a) Third order subharmonic

Figure 6.16 Third order subharmonic waveforms

Solutions for the regions of subharmonic oscillations can also be obtained by the parameter plane approach of reference 13 or using the DF procedure in which case equation (6.63) has

$$K_1(\beta_1) = \pi \cos m\beta_1/2h \sin 2\beta_1,$$
$$K_2(\beta_1) = \pi \sin m\beta_1/4h \sin^2 \beta_1.$$

Similar procedures[14] can be used to investigate subharmonics of all orders which occur in the asymmetrical operation of relay systems.

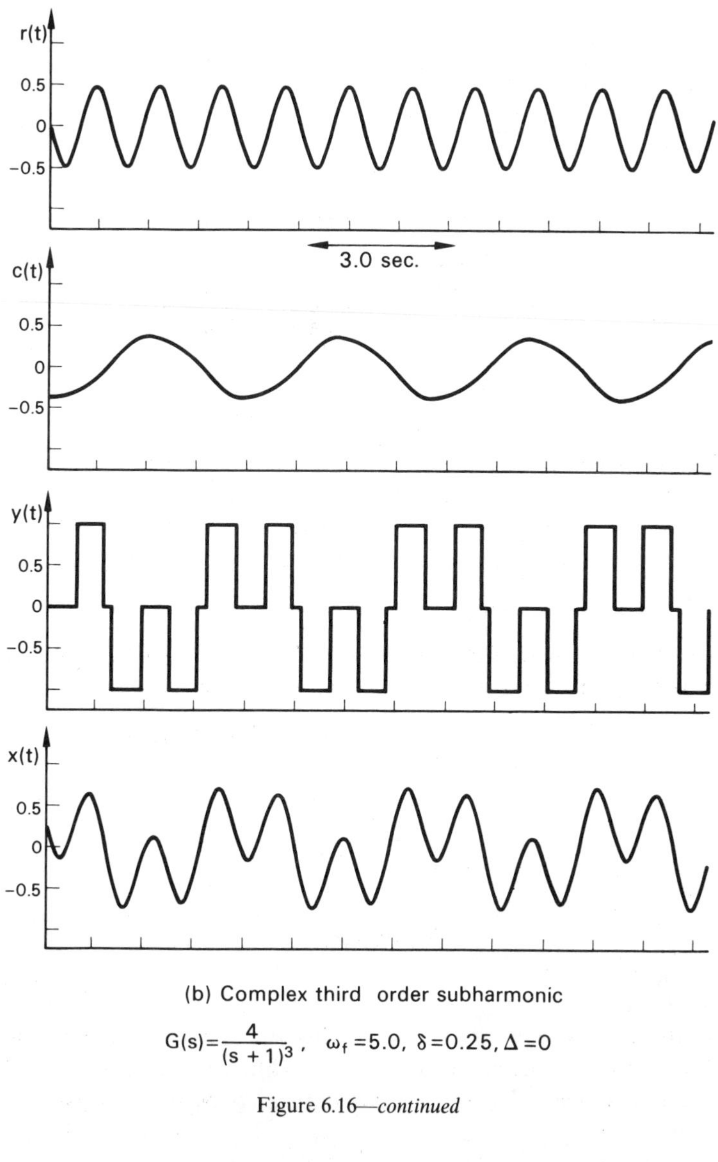

(b) Complex third order subharmonic

$G(s)=\dfrac{4}{(s+1)^3}$, $\omega_f = 5.0$, $\delta = 0.25$, $\Delta = 0$

Figure 6.16—*continued*

6.9.2 *Stability of forced oscillations*

Tsypkin[4,8] has derived a stability criterion for forced oscillations by investigating the effect of a small perturbation of the relay input $x(t)$. Consider the situation shown in figure 6.17 where the input to the ideal relay $x(t) = x_T(t) + x_d(t)$, with $x_d(t)$ the deviation of the relay input from its value $x_T(t)$ under periodic conditions. Switching of the relay output will take place at times t_1, t_2, etc. which are solutions of $x_T(t) + x_d(t) = 0$.

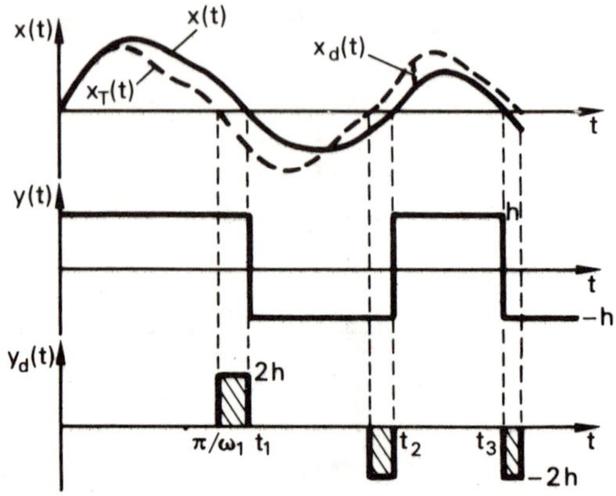

Figure 6.17 Perturbed waveforms at the relay

The difference in the relay output from its value in the periodic state is $y_d(t)$, which as shown in figure 6.17, consists of pulses occurring at times $t = \pi/\omega_1$, $2\pi/\omega_1 \ldots$ of duration $(t_1 - \pi/\omega_1)$, $(t_2 - 2\pi/\omega_1) \ldots$. Assuming that $x_d(t)$ is small and does not vary rapidly the pulse occurring at time $t = n\pi/\omega_1$,

$$2h\{u(t - n\pi/\omega_1) - u(t - t_n)\}$$

may be approximated by the impulse $(-1)^{n-1}2h(t_n - n\pi/\omega_1)\delta(t - n\pi/\omega_1)$. Now in the limit as $x_d(t)$ tends to zero, it is seen from figure 6.18, that

$$(-1)^{n-1}(t_n - n\pi/\omega_1) \approx x_d(n\pi/\omega_1)/|\dot{x}_T(n\pi/\omega_1)|.$$

Thus

$$y_d(t) = 2h \sum_{n=0}^{\infty} \frac{x_d(n\pi/\omega_1)}{|\dot{x}_T(n\pi/\omega_1)|}\delta(t - n\pi/\omega_1)$$

$$= \frac{2h}{|\dot{x}_T(\pi/\omega_1)|} \sum_{n=0}^{\infty} x_d(n\pi/\omega_1)\delta(t - n\pi/\omega_1), \qquad (6.64)$$

as $\dot{x}_T(t)$ is periodic.

Equation (6.64) is seen to correspond to the output of an ideal impulse sampler,[15] which samples the input $x_d(t)$ at a frequency ω_s equal to $2\omega_1$, followed by a gain of $2h/|\dot{x}_T(\pi/\omega_1)|$. Thus the stability of the perturbed system can be

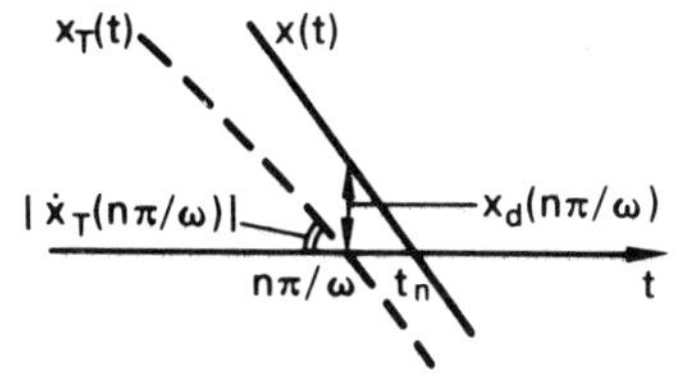

Figure 6.18 Impulse magnitude determination

determined by studying the behaviour of the open loop pulse transfer function $[\{2h/|\dot{x}_T(\pi/\omega_1)|\}G^*(j\omega)]$. Now a pulse transfer function can be expressed as a summation,[15] that is

$$G^*(j\omega) = \left\{(\omega_s/2\pi) \sum_{n=-\infty}^{\infty} G(j\omega + jn\omega_s)\right\} + g(0^+)/2,$$

so that

$$\begin{aligned} G^*(j\omega_1) &= \left\{(\omega_1/\pi) \sum_{n=-\infty}^{\infty} G(j\omega_1 + jn2\omega_1)\right\}, \quad \text{with } g(0^+) = 0 \\ &= \left\{(\omega_1/\pi) \sum_{n=-\infty}^{\infty} G\{j(2n+1)\omega_1\}\right\} \\ &= \left\{(\omega_1/\pi) \sum_{n=-\infty}^{\infty} U_G\{(2n+1)\omega_1\} + jV_G\{(2n+1)\omega_1\}\right\}. \end{aligned}$$

But U_G is an even function and V_G an odd function of ω, so that

$$\begin{aligned} G^*(j\omega_1) &= (2\omega_1/\pi) \sum_{n=1(2)}^{\infty} U(n\omega_1)\dagger \\ &= (\omega_1/2h)\,\mathrm{Re}\{\Lambda(\omega_1)\}. \end{aligned}$$

Thus

$$\left.\frac{2hG^*(j\omega)}{|\dot{x}_T(\pi/\omega_1)|}\right|_{\omega=\omega_1} = \frac{\omega_1\,\mathrm{Re}\{\Lambda(\omega_1)\}}{|\dot{x}_T(\pi/\omega_1)|} = \frac{\mathrm{Re}\{\Lambda(\omega_1)\}}{|\mathrm{Re}\{\Lambda(\omega_1) + R(\pi + \phi)\}|}.$$

Applying the Nyquist criterion to the pulse transfer function the system will be stable provided

$$\{2h/|\dot{x}_T(\pi/\omega_1)|\}G^*(j\omega)|_{\omega=\omega_1} \gtrless -1,$$

where the inequality depends upon the form of $G^*(j\omega)$. For example, if $G(s) = K/s(s+a)(s+b)$, which is the form of open loop transfer function for which $\Lambda(\omega)$ is sketched in figure 6.11, the greater than sign applies and the criterion

† Note this equation can be written $G^*(j\omega_1) = (\omega_1/2h)\,\mathrm{Re}\{\Lambda(\omega_1)\}$ only if $g(0^+) = 0$. It can also be shown that[16] $H^*(j\omega_1) = (1/2h)\,\mathrm{Im}\{\Lambda(\omega_1)\}$ where $H(s) = G(s)/s$ and $G(\infty) = 0$. By definition[15] $G^*(j\omega_1) = G(z)|_{z\to-1,\,T=\pi/\omega_1}$, so that z transforms may be used to sum the series involved in $\Lambda(\omega)$.

becomes

$$\mathrm{Re}\{\Lambda(\omega)\} > -|\mathrm{Re}\{R(\pi+\phi)+\Lambda(\omega)\}|.$$

This can be interpreted as a solution to the left of a vertical line through the centre of the circular $R(\pi+\phi)$ locus is stable and to the right unstable. That is A, for which $\mathrm{Re}\{(R(\pi+\phi)\} < 0$, or equivalently $\dot{r}(\pi/\omega_1) < 0$, is stable and B for which $\dot{r}(\pi/\omega_1) > 0$ is unstable. The condition for stability in this case may alternatively be stated as the stable solution is that for which the slope of the input $r(t)$ at the switching instant is the same as the direction of switching. The criterion obviously applies in a similar manner for the subharmonic oscillation solutions of figure 6.14.

In the case of a self-oscillation $\omega_1 = \omega_0$ and the open loop pulse transfer function of the perturbed system $\{2h/|\dot{x}_T(\pi/\omega_1)|\}G^*(j\omega)$ passes through the $(-1, j0)$ point for $\omega = \omega_0$. To obtain the limit cycle stability criterion stated in section 6.8 it is necessary to study the behaviour of the open loop pulse transfer function divided by $(s - j\omega_0)$.

With reference to figure 6.11 it can be seen that if R is increased by a small amount, $|\dot{r}(\pi^-/\omega)|$ increases at both solutions whilst $|\dot{x}(\pi^-/\omega)|$ increases at A and decreases at B. If x were sinusoidal then this would correspond to $\mathrm{d}R/\mathrm{d}x > 0$ at A and $\mathrm{d}R/\mathrm{d}x < 0$ at B, in agreement with the result obtained by the DF method (see equation (5.62)). However for a relay with no dead zone $y(t)$ is always a square wave so that at a given frequency the contribution of the distortion terms to $\dot{x}(\pi^-/\omega)$ remains constant and the change in $|\dot{x}(\pi^-/\omega)|$ is due to the fundamental alone. Thus for a stable forced oscillation Tsypkin's criterion is equivalent to $\mathrm{d}R/\mathrm{d}X_1 > 0$, where X_1 is the magnitude of the fundamental component in x. Similarly for a stable subharmonic the criterion becomes $\mathrm{d}R/\mathrm{d}X_m > 0$, where X_m is the magnitude of the mth harmonic component (the same frequency as R) in x.

In a similar manner to the above Tsypkin[4] derived necessary conditions for the stability of a forced oscillation in a system containing a relay with dead zone as

$$\mathrm{Re}\{R(\pi+\phi)\}\,\mathrm{Re}[\{R(\rho\pi+\phi)+\Lambda'(\omega_1)\}+\{\Lambda'(\omega_1)-\Lambda_1(\omega_1)\}]$$
$$+\,\mathrm{Re}\{R(\rho\pi+\phi)\}\,\mathrm{Re}[\{R(\pi+\phi)+\Lambda(\omega_1)\}+\{\Lambda(\omega_1)-\Lambda_1(\omega_1)\}] > 0,$$

where

$$\Lambda_1(\omega_1) = \Lambda(\omega_1)|_{\rho=1} = \Lambda'(\omega_1)|_{\rho=1} \quad \text{and} \quad \lim_{s\to\infty} sG(s) = 0.$$

Experimental results for several examples have confirmed the intuitive criterion,[7] similar to the above for the relay with no dead zone, that for a relay with dead zone stability requires $\mathrm{d}R/\mathrm{d}\beta > 0$, where β is the angle given in the notation of figure 6.9. This conjecture is extremely easy to apply.

6.9.3 *D.c. incremental gain*

The Tsypkin locus sketched in figure 6.11 is typical of that for a transfer function of the form $G(s) = K/s(s+a)(s+b)$. In section 5.11.2 it has been shown that the feedback loop of figure 6.1 with a transfer function of the above form and an ideal relay oscillates in the autonomous state and a finite magnitude high frequency sinusoidal input is required to quench the oscillation. The region of synchronised forced oscillations obtained by Tsypkin's method including the necessary stability condition, as shown in figure 6.15, indicates, in contrast, that for a high frequency input a near zero input amplitude produces synchronisation.

The Tsypkin necessary criterion for a stable forced oscillation, as will be appreciated from the previous comments, is equivalent to the dR/dX criterion of the DF method. In the DF method an additional condition, which ensures that no limit cycle can build up at the forced oscillation solution, must also be checked. Tsypkin shows how such a check can be made, however, an alternative approach with a simple graphical interpretation is given below.[17]

Consider the system of figure 6.1 to contain a relay with no dead zone. Figure 6.19 shows the relay input and output when a small additional bias signal

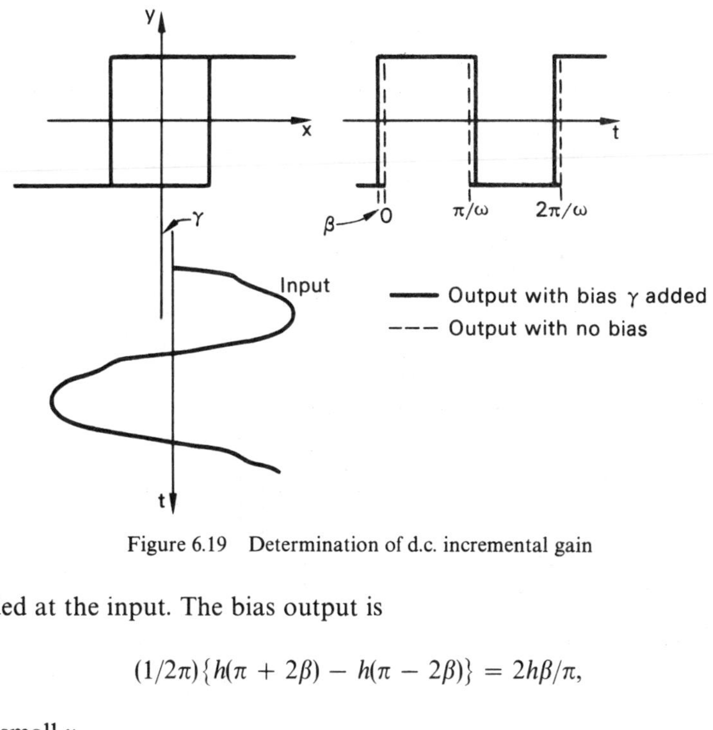

Figure 6.19 Determination of d.c. incremental gain

γ is added at the input. The bias output is

$$(1/2\pi)\{h(\pi + 2\beta) - h(\pi - 2\beta)\} = 2h\beta/\pi,$$

but for small γ,

$$\dot{x}(0) = -\dot{x}(\pi^-/\omega) = \omega\gamma/\beta,$$

so that the d.c. incremental gain is

$$(2h\beta/\pi\gamma) = (2h/\pi)\left\{\frac{-\omega}{\dot{x}(\pi^-/\omega)}\right\}. \tag{6.65}$$

If $x(t)$ is sinusoidal, that is

$$x(t) = a\sin(\omega t + \phi),$$
$$\dot{x}(t) = a\omega\cos(\omega t + \phi),$$

then

$$-\dot{x}(\pi^-/\omega)/\omega = -a\cos(\pi + \phi) = a\cos\phi = a^2/(a^2 - \Delta^2)^{1/2},$$

and the d.c. incremental gain is $2h(a^2 - \Delta^2)^{1/2}/a^2\pi$ in agreement with the result of section 5.3.4.

For the forced oscillation predicted by the Tsypkin method to exist the d.c. incremental gain must be less than the critical gain K_c given by the Routh criterion for the linear elements in the loop otherwise a self-oscillation can exist. Thus for stability

$$(2h/\pi)\{-\omega/\dot{x}(\pi^-/\omega)\} < K_c,$$

that is

$$(2h/\pi K_c) < -\dot{x}(\pi^-/\omega)/\omega = -\mathrm{Re}\{\Lambda(\omega) + R(\pi + \phi)\}.$$

The criterion is easily interpreted on the forced oscillation solution diagram since $\mathrm{Re}\{\Lambda(\omega) + R(\pi + \phi)\}$ is known at the solution point. As shown in figure 6.20 if the solution lies to the left of the vertical line passing through $-2h/\pi K_c$ it is stable and if it lies to the right unstable.

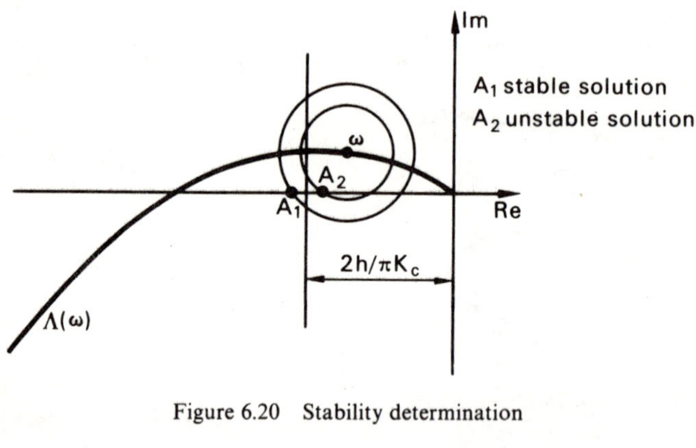

Figure 6.20 Stability determination

When the relay has dead zone it is similarly easily shown that the d.c. incremental gain is

$$-\frac{h}{\pi}\left\{\frac{\omega}{\dot{x}(\pi^-/\omega)} + \frac{\omega}{\dot{x}(\rho\pi^-/\omega)}\right\}, \tag{6.66}$$

and the condition for stability

$$\frac{h}{\pi K_c} > \frac{-1}{\{\omega/\dot{x}(\pi^-/\omega)\} + \{\omega/\dot{x}(\rho\pi^-/\omega)\}}.$$

$$\dot{x}(\rho\pi^-/\omega)/\omega = \mathrm{Re}\{\Lambda'(\omega) + R(\rho\pi + \phi)\}$$

and

$$\dot{x}(\pi^{-}/\omega)/\omega = \mathrm{Re}\{\Lambda(\omega) + R(\pi + \phi)\}$$

for $\lim_{s\to\infty} sG(s) = 0$, are the real coordinates at the forced oscillation solution points so that the above criterion is easily applied.

It should be remembered that the above criterion gives the condition for the induction of a limit cycle. As discussed previously when a limit cycle can exist in a relay system a different value of R to that for induction may be required for quenching. The value of R for quenching cannot be found by Tsypkin's method since prior to quenching the relay output will be a complex waveform produced by approximately sinusoidal signals of different frequency. R can however be found approximately as described in section 5.11.2 using the SSDF.

Example

Example 1 of section 5.11.2 is again considered. Figure 6.21 shows the $G(j\omega)$ and $A_G(0, \omega)$ loci for the transfer function $2/s(s + 1)^2$. The characteristic equation of the linear system is

$$s^3 + 2s^2 + s + 2K = 0,$$

for which the critical gain K_c for K is 1·0. The stability line when $A_G(0, \omega)$ is

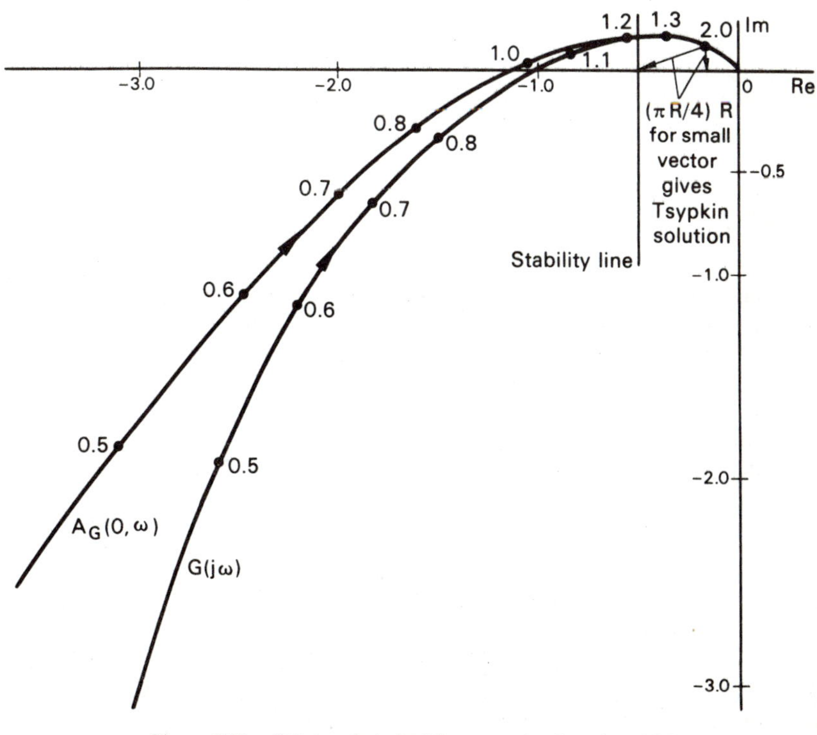

Figure 6.21 $G(j\omega)$ and A_G loci for example of section 6.9.3

plotted passes through $-(\pi/4h)(2h/\pi K_c) = -1/2K_c = -1/2$ as shown in the diagram. The circles for the input have radii of $(\pi R/4h) = (\pi R/4)$.

Figure 6.22 shows theoretical results obtained by the exact methods described above and approximate solutions from the complete DF and universal chart as described in section 5.11.2, example 1. The exact d.c. incremental gain curve (b)

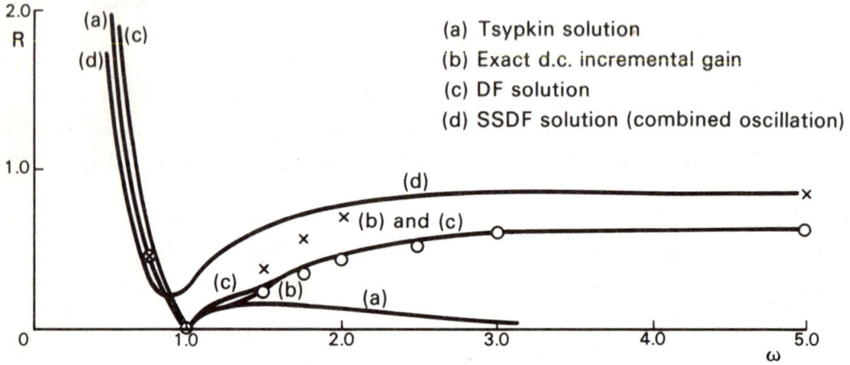

Figure 6.22 Frequency response results for example of section 6.9.3

breaks away from the Tsypkin curve (a), obtained using the necessary stability condition, for ω slightly greater than 1·3 and almost immediately becomes coincident with the DF solution curve (c) showing where a synchronised forced oscillation can no longer be maintained when R is reduced. A solution for a point on this curve is given by curve (ii) in figure 5.27. Curve (d) shows the SSDF solution for the boundary where synchronisation takes place as R is increased, a boundary which cannot be found by an exact method. A point on this boundary is given by curve (iv) in figure 5.27.

Also shown in the diagram are some simulation results obtained at various fixed frequencies by first increasing and then decreasing R. As expected agreement with the exact solution boundary (a) for $0 < \omega < 1{\cdot}3$ and (b) $\omega > 1{\cdot}3$ is good but the values for synchronisation on increasing R differ appreciably in places from the approximate solution boundary (d). This, as explained in section 5.11.2, is due to the assumption of a two frequency input to the nonlinearity being overly optimistic. Further, as expected at other frequencies, such as $\omega \approx 3{\cdot}0$ for example, subharmonic behaviour was encountered and synchronisation was delayed until R reached a higher value than that given by (d).

REFERENCES

1. Hamel, B.: 'Contribution a l'étude mathematique des systèmes de réglage par tout-ou-rien', *C.E.M.V., Service Technique Aeronautique*, 1949, 17.
2. Bohn, E. V.: 'Stability margins and steady state oscillations in on-off feedback systems', *I.R.E. Trans. Circuit Theory*, 1961, CT-8, pp. 127–130.
3. Chung, J. K-C. and Atherton, D. P.: 'The determination of periodic modes in relay systems using the state space approach', *Int. J. of Control*, 1966, 4, pp. 105–126.

4. Tsypkin, J. A.: *Theorie der relais systeme der automatischen regelung* (R. Oldenbourg-Verlag, Munich, 1958).
5. Kovatch, G.: 'A method for the computation of self-sustained oscillations in systems with piecewise linear elements', *I.E.E.E. Trans.*, 1963, AC-8, p. 358.
6. Leach, B. W.: 'An evaluation of estimates of stability of oscillations in nonlinear systems', *Ph.D. Thesis, University of Manitoba*, 1972.
7. McCardle, J. C.: 'Forced oscillations in relay control systems', *M.Sc.E. Thesis, University of New Brunswick, Canada*, 1968.
8. Gille, J. C., Pelegrin, M. J. and Decaulne, P.: *Feedback control systems* (McGraw-Hill, New York, 1959, Chapter 26).
9. Jolley, L. B. W.: *Summation of series* (Dover, New York, 1961, p. 104).
10. Atherton, D. P.: 'Conditions for periodicity in control systems containing several relays', *Proc. I.F.A.C.*, Paper 28E, London, 1966.
11. Carnahan, B., Luther, H. A. and Wilkes, J. O.: *Applied numerical methods* (Wiley, New York, 1964, pp. 421–429).
12. Turnbull, G. F., Atherton, D. P. and Townsend, J. M.: 'A method for the theoretical analysis of relay amplifiers', *Proc. I.E.E.*, 1965, 112, pp. 1039–1055.
13. LeMaitre, J.-F., Paquet, J.-G. and Gille, J.-C.: 'A general approach for on-off control system oscillations, *Automatica*, 1970, 6, pp. 609–613.
14. Sakawa, Y.: 'Subharmonic oscillations in relay control systems', *Proc. I.F.A.C.*, Moscow, 1960, pp. 404–409.
15. Jury, E.: *Sampled-data control systems* (Wiley, New York, 1958).
16. Hsu, J. C. and Meyer, A. U.: *Modern control principles and applications* (McGraw-Hill, New York, 1968, pp. 304–306).
17. Atherton, D. P.: 'The stability of oscillations', *Automatic Control Theory and Applications*, 1974, 2, pp. 47–49.

BIBLIOGRAPHY

Bauer, W. L.: 'Analytische berechnung von dauerschwingungen in mehrpunktregelkreisen nach dem zypkinverfahren', *Regelungstechnik und Proceß-Datenverarbeitung*, 1973, 21, pp. 110–118.

Bonenn, Z.: 'Frequency response of feedback relay amplifiiers', *Proc. I.E.E.*, 1961, 108C, pp. 287–295.

Brookes, B. E. and Johnson, R. A.: 'Limit cycles in time-delay relay systems', *Int. J. Control*, 1969, Vol. 9, No. 4, pp. 387–398.

Dolgolenko, Y. V.: 'Stability and self oscillations of an on-off servomechanism with time delay' (Russian), *Automatika i Telemekhanika*, 1952, 12, pp. 109–120.

Fel'dbaum, A. A.: 'Elementary on-off servomechanisms' (Russian), *Automatika i Telemekhanika*, 1949, 10, pp. 249–266.

Gille, J.-C. and Paquet, J.-G.: 'Subharmonic oscillations in on-off control systems', *A.I.E.E.*, 1962, 81, pt. 2, pp. 210–216.

Gille, J.-C., Le Maitre, J.-F. and Paquet, J.-G.: 'Application de la méthode d'Andronov et Vitt aux oscillations forcées des asservissements par plus-ou-moins', *Annales des Télécommunications*, 1967, 22, pp. 188–191.

Gille, J.-C., Paquet, J.-G. and Le Maitre, J.-F.: 'Synchronisation des asservissements par plus-ou-moins', *C.R. Acad. Sc. Paris*, 1968, 267, pp. 238–240.

Gille, J.-C., Paquet, J.-G. and Le Maitre, J.-F.: 'Oscillations libres des systèmes asservis par plus-ou-moins', *C.R. Acad. Sc. Paris*, 1968, 267, pp. 335–336.

Hudon, R., Paquet, J.-G. and Gille, J.-C.: 'On the hysteresis of on-off elements', *I.E.E.E. Trans.*, 1969, AC-14, pp. 52–55.

Jumarie, G.: 'Generalization of the theory of Ya. Z. Tsypkin for relay automatic control system', *Automation and Remote Control*, 1969, 12, pp. 1929–1937.

Lambert, J. C.: 'Etude des oscillations complexes d'ordre 4 dans les systèmes asservis par plus-ou-moins', *Thèse de Maîtrise, Université Laval*, 1966.

Le Maitre, J.-F., Champenois, A., Paquet, J.-G. and Gille, J.-C.: 'Oscillations combinées des systèmes par plus-ou-moins', *Annales des Télécommunications*, 1968, 23.

Le Pourhiet, A.: 'Contribution à l'étude des oscillations forcées dans les asservissements par plus-ou-moins', *Thèse de Maîtrise, Université Laval*, 1969.

Mohammed, A.: 'Steady state oscillations and stability of on-off feedback systems', *Ph.D. Thesis, Electrical Engineering Dept.*, University of British Columbia, 1965.

Neimark, U. I.: 'On periodic states and the stability of relay systems' (Russian), *Automatika i Telemekhanika*, 1953, 14, 5, pp..556–569.

Neimark, U. I. and Kublanov, I. M.: 'Study of periodic motions in a simple on-off temperature control system' (Russian), *Automatika i Telemekhanika*, 1953, 14, pp. 34–43.

Neimark, U. I.: 'Self and forced oscillations of on-off systems with time delay' (Russian), *Automatika i Telemekhanika*, 1955, 16, pp. 225–232.

Paquet, J.-G.: 'Oscillations sous-harmoniques dans les asservissements par plus-ou-moins', *Thèse de Doctorat, Université Laval*, 1962.

Paquet, J.-G., Le Maitre, J.-F. and Gille, J.-C.: 'La méthode de Hamel-Cypkin et son application aux fonctions de transfert particulières', *Automatisme*, 1966, 11, pp. 3–5.

Paquet, J.-G. and Le Maitre, J.-F.: *Méthodes pratiques d'etude des oscillations nonlinéaires* (Gordon & Breach, Paris, 1970).

Roots, W. K. and Walker, F.: 'Dynamic equilibrium cycling in discontinuously controlled electric process heating', *I.E.E.E. Trans.*, 1967, IGA-3, pp. 131–140.

Roots, W. K. and Walker, F.: 'Secondary feedback in on-off electric heating processes', *I.E.E.E. Trans.*, 1967, IGA-3, pp. 401–420.

Roots, W. K., Woods, J. T., Wells, F. M. and Tull, R. H.: 'Mode dependent time constants in electric heating processes', *I.E.E.E. Trans.*, 1967, IGA-3, pp. 315–322.

Roots, W. K., Shridhar, M., Pfafflin, J. R. and Tull, R. H.: 'Mode dependent time constants in three forms of space heating', Paper presented at the *I.E.E.E., IGA Technical Conference*, May 1970, San Francisco.

Tsypkin, J. A.: 'Forced oscillations in on-off servomechanisms' (Russian), *Automatika i Telemekhanika*, 1952, 13, pp. 501–524.

Tsypkin, J. A.: 'Stability of periodic motions in on-off servomechanisms' (Russian), *Automatika i Telemekhanika*, 1953, 14, pp. 638–646.

Turnbull, G. F. and Townsend, J. M.: 'A feedback pulse with modulated audio amplifier', *Wireless World*, 1965, 71, pp. 160–168.

PROBLEMS

P.6.1 Use Tsypkin's method to determine the frequency of the limit cycle for the system of example 4, section 2.11.

P.6.2 Determine by Tsypkin's method the exact solutions for the amplitude and frequency of problem P.4.2.

P.6.3 Obtain exact solutions for problem P.4.3.

P.6.4 Determine the exact condition for stability of the system of problem P.4.9.

P.6.5 A relay system of the form shown in figure 6.1 has an ideal relay ($\delta = \Delta = 0$) and $G(s) = e^{-s/2}/s(s+1)$. Determine, with $r(t) = 0$, the exact frequency of oscillation and compare its value with the DF solution. If $r(t) = R \sin t$ what is the minimum value of R which will cause the system to synchronise with the input frequency?

P.6.6 Find the conditions for stability of the relay system of figure 6.1 if the relay has no hysteresis ($\Delta = 0$) and $G(s) = K(1 - sT)/(1 + sT)^2$.

P.6.7 If the relay system of figure 6.1 contains an ideal relay ($\delta = \Delta = 0$) and $G(s) = K/(1 + sT)^3$, determine the frequency of oscillation of the system as a function of a constant input $r(t) = R$ and, in addition, the modified nonlinearity in the presence of the oscillation.

P.6.8 What is the closed loop transfer function for the system of P.6.7 for a small input $r(t)$. Compare the result obtained using (a) the exact incremental describing function and (b) the IDF.

P.6.9 Determine for the system of P.6.7 the smallest high frequency sinusoidal input which will produce a synchronised forced response.

P.6.10 Determine the maximum value of K for which the system of figure P.6.1 will be stable.

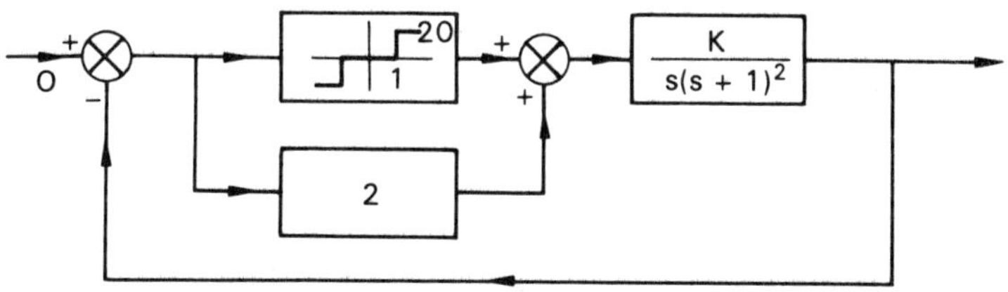

Figure P.6.1

P.6.11 What is the maximum value of K for which the system of figure P.6.2 will be stable?

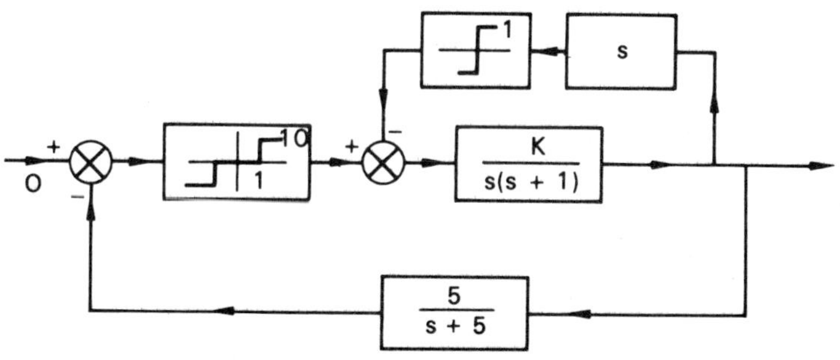

Figure P.6.2

P.6.12 A relay type controller has a symmetrical odd characteristic which for positive inputs is given by

$$n(x) = \begin{cases} 0 & x < 1 \\ 1 & 1 < x < 2. \\ 2 & 2 < x \end{cases}$$

Find the exact period of the limit cycle if the linear elements of the feedback loop have a transfer function $G(s) = 64/s(s+2)^2$. {Hint: Represent the characteristic as two relays with dead zone in parallel.} Compare the result with the approximate DF solution.

P.6.13 A relay type controller has a symmetrical odd characteristic which for positive inputs is given by

$$n(x) = \begin{cases} 0 & x < 1 \\ 1 & 1 < x < 3 \\ 2 & 3 < x < 5 \\ 3 & 5 < x \end{cases}$$

Find the exact period of the limit cycle if the linear elements of the feedback loop have a transfer function $G(s) = 20/s(s + 1)^2$. Compare the result with the approximate DF solution.

Chapter 7

Random Signal Analysis

7.1 Introduction

In many control systems the signals which occur cannot be described explicitly as functions of time although various characteristics of the signal may be known. Typically these may be limits on the amplitude variations, the mean squared value, the frequency spectrum and so on. Such a probabilistic, rather than deterministic, signal is known as a random or stochastic signal. A random signal may enter a control system as unwanted noise or may provide a more accurate description of the expected system input. The term unwanted noise may be used to describe actual disturbance signals, the inaccuracy of measurements, or drift and noise generated in the system components.

The definitions of statistics are based on the outcome of a large number of identical experiments, so that to obtain a statistical description of a random variable X its value in all the experiments should be measured. A random process is the term given to the ensemble of time functions, figure 7.1, showing the variations of the random variable as a function of time in all the experiments. The time functions $x_i(t)$, for example, may be the output signals from a large number of noise generators. A particular signal in the random process, for example $x_1(t)$, is known as a member function. Statistical data for the random variable X can be obtained by taking averages over the ensemble at any instant of time. If this data is independent of the origin of the time axis the process is said to be stationary.

In a practical situation one is usually required to assess the statistical behaviour of a system when it is subject to an input signal, which may be any member function $x_i(t)$ of the ensemble, from measurements taken with that input alone. This leads to the intuitive assumption that the statistical properties obtained from a time average of any member function of an ensemble belonging to a stationary random process will be equal to those obtained by taking an ensemble average. This assumption is known as the ergodic hypothesis. This chapter considers in detail the properties of ergodic processes and the methods used in the analysis of linear systems with such inputs. We start by presenting basic probability theory for one or more continuous random variables.

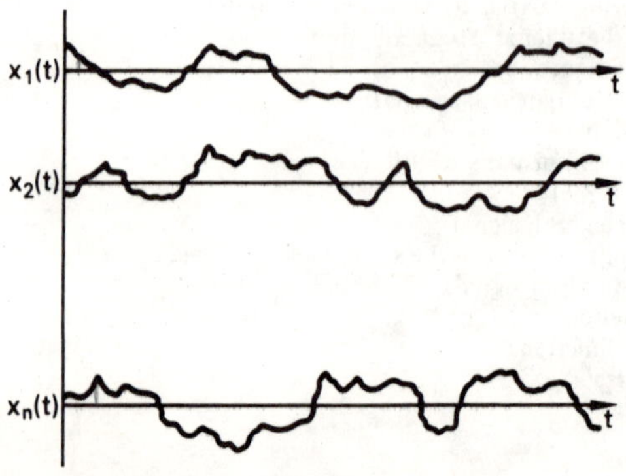

Figure 7.1 A random process, $x(t)$

7.2 Probability theory for one random variable

A random variable, X, may be defined as any quantity whose value cannot be predicted accurately but may be estimated by statistical analysis. Assuming that at any particular measurement the random variable assumes a certain value, x, the sample space is that space which can accommodate all possible measurements of the random variable. For example in throwing a single die the only possible results are the integers 1 to 6, the random variable, x, is thus discontinuous having only specific discrete values and the sample space is finite. When the random variable is continuous the sample space becomes infinite.

7.2.1 *The probability density function*

The distribution function $P(x)$, associated with the random variable X, which may take any value from $-\infty$ to ∞, is defined as the probability of $X \leq x$. The event $X \leq x$ is a subset of the totality of possible events $X \leq \infty$ and

$$P(x) = \text{Prob}[X \leq x]. \tag{7.1}$$

When X is a continuous random variable, $P(x)$ is a continuous function, but when X is discontinuous (i.e. discrete) so also is $P(x)$. Clearly $P(x)$ is nondecreasing, so that

$$P(a) \leq P(b) \quad \text{if } a \leq b,$$

$$P(-\infty) = 0, \quad \text{and} \quad P(\infty) = 1,$$

since a probability must fall between zero and one. Also

$$\text{Prob}[X > x] = 1 - \text{Prob}[X \leq x] = 1 - P(x).$$

Assuming X is continuous or possesses finite discontinuities then $P(x)$ will be differentiable and the probability density function of the distribution is defined by

$$p(x) = \frac{\mathrm{d}P(x)}{\mathrm{d}x}, \tag{7.2}$$

where, since $P(x)$ is monotonically nondecreasing, $p(x) \geq 0$ for all x. The probability of X occurring in the interval x to $x + \delta x$ is $p(x)\,\delta x$, and

$$\int_a^b p(x)\,\mathrm{d}x = \int_a^b \mathrm{d}[P(x)] = P(b) - P(a),$$

$$= \text{Prob}[a < X \leq b]. \tag{7.3}$$

The probability of any particular value, a, of the continuous random variable X is thus zero. Also

$$\int_{-\infty}^{x_0} p(x)\,\mathrm{d}x = P(x_0) - P(-\infty) = P(x_0),$$

and

$$\int_{-\infty}^{\infty} p(x)\,\mathrm{d}x = P(\infty) - P(-\infty) = 1.$$

7.2.2 *The expected values and moments of X*

From above

$$\text{Prob}[x_i < X \leq x_i + \delta x] = \int_{x_i}^{x_i+\delta x} p(x)\,\mathrm{d}x \simeq p(x_i)\,\delta x, \tag{7.4}$$

for small δx. Now, the expected or mean value of X is the limit of the sum of the possible values of X when each value is multiplied by its probability of occurrence, that is

$$\mu_1 = E[X] = \lim_{N\to\infty} \sum_{i=1}^{N} x_i p(x_i)\,\delta x = \int_{-\infty}^{\infty} xp(x)\,\mathrm{d}x. \tag{7.5}$$

Similarly, the expected value of any real single valued function $f(X)$ of X is

$$E[f(X)] = \int_{-\infty}^{\infty} f(x)p(x)\,\mathrm{d}x. \tag{7.6}$$

In particular, the mean squared value

$$E[X^2] = \int_{-\infty}^{\infty} x^2 p(x)\,\mathrm{d}x, \tag{7.7}$$

and the mean squared value about the mean, or variance,

$$E[(X - E[X])^2] = \int_{-\infty}^{\infty} (x - \mu_1)^2 p(x)\,\mathrm{d}x. \tag{7.8}$$

The standard deviation of X is defined as the positive square root of the variance.

Since

$$E[(X - E[X])^2] = E[X^2 - 2X\mu_1 + \mu_1^2],$$
$$= E[X^2] - \mu_1^2,$$

the variance is the mean squared value minus the mean value squared. The nth moment, μ_n, of X is defined as the expected value of X^n

$$\mu_n = E[X^n] = \int_{-\infty}^{\infty} x^n p(x)\,\mathrm{d}x, \tag{7.9}$$

and the nth central moment, ψ_n, is obtained by subtracting the mean value before averaging,

$$\psi_n = E[(X - E[X])^n] = \int_{-\infty}^{\infty} (x - \mu_1)^n p(x)\,\mathrm{d}x. \tag{7.10}$$

7.2.3 *The characteristic and moment generating functions*

The characteristic function,† $M(j\omega)$ is defined as the Fourier transform of the probability density function, that is,

$$M(j\omega) = E[e^{-j\omega X}] = \int_{-\infty}^{\infty} e^{-j\omega x} p(x)\,dx. \tag{7.11}$$

In addition, using the inverse transform one may write

$$p(x) = (1/2\pi) \int_{-\infty}^{\infty} M(j\omega) e^{j\omega x}\,d\omega. \tag{7.12}$$

The moment generating function $m(t)$, defined as the expected value of e^{tX}, is often used. Thus

$$m(t) = E[e^{tX}] = \int_{-\infty}^{\infty} e^{tx} p(x)\,dx = M(j\omega)\Big|_{\omega = -t/j} \tag{7.13}$$

Successive differentiation with respect to t, yields

$$\frac{dm(t)}{dt} = \int_{-\infty}^{\infty} x e^{tx} p(x)\,dx,$$

$$\frac{d^2 m(t)}{dt^2} = \int_{-\infty}^{\infty} x^2 e^{tx} p(x)\,dx, \tag{7.14}$$

and

$$\frac{d^n m(t)}{dt^n} = \int_{-\infty}^{\infty} x^n e^{tx} p(x)\,dx.$$

Thus putting $t = 0$

$$\frac{dm(t)}{dt}\bigg|_{t=0} = \int_{-\infty}^{\infty} x p(x)\,dx = \mu_1,$$

$$\frac{d^2 m(t)}{dt^2}\bigg|_{t=0} = \int_{-\infty}^{\infty} x^2 p(x)\,dx = \mu_2$$

and

$$\frac{d^n m(t)}{dt^n}\bigg|_{t=0} = \int_{-\infty}^{\infty} x^n p(x)\,dx = \mu_n.$$

7.2.4 *The cumulant function*

The cumulant function $K(t)$ of the distribution is the logarithm of the moment generating function, that is

$$K(t) = \log m(t), \tag{7.15}$$

† This function is also defined as $E[e^{j\zeta X}]$, that is with $\zeta = -\omega$.

and the nth order cumulant λ_n is given by

$$\lambda_n = \left.\frac{\mathrm{d}^n K(t)}{\mathrm{d}t^n}\right|_{t=0}. \tag{7.16}$$

Successive differentiation with respect to t of the relationship

$$m(t) = \mathrm{e}^{K(t)},$$

and substitution of $t = 0$ gives the moments in terms of the cumulants of the distribution. The first few relationships are

$$\mu_1 = \lambda_1,$$
$$\mu_2 = \lambda_2 + \lambda_1^2,$$
$$\mu_3 = \lambda_3 + 3\lambda_1\lambda_2 + \lambda_1^3,$$
$$\mu_4 = \lambda_4 + 3\lambda_2^2 + 4\lambda_1\lambda_3 + 6\lambda_1^2\lambda_2 + \lambda_1^4,$$

and for the cumulants in terms of the moments

$$\lambda_1 = \mu_1,$$
$$\lambda_2 = \mu_2 - \mu_1^2,$$
$$\lambda_3 = \mu_3 - 3\mu_1\mu_2 + 2\mu_1^3,$$
$$\lambda_4 = \mu_4 - 4\mu_3\mu_1 - 3\mu_2^2 + 12\mu_2\mu_1^2 - 6\mu_1^4.$$

7.3 Probability theory for two random variables

Two continuous random variables X and Y are considered, whose values obtained at any particular measurement are x and y respectively. The totality of possible values for X and Y are $-\infty \leq X \leq \infty$ and $-\infty \leq Y \leq \infty$, the sample space is thus the infinite area of the XY plane.

7.3.1 *The joint probability density function*

The joint distribution function, $P(x, y)$, associated with the continuous random variables X and Y is defined as the probability of the combined events $X \leq x$ and $Y \leq y$.

$$P(x, y) = \text{Prob}[X \leq x \text{ and } Y \leq y]. \tag{7.17}$$

Both inequalities must be satisfied simultaneously. This combined event is a subset of the totality of possible combined events $X \leq \infty$ and $Y \leq \infty$. Clearly

$$P(-\infty, y) = P(x, -\infty) = 0 \quad \text{and} \quad P(\infty, \infty) = 1.$$

Also

$$\begin{aligned}\text{Prob}[X > x \text{ and } Y > y] &= 1 - \text{Prob}[X \leq x \text{ and } Y \leq y],\\ &= 1 - P(x, y).\end{aligned}$$

Assuming that the partial derivatives of $P(x, y)$ exist, the joint or second probability density function $p(x, y)$ is defined by

$$p(x, y) = \frac{\partial^2 P(x, y)}{\partial x \, \partial y}. \tag{7.18}$$

Therefore if $a < b$ and $c < d$,

$$\int_a^b \int_c^d p(x, y) \, dy \, dx = \int_a^b [p'(x, d) - p'(x, c)] \, dx,$$

where $p'(x, y) = \int_0^y p(x, \xi) \, d\xi$. Thus

$$\begin{aligned}\int_a^b \int_c^d p(x, y) \, dy \, dx &= P(b, d) - P(a, d) - P(b, c) + P(a, c), \\ &= \text{Prob}[a < X \leq b \text{ and } Y \leq d] \\ &\quad - \text{Prob}[a < X \leq b \text{ and } Y \leq c], \\ &= \text{Prob}[a < X \leq b \text{ and } c < Y \leq d].\end{aligned} \tag{7.19}$$

Thus it also follows that

$$\begin{aligned}\int_{-\infty}^{x_0} \int_{-\infty}^{y_0} p(x, y) \, dy \, dx &= P(x_0, y_0) - P(-\infty, y_0) - P(x_0, -\infty) + P(-\infty, -\infty) \\ &= P(x_0, y_0),\end{aligned}$$

and

$$\int_{-\infty}^{\infty} \int_{-\infty}^{\infty} p(x, y) \, dy \, dx = P(\infty, \infty) = 1.$$

7.3.2 *The expected values and moments of X and Y*

For the two variables

$$\text{Prob}[x_i < X \leq x_i + \delta x \text{ and } y_i < Y \leq y_i + \delta y] \simeq p(x_i, y_i) \, \delta x \, \delta y. \tag{7.20}$$

The expected or mean value of $f(X, Y)$, a function of X and Y, is the limit of the sum of the possible values of $f(X, Y)$ multiplied by their probability of occurrence, that is

$$\begin{aligned}E[f(X, Y)] &= \lim_{\substack{N \to \infty, \\ \delta x \to 0, \\ \delta y \to 0}} \sum_{i=1}^{N} f(x_i, y_i) p(x_i, y_i) \, \delta x \, \delta y, \\ &= \int_{-\infty}^{\infty} \int_{-\infty}^{\infty} f(x, y) p(x, y) \, dx \, dy.\end{aligned} \tag{7.21}$$

The variables X and Y in $f(X, Y)$ may in many cases be separable, so that if

$$p(x) = \frac{dP(x)}{dx} \quad \text{is the probability density function for } x \text{ alone,}$$

and

$$q(y) = \frac{dQ(y)}{dy} \quad \text{is the probability density function for } y \text{ alone,}$$

by definition

$$P(x) = \text{Prob}[X \le x \text{ and } Y \le \infty] = \int_{-\infty}^{x} \int_{-\infty}^{\infty} p(x, y)\, dy\, dx.$$

therefore

$$p(x) = \frac{dP(x)}{dx} = \int_{-\infty}^{\infty} p(x, y)\, dy, \tag{7.22}$$

and similarly

$$q(y) = \int_{-\infty}^{\infty} p(x, y)\, dx.$$

Thus if

$$f(X, Y) = f_1(X) + f_2(Y),$$

$$\begin{aligned} E[f(X, Y)] &= \int_{-\infty}^{\infty} \int_{-\infty}^{\infty} [f_1(x) + f_2(y)]p(x, y)\, dy\, dx, \\ &= \int_{-\infty}^{\infty} f_1(x)p(x)\, dx + \int_{-\infty}^{\infty} f_2(y)q(y)\, dy, \\ &= E[f_1(X)] + E[f_2(Y)]. \end{aligned}$$

For the particular case when X and Y are mutually independent random variables, by the definition of independence,

$$\text{Prob}[X \le x \text{ and } Y \le y] = \{\text{Prob}[X \le x]\}\{\text{Prob}[Y \le y]\}. \tag{7.23}$$

Hence

$$P(x, y) = P(x)Q(y), \tag{7.24}$$

and

$$p(x, y) = p(x)q(y). \tag{7.25}$$

Therefore if for example $f(X, Y) = f_1(X)f_2(Y)$

$$E[f(X, Y)] = E[f_1(X)]E[f_2(Y)],$$

when X and Y are independent. The set of possible second order moments μ_{jk} are given by

$$\mu_{jk} = E[X^j Y^k] = \int_{-\infty}^{\infty} \int_{-\infty}^{\infty} x^j y^k p(x, y)\, dy\, dx, \tag{7.26}$$

and the corresponding central moments are

$$\begin{aligned}\psi_{jk} &= E[(X - \mu_{10})^j(Y - \mu_{01})^k] \\ &= \int_{-\infty}^{\infty}\int_{-\infty}^{\infty} (x - \mu_{10})^j(y - \mu_{01})^k p(x, y)\,dy\,dx.\end{aligned} \tag{7.27}$$

In particular

$$\psi_{20} = E[(X - \mu_{10})^2] = \int_{-\infty}^{\infty} (x - \mu_{10})^2 p(x)\,dx = \mu_{20} - \mu_{10}^2,$$

$$\psi_{02} = E[(Y - \mu_{01})^2] = \int_{-\infty}^{\infty} (y - \mu_{01})^2 q(y)\,dy = \mu_{02} - \mu_{01}^2,$$

and

$$\begin{aligned}\psi_{11} &= E[(X - \mu_{10})(Y - \mu_{01})], \\ &= \int_{-\infty}^{\infty}\int_{-\infty}^{\infty} (x - \mu_{10})(y - \mu_{01})p(x, y)\,dy\,dx, \\ &= \mu_{11} - \mu_{10}\mu_{01}.\end{aligned}$$

The term $\mu_{11} = E[XY]$ is called the crosscorrelation of X and Y, and $\psi_{11} = E[(X - \mu_{10})(Y - \mu_{01})]$ is the covariance of X and Y. The terms covariance and crosscorrelation are often used synonymously, particularly in system analysis, as most distributions are assumed to have a zero mean. A distribution having a non-zero mean can, of course, always be replaced by a constant (mean) value and a distribution with a zero mean. If $E[XY] = E[X]E[Y]$ the variables X and Y are uncorrelated and if $\psi_{11} = 0$ they are said to be orthogonal.

7.3.3 *The joint characteristic and moment generating functions*

The joint characteristic function, $M(ju, jv)$, is defined as the double Fourier transform of the joint probability density function or as $E[e^{-j(uX+vY)}]$, so that

$$M(ju, jv) = \int_{-\infty}^{\infty}\int_{-\infty}^{\infty} p(x, y)e^{-j(ux+vy)}\,dy\,dx, \tag{7.28}$$

and by the inverse transformation

$$p(x, y) = (1/4\pi^2)\int_{-\infty}^{\infty}\int_{-\infty}^{\infty} M(ju, jv)e^{j(ux+vy)}\,dv\,du. \tag{7.29}$$

Similarly the joint moment generating function $m(s, t)$ is defined as

$$E[e^{sX+tY}],$$

so that

$$m(s, t) = \int_{-\infty}^{\infty}\int_{-\infty}^{\infty} p(x, y)e^{sx+ty}\,dy\,dx. \tag{7.30}$$

The partial derivatives evaluated at $s = t = 0$ yield the moments of the distribution, thus

$$\left.\frac{\partial^{j+k}m}{\partial s^j\,\partial t^k}\right|_{s,t=0} = \int_{-\infty}^{\infty}\int_{-\infty}^{\infty} x^j y^k p(x, y)\,\mathrm{d}y\,\mathrm{d}x = E[X^j Y^k] = \mu_{jk}. \tag{7.31}$$

7.3.4 *The joint cumulant function*

The joint cumulant function $K(s, t)$ is the logarithm of the joint moment generating function, that is

$$K(s, t) = \log m(s, t), \tag{7.32}$$

and the corresponding cumulants are given by

$$\lambda_{jk} = \left.\frac{\partial^{j+k}K(s, t)}{\partial s^j\,\partial t^k}\right|_{s,t=0}. \tag{7.33}$$

Partial derivatives with respect to s and t of the relationship

$$m(s, t) = \mathrm{e}^{K(s,t)},$$

evaluated at $s = t = 0$, give the relationship between the joint moments and cumulants.

7.3.5 *Conditional probability*

The conditional probability density denoted by $p(x|y)$ is defined as

$$p(x|y) = \frac{p(x, y)}{q(y)}. \tag{7.34}$$

Thus $p(x|y)\,\delta x$ is the probability that the continuous random variable X lies between x and $x + \delta x$ when it is known that Y lies between y and $y + \delta y$ and is the probability that both X and Y are in the required range divided by the probability of Y being in the range. Correspondingly the conditional dependence of Y on X is given by

$$p(y|x) = \frac{p(x, y)}{p(x)}. \tag{7.35}$$

If X and Y are statistically independent then a knowledge of Y will provide no additional information about X, so that

$$p(x|y) = p(x),$$

since $p(x, y) = p(x)q(y)$.

In a similar manner the conditional expectation of the function $f(y)$ of y given $X = x$ is given by

$$\begin{aligned} E[f(y)|x] &= \int_{-\infty}^{\infty} f(y)p(y|x)\,\mathrm{d}y, \\ &= \int_{-\infty}^{\infty} f(y)\frac{p(x, y)}{p(x)}\,\mathrm{d}y = g(x), \end{aligned} \tag{7.36}$$

say, which in general is a function of x. In particular if $f(y) = y$ the resulting $g(x)$ is called the regression of y on x and if this is linear so that $g(x) = ax + b$, where a and b are constants, then $E[y|x] = ax + b$ and the random variable Y is said to be separable with respect to the random variable X. Higher orders of separability have been defined so that if

$$E[y|x] = \sum_{i=0}^{n} a_i x^i,$$

then Y is said to be separable of order n with respect to X.

7.4 Stochastic processes

Considering again the random process consisting of the ensemble of signals in figure 7.1, then at any instant of time t_1 the values of all these signals yield a random variable $X(t_1)$. Thus we can obtain a probability density function for the random variable $X(t_1)$; a joint density function for the random variables $X(t_1)$, $X(t_2)$ and so on, where the density functions and properties for more than two random variables follow as a logical extension of the two variable situation. Similarly for two stochastic processes whose ensembles consist of the signals $x_i(t)$, $y_i(t)$ we can evaluate a joint probability density function for the random variables $X(t_1)$ and $Y(t_2)$. In general these probability density functions will depend upon the times chosen, for example $p[x(t_1)]$ on t_1 and $p[x(t_1), y(t_2)]$ on t_1 and t_2.

The mean value of a random process at a particular instant of time is $E[X(t_1)]$, where

$$E[X(t_1)] = \int_{-\infty}^{\infty} xp(x, t_1)\, dx, \tag{7.37}$$

and the autocorrelation at times t_1 and t_2 is $E[X(t_1)X(t_2)]$ where

$$[E[X(t_1)X(t_2)] = R_{x_1x_2}(t_1, t_2) = \int_{-\infty}^{\infty}\int_{-\infty}^{\infty} x_1x_2p(x_1, x_2, t_1, t_2)\, dx_1\, dx_2. \tag{7.38}$$

For two processes the crosscorrelation is $E[X(t_1)Y(t_2)]$ where

$$E[X(t_1)Y(t_2)] = R_{xy}(t_1, t_2) = \int_{-\infty}^{\infty}\int_{-\infty}^{\infty} xyp(x, y, t_1, t_2)\, dx\, dy. \tag{7.39}$$

7.4.1 *Stationary processes*

A process is completely stationary when the probability density functions depend only on the difference in time at which the measurements are taken, that is, they are invariant with respect to time translation. Thus, for a completely stationary process the (first) probability density function is independent of time, the second, or joint, probability density function is dependent only on the time difference, $\tau = t_2 - t_1$, and so on. Mathematically

$$p(x, t_i) = p(x),$$

and

$$p(x, y, t_i, t_i + \tau) = p(x, y, \tau),$$

etc., where t_i is any instant of time.

In many practical situations one is usually only interested in a limited number of statistical properties, for example the mean values and correlation functions of the variables. When the random process $X(t)$ is such that its mean value $E[X(t)]$ is independent of the time t and the correlation $E[X(t_1)X(t_2)]$ depends only on the time difference $(t_2 - t_1)$ the random process is said to be weakly stationary. For the examples considered in this book only weak stationarity is implied.

7.4.2 *Ergodic processes*

A random process is said to be ergodic if time averages of member functions of the process can be assumed equivalent to the corresponding ensemble averages. Mathematically for an ergodic process

$$E[f\{X(t)\}] = \lim_{T\to\infty} (1/2T) \int_{-T}^{T} f[x(t)]\,\mathrm{d}t$$

for all functions f. For example, the random process given by $x_i(t) = c_i$, where c_i is a constant and i has values $1,2,3,\ldots N$, is stationary but only ergodic when $c_i = c$ for all i. Similarly the random process $x_i(t) = c_i \sin(\omega t + \phi_i)$ with the phase angle ϕ_i uniformly distributed in the range $0 \leq \phi \leq 2\pi$, and the amplitude c_i a random variable is again stationary only, but stationary and ergodic if $c_i = c$ for all i. Stationarity is thus a necessary condition for ergodicity.

The probability density function for an ergodic process can be obtained from any member function $x_i(t)$ by using the relationship

$$p(x) = \lim_{\substack{T\to\infty \\ \delta x\to 0}} \frac{\delta T(x, \delta x)}{T\delta x}, \tag{7.40}$$

where $T(x, \delta x)$ is the time spent by $x_i(t)$ in the range x to $x + \delta x$, and the second probability density function from

$$p(x_1, x_2) = \lim_{\substack{T\to\infty, \\ \delta x_1\to 0 \\ \delta x_2\to 0}} \frac{\delta T(x_1, x_2, \delta x_1, \delta x_2, \tau)}{T\delta x_1\,\delta x_2}, \tag{7.41}$$

where $\delta T(x_1, x_2, \delta x_1, \delta x_2, \tau)$ is the total time for which the signal $x_i(t)$ lies between x_1 and $x_1 + \delta x_1$ and also between x_2 and $x_2 + \delta x_2$ τ seconds later.

Therefore for an ergodic process

$$E[f\{X(t)\}] = \int_{-\infty}^{\infty} f(x)p(x)\,\mathrm{d}x = \lim_{T\to\infty} (1/2T) \int_{-T}^{T} f\{x(t)\}\,\mathrm{d}t, \tag{7.42}$$

and in particular the mean value is

$$E[X(t)] = \int_{-\infty}^{\infty} xp(x)\,\mathrm{d}x = \lim_{T\to\infty} (1/2T) \int_{-T}^{T} x(t)\,\mathrm{d}t. \tag{7.43}$$

Similarly the autocorrelation, which being a function of τ is usually called the autocorrelation function, is

$$E[X(t)X(t+\tau)] = \int_{-\infty}^{\infty}\int_{-\infty}^{\infty} x_1 x_2 p(x_1, x_2, \tau)\,\mathrm{d}x_1\,\mathrm{d}x_2,$$

$$= \lim_{T\to\infty} (1/2T)\int_{-T}^{T} x(t)x(t+\tau)\,\mathrm{d}t, \tag{7.44}$$

and the crosscorrelation function between two ergodic processes is

$$E[X(t)Y(t+\tau)] = \int_{-\infty}^{\infty}\int_{-\infty}^{\infty} xyp(x, y, \tau)\,\mathrm{d}x\,\mathrm{d}y,$$

$$= \lim_{T\to\infty} (1/2T)\int_{-T}^{T} x(t)y(t+\tau)\,\mathrm{d}t. \tag{7.45}$$

7.5 Correlation functions

Since the autocorrelation and crosscorrelation functions, defined in equations (7.44) and (7.45) are extremely important in the consideration of random signals some of their major properties are reviewed below.

(1) The autocorrelation function is an even function of τ.

By definition

$$R_{xx}(\tau) = \lim_{T\to\infty} (1/2T)\int_{-T}^{T} x(t)x(t+\tau)\,\mathrm{d}t,$$

therefore putting $t' = t + \tau$ gives

$$R_{xx}(\tau) = \lim_{T\to\infty} (1/2T)\int_{-T+\tau}^{T+\tau} x(t'-\tau)x(t')\,\mathrm{d}t',$$

i.e.

$$R_{xx}(\tau) = R_{xx}(-\tau). \tag{7.46}$$

In a similar manner it can be shown for the crosscorrelation function, which in general is not an even function of τ, that

$$R_{xy}(\tau) = R_{yx}(-\tau). \tag{7.47}$$

(2) The autocorrelation function for zero argument is equal to the mean squared value of the signal, that is

$$R_{xx}(0) = \lim_{T\to\infty} (1/2T)\int_{-\infty}^{T} x^2(t)\,\mathrm{d}t = \int_{-\infty}^{\infty} x^2 p(x)\,\mathrm{d}x = \overline{x^2(t)}. \tag{7.48}$$

$\overline{x^2(t)}$, where the bar denotes a time average, is also referred to as the variance, when $\overline{x(t)} = 0$, or the power of the signal $x(t)$. The latter term is used since the power dissipated in a unit resistance to which a voltage $x(t)$ is applied is $\overline{x^2(t)}$.

The crosscorrelation function for zero argument $R_{xy}(0)$ may also be referred to as the covariance or cross power, when $\overline{x(t)} = \overline{y(t)} = 0$.

(3) The magnitude of the autocorrelation function for all τ is never greater than the value for $\tau = 0$. Since

$$\pm x(t)x(t+\tau) = (1/2)[\{x(t) \pm x(t+\tau)\}^2 - x^2(t) - x^2(t+\tau)],$$

averaging over the time from $-T$ to T as $T \to \infty$,

$$\pm R_{xx}(\tau) = (1/2)\overline{\{x(t) \pm x(t+\tau)\}^2} - (1/2)\overline{x^2(t)} - (1/2)\overline{x^2(t+\tau)},$$

that is

$$\mp R_{xx}(\tau) = R_{xx}(0) - (1/2)\overline{\{x(t) \pm x(t+\tau)\}^2}.$$

Hence

$$|R_{xx}(\tau)| \leq R_{xx}(0), \tag{7.49}$$

as

$$\overline{\{x(t) \pm x(t+\tau)\}^2} \geq 0.$$

(4) The autocorrelation function of a signal containing periodic components or an average value also contains periodic components of the same frequency or an average value. This is easily shown by representing the signal as the sum of two uncorrelated signals. The autocorrelation of the sum is the sum of the autocorrelations.

(5) The autocorrelation function of a random signal with zero mean and no periodic component tends to zero as $\tau \to \infty$. This implies that the value of a random signal at a particular instant of time is independent of the value of the signal τ seconds previously as τ becomes large. The rate at which the function tends to zero depends upon the frequency content or spectrum of the signal, which is considered in the next section.

(6) Convolution of the autocorrelation function of a signal applied to a linear network and the network impulse response gives the crosscorrelation function between the signal and the network output. Taking the network input as $x(t)$ and the corresponding output as $y(t)$, then

$$y(t) = \int_{-\infty}^{\infty} g(u)x(t-u)\,du,$$

where $G(s) = \mathscr{L}[g(u)]$ is the transfer function of the element. But

$$R_{xy}(\tau) = \lim_{T\to\infty} (1/2T)\int_{-T}^{T} x(t)y(t+\tau)\,dt;$$

therefore

$$R_{xy}(\tau) = \lim_{T\to\infty} (1/2T)\int_{-T}^{T} x(t)\left\{\int_{-\infty}^{\infty} g(u)x(t+\tau-u)\,du\right\} dt.$$

Interchanging the order of integration

$$R_{xy}(\tau) = \int_{-\infty}^{\infty} g(u)\left\{\lim_{T\to\infty} (1/2T)\int_{-T}^{T} x(t)x(t+\tau-u)\,\mathrm{d}t\right\} \mathrm{d}u,$$

that is

$$R_{xy}(\tau) = \int_{-\infty}^{\infty} g(u)R_{xx}(\tau - u)\,\mathrm{d}u. \tag{7.50}$$

Thus the input–output crosscorrelation is the convolution of the system impulse response and the autocorrelation of the input. In other words if the input to the system were the input autocorrelation function $R_{xx}(t)$ the output would be the crosscorrelation $R_{xy}(t)$. In a similar manner it can be shown that

$$R_{yy}(\tau) = \int_{-\infty}^{\infty} g(u)R_{yx}(\tau - u)\,\mathrm{d}u, \tag{7.51}$$

or alternatively

$$R_{yy}(\tau) = \int_{-\infty}^{\infty} g(u)R_{xy}(u - \tau)\,\mathrm{d}u.$$

Hence using equation (7.50) to substitute for $R_{xy}(u - \tau)$ gives

$$R_{yy}(\tau) = \int_{-\infty}^{\infty} g(u)\left\{\int_{-\infty}^{\infty} g(v)R_{xx}(u-\tau-v)\,\mathrm{d}v\right\} \mathrm{d}u,$$

which becomes

$$R_{yy}(\tau) = \int_{-\infty}^{\infty}\int_{-\infty}^{\infty} g(u)g(v)R_{xx}(\tau + v - u)\,\mathrm{d}v\,\mathrm{d}u. \tag{7.52}$$

7.6 The power density spectrum

Consider a representative member signal $x(t)$ of a stationary random process which exists for $-\infty < t < \infty$. Define a signal $x_T(t)$ such that

$$\begin{aligned} x_T(t) &= x(t) \qquad |t| \leq T, \\ x_T(t) &= 0 \qquad |t| > T. \end{aligned} \tag{7.53}$$

This artifice, which represents a practical situation as only a finite length of the signal will be measured, enables the Fourier transform $x_T(j\omega)$ of $x_T(t)$ to be defined since $\int_{-\infty}^{\infty} x_T(t)\,\mathrm{d}t$ is finite. Relationships for $x(t)$ can be found by letting $T \to \infty$, as

$$x(t) = \lim_{T\to\infty} x_T(t).$$

By definition

$$X_T(j\omega) = \int_{-\infty}^{\infty} x_T(t)\mathrm{e}^{-j\omega t}\,\mathrm{d}t, \tag{7.54}$$

and

$$x_T(t) = (1/2\pi)\int_{-\infty}^{\infty} X_T(j\omega)e^{j\omega t}\,d\omega. \tag{7.55}$$

The mean squared value, or average power of $x_T(t)$ in the interval $-T$ to T is

$$\overline{x_T^2(t)} = (1/2T)\int_{-\infty}^{\infty} x_T^2(t)\,dt.$$

Using equation (7.55)

$$\overline{x_T^2(t)} = (1/2T)\int_{-\infty}^{\infty} x_T(t)\left\{(1/2\pi)\int_{-\infty}^{\infty} X_T(j\omega)e^{j\omega t}\,d\omega\right\}dt,$$

which on changing the order of integration and using equation (7.54) gives

$$\overline{x_T^2(t)} = (1/4\pi T)\int_{-\infty}^{\infty} X_T(j\omega)X_T(-j\omega)\,d\omega,$$

$$= (1/4\pi T)\int_{-\infty}^{\infty} |X_T(j\omega)|^2\,d\omega.$$

Thus

$$(1/2T)\int_{-T}^{T} x_T^2(t)\,dt = (1/4\pi T)\int_{-\infty}^{\infty} |X_T(j\omega)|^2\,d\omega, \tag{7.56}$$

a result which could be written down immediately using Parseval's theorem.(1) It is not possible at this stage to let T approach infinity in the above equation as $X_T(j\omega)$ does not exist in the limit. Since $X_T(j\omega)$ is a random variable with respect to the ensemble of sample functions from which $x(t)$ was taken it is reasonable to suppose, and can in fact be proved under certain general conditions,(1) that the limit of the expected value of $(1/T)|X_T(j\omega)|^2$ does exist. Therefore, taking the expected value of both sides of equation (7.56), interchanging the order of the expectation and integration, and taking the limit as $T \to \infty$ we obtain

$$\lim_{T\to\infty}(1/2T)\int_{-T}^{T} E\{x^2(t)\}\,dt = \lim_{T\to\infty}(1/4\pi T)\int_{-\infty}^{\infty} E\{|X_T(j\omega)|^2\}\,d\omega,$$

which may be written

$$E\{\overline{x^2(t)}\} = (1/2\pi)\int_{-\infty}^{\infty} S_{xx}(\omega)\,d\omega,$$

where

$$S_{xx}(\omega) = \lim_{T\to\infty} E\{X_T(j\omega)X_T(-j\omega)\}/2T. \tag{7.57}$$

$S_{xx}(\omega)$ is known as the power spectral density of the random process and is seen to be both a real and even function of ω. If, in addition, the process is ergodic then the mean squared value, or average power P_{av}, is given by

$$P_{av} = \overline{x^2(t)} = E\{\overline{x^2(t)}\} = (1/2\pi)\int_{-\infty}^{\infty} S_{xx}(\omega)\, d\omega$$

which, since $S_{xx}(\omega)$ is an even function of ω, may also be written

$$\overline{x^2(t)} = (1/\pi)\int_{0}^{\infty} S_{xx}(\omega)\, d\omega.$$

The power contained between angular frequencies ω_1 and ω_2 of $x(t)$ is then P'_{av} where

$$P'_{av} = (1/\pi)\int_{\omega_1}^{\omega_2} S_{xx}(\omega)\, d\omega. \tag{7.58}$$

The unit of power is watts so that $S_{xx}(\omega)$ has units of watts per hertz.† When $x(t)$ is the input to a network of transfer function $G(s)$, the Fourier transform of the output $y(t)$ is given by

$$Y(j\omega) = G(j\omega)X(j\omega). \tag{7.59}$$

Therefore

$$\begin{aligned} S_{yy}(\omega) &= \lim_{T\to\infty} E\{Y_T(j\omega)Y_T(j\omega)\}/2T, \\ &= G(j\omega)G(-j\omega)\lim_{T\to\infty} E\{X_T(j\omega)X_T(-j\omega)\}/2T, \end{aligned}$$

so that

$$S_{yy}(\omega) = |G(j\omega)|^2 S_{xx}(\omega). \tag{7.60}$$

The cross power spectral density $S_{xy}(j\omega)$ of $x(t)$ and $y(t)$ is defined similarly to equation (7.57) by

$$S_{xy}(j\omega) = \lim_{T\to\infty} E\{X_T(-j\omega)Y_T(j\omega)\}/2T, \tag{7.61}$$

giving, when $x(t)$ and $y(t)$ are related by equation (7.59)

$$S_{xy}(j\omega) = G(j\omega)S_{xx}(\omega). \tag{7.62}$$

The cross spectral density is thus seen to be neither a real nor an even function of ω. Also

$$S_{yx}(j\omega) = G(-j\omega)S_{xx}(\omega) = S_{xy}(-j\omega), \tag{7.63}$$

since $S_{xx}(\omega)$ is an even function of ω.

† Equation (7.58) can be written $P'_{av} = \int_{\omega_1}^{\omega_2} S'_{xx}(\omega)\, d\omega$ if $S'_{xx}(\omega)$ is in watts per radian per second. The notation given has been used so that the power spectral density and correlation function are Fourier transform pairs.

7.6.1 *Relationship between the correlation function and the power spectral density*

Comparison of equations (7.50) and (7.62) suggests that $R_{xx}(\tau)$ and $S_{xx}(\omega)$ are Fourier transform pairs. This is easily proved starting from the definition of the power spectral density given in equation (7.57). Substituting in equation (7.57) for $X_T(j\omega)$ from equation (7.54) gives

$$S_{xx}(\omega) = \lim_{T\to\infty} (1/2T)E\left\{\int_{-T}^{T} x_T(t_1)\mathrm{e}^{-j\omega t_1}\,\mathrm{d}t_1 \int_{-T}^{T} x_T(t_2)\mathrm{e}^{-j\omega t_2}\,\mathrm{d}t_2\right\}.$$

Changing the order of the expectation and integration and substituting $t_2 - t_1 = \tau$ we obtain

$$S_{xx}(\omega) = \lim_{T\to\infty} (1/2T)\int_{-T-t_1}^{T-t_1} \mathrm{d}\tau \int_{-T}^{T} \mathrm{e}^{-j\omega\tau}E\{x_T(t_1)x_T(t_1+\tau)\}\,\mathrm{d}t_1,$$

$$= \lim_{T\to\infty} (1/2T)\int_{-T-t_1}^{T-t_1} \mathrm{d}\tau \int_{-T}^{T} \mathrm{e}^{-j\omega\tau}R_{xx}(t_1, t_1+\tau)\,\mathrm{d}t_1,$$

$$= \int_{-\infty}^{\infty}\left\{\lim_{T\to\infty} (1/2T)\int_{-T}^{T} R_{xx}(t_1, t_1+\tau)\,\mathrm{d}t_1\right\}\mathrm{e}^{-j\omega\tau}\,\mathrm{d}\tau.$$

For a stationary process, the autocorrelation function depends only on the time difference, thus

$$\overline{R_{xx}(t_1, t_1+\tau)} = R_{xx}(\tau),$$

so that

$$S_{xx}(\omega) = \int_{-\infty}^{\infty} R_{xx}(\tau)\mathrm{e}^{-j\omega\tau}\,\mathrm{d}\tau. \tag{7.64}$$

The autocorrelation function can thus be obtained from the power spectrum by the inverse Fourier transform relationship, that is

$$R_{xx}(\omega) = (1/2\pi)\int_{-\infty}^{\infty} S_{xx}(\omega)\mathrm{e}^{j\omega\tau}\,\mathrm{d}\omega. \tag{7.65}$$

Since $S_{xx}(\omega)$ and $R_{xx}(\omega)$ are even functions we may also write

$$R_{xx}(\tau) = (1/\pi)\int_{0}^{\infty} S_{xx}(\omega)\cos\omega\tau\,\mathrm{d}\omega, \tag{7.66}$$

and

$$S_{xx}(\omega) = 2\int_{0}^{\infty} R_{xx}(\tau)\cos\omega\tau\,\mathrm{d}\tau. \tag{7.67}$$

It is worth noting here that equations (7.57) and (7.67) are useful for the numerical computation of power spectra from finite time length signal and autocorrelation function records respectively.

Similarly it may be shown that $S_{xy}(j\omega)$ and $R_{xy}(\tau)$ are also transform pairs.

7.6.2 *The evaluation of mean squared values*

The mean squared value of a random signal can be evaluated from the spectral density using the relationship

$$\overline{x^2(t)} = (1/2\pi)\int_{-\infty}^{\infty} S_{xx}(\omega)\,\mathrm{d}\omega. \tag{7.68}$$

Since $S_{xx}(\omega)$ is an even function of ω this may be written

$$\overline{x^2(t)} = (1/2\pi)\int_{-\infty}^{\infty} \frac{c(j\omega)c(-j\omega)}{d(j\omega)\,d(-j\omega)}\,\mathrm{d}\omega,$$

that is

$$\overline{x^2(t)} = \frac{1}{2\pi j}\int_{-j\infty}^{j\infty} \frac{c(s)c(-s)}{d(s)\,d(-s)}\,\mathrm{d}s, \tag{7.69}$$

where in general,†

$$c(s) = \sum_{j=0}^{n-1} c_j s^j,$$

and

$$d(s) = \sum_{j=0}^{n} d_j s^j,$$

and $d(s)$ has no right half plane zeros. This integral has been tabulated[(2)] up to a value of $n = 10$; a shortened version is given in table 7.1 for values up to $n = 6$.

7.6.3 *White noise applied to a network*

White noise is a random signal having a constant spectral density at all frequencies. Thus, if $x(t)$ is a white noise signal

$$S_{xx}(\omega) = S \quad \text{a constant,}$$

giving

$$R_{xx}(\tau) = (1/\pi)\int_{0}^{\infty} S\cos\omega\tau\,\mathrm{d}\omega = S\delta(\tau),\ddagger \tag{7.70}$$

where $\delta(\tau)$ is a unit impulse at $\tau = 0$. Ideal white noise which is physically unrealisable since it contains infinite power is seen to have the further properties:

(1) When applied to a network with S equal to unity the crosscorrelation function between input and output by virtue of equation (7.50) is the network impulse response.

---

† Most transfer functions of physical systems are rational functions of s. One common exception is the pure time delay (dead time) e^{-sT}, which can however be approximated by the ratio of polynomials in s by using the Pade Table.[(3)]

‡ This result is not obvious (see appendix B), although it is easily shown that the Fourier transform of $S\delta(\tau)$ is S.

TABLE 7.1

Table of integrals

$$I_n = \frac{1}{2\pi j}\int_{-j\infty}^{j\infty} \frac{c(s)c(-s)}{d(s)d(-s)} ds$$

where

$$c(s) = \sum_{j=0}^{n-1} c_j s^j \quad \text{and} \quad d(s) = \sum_{j=0}^{n} d_j s^j$$

$$I_1 = \frac{c_0^2}{2d_0 d_1}$$

$$I_2 = \frac{c_1^2 d_0 + c_0^2 d_2}{2d_0 d_1 d_2}$$

$$I_3 = \frac{c_2^2 d_0 d_1 + (c_1^2 - 2c_0c_2)d_0 d_3 + c_0^2 d_2 d_3}{2d_0 d_3(-d_0 d_3 + d_1 d_2)}$$

$$I_4 = \frac{c_3^2(-d_0^2 d_3 + d_0 d_1 d_2) + (c_2^2 - 2c_1c_3)d_0 d_1 d_4 + (c_1^2 - 2c_0c_2)d_0 d_3 d_4 + c_0^2(-d_1 d_4^2 + d_2 d_3 d_4)}{2d_0 d_4(-d_0 d_3^2 - d_1^2 d_4 + d_1 d_2 d_3)}$$

$$I_5 = \frac{1}{2\Delta_5}[c_4^2 m_0 + (c_3^2 - 2c_2c_4)m_1 + (c_2^2 - 2c_1c_3 + 2c_0c_4)m_2 + (c_1^2 - 2c_0c_2)m_3 + c_0^2 m_4]$$

where

$$m_0 = \frac{1}{d_5}(d_3 m_1 - d_1 m_2) \qquad m_3 = \frac{1}{d_0}(d_2 m_2 - d_4 m_1)$$

$$m_1 = -d_0 d_3 + d_1 d_2 \qquad m_4 = \frac{1}{d_0}(d_2 m_3 - d_4 m_2)$$

$$m_2 = -d_0 d_5 + d_1 d_4 \qquad \Delta_5 = d_0(d_1 m_4 - d_3 m_3 + d_5 m_2)$$

$$I_6 = \frac{1}{2\Delta_6}[c_5^2 m_0 + (c_4^2 - 2c_3c_5)m_1 + (c_3^2 - 2c_2c_4 + 2c_1c_5)m_2 + (c_2^2 - 2c_1c_3 + 2c_0c_4)m_3 + (c_1^2 - 2c_0c_2)m_4 + c_0^2 m_5]$$

where

$$m_0 = \frac{1}{d_6}(d_4 m_1 - d_2 m_2 + d_0 m_3) \qquad m_4 = \frac{1}{d_0}(d_2 m_3 - d_4 m_2 + d_6 m_1)$$

$$m_1 = -d_0 d_1 d_5 + d_0 d_3^2 + d_1^2 d_4 - d_1 d_2 d_3 \qquad m_5 = \frac{1}{d_0}(d_2 m_4 - d_4 m_3 + d_6 m_2)$$

$$m_2 = d_0 d_3 d_5 + d_1^2 d_6 - d_1 d_2 d_5 \qquad \Delta_6 = d_0(d_1 m_5 - d_3 m_4 + d_5 m_3)$$

$$m_3 = d_0 d_5^2 + d_1 d_3 d_6 - d_1 d_4 d_5$$

Reproduced by permission of John Wiley, New York.[2]

(2) As the autocorrelation is zero except for $\tau = 0$ the amplitude at any particular instant of time is completely independent of all previous amplitudes.

A random signal having a flat spectrum over a limited frequency range say from $-\omega_1$ to ω_1 is referred to as band limited white noise and has an autocorrelation function

$$R_{xx}(\tau) = (1/2\pi)\int_{-\omega_1}^{\omega_1} Se^{j\omega\tau}\,d\omega = (S/2\pi)(e^{j\omega_1\tau} - e^{-j\omega_1\tau})/j\tau,$$

$$= (S\sin\omega_1\tau)/\pi\tau, \tag{7.71}$$

and a mean squared value $R_{xx}(0) = S\omega_1/\pi$. Note that $\lim_{\omega_1\to\infty}(\sin\omega_1\tau)/\pi\tau = \delta(\tau)$ which is in agreement with the above result for white noise. Application of a white noise signal $x(t)$ of spectral density S to a low pass filter with transfer function $G(s) = a/(s + a)$ results in an output $y(t)$ having a spectral density

$$S_{yy}(\omega) = a^2S/(a^2 + \omega^2), \tag{7.72}$$

and

$$\overline{y^2(t)} = (S/2\pi j)\int_{-j\infty}^{j\infty} \{a^2/(a^2 - s^2)\}\,ds = (S/2\pi)\int_{-\infty}^{\infty} \{a^2/(a^2 + \omega^2)\}\,d\omega,$$

which, using table 7.1, or integrating directly gives

$$\overline{y^2(t)} = (Sa/2\pi)\tan^{-1}(\omega/a)|_{-\infty}^{\infty} = Sa/2. \tag{7.73}$$

Also

$$R_{yy}(\tau) = (S/2\pi j)\int_{-j\infty}^{j\infty} \{(a^2e^{st})/(a^2 - s^2)\}\,ds,$$

which when put into partial fractions becomes

$$R_{yy}(\tau) = \frac{Sa}{4\pi j}\int_{-j\infty}^{j\infty}\left\{\frac{1}{a+s} + \frac{1}{a-s}\right\}e^{s\tau}\,ds.$$

This is the inverse relationship for the double sided Laplace transform (see appendix **B**), where the contribution to the solution from the integrand for the term with the positive pole is to be interpreted as giving a bounded solution for $\tau < 0$ and not an unbounded solution for $\tau > 0$. Obviously, since the autocorrelation function is an even function of τ, it is only necessary to evaluate the contribution from the negative pole, that is for $\tau > 0$. The solution for $\tau < 0$ can then be written down from symmetry. Thus

$$R_{yy}(\tau) = (Sa/2)\{e^{-a\tau}u(\tau) + e^{a\tau}u(-\tau)\},$$

where $u(\tau)$† is a unit step at $\tau = 0$, or

$$R_{yy}(\tau) = (Sa/2)e^{-a|\tau|} \quad \text{for all } \tau. \tag{7.74}$$

† Note that (appendix **B**) the unit step $u(\tau)$ is defined by

$$u(\tau) = \begin{cases} 0, t < 0 \\ \frac{1}{2}, t = 0. \\ 1, t > 0 \end{cases}$$

The rapidity with which $R_{yy}(\tau) \to 0$ as $|\tau| \to \infty$ depends upon the time constant a^{-1} of $G(s)$, the larger the time constant (i.e. the smaller a) the slower $R_{yy}(\tau) \to 0$.

A physical explanation for the behaviour of $R_{yy}(\tau)$ from a consideration of the second amplitude probability density function is helpful. Consider, for example, a signal which is known to have a maximum rate of change V_m and has a value x_1 at time t_1, then at a time $t_1 + \tau$ its value x_2 must lie within the range $x_1 \pm V_m\tau$. Therefore as V_m is increased, corresponding to a signal with a higher frequency content, the range of possible values of x_2 becomes larger and less dependent on the value x_1, such that the autocorrelation function tends more rapidly to zero. Obviously, a white noise signal has a possible unlimited (i.e. infinite) value of V_m.

7.7 Optimisation using a mean squared error criterion for a fixed configuration

Representation of a random signal by its power density spectrum or autocorrelation function has been seen to be a sufficient description for evaluation of the mean squared value of the signal. Therefore if the 'goodness' of a control system is defined in terms of the minimisation of the mean squared value of the error signal, it is theoretically possible to evaluate optimum values of the system parameters for stationary random inputs. To illustrate the procedure several examples are considered. Simple illustrations have been selected as the numerical calculations become particularly tedious and unwieldy for high order systems.

7.7.1 *System with a signal and noise input*

The block diagram of the system to be considered is shown in figure 7.2.

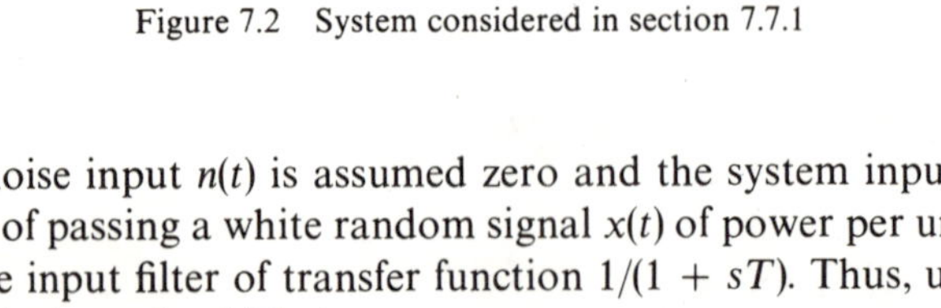

Figure 7.2 System considered in section 7.7.1

Initially the noise input $n(t)$ is assumed zero and the system input is the signal $r(t)$, the result of passing a white random signal $x(t)$ of power per unit bandwidth S_s through the input filter of transfer function $1/(1 + sT)$. Thus, using equation (7.73) and replacing a by $1/T$ gives

$$\overline{r^2(t)} = S_s/2T.$$

Also

$$C(s) = \frac{KX(s)}{(1 + sT)(s + K)}.$$

Therefore

$$\overline{c^2(t)} = \frac{S_s}{2\pi j}\int_{-j\infty}^{j\infty}\left\{\frac{K}{(1+sT)(s+K)}\right\}\left\{\frac{K}{(1-sT)(K-s)}\right\}\mathrm{d}s,$$

which using table 7.1 gives

$$\overline{c^2(t)} = S_sK/2(1+KT). \tag{7.75}$$

Now

$$\overline{e_s^2(t)} = \overline{\{r(t)-c(t)\}^2} = \overline{r^2(t)} + \overline{c^2(t)} - \overline{2r(t)c(t)},$$

where

$$\overline{r(t)c(t)} = (1/2\pi j)\int_{-j\infty}^{j\infty} S_{rc}(s)\,\mathrm{d}s \quad \text{with } s = j\omega.$$

But

$$S_{rc}(s) = \frac{KS_s}{(s+K)(1-sT)(1+sT)},$$

giving

$$\overline{r(t)c(t)} = \frac{S_s}{2\pi j}\int_{-j\infty}^{j\infty}\frac{K}{(s+K)(1+sT)(1-sT)}\mathrm{d}s. \tag{7.76}$$

Thus

$$\overline{r(t)c(t)} = S_sK\lim_{\tau\to 0}\mathscr{L}_D^{-1}\left\{\frac{1}{(s+K)(1+sT)(1-sT)}\right\},$$

where $\mathscr{L}_D{}^1$ denotes the inverse double sided Laplace transform with respect to the time parameter τ.

Now

$$\begin{aligned}\mathscr{L}_D^{-1}\frac{1}{(s+K)(1+sT)(1-sT)} &= \mathscr{L}_D^{-1}\left\{\frac{1/(1-T^2K^2)}{s+K} - \frac{T/2(1-TK)}{1+sT}\right.\\ &\qquad\left. + \frac{T/2(1+KT)}{1-sT}\right\},\\ &= \left\{\frac{e^{-K\tau}}{1-T^2K^2} - \frac{e^{-\tau/T}}{2(1-TK)}\right\}u(\tau)\\ &\qquad + \frac{e^{\tau/T}}{2(1+TK)}u(-\tau),\end{aligned}$$

giving as $\tau \to 0$

$$\frac{1}{2(1+KT)}\lim_{\tau\to 0}u(\tau) + \frac{1}{2(1+TK)}\lim_{\tau\to 0}u(-\tau) = 1/2(1+KT).$$

Hence

$$\overline{r(t)c(t)} = S_s K/2(1 + KT),$$

and

$$\overline{e_s^2(t)} = \frac{S_s}{2(1 + KT)} + \frac{S_s}{2T} - \frac{S_s K}{(1 + KT)},$$

$$= S_s(1 + KT + KT - 2KT)/2T(1 + KT) = S_s/2T(1 + KT).$$

In the above procedure evaluation of the cross power $\overline{r(t)c(t)}$ was complicated by the fact that the integrand of equation (7.76) was not of the form listed in table 7.1; a normal situation in cross power calculations. It is, however, possible to make use of the table in the following manner:

(1) Write the integral in the form

$$J = \frac{1}{2\pi j}\int_{-j\infty}^{j\infty} \frac{f(s)}{d(s)\,d(-s)}\,ds.$$

(2) Since the solution for J is real (the imaginary contribution between $-j\infty$ and 0 is equal and opposite to that between 0 and $j\infty$) only the even powers of s in $f(s)$ need be considered.

(3) Each even power may be identified with one term in the numerator $c(s)c(-s)$ of the standard integral, I_n.

Proceeding in this way for equation (7.76),

$$d(s) = s^2T + s(1 + KT) + K,$$

giving

$$d_2 = T, \qquad d_1 = 1 + KT \quad \text{and} \quad d_0 = K,$$

even $\{f(s)\}$ = even $\{K(K - s)\} = K^2$, so that $c_0 = K$. Using the table this gives

$$\overline{r(t)c(t)} = S_s K^2 T/2KT(1 + KT) = S_s K/2(1 + KT),$$

as before.

When we do not need to know the cross power but only the mean squared error, this can be evaluated more easily from the expression for $E_s(s)$ in terms of $X(s)$, that is

$$E_s(s) = \frac{sX(s)}{(s + K)(1 + sT)},$$

giving

$$\overline{e_s^2(t)} = \frac{S_s}{2\pi j}\int_{-j\infty}^{j\infty} \left\{\frac{s}{(1 + sT)(s + K)}\right\}\left\{\frac{-s}{(1 - sT)(K - s)}\right\} ds,$$

which using table 7.1 gives the desired solution

$$\overline{e_s^2(t)} = S_s/2T(1 + KT). \tag{7.77}$$

From inspection of equation (7.77) it is seen that the mean squared error decreases as K increases for a given value of T. The optimum system thus has infinite open loop gain, a physically unrealistic situation. If, however, the system input also contains a noise signal $n(t)$ as well as the signal input $r(t)$ an optimum value of K can be obtained. Before doing this it is necessary to consider the concept of error when a disturbance signal $n(t)$ is present in the system. The block diagram is redrawn in figure 7.3 with the closed loop system replaced by the

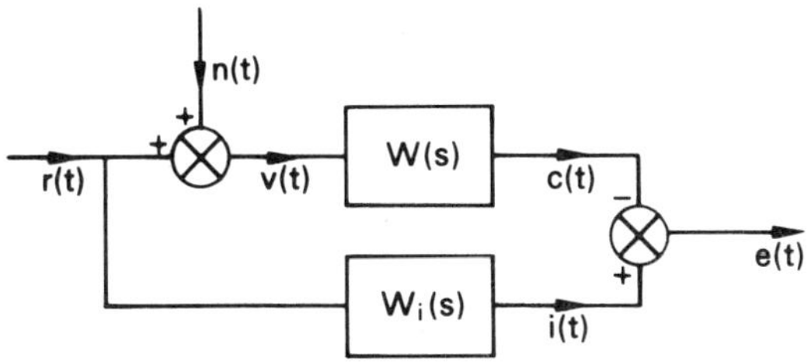

Figure 7.3 Reconfiguration of figure 7.2 with additional noise input

transfer function $W(s)$ and with a transfer function $W_i(s)$ relating the signal input $r(t)$ to the desired or ideal system output $i(t)$. The error signal $e(t)$ by definition is then given by $e(t) = i(t) - c(t)$, giving

$$e^2(t) = \{i(t) - c(t)\}^2 = i^2(t) + c^2(t) - 2i(t)c(t). \tag{7.78}$$

Since

$$I(s) = W_i(s)R(s),$$

and

$$C(s) = W(s)R(s) + W(s)N(s),$$

$$E(s) = \{W_i(s) - W(s)\}R(s) - W(s)N(s).$$

Writing $W_e(s) = W_i(s) - W(s)$ gives

$$E(s)E(-s) = W_e(s)W_e(-s)R(s)R(-s) + W(s)W(-s)N(s)N(-s)$$
$$- W_e(s)W(-s)R(s)N(-s) - W_e(-s)W(s)R(-s)N(s),$$

or, in terms of power density spectra,

$$S_{ee}(s) = W_e(s)W_e(-s)S_{rr}(s) + W(s)W(-s)S_{nn}(s) - W_e(s)W(-s)S_{nr}(s)$$
$$- W_e(-s)W(s)S_{rn}(s), \tag{7.79}$$

from which the mean squared value of the error, $e(t)$, can be evaluated.

In a control system with a feedforward transfer function $G(s)$ and a feedback transfer function $H(s)$ the desired output is normally related to the reference input by

$$\frac{I(s)}{R(s)} = \frac{1}{H(s)} = W_i(s),$$

and

$$W(s) = \frac{G(s)}{1 + G(s)H(s)}$$

so that

$$W_e(s) = \frac{1}{H(s)} - \frac{G(s)}{1 + G(s)H(s)} = \frac{1}{1 + G(s)H(s)}.$$

When the signal and noise are uncorrelated, $S_{nr}(s) = S_{rn}(s) = 0$ giving

$$S_{ee}(s) = W_e(s)W_e(-s)S_{rr}(s) + W(s)W(-s)S_{nn}(s),$$

and on integrating

$$\overline{e^2(t)} = \overline{e_s^2(t)} + \overline{c_n^2(t)},$$

where $\overline{e^2(t)}$ is the total mean squared error, $\overline{e_s^2(t)}$ is the mean squared error due to the signal only and $\overline{c_n^2(t)}$ is the output of the system due to the noise only.

Therefore, for the particular example,

$$\overline{c_n^2(t)} = \frac{S_n}{2\pi j}\int_{-j\infty}^{j\infty} \left\{\frac{K}{(s + K)}\right\}\left\{\frac{K}{(K - s)}\right\} \mathrm{d}s = S_n K/2,$$

giving

$$\overline{e^2(t)} = \frac{S_s}{2T(1 + TK)} + \frac{S_n K}{2}. \tag{7.80}$$

This has a minimum value when

$$\frac{\mathrm{d}\{\overline{e^2(t)}\}}{\mathrm{d}K} = 0 = \frac{S_n}{2} - \frac{S_s}{2T}\left\{\frac{T}{(1 + TK)^2}\right\},$$

that is

$$K = (1/T)\{(S_s/S_n)^{1/2} - 1\},$$

an acceptable value for K provided $S_s > S_n$.

7.7.2 *System with a constraint on another variable*

The system considered is shown in the block diagram of figure 7.4. It gives rise to the conventional second order characteristic equation

$$\frac{C(s)}{R(s)} = \frac{\omega_0^2}{s^2 + 2\zeta s\omega_0 + \omega_0^2}.$$

The system input is the random signal $r(t)$ having the power density spectrum obtained by passing white noise of unit power per unit bandwidth through the transfer function $1/(1 + sT)$. Initially it is assumed that all the parameters except ζ are fixed and the value of ζ which minimises the mean squared error is required.

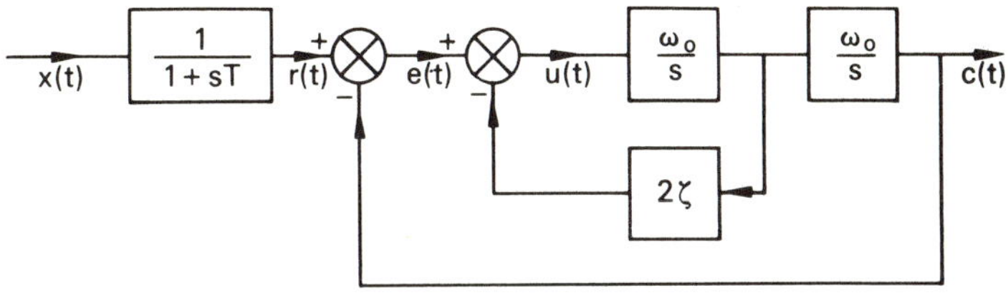

Figure 7.4 System considered in section 7.7.2

From the block diagram

$$\frac{E(s)}{R(s)} = \frac{s^2 + 2\zeta s\omega_0}{s^2 + 2\zeta s\omega_0 + \omega_0^2},$$

giving

$$\overline{e^2(t)} = (1/2\pi j)\int_{-j\infty}^{j\infty} A(s)A(-s)\,\mathrm{d}s,$$

where

$$A(s) = \frac{s^2 + 2\zeta s\omega_0}{(s^2 + 2\zeta s\omega_0 + \omega_0^2)(1 + sT)}.$$

Using table 7.1 the solution is

$$\overline{e^2(t)} = \frac{2\zeta + K + 4\zeta^2 K}{4T\zeta(1 + 2\zeta K + K^2)}, \tag{7.81}$$

where $K = T\omega_0$. Differentiating and setting $\mathrm{d}\overline{e^2(t)}/\mathrm{d}\zeta = 0$ to find the minimum with respect to ζ gives

$$(2 + 8\zeta K)(\zeta K^2 + 2\zeta^2 K + \zeta) = (2\zeta + K + 4\zeta^2 K)(1 + 4\zeta K + K^2),$$

that is

$$4\zeta^2 K^2 - 4\zeta K - K^2 - 1 = 0.$$

Solving the quadratic the permissible solution is

$$\zeta = \{1 + (K^2 + 2)^{1/2}\}/2K. \tag{7.82}$$

Thus for large K, that is when the system bandwidth is much greater than the bandwidth of the random input signal, the optimum value of ζ is 0·5. Again, it is seen from equation (7.81), that the optimum value of the loop gain ω_0^2 is infinite.

It may be desirable when designing a system to limit the value of a particular signal. If an analytical expression can be obtained for the signal which it is desired to limit then the technique of Lagrange can be used to minimise the mean squared error subject to the desired constraint on this signal. Usually, since it is only possible to evaluate analytically the mean squared value of a signal in a system, the constraint must be applied to the mean squared value. However, if the amplitude probability density of the signal is known a constraint on the mean squared value can be made so as to make the probability of exceeding a particular value as small as desired. Again, in many cases, a mean squared (power) constraint may be desirable to avoid say, overheating of a component.

To illustrate the application of Lagrange's method the above minimisation of the mean squared error will be repeated subject to a constraint on the mean squared value of $u(t)$.

From figure 7.4

$$\frac{U(s)}{R(s)} = \frac{s^2}{s^2 + 2\zeta s\omega_0 + \omega_0^2},$$

giving

$$\overline{u^2(t)} = (S/2\pi j)\int_{-j\infty}^{j\infty} B(s)B(-s)\,\mathrm{d}s,$$

where

$$B(s) = \frac{s^2}{(s^2 + 2\zeta s\omega_0 + \omega_0^2)(1 + sT)}.$$

The solution, using table 7.1, is

$$\overline{u^2(t)} = \frac{2\zeta + K}{4T\zeta(1 + 2\zeta K + K^2)}. \tag{7.83}$$

Optimum values of ζ and K can be obtained theoretically which minimise $\overline{e^2(t)}$ and constrain $\overline{u^2(t)}$ to be less than or equal to some positive value U_0. Since the expressions for $\overline{e^2(t)}$ and $\overline{u^2(t)}$ however both tend to zero as K tends to infinity, the optimum value of K is infinite. On the other hand, if the maximum value of K is limited so that $K \leq K_m$, it is possible to find optimum values of K and ζ for a given choice of the constraint value U_0.

This problem can be studied using Lagrange's method of undetermined multipliers. In this method the synthetic function, F, where

$$F = \overline{e^2(t)} + \rho\overline{u^2(t)} \tag{7.84}$$

is minimised with respect to ζ and K. The resulting relationships from $\partial F/\partial\zeta = 0$ and $\partial F/\partial K = 0$ plus the constraint $\overline{u^2(t)} = U_0$ give three equations which can be solved for the three unknowns ζ, K and ρ the undetermined multiplier.

Thus for this example

$$F = \frac{2\zeta + K + 4\zeta^2 K + \rho(2\zeta + K)}{4T\zeta(1 + 2\zeta K + K^2)},$$

which gives, for $\partial F/\partial K = 0$,

$$\rho(K^2 + 4\zeta^2 + 4\zeta K - 1) = 1 - K^2 - 4\zeta K - 4\zeta^2 K^2,$$

and, for $\partial F/\partial\zeta = 0$,

$$\rho(4\zeta^2 + 4\zeta K + K^2 + 1) = 4\zeta^2 K^2 - 4\zeta K - K^2 - 1.$$

Eliminating ρ from these two expressions,

$$(4\zeta^2K^2 - 4\zeta K - K^2 - 1)(K^2 + 4\zeta^2 + 4\zeta K - 1)$$
$$= (1 - K^2 - 4\zeta K - 4\zeta^2K^2)(4\zeta^2 + 4\zeta K + K^2 + 1),$$

which reduces to

$$4\zeta^2K^2 + 4\zeta K^3 + K^4 - 1 = 0,$$

that is

$$\zeta = (1 - K^2)/2K, \tag{7.85}$$

as the only positive solution for ζ provided K is less than unity.

Substituting this value of ζ in equation (7.83) with the constraint $\overline{u^2(t)} = U_0$ results in

$$TU_0 \geq \frac{1}{4(1 - K^2)}. \tag{7.86}$$

The right hand side of equation (7.86) is negative for $K > 1$, and in consequence if the specified K_m is greater than unity, that is $\omega_0 > 1/T$, the required value of ζ is obtained from equation (7.82) and the minimum possible value of $\overline{u^2(t)}$ is realised using $K = K_m$ and this value of ζ. If the value of $\overline{u^2(t)}$ is greater than the desired U_0 then it can only be decreased by increasing K_m.

Alternatively, if $\omega_0 < 1/T$, the above solution using Lagrange's method is applicable. The value of K required is obtained from equation (7.86) using the given value of U_0, and the corresponding value of ζ is then found from equation (7.85). The optimum value of K may now be less than K_m.

For example, to satisfy the constraint $U_0 = 0{\cdot}50/T$ with $K < 1$, equation (7.86) givcs

$$\frac{1}{4(1 - K^2)} = 0{\cdot}5, \qquad K = 0{\cdot}707,$$

and using this value in equation (7.85) gives

$$\zeta = \frac{1 - (0{\cdot}707)^2}{2(0{\cdot}707)} = 0{\cdot}353.$$

Thus a minimum mean squared error is obtained with $\zeta = 0{\cdot}353$ and $K = 0{\cdot}707$ with the constraints $U_0 = 0{\cdot}50/T$ and $0 < K_m < 1$. The magnitude of the mean squared error under these conditions, however, is somewhat greater than the mean squared value of the input. This unsatisfactory design is of course due to the unsuitable constraint imposed on the magnitude of K.

7.8 Minimisation of mean squared error with a free configuration

In the previous sections we have been concerned with optimising some variable parameters of the system transfer function $W(s)$ of figure 7.3. Here we consider the more general problem of finding the transfer function $W(s)$ which minimises the mean squared error $\overline{e^2(t)}$ in figure 7.3 given the system input $v(t)$ and desired output $i(t)$.

From figure 7.3 the convolution integral gives

$$c(t) = \int_{-\infty}^{\infty} w(t_1)v(t - t_1)\,dt_1,$$

and, using this expression in equation (7.78),

$$e^2(t) = i^2(t) - 2i(t)\int_{-\infty}^{\infty} w(t_1)v(t - t_1)\,dt_1 + \left\{\int_{-\infty}^{\infty} w(t_1)v(t - t_1)\,dt_1\right\}\left\{\int_{-\infty}^{\infty} w(t_2)v(t - t_2)\,dt_2\right\},$$

where the additional variable t_2 has been introduced to avoid confusion in later manipulations.

Averaging with respect to time gives

$$\overline{e^2(t)} = \overline{i^2(t)} - 2\lim_{T\to\infty}(1/2T)\int_{-T}^{T}\int_{-\infty}^{\infty} w(t_1)v(t - t_1)i(t)\,dt_1\,dt + \lim_{T\to\infty}(1/2T)\int_{-T}^{T}\int_{-\infty}^{\infty}\int_{-\infty}^{\infty} w(t_1)v(t - t_1)w(t_2)v(t - t_2)\,dt_1\,dt_2\,dt.$$

Changing the order of integration and integrating with respect to t first the expression reduces to

$$\overline{e^2(t)} = R_{ii}(0) - 2\int_{-\infty}^{\infty} w(t_1)R_{vi}(t_1)\,dt_1 + \int_{-\infty}^{\infty}\int_{-\infty}^{\infty} w(t_1)w(t_2) R_{vv}(t_1 - t_2)\,dt_1\,dt_2. \tag{7.87}$$

This equation gives an expression for the mean squared error as a functional of the system weighting function $w(t)$. We now have to determine the function $w(t)$ which minimises the value of $\overline{e^2(t)}$. This can be done using the calculus of variations by assuming a solution exists for $w(t)$ equal to $w_0(t)$ and considering the weighting function

$$w(t) = w_0(t) + \varepsilon w_\varepsilon(t), \tag{7.88}$$

where ε is a variable parameter and $\varepsilon w_\varepsilon(t)$ the variation of $w(t)$. Substituting the expression for $w(t)$ from equation (7.88) in equation (7.87) and differentiating with respect to ε, yields

$$\frac{d\overline{e^2(t)}}{d\varepsilon} = -2\int_{-\infty}^{\infty} R_{vi}(t_1)w_\varepsilon(t_1)\,dt_1 + \int_{-\infty}^{\infty}\int_{-\infty}^{\infty} w_0(t_1)w_\varepsilon(t_2)R_{vv}(t_1 - t_2)\,dt_1\,dt_2 + \int_{-\infty}^{\infty}\int_{-\infty}^{\infty} w_\varepsilon(t_1)w_0(t_2)R_{vv}(t_1 - t_2)\,dt_1\,dt_2 + 2\varepsilon\int_{-\infty}^{\infty}\int_{-\infty}^{\infty} w_\varepsilon(t_1)w_\varepsilon(t_2)R_{vv}(t_1 - t_2)\,dt_1\,dt_2. \tag{7.89}$$

Since the autocorrelation function is even, by changing the variables t_1 and t_2 in one double integral we can show that

$$\int_{-\infty}^{\infty}\int_{-\infty}^{\infty} w_0(t_1)w_\varepsilon(t_2)R_{vv}(t_1 - t_2)\,dt_1\,dt_2$$

$$= \int_{-\infty}^{\infty}\int_{-\infty}^{\infty} w_\varepsilon(t_1)w_0(t_2)R_{vv}(t_1 - t_2)\,dt_1\,dt_2.$$

Thus setting

$$d\overline{e^2(t)}/d\varepsilon = 0 \quad \text{at } \varepsilon = 0,$$

equation (7.89) gives

$$2\int_{-\infty}^{\infty}\left\{\int_{-\infty}^{\infty} w_0(t_2)R_{vv}(t_1 - t_2)\,dt_2 - R_{vi}(t_1)\right\}w_\varepsilon(t_1)\,dt_1 = 0. \tag{7.90}$$

Now $w_\varepsilon(t)$ is an arbitrary realisable weighting function and therefore must be zero for t less than zero. Thus the only way in which equation (7.90) can be satisfied for values of t_1 equal to or greater than zero is if the condition

$$\int_{-\infty}^{\infty} w_0(t_2)R_{vv}(t_1 - t_2)\,dt_2 - R_{vi}(t_1) = 0 \quad \text{for } t_1 \geq 0$$

is met. Evaluation of $d^2\overline{e^2(t)}/d\varepsilon^2$ shows that the stationary value of $w_0(t)$ satisfying the above integral equation corresponds to a minimum. The weighting function $w(t)$ which minimises the mean squared error is thus the solution of the integral equation, usually known as the Wiener–Hopf equation

$$\int_{0}^{\infty} w(t)R_{vv}(\tau - t)\,dt - R_{vi}(\tau) = 0 \quad \text{for } \tau \geq 0, \tag{7.91}$$

where since $w(t) = 0$ for $t < 0$ the lower limit has been changed to zero.

7.8.1 *Solution of the Wiener–Hopf equation*

Although it is possible to solve the integral equation (7.91) in a direct manner for certain special cases the more general procedure is to make use of transform methods. Defining the left hand side of equation (7.91) by $q(\tau)$, that is

$$q(\tau) = \int_{0}^{\infty} w(t)R_{vv}(\tau - t)\,dt - R_{vi}(\tau), \tag{7.92}$$

then the Wiener–Hopf equation requires

$$q(\tau) = 0 \quad \text{for all } \tau \geq 0. \tag{7.93}$$

Taking the double sided Laplace transform of equation (7.92) gives

$$Q(s) = W(s)S_{vv}(s) - S_{vi}(s). \tag{7.94}$$

Now to satisfy equation (7.93) $Q(s)$ must have no left half plane poles and also, since $w(t)$ is a physically realisable impulse response, it must have no right half plane poles. The solution we obtain from equation (7.94) for $W(s)$ must therefore satisfy these constraints. The procedure for doing this is first to perform a spectrum factorisation of $S_{vv}(s)$, that is it is written

$$S_{vv}(s) = S_{vv}(s)_R S_{vv}(s)_L, \tag{7.95}$$

where $S_{vv}(s)_R$ and $S_{vv}(s)_L$ have both their poles and zeros in the right and left half of the s plane respectively. Obviously from previous considerations the form of the expressions for power density spectra of systems with rational transfer functions allows this factorisation. Equation (7.94) with (7.95) substituted may be written

$$Q(s) = S_{vv}(s)_R \left\{ W(s)S_{vv}(s)_L - \frac{S_{vi}(s)}{S_{vv}(s)_R} \right\},$$

and on putting the ratio $S_{vi}(s)/S_{vv}(s)_R$ into partial fractions it yields

$$Q(s) = S_{vv}(s)_R \left\{ W(s)S_{vv}(s)_L - \left[\frac{S_{vi}(s)}{S_{vv}(s)_R}\right]_{Lp} - \left[\frac{S_{vi}(s)}{S_{vv}(s)_R}\right]_{Rp} \right\},$$

where the subscripts Lp and Rp are used to denote all those terms in the partial fraction expansion with left half and right half plane poles respectively. Thus for $Q(s)$ to have right half plane poles only we require the first two terms within the brackets to add to zero. That is

$$W(s) = \frac{1}{S_{vv}(s)_L} \left[\frac{S_{vi}(s)}{S_{vv}(s)_R}\right]_{Lp}, \tag{7.96}$$

which is the required solution of the Wiener–Hopf equation.

7.8.2 *Semi-free configuration*

To apply the Wiener–Hopf solution to a control system design normally requires a modification to account for the fact that the closed loop transfer function $W(s)$ is not completely arbitrary. Figure 7.5(a) shows a common situation where

$$W(s) = \frac{C(s)}{V(s)} = \frac{G_c'(s)G(s)}{1 + G_c'(s)G(s)H(s)},$$

the plant has a fixed transfer function $G(s)$, the feedback transfer function $H(s)$ may be open to some choice and the optimum compensator $G_c'(s)$ is to be found. It is shown in reference 2 that without any loss in design freedom the feedback configuration for $W(s)$ may be replaced by the equivalent cascade combination of a compensator $G_c(s)$ and the plant $G(s)$ as shown in figure 7.5(b). Thus

$$G_c(s)G(s) = \frac{G_c'(s)G(s)}{1 + G_c'(s)G(s)H(s)},$$

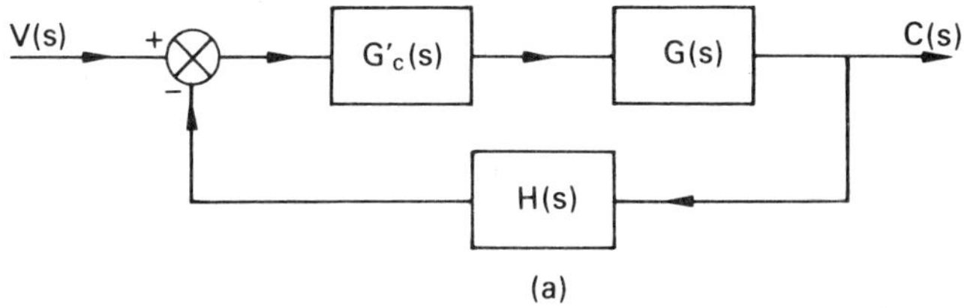

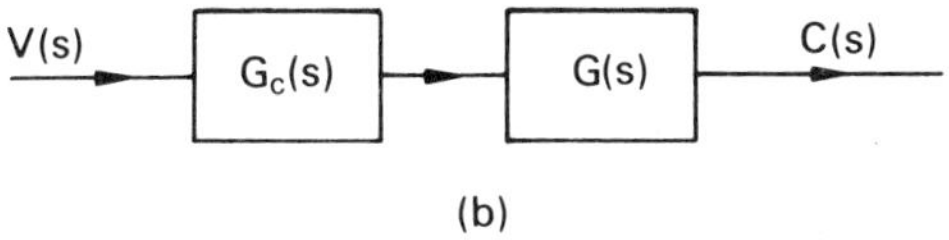

Figure 7.5 Feedback system rearrangement for semifree configuration design

giving

$$G_c'(s) = \frac{G_c(s)}{1 - G_c(s)G(s)H(s)}. \tag{7.97}$$

The design procedure for this situation, which is known as the semifree configuration, thus consists of determining the optimum compensator $G_c(s)$, with $W(s)$ equal to $G_c(s)G(s)$, and then using equation (7.97) to find the corresponding compensator $G_c'(s)$ required in the actual closed loop system.

When the plant transfer function $G(s)$ has right half plane zeros direct use of equation (7.96) is not possible since the solution for $G_c(s)$ will be $W(s)/G(s)$ an unstable transfer function. The determination of a valid solution in this case requires a logical extension of that given for the free configuration. By interchanging $G_c(s)$ and $G(s)$ in figure 7.5(b), the problem becomes one of finding the optimum $G_c(s)$, equivalent to $W(s)$ in the free configuration, with an input $X(s) = G(s)V(s)$ rather than $V(s)$. Making such a substitution in equation (7.96) gives

$$G_c(s) = \frac{1}{S_{xx}(s)_L}\left[\frac{S_{xi}(s)}{S_{xx}(s)}\right]_{Lp}.$$

But

$$S_{xx}(s) = G(s)G(-s)S_{vv}(s),$$

$$S_{xi}(s) = G(-s)S_{vi}(s),$$

so that the solution becomes

$$G_c(s) = \frac{1}{\{G(s)G(-s)\}_L S_{vv}(s)_L}\left[\frac{G(-s)S_{vi}(s)}{\{G(s)G(-s)\}_R S_{vv}(s)_R}\right]_{Lp}. \tag{7.98}$$

7.9 Statistical properties of particular signals

The statistical properties of the more important signals needed in control systems analysis are discussed below. Some deterministic signals are also considered since they are often best described for analytical purposes in a probabilistic manner particularly when they occur in the presence of random signals.

7.9.1 *Sinusoidal signal*

Consider the random process $x(t) = a\cos(\theta + \phi)$ where $\theta = \omega_0 t$, a is a constant and ϕ is uniformly distributed between 0 and 2π. As ϕ goes from 0 to 2π, x twice takes all the possible values between $+a$ and $-a$. The probability $r(x_1)\,\delta x$ of x lying between x_1 and $x_1 + \delta x$ is equal to twice the probability of $p(\phi_1)\,\delta\phi$ of ϕ lying between ϕ_1 and $\phi_1 + \delta\phi$ where $x_1 = a\cos(\theta + \phi_1)$. Hence

$$r(x)\,\delta x = 2p(\phi)\,\delta\phi,$$

giving

$$r(x) = 2p(\phi)(\mathrm{d}\phi/\mathrm{d}x).$$

But

$$p(\phi) = 1/2\pi \quad \text{and} \quad \mathrm{d}\phi/\mathrm{d}x = (a^2 - x^2)^{-1/2},$$

thus

$$r(x) = 1/\pi(a^2 - x^2)^{1/2}.$$

The amplitude probability for the sinusoidal signal is thus

$$\begin{aligned} r(x) &= 1/\pi(a^2 - x^2)^{1/2} && \text{for } |x| \le a, \\ &0 && \text{for } |x| > a, \end{aligned} \tag{7.99}$$

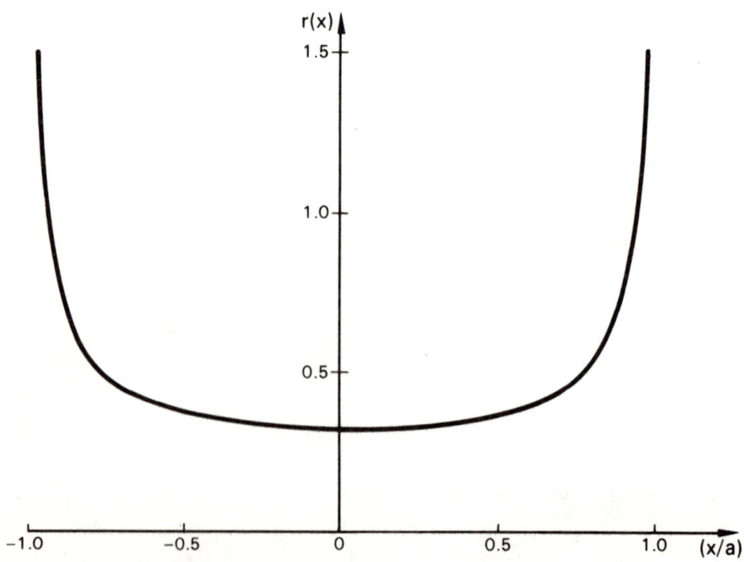

Figure 7.6 Graph of amplitude probability density $r(x)$ for a sinusoid

and is shown plotted in figure 7.6. The characteristic function $M(j\omega)$ by definition, is

$$M(j\omega) = E\{e^{j\omega x}\} = E\{e^{j\omega a\cos(\theta+\phi)}\}.$$

The exponential can be expanded using thc Jacobi–Anger formula[4]

$$e^{z\cos\theta} = \sum_{s=0}^{\infty} \varepsilon_s I_s(z)\cos s\theta = \sum_{s=0}^{\infty} \varepsilon_s j^{-s} J_s(jz)\cos s\theta, \tag{7.100}$$

where $I_s(z)$ and $J_s(z)$ are respectively the modified Bessel and Bessel functions of order s, and ε_s the Neumann factor (see section 3.2). Therefore

$$M(j\omega) = E\left\{\sum_{s=0}^{\infty} \varepsilon_s j^s J_s(a\omega)\cos s(\theta+\phi)\right\} = J_0(a\omega), \tag{7.101}$$

since $E\{\cos s\phi\}$ is zero for $s \neq 0$. Similarly, the moment generating function is

$$m(t) = J_0(ajt),$$

and the nth moment of the distribution is

$$\begin{aligned} \mu_n = \left.\frac{d^n m(t)}{dt^n}\right|_{t=0} &= \int_{-a}^{a} x^n r(x)\,dx \\ &= \frac{1\times 3\times 5\ldots(n-1)}{2\times 4\times 6\ldots n}a^n = \frac{2\Gamma\{(n+1)/2\}}{\pi^{1/2}\Gamma\{(n+2)/2\}}a^n, \quad \text{for } n \text{ even}, \\ &\quad 0 \qquad\qquad\qquad\qquad\qquad\qquad\qquad \text{for } n \text{ odd}. \end{aligned} \tag{7.102}$$

The autocorrelation function of the signal is

$$\begin{aligned} R_{xx}(\tau) &= E\{x(t)x(t+\tau)\}, \\ &= E\{a^2\cos(\theta+\phi)\cos(\theta+\phi+\omega_0\tau)\}, \\ &= E\{(a^2/2)[\cos\omega_0\tau + \cos(2\theta+2\phi+\omega_0\tau)]\}. \end{aligned}$$

Thus

$$R_{xx}(\tau) = (a^2/2)\cos\omega_0\tau, \tag{7.103}$$

which as expected depends on the time shift, τ, and in particular, $R_{xx}(0)$, the mean squared value is equal to $a^2/2$. The spectral density, $S(\omega)$, being the Fourier transform of the autocorrelation function is given by

$$\begin{aligned} S_{xx}(\omega) &= \int_{-\infty}^{\infty} (a^2/2)\cos\omega_0\tau\, e^{-j\omega\tau}\,d\tau, \\ &= \int_0^{\infty} a^2\cos\omega_0\tau\cos\omega\tau\,d\tau, \\ &= (a^2/2)\int_0^{\infty} \{\cos(\omega-\omega_0)\tau + \cos(\omega+\omega_0)\tau\}\,d\tau, \\ &= (a^2\pi/2)\{\delta(\omega-\omega_0) + \delta(\omega+\omega_0)\}, \end{aligned}$$

so that

$$S_{xx}(\omega) = (a^2\pi/2)\,\delta(|\omega| - \omega_0) \tag{7.104}$$

Since the process is ergodic the above statistical properties could also have been found from time averages of a member signal. To illustrate the time average approach we obtain the joint characteristic function from a time average on the member signals $x(t)$ and $y(t)$ of two random processes, where

$$x(t) = a \cos \omega_0(t + \tau) \quad \text{and} \quad y(t) = b \cos \omega_1 t.$$

Using E_t rather than a bar to denote the time average, by definition the joint characteristic function is

$$M(ju, jv, \tau) = E_t\{\exp j(au \cos \omega_0(t + \tau) + bv \cos \omega_1 t)\}, \tag{7.105}$$

which on using the Jacobi–Anger expansion yields

$$M(ju, jv, \tau) = E_t\left\{\sum_{s=0}^{\infty}\sum_{k=0}^{\infty} j^{s+k}\varepsilon_s\varepsilon_k J_s(au)J_k(bv)\cos s\omega_0(t + \tau)\cos k\omega_1 t\right\}$$

$$= \sum_{s=0}^{\infty}\sum_{k=0}^{\infty} j^{s+k}\varepsilon_s\varepsilon_k J_s(au)J_k(bv)E_t\{\cos s\omega_0(t + \tau)\cos k\omega_1 t\}.$$

The expectation has different solutions depending upon whether the frequencies ω_0 and ω_1 are commensurate or not. We consider two particular cases, the commensurate case when the frequencies are equal and the incommensurate case. For the latter case the solution is

$$M(ju, jv, \tau) = J_0(au)J_0(bv)$$

for ω_1/ω_0 not a ratio of integers, and for the former case

$$M(ju, jv, \tau) = \sum_{s=0}^{\infty} (-1)^s\varepsilon_s J_s(au)J_s(bv)\cos s\omega_0\tau \quad \text{for } \omega_1 = \omega_0. \tag{7.106}$$

An alternative form for this latter result can be obtained if the two cosine functions in equation (7.105), which are now of the same frequency, are added vectorially to yield

$$M(ju, jv, \tau) = E_t\{\exp j(a^2u^2 + b^2v^2 + 2abuv \cos \omega_0\tau)^{1/2}\cos(\omega_0 t + \psi)\},$$

which reduces to

$$M(ju, jv, \tau) = J_0\{(a^2u^2 + b^2v^2 + 2aubv \cos \omega_0\tau)^{1/2}\}.$$

The joint density function for the two processes can be found from the inverse double Fourier transform of the joint characteristic function. In particular if $\omega_1 = \omega_0$

$$r(x, y, \tau) = (1/4\pi^2)\int_{-\infty}^{\infty}\int_{-\infty}^{\infty} M(ju, jv, \tau)e^{j(ux+vy)}\,du\,dv,$$

which on substituting for $M(ju, jv, \tau)$ from equation (7.106) gives

$$r(x, y, \tau) = (1/4\pi^2) \sum_{s=0}^{\infty} \varepsilon_s \cos s\omega_0\tau \int_{-\infty}^{\infty} \int_{-\infty}^{\infty} (-1)^s J_s(au) J_s(bv) e^{j(ux+vy)}\, du\, dv.$$

Now[5]

$$(1/2\pi) \int_{-\infty}^{\infty} j^s J_s(au) e^{jux}\, du = (-1)^s T_s(x/a) r(x),$$

where $T_s(x)$ is the Chebyshev polynomial of order s (see section 3.2.2). Thus on separating the double integral the solution becomes

$$r(x, y, \tau) = \sum_{s=0}^{\infty} \varepsilon_s T_s(x/a) T_s(y/b) r(x) r(y) \cos s\omega_0\tau. \qquad (7.107)$$

Using the relationship

$$R_{xy}(\tau) = \int_{-\infty}^{\infty} \int_{-\infty}^{\infty} xy r(x, y, \tau)\, dx\, dy,$$

it is easy to show that on substituting equation (7.107) for $r(x, y, \tau)$ the known result $R_{xy}(\tau) = (ab/2)\cos \omega_0\tau$ is obtained.

7.9.2 *Signal with a uniform distribution*

For a signal $z(t)$ with a uniform amplitude probability density, $p(z)$, between the amplitude levels $\pm a$ we have

$$p(z) = (1/2a).$$

The characteristic function of the signal is

$$M(j\omega) = (1/2a) \int_{-a}^{a} e^{-j\omega z}\, dz = (\sin a\omega)/a\omega,$$

and its moments are

$$\mu_n = \begin{cases} 2a^{n+1}/(n+1) & \text{for } n \text{ even,} \\ 0 & \text{for } n \text{ odd.} \end{cases}$$

To obtain other properties of the signal, which are dependent upon the second density function, more information is required. For example, if the signal is periodic then these properties can be obtained from a Fourier series expansion for the signal. The triangular waveform $z(t)$ of period $2\pi/\omega_0$ can be represented by the series

$$z(t) = \sum_{n=1(2)}^{\infty} a_n \cos n\omega_0 t,$$

where

$$a_n = 8a/n^2\pi^2.$$

Hence, using the results of section 7.9.1, the autocorrelation function is

$$R_{zz}(\tau) = \sum_{n=1(2)}^{\infty} (a_n^2/2) \cos n\omega_0\tau,$$

and the spectral density is

$$S_{zz}(\omega) = (\pi/2) \sum_{n=1(2)}^{\infty} a_n^2\, \delta(|\omega| - n\omega_0).$$

7.9.3 *A Gaussian signal*

Consider the random process $x(t)$ of figure 7.1, then over the range 0 to T it may be represented by the Fourier series

$$x(t) = \sum_{n=0}^{\infty} a_n \cos \omega_n t + b_n \sin \omega_n t, \tag{7.108}$$

or alternatively

$$x(t) = \sum_{n=0}^{\infty} c_n \cos(\omega_n t - \phi_n), \tag{7.109}$$

where

$$a_0 = (1/T) \int_0^T x(t)\, \mathrm{d}t,$$

$$a_n = (2/T) \int_0^T x(t) \cos \omega_n t\, \mathrm{d}t,$$

$$b_n = (2/T) \int_0^T x(t) \sin \omega_n t\, \mathrm{d}t,$$

$$c_n = (a_n^2 + b_n^2)^{1/2},$$

$$\phi_n = \tan^{-1}(b_n/a_n),$$

and

$$\omega_n = 2\pi n/T = n\omega_1.$$

In the limit as $T \to \infty$ the spacing between each frequency component, $\delta\omega$, which is equal to $2\pi/T$ tends to zero. The coefficients a_n and b_n will be random variables which can be shown[1] to be approximately uncorrelated for any process whose spectral density is approximately constant over large intervals compared to $1/T$.

Further it can be shown[1] that a linear transformation on a set of normally (Gaussian) distributed random variables is another set of normally distributed random variables. Thus if a Gaussian input is applied to a linear network the output is Gaussian. Here, therefore, if the process $x(t)$ is Gaussian with zero mean, since the coefficients a_0, a_n and b_n are found by the above linear transformations, a_0 will be zero and a_n and b_n will be normally distributed. From the theory of nonlinear transformations it may be shown that c_n has a Rayleigh distribution (see section 8,10) and ϕ_n is uniformly distributed between 0 and 2π.

When a random signal is passed through a narrow band filter the observed output is a sinusoid of slowly varying amplitude and phase. This suggests that such a narrow band waveform can be represented by

$$x(t) = R(t)\cos\{\omega_c t + \delta(t)\}, \tag{7.110}$$

where ω_c is the centre frequency and $R(t)$ and $\delta(t)$ are random processes called the envelope function and phase function respectively.[6] From a comparison of the above representation for a narrow band signal with that given in equation (7.108) it can again be shown that $R(t)$ has a Rayleigh distribution and that $\delta(t)$ is uniformly distributed between 0 and 2π if the signal is Gaussian.

The large frequency of occurrence of the normal distribution is a consequence of the central limit theorem which states that if a signal consists of a large number of linearly independent components none of which contributes significantly to the sum then the signal will have a Gaussian distribution. To show this consider the random variables $Y = y(t)$ and $X_n = x_n(t)$ where the X_n are independent. Then if

$$y(t) = \sum_{n=1}^{\infty} x_n(t),$$

we know that (see section 7.10)

$$K_y(t) = \sum_{n=1}^{\infty} K_{x_n}(t),$$

where $K_y(t)$ and $K_{x_n}(t)$ are the cumulant functions of Y and X_n respectively. Therefore

$$K_y(t) = \sum_{n=1}^{\infty} \log E_t(e^{X_n t}),$$

which on expanding the exponential gives

$$K_y(t) = \sum_{n=1}^{\infty} \log E_t\left(1 + \frac{tX_n}{1!} + \frac{t^2 X_n^2}{2!} + \cdots + \frac{t^i X_n^i}{i!} + \cdots\right).$$

Since the contribution of each X_n to Y is small this reduces to

$$K_y(t) \approx E_t\left(\sum_{n=1}^{\infty} X_n\right)t + E_t\left(\sum_{n=1}^{\infty} X_n^2\right)t^2/2!,$$

that is

$$K_y(t) = \mu t + \sigma^2 t^2/2, \tag{7.111}$$

where

$$\mu = E_t\left(\sum_{n=1}^{\infty} X_n\right) \quad \text{and} \quad \sigma^2 = E_t\left(\sum_{n=1}^{\infty} X_n^2\right).$$

The corresponding moment generating and characteristic functions are

$$m_y(t) = \exp\{\mu t + \sigma^2 t^2/2\},$$

and

$$M_y(j\omega) = \exp\{-j\mu\omega - \sigma^2\omega^2/2\}. \tag{7.112}$$

Taking the inverse Fourier transform of $M_y(j\omega)$ gives for the amplitude probability density of y,

$$q(y) = \frac{1}{\sigma\sqrt{2\pi}} \exp\{-(y-\mu)^2/2\sigma^2\},$$

which is the density for a Gaussian distribution with mean value μ and variance σ^2. Since in signal analysis problems it is usually most convenient to consider signals with a zero mean and take any existing mean value as a separate bias signal we let $q(y)$ from hereon represent the amplitude probability density of a Gaussian signal with zero mean, that is

$$q(y) = (1/\sigma\sqrt{2\pi})e^{-y^2/2\sigma^2}. \tag{7.113}$$

Figure 7.7 shows graphs of $q(y)$ for various values of σ. The nth moment of the distribution is given by

$$\begin{aligned} \mu_n = {} & \Gamma(n)/2^{(n-2)/2}\Gamma(n/2) && \text{for } n \text{ even}, \\ & 0 && \text{for } n \text{ odd}, \end{aligned} \tag{7.114}$$

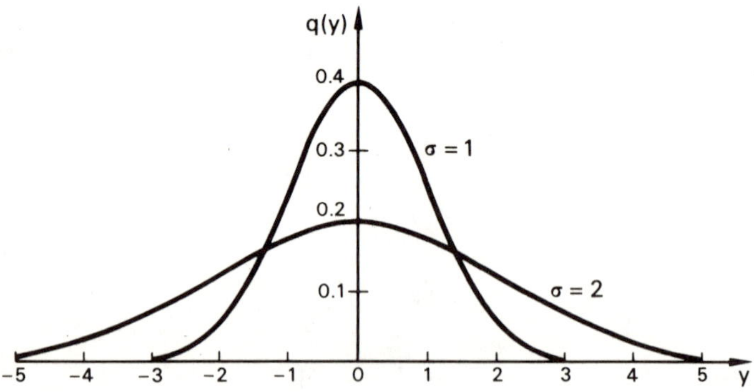

Figure 7.7 Graph of amplitude probability density $q(y)$ for a Gaussian signal

and the cumulants are

$$\begin{aligned} \lambda_n = {} & 0 && \text{for } n \neq 2, \\ & \sigma^2 && \text{for } n = 2. \end{aligned}$$

A multidimensional central limit theorem also exists which states that if there are n signals $y_n(t)$ (i.e. Y_n random variables) of zero mean, each of which consists of a large number of small independent components, then the nth order probability density function will be normally distributed. The nth order normal probability density function is given by

$$q(y_1, y_2 \dots y_n) = (2\pi)^{-n/2}|M|^{-1/2} \exp\left\{(-1/2|M|) \sum_{i,j=1}^{n} M_{ij}y_iy_j\right\}, \quad (7.115)$$

where M is the matrix

$$\begin{pmatrix} \mu_{11} & \mu_{12} & \cdots & \mu_{1n} \\ \mu_{21} & \mu_{22} & \cdots & \mu_{2n} \\ \vdots & \vdots & \vdots & \vdots \\ \mu_{n1} & \mu_{n2} & \cdots & \mu_{nn} \end{pmatrix},$$

$$|M| = \text{determinant of matrix } M,$$

$$\mu_{ij} = E(Y_i Y_j),$$

and

$$M_{ij} = \text{the cofactor of } \mu_{ij} \text{ in the determinant } |M|.$$

In particular for $n = 1$, equation (7.115) reduces to

$$q(y_1) = (2\pi|M|)^{-1/2} \exp(-M_{11}y_1^2/2|M|),$$

where $M = \mu_{11}, |M| = \mu_{11}$,

$$M_{11} = 1, \quad \text{and} \quad \mu_{11} = E(Y_1^2) = \sigma^2,$$

giving

$$q(y_1) = (1/\sigma\sqrt{2\pi})e^{-y_1^2/2\sigma^2},$$

as previously derived. Again, for $n = 2$ equation (7.115) becomes

$$q(y_1, y_2) = (2\pi)^{-1}|M|^{-1/2} \exp\{(-1/2|M|)(M_{11}y_1^2 + 2M_{12}y_2y_1 + M_{22}y_2^2)\},$$

where

$$M = \begin{pmatrix} \mu_{11} & \mu_{12} \\ \mu_{21} & \mu_{22} \end{pmatrix}, \qquad |M| = \mu_{11}\mu_{22} - \mu_{12}\mu_{21},$$

$$\mu_{11} = E(Y_1^2), \qquad \mu_{12} = E(Y_1 Y_2) = E(Y_2 Y_1) = \mu_{21}, \qquad \mu_{22} = E(Y_2^2),$$

and

$$M_{11} = \mu_{22}, \qquad M_{12} = -\mu_{21} = -\mu_{12} = M_{21}, \qquad M_{22} = \mu_{11}.$$

If the random variables Y_1 and Y_2 are the Gaussian signals $y_1(t)$ and $y_2(t + \tau)$ of variances σ_1^2 and σ_2^2 respectively, then

$$E(Y_1^2) = \sigma_1^2, \qquad E(Y_2^2) = \sigma_2^2, \quad \text{and} \quad E(Y_1 Y_2) = R_{y_1y_2}(\tau) = \rho(\tau)\sigma_1\sigma_2,$$

where $\rho(\tau)$ is the normalised crosscorrelation function. Hence

$$q(y_1, y_2, \tau) = \frac{1}{2\pi\sigma_1\sigma_2\{1 - \rho^2(\tau)\}^{1/2}} \exp\left[\frac{-1}{2\{1 - \rho^2(t)\}}\{(y_1/\sigma_1)^2 - \{2\rho(\tau)y_1y_2/\sigma_1\sigma_2\} + (y_2/\sigma_2)^2\}\right]. \tag{7.116}$$

Using Mehler's expansion,[7] equation (7.116) can be written as

$$q(y_1, y_2, \tau) = \sum_{k=0}^{\infty} \{\rho^k(\tau)/k!\} H_k(y_1/\sigma_1) H_k(y_2/\sigma_2) q(y_1) q(y_2), \tag{7.117}$$

where $H_k(y)$ is the Hermite polynomial defined by

$$H_k(y) = (-1)^k e^{y^2/2} \frac{d^k}{dy^k}\{e^{-y^2/2}\}. \tag{7.118}$$

Taking the double Fourier transform of equation (7.117) gives the characteristic function

$$M(ju, jv, \tau) = \exp\{(-1/2)(\sigma_1^2 u^2 + 2\sigma_1\sigma_2 uv\rho(\tau) + \sigma_2^2 v^2)\}, \tag{7.119}$$

since[5]

$$\int_{-\infty}^{\infty} H_k(y/\sigma) q(y) e^{-j\omega y} \, dy = (j\omega\sigma)^k e^{-\sigma^2\omega^2/2}.$$

7.9.4 *A random telegraph signal*

A random telegraph signal $x(t)$ is a two state signal with the probability of the signal switching from one value to the other being Poisson distributed. To satisfy the Poisson distribution the number of zero crossings (i.e. events) must comply with the following conditions:

(i) The occurrence of an event at any particular instant of time is independent of the previous history.

(ii) As the time of observation δT approaches zero the probability of observing one event tends to $\mu\,\delta T$, and the probability of observing more than one event tends to zero more rapidly than δT, where μ is the mean rate of occurrence of the events per unit time.

To determine the probability distribution of the zero crossings, the interval of time T is assumed divided into n small intervals δT. The probability of an event in δT is thus $\mu\,\delta T$ and the probability of no event is $1 - \mu\,\delta T$, since as $\delta T \to 0$, the probability of more than one event is negligible. Hence in the time T, since the probability $p(r, T)$ of the occurrence of r events is the number of ways of selecting the probability of r events and $(n - r)$ no events

$$p(r, T) = \lim_{\substack{\delta T \to 0 \\ n \to \infty}} {}_nC_r(\mu\,\delta T)^r(1 - \mu\,\delta T)^{n-r},$$

with $n\,\delta T = T$. Therefore the probability of occurrence of r events in the time T is

$$p(r, T) = \lim_{n\to\infty} \frac{n(n-1)\cdots(n-r+1)(\mu T)^r\{1-(\mu T/n)\}^n}{r!n!\{1-(\mu T/n)\}^r}$$

$$= \{(\mu T)^r/r!\} \lim_{n\to\infty} \frac{\{1-(1/n)\}\{1-(2/n)\}\cdots\{1-[(r-1)/n]\}\{1-(\mu T/n)\}^n}{\{1-(\mu T/n)\}^r}$$

$$= \{(\mu T)^r/r!\}e^{-\mu T}. \quad (7.120)$$

Since the average rate of zero crossings is constant the signal must spend equal amounts of time at the two levels of $\pm a$. Hence the amplitude probability density, $p(z)$, of the signal is given by

$$p(z) = (1/2)\{\delta(z-a) + \delta(z+a)\}. \quad (7.121)$$

The corresponding characteristic function is

$$M(j\omega) = \int_{-\infty}^{\infty} (1/2)\{\delta(z-a) + \delta(z+a)\}e^{-j\omega z}\,dz = \cos a\omega. \quad (7.122)$$

The second amplitude probability density distribution must obviously be of the form

$$p(z_1, z_2, \tau) = p_{++}(\tau)\delta(z_1 - a)\delta(z_2 - a) + p_{+-}(\tau)\delta(z_1 - a)\delta(z_2 + a)$$
$$+ p_{-+}(\tau)\delta(z_1 + a)\delta(z_2 - a) + p_{--}(\tau)\delta(z_1 + a)\delta(z_2 + a),$$

where p_{++} is the probability that $z_1(t) = z(t)$ is at $+a$ and $z_2(t) = z(t+\tau)$ is at $+a$ and so on. Clearly, this probability is given by

$$p_{++}(\tau) = \{\text{Probability that } z(t) = +a\}\{\text{Probability that } z(t) \text{ has an even number of zero crossings in } t_i < t < t_i + \tau\},$$

and since the probability of $z(t) = +a$ or $-a$ is 1/2,

$$p_{++}(\tau) = p_{--}(\tau) \quad \text{and} \quad p_{+-}(\tau) = p_{-+}(\tau).$$

Now the probability that $z(t)$ has an even number of zero crossings in the interval τ is, using equation (7.120),

$$\sum_{r=0(2)}^{\infty} p(r, \tau) = e^{-\mu\tau}[1 + \{(\mu\tau)^2/2!\} + \{(\mu\tau)^4/4!\} + \cdots]$$
$$= e^{-\mu\tau}(e^{\mu\tau} + e^{-\mu\tau})/2 = (1/2)(1 + e^{-2\mu\tau}),$$

and the probability that $z(t)$ has an odd number of zero crossings in the interval τ is

$$1 - (1/2)(1 + e^{-2\mu\tau}) = (1/2)(1 - e^{-2\mu\tau}).$$

Hence

$$p_{++}(\tau) = p_{--}(\tau) = (1/4)(1 + e^{-2\mu\tau}),$$

and

$$p_{+-}(\tau) = p_{-+}(\tau) = (1/4)(1 - e^{-2\mu\tau}).$$

Therefore

$$p(z_1, z_2, \tau) = (1/4)(1 + e^{-2\mu\tau})\{\delta(z_1 - a)\delta(z_2 - a) + \delta(z_1 + a)\delta(z_2 + a)\}$$
$$+ (1/4)(1 - e^{-2\mu\tau})\{\delta(z_1 - a)\delta(z_2 + a) + \delta(z_1 + a)\delta(z_2 - a)\}. \tag{7.123}$$

Thus the autocorrelation function of the signal is

$$R_{zz}(\tau) = \int_{-\infty}^{\infty}\int_{-\infty}^{\infty} z_1 z_2 p(z_1, z_2, \tau)\,dz_1\,dz_2$$
$$= 2a^2(1/4)(1 + e^{-2\mu\tau}) - 2a^2(1/4)(1 - e^{-2\mu\tau})$$
$$= a^2 e^{-2\mu\tau} \quad \text{for } \tau \text{ positive.}$$

Since the autocorrelation function has even symmetry the expression becomes for all τ,

$$R_{zz}(\tau) = a^2 e^{-2\mu|\tau|}, \tag{7.124}$$

and the corresponding power spectral density is

$$S(\omega) = 2a^2 T_1/(1 + \omega^2 T_1^2),$$

where $T_1 = 1/2\mu$. The joint characteristic function of the signal is

$$M(ju, jv, \tau) = \int_{-\infty}^{\infty}\int_{-\infty}^{\infty} p(z_1, z_2, \tau)e^{-j(uz_1 + vz_2)}\,dz_1\,dz_2,$$

which on substituting for $p(z_1, z_2, \tau)$ from equation (7.123) gives

$$M(ju, jv, \tau) = \cos au \cos av - e^{-2\mu\tau} \sin au \sin av. \tag{7.125}$$

7.10 Amplitude probability density of the sum of several signals

In analysing systems with multiple inputs one often needs to determine the statistics of a signal which is a function of several other signals. Here we consider the most important case only; when the signal is the sum of several other signals. For other situations the reader is referred to more specialised texts.[8]

Consider first the signal $z(t)$ where

$$z(t) = x(t) + y(t);$$

then to determine the distribution function $P(z)$ we note that the region in the $x - y$ plane for which

$$x + y \leq z$$

contains all possible values of x and y for which the sum is less than or equal to z.

Therefore, integrating over suitable strips in the plane

$$P(z) = \int_{-\infty}^{\infty} \int_{-\infty}^{z-y} p(x, y)\, dx\, dy.$$

Differentiating with respect to z we obtain

$$p(z) = \int_{-\infty}^{\infty} p(z - y, y)\, dy.$$

When the variables x and y are independent with density functions $r(x)$ and $q(y)$ respectively,

$$p(x, y) = r(x)q(y),$$

and the above equation becomes

$$p(z) = \int_{-\infty}^{\infty} r(z - y)q(y)\, dy, \tag{7.126}$$

or alternatively

$$p(z) = \int_{-\infty}^{\infty} q(z - x)r(x)\, dx.$$

Taking Fourier transforms of either of the above equations gives

$$M_z(j\omega) = M_x(j\omega)M_y(j\omega). \tag{7.127}$$

The characteristic function of the amplitude probability density of the sum is thus the product of the separate characteristic functions. The corresponding relationships between the moment generating and cumulant functions are

$$m_z(t) = m_x(t)m_y(t), \tag{7.128}$$

and

$$K_z(t) = K_x(t) + K_y(t). \tag{7.129}$$

In particular, equation (7.129) shows that the cumulants of the sum are equal to the sum of the cumulants, that is

$$\lambda_{n_z} = \lambda_{n_x} + \lambda_{n_y}.$$

In a similar manner to the above, it can also be shown that if $p(z_1, z_2, \tau)$ is the second amplitude probability density function of z, then

$$p(z_1, z_2, \tau) = \int_{-\infty}^{\infty} \int_{-\infty}^{\infty} r(x_1, x_2, \tau)q(z_1 - x_1, z_2 - x_2, \tau)\, dx_1\, dx_2,$$

or

$$p(z_1, z_2, \tau) = \int_{-\infty}^{\infty} \int_{-\infty}^{\infty} r(z_1 - y_1, z_2 - y_2, \tau)q(y_1, y_2, \tau)\, dy_1\, dy_2,$$

and after taking double Fourier Transforms that

$$M_z(ju, jv, \tau) = M_x(ju, jv, \tau)M_y(ju, jv, \tau),$$

$$m_z(s, t, \tau) = m_x(s, t, \tau)m_y(s, t, \tau),$$

and

$$K_z(s, t, \tau) = K_x(s, t, \tau) + K_y(s, t, \tau).$$

Obviously the above approach can be extended to cover the case of several summed signals.

7.11 Measurements on random signals

It is appropriate to conclude this chapter with a few comments on experimental methods of random signal analysis. In recent years, particularly with the increased use of digital methods, a large amount of material has been written on this subject and for detailed information the interested reader is referred to the more specialised works on the subject.(9,10,11) Here our concern is with a discussion of some analog measurement techniques for true random signals. This excludes considerations relating to the generation and measurement methods of pseudo random signals, such as m sequences, which have received considerable attention in recent years.(12)

Random signals with an approximately Gaussian distribution and a spectrum from a few hertz to several kilohertz can be obtained from noisy diodes, zener diodes, photocells or thyratrons.(10) To achieve a power spectrum which is approximately flat down to d.c. from these noise sources, as is often required for control applications or simulations either nonlinear(13) or frequency shifting methods can be used.(14,15) A random telegraph signal can be obtained from a radioactive source and Geiger counter(16) or from a binary operation on a Gaussian signal.(17)

7.11.1 *Density functions*

In principle analog measurement of the probability density functions of ergodic random signals is straightforward, however the complexity of practical schemes often limits their implementation to the first density function only. To determine the fraction of the total time a signal spends between the level x_n and $x_n + \delta x$, that is $p_n(x)\,\delta x$ the signal can be passed through an amplitude window of width δx which only gives an output of magnitude, h, when the input is within the above range. If the output pulses from this amplitude window are fed to an integrator of gain K for a time T then the integrator output divided by hKT gives the value of $p(x_n)\,\delta x$. By repeating the measurement N times for different x_n so that $N\,\delta x$ covers the entire range of input amplitudes a stepped approximation to $p(x)$ can be obtained. The measurements may be taken on the same sample of signal, which is stored and replayed, or measured using N channels to give simultaneous readings, or on different sample lengths of duration T of the signal. A scheme for the automatic plotting of the density function histogram using an analog computer with parallel digital logic is described in reference 18.

A similar procedure can be used for finding $p(x, y, \tau)$ for the two signals $x(t)$ and $y(t)$. In this case two amplitude windows of width δx and δy respectively are required and a pulse is now only applied to the integrator input when $x(t)$ and $y(t + \tau)$ lie simultaneously within the intervals x_n to $x_n + \delta x$ and y_n to $y_n + \delta y$ respectively. N^2 measurements are thus required to obtain the density function for only one value of τ. As results are usually required for various values of τ it is normally only possible to obtain the measurements from recorded data.

7.11.2 *Correlation functions*

The correlation functions of random signals are probably measured and used more than any other statistical property. On line correlators, that is real time correlation measurements are used for the determination of system transfer functions and for providing correlation function criteria for adaptive systems. Off line correlators require a variable time shift τ, a multiplier and an integrator to evaluate $R_{xy}(\tau)$ as given by equation (7.45), since the signals can be played back again for each required value of τ. When magnetic tape is used as the storage medium a variable τ is easily obtained with a movable reading head. However, for on line measurement, evaluation of the correlation function for N values of τ requires N multipliers and integrators in addition to $(N - 1)$ reading heads† if magnetic tape is the storage medium providing the delays.

Various methods for measuring correlation functions using simplified multiplying circuitry and/or time delay schemes, with on line applications in mind, have thus been suggested. All these methods involve amplitude quantisation of one or both the signals $x(t)$ and $y(t)$. For example, if the signal $y(t)$ in one channel is converted to a two state signal with amplitudes ± 1, the product $x(t)y(t)$ becomes $x(t) \operatorname{sgn} y(t)$, which is easily obtained using a simple gate circuit. Block diagrams of a true correlator, a coincidence correlator, a relay correlator and a linearised relay correlator are shown in figure 7.8.

The correlation functions $R'_{xy}(\tau)$ given by the coincidence and relay correlators can be shown to be simply related to the true correlation function $R_{xy}(\tau)$ when the signal x is separable with respect to y (see equation (10.78)). If x and y are jointly Gaussian then for the coincidence correlator $R'_{xy}(\tau) = (2h^2/\pi) \sin^{-1} \{R_{xy}(\tau)/R_{xy}(0)\}$ and for the relay correlator the ratio $R'_{xy}(\tau)/R_{xy}(\tau)$ is the random describing function, RDF, of the relay (see example P.8.15). The linearised relay correlator is based on the signal modification procedure, which produces a 'linear region in the relay', discussed in section 5.12. When $z(t)$ is uniformly distributed over an amplitude range greater than that of $x(t)$ and is independent of $x(t)$ and $y(t)$ the output correlation function is identical with the result which would be obtained if the $x(t)$ channel were linear with a gain dependent upon the amplitude range of $z(t)$.

---

† This assumes one value of τ is zero, for which no storage is required.

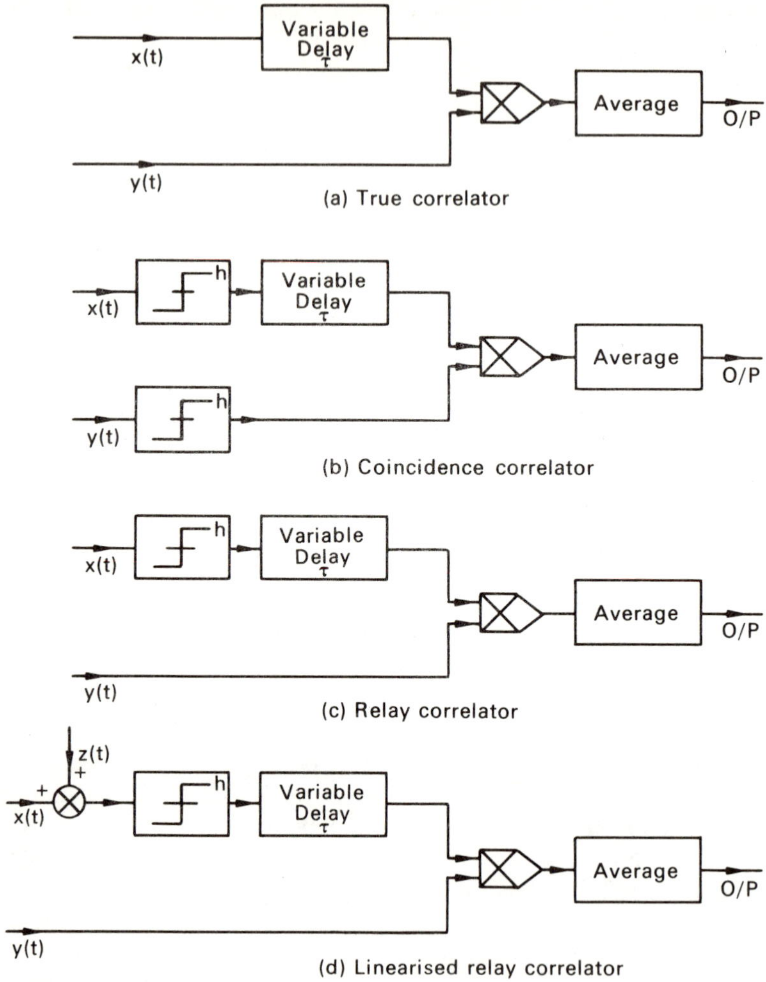

Figure 7.8 Types of analog correlators

7.11.3 *Power spectra*

The power density spectrum of a random signal can be measured using the same type of wave analyser as is employed for discrete spectrum analysis. Two basic types of wave analyser are used, namely, the 'constant-Q' and 'constant bandwidth' analysers. In the 'constant-Q' analyser the unknown signal is passed through a narrow band-pass tuned filter whose centre frequency may be adjusted to cover the required range of the frequency analysis, the mid-band gain remaining constant as the centre frequency is adjusted and the filter bandwidth increasing in proportion to the centre frequency. The 'constant bandwidth' analyser, however, gives both a constant mid-band gain and a constant bandwidth as the frequency is adjusted. This characteristic is usually achieved using the superheterodyne principle. The output from both types of analyser is normally measured on a rectified moving coil instrument.

When the analyser has a single tuned filter with transfer function $G(s)$, where

$$G(s) = \frac{G}{1 + Q\{(s/\omega_0) + (\omega_0/s)\}},$$

and the input to the analyser has a constant spectral density of S volts^2/hertz around the tuned frequency, f_0; the mean squared value of the filter output is

$$(S/4\pi j)\int_{-j\infty}^{j\infty} G(s)G(-s)\,\mathrm{d}s = \omega_0 G^2 S/4Q.$$

The equivalent bandwidth, f_B, of the filter is defined as the bandwidth of an ideal filter of constant gain G over the bandwidth, f_B, and zero gain elsewhere which will give the same mean squared output, hence

$$SG^2 f_B = SG^2\omega_0/4Q,$$

giving

$$f_B = \omega_0/4Q = (\pi/2)\,\delta f_0 \text{ hertz},$$

where

$$\delta f_0 \approx f_0/Q$$

is the filter bandwidth. The reading, M, of the analyser meter is given by

$$M - FG(\omega_0 S/4Q)^{1/2} \text{ volts},$$

where F is the form factor of the filter output signal. Even when the zero mean random input signal to the analyser is not Gaussian, the filter output will be nearly Gaussian, and it may be assumed that $F = (\pi/2)^{1/2}$ giving

$$M = G(\pi S\omega_0/8Q)^{1/2} = (G\pi/2)(S\,\delta f_0)^{1/2}. \qquad (7.130)$$

Thus for a random input with a constant spectral density the meter reading is proportional to the square root of the tuned frequency, f_0, for the 'constant-Q' analyser and is constant for the 'constant bandwidth' analyser. The 'constant bandwidth' analyser is thus generally preferred since, in addition, it also gives better separation of the high order harmonics when used with periodic signals.

Ideally the narrower the bandwidth of the wave analyser the more accurate will be the measurement of the spectral density of the random input. But the narrower the bandwidth the larger the magnitude of the very low frequency components in the signal through the meter coil produced by the nonlinear rectifier characteristic. In consequence a much longer measuring time is required to obtain a more accurate assessment of the meter average value.[9]

REFERENCES

1. Thomas, J. B.: *An introduction to statistical communication theory* (Wiley, New York, 1969, Chapters 2 to 4).
2. Newton, G. C., Gould, L. A. and Kaiser, J. F.: *Analytical design of linear feedback controls* (Wiley, New York, 1957).
3. Truxal, J. G.: *Control system synthesis* (McGraw-Hill, New York, 1955, Chapter 9).
4. Magnus, W. and Oberhettinger, F.: *Special functions of mathematical physics* (Chelsea, New York, 1949, Chapter 3).
5. Erdelyi et al.: *Tables of integral transforms* (McGraw-Hill, New York, 1954, Vol. 1).
6. Rice, S. O.: 'Mathematical analysis of random noise', N. Wax, *Selected papers on noise and stochastic processes* (Dover, New York, 1954, pp. 133–294).
7. Barrett, J. F. and Coales, J. F.: 'An introduction to the analysis of non-linear control systems with random inputs', *Proc. I.E.E.*, 1956, 103C, pp. 190–199.
8. Papoulis, A.: *Probability, random variables, and stochastic processes* (McGraw-Hill, New York, 1965).
9. Bendat, J. S. and Piersol, A. G.: *Measurement and analysis of random data* (Wiley, New York, 1966).
10. Korn, G. A.: *Random-process simulation and measurements* (McGraw-Hill, New York, 1966).
11. Jenkins, G. M. and Watts, D. G.: *Spectral analysis and its applications* (Holden Day, San Francisco, 1968).
12. Davies, W. D. T.: *System identification for self-adaptive control* (Wiley, New York, 1970).
13. Douce, J. L. and Shackleton, J. M.: 'L.f. random-signal generator', *Electronic Radio Engr.*, 1958, 35, Aug., pp. 295–297.
14. Bennett, R. R. and Fulton, A. S.: 'The generation and measurement of low frequency random noise', *J. Appl. Phys.*, 1951, 22, pp. 1187–1191.
15. Beecher, D. E., Bennett, R. R. and Low, H.: 'Stabilized noise source for air weapons design', *Electronics*, 1954, 27, July, pp. 163–165.
16. Manelis, J. B.: 'Generating random noise with radioactive sources', *Electronics*, 1961, 34, Sept., pp. 66–69.
17. Douce, J. L. and Leary, B. G.: 'A random pulse generator with variable mean rate', *Proc. I.E.E.*, 1961, 108B, pp. 524–528.
18. Kavanagh, R. J.: 'Automatic measurement and plotting of probability histograms using an analogue computer with digital logic', *Simulation*, 1969, 13, pp. 285–288.

BIBLIOGRAPHY

Beckmann, P.: *Probability in communications engineering* (Harcourt, Brace and World, New York, 1967).

Bekey, G. A. and Karplus, W. J.: *Hybrid computation* (Wiley, New York, 1968, Chapter 11).

Bell, D. A.: *Electrical noise* (Van Nostrand, Princeton, N.J., 1960).

Betts, J. A.: *Signal processing, modulation and noise* (E.U.P., London, 1970).

Bendat, J. S.: *Principles and applications of random noise theory* (Wiley, New York, 1958).

Cooper, G. R. and McGillem, C. D.: *Probabilistic methods of signal and system analysis* (Holt, Rinehart and Winston, New York, 1971).

Cramer, H.: *Mathematical methods of statistics* (Princeton Univ. Press, Princeton, N.J., 1946).

Cramer, H.: *The elements of probability theory and some of its applications* (Wiley, New York, 1955).

Davenport, W. B.: *Probability and random processes; an introduction for applied scientists and engineers* (McGraw-Hill, New York, 1970).

Davenport, W. B. and Root, W. L.: *An introduction to the theory of random signals and noise* (McGraw-Hill, New York, 1958).

Doob, J. L.: *Stochastic processes* (Wiley, New York, 1953).

James, H. M., Nichols, N. B. and Phillips, R. S.: *Theory of servomechanisms* (McGraw-Hill, New York, 1947).

Kolmogorov, A. A.: *Foundation of the theory of probability* (Chelsea, New York, 1954).

Lanning, J. H. and Battin, R. H.: *Random processes in automatic control* (McGraw-Hill, New York, 1956).

Lathi, B. P.: *An introduction to random signals and communication theory* (International Textbook Co., Scranton, Penn., 1968).

Lee, Y. W.: *Statistical theory of communication* (Wiley, New York, 1960).
Loeve, M.: *Probability theory* (Van Nostrand, Princeton, N.J., 1963).
Melsa, J. L. and Sage, A. P.: *An Introduction to probability and stochastic processes* (Prentice-Hall, New Jersey, 1973).
Merriam, C. W.: *Optimization theory and the design of feedback control systems* (McGraw-Hill, New York, 1964).
Middleton, D.: *Introduction to statistical communication theory* (McGraw-Hill, New York, 1960).
Mix, D. F.: *Random signal analysis* (Addison-Wesley, Reading, Mass., 1969).
Parzen, E.: *Modern probability theory and its applications* (Wiley, New York, 1960).
Schwarz, R. J. and Friedland, B.: *Linear systems* (McGraw-Hill, New York, 1965).
Solodovnikov, V. V.: *Statistical dynamics of linear automatic control systems*, 2nd edn. (Van Nostrand, London, 1965).
Wadsworth, G. P. and Bryan, J. G.: *Introduction to probability and random variables* (McGraw-Hill, New York, 1960).

PROBLEMS

P.7.1 Show that the form factor, F, which is defined as the ratio of the rectified average to the root mean squared value is
(a) $\pi/2\sqrt{2}$ for a sinusoidal signal
(b) $(\pi/2)^{1/2}$ for a Gaussian signal
(c) $(\pi/8)^{1/2}$ for a Rayleigh distributed signal for which the density function

$$p(R) = (R/\sigma^2)e^{-R^2/2\sigma^2} \qquad R > 0$$
$$= 0 \qquad R < 0.$$

P.7.2 If two random processes $x(t)$ and $y(t)$ are independent show that
(a) they are also uncorrelated, and
(b) the two processes $g\{x(t)\}$ and $h\{y(t)\}$ are independent.

P.7.3 If the random processes $x(t)$ and $y(t)$ are jointly normal and uncorrelated show that
(a) they are independent, and
(b) the autocorrelation function of their sum is equal to the sum of their autocorrelation functions.

P.7.4 If for all t the random process $x(t)$ satisfies $a < x(t) < b$, show that
(a) $a < E(x) < b$
(b) $E[x^2 - \{E(x)\}^2] \leq (b - a)^2/4$.

P.7.5 Show that if a random process $x(t)$ has a Cauchy density, namely $p(x) = \alpha/\pi(\alpha^2 + x^2)$, its characteristic function is given by $M(j\omega) = e^{-\alpha|\omega|}$.

P.7.6 If $z(t)$ is the product of the two signals $x(t)$ and $y(t)$ show that the probability of $z(t)$ being less than zero, that is $P(0) = \text{Prob}(z \leq 0)$, is given by

$$P(0) = \int_0^{\infty}\int_{-\infty}^{0} xyp(x, y)\,dx\,dy + \int_{-\infty}^{0}\int_0^{\infty} xyp(x, y)\,dx\,dy.$$

Show also that if $x(t)$ and $y(t)$ are both Gaussian with means of μ_x and μ_y and variances σ_x^2 and σ_y^2 then

$$P(0) = (1/2) - 2\,\text{erf}(\mu_x/\sigma_x)\,\text{erf}(\mu_y/\sigma_y).$$

P.7.7 A stochastic process $x(t)$ is defined such that $x_i(t)$ consists of a series of rectangular voltage pulses as shown in figure P.7.1. The pulses have random heights but a fixed duration and alternate between positive and negative values at the rate of 20 pulses per second. The probability density of the magnitude of the pulses is $p(x) = cxe^{-x}$ for $x > 0$, where c is a constant. Find the autocorrelation function and power spectrum of $x(t)$. If $x(t)$ is the input to the network, which is also shown in the figure, evaluate the output power spectrum if $RC = 1$.

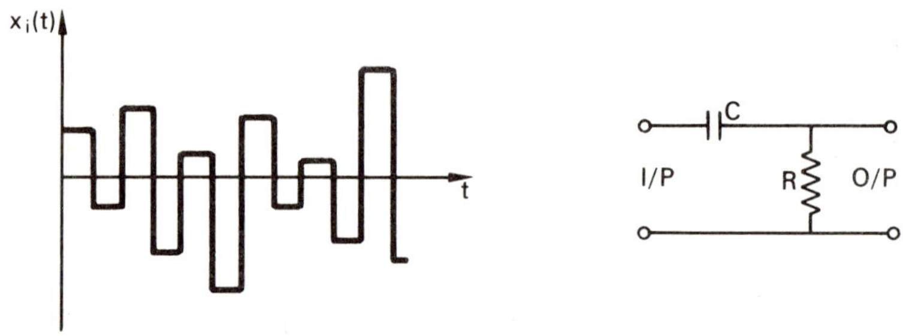

Figure P.7.1

P.7.8 A stationary signal consists of a train of current impulses with random sign (+ or − equally likely). Each carries a charge of 10^{-12} coulombs and the impulses are Poisson distributed in time with an average rate of occurrence of 1000 per second. The current impulses are applied to the parallel combination of a resistance R of $10^9\ \Omega$ and capacitance C of 100 pF. If $v(t)$ is the voltage across R and C find its average and mean squared value and its autocorrelation function $R_{vv}(\tau)$.

P.7.9 The input to a linear system $w(t)$ is a stationary process $x(t)$ with power spectrum, $S_{xx}(\omega)$. Show that if $R_{xx}(\tau) = 0$ for $|\tau| > \varepsilon$ where ε is sufficiently small then

$$R_{yx}(\tau) \approx w(t)S_{xx}(0) - (1/2)\left\{\frac{\mathrm{d}^2 S_{xx}(\omega)}{\mathrm{d}\omega^2}\bigg|_{\omega=0}\right\}\left\{\frac{\mathrm{d}^2 w(\tau)}{\mathrm{d}\tau^2}\right\}$$

where $y(t)$ is the system output.

P.7.10 Show that, if the input $x(t)$ to a realisable linear system $w(t)$ is white noise with $S_{xx}(\omega) = K$, then the mean squared value of the output is $K\int_0^\infty w^2(t)\,\mathrm{d}t$.

P.7.11 In the circuit of figure P.7.2, $v(t)$ is a Gaussian voltage with $R_{vv}(\tau) = 10e^{-|\tau|} + 10$. Assuming that the diodes are ideal, determine the mean value of the voltage between A and B.

P.7.12 A Gaussian signal, $x(t)$ with power spectrum $S_{xx}(\omega) = 1/(1 + \omega^2)^2$ is applied to a network with weighting function $w(t) = e^{-2t}$. Find the probability that at any specific time the network output will exceed unity.

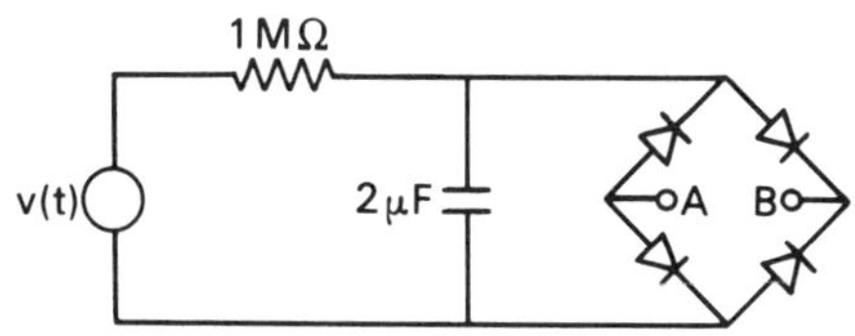

Figure P.7.2

P.7.13 A physical system is subjected to a test input, $x(t)$, which is a stationary random process with autocorrelation function

$$R_{xx}(\tau) = 2\pi\delta(\tau) + 3\pi e^{-|\tau|}.$$

The crosscorrelation between the system input and output, $y(t)$, is found to be $R_{xy}(\tau) = 2\pi e^{-|\tau|}$. Determine a linear model for the system which is best in the minimum mean squared error sense.

P.7.14 A linear system has a weighting function $w(t)$, input $x(t)$ and output $y(t)$. A second linear system has a weighting function $g(t)$, input $u(t)$ and output $v(t)$. Derive an expression for the cross power spectrum $S_{yv}(\omega)$ in terms of the cross power spectrum $S_{xu}(\omega)$.

P.7.15 A low pass filter is frequently used as a means of estimating the mean value of a random signal. The function of the filter can be regarded as that of reducing the mean squared value of the random component of the signal relative to that of the zero frequency component. Ideally the mean squared value of the random component would be reduced to zero.

Suppose that the input to a low pass filter of transfer function $1/(1 + sT)$ is a random signal $x(t)$ with power density spectrum

$$S_{xx}(\omega) = A\delta(\omega) + B/(1 + a^2\omega^2).$$

Find the ratio of the squared value of the average component to the mean squared value of the random component for both the input $x(t)$ and filter output $y(t)$. If these ratios are denoted by R_x and R_y respectively, show that as the ratio T/a becomes large, R_y/R_x approaches T/a.

P.7.16 In the position control system of figure P.7.3 tachometer feedback compensation is used. The system input $r(t)$ is an ergodic random process with autocorrelation function $R_{rr}(\tau) = e^{-|\tau|} \cos 2\pi\tau$. Determine the value of K which gives a minimum mean squared value of the error, e.

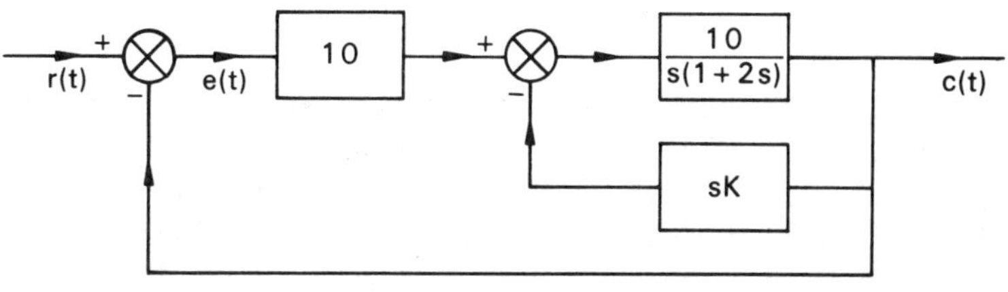

Figure P.7.3

P.7.17 A position control system with negligible friction has an undamped natural frequency of 1 rad/sec. It is stabilised by rate feedback. Find

(a) the gain of the rate feedback loop to give a minimum mean squared error, and

(b) an expression for the mean squared torque and the gain of the rate feedback loop which minimises the mean squared error subject to the constraint that the mean squared value of torque shall not exceed 0·5 unit^2.

The power spectrum of the input signal $S_{rr}(\omega) = 1/(1 + \omega^2)$.

P.7.18 A signal has a power density spectrum $S_{rr}(\omega) = \omega^2/(1 + \omega^2)^2$. It is mixed additively with a noise whose power density spectrum is $S_{nn}(\omega) = 1/16$. Determine the transfer function for the minimum mean squared linear filter if the desired output is the signal.

P.7.19 The input to a filter of transfer function $1/(s + \alpha)$ is the sum of a sinusoidal signal $A \sin \omega t$ and a stationary noise $n(t)$ with $S_{nn}(\omega) = K$. Show that the noise to signal ratio at the output is a minimum when $\alpha = \omega$.

P.7.20 We have shown that for the system of figure 7.3 the optimum realisable linear filter to minimise $\overline{e^2(t)}$ is $W(s)$, where

$$W(s) = \frac{1}{S_{vv}(s)_L}\left[\frac{S_{vi}(s)}{S_{vv}(s)_R}\right]_{Lp}.$$

If the signals $v(t)$ and $i(t)$ are jointly normal processes show that the above transfer function continues to be that of the optimum realisable linear filter when the criterion of optimality is changed to $\overline{f\{e(t)\}}$ is a minimum if (a) $f\{e(t)\}$ is an even function of $e(t)$ and (b) $f\{e(t)\}$ is non-decreasing.

P.7.21 A position control system is described by the following transform equations

$$(1 + 0{\cdot}1s)T(s) = KE(s)$$
$$s(1 + 0{\cdot}5s)C(s) = T(s)$$
$$E(s) = R(s) - C(s)$$

where R is the input, C the output, E the error and T the torque. The system input is an ergodic zero mean Gaussian signal with power density spectrum $S_{rr}(\omega) = (195{\cdot}5/\pi)\{1/(1 + 12{\cdot}5^2\omega^2)\}$ unit^2/rad/sec. Determine the value of K for which the mean squared error will be a minimum with (a) no constraints and (b) a constraint that the torque T may only exceed a magnitude of 0·78 unit for 5 per cent of the time.

P.7.22 For the system shown in figure P.7.4 $i(t) = r(t)$ and the following data applies

$$G_f(s) = \frac{\omega_0^2(1 + sT)}{s(s^2 + 2\zeta s\omega_0 + \omega_0^2)}, \qquad G_c(s) = K_v, \qquad H_f(s) = 1,$$

$$S_{rr}(\omega) = \frac{9}{\omega^2(\omega^2 + 0{\cdot}25)}, \qquad S_{nn}(\omega) = 0{\cdot}2, \qquad S_{nr}(\omega) = 0,$$

$$\omega_0 = 50, \qquad \zeta = 0{\cdot}5 \quad \text{and} \quad T = 0{\cdot}005.$$

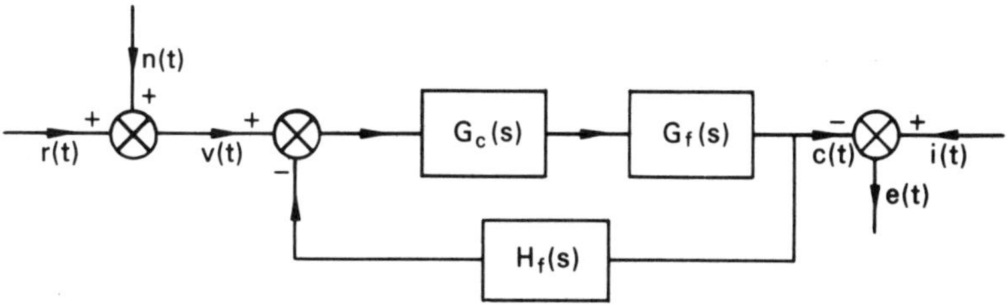

Figure P.7.4

Find the value of K_v which minimises the mean squared error and the corresponding minimum value of this error. If the compensator $G_c(s)$ is changed to $K_v(1 + sT_c)$, calculate the values of K_v and T_c which minimise the mean squared error and the new minimum value.

P.7.23 If in the system shown in figure P.7.4

$$S_{rr}(\omega) = \frac{K}{\omega^2(\omega^2 + 0{\cdot}01)}, \qquad K = 10{\cdot}12, \qquad S_{nn}(\omega) = 2,$$

$$S_{rn}(\omega) = 0, \qquad G_f(s) = 1 \quad \text{and} \quad i(t) = \frac{dr(t)}{dt},$$

find the compensator $G_c(s)$ which minimises the mean squared error and evaluate its magnitude.

P.7.24 For the system of figure P.7.4 the following data are given:

$$G_f(s) = \frac{1 - s}{1 + s}, \qquad S_{rr}(\omega) = \frac{20}{\omega^2 + 0{\cdot}01}, \qquad S_{nn}(\omega) = 2,$$

$$S_{rn}(\omega) = 0 \quad \text{and} \quad i(t) = r(t).$$

Find the compensator $G_c(s)$ which minimises the mean squared error and determine its magnitude.

P.7.25 The system shown in figure P.7.5 is required to integrate its input $r(t)$, that is $i(t) = \int_{-\infty}^{t} r(u)\,du$. If $G_f(s) = 1/(s + 10)$ $S_{rr}(\omega) = 4/(\omega^2 + 0{\cdot}25)$, find the compensator $G_c(s)$ that minimises the mean squared value of the error subject to the constraint that the mean squared value of the rate of

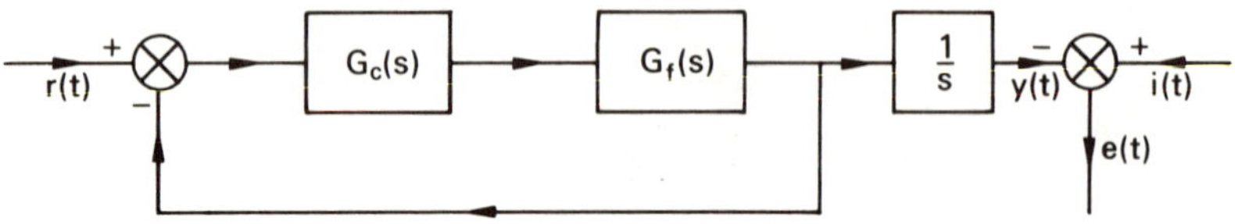

Figure P.7.5

change of $y(t)$, $dy(t)/dt$, is limited to 2·5. What is the magnitude of this mean squared error?

P.7.26 Show that if both equations (7.108) and (7.110) are to represent a narrow band Gaussian process the relationships

$$R(t)\cos\delta(t) = \sum_{k=1}^{\infty} a_k \cos\omega_k t + b_k \sin\omega_k t$$

$$R(t)\sin\delta(t) = \sum_{k=1}^{\infty} a_k \sin\omega_k t - b_k \cos\omega_k t$$

where $\omega_k = k\omega_1 - \omega_c$, must hold.

P.7.27 To illustrate the effect of the width of the amplitude window, calculate the values of the probability density function which would be obtained by an analog probability density function analyser for the following signals:

(a) A sine wave of unit peak value.

(b) A Gaussian signal with zero mean and unit mean squared value.

Calculate values of $p(x)$ for an appropriate range of values of x for values of window width equal to 0·1 and 0·25 and compare with the true values. Assume that an ideal averaging device is used.

Chapter 8

RDFs and Nonlinear System Analysis

8.1 Introduction

The aim of this chapter is to discuss the DF, or quasilinear, analysis of nonlinear systems with random inputs. Before this can be done however, it is necessary to investigate the response of a nonlinearity to a random input. To avoid some repetition we start straight away by considering a random plus bias input and the determination of the random plus bias describing function, the RBDF. Modelling the SVNL by two linear gains one for the bias and one for the random signal, the optimum values of these gains, the components of the RBDF, are again found by minimising the mean squared error between the nonlinearity output and that from the linear gains. Once more it is shown that the gains may also be found from variance and covariance relationships. It is further shown that the equivalent gain model is the best linear approximation if the random signal satisfies certain conditions, in which case the random signal is known as a separable process. The Gaussian signal is shown to belong to this class of processes.

Throughout the remainder of the chapter the random signal is assumed to be Gaussian and several additional results are first derived for such a signal. In particular the random signal component of the RBDF is shown to be equal to the d.c. incremental gain for the random plus bias input. For a known input spectrum the complete output autocorrelation function of the nonlinearity is evaluated. The series expression obtained is seen to be quite similar to that for a sinusoidal input, and it is shown that the random signal component of the RBDF, or equivalent gain, determines that part of the output correlated with the input. An interesting property is that the output spectrum may contain signals at a lower frequency than the input, a fact which has a direct bearing upon the accuracy of the quasilinear analysis of nonlinear feedback systems with Gaussian inputs.

Exact and approximate analytical solutions for Gaussian modified nonlinearities are considered. The approximate methods given are similar to those presented earlier for a sine plus bias input. The effect of nonlinearity on the statistical properties of the signal at its output for a Gaussian input signal is examined. Such a transformation is useful for producing random signals with different statistical properties.

The describing function analysis and synthesis of nonlinear feedback systems is then considered. The initial studies cover the analysis of those situations where the nonlinearity input may be modelled as a random signal plus, in some cases, a bias. The accuracy of the DF method for random signals is discussed and it is followed by consideration of the synthesis of optimum compensators for such systems. Finally, the modified nonlinearity approach for determining a describing function for a random plus sinusoidal input, the RSDF, is presented and some examples of its application in feedback systems are given.

8.2 The RBDF

To analyse a nonlinear system by DF methods requires replacement of the nonlinearity by its best linear approximation. Since mean squared values are commonly used in linear systems with a Gaussian input, Booton's[1] suggestion of using an equivalent gain found by minimising the mean squared error between the output from the nonlinearity and the output from the linear gain appears logical, particularly when it agrees with the DF for a sinusoid.

Consider a nonlinearity, n, with a bias plus random input,[2] $\gamma + x(t)$, as shown in figure 8.1. The error signal, $e(t)$, given by

$$\begin{aligned} e(t) &= y(t) - K_r x(t) - K_\gamma \gamma \\ &= n\{\gamma + x(t)\} - K_r x(t) - K_\gamma \gamma, \end{aligned}$$

has a mean squared value

$$E(e^2) = \int_{-\infty}^{\infty} \{n(\gamma + x) - K_r x - K_\gamma \gamma\}^2 p(x)\,\mathrm{d}x,$$

where $p(x)$ is the amplitude probability distribution of the zero mean random signal.

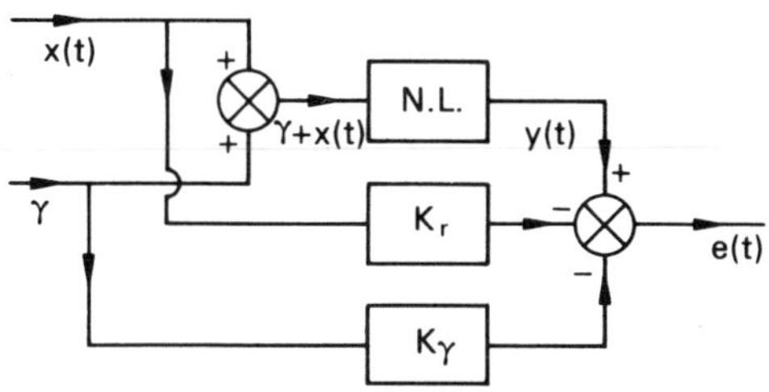

Figure 8.1 Equivalent gain model

Now $E(e^2)$ is a minimum when

$$\frac{\partial}{\partial K_r}\{E(e^2)\} = \frac{\partial}{\partial K_\gamma}\{E(e^2)\} = 0.$$

Thus, first differentiating with respect to K_r, we obtain

$$\int_{-\infty}^{\infty} 2\{n(\gamma + x) - K_r x - K_\gamma \gamma\} x p(x)\,\mathrm{d}x = 0,$$

which gives

$$K_r = \{1/R_{xx}(0)\} \int_{-\infty}^{\infty} x n(x + \gamma) p(x)\,\mathrm{d}x, \tag{8.1}$$

where $R_{xx}(0)$ is the variance of x.

Similarly, differentiating with respect to K_γ, gives

$$K_\gamma = (1/\gamma) \int_{-\infty}^{\infty} n(x + \gamma) p(x)\,\mathrm{d}x. \tag{8.2}$$

The results are seen to agree with those of section 5.2 when $x(t)$ is a sinusoid.

8.2.1 *Correlation property*

The crosscorrelation between x and y in figure 8.1 is

$$R_{xy}(\tau) = \int_{-\infty}^{\infty}\int_{-\infty}^{\infty} xyp(x, y, \tau)\,\mathrm{d}x\,\mathrm{d}y,$$

and the covariance

$$R_{xy}(0) = \int_{-\infty}^{\infty}\int_{-\infty}^{\infty} xyp(x, y, 0)\,\mathrm{d}x\,\mathrm{d}y.$$

Substituting for y gives

$$R_{xy}(0) = \int_{-\infty}^{\infty}\int_{-\infty}^{\infty} xn(x + \gamma)p(x, y)\,\mathrm{d}x\,\mathrm{d}y$$

$$= \int_{-\infty}^{\infty} xn(x + \gamma)p(x)\,\mathrm{d}x = K_r R_{xx}(0),$$

since

$$\int_{-\infty}^{\infty} p(x, y)\,\mathrm{d}y = p(x).$$

Thus

$$K_r = R_{xy}(0)/R_{xx}(0). \tag{8.3}$$

Similarly

$$R_{\gamma y}(0) = \int_{-\infty}^{\infty} \gamma n(x + \gamma)p(x)\,\mathrm{d}x = \gamma^2 K_\gamma,$$

so that

$$K_\gamma = R_{\gamma y}(0)/R_{\gamma\gamma}(0). \tag{8.4}$$

Again, therefore, the gains may be defined as the covariance of the output and the input signal whose gain is required divided by the variance of that signal.

8.2.2 *Other quasilinearisation criteria*

As discussed in section 3.9 for the sinusoidal DF other definitions for a DF have been used. The most common alternative description for a random signal is the r.m.s. DF,[3] $K_{\mathrm{r.m.s.}}$, where

$$K_{\mathrm{r.m.s.}} = \frac{\text{Variance of random signal output}}{\text{Variance of random signal input}}$$

$$= \left[\frac{\int_{-\infty}^{\infty} \{n(x + \gamma)\}^2 p(x)\,\mathrm{d}x - \{\int_{-\infty}^{\infty} n(x + \gamma)p(x)\,\mathrm{d}x\}^2}{\int_{-\infty}^{\infty} x^2 p(x)\,\mathrm{d}x}\right]^{1/2}. \tag{8.5}$$

The bias gain, defined on a similar r.m.s. basis, obviously remains equal to K_γ. In general analytical solutions for $K_{\mathrm{r.m.s.}}$ are more complex than those for K_r.

Extensive investigations[4,5] suggest that in the analysis of nonlinear feedback systems with random inputs no better results can be obtained using alternative RDFs such as the above.

8.3 The optimum linear model

An obvious question that arises with respect to the linear gain model of figure 8.1 is: can a lower mean squared error be obtained if a linear dynamic element is used as the model rather than the linear gain? Consider therefore the situation of figure 8.1 with $\gamma = 0$ and K_r replaced by a linear network with impulse response $h(t)$. The necessary and sufficient condition for $h(t)$ to be the impulse response of the optimum realisable linear model is given by the Wiener–Hopf equation (7.91), that is

$$R_{xy}(\tau) - \int_{-\infty}^{\infty} R_{xx}(\tau - t)h(t)\,dt = 0 \quad \text{for } \tau \geq 0. \tag{8.6}$$

Assuming the optimum linear filter is a gain K_r so that $h(t) = K_r\,\delta(t)$, where $\delta(t)$ is the Dirac delta function, equation (8.6) becomes

$$R_{xy}(\tau) - K_r R_{xx}(\tau) = 0,$$

or

$$R_{xy}(\tau) = K_r R_{xx}(\tau). \tag{8.7}$$

But from the figure $e = y - K_r x$ so that

$$R_{xe}(\tau) = R_{xy}(\tau) - K_r R_{xx}(\tau)$$

and the necessary condition for the linear filter to be the gain K_r is

$$R_{xe}(\tau) = 0. \tag{8.8}$$

Now

$$R_{xe}(\tau) = \int_{-\infty}^{\infty}\int_{-\infty}^{\infty} x_1\{n(x_2) - K_r x_2\}p(x_1, x_2, \tau)\,dx_1\,dx_2, \tag{8.9}$$

but from equation (8.1), with $\gamma = 0$,

$$\int_{-\infty}^{\infty} \{n(x_2) - K_r x_2\}x_2 p(x_2)\,dx_2 = 0,$$

so that for equations (8.8) and (8.9) to be satisfied for all n and τ,

$$\int_{-\infty}^{\infty} x_1 p(x_1, x_2, \tau)\,dx_1 = x_2 p(x_2) f(\tau). \tag{8.10}$$

However

$$R_{xx}(\tau) = \int_{-\infty}^{\infty}\int_{-\infty}^{\infty} x_1 x_2 p(x_1, x_2, \tau)\,dx_1\,dx_2,$$

so that if the process satisfies equation (8.10),

$$R_{xx}(\tau) = \int_{-\infty}^{\infty} x_2^2 p(x_2) f(\tau)\, dx_2 = R_{xx}(0) f(\tau),$$

giving

$$f(\tau) = R_{xx}(\tau)/R_{xx}(0).$$

Thus a necessary, and as may also be shown sufficient,[6] condition for the linear gain K_r to be the optimum linear model is that the random input satisfy

$$\int_{-\infty}^{\infty} x_1 p(x_1, x_2, \tau)\, dx_1 = x_2 p(x_2) R_{xx}(\tau)/R_{xx}(0). \tag{8.11}$$

A random process satisfying this condition was defined by Nuttall[6] as a separable random process. The term factorable process[7] is also used.

In addition for a separable process

$$R_{yy}(\tau) = \int_{-\infty}^{\infty} \int_{-\infty}^{\infty} (e_1 + K_r x_1)(e_2 + K_r x_2) p(x_1, x_2, \tau)\, dx_1 dx_2,$$

giving

$$R_{yy}(\tau) = R_{ee}(\tau) + K_r^2 R_{xx}(\tau). \tag{8.12}$$

8.3.1 *Separable processes*

Are there any random processes which satisfy the condition of separability given by equation (8.11)? It has been shown in section 3.3.2 that for a sinusoid, $N(a)$ satisfies equation (8.7), that is

$$N(a) = R_{xy}(\tau)/R_{xx}(\tau).$$

Thus the sinusoidal process, that is the process for which the member signals are all sinusoids of the same amplitude and frequency but with phase distributed uniformly, is separable. This may also be shown by evaluating the left hand side of equation (8.11) with the expression for $p(x_1, x_2, \tau)$ substituted from equation (7.107). That is

$$\int_{-\infty}^{\infty} x_1 p(x_1, x_2, \tau)\, dx_1 = \int_{-\infty}^{\infty} x_1 \sum_{s=0}^{\infty} \varepsilon_s T_s(x_1/a) T_s(x_2/a) r(x_1) r(x_2) \cos s w_0 \tau\, dx_1.$$

Changing the order of the summation and integration, and using the orthogonal relationship for $T_s(x/a)$, the expression reduces to

$$\int_{-\infty}^{\infty} x_1 p(x_1, x_2, \tau)\, dx_1 = x_2 r(x_2) \cos s w_0 \tau,$$

which satisfies equation (8.11). Similarly for the Gaussian process, using the notation of the previous chapter,

$$\int_{-\infty}^{\infty} y_1 q(y_1, y_2, \tau)\, dy_1 = \int_{-\infty}^{\infty} y_1 \sum_{k=0}^{\infty} H_k(y_1/\sigma) H_k(y_2/\sigma) \{\rho^k(\tau)/k!\} q(y_1) q(y_2)\, dy_1,$$

which on using the orthogonal property of the Hermite polynomials, namely

$$\int_{-\infty}^{\infty} H_j(y/\sigma)H_k(y/\sigma)q(y)\,\mathrm{d}y = k!\delta_{jk},$$

gives

$$\int_{-\infty}^{\infty} y_1 q(y_1, y_2, \tau)\,\mathrm{d}y_1 = y_2 q(y_2)\rho(\tau).$$

Since this expression is in agreement with equation (8.11), the Gaussian process is also separable.

8.4 The response of a nonlinearity to a Gaussian plus bias input

In this section we obtain additional results for the nonlinearity output which are only valid when the random input signal is Gaussian. The RBDF components K_r and K_γ will also be denoted by $N_\sigma(\sigma, \gamma)$ and $N_\gamma(\sigma, \gamma)$ in agreement with the notation used for a sinusoidal plus bias input.

8.4.1 *The output autocorrelation function*

As the Gaussian signal is denoted by y, let the nonlinearity output be z, then

$$\begin{aligned} R_{zz}(\tau) &= \int_{-\infty}^{\infty}\int_{-\infty}^{\infty} z_1 z_2 p(z_1, z_2, \tau)\,\mathrm{d}z_1\,\mathrm{d}z_2 \\ &= \int_{-\infty}^{\infty}\int_{-\infty}^{\infty} n(y_1+\gamma)n(y_2+\gamma)q(y_1, y_2, \tau)\,\mathrm{d}y_1\,\mathrm{d}y_2. \end{aligned}$$

Substituting for $q(y_1, y_2, \tau)$ from equation (7.117) gives

$$\begin{aligned} R_{zz}(\tau) = \int_{-\infty}^{\infty}\int_{-\infty}^{\infty} & n(y_1+\gamma)n(y_2+\gamma)\sum_{k=0}^{\infty} H_k(y_1/\sigma)H_k(y_2/\sigma)q(y_1)q(y_2) \\ & \times \{\rho^k(\tau)/k!\}\,\mathrm{d}y_1\,\mathrm{d}y_2, \end{aligned}$$

which on separating the integrals gives

$$R_{zz}(\tau) = \sum_{k=0}^{\infty} \alpha_k^2\{R_{yy}^k(\tau)/k!\} = \sum_{k=0}^{\infty} (\alpha_k')^2\{\rho^k(\tau)/k!\}, \tag{8.13}$$

where $\rho(\tau) = R_{yy}(\tau)/\sigma^2$ the normalised input autocorrelation function and

$$\alpha_k' = \sigma^k\alpha_k = \int_{-\infty}^{\infty} n(y+\gamma)H_k(y/\sigma)q(y)\,\mathrm{d}y, \tag{8.14}$$

from which we see that $\alpha_1 = N_\sigma(\sigma, \gamma) = \alpha_1'/\sigma$. Equation (8.13) may also be written

$$R_{zz}(\tau) = \alpha_0^2 + \alpha_1^2 R_{yy}(\tau) + \sum_{k=2}^{\infty} \alpha_k^2 R_{yy}^k(\tau)/k!$$

which shows that $\{N_\sigma(\sigma, \gamma)\}^2$ is the portion of the output autocorrelation function with the same 'form' as the input autocorrelation function and that α_0 is the bias output $n(\sigma, \gamma)$. Another particularly useful expression for α_k can be obtained by substituting in equation (8.14) for $q(y)$ and $H_k(y/\sigma)$. This gives

$$\alpha_k = \int_{-\infty}^{\infty} n(y+\gamma)(-1)^k e^{y^2/2\sigma^2} \frac{d^k}{dy^k}\{e^{-y^2/2\sigma^2}\}(1/\sigma\sqrt{2\pi})e^{-y^2/2\sigma^2}\, dy$$

$$= (-1)^k(1/\sigma\sqrt{2\pi}) \int_{-\infty}^{\infty} n(y+\gamma)\frac{d}{dy}\left\{\frac{d^{k-1}}{dy^{k-1}}(e^{-y^2/2\sigma^2})\right\} dy.$$

Integrating by parts,

$$\alpha_k = \left\{(-1)^k(1/\sigma\sqrt{2\pi})n(y+\gamma)\frac{d^{k-1}}{dy^{k-1}}(e^{-y^2/2\sigma^2})\right\}_{-\infty}^{\infty}$$

$$-(-1)^k(1/\sigma\sqrt{2\pi}) \int_{-\infty}^{\infty} \frac{\partial}{\partial y}\{n(y+\gamma)\}\frac{d^{k-1}}{dy^{k-1}}(e^{-y^2/2\sigma^2})\, dy.$$

The first term vanishes at both limits so that

$$\alpha_k = (1/\sigma^{k-1}) \int_{-\infty}^{\infty} \frac{\partial}{\partial y}\{n(y+\gamma)\}H_{k-1}(y/\sigma)q(y)\, dy.$$

Continued integration by parts gives

$$\alpha_k = \int_{-\infty}^{\infty} \frac{\partial^k}{\partial y^k}\{n(y+\gamma)\}H_0(y/\sigma)q(y)\, dy,$$

that is

$$\alpha_k = \int_{-\infty}^{\infty} \frac{\partial^k}{\partial y^k}\{n(y+\gamma)\}q(y)\, dy, \tag{8.15}$$

a result which may also be derived from Price's theorem.[(8)] In particular

$$\alpha_0 = \gamma N_\gamma(\sigma, \gamma) = \int_{-\infty}^{\infty} n(y+\gamma)q(y)\, dy,$$

and

$$\alpha_1 = N_\sigma(\sigma, \gamma) = \int_{-\infty}^{\infty} \frac{\partial}{\partial y}\{n(y+\gamma)\}q(y)\, dy.$$

Since $(\partial/\partial y)\{n(y+\gamma)\} = (\partial/\partial \gamma)\{n(y+\gamma)\}$ the expression for α_1 may also be written

$$\alpha_1 = \frac{\partial \alpha_0}{\partial \gamma} = g(\sigma, \gamma), \tag{8.16}$$

where $g(\sigma, \gamma)$ is the incremental gain. We thus have the important result that for a Gaussian plus bias input the DF for the Gaussian signal is equal to the d.c. incremental gain.

8.4.2 *Modified nonlinearities*

The concept of a modified nonlinearity has been seen to be particularly useful in the analysis of systems with multiple inputs. For a Gaussian plus bias input we define the kth modified nonlinearity by

$$n(\sigma,\gamma)_k = \alpha'_k = \int_{-\infty}^{\infty} n(y+\gamma)H_k(y/\sigma)q(y)\,\mathrm{d}y.$$

Using equation (8.15) and noting that

$$\frac{\partial^k}{\partial y^k}\{n(y+\gamma)\} = \frac{\partial^k}{\partial \gamma^k}\{n(y+\gamma)\},$$

the expression for the modified nonlinearity may also be written as

$$n(\sigma,\gamma)_k = \sigma^k \int_{-\infty}^{\infty} \frac{\partial^k}{\partial \gamma^k}\{n(y+\gamma)\}q(y)\,\mathrm{d}y,$$

so that

$$n(\sigma,\gamma)_k = \sigma^k \frac{\partial^k}{\partial \gamma^k}\{n(\sigma,\gamma)\}, \tag{8.17}$$

where $n(\sigma,\gamma)_0$ is denoted by $n(\sigma,\gamma)$ and referred to as the modified nonlinearity. Thus for a Gaussian input the higher order modified nonlinearities can be obtained by differentiation of the modified nonlinearity.

8.4.3 *The components of the output spectrum*

Transforming equation (8.13) gives

$$S_{zz}(\omega) = \sum_{k=0}^{\infty} (\alpha_k^2/k!)S_k(\omega), \tag{8.18}$$

where

$$S_k(\omega) = \int_{-\infty}^{\infty} R_{yy}^k(\tau)\mathrm{e}^{-j\omega\tau}\,\mathrm{d}\tau, \tag{8.19}$$

so that

$$S_1(\omega) = S_{yy}(\omega)$$

the spectrum of the input signal y. However, equation (8.19) can be written

$$S_k(\omega) = \int_{-\infty}^{\infty} R_{yy}^{k-1}(\tau)R_{yy}(\tau)\mathrm{e}^{-j\omega\tau}\,\mathrm{d}\tau,$$

and substituting

$$R_{yy}(\tau) = (1/2\pi)\int_{-\infty}^{\infty} S_1(\omega')\mathrm{e}^{j\omega'\tau}\,\mathrm{d}\omega'$$

gives, after changing the order of integration,

$$S_k(\omega) = (1/2\pi)\int_{-\infty}^{\infty} S_1(\omega')\left\{\int_{-\infty}^{\infty} R_{yy}^{k-1}(\tau)e^{-j\tau(\omega-\omega')}\,d\tau\right\}d\omega',$$

that is

$$S_k(\omega) = (1/2\pi)\int_{-\infty}^{\infty} S_1(\omega')S_{k-1}(\omega-\omega')\,d\omega'. \tag{8.20}$$

Thus by repeated convolution, expressed by the above equation, starting with $k = 2$ and incrementing k by one each time the output spectral components $S_k(\omega)$ for any k can be obtained. We consider this result in more detail for various forms of input spectra $S_1(\omega)$.

A narrow band input

Suppose that the spectrum $S_1(\omega)$ of the Gaussian input is narrow band of width $\delta\omega_c$ about $\pm\omega_c$, as shown in figure 8.2. Neglecting the bias output term, that is

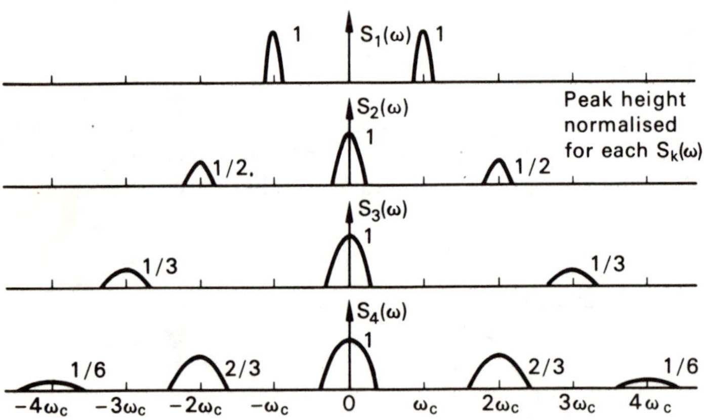

Figure 8.2 Output spectral components

the term involving α_0, the next output spectral density uncorrelated with the input is $S_2(\omega)$, the convolution of $S_1(\omega)$ with itself. The result is a spectrum with bands of width $2\,\delta\omega_c$ about $\pm 2\omega_c$ and 0. Convolving this spectrum and $S_1(\omega)$ gives the spectrum, $S_3(\omega)$, with bands of width $3\,\delta\omega_c$ about $\pm 3\omega_c$ and $\pm\omega_c$. In general, therefore, $S_k(\omega)$ has bands of width $k\,\delta\omega_c$ about $\pm\omega_c\sum_{j=0(2)}^{k}(k-j)$. These spectra are shown in figure 8.2, the magnitudes of the various bands being normalised with respect to the largest band in each $S_k(\omega)$. It is of particular interest to note that the output spectral components neglected in the quasi-linearisation, that is $S_k(\omega)$ for $k \neq 1$, may contain lower frequencies than the input spectrum $S_1(\omega)$.

Alternatively if we use Rice's representation,

$$y(t) = R(t)\cos\{\omega_c t + \delta(t)\},$$

to describe the narrow band process, the effect of power law terms in a nonlinear characteristic in producing spectral bands of the type shown is obvious.

A low pass band input

Assume the input spectrum is given by

$$S_1(\omega) = \frac{S}{1 + (\omega/\omega_0)^2}. \tag{8.21}$$

From the convolution relationship of equation (8.20) it is seen immediately that the output spectrum component $S_k(\omega)$ will be a low fequency spectrum with a bandwidth which increases as k increases.

Now the autocorrelation function of the input is

$$R_{yy}(\tau) = (1/2\pi) \int_{-\infty}^{\infty} S_1(\omega) e^{j\omega\tau}\,d\omega = (S\omega_0/2)e^{-\omega_0|\tau|}.$$

Hence

$$R_{yy}^k(\tau) = (S\omega_0/2)^k e^{-k\omega_0|\tau|},$$

giving

$$S_k(\omega) = \frac{(S\omega_0/2)^k(2/k\omega_0)}{1 + (\omega/k\omega_0)^2}.$$

The normalised spectrum, $S_k'(\omega)$, is defined by

$$S_k'(\omega) = \int_{-\infty}^{\infty} \rho^k(\tau)e^{-j\omega\tau}\,d\tau = S_k(\omega)/R_{yy}^k(0),$$

so that

$$S_k'(\omega) = \frac{(2/k\omega_0)}{1 + (\omega/k\omega_0)^2},$$

which has a low frequency magnitude inversely proportional to k and a bandwidth proportional to k.

Thus using the normalised form of equation (8.18), the output spectrum is

$$S_{zz}(\omega) = \sum_{k=0}^{\infty} \frac{(\alpha_k'/k!)(2/k\omega_0)}{1 + (\omega/k\omega_0)^2},$$

and its magnitude at low frequencies is

$$\{S_{zz}(\omega)\}_{\omega\to 0} = (2/\omega_0) \sum_{k=0}^{\infty} (\alpha_k'^2/k \times k!),$$

of which a significant amount may be from the distortion terms, that is

$$\{S_{ee}(\omega)\}_{\omega\to 0} = (2/\omega_0)\left\{\alpha_0'^2 + \sum_{k=2}^{\infty} (\alpha_k'^2/k \times k!)\right\}.$$

A limited bandwidth input

Consider an input spectrum of the form

$$S_1(\omega) = S\left\{\frac{1}{1 + (\omega/n\omega_0)^2} - \frac{1}{1 + (\omega/\omega_0)^2}\right\},$$

where $n > 1$. This corresponds to white noise with a power per hertz of S passed through a filter with the transfer function

$$\frac{s(n^2 - 1)^{1/2}}{n\omega_0\{1 + (s/n\omega_0)\}\{1 + (s/\omega_0)\}}.$$

At low frequencies, where $\omega \ll \omega_0$,

$$S_1(\omega) \to S\omega(n^2 - 1)^{1/2}/n\omega_0,$$

which tends to zero as ω tends to zero. The input autocorrelation function, $R_{yy}(\tau)$ is given by

$$R_{yy}(\tau) = (Sn\omega_0/2)\mathrm{e}^{-n\omega_0|\tau|} - (S\omega_0/2)\mathrm{e}^{-\omega_0|\tau|},$$

so that

$$R_{yy}(0) = (n - 1)S\omega_0/2,$$

and

$$\rho(\tau) = (ne^{-n\omega_0|\tau|} - e^{-\omega_0|\tau|})/(n - 1).$$

The output correlation function is thus

$$R_{zz}(\tau) = \sum_{k=0}^{\infty} \frac{\alpha_k'^2\{n\mathrm{e}^{-n\omega_0|\tau|} - \mathrm{e}^{-\omega_0|\tau|}\}^k}{k!(n - 1)^k},$$

and the kth normalised spectral component,

$$S_k'(\omega) = \frac{1}{(n - 1)^k}\int_{-\infty}^{\infty} \{n\mathrm{e}^{-n\omega_0|\tau|} - \mathrm{e}^{-\omega_0|\tau|}\}^k \mathrm{e}^{-j\omega\tau}\, \mathrm{d}\tau,$$

which can be evaluated term by term after the expression in the brackets has been multiplied by itself $(k - 1)$ times.

For example, assume that the nonlinearity is such that $\alpha_k'^2$ is negligible when $k \neq 1$ and 3. Then

$$S_3'(\omega) = \frac{2}{\omega_0(n - 1)^3}\left[\frac{n^2}{3\{1 + (\omega/n\omega_0)^2\}} - \frac{3n^2}{(2n + 1)[1 + \omega^2/\{(2n + 1)\omega_0\}^2]} + \frac{3n}{(n + 2)[1 + \omega^2/\{(n + 2)\omega_0\}^2]} - \frac{1}{3\{1 + (\omega/3\omega_0)^2\}}\right],$$

which at low frequencies becomes

$$S_3'(\omega)_{\omega\to 0} = \frac{2}{\omega_0(n-1)^3}\left\{\frac{n^2}{3} - \frac{3n^2}{2n+1} + \frac{3n}{n+2} - \frac{1}{3}\right\}$$
$$= \frac{4(n+1)}{3\omega_0(n+2)(2n+1)}.$$

Thus there is a finite output power as $\omega \to 0$, although the input power in this region is negligible. To see how this low frequency distortion output compares with the output $S_1'(\omega)$ which is correlated with the input it is seen that

$$S_1'(\omega) = \frac{2}{\omega_0(n-1)}\left\{\frac{1}{1+(\omega/n\omega_0)^2} - \frac{1}{1+(\omega/\omega_0)^2}\right\}$$

has a maximum value at $\omega = \omega_0\sqrt{n}$, equal to

$$S_1'(\omega)|_{\max} = 2/\{\omega_0(n+1)\}.$$

Hence the ratio

$$\frac{\text{Low frequency output distortion power per unit bandwidth}}{\text{Maximum power per unit bandwidth of the correlated output}}$$
$$= \frac{2\alpha_3'^2(n+1)^2}{3!3\alpha_1'^2(n+2)(2n+1)}.$$

The actual magnitude of this ratio depends upon α_1', α_3' and n, although the latter parameter has only a small effect, since the ratio varies less than 10 per cent for $1 < n < \infty$.

8.5 Analytical solutions for the RBDF

Analytical solutions for the RBDF and the output coefficients α_k are obtained below for some common nonlinearities. The results given for the integer power law and the harmonic nonlinearity can be used to obtain approximate solutions for other characteristics. Additional results are tabulated for the RDF in appendix A5 and the RBDF in appendix A6.

8.5.1 *Integer power law*

For the characteristic $n(x) = x^m$, using equation (8.14) gives

$$n(\sigma, \gamma)_k = \alpha_k\sigma^k = \int_{-\infty}^{\infty} (y+\gamma)^m H_k(y/\sigma)q(y)\,dy.$$

Expanding $(y+\gamma)^m$ gives

$$n(\sigma, \gamma)_k = \sum_{j=0}^{m} {}_mC_j\gamma^j \int_{-\infty}^{\infty} y^{m-j}H_k(y/\sigma)q(y)\,dy.$$

Using the integral,

$$\int_{-\infty}^{\infty} y^m H_k(y/\sigma) q(y)\,\mathrm{d}y = \begin{cases} m(m-1)\cdots(m-k+1)\sigma^k \mu_{m-k} & \text{for } k < m \\ 0 & \text{for } k > m \end{cases}, \tag{8.23}$$

where μ_n is the nth moment of the distribution $q(y)$, gives the solution

$$n(\sigma, \gamma)_k = \sum_{j=0}^{m} {}_mC_j \gamma^j (m-j)(m-j-1)\cdots(m-j-k+1)\sigma^k \mu_{m-j-k}, \tag{8.24}$$

and in particular, the modified nonlinearity,

$$n(\sigma, \gamma) = \sum_{j=0}^{m} {}_mC_j \mu_{m-j} \gamma^j. \tag{8.25}$$

Thus if $n(x)$ is odd, $n(\sigma, \gamma)_k$ is odd if k is even and even if k is odd, and vice versa for $n(x)$ even.

If for example $n(x) = x + Ax^3$, from equation (8.14) with $k = 0$,

$$n(\sigma, \gamma) = (1 + 3A\sigma^2)\gamma + A\gamma^3,$$

using equation (8.17), by successive differentiation we have

$$n(\sigma, \gamma)_1 = (\sigma + 3A\sigma^3) + 3A\sigma\gamma^2,$$
$$n(\sigma, \gamma)_2 = 6A\sigma^2\gamma,$$
$$n(\sigma, \gamma)_3 = 6A\sigma^3,$$

and

$$n(\sigma, \gamma)_k = 0 \quad \text{for } k > 3.$$

8.5.2 *Harmonic nonlinearity*

If $n(x) = \sin mx$,

$$n(\sigma, \gamma)_k = \sigma^k \alpha_k = \int_{-\infty}^{\infty} \sin(my + m\gamma) H_k(y/\sigma) q(y)\,\mathrm{d}y$$
$$= \cos m\gamma \int_{-\infty}^{\infty} \sin my H_k(y/\sigma) q(y)\,\mathrm{d}y$$
$$+ \sin m\gamma \int_{-\infty}^{\infty} \cos my H_k(y/\sigma) q(y)\,\mathrm{d}y.$$

Now[9]

$$\int_{-\infty}^{\infty} H_k(y/\sigma) q(y) \mathrm{e}^{jmy}\,\mathrm{d}y = (jm\sigma)^k \mathrm{e}^{-\sigma^2 m^2/2},$$

so that

$$\int_{-\infty}^{\infty} H_k(y/\sigma)q(y)\cos my\,\mathrm{d}y = (-1)^{k/2}(m\sigma)^k \mathrm{e}^{-\sigma^2 m^2/2} \quad \text{for } k \text{ even,}$$

and

$$\int_{-\infty}^{\infty} H_k(y/\sigma)q(y)\sin my\,\mathrm{d}y = (-1)^{(k-1)/2}(m\sigma)^k \mathrm{e}^{-\sigma^2 m^2/2} \quad \text{for } k \text{ odd.}$$

Thus

$$\begin{aligned} n(\sigma, \gamma)_k &= (-1)^{(k-1)/2}(m\sigma)^k \mathrm{e}^{-\sigma^2 m^2/2}\cos m\gamma \quad &\text{for } k \text{ odd,}\\ &\ (-1)^{k/2}(m\sigma)^k \mathrm{e}^{-\sigma^2 m^2/2}\sin m\gamma \quad &\text{for } k \text{ even,}\end{aligned}$$

or in one expression

$$n(\sigma, \gamma)_k = (-1)^k(m\sigma)^k \mathrm{e}^{-\sigma^2 m^2/2}\sin(m\gamma - k\pi/2). \tag{8.26}$$

Similarly for $n(x) = \cos mx$,

$$n(\sigma, \gamma)_k = (-1)^k(m\sigma)^k \mathrm{e}^{-\sigma^2 m^2/2}\cos(m\gamma - k\pi/2). \tag{8.27}$$

Hence Gaussian modified harmonic nonlinearities remain harmonic. In fact for any signal z with an even amplitude probability distribution $p(z)$ the modified nonlinearity $n(z, \gamma)$ of $\sin mx$ is

$$\begin{aligned} n(z, \gamma) &= \int_{-\infty}^{\infty} \sin(mz + m\gamma)p(z)\,\mathrm{d}z \\ &= M(jm)\sin m\gamma, \end{aligned}$$

where $M(jm)$ is the characteristic function of $p(z)$. Similarly for $\cos mx$, $n(z, \gamma)$ is $M(jm)\cos m\gamma$.

8.5.3 *Ideal saturation*

Here

$$\begin{aligned} n(y + \gamma) = m(y + \gamma), &\qquad -(\delta + \gamma) < y < \delta - \gamma,\\ m\delta, &\qquad y > \delta - \gamma,\\ -m\delta, &\qquad y < -(\delta + \gamma), \end{aligned}$$

so that

$$\begin{aligned} n(\sigma, \gamma)_k = &\int_{-\infty}^{-(\delta+\gamma)} -m\,\delta H_k(y/\sigma)q(y)\,\mathrm{d}y \\ &+ \int_{-(\delta+\gamma)}^{\delta-\gamma} m(y + \gamma)H_k(y/\sigma)q(y)\,\mathrm{d}y + \int_{\delta-\gamma}^{\infty} m\,\delta H_k(y/\sigma)q(y)\,\mathrm{d}y. \end{aligned}$$

Using the integrals

$$\int y^m H_k(y/\sigma)q(y)\,\mathrm{d}y = -q(y)\sum_{j=0}^{m}\sigma^{j+1}H_{k-j-1}(y/\sigma)\frac{\mathrm{d}^j(y^m)}{\mathrm{d}y^j} \quad \text{for } k > m+1, \tag{8.28}$$

$$\int H_0(y/\sigma)q(y)\,\mathrm{d}y = (1/2)\{1 + \operatorname{erf}(y/\sigma\sqrt{2})\},$$

$$\int yH_0(y/\sigma)q(y)\,\mathrm{d}y = -\sigma^2 q(y),$$

$$\int yH_1(y/\sigma)q(y)\,\mathrm{d}y = -\sigma y q(y) + (\sigma/2)\{1 + \operatorname{erf}(y/\sigma\sqrt{2})\},$$

where

$$\operatorname{erf} x = (2/\sqrt{\pi})\int_0^x \mathrm{e}^{-u^2}\,\mathrm{d}u, \tag{8.29}$$

the solution in terms of the normalised parameters

$$\gamma_N = \gamma/\delta \quad \text{and} \quad \sigma_N = \sigma\sqrt{2}/\delta$$

is

$$n(\sigma_N, \delta_N)_k = (m\sigma_N/2\sqrt{\pi})\{e^{-\alpha^2}H_{k-2}(-\alpha\sqrt{2}) - e^{-\beta^2}H_{k-2}(\beta\sqrt{2})\} \quad \text{for } k \geq 2, \tag{8.30}$$

$$n(\sigma_N, \gamma_N)_1 = (m\sigma_N/2\sqrt{2})\{\operatorname{erf}\alpha + \operatorname{erf}\beta\}, \tag{8.31}$$

and

$$n(\sigma_N, \gamma_N) = (m/2\sqrt{\pi})\{\sqrt{\pi}(\operatorname{erf}\alpha - \operatorname{erf}\beta) + \gamma_N\sqrt{\pi}(\operatorname{erf}\alpha + \operatorname{erf}\beta) - \sigma_N(\mathrm{e}^{-\beta^2} - \mathrm{e}^{-\alpha^2})\}, \tag{8.32}$$

where

$$\alpha = (1 + \gamma_N)/\sigma_N \quad \text{and} \quad \beta = (1 - \gamma_N)/\sigma_N.$$

Figure 8.3 shows graphs of some of the modified characteristics, which are symmetrical about the origin being odd for k even and even for k odd, for a normalised input magnitude σ_N of unity.

8.5.4 *On-off relay with dead zone*

In this case

$$n(y+\gamma) = \begin{cases} 0, & -(\delta+\gamma) < y < \delta - \gamma, \\ h, & y > \delta - \gamma, \\ -h, & y < -(\delta+\gamma), \end{cases}$$

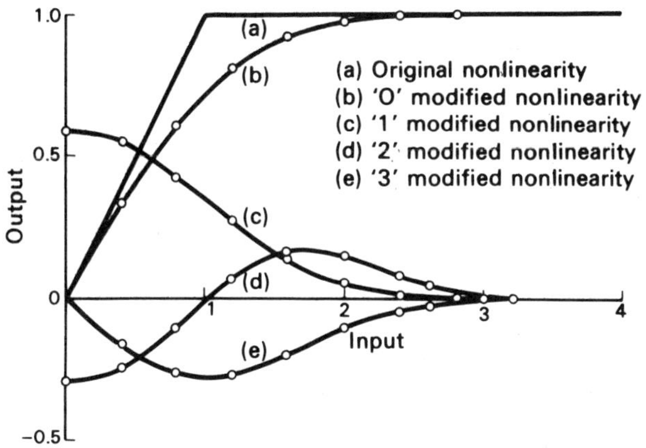

Figure 8.3 Gaussian modified saturation characteristics

so that

$$n(\sigma, \gamma)_k = \int_{-\infty}^{-(\delta+\gamma)} -hH_k(y/\sigma)q(y)\,\mathrm{d}y + \int_{\delta-\gamma}^{\infty} hH_k(y/\sigma)q(y)\,\mathrm{d}y.$$

Using the integral of equation (8.28) gives

$$n(\sigma, \gamma)_k = h\sigma[H_{k-1}\{-(\delta+\gamma)/\sigma\}q\{-(\delta+\gamma)\} + H_{k-1}\{(\delta-\gamma)/\sigma\}q(\delta-\gamma)] \quad \text{for } k \geq 1, \qquad (8.33)$$

and

$$n(\sigma, \gamma) = (h/2)[\mathrm{erf}\{(\delta+\gamma)/\sigma\sqrt{2}\} - \mathrm{erf}\{(\delta-\gamma)/\sigma\sqrt{2}\}]. \qquad (8.34)$$

8.6 Approximate solutions for Gaussian modified nonlinearities

Often RBDFs are required for asymmetrical nonlinear characteristics. The best procedure for this case is probably that described in section 5.4 where the nonlinearity is considered divided into two characteristics $n_+(y)$ and $n_-(y)$ about a new origin 0_γ the point $(\gamma, 0)$. Then

$$\alpha_0 = \alpha_{+0} + \alpha_{-0},$$

and

$$\sigma^k\alpha_k = n(\sigma, \gamma)_k = \alpha'_k = \alpha'_{+k} + \alpha'_{-k}.$$

We therefore consider below only single sided characteristics and show how approximate solutions for the coefficients α_{+k} can be obtained from a quantised approximation and by graphical integration.

8.6.1 *Quantised approximation*

In this case

$$\alpha'_{+k} = \int_{\delta_1}^{\delta_2} h_1 H_k(y/\sigma)q(y)\,\mathrm{d}y + \int_{\delta_2}^{\delta_3} (h_1 + h_2)H_k(y/\sigma)q(y)\,\mathrm{d}y$$

$$+ \int_{\delta_m}^{\infty} (h_1 + h_2 + \cdots h_m)H_k(y/\sigma)q(y)\,\mathrm{d}y,$$

which on integrating and cancelling terms reduces to

$$\alpha'_{+k} = \sum_{m=1}^{M} h_m H_{k-1}(\delta_m/\sigma)q(\delta_m) \quad \text{for } k \geq 1,$$

and in particular

$$\alpha_{+1} = (\alpha'_{+1}/\sigma) = \sum_{m=1}^{M} h_m q(\delta_m).$$

The solution for α_{+0} is

$$\alpha_{+0} = \sum_{m=1}^{M} (h_m/2)\{1 - \mathrm{erf}(\delta_m/\sigma\sqrt{2})\}.$$

Again the solutions are seen to be simple expressions which can easily be used to obtain approximate results for any nonlinearity. Indications are that for nonlinearities approximated with ten equal increments $(\delta_m - \delta_{m-1})$ up to an input magnitude of $2\sqrt{2}\sigma$ the values obtained for α_{+0} and α_{+1} will be in error by less than 1 per cent.[10]

8.6.2 *Graphical integration*

Equation (8.14) can be written

$$\alpha'_k = n(\sigma, \gamma)_k = \sigma^k \alpha_k = \int_{-\infty}^{\infty} n(y + \gamma)\,\mathrm{d}\{A_k(y)\},$$

where

$$A_k(y) = \int H_k(y/\sigma)q(y)\,\mathrm{d}y = -\sigma H_{k-1}(y/\sigma)q(y) \quad \text{for } k \geq 1. \tag{8.35}$$

The function $H_{k-1}(y/\sigma)q(y)$ possesses $(k-2)$ turning points, but unlike $D_s(x/a)$ and $T_s(x/a)$, the values of the function at these points for any particular value of k are different. Graphical integration with the above expression for $A_k(y)$ is therefore impracticable. However, the disadvantage can be overcome by expanding the polynomial $H_k(y/\sigma)$ and performing the graphical integration for equation (8.35) term by term.

Consider, for example, $A_3(y)$ where

$$A_3(y) = \int H_3(y/\sigma)q(y)\,\mathrm{d}y = \int \left(\frac{y^3}{\sigma^3} - \frac{3y}{\sigma}\right)\frac{1}{\sigma\sqrt{2\pi}}\mathrm{e}^{-y^2/2\sigma^2}\,\mathrm{d}y,$$

putting $u = y/\sigma\sqrt{2}$, and integrating

$$A_3(y) = -\frac{2}{\sqrt{2\pi}}(1 + u^2)e^{-u^2} + \frac{3}{\sqrt{2\pi}}e^{-u^2},$$

that is

$$A_3(y) = \frac{B_3(u)}{c_3} - \frac{3B_1(u)}{c_1},$$

where

$$B_3(u) = -(1 + u^2)e^{-u^2}, \qquad B_1(u) = -e^{-u^2}, \qquad c_3 = F, \qquad c_1 = 2F$$

and $F = \sqrt{\pi/2}$ the form factor for a Gaussian signal. Similarly functions $B_i(u)$ can be obtained from

$$B_i(u) = \left[\int c_i(y/\sigma)^i q(y)\,dy\right]_{y=u\sigma\sqrt{2}}, \tag{8.36}$$

with c_i chosen so that the value of the integral from 0 to ∞ is unity. These functions[11] for several values of i are shown in figure 8.4. The required cursor

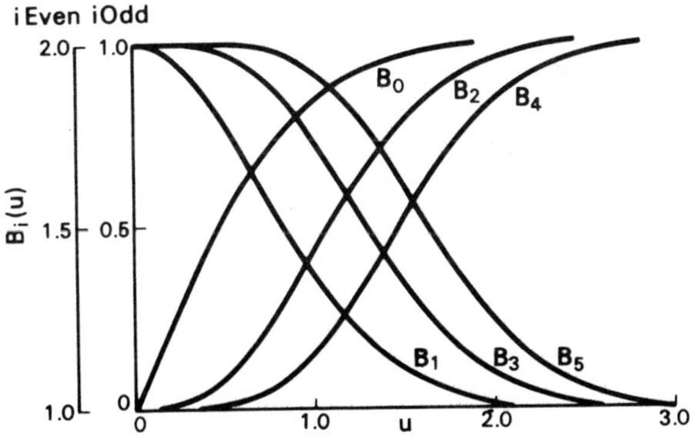

Figure 8.4 Graph of the functions $B_i(u)$

points, for positive values of u, and for intervals of 0·1 in the functions $B_i(u)$ are given in table 8.1 together with the constants c_i. The solutions for α'_k are then obtained as indicated in table 8.2, where β_i denotes the mean value of the points on the nonlinearity $n(y + \gamma)$ at the values of u given for $B_i(u)$ in table 8.1.

8.7 Nonlinearity identification

In section 3.8 a method was given for the identification of a nonlinear element from its harmonic response. A similar procedure is available based on the response of the nonlinearity to a Gaussian input. It can be shown that the best

TABLE 8.1
Cursor points for intervals of 0·1 in $B_i(u)$

$B_i(u)$	c_i	Values of u									
$B_0(u) = 1 + \operatorname{erf} u$	2	0·044	0·134	0·225	0·321	0·423	0·534	0·661	0·813	1·018	1·386
$B_1(u) = e^{-u^2}$	$2F$	0·231	0·403	0·536	0·658	0·773	0·893	1·025	1·177	1·379	1·733
$B_2(u) = 1 + \operatorname{erf} u - \frac{u}{\sqrt{\pi}} e^{-u^2}$	2	0·420	0·632	0·779	0·906	1·027	1·150	1·282	1·434	1·631	1·977
$B_3(u) = (1 + u^2)e^{-u^2}$	F	0·596	0·828	0·980	1·110	1·235	1·358	1·490	1·641	1·837	2·175
$B_4(u) = 1 + \operatorname{erf} u - \frac{2u}{\sqrt{\pi}} e^{-u^2} - \frac{4u^3}{3\sqrt{\pi}} e^{-u^2}$	$\frac{2}{3}$	0·757	0·998	1·157	1·289	1·413	1·537	1·669	1·820	2·014	2·353
$B_5(u) = \left(1 + u^2 + \frac{u^4}{2}\right) e^{-u^2}$	$\frac{F}{4}$	0·904	1·153	1·314	1·449	1·574	1·700	1·830	1·980	2·174	2·509

TABLE 8.2

Values of α'_k from $B_i(u)$ cursors

$\alpha'_0 = \beta_0$
$\alpha'_1 = \dfrac{\beta_1}{F}$
$\alpha'_2 = \beta_2 - \beta_0$
$\alpha'_3 = \dfrac{2\beta_3 - 3\beta_1}{F}$
$\alpha'_4 = 3\beta_4 - 6\beta_2 + 3\beta_0$
$\alpha'_5 = \dfrac{8\beta_5 - 20\beta_3 + 15\beta_1}{F}$

mean squared error estimate of the nonlinearity is

$$n(y) = \sum_{k=0}^{M} \alpha'_k H_k(y/\sigma), \tag{8.37}$$

where α'_k are the coefficients associated with the output autocorrelation function series

$$R_{zz}(\tau) = \sum_{k=0}^{\infty} \alpha'_k \{\rho^k(\tau)/k!\}.$$

Experimental determination of the α'_k from a measurement of $R_{zz}(\tau)$ or $S_{zz}(\omega)$ is however not as simple, except possibly for a narrow band input, as the equivalent sine input case. Use of the method is thus not very attractive.

The identification of a symmetrical single valued nonlinearity which yields a known RDF, namely the inverse RDF problem, is straightforward. From equation (8.14) with $\gamma = 0$

$$N_+(\sigma) = \alpha_{+1} = \frac{1}{\sigma^3\sqrt{2\pi}} \int_0^{\infty} y n_+(y) \mathrm{e}^{-y^2/2\sigma^2} \, \mathrm{d}y.$$

Substituting $u = y^2/2$ gives

$$N_+(\sigma) = \frac{1}{\sigma^3\sqrt{2\pi}} \int_0^{\infty} n_+\{(2u)^{1/2}\} \mathrm{e}^{-u/\sigma^2} \, \mathrm{d}u,$$

and writing $s = 1/\sigma^2$,

$$N_+(s^{-1/2}) = \frac{s^{3/2}}{\sqrt{2\pi}} \int_0^{\infty} n_+\{(2u)^{1/2}\} \mathrm{e}^{-su} \, \mathrm{d}u. \tag{8.38}$$

Determination of the RDF can thus be interpreted as a Laplace transform and the inverse expression is

$$n_+(x) = \sqrt{2\pi}[\mathcal{L}^{-1}\{s^{-3/2} N_+(s^{-1/2})\}]_{t=x^2/2}. \tag{8.39}$$

Thus for the odd symmetrical nonlinearity for $x > 0$,

$$n(x) = (\pi/2)^{1/2}[\mathscr{L}^{-1}\{s^{-3/2}N(s^{-1/2})\}]_{t=x^2/2}. \tag{8.40}$$

Example

For the relay with dead zone,

$$N(\sigma) = (2/\pi)^{1/2}(h/\sigma)e^{-\delta^2/2\sigma^2}.$$

Thus from equation (8.40),

$$\begin{aligned} n(x) &= h[\mathscr{L}^{-1}\{s^{-3/2}s^{1/2}e^{-\delta^2 s/2}\}] \\ &= [\mathscr{L}^{-1}\{(h/s)e^{-s\delta^2/2}\}]_{t=x^2/2}. \end{aligned}$$

Therefore

$$\begin{aligned} n(t) &= 0, \qquad t < \delta^2/2, \\ &\;\;\; h, \qquad t > \delta^2/2, \end{aligned}$$

giving

$$\begin{aligned} n(x) &= 0, \qquad x < \delta, \\ &\;\;\; h, \qquad x > \delta, \quad \text{as required.} \end{aligned}$$

The approximate techniques of the inverse DF problem described in section 3.8.2 may also be used for the RDF.

Polynomial expansion

Since the RDF for a polynomial characteristic is also a polynomial, if it is assumed that a polynomial curve-fit for $N(\sigma)$ has been found then the coefficients of the corresponding $n(x)$ polynomial are easily obtained.

Let

$$N(\sigma) = \sum_{m=0}^{M} c_m \sigma^m;$$

then comparing coefficients with the RDF of

$$n(y) = \sum_{m=0}^{M+1} A_m y^m$$

gives

$$A_m = c_{m-1}\sqrt{\pi}/[2^{(m+1)/2}\Gamma\{(2+m)/2\}].$$

Quantised nonlinearity

Consider a quantised nonlinearity to have steps of height h_m at the points $\{(2m-1)3\sigma\}/2M$, $m = 1$ to M, in the distance 3σ. Then knowing M points on the RDF, $N(\sigma)$, the values of h_m and therefore the nonlinearity which yields a RDF passing through the M points can be determined.

For the quantised nonlinearity,

$$N(\sigma) = (\sqrt{2}/\pi\sigma) \sum_{m=1}^{M} h_m e^{-\delta_m^2/2\sigma^2},$$

so that

$$N(3k\sigma/M) = (M\sqrt{2}/3k\sigma\sqrt{\pi}) \sum_{m=1}^{k} h_m e^{-\{(2m-1)/2k\sqrt{2}\}^2}.$$

Taking values of k from 1 to M gives

$$N(3\sigma/M) = (M/3\sigma F)h_1 e^{-1/8},$$

$$N(6\sigma/M) = (M/6\sigma F)\{h_1 e^{-1/32} + h_2 e^{-9/32}\},$$

$$N(3\sigma) = (1/3\sigma F) \sum_{m=1}^{M} h_m e^{-\{(2m-1)/2M\sqrt{2}\}^2},$$

from which the values of h_1 to h_m can be found by successive elimination.

8.8 Measurement of the RBDF

The equivalent gain block diagram model of figure 8.1 indicates a direct approach to the determination of K_r and K_γ, that is K_r and K_γ are successively adjusted until the mean squared value of $e(t)$ is a minimum. Since the gains K_r and K_γ arc orthogonal there is no problem of interaction in the adjustments and the procedure is straightforward. One disadvantage of the method is that in most cases the minimum mean squared value of $e(t)$ is badly defined, changes of K_r and K_γ about their optimum values producing little change in $\overline{e^2(t)}$. Accurate measurements of $\overline{e^2(t)}$ for several values of K_r and K_γ near to the optimum are therefore needed to obtain satisfactory results. It is often therefore quicker and more accurate to measure the covariance functions $R_{xy}(0)$, $R_{xx}(0)$ plus the bias input and output signals and use equations (8.3) and (8.4).

It has been shown that the DF for a Gaussian input signal is equal to the d.c. incremental gain, so that measurement of this parameter provides a suitable means for obtaining the DF. This approach is also applicable if the input to the nonlinearity contains other signals independent of the Gaussian input as the d.c. incremental gain result still applies. A particular example is the measurement of the RSDF, the describing function for a sine plus Gaussian input which is introduced in section 8.13, where the DF for the sinusoid can be found using a wave analyser as described in section 3.7 and the DF for the Gaussian signal from the measurement of the d.c. incremental gain.

8.9 Amplitude probability density of nonlinearity output

An amplitude dependent nonlinearity distorts the input signal, so that the amplitude probability density of the output is different from that of the input.

Consider the nonlinear characteristic

$$z = n(y), \tag{8.41}$$

with an input signal $y(t)$ of amplitude probability density $q(y)$.

The probability that y lies in the range $(y_1, y_1 + \delta y)$ is $q(y_1)\,\delta y$ and the probability that z lies in the range $(z_1, z_1 + \delta z)$ is $p(z_1)\,\delta z$ where the amplitude probability density of z is $p(z)$. If the relationship between z and y given by equation (8.41) is single valued so that z can only have the value z_1 when y is equal to y_1, the above probabilities must be equal, that is

$$q(y_1)\,\delta y = p(z_1)\,\delta z,$$

giving

$$p(z_1) = q(y_1)/\{|\mathrm{d}z/\mathrm{d}y|\}_{y=y_1}.$$

Hence

$$p(z) = q(y)/\{|\mathrm{d}z/\mathrm{d}y|\} = q(y)/\{|n'(y)|\}, \tag{8.42}$$

assuming that $n'(y)$ exists for all y. Since $p(z)$ and $q(y)$ are always positive equation (8.42) relates only to magnitudes so that the modulus of the derivative is used. If the inverse relationship of equation (8.41) is

$$y = m(z), \tag{8.43}$$

then the required density is

$$p(z) = [q(y)/|n'(y)|]_{y=m(z)}. \tag{8.44}$$

When each value of z corresponds to N values of y, that is equation (8.43) is multivalued,

$$p(z) = \sum_{i=1}^{N} [q(y_i)/|n'(y_i)|]_{y_i=m(z)}. \tag{8.45}$$

For example, consider the nonlinearity $z = y^2$. Then $n'(y) = 2y$ and $y = \pm\sqrt{z}$ is double valued so that

$$p(z) = \{q(\sqrt{z}) + q(-\sqrt{z})\}/2\sqrt{z}.$$

In particular, for the input y Gaussian,

$$p(z) = (1/\sigma\sqrt{2\pi z})\mathrm{e}^{-z/2\sigma^2} \quad \text{for } z > 0.$$

In many practical problems the characteristic $n(y)$ may not be known analytically or the determination of the inverse relationship $y = m(z)$ may be difficult in which case a graphical solution can be used. The procedure is illustrated in figure 8.5(a) and (b) for two specific nonlinear characteristics. Diagram (i) shows the assumed input distribution $q(y)$, diagram (ii) the nonlinearity $n(y)$ and (iii) the slope of the nonlinearity derived from (ii). The ratio $q(y)/|n'(y)|$ is shown in diagram (iv). To obtain the distribution with respect to the output variable z the horizontal scale of $q(y)/|n'(y)|$ must be adjusted according to the nonlinear

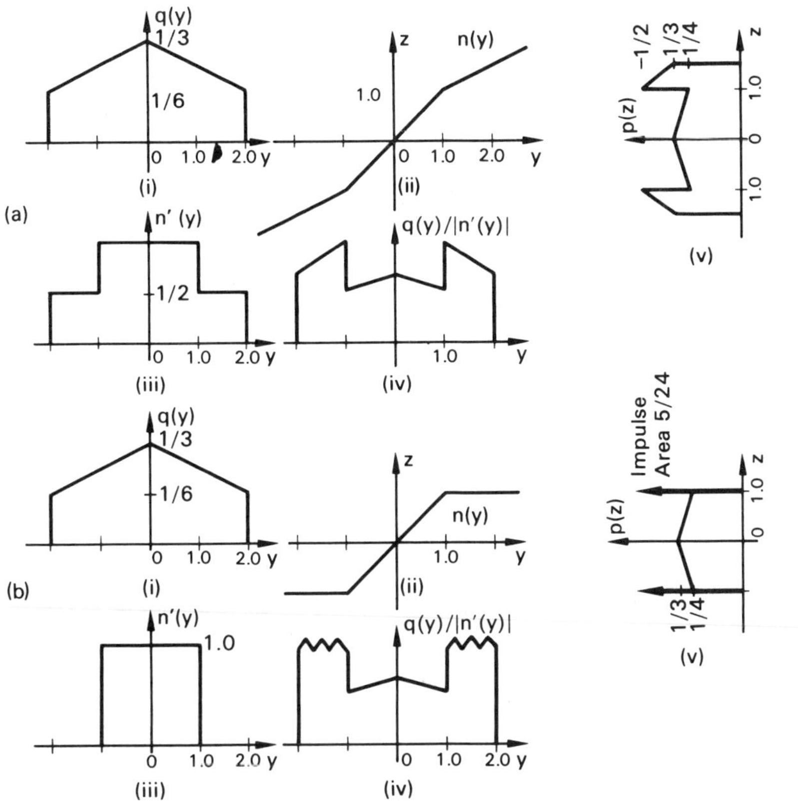

Figure 8.5 Graphical determination of amplitude probability density of nonlinearity output

relationship $z = n(y)$, the resulting distribution $p(z)$ is shown turned on its side in diagram (v). The area under the curve $p(z)$, but not under the curve $q(y)/|n'(y)|$, is unity. As the value of $n'(y)$ is zero for $|y| > 1$ in example (b) the output distribution $p(z)$ contains impulses at $n(\pm 1)$ each of area A where

$$A = \int_1^\infty q(y)\,\mathrm{d}y = (1/6) + (1/2)(1/12) = 5/24.$$

8.9.1 *The characteristic function and moments*

The characteristic function $M_z(j\omega)$ of the output signal can be evaluated from the definition

$$M_z(j\omega) = \int_{-\infty}^{\infty} p(z)\mathrm{e}^{-j\omega z}\,\mathrm{d}z$$

when $p(z)$ is known. It is, however, usually more easily evaluated directly from the input signal y since substituting

$$p(z)\,\mathrm{d}z = q(y)\,\mathrm{d}y,$$

and

$$z = n(y)$$

in the above integral gives

$$M_z(j\omega) = \int_{-\infty}^{\infty} q(y)\mathrm{e}^{-j\omega n(y)}\,\mathrm{d}y. \tag{8.46}$$

Similarly the nth moment μ_n of the output z is given by

$$\mu_n = \int_{-\infty}^{\infty} z^n p(z)\,\mathrm{d}z = \int_{-\infty}^{\infty} [n(y)]^n q(y)\,\mathrm{d}y. \tag{8.47}$$

For example, for the characteristic $z = y^2$ with a Gaussian input $y(t)$

$$M_z(j\omega) = (1/\sigma\sqrt{2\pi})\int_{-\infty}^{\infty} \mathrm{e}^{-y^2/2\sigma^2}\mathrm{e}^{-j\omega y^2}\,\mathrm{d}y,$$

and substituting $u = y\{(1 + 2\sigma^2 j\omega)^{1/2}/2\sigma^2\}$, gives

$$M_{y^2}(j\omega) = M_z(j\omega) = \int_{-\infty}^{\infty} \frac{\mathrm{e}^{-u^2}}{\{\pi(1 + 2j\omega\sigma^2)\}^{1/2}}\,\mathrm{d}u = \frac{1}{(1 + 2j\omega\sigma^2)^{1/2}}. \tag{8.48}$$

8.9.2 *Distributions generated from a Gaussian one*

Various signals with probability density distributions that are frequently used or occur in physical situations can be generated from one or more Gaussian distributions. The chi-squared, χ^2, distribution is the distribution of the sum of the squares of two or more independent random variables, each having a Gaussian distribution. Thus if

$$z = y_1^2 + y_2^2, \tag{8.49}$$

then $p(z)$ has a χ^2 distribution when y_1 and y_2 are Gaussian. From equation (8.48) above, if $y_1(t)$ and $y_2(t)$ have equal mean squared values σ^2

$$M_{y_1^2}(j\omega) = M_{y_2^2}(j\omega) = \frac{1}{(1 + 2j\omega\sigma^2)^{1/2}},$$

giving

$$M_z(j\omega) = M_{y_1^2}(j\omega)M_{y_2^2}(j\omega) = \frac{1}{(1 + 2j\omega\sigma^2)}, \tag{8.50}$$

and

$$p(z) = (1/2\pi)\int_{-\infty}^{\infty} \frac{\mathrm{e}^{j\omega z}}{1 + 2j\omega\sigma^2}\,\mathrm{d}\omega = (1/2\sigma^2)\mathrm{e}^{-z/2\sigma^2}, \quad \text{for } z > 0. \tag{8.51}$$

In the general case where

$$z = \sum_{i=1}^{f} y_i^2,$$

and

$$\overline{y_i^2(t)} = 1,$$

$$M_z(j\omega) = \frac{1}{(1 + 2j\omega)^{f/2}}, \tag{8.52}$$

and

$$\begin{aligned} p(z) &= z^{(f-2)/2}e^{-z/2}/(2^{f/2})\Gamma(f/2) \quad \text{for } z > 0, \\ &= 0 \quad \text{for } z < 0. \end{aligned} \tag{8.53}$$

The variable f is referred to as the degrees of freedom of the variable z. The χ^2 distribution which approaches the Gaussian distribution for large values of f, has a mean value of f and a variance of $2f$.

When z in equation (8.49) is replaced by u^2, the density $q(u)$ of u is given by

$$q(u) = p(z)(dz/du) = 2up(u^2),$$

that is

$$q(u) = (u/\sigma^2)e^{-u^2/2\sigma^2} \quad \text{for } u > 0. \tag{8.54}$$

This density, $q(u)$, is known as the Rayleigh probability density and has a mean value of $\sigma\sqrt{\pi/2}$ and a variance of $2\sigma^2$. The characteristic function for the Rayleigh distribution is

$$M_u(j\omega) = \{1 - (j\omega\sigma\sqrt{\pi})/2\}e^{-\sigma^2\omega^2/2}. \tag{8.55}$$

8.10 Higher order distribution functions of nonlinearity output

Consider the general case where two random variables u and v are related to two other random variables y and z by

$$u = u(y, z),$$

$$v = v(y, z),$$

and u and v are single valued in their arguments. Assuming that the joint probability density function of y and z is $q(y, z)$ and the inverse relationships

$$y = y(u, v),$$

$$z = z(u, v),$$

are single valued, as y and z move about some area $\delta y\, \delta z$ in the y, z plane u and v move about an area $\delta u\, \delta v$ in the u, v plane where

$$u_1 = u(y_1, z_1) \quad \text{and} \quad v_1 = v(y_1, z_1).$$

Thus

$$q(y, z)\, dy\, dz = p(u, v)\, du\, dv.$$

The relationship between elements of area in the two planes is determined from the Jacobian transformation of coordinates, namely

$$\mathrm{d}u\,\mathrm{d}v = |J|\,\mathrm{d}y\,\mathrm{d}z,$$

where

$$J = \left|\frac{\partial(u, v)}{\partial(y, z)}\right| = \begin{vmatrix} \dfrac{\partial u}{\partial y} & \dfrac{\partial u}{\partial z} \\ \dfrac{\partial v}{\partial y} & \dfrac{\partial v}{\partial z} \end{vmatrix} = \frac{\partial u}{\partial y}\frac{\partial v}{\partial z} - \frac{\partial u}{\partial z}\frac{\partial v}{\partial y}; \tag{8.56}$$

hence

$$p(u, v) = \frac{q(y, z)}{|J|} = \frac{q\{y(u, v), z(u, v)\}}{|J|}. \tag{8.57}$$

Similar equations can be written down for three or more variables. When a random signal $y(t)$ with a second amplitude probability density function $q(y_1, y_2, \tau)$ is applied to the nonlinear characteristic $z = n(y)$, the Jacobian matrix is diagonal as $z_1 = n(y_1)$ and $z_2 = n(y_2)$, thus

$$p(z_1, z_2, \tau) = \left\{\frac{q(y_1, y_2, \tau)}{|n'(y_1)||n'(y_2)|}\right\}_{y_1 = m(z_1),\, y_2 = m(z_2)}, \tag{8.58}$$

where $y = m(z)$ is the inverse relationship for the nonlinearity. Thus for the characteristic $z = y^2$,

$$p(z_1, z_2, \tau) = \{q(\sqrt{z_1}, \sqrt{z_2}, \tau)\}/\sqrt{z_1 z_2} \quad \text{for } z_1, z_2 > 0,$$

as $y = \pm\sqrt{z}$ is double valued. In particular when $q(y_1, y_2, \tau)$ is Gaussian,

$$p(z_1, z_2, \tau) = \sum_{k=0}^{\infty} H_k(\sqrt{z_1}/\sigma)q(\sqrt{z_1})H_k(\sqrt{z_2}/\sigma)q(\sqrt{z_2})\rho^k(\tau)/\sqrt{z_1 z_2}k!,$$

where $q(z^{1/2}) = (1/\sigma\sqrt{2\pi})\mathrm{e}^{-z/2\sigma^2}$ and $\rho(\tau)$ is the normalised correlation function of the Gaussian input.

8.10.1 *The joint characteristic function*

The joint characteristic function of the output signal, $M_z(ju, jv, \tau)$ by definition is given by

$$M_z(ju, jv, \tau) = \int_{-\infty}^{\infty}\int_{-\infty}^{\infty} p(z_1, z_2, \tau)\mathrm{e}^{-(uz_1 + vz_2)}\,\mathrm{d}z_1\,\mathrm{d}z_2.$$

Substituting $z = n(y)$ and $p(z_1, z_2, \tau)\,\mathrm{d}z_1\,\mathrm{d}z_2 = q(y_1, y_2, \tau)\,\mathrm{d}y_1\,\mathrm{d}y_2$ gives

$$M_z(ju, jv, \tau) = \int_{-\infty}^{\infty}\int_{-\infty}^{\infty} q(y_1, y_2, \tau)\mathrm{e}^{-j\{(un(y_1) + vn(y_2)\}}\,\mathrm{d}y_1\,\mathrm{d}y_2. \tag{8.59}$$

In particular when $q(y_1, y_2, \tau)$ is Gaussian and $n(y)$ is an ideal on-off characteristic with levels $\pm h$

$$M_z(ju, jv, \tau) = \sum_{k=0}^{\infty} \{\rho^k(\tau)/k!\} \int_{-\infty}^{\infty} H_k(y_1/\sigma)q(y_1)e^{-jun(y_1)}\,dy_1 \int_{-\infty}^{\infty} H_k(y_2/\sigma)q(y_2)e^{-jvn(y_2)}\,dy_2.$$

Now

$$\int_{-\infty}^{\infty} H_k(y_1/\sigma)q(y_1)e^{-jun(y_1)}\,dy_1$$

$$= \int_{-\infty}^{0} H_k(y_1/\sigma)q(y_1)e^{+juh}\,dy_1 + \int_{0}^{\infty} H_k(y_1/\sigma)q(y_1)e^{-juh}\,dy_1$$

$$= -\{2jH_{k-1}(0)/\sqrt{2\pi}\}\sin uh \quad \text{for } k \geq 1,$$

$$\cos uh \quad \text{for } k = 0.$$

Thus

$$M_z(ju, jv, \tau) = -(2/\pi)\sin uh \sin vh \sum_{k=1}^{\infty} \{H_{k-1}(0)\}^2\{\rho^k(\tau)/k!\} + \cos uh \cos vh,$$

that is

$$M_z(ju, jv, \tau) = -(2/\pi)\sin uh \sin vh \sin^{-1}\rho(\tau) + \cos uh \cos vh. \tag{8.60}$$

The corresponding value of $p(z_1, z_2, \tau)$ is

$$p(z_1, z_2, \tau) = (2/\pi)p(z_1)p(z_2)\,\text{sgn}\, z_1\, \text{sgn}\, z_2 \sin^{-1}\rho(\tau) + p(z_1)p(z_2), \tag{8.61}$$

where

$$p(z) = (1/2)\{\delta(z - h) + \delta(z + h)\}.$$

The autocorrelation function of the output signal can be evaluated from

$$R_{zz}(\tau) = \int_{-\infty}^{\infty}\int_{-\infty}^{\infty} z_1 z_2 p(z_1, z_2, \tau)\,dz_1\,dz_2,$$

or alternatively from the integral with respect to y_1 and y_2 to give the result

$$R_{zz}(\tau) = (2/\pi)h^2 \sum_{k=1}^{\infty} \{H_{k-1}(0)\}^2\{\rho^k(\tau)/k!\} = (2/\pi)h^2 \sin^{-1}\rho(\tau). \tag{8.62}$$

Comparison with equation (7.124), with $a = h$, shows that the output signal $z(t)$ will be a random telegraph signal[12] if

$$(2/\pi)\sin^{-1}\rho(\tau) = e^{-2\mu|\tau|},$$

that is

$$\rho(\tau) = \sin\{(\pi/2)e^{-2\mu|\tau|}\}. \tag{8.63}$$

Therefore the second amplitude probability density function of the random telegraph signal is

$$p(z_1, z_2, \tau) = p(z_1)p(z_2)e^{-2\mu|\tau|}\,\text{sgn}\, z_1\, \text{sgn}\, z_2 + p(z_1)p(z_2). \tag{8.64}$$

8.11 Closed loop systems with random inputs

Here the analysis of those systems where the input to the nonlinearity may be approximated as a Gaussian signal with or without bias is considered. This will include those situations where the system inputs, either intentional or disturbance signals, are Gaussian random and bias signals. Since the sum of two Gaussian signals is also Gaussian the same analysis can also be used for multiple Gaussian inputs.

A large number of problems exist where input signals can only be described adequately in statistical terms so that the study of nonlinear systems with random inputs is a field of growing importance in many branches of engineering.(13,14) The basic assumption in the analysis is that the nonlinearity input is Gaussian or equivalently for Gaussian system inputs the distortion spectral components produced by the nonlinearity are negligible when fedback to the nonlinearity. In support of this assumption it can be shown that a random signal becomes more Gaussian when passed through a low pass filter,(15) a characteristic of most control system plants. The analysis closely parallels that on the frequency response of nonlinear systems using the DF and SBDF as done in chapters 4 and 5. The same requirement regarding the absence of autonomous or induced limit cycles again applies.

When no bias signals exist within the loop the mean squared value of the input signal at the nonlinearity can be evaluated from two nonlinear algebraic equations, one describing the quasilinear behaviour of the nonlinearity and the other, obtained from the linearised system, giving the relationship between the mean squared value of the nonlinearity input and its equivalent gain. If a bias signal exists at the nonlinearity input four nonlinear equations normally result, two describing the behaviour of the nonlinearity and two from the separate analysis of the linearised system for the bias and random signals. The bias and mean squared value of the random signal at the nonlinearity input can be obtained from these equations. The determination of the magnitudes of signals elsewhere in the system then follows, although it should be remembered that the quasilinear analysis only attempts to model the signals at points removed from the nonlinearity output by a low pass filter. Should, for example, the mean squared value of the nonlinearity output be required then it is best done by direct evaluation (equation (8.5)) using the available solution for the nonlinearity input. Simulation studies(16) show good agreement between signals at the filtered points in a nonlinear system and its quasilinear equivalent.

Instances do occur where for certain system parameters the above equations yield more than one solution for the nonlinearity input; a situation analogous to the jump phenomenon found with sinusoidal inputs. This behaviour was first predicted and observed by Booton(1) who referred to it as dual-mode operation; a term which appears to have been superseded by the designation jump resonance or jump phenomena for a random input. As one might expect for a random input, a system with two possible stable theoretical solution modes tends to move in a random manner between the two states, where based on short time averages the nonlinearity input signal first has one solution then the other. When averages of a sufficient length to satisfy the statistical requirements for good accuracy[17] are taken the curve showing say the mean squared input to the nonlinearity versus the mean squared system input has a steep slope in the jump region.

Two specific examples are considered to illustrate the above method and are followed by some comments on its accuracy.

8.11.1 *A Gaussian input*

Consider the block diagram of figure 8.6 which shows the system studied in section 7.7.2 with the addition of a saturation nonlinearity. The system is required

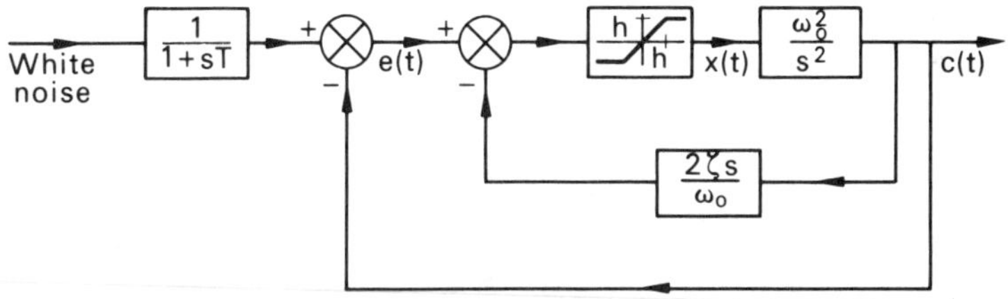

Figure 8.6 Block diagram for example of section 8.11.1

to follow the Gaussian input signal with a minimum error. Replacing the nonlinearity by the equivalent gain, $N(\sigma)$, the linearised system gives

$$\frac{E(s)}{R(s)} = \frac{s^2 + 2\zeta s\omega_0 N(\sigma)}{s^2 + 2\zeta s\omega_0 N(\sigma) + \omega_0^2 N(\sigma)},$$

and

$$\frac{X(s)}{R(s)} = \frac{s^2}{s^2 + 2\zeta s\omega_0 N(\sigma) + \omega_0^2 N(\sigma)}.$$

When the input to the filter of transfer function $1/(1 + sT)$ is white noise of S volts^2/Hz evaluation of the relevant integrals (see section 7.6.2) gives mean squared values for $e(t)$ and $x(t)$ of

$$\overline{e^2(t)} = \frac{\{2\zeta + k + 4\zeta^2 kN(\sigma)\}S}{4T\zeta\{1 + k^2N(\sigma) + 2\zeta kN(\sigma)\}}, \tag{8.65}$$

and

$$\overline{x^2(t)} = \frac{(2\zeta + k)S}{4T\zeta\{1 + k^2N(\sigma) + 2\zeta kN(\sigma)\}}, \tag{8.66}$$

where $k = w_0 T$. Writing $\overline{x^2(t)} = \sigma_x^2$, $\sigma_{xn} = \sigma_x/h$ and $(S/T) = n^2h^2$ where $\pm h$ are the limit levels of the saturation and n is a constant, equation (8.66) becomes

$$\frac{\sigma_{xn}}{n} = \left\{\frac{2\zeta + k}{4\zeta\{1 + k^2N(\sigma) + 2\zeta kN(\sigma)\}}\right\}^{1/2}. \tag{8.67}$$

In addition for the saturation nonlinearity, from equation (8.31) with $\gamma_N = 0$

$$N(\sigma) = \mathrm{erf}(1/\sigma_{xn}\sqrt{2}). \tag{8.68}$$

Thus if the values of k, ζ and n are given equations (8.67) and (8.68) can be solved for $N(\sigma)$ and σ_{xn}. This is done graphically in figure 8.7 where curves of (σ_{xn}/n)

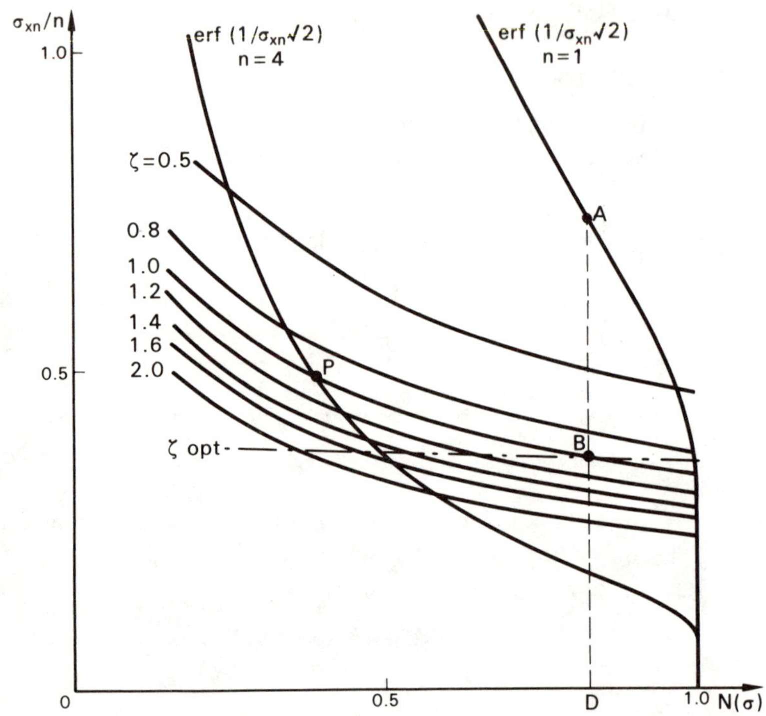

Figure 8.7 Solution loci for system of figure 8.6

versus $N(\sigma)$ are plotted from equation (8.67) for several values of ζ with $k = 2{\cdot}0$ and from equation (8.68) for values of n equal to 1·0 and 4·0. For $n = 4{\cdot}0$ and $\zeta = 1{\cdot}0$ the solution is at the point P where $N(\sigma) = 0{\cdot}39$ and $(\sigma_{xn}/n) = 0{\cdot}49$. On substituting for $N(\sigma)$ in equation (8.65) this gives

$$\overline{e^2(t)} = 6{\cdot}92h^2.$$

Since the system is nonlinear the optimum value of ζ to minimise the mean squared error will depend upon the power, S, as well as the relative bandwidth factor k. Assuming k is fixed, differentiating equation (8.65) gives for the optimum value of ζ,

$$\zeta_{opt} = [1 + \{2 + k^2N(\sigma)\}^{1/2}]/2kN(\sigma),$$

which for $k = 2{\cdot}0$ becomes

$$\zeta_{opt} = [1 + \{2 + 4N(\sigma)\}^{1/2}]/4N(\sigma).$$

The points on the loci of equation (8.67) for various values of ζ which satisfy the above equation for ζ_{opt} are shown by the dotted line in figure 8.7. Thus for $\zeta = 1{\cdot}0$ the operation will be optimum when n is such that the curve for the equivalent gain of the nonlinearity passes through B, that is when $n = AD/BD = 2{\cdot}05$. Figure 8.8 shows the required relationship between ζ and n evaluated for the optimum condition of a minimum mean squared error in the above manner.

With $k = 2{\cdot}0$ and $n = 4{\cdot}0$, the optimum value of ζ is 1·52, $N(\sigma) = 0{\cdot}495$ and the corresponding mean squared error is

$$\overline{e^2(t)} = 6{\cdot}20h^2.$$

For the same input the linear system with optimum damping of $\zeta = 0{\cdot}863$ has a mean squared error

$$\overline{e^2(t)} = 5{\cdot}32h^2.$$

The optimum nonlinear system thus has a mean squared error 16·6 per cent greater than that of the optimum linear system. Experimental measurements on a simulated system gave a slightly higher value for the mean squared error.

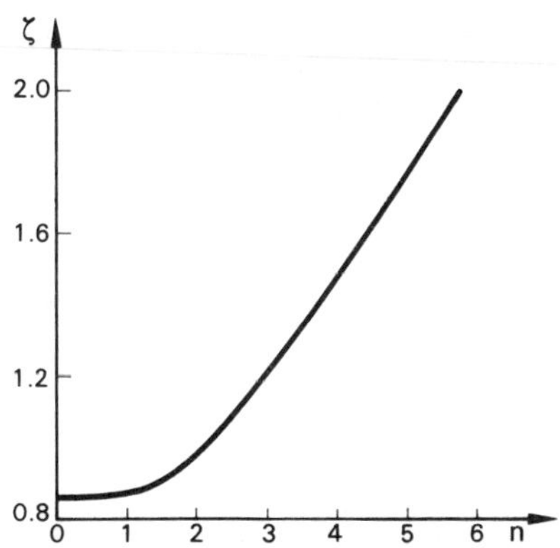

Figure 8.8 Optimum damping as a function of input power for system of figure 8.6

8.11.2 *Gaussian and bias inputs*

The block diagram of the system considered is shown in figure 8.9. The signal $z(t)$, assumed to have a Gaussian distribution and white spectrum of power

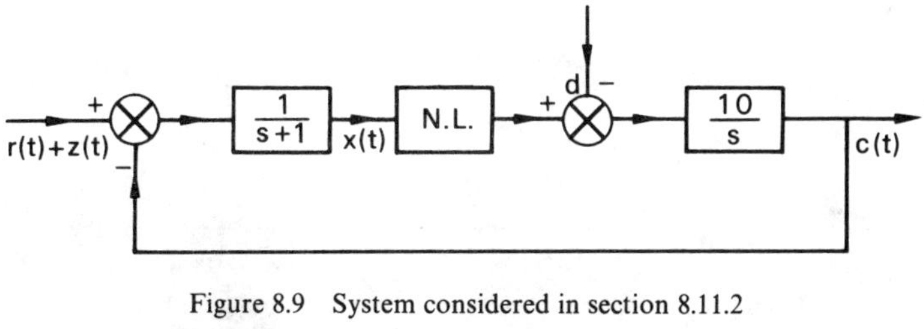

Figure 8.9 System considered in section 8.11.2

per unit bandwidth S, is a disturbance signal contaminating the constant reference input $r(t) = r$. In addition, a further constant disturbance, d, exists at the nonlinearity output. We wish to determine how the system error is affected by d, when $S = 0{\cdot}1$ and the nonlinearity is defined by $n(x) = x + x^3$.

Provided the system is stable the bias input to the integrator in the steady state must be zero, so that the bias output from the nonlinearity is equal to d. Assuming that the random signal has a mean squared value of σ_x^2 and the bias signal a value γ at the input to the nonlinearity, then the equivalent gain expressions for the nonlinearity yield

$$(1 + 3\sigma_x^2)\gamma + \gamma^3 = d, \tag{8.69}$$

$$(1 + 3\sigma_x^2) + 3\gamma^2 = N_\sigma(\sigma, \gamma) = N_\sigma. \tag{8.70}$$

In addition for the linearised system

$$\frac{X(s)}{Z(s)} = \frac{s}{s^2 + s + 10N_\sigma},$$

which gives on evaluating the mean squared value integral

$$\sigma_x^2 = S/2 = 0{\cdot}05.$$

Substituting this value of σ_x^2 in equations (8.69) and (8.70) gives the two equations

$$\gamma^3 + 1{\cdot}15\gamma = d, \tag{8.71}$$

$$3\gamma^2 + 1{\cdot}15 = N_\sigma. \tag{8.72}$$

Figure 8.10 shows the solution of equations (8.71) and (8.72) for values of d between 0 and 2·0. The mean squared value of the random noise signal at the

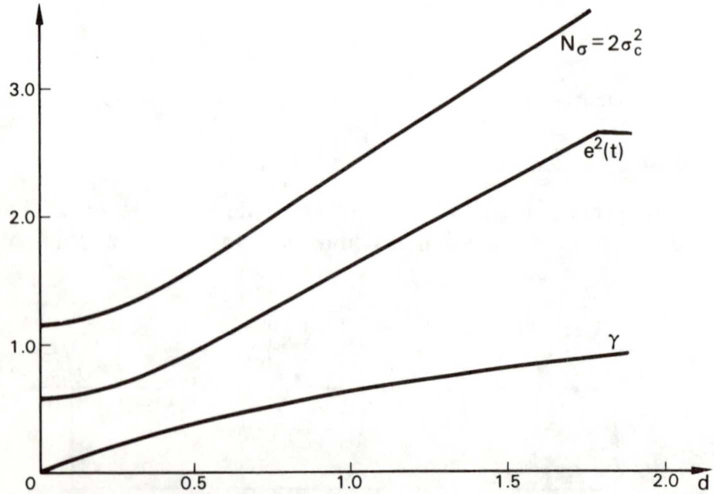

Figure 8.10 Mean squared values in system of figure 8.9

system output is given by

$$\sigma_c^2 = \frac{S}{2\pi j}\int_{-j\infty}^{j\infty} \frac{100N_\sigma^2}{(s^2 + s + 10N_\sigma)(s^2 - s + 10N_\sigma)}\,ds,$$

that is

$$\sigma_c^2 = 5SN_\sigma = 0{\cdot}5N_\sigma.$$

In addition the bias value at the output $c_b(t)$ is given by

$$c_b(t) = r - \gamma,$$

so that the total mean squared error $\overline{e^2(t)}$ between $c(t)$ and the desired output r is

$$\overline{e^2(t)} = \gamma^2 + 0{\cdot}5N_\sigma = 2{\cdot}5\gamma^2 + 0{\cdot}575.$$

The value of $\overline{e^2(t)}$ is also shown in figure 8.10.

8.11.3 *Accuracy*

Attempts have been made by various authors(18,19,4,5,20) to find methods for estimating the accuracy of quasilinear analysis of nonlinear systems with random inputs but no truly satisfactory solution has resulted. Obviously the accuracy is in some way dependent on the amount of distortion produced by the nonlinearity and the effectiveness of the low pass plant in filtering the signal and producing a near Gaussian input to the nonlinearity.

Probably the most useful work is that of Smith(4,5) who appears to be the only investigator to have undertaken any detailed experimental work. His main theoretical contribution was to develop an alternative approximate nonlinear method of analysis based on functional methods.(21) The method, which assumes the nonlinearity input to be Gaussian and the distortion at the nonlinearity output to be white noise, can be used to calculate the mean squared value of the nonlinearity input signal.

For the particular case of a first order system with a nonlinearity and integrator in the feedforward path and unity feedback, exact solutions can be obtained from the Fokker–Planck equation(22) for the first and second order probability densities of the nonlinearity input when the input signal is integrated white noise. Smith gives theoretical curves showing the r.m.s. nonlinearity input (error) versus the r.m.s. system input evaluated by the exact, quasilinear and approximate nonlinear method for the above system with three different nonlinearities. In all cases the accuracy of the approximate nonlinear method was found to be worse than the quasilinear, but where the effects of distortion could be estimated as large due to a considerable difference between the results of the two approximate methods, both were considerably in error. Conversely when the two methods gave nearly identical results the accuracy obtained was good. It was also noted that there was a strong correlation between the magnitude of the deviation from the Gaussian distribution at the nonlinearity input and the inaccuracies of the approximate solutions. Unfortunately a theoretical estimate of this deviation for any system, say by assuming a Gaussian input to the nonlinearity and calculating the distribution of the fedback signal, is not in general possible.

For the systems of figure 8.11 r.m.s. error versus r.m.s. input curves were calculated by the quasilinear and approximate nonlinear method and compared

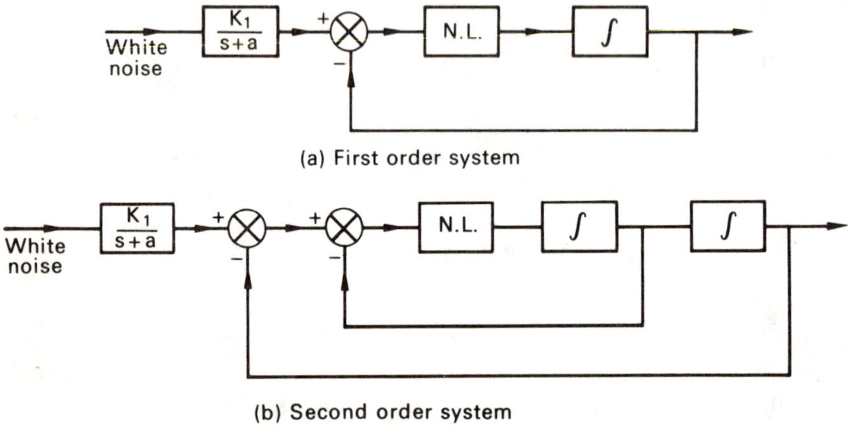

Figure 8.11 Systems studied by Smith[4]

with simulation measurements using cubic, relay with dead zone and ideal limiter nonlinearities. The amplitude probability density function of the error was also measured. Smith's results may be summarised as follows:

(1) Wherever the r.m.s. errors calculated by the quasilinear and approximate nonlinear methods differed appreciably, the experimental results showed both methods to be in error. The measured r.m.s. errors were normally higher than the calculated ones.

(2) The accuracy of the approximate analyses was better for inputs having wide bandwidth than for narrow band inputs and better for the second order system than the first.

(3) In regions where the approximate analyses gave good agreement with experiment the error signal was Gaussian to a high degree of accuracy. Conversely, where agreement was poor, the amplitude distribution of the signal was characteristically non-Gaussian.

(4) The nature of the non-Gaussian distributions was such that they would yield lower values of equivalent gain for the nonlinearities; hence a higher r.m.s. error in agreement with (1).

(5) R.m.s. errors of better than 5 per cent usually resulted when the Gaussian assumption was valid. For the first order systems errors of up to 25 per cent were found from the simulation and even higher for the system amenable to exact analysis.

As a result of these investigations Smith's main conclusions on the accuracy of quasilinear analysis were:

(1) When quasilinear methods give inaccurate predictions of the performance of a system subject to a random input, they do so because the assumption of a Gaussian distribution at the nonlinearity input does not hold.

(2) The approximate nonlinear method of taking into account distortion allows the occurrence of the above to be predicted and gives a good test of the accuracy of quasilinear results.

(3) Any quasilinear model which attempts to take into account distortion effects while retaining the Gaussian input assumption[23] ignores the real cause of difficulty. The crosscorrelation model for the nonlinearity is preferable to any modification of it proposed by other writers.[3,24,25]

(4) The quasilinear method of analysis will yield results of accuracy sufficient for most engineering purposes if the power spectral density of the input signal to the nonlinear element shows a peak at a frequency outside the effective bandwidth of the system, both quantities being found by quasilinear methods. In such cases the predicted r.m.s. level of signals is unlikely to be more than 5 to 10 per cent in error.

(5) Results obtained using quasilinear methods in the analysis of low order nonlinear feedback systems with narrow band inputs should be regarded as subject to considerable error.

In reference 23 the distortion $d(t)$ is taken into account by assuming the model shown in figure 8.12 for the nonlinearity. The signal $d(t)$ is taken as white noise

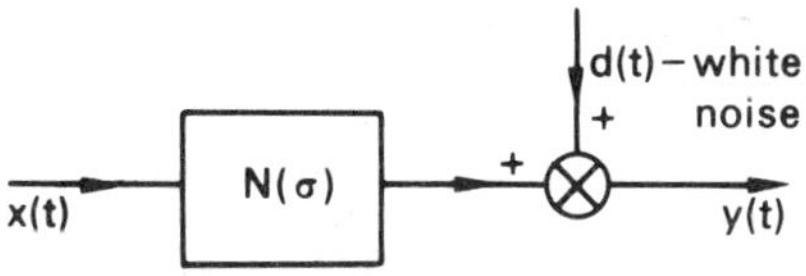

Figure 8.12 Nonlinearity model of reference 23

with power per unit bandwidth equal to the value obtained at low frequencies at the nonlinearity output assuming a nonlinearity input spectrum of the form given by a quasilinear analysis. The calculation of the power is done according to the procedure of section 8.4.3. Although Smith concluded that this method does not improve the accuracy of mean squared error calculations, analysis using the above model predicts a low frequency error signal as found in practice.

8.12 Optimum systems

In the previous sections we have primarily been concerned with analysing the behaviour of nonlinear systems having random inputs. The system designer, however, is normally faced with the problem of synthesising a system to meet given performance criteria. Two situations usually occur where the designer has limited flexibility and the optimum design entails selecting the best values for certain variable system parameters or, alternatively, where only some portion of the system is fixed and the designer has considerable choice with regard to the system configuration and compensation. These two situations are referred to as fixed and semi-free configurations respectively in reference 26. The first case is straightforward as exemplified by the problem studied in section 8.11.1 where the optimum value of ζ for a minimum mean squared error was determined. The classic example of the second category is the Wiener problem, considered in section 7.8, of the separation of a signal from noise with a linear filter. As described in its basic form it is a free configuration problem in that no fixed elements are involved, however it can readily be extended as done in section 7.8.2 to cover the semi-free configuration which arises in control problems.

The further extension of the Wiener problem to the design of the optimum linear compensator for a system with fixed linear and nonlinear parts is in principle not difficult using an equivalent gain model for the nonlinearity. Smith[4] has described and illustrated the appropriate procedure which is given in the next section. The resulting system, due to its inherent nonlinearity, is only optimum for the given input conditions. Because of this property of a linear compensated nonlinear system Douce and King[27] adopted a somewhat different approach and presented a method for designing a nonlinear compensator to give a minimum mean squared error for various magnitudes of random input. Their method is considered in section 8.12.2.

8.12.1 *Optimum linear compensation*

The description of the following method of Smith[4] closely follows its presentation in reference 28. The conditions of the problem are:

(1) The system responds to an input which is the sum of a signal and noise, both of which are considered stationary Gaussian random processes with given power spectral density functions.

(2) The desired output of the system is a function derived from the signal component of the input by some linear invariant operation, not necessarily physically realisable.

(3) The criterion for optimality is minimum steady state mean squared error and is the difference between the actual output of the system and that desired.

(4) The system contains a stable linear part and a single valued nonlinearity both of which are fixed. The optimum linear physically realisable compensation is to be designed for the system.

The system is shown in the block diagram of figure 8.13, where N and $G(s)$ characterise the fixed nonlinear and linear parts and $G_c'(s)$ is the compensator to be designed. The feedback system configuration is also shown redrawn in a single path configuration which is the form, apart from N, to which the filter theory of section 7.8.2 is applicable. The new cascade compensator $G_c(s)$ and the original $G_c'(s)$, assuming quasilinear analysis, are related by

$$G_c = \frac{G_c'}{1 + G_c' N_\sigma G}, \quad \text{or} \quad G_c' = \frac{G_c}{1 - G_c N_\sigma G}. \tag{8.73}$$

The additional difficulty caused by the nonlinearity is that its RDF depends upon σ_x^2 which is not known until $G_c(s)$ is found. To overcome this problem a

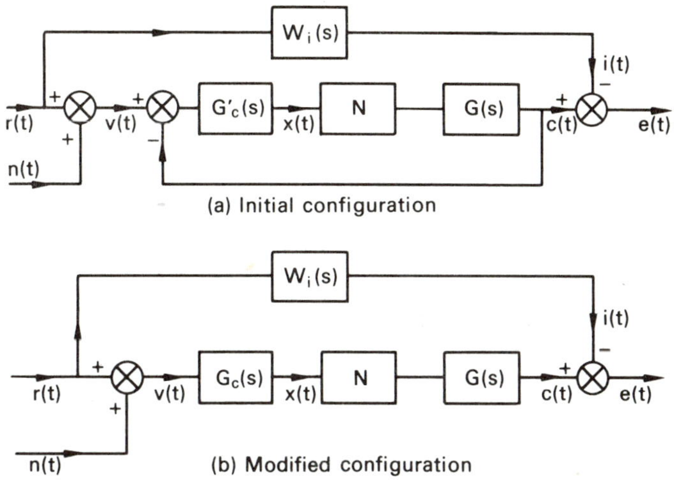

Figure 8.13 Illustrations for design of optimum linear compensator

value of σ_x^2 is assumed and the optimum design, which minimises $\sigma_e^2 + \lambda\sigma_x^2$ where λ is a Lagrange multiplier, carried out with the constraint on σ_x^2. By repeating the procedure for different values of σ_x^2 and recording the mean squared error, σ_e^2, for each solution, the optimum solution yielding the smallest mean squared error can be found.

The solution to the constrained optimum design problem for the linear semi-free configuration of figure 8.13 is given in section 7.8.2. Taking account of the nonlinearity it proceeds as follows:

(1) Choose a value of σ_x and calculate the corresponding value of N_σ.

(2) Form the function

$$H(s) = N_\sigma^2 G(s)G(-s) + \lambda,$$

and factor it into

$$H(s) = H(s)_L H(s)_R,$$

where $H(s)_L$ contains all the factors of $H(s)$ which define poles or zeros in the left half plane and $H(s)_R$ those in the right. Any poles or zeros on the $j\omega$ axis are moved just off the axis by the addition of a real part, ε. For example if $H(s)$ has a factor $1/(s)(-s)$ it is replaced by $1/(s+\varepsilon)(-s+\varepsilon)$.

(3) Form the power spectral density functions

$$S_{vr}(s) = S_{rr}(s) + S_{nr}(s)$$

$$S_{vv}(s) = S_{rr}(s) + S_{nr}(s) + S_{rn}(s) + S_{nn}(s),$$

where

$$S(s) = S(\omega)|_{\omega^2=-s^2} \quad \text{and} \quad v(t) = r(t) + n(t).$$

Factor the input spectrum

$$S_{vv}(s) = S_{vv}(s)_L S_{vv}(s)_R$$

as for $H(s)$ in (2).

(4) Form the function

$$M(s) = \frac{W_i(s)N_\sigma G(-s)S_{vr}(s)}{H(s)_R S_{vv}(s)_R}$$

and separate into

$$M(s) = M(s)_{Lp} + M(s)_{Rp},$$

where $M(s)_{Lp}$ and $M(s)_{Rp}$ have their poles only in the left and right half plane respectively. $M(s)_{Lp}$ only need be evaluated explicitly.

(5) The optimum cascade compensation function is

$$G_c(s) = \frac{M(s)_{Lp}}{H(s)_L S_{vv}(s)_L},$$

which is a function of the Lagrange multiplier λ. The above expression for the optimum compensator if there is no constraint, that is $\lambda = 0$, is of course identical with that derived in equation (7.98) when $G(s)$ is replaced by $N_\sigma G(s)$.

(6) Determine σ_x from

$$\sigma_x^2 = \frac{1}{2\pi j}\int_{-j\infty}^{j\infty} S_{rr}(s)G_c(s)G_c(-s)\,ds,$$

and, equating it to the assumed value in (1), find λ. Using this value of λ find $G_c(s)$ from (5).

(7) Calculate the mean squared error from

$$\overline{e^2(t)} = \frac{1}{2\pi j}\int_{-j\infty}^{j\infty} S_{ee}(s)\,ds,$$

where

$$S_{ee}(s) = E(s)E(-s)S_{rr}(s) + N_\sigma^2 G_c(s)G(s)G_c(-s)G(-s)S_{nn}(s)$$
$$+ N_\sigma G_c(s)G(s)E(-s)S_{rn}(s) + N_\sigma G_c(-s)G(-s)E(s)S_{nr}(s),$$

and

$$E(s) = N_\sigma G_c(s)G(s) - W_i(s).$$

The first two terms and the sum of the last two in the expression for $S_{ee}(s)$ are of the form required for the standard integral of table 7.1.

(8) Repeat for several values of σ_x to find that which yields the minimum $\overline{e^2(t)}$. Use the corresponding value of $G_c(s)$ in equation (8.73) to give $G_c'(s)$.

Obviously the solution of the above synthesis problem is quite tedious but it does provide a satisfactory solution to a rather complex problem. The big advantage of having an optimum design procedure available is that though it may not be implemented in practice it serves as both a guide and standard for alternative designs.

Example

For the system shown in figure 8.14 determine the optimum linear compensator $G_c'(s)$ and the corresponding minimum mean squared error. Given

$$S_{rr}(s) = 24/(1 - s^2), \qquad S_{nn}(s) = 1, \qquad S_{rn}(s) = S_{nr}(s) = 0$$

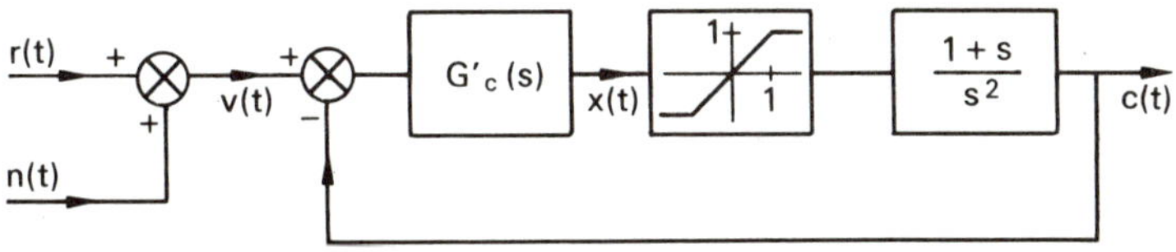

Figure 8.14 System considered in example of section 8.12.1

and that the desired output is the signal component of the input. Proceeding as described above

(1) $N_\sigma = \text{erf}(1/\sigma_x\sqrt{2})$.

(2) $H(s) = \{N_\sigma^2(1 + s)(1 - s^2)/s^4\} + \lambda$,

$H(s) = N_\sigma^2\{\rho^2 s^2 + as + 1)(\rho^2 s^2 - as + 1)\}/s^4$,

where $\rho^4 = \lambda/N_\sigma^2$ and $a = (1 + 2\rho^2)^{1/2}$.

(3) $$S_{vv}(s) = \frac{24}{1 - s^2} + 1 = \left(\frac{5 + s}{1 + s}\right)\left(\frac{5 - s}{1 - s}\right).$$

(4) $$M(s) = N_\sigma\left(\frac{1 - s}{s^2}\right)\left(\frac{24}{1 - s^2}\right)\left(\frac{1 - s}{5 - s}\right)\left\{\frac{s^2}{(\rho^2 s^2 - as + 1)N_\sigma}\right\}$$

$$= \frac{24(1 - s)}{(1 + s)(5 - s)(\rho^2 s^2 - as + 1)},$$

since $W_i(s) = 1$. Thus $M(s)_L = 8/(\rho^2 + a + 1)(1 + s)$.

(5) $$G_c(s) = \left\{\frac{8}{(\rho^2 + a + 1)(1 + s)}\right\}\left\{\frac{s^2}{(\rho^2 s^2 + as + 1)N_\sigma}\right\}\left(\frac{1 + s}{5 + s}\right)$$

$$= \frac{8s^2}{(\rho^2 + a + 1)N_\sigma(5 + s)(\rho^2 s^2 + as + 1)}.$$

(6) The mean squared input to the nonlinearity

$$\sigma_x^2 = \frac{K}{2\pi j}\int_{-j\infty}^{j\infty}\left\{\frac{s^2}{(1 + s)(\rho^2 s^2 + as + 1)}\right\}\left\{\frac{s^2}{(1 - s)(\rho^2 s^2 - as + 1)}\right\} ds,$$

where $K = 64/N_\sigma^2(\rho^2 + a + 1)^2$. Evaluating the integral gives

$$\sigma_x^2 = 32(1 + a)/\rho^2 N_\sigma^2 a(\rho^2 + a + 1)^3.$$

Assuming values for σ_x and using the above expression together with that given in (1) the relationship between λ and σ_x shown in figure 8.15 is obtained.

(7) Since $S_{rn}(s) = S_{nr}(s) = 0$,

$$S_{ee}(s) = E(s)E(-s)S_{rr}(s) + N_\sigma^2 G_c(s)G(s)G_c(-s)G(-s)S_{nn}(s),$$

with

$$E(s) = N_\sigma G_c(s)G(s) - 1$$

$$= \frac{8(s+1) - (\rho^2 + a + 1)(5+s)(\rho^2 s^2 + as + 1)}{(\rho^2 + a + 1)(5+s)(\rho^2 s^2 + as + 1)},$$

as

$$N_\sigma G_c(s)G(s) = \frac{8(s+1)}{(\rho^2 + a + 1)(5+s)(\rho^2 s^2 + as + 1)}.$$

The result of evaluating

$$\sigma_e^2 = \overline{e^2(t)} = \frac{1}{2\pi j}\int_{-j\infty}^{j\infty} S_{ee}(s)\,\mathrm{d}s$$

is also shown in figure 8.15 as a function of σ_x.

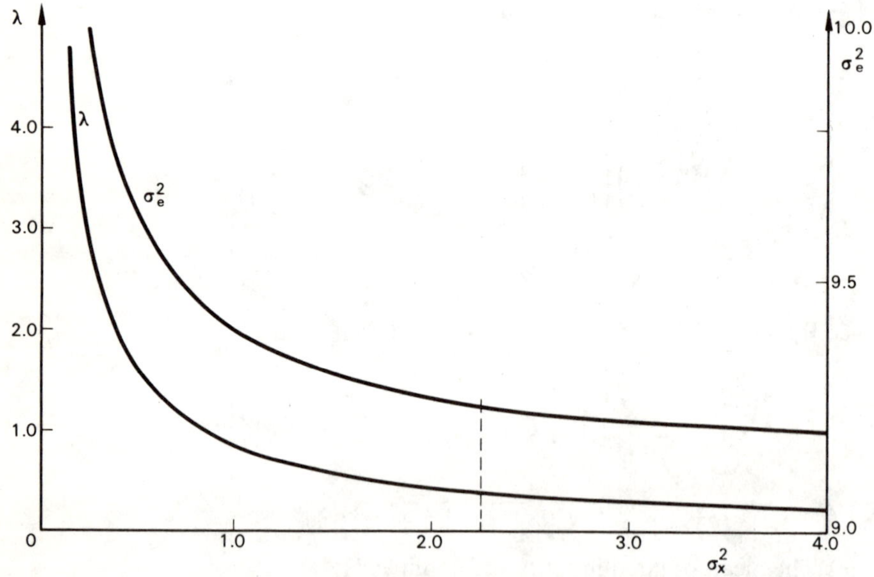

Figure 8.15 Solution curves for example of figure 8.14

(8) From figure 8.15 it is seen that there is no minimum value of σ_e for finite σ_x, however, the improvement for σ_x say greater than 1·5 is negligible. The optimum solution is therefore taken as $\sigma_x = 1{\cdot}5$, $\sigma_e^2 = 9{\cdot}25$ and $\lambda = 0{\cdot}388$. A value of σ_x which gives $N_\sigma = 0{\cdot}495$. Using these values in the expression for $G_c(s)$ gives

$$G_c(s) = \frac{3{\cdot}06s^2}{(s^2 + 1{\cdot}5s + 0{\cdot}8)(s+5)},$$

and from

$$G_c'(s) = \frac{G_c(s)}{1 - N_\sigma G_c(s)G(s)},$$

$$G_c'(s) = \frac{3s^2}{s^3 + 6{\cdot}5s^2 + 6{\cdot}8s + 2{\cdot}5},$$

the required optimum compensator.

8.12.2 *Optimum nonlinear compensation*

A block diagram of the type of system considered by Douce and King[27] is shown in figure 8.16. The fixed part of the system contains the nonlinearity N_1,

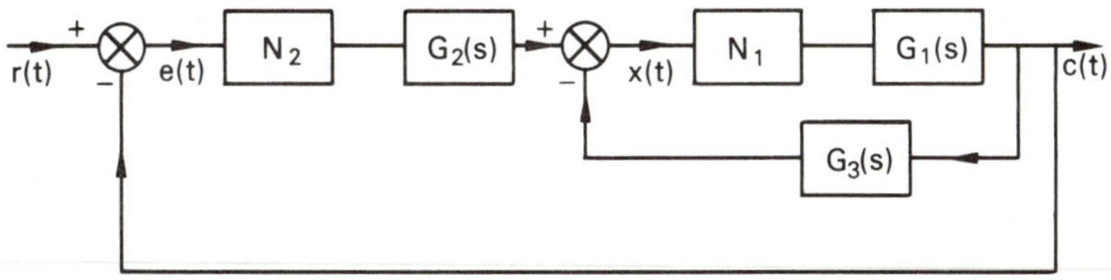

Figure 8.16 Block diagram of system for optimum nonlinear compensator design

the plant with transfer function $G_1(s)$ and any linear cascade or feedback compensators such as $G_2(s)$ and $G_3(s)$. The input signal is assumed to be Gaussian with a known spectral shape but only short time stationary in that its mean squared value changes very slowly compared to the time constants of the feedback system. The static nonlinear compensator N_2 is to be designed so that the mean squared value of the error $e(t)$ remains a minimum for variations in the input power. The procedure if both nonlinearities are replaced by their RDFs is relatively straightforward although again somewhat laborious.

The approximation of the inherent system nonlinearity N_2 by its RDF is certainly questionable since in the majority of cases $G_2(s)$ will not be a low pass filter. The authors argued that RDFs for many nonlinearities, particularly the ideal saturation characteristic which was the inherent system nonlinearity of the problem, are not strongly dependent on the distribution of the input. Whilst this could be true in some instances it is difficult to reconcile with Smith's arguments on the accuracy of quasilinearisation. With the understanding therefore that the accuracy of the proposed method may not be entirely satisfactory, the procedure is as follows:

(1) Choose values for $N_2(\sigma)$, the RDF of the compensator, and for σ_x the r.m.s. value at the input to N_1. Evaluate σ_r and σ_e the r.m.s. values of the system input and error.

(2) Repeat the procedure for enough values of $N_2(\sigma)$ and σ_x to obtain the graphs of figure 8.17.

(3) For a chosen value of σ_r determine the corresponding values of $N_2(\sigma)$ and σ_e from the curves of figure 8.17. Repeat for several values of σ_r to obtain the σ_e versus $N_2(\sigma)$ curves of figure 8.18.

(4) Determine from figure 8.18 the values of $N_2(\sigma)$ which produce minimum values of σ_e. Such points will normally exist unless the solution is trivial. Plot these on the $N_2(\sigma)$ versus σ_e plane, figure 8.19, to obtain the RDF required for the optimum nonlinear compensator.

(5) Find the nonlinearity from its RDF (see section (8.7).

The system considered in reference 27 had $G_1(s) = 1/s^2T^2$, $G_2(s) = 1$, $G_3(s) = 2\zeta sT$ and an ideal saturation characteristic for N_1. The input signal was white noise passed through a filter with transfer function $K/(1 + s\mu T)$. Theoretical results for N_2 were found for $\mu = 2, 4, 8$ and ∞, where the latter case corresponding to a very low frequency input, N_2 approached the form required for an optimum step response (see section 2.12). Simulation results for the system containing the optimum compensator nonlinearity for a particular input band-

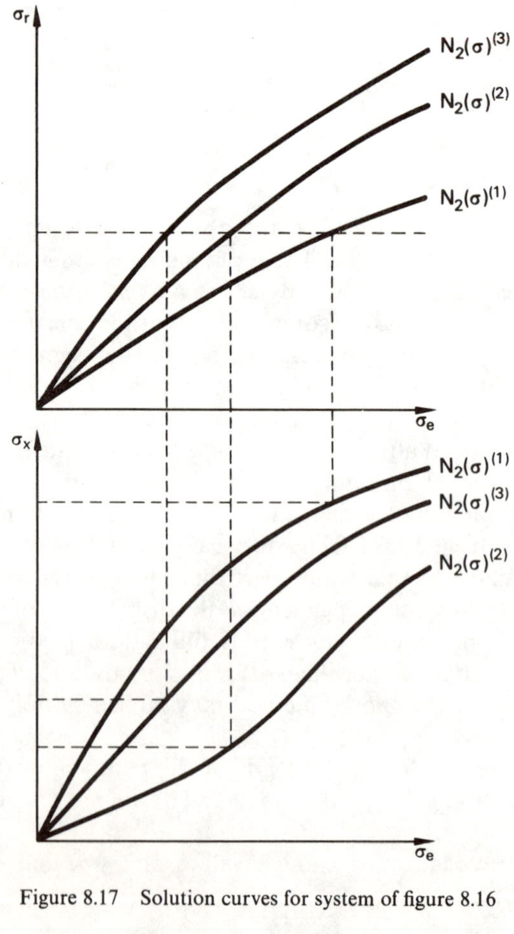

Figure 8.17 Solution curves for system of figure 8.16

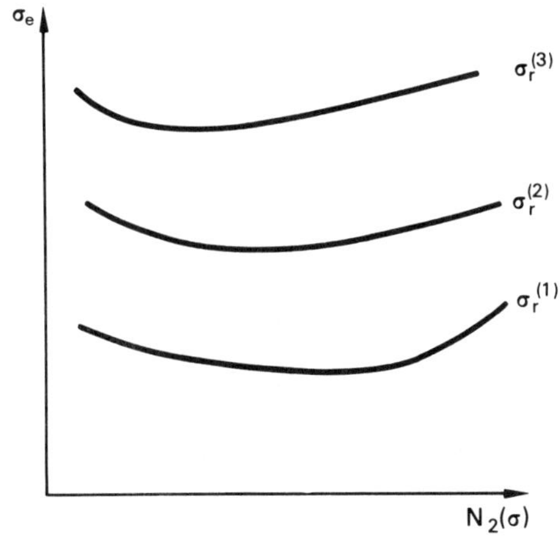

Figure 8.18 Relationships obtained from figure 8.17

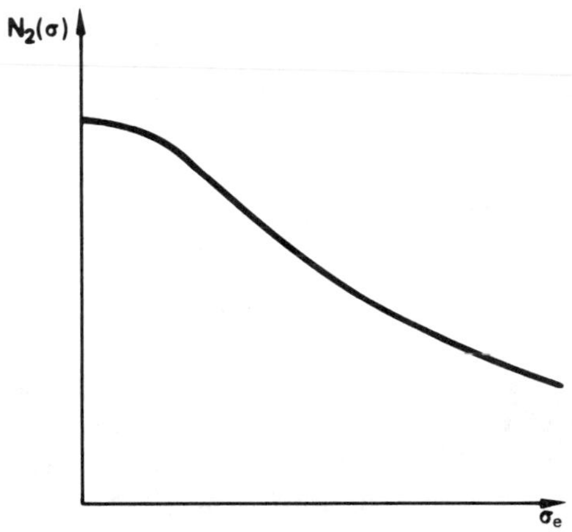

Figure 8.19 Required RDF of compensator from minima of figure 8.18

width confirmed the theory. In all cases lower mean squared errors were obtained with the compensator present, with a maximum improvement of 3:1 for a large input power with $\mu = 8$. Despite the coarse approximation the experimental results, as witnessed by those reproduced in figure 8.20 for this case, showed reasonable agreement with the theory.

8.13 The RSDF

In this section we are concerned with the analysis of those problems where the input to the nonlinearity may be modelled by a sinusoidal plus Gaussian signal. These situations exist in a nonlinear system with sinusoidal and Gaussian

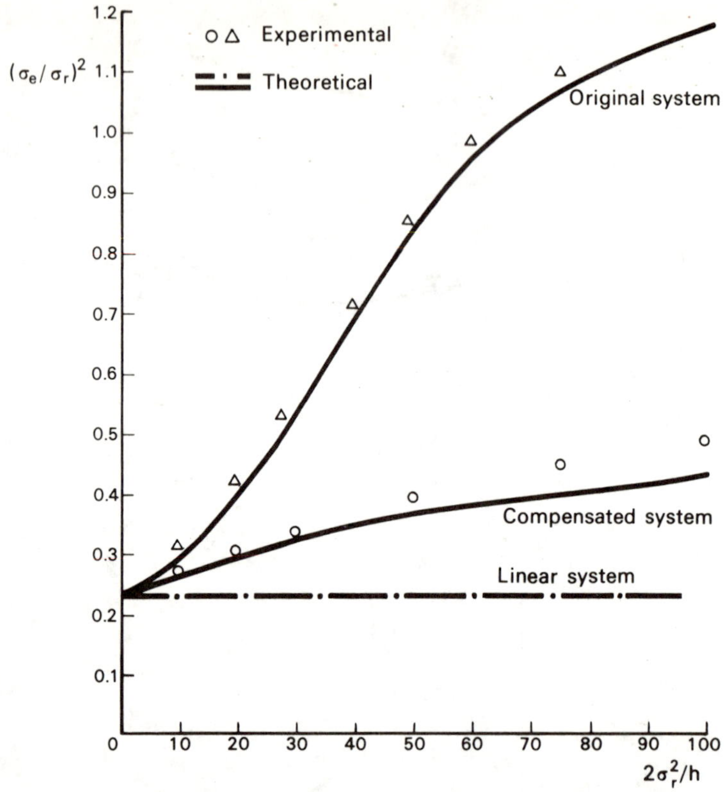

Figure 8.20 Results given in reference 27

inputs or where only a Gaussian input is applied and a limit cycle exists or is induced in the system. Chapter 10 discusses in some detail various methods for evaluating multiple input describing functions; however in this section, since our concern is directed more towards an examination of system behaviour, the validity of the modified nonlinearity approach will again be assumed and be sufficient for our investigations.

The general method of analysis assumes values for the sinusoidal and Gaussian components at the nonlinearity input and the resulting four nonlinear equations, two from the RSDF defining the quasilinear behaviour of the nonlinearity and two from separate considerations for the sinusoidal and Gaussian signals around the quasilinear loop, are solved. The actual procedure adopted for solving the equations is often dependent upon the form of the system under study. In particular, for analysing systems with a Gaussian input in which a limit cycle exists an approach similar to that of section 5.11.2 is convenient. The DF for the Gaussian signal is evaluated for those values of the sinusoidal DF which maintain the limit cycle.

The stability of solutions obtained for the above types of systems can be investigated using the incremental describing function in the presence of the sinusoidal and random signals, namely the IRSDF. Its value is easily obtained using the modified nonlinearity approach as it is simply the IDF of the nonlinearity modified by the random signal. Thus for complete stability, analogous to previous considerations with the IDF, the IRSDF must be obtained for incremental inputs both unrelated, as the bias γ, or related to the sinusoidal signal.

Applications of the RSDF are illustrated by way of two examples, one dealing with signal stabilisation and the other with three external inputs applied to a system. It should be emphasised however that many other applications of the RSDF are possible which deal with other facets of nonlinear behaviour such as the induction of limit cycles or the effect of a random signal on jump resonance. Again the three input describing function, the RSSDF, can be used to find limit cycles in a system with random and sinusoidal inputs or the effect of a random signal on subharmonic oscillations. All these possibilities as well as others which exist when bias signals are involved should be clear to the reader, as in addition should be the way in which the required DFs can be found using the modified nonlinearity approach.

8.13.1 *Signal stabilisation*

The system shown in figure 8.21 is considered. It is identical to example 1 of section 5.11.2 apart from the increased gain of the linear plant. The input $r(t)$

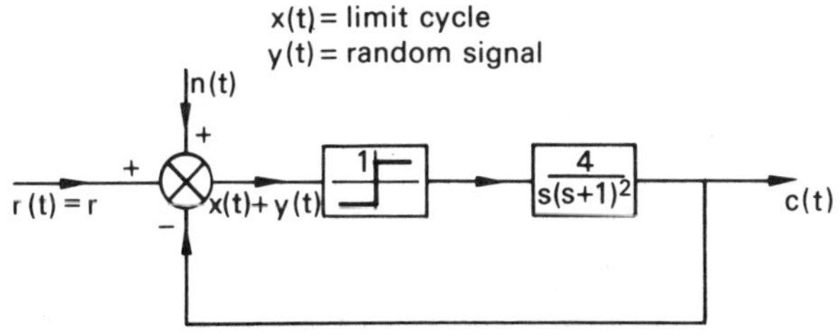

Figure 8.21 System considered in section 8.13.1

is a constant reference r and when the random input $n(t)$ is zero the system has a limit cycle of frequency 1 rad/sec. It is assumed that $n(t)$ has a frequency spectrum sufficiently high to neglect feedback of the random signal, so that $y(t) = n(t)$. It should be stressed here however that due to the production of low frequency intermodulation products by the nonlinearity some distortion will be fedback. The distortion component also causes undesirable low frequency variations in the output level and for this reason signal stabilisation is preferably done using a periodic signal.

For the ideal relay with a random input, y,

$$n(\sigma, \gamma) = 2h \int_0^{\gamma} q(y)\,\mathrm{d}y,$$

for $q(y)$ an even function about $y = 0$. From this expression it can be seen that if $q(y)$ has a negative slope for $y > 0$, the d.c. incremental gain $g(\sigma, \gamma)$ being the slope of $n(\sigma, \gamma)$ has its maximum value at $\gamma = 0$. Thus for such a case the oscillation will be slowly quenched, dying out when $g(\sigma, 0) = 0{\cdot}5$. In particular (i) if y is uniformly distributed between $\pm a$ the oscillation ceases for $a > 2h$ and (ii) if y is Gaussian, since $g(\sigma, 0) = N(\sigma) = (2/\pi)^{1/1}(h/\sigma)$, the oscillation ceases for $\sigma > 2h(2/\pi)^{1/2} = 1{\cdot}596h$.

When y has a sinusoidal distribution, since the slope of $q(y)$ for $y > 0$ is positive, $g(\sigma, \gamma)$ does not have its maximum value at $\gamma = 0$, and the oscillation as seen in section 5.11.2 for a deterministic sinusoidal input suddenly dies away from a finite amplitude as y is increased.

To determine the operating conditions for a magnitude of y smaller than that required to quench the limit cycle the RSDF is required. Assuming y is Gaussian then

$$n(\sigma, \gamma) = (2h/\sigma\sqrt{2\pi}) \int_0^\gamma e^{-y^2/2\sigma^2}\, dy = h \operatorname{erf}(\gamma/\sigma\sqrt{2}),$$

so that

$$N_a(a, \sigma) = (4h/a^2\pi) \int_0^a x(a^2 - x^2)^{-1/2} \operatorname{erf}(x/\sigma\sqrt{2})\, dx,$$

which on integrating by parts gives

$$N_a(a, \sigma) = (4\sqrt{2}h/a^2\sigma\pi^{3/2}) \int_0^a (a^2 - x^2)^{1/2} e^{-x^2/2\sigma^2}\, dx,$$

and on substituting $x = au^{1/2}$ becomes

$$N_a(a, \sigma) = (2\sqrt{2}h/\sigma\pi^{3/2}) \int_0^1 e^{-\rho^2 u} u^{-1/2}(1 - u)^{1/2}\, du. \tag{8.74}$$

Now[29]

$$\{\Gamma(\gamma)/\Gamma(\alpha)\Gamma(\gamma - \alpha)\} \int_0^1 e^{xu} u^{\alpha - 1}(1 - u)^{\gamma - \alpha - 1}\, du = {}_1F_1(\alpha; \gamma; x) \tag{8.75}$$

for $\mathrm{Re}\,\gamma > \mathrm{Re}\,\alpha > 0$, where ${}_1F_1(\alpha; \gamma; x)$ is the confluent hypergeometric function defined by[29]

$${}_1F_1(\alpha; \gamma; x) = 1 + \frac{\alpha x}{\gamma} + \frac{\alpha(\alpha + 1)}{\gamma(\gamma + 1)} \frac{x^2}{2!} + \cdots. \tag{8.76}$$

Thus on substituting for the integral in equation (8.74) the solution is

$$N_a(a, \sigma) = (2/\pi)^{1/2}(h/\sigma)\,{}_1F_1(1/2; 2; -\rho^2).$$

Similarly to find the gain $N_\sigma(a, \sigma)$ we know that

$$n(\sigma, \gamma)_1 = (1/\sigma^2\sqrt{2\pi}) \int_{-\infty}^{\infty} n(y + \gamma) y e^{-y^2/2\sigma^2}\, dy$$

$$= (2/\pi)^{1/2} h e^{-\gamma^2/2\sigma^2},$$

so that

$$N_\sigma(a, \sigma) = (2/\pi)^{3/2}(h/\sigma) \int_0^a e^{-x^2/2\sigma^2}(a^2 - x^2)^{-1/2}\,dx,$$

which on substituting $x = au^{1/2}$ becomes

$$N_\sigma(a, \sigma) = (2/\pi)^{3/2}(h/\sigma) \int_0^1 e^{-\rho^2 u}(1 - u)^{-1/2}u^{-1/2}\,du.$$

Using equation (8.75) the solution is

$$N_\sigma(a, \sigma) = (2/\pi)^{1/2}(h/\sigma)\,{}_1F_1(1/2\,;1\,;-\rho^2).$$

Figure 8.22 shows graphs of the functions $\sigma N_a(a, \sigma)/h$ and $\sigma N_\sigma(a, \sigma)/h$ versus ρ.

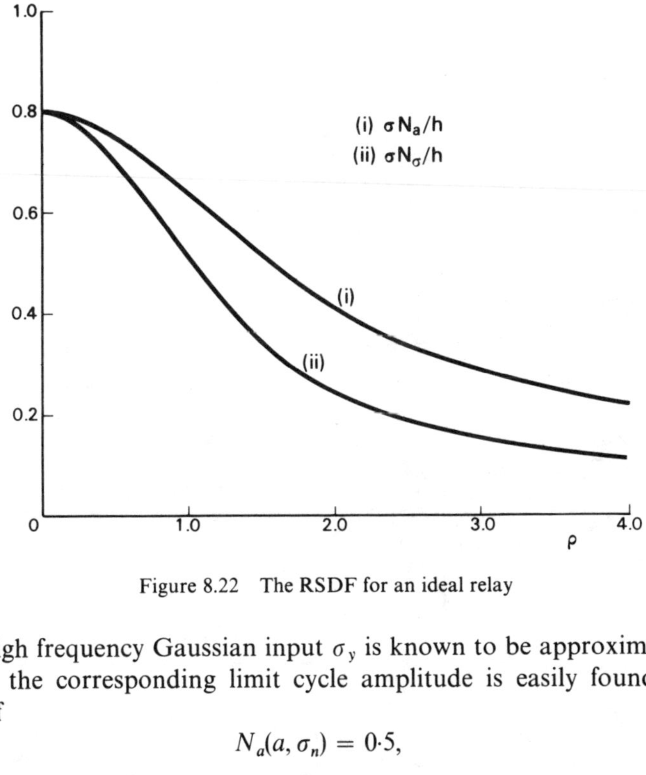

Figure 8.22 The RSDF for an ideal relay

For a high frequency Gaussian input σ_y is known to be approximately equal to σ_n and the corresponding limit cycle amplitude is easily found from the solution of

$$N_a(a, \sigma_n) = 0{\cdot}5,$$

which becomes $(2/\pi)^{1/2}\,{}_1F_1(1/2\,;2\,;-\rho^2) = \sigma_n/2$.

When $n(t)$ does not have a high frequency spectrum the operating point can be found by determining $N_\sigma(a, \sigma)$ for those values of a and σ which satisfy the constraint $N_a(a, \sigma) = 0{\cdot}5$. The complete DF N_σ for the Gaussian input, that is $N_\sigma(a, \sigma)$ versus σ for the range of existence of the limit cycle and $N(\sigma)$ versus σ outside that range, can then be drawn as shown in figure 8.23. For the linearised system σ can also be calculated as a function of N_σ for a given input

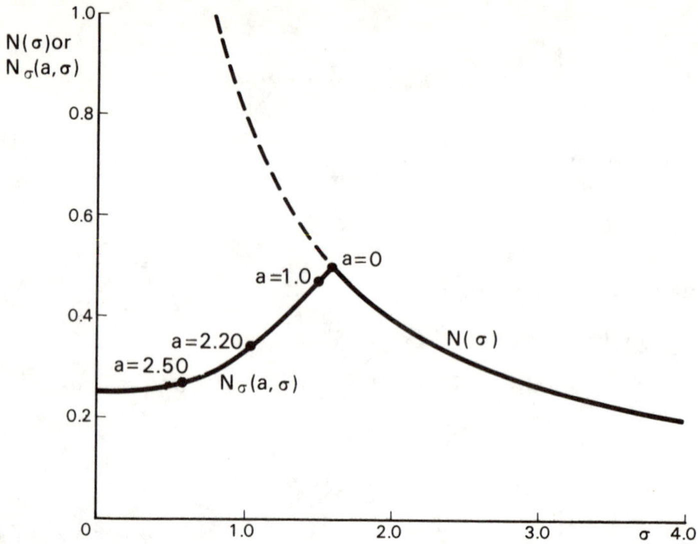

Figure 8.23 Complete DF for system of figure 8.21

spectrum $S_n(\omega)$, and the intersection of the curve with that of figure 8.23 gives the operating point, which may be within or outside the limit cycle region.

8.13.2 *A system with three inputs*

Figure 8.24 shows the system considered in reference 2. It appears to be the first example given on the multi-gain analysis of a nonlinear system with several

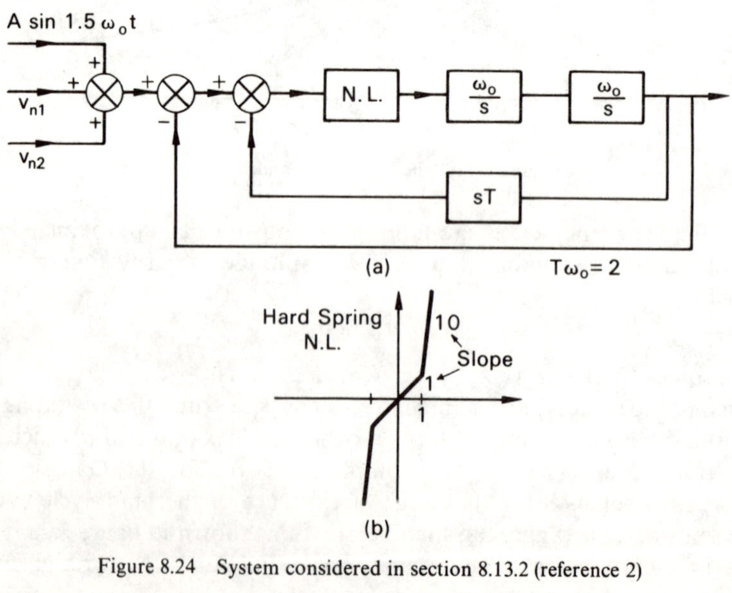

Figure 8.24 System considered in section 8.13.2 (reference 2)

inputs, although a similar problem was studied in a previous work[30] where a single equivalent gain, dependent upon the statistical properties of a sine plus Gaussian signal, was used. Such a quasilinear model is shown in section 10.3.1 to be non optimum.

The system of figure 8.24 contains a hard spring nonlinear characteristic and three inputs; a sinusoid of amplitude A at a frequency $1{\cdot}5\omega_0$, and two Gaussian signals v_{n1} and v_{n2}. Both the noise signals are of limited bandwidth having 6 dB/octave attenuation at low and high frequencies. The break points for v_{n1} being at $\omega_0/8$ and $\omega_0/4$ and for v_{n2} at ω_0 and $2\omega_0$ respectively. The noise signals are present in a power ratio $v_{n1}/v_{n2} = 4$. The expression for the spectrum of the total input noise, since the two components have different break points, is such that the two noise signals are best considered separately through the linear part of the system.

For the input sinusoid the corresponding magnitude X at the input to the nonlinearity is given by

$$X = A\left\{\frac{2{\cdot}25}{(1{\cdot}09N_a^2 - 4{\cdot}5N_a + 5{\cdot}06)^{1/2}}\right\}, \tag{8.77}$$

where $N_a = N_a(a, \sigma)$. The ratio $R = X/A$ is plotted in figure 8.25, curve (a), for values of N_a from 1 to 5.

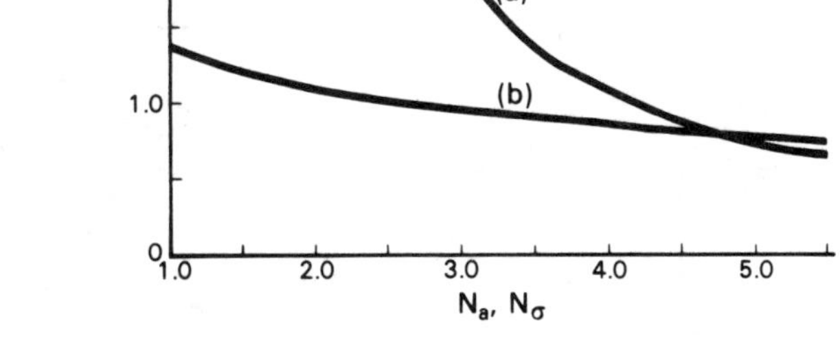

Figure 8.25 Magnitudes of signal and noise at nonlinearity input

The power spectrum $S(\omega)$ for each of the input noise signals is given by

$$S(\omega) = \left|\frac{2(\omega_1 + \omega_2)sv^2}{(s + \omega_1)(s + \omega_2)}\right|_{s=j\omega},$$

where ω_1 and ω_2 are respectively the frequencies at which the low and high frequency attenuations of 6 dB/octave begin, and v^2 is a constant. The transfer function for the noise linearised system from the system input to the nonlinearity input, $H(s)$, is given by

$$H(s) = \frac{s^2}{s^2 + 0{\cdot}2N_\sigma s\omega_0 + N_\sigma \omega_0^2},$$

where the equivalent gain to the noise $N_\sigma = N_\sigma(a, \sigma)$ is the same for both noise signals. Thus the mean squared value of a noise signal at the nonlinearity input is given by

$$\sigma^2 = (1/2\pi)\int_{-\infty}^{\infty} S(\omega)H(j\omega)H(-j\omega)\,d\omega.$$

Evaluating this integral for the two noise signals gives

$$\sigma_{n1}^2 = \frac{(0{\cdot}02 + 413{\cdot}3N_\sigma)v_{n1}^2}{0{\cdot}2 + 16{\cdot}48N_\sigma + 220{\cdot}4N_\sigma^2},$$

and

$$\sigma_{n2}^2 = \frac{(1 + 6{\cdot}30N_\sigma)v_{n2}^2}{1 + 1{\cdot}55N_\sigma + 0{\cdot}42N_\sigma^2}.$$

The total noise input power is made from v_{n1} and v_{n2} in the ratio 4:1 so that

$$\sigma_n^2 = \sigma_{n1}^2 + \sigma_{n2}^2 = v_n^2\left\{\frac{0{\cdot}8(0{\cdot}02 + 413{\cdot}3N_\sigma)}{0{\cdot}2 + 16{\cdot}48N_\sigma + 220{\cdot}4N_\sigma^2} + \frac{0{\cdot}2(1 + 6{\cdot}30N_\sigma)}{1 + 1{\cdot}55N_\sigma + 0{\cdot}42N_\sigma^2}\right\}.$$

The ratio $r = \sigma_n/v_n$ from the above expression is also shown plotted in figure 8.25, curve (b), for values of N_σ from 1 to 5.

The RSDF or equivalent gain curves for the nonlinearity, which were evaluated using cursors, are shown in figure 8.26. Using the curves for N_a from figure 8.26,

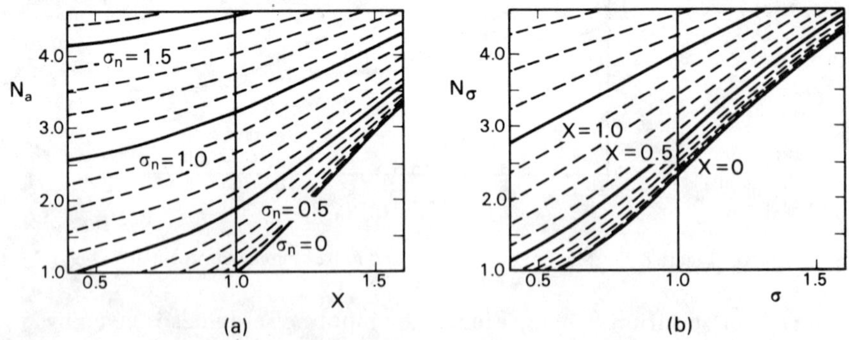

Figure 8.26 RSDF curves for the nonlinearity

curve (a) of figure 8.25 showing the variation of X with N_a can be converted to a curve of X against σ_n', the noise required at the input to the nonlinearity together with X to give the gain N_a to X. Similarly the curves for N_σ from figure 8.26 can be used to convert curve (b) of figure 8.25 to a curve of σ_n versus X'.

The curves X versus σ_n' and X' versus σ_n are plotted together in figure 8.27 for several values of X and σ_n. Intersections of these curves for the particular values of A and v_n give the operating conditions at the nonlinearity input. The operating points obtained in a simulation study are also shown in the figure. For $A = 0{\cdot}5$ and v_n slightly less than 0·25 it will be seen that three intersections can occur indicating that a jump in operating conditions is possible.

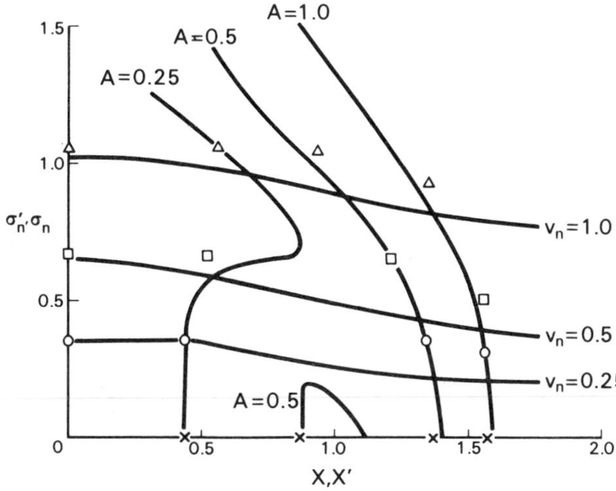

Figure 8.27 Determination of operating point

REFERENCES

1. Booton, R. C., Jr.: 'Nonlinear control systems with random inputs', *I.R.E. Trans.*, 1954, CT-1, pp. 9–17.
2. Somerville, M. J. and Atherton, D. P.: 'Multi-gain representation for a single-valued nonlinearity with several inputs, and the evaluations of the equivalent gains by a cursor method', *Proc. I.E.E.*, 1958, 105C, pp. 537–549.
3. Axelby, G. S.: 'Random noise with bias signals in nonlinear devices', *I.R.E. Trans.*, 1959, AC-4, pp. 167–172.
4. Smith, H. W.: *Approximate analysis of randomly excited nonlinear controls, Research Monograph 34, M.I.T. Press, Cambridge, Mass.*, 1966.
5. Smith, H. W.: 'The applicability of quasi-linear methods to nonlinear feedback systems with random inputs', *Proc. II I.F.A.C. Congress*, Basel, Switzerland, 1963.
6. Nuttall, A. H.: 'Theory and application of the separable class of random processes', *M.I.T. Res. Lab. Electronics*, 1958, Rept. 343.
7. Deutsch, R.: *System analysis techniques* (Prentice-Hall, Englewood Cliffs, N.J., 1969, Chapter 5).
8. Price, R.: 'A useful theorem for nonlinear devices having Gaussian inputs', *I.R.E. Trans.*, 1958, IT-4, pp. 69–72.
9. Erdelyi, A. et al.: *Tables of integral transforms* (McGraw-Hill, New York, 1954, 1).
10. Atherton, D. P.: 'Rapid evaluation of the auto-correlation function of the output of single-valued nonlinearities in response to sinusoidal and Gaussian signals', *Proc. I.E.E.*, 1962, 109C, pp. 656–664.
11. Atherton, D. P.: 'The evaluation of the response of single-valued nonlinearities to several inputs', *Proc. I.E.E.*, 1962, 109C, pp. 146–157.
12. Douce, J. L. and Leary, B. G.: 'A random pulse generator with variable mean rate', *Proc. I.E.E.*, 1961, 108B, pp. 524–528.

13. Thompson, A. G.: 'Optimum damping in a randomly excited nonlinear suspension', *Proc. I. Mech. E.*, 1969, 184, 2A, pp. 169–184.
14. Crandall, S. H.: 'Random vibration of systems with nonlinear restoring forces', *Proc. Int. Symposium on Nonlinear Vibrations*, Kiev, 1963, Vol. 1, pp. 306–314.
15. Middleton, D.: *Introduction to statistical communication theory* (McGraw-Hill, New York, 1960, Chapter 17).
16. Gibson, J. E. and Sridhar, R.: 'The response of nonlinear closed loop systems to random inputs', *Trans. A.S.M.E., J. Basic Eng.*, 1964, D-86, pp. 132–138.
17. Bendat, J. S. and Piersol, A. G.: *Measurement and analysis of random data* (Wiley, New York, 1966, pp. 185–186).
18. Sawaragi, Y. and Sugai, N.: *Accuracy considerations of the equivalent linearization technique for the analysis of a nonlinear control system with a Gaussian random input* (Memoirs of the Faculty of Engineering, University of Kyoto, 1961, 23).
19. Sawaragi, Y. et al.: 'The discrepancy from the normal distribution of the probability of the response of nonlinear control systems subjected to a Gaussian random input', *Engineering Research Institute*, University of Kyoto, 1961, 79.
20. Kolovskii, M. Z.: 'Estimating the accuracy of solutions obtained by the method of statistical linearization', *Automatica*, pp. 1962–1701. Translated from *Automatika i Telemekhanika*, 1966, 27, pp. 430–453.
21. George, D. A.: *Continuous nonlinear systems* (Sc.D. Thesis, Dept. of Elec. Eng., M.I.T., Cambridge, Mass., 1958).
22. Wang, M. C. and Uhlenbeck, G. E.: 'On the theory of the Brownian Motion', II, *Rev. Mod. Phys.*, 1945, 17.
23. West, J. C., Douce, J. L. and Leary, B. G.: 'Frequency spectrum distortion of random signals in nonlinear feedback systems', *Proc. I.E.E.*, 1961, 108C, pp. 259–264.
24. Caron, J.-Y.: 'Errors in the quasi-linearization technique for nonlinear systems', *S.M. Thesis, Dept. of Elec. Eng., M.I.T., Cambridge, Mass.*, 1955.
25. Pupkov, K. A.: 'Method of investigating the accuracy of essentially nonlinear automatic control systems by means of equivalent transfer functions', *Automatika i Telemekhanika*, 1960, 21; Translation *Automation and Remote Control*, 1960, 21.
26. Newton, G. C., Gould, L. A. and Kaiser, J. F.: *Analytic design of linear feedback controls* (Wiley, New York, 1957).
27. Douce, J. L. and King, R. E.: 'The effect of an additional nonlinearity on the performance of torque-limited control systems subjected to random inputs', *Proc. I.E.E.*, 1960, 107C, pp. 190–197.
28. Gelb, A. and Vander Velde, W. E.: *Multiple-input describing functions and nonlinear system design* (McGraw-Hill, New York, 1968, Chapter 7).
29. Abramowitz, M. and Stegun, I. A.: *Handbook of mathematical functions* (Dover, New York, 1965, Chapter 13).
30. Nikiforuk, P. N. and West, J. C.: 'The describing function analysis of a nonlinear servo mechanism subject to stochastic signals and noise', *Proc. I.E.E.*, 1957, 104C, pp. 193–203.

BIBLIOGRAPHY

Atherton, D. P.: 'Equivalent gain of single-valued non-linearities with multiple inputs', *Trans. I.E.E.E.*, 1963, AC-8, pp. 385–386.

Barrett, J. F. and Coales, J. F.: 'An introduction to the analysis of non-linear control systems with random inputs', *Proc. I.E.E.*, 1955, 103C, pp. 190–199.

Barrett, J. F. and Lampard, D. G.: 'An expansion for some second order probability distributions and its application to noise problems', *Trans. I.R.E.*, 1955, IT-1, pp. 10–15.

Brown, J. L.: 'Application of the theory of orthogonal polynomials in two variables to a multigain equivalent linearization problem', *Proc. I.E.E.*, 1961, 108C, pp. 115–118.

Brown, J. L.: 'On a cross correlation property for stationary random processes', *Trans. I.R.E.*, 1957, IT-3, pp. 28–31.

Chuang, K. and Kazda, L. F.: 'A study of nonlinear systems with random inputs', *Trans. A.I.E.E.*, 1959, 78, pt. II, pp. 100–105.

Crandall, S. H.: *Correlations and spectra of nonlinear response*, Zagadnienia Drgan Nieliniowych (Poland), 1973, 14, pp. 39–53.

Deutsch, R.: *Nonlinear transformations of random processes* (Prentice-Hall, Englewood Cliffs, N.J., 1962).

Douce, J. L. and Roberts, P. D.: 'Effect of distortion on the performance of nonlinear control systems', *Proc. I.E.E.*, 1963, 110, pp. 1497–1502.

Gelb, A. and Warren, R. S.: 'Direct statistical analysis of nonlinear systems: cadet', *A.I.A.A. Journal*, 1973, 11, pp. 689–694.
George, K. R.: 'Accuracy considerations in the analysis of nonlinear control systems subjected to random inputs', *M.Sc. Thesis, U.N.B.*, May 1973.
Gibson, J. E.: *Nonlinear automatic control* (McGraw-Hill, New York, 1963, Chapter 9).
Graham, D. and McRuer, D.: *Analysis of nonlinear control systems* (Wiley, New York, 1961, Chapter 6).
Gujar, U. G. and Kavanagh, R. J.: 'Generation of random signals with specified probability density functions and power density spectra', *Trans. I.E.E.E.*, 1968, AC-13, pp. 716–719.
Gusev, M. I.: 'Determining the effect of regular signal dynamics in the statistical linearization method', *Auto. and Remote Control*, 1961, 21, pp. 1093–1100.
Heinen, J. A. and Wu, S. H.: 'Modified equivalent gain analysis of nonlinear systems with Gaussian noise', *Int. J. Control*, 1968, 7, pp. 63–71.
Kazakov, I. Y.: 'Approximate probabilistic analysis of the operational accuracy of essentially nonlinear automatic systems', *Auto. and Remote Control*, 1956, 17, pp. 385–409.
Kazakov, I. Y.: 'Problems of the theory of statistical linearization and its applications', *Proc. 1st I.F.A.C., Moscow*, 1960, pp. 1968–1972, Butterworth, London.
Kislov, B. D.: 'Reduced equivalent amplification factor of a nonlinear element in the presence of noise', *Auto. and Remote Control*, 1961, 21, pp. 804–809.
Lubbock, J. K.: 'The optimization of a class of non-linear filters', *Proc. I.E.E.*, 1960, 107C, pp. 60–74.
Merchav, S. J.: 'Equivalent gain of single-valued nonlinearities with multiple inputs', *Trans. I.R.E.*, 1962, AC-7, pp. 122–124.
Natali, F. D.: 'An equivalent gain for nonlinear elements different from that proposed by Booton', *Proc. I.E.E.E.*, 1967, 55, pp. 414–415.
Nikiforuk, P. N.: 'Response of a particular nonlinear control system to random signals', *Trans. A.I.E.E.*, 1957, 75 pt. II, pp. 419–422.
Ohta, M. and Yamaguchi, S.: 'Statistical approximation of even-function nonlinear elements by an equivalent squaring operation', *Electrical Engineering in Japan*, 1972, 92-C, pp. 122–131.
Oldenburger, R. and Sridhar, R.: 'Signal stabilization of a control system with random inputs', *Trans. A.I.E.E.*, 1961, 80 pt. II, pp. 260–267.
Pastel, M. P.: 'Analyzing nonlinear systems with random inputs', *Control Eng.*, 1962, 9, pp. 113–117.
Pervozvanskii, A. A.: *Random processes in nonlinear control systems* (Academic Press, New York, 1965).
Popov, E. P.: 'On evaluating the quality of nonlinear automatic systems with random interference', *Auto. and Remote Control*, 1960, 20, pp. 1281–1287.
Pyatnitskii, G. I.: 'The effect of stationary random processes on automatic control systems containing essentially nonlinear elements', *Auto. and Remote Control*, 1960, 21, pp. 328–332.
Sawaragi, Y. and Takahashi, S.: 'Reponse of control systems containing zero-memory non-linearity to sinusoidal and Gaussian inputs', *Proc. Heidelberg Conference on Automatic Control*, 1956, pp. 271–274.
Sridhar, R. and Oldenburger, R.: 'Stability of a nonlinear feedback system in the presence of Gaussian noise', *Trans. A.S.M.E., J. Basic. Eng.*, 1962, 84, pp. 61–70.
Thaler, G. J. and Pastel, M. P.: *Analysis and design of nonlinear feedback control systems* (McGraw-Hill, New York, 1962, Chapter 5).
Thompson, W. E.: 'The response of a non-linear system to random noise', *Proc. I.E.E.*, 1954, 102C, pp. 46–48.
Watts, D. G.: 'A general theory of amplitude quantization with applications to correlation determinations', *Proc. I.E.E.*, 1962, 109C, pp. 209–218.
West, J. C. and Nikiforuk, P. N.: 'The response of remote-position control systems with hard-spring non-linear characteristics to step functions and random inputs', *Proc. I.E.E.*, 1954, 102B, pp. 575–593.
West, J. C.: *Analytical techniques for non-linear control systems* (E.U.P., London, 1960, Chapter 13).

PROBLEMS

P.8.1 If the random variable X has a distribution function $P(x)$ and density function $p(x)$ show that the constant m which minimises $E[|X - m|]$ is such that $P(m) = 0{\cdot}5$. With m this value and α an arbitrary constant show that $E[|X - \alpha|] = E[|X - m|] + 2\int_{\alpha}^{m} (x - \alpha)p(x)\,dx$.

P.8.2 The equivalent gain, K_{mod}, for a random input, $x(t)$, to a nonlinearity is defined as the value of the gain which minimises the modulus of the error between the nonlinearity output, $n(x)$, and the linear gain output. Show that if the slope of the nonlinearity, $n(x)$, is monotonic then (i) $K_{mod} = n(0{\cdot}982\sigma)/0{\cdot}982\sigma$ for $x(t)$ a Gaussian signal and (ii) $K_{mod} = n(0{\cdot}866a)/0{\cdot}866a$ for $x(t)$ a sinusoid.

P.8.3 A Gaussian signal with zero mean is applied to a nonlinear characteristic $n(y)$, determine the amplitude probability density of the output signal if (a) $n(y) = y^n$ where n is an integer and (b)

$$n(y) = \begin{cases} y^\alpha & \text{for } y > 0 \\ 0 & \text{for } y < 0 \end{cases}$$

where α is a real number.

P.8.4 A random voltage, which has a uniform probability density over the range -5 to $+5$ and zero elsewhere, is applied to an ideal saturation characteristic with unit gain in the linear region and saturation levels of ± 4 volts. Determine the probability density of the output signal.

P.8.5 A randomly varying voltage has a probability density which is uniform over the range -5 to $+5$ and zero elsewhere. It is applied to a variable diode function generator on an analog computer which approximates a square law characteristic. The function generator has even symmetry and is linear between break point values of (0, 0), (1, 1), (2, 4), (3, 9), (4, 16) and (5, 25). Calculate the error which will exist between the mean value of the output of the function generator and that of an ideal squarer.

P.8.6 A Gaussian signal of mean value μ and variance σ is the input to a full wave rectifier. Show that the mean value of the output is $(2/\pi)^{1/2}\sigma e^{-\mu^2/2\sigma^2} + 2\mu \operatorname{erf}(\mu/\sigma)$ and find the RDF.

P.8.7 Show that the product of a sinusoidal signal of amplitude a and a signal with a Rayleigh distribution (see equation (8.54)) is a Gaussian signal with zero mean and variance $a^2\sigma^2$.

P.8.8 Two random signals $x(t)$ and $y(t)$ have amplitude probability densities $p(x) = e^{-x}u(x)$ and $p(y) = e^{-y}u(y)$ where $u(\alpha)$ denotes the unit step at $\alpha = 0$. The signals $z(t)$ and $w(t)$ are related to $x(t)$ and $y(t)$ by the expressions

$$z(t) = x(t) + y(t); \qquad w(t) = x(t)/\{x(t) + y(t)\}.$$

Show that $z(t)$ and $w(t)$ are independent, that $p(z) = ze^{-z}u(z)$ and that $w(t)$ is uniformly distributed in the interval zero to unity.

P.8.9 A random signal $x(t)$ with a uniform distribution in the range $-4{\cdot}5$ to $+4{\cdot}5$ is applied to a symmetrical odd uniform quantiser with steps of height h at intervals 0·5, 1·5, 2·5, etc. Prove that the moment generating function for the output signal is given by $m(t) = (e^{-4ht} - e^{5ht})/9(1 - e^{ht})$.

P.8.10 It is required to produce a random signal uniformly distributed between 0 and 1 from a random signal $x(t)$ with distribution function $P(x)$. Show that this can be accomplished by passing the signal $x(t)$ through a nonlinear characteristic $P(x)$.

P.8.11 In the block diagram of figure P.8.1 the input $x(t)$ is a strictly Gaussian ergodic process with power spectrum $S_{xx}(\omega) = 1/(\omega^2 + 1)$. The gain K has the value $R_{x_1x_2}(0)/R_{x_1x_1}(0)$

(a) Find the joint probability density $p(x_1, x_2)$.
(b) Find the amplitude probability density $p(y_1)$.
(c) If $y_2(t) = f_1\{x_1(t)\}$ and $y_3(t) = f_2\{z(t)\}$, show that

$$R_{y_2y_3}(\tau) = \sum_{n=0}^{\infty} a_n R^n_{x_1z}(\tau);$$

and if f_1 is odd and f_2 even a_n is zero for all n.

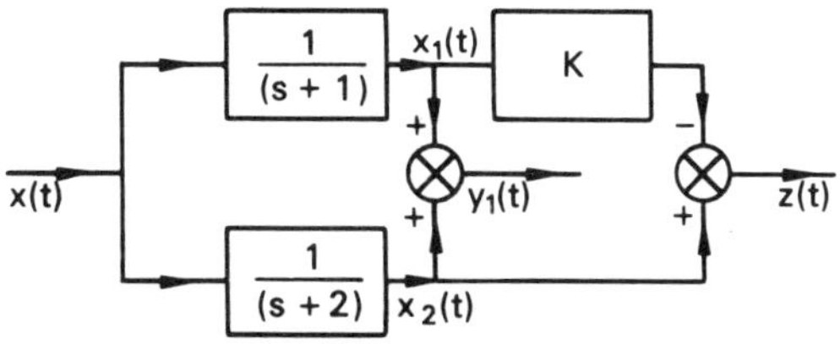

Figure P.8.1

P.8.12 A Gaussian signal with zero mean $y(t)$ is passed through a full wave rectifier characteristic and a filter of transfer function $G_1(s)$ to produce the signal $v_1(t)$. $y(t)$ is also passed through an ideal saturation characteristic of unit slope and limit levels and then through the filter $G_2(s)$ to give the signal $v_2(t)$. Show that $R_{v_1v_2}(\tau)$ is zero for all τ and determine expressions for $R_{v_1v_1}(\tau)$ and $R_{v_2v_2}(\tau)$ in terms of $R_{yy}(\tau)$ and the filters $G_1(s)$ and $G_2(s)$.

P.8.13 For a signal with a Rayleigh distribution

$$p(R) = \begin{cases} (R/\sigma^2)e^{-R^2/2\sigma^2} & \text{for } R > 0 \\ 0 & \text{for } R < 0 \end{cases}$$

Show that the polynomials, $t_i(R)$, which form an orthogonal set with respect to the weighting function $p(R)$, that is satisfy $\int_0^\infty t_i(R)t_j(R)p(R)\,dR = \delta_{ij}$ where δ_{ij} is the Kronecker delta, are given by

$$t_i(R) = \left\{\frac{\sigma\Gamma(n+1)}{\sqrt{2}\Gamma(n+3/2)}\right\}^{1/2} L_n^{(1/2)}(x^2/2\sigma^2)$$

where $L_n^{(\alpha)}(x)$ is the generalised Laguerre polynomial of order n.

P.8.14 Determine the nonlinearities whose RDFs are (a) $(\sigma^2 - 1)^n$ and (b) $(1 - \alpha^2\sigma^2)^{-1}$.

P.8.15 Show that for the coincidence and relay correlators given in figure 7.8 that when the inputs $x(t)$ and $y(t)$ are jointly Gaussian the measured correlation functions are $(2h^2/\pi)\sin^{-1}\{R_{xy}(\tau)/R_{xy}(0)\}$ and $\{2/\pi R_{xy}(0)\}^{1/2} \times hR_{xy}(\tau)$ respectively.

P.8.16 What will be the measured correlation functions of the coincidence and relay correlators of figure 7.8 if (i) $x(t) = a\cos\omega t$, $y(t) = b\cos(\omega t + \phi)$ and (ii) $x(t) = a\cos\omega t$, $y(t) = b\,\mathrm{sgn}\{a\cos(\omega t + \phi)\}$.

P.8.17 The Jacobian transformation

$$J = \left|\frac{\partial(u, v)}{\partial(y, z)}\right|$$

is defined in equation (8.56). Show that if

$$J' = \left|\frac{\partial(y, z)}{\partial(u, v)}\right|$$

then $JJ' = 1$. Check your result for $y = u\cos v$, $z = u\sin v$.

P.8.18 Determine an expression for the DF of x for transmission through the block diagram of figure P.8.2. If n_1 is a saturation characteristic with unit linear gain and limit levels at $\pm h$, n_2 a cubic, y a Gaussian signal of variance σ^2 and x a sinusoid of amplitude a evaluate the DF and IDF for x.

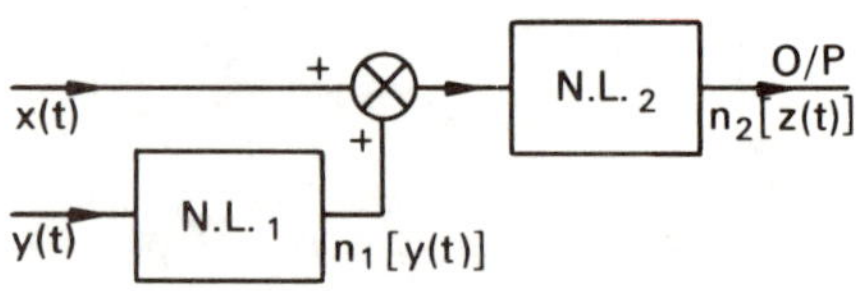

Figure P.8.2

P.8.19 For the block diagram of figure P.8.3 determine the RDF, $N(\sigma)$, for y if y is a Gaussian signal.

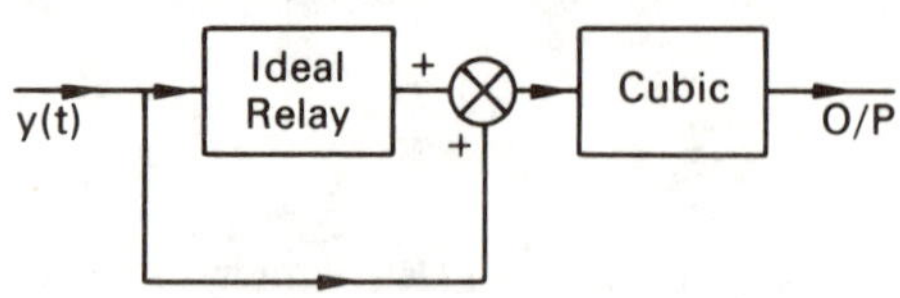

Figure P.8.3

P.8.20 Evaluate the r.m.s. DF, $K_{\text{r.m.s.}}$, (equation (8.5)) for a Gaussian input to (a) a cubic characteristic and (b) an ideal relay. Compare the results with those obtained for the RDF, $N(\sigma)$.

P.8.21 Determine the modified nonlinearity with Gaussian input $n(\sigma, \gamma)$ for (a) an ideal relay and (b) a cubic. Making use of these results in equation (8.17) determine $N(\sigma)$ for the two characteristics.

P.8.22 Determine the modified nonlinearity with a sinusoidal plus Gaussian input $n(a, \sigma, \gamma)$ for (a) an ideal relay and (b) a cubic. Explain how the RSDF, $N_\sigma(a, \sigma)$, can be obtained from $n(a, \sigma, \gamma)$ and find its value for the above two characteristics.

P.8.23 Evaluate the mean squared error for the system shown in figure P.8.4 if the input is Gaussian with spectral density $4/(1 + 16\omega^2)$.

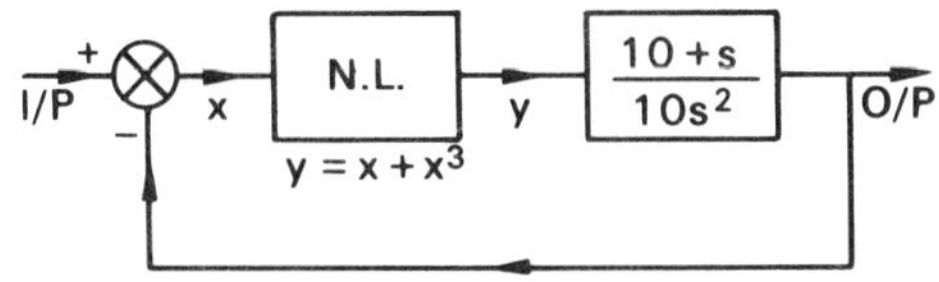

Figure P.8.4

P.8.24 Evaluate the mean squared error for the system with a nonlinear proportional plus derivative controller shown in figure P.8.5 if the input is Gaussian with a spectral density $\delta(\omega) + 9/(1 + 9\omega^2)$.

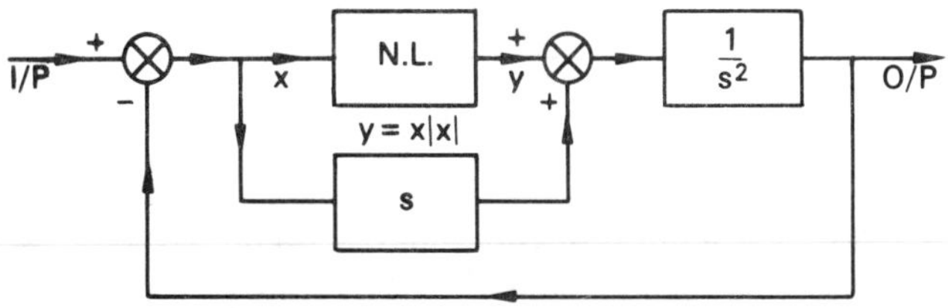

Figure P.8.5

P.8.25 For the system shown in figure P.8.6 the input is white noise of unit power per unit bandwidth. Determine the value of K which just quenches the system limit cycle and the corresponding mean squared value of the random input to the nonlinearity. What is the mean squared value of the random input to the nonlinearity if K has (a) half the above value and (b) twice the above value.

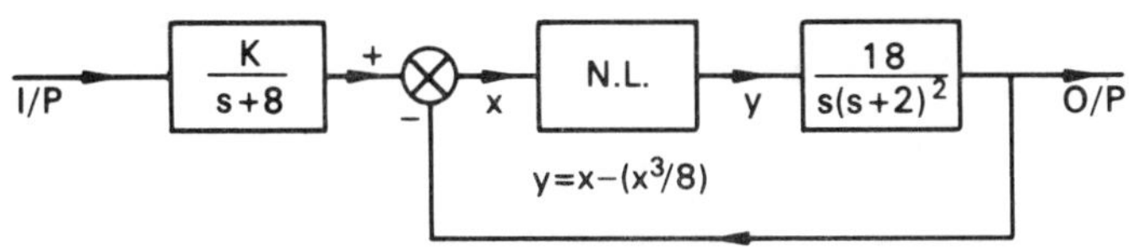

Figure P.8.6

P.8.26 Determine for the system shown in figure P.8.7, the value of K which will ensure that subharmonic oscillations cannot take place in the system for input frequencies greater than 0·02 rads/sec. $n(t)$ is white Gaussian noise of unit power per unit bandwidth.

P.8.27 Determine the minimum value of the variance of a Gaussian dither which assures that the system of figure P.8.8 will not limit cycle in the absence of other inputs.

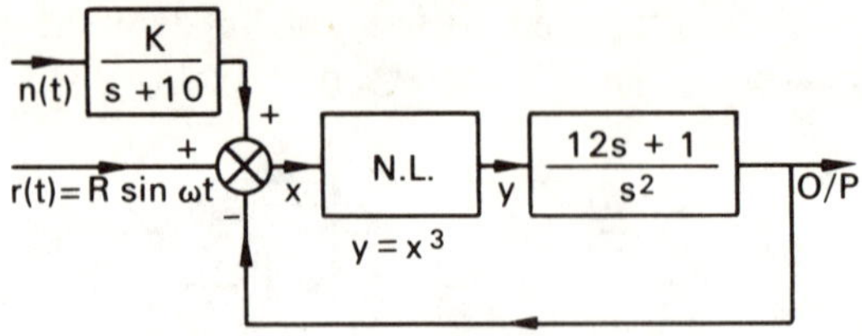

Figure P.8.7

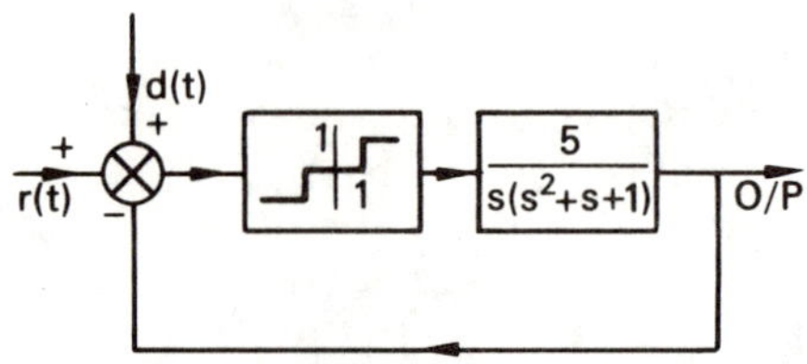

Figure P.8.8

P.8.28 Using the representation of equation (7.110) for a Gaussian signal show that the equivalent gain of a nonlinearity is given by

$$N(\sigma) = (1/\sigma^2)\int_0^{\infty}\int_0^{2\pi} R\cos\theta\, n(R\cos\theta)p(R)p(\theta)\,\mathrm{d}\theta\,\mathrm{d}R,$$

where $p(\theta) = 1/2\pi$ and $p(R)$ has a Rayleigh distribution (see equation (8.54)). Use this expression to evaluate $N(\sigma)$ for (a) an ideal relay and (b) a cubic.

Chapter 9

Describing Function Analysis of Transient Behaviour

9.1 Introduction

The purpose of this chapter is to examine the application of describing function methods for the analysis of transients in nonlinear systems. The first and most obvious situation of interest here is the determination of the step response, or more generally the response from some non-equilibrium initial state, of a stable nonlinear system. Obviously for a transient situation the basic assumption to be made for a describing function analysis in the sense previously considered, namely a description for the nonlinearity input waveform, is not clear.

It is perhaps for this reason, therefore, that one of the first approaches to the problem[1,2,3] used a so-called transient input describing function which avoids the above problem and thus does not fit the normal describing function concept. In this work, the nonlinearity is replaced by a constant gain equal to its initial slope together with a residual signal subtracted from the gain output. The residual signal is obtained from the system step input via a linear transfer function $L_r(s)$. The major difficulty with the procedure, which can only be used for piecewise linear characteristics situated at the error position, is the synthesis of a suitable $L_r(s)$ which is a function of both the nonlinearity and the input step amplitude.

Despite the fact that the nonlinearity input is not sinusoidal during a transient, investigations of the problem using the sinusoidal describing function were inevitable. Grensted's[4] work using this description shows, as one would anticipate, that the accuracy is poor for other than lightly damped transients where the nonlinearity input in fact approaches a sinusoid. A better approach for such situations, however, still using a sinusoidal describing function to characterise the nonlinearity, is to assume that the describing function amplitude and phase are not constant but vary slowly with time. This is in agreement with the nonlinearity input for such cases, which may be called a transient oscillation. Transient oscillations are of considerable interest in situations other than the step response of lightly damped systems. Instability often takes place in the form of a rising transient oscillation and the changes occurring in limit cycling systems due to variations in set point or a system parameter involve transients of this type. Considerations such as these form a further aspect of the material covered in this chapter. This dynamic sinusoidal describing function approach has been used by several authors[5,6,7,8,9] both in the study of step responses and limit cycle dynamics. Good results have been obtained where the initial assumption that the amplitude and frequency of the transient oscillation at the input to the nonlinearity is 'slowly' varying is justified.

For a system with heavy damping the nonlinearity input may approach an exponential waveform so that the use of an exponential describing function may give satisfactory results. Such a describing function, an EDF, has been defined and used in reference 10. This approach is therefore considered together with an extension for use with oscillatory systems where the input is taken as an exponentially damped sinusoid. A more recent development[11,12,13,14] is the use of an exponentially damped sinusoidal describing function defined over a portion, rather than the complete exponential response. The describing function representation for the nonlinearity is then changed at intervals, usually every half period, throughout the transient oscillation rather than using one describing function representation for the complete response.

The evaluation and use of these various describing functions, which fit our general quasilinear model concept of a nonlinearity, are investigated in this chapter.

9.2 The dynamic DF

Here we consider the application of the sinusoidal DF in the analysis of transient oscillations. Assuming that the nonlinearity input $x(t)$ is a sinusoid with slowly varying amplitude and frequency, that is

$$x(t) = a(t)\cos\omega(t)t = \mathrm{Re}\{a(t)e^{j\theta(t)}\},$$

where the frequency $\omega(t)$ is equal to $\dot{\theta}(t)$ the instantaneous rate of change of phase. For mathematical convenience we take

$$x(t) = a(t)e^{j\theta(t)}, \tag{9.1}$$

where the physical variable is the real or imaginary part of the complex function for $x(t)$. The DF for a general nonlinear element with this input is then

$$N(t) = M(t)e^{j\psi(t)}, \tag{9.2}$$

where both the magnitude and phase of $N(t)$ are slowly varying functions of time.

Consider now the autonomous feedback system shown in figure 9.1, the response of which is required for some specified initial conditions. Since the

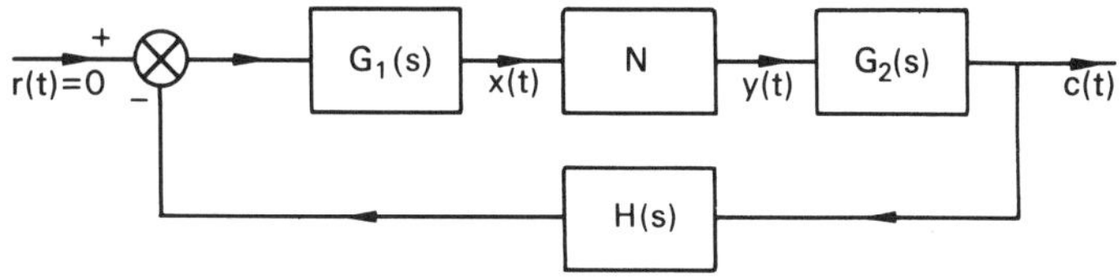

Figure 9.1 Nonlinear feedback system

assumed form for $x(t)$ just involves the two parameters $a(t)$ and $\theta(t)$ only two initial conditions on $x(t)$ and its derivatives can be satisfied. $a(0)$ and $\theta(0)$ will thus normally be chosen so that $x(0)$ and $\dot{x}(0)$ are correct. As is usual in DF problems we look for a solution for the nonlinearity input $x(t)$. Once obtained it may then be used, if required, to determine by linear theory the signals at other points in the loop, which will in general have both instantaneous amplitudes and frequencies different from those of $x(t)$. From figure 9.1 the two expressions relating the nonlinearity input and output are

$$X(s) = -G_2(s)H(s)G_1(s)Y(s) = -G(s)Y(s) \tag{9.3}$$

for the linear elements and

$$y(t) = N(t)x(t), \tag{9.4}$$

the quasilinearised representation for the nonlinearity. If the transfer function $G(s)$ has the form

$$G(s) = \frac{b_0 + b_1 s + \cdots + b_m s^m}{a_0 + a_1 s + \cdots + a_n s^n},$$

then writing equation (9.3) in differential equation form gives

$$a_n \frac{d^n x}{dt^n} + \cdots + a_1 \frac{dx}{dt} + a_0 x = -\left\{b_0 y + b_1 \frac{dy}{dt} + \cdots + b_m \frac{d^m y}{dt^m}\right\}. \tag{9.5}$$

Now from equation (9.1)

$$x = a(t)e^{j\theta(t)},$$

so that

$$\begin{aligned}\dot{x} &= \dot{a}e^{j\theta} + ja\dot{\theta}e^{j\theta}\\ &= \{(\dot{a}/a) + j\dot{\theta}\}ae^{j\theta} = \{(\dot{a}/a) + j\omega\}x.\end{aligned}$$

By analogy with the standard description of an exponentially varying sinusoid we define the relative rate of change of amplitude to be σ, so that σ is negative for a decaying exponential, then

$$\sigma = \dot{a}/a, \tag{9.6}$$

and

$$\dot{x} = (\sigma + j\omega)x = sx, \tag{9.7}$$

where the variable s, which has no interpretation in terms of an integral transform, is chosen equal to $\sigma + j\omega$. Since σ, ω and therefore s are functions of time further differentiation of $\dot{x}$ gives the results

$$\ddot{x} = \dot{s}x + s\dot{x} = \dot{s}x + s^2x = (s^2 + \dot{s})x,$$

$$\dddot{x} = (s^3 + 3s\dot{s} + \ddot{s})x,$$

$$\ddddot{x} = (s^4 + 6s^2\dot{s} + 3\dot{s}^2 + 4s\ddot{s} + \dddot{s})x,$$

and so on. Similarly successive differentiation of equation (9.4) gives

$$\dot{y} = \dot{N}x + N\dot{x} = (\dot{N} + Ns)x,$$

$$\ddot{y} = \{N(s^2 + \dot{s}) + 2\dot{N}s + \ddot{N}\}x,$$

$$\dddot{y} = \{N(s^3 + 3s\dot{s} + \ddot{s}) + 3\dot{N}(s^2 + \dot{s}) + 3\ddot{N}s + \dddot{N}\}x,$$

and so on.

It can be shown[9] that the derivatives of x and y may be obtained from the recursion relationships

$$D^{n+1}x = \sum_{k=0}^{n} {}_nC_k D^{n-k}sD^k_x \quad \text{with } D^0x = x,$$

and

$$D^{n+1}y = \sum_{k=0}^{n} {}_nC_k(D^{n-k}sD^ky + D^{n-k+1}ND^kx), \quad \text{with } D^0y = N(a)x,$$

where ${}_nC_k = n!/(n-k)!k!$.

If these expressions for x, y and their derivatives are substituted in equation (9.5) the result is an equation involving s, N and their derivatives. Now N is a function of a only for a SVNL and a complex function of a and ω for a frequency dependent nonlinearity. Thus since s is a complex function of a, $\dot{a}$ and $\dot{\theta}$ substitution for s, N and their derivatives in this differential equation gives a complex differential equation in a, $\dot{\theta}$ and their derivatives. Separation of the differential equation into its real and imaginary parts then produces two nonlinear differential equations in a and $\dot{\theta}$ which can in principle be solved for $a(t)$ and $\theta(t)$ to give the required signal $x(t)$. Being nonlinear, these differential equations do not normally permit an explicit solution. However, since our aim in formulating the problem was to obtain results for signals with slowly varying amplitude and phase, solutions of the differential equations neglecting some carefully selected terms, such as the higher derivatives of a and ω may yield satisfactory answers.

Two possible procedures appear to be generally mathematically tractable, namely (a) the quasi-static solution and (b) the small σ solution.

(a) In the quasi-static solution all derivatives of a and ω are neglected. This, of course, is simply the procedure of replacing the nonlinearity by its DF, $N(a)$, with a constant. The method only produces satisfactory solutions for extremely lightly damped transient oscillations.[4,15]

(b) The procedure for the small σ solution is as follows[9]:

(i) In the differential equation formed from the real terms, drop derivatives of σ and ω. Powers of σ greater than 2 can also usually be neglected. Solve for ω, denoted $\omega(\sigma, a)$, since it will be a function of σ and a.

(ii) Differentiate the solution of (i) dropping the derivatives of σ to obtain $\dot{\omega}(\sigma, a)$.

(iii) In the differential equation obtained from the imaginary terms drop derivatives of σ and derivatives of ω higher than the first. Use $\omega(\sigma, a)$ from (i) and $\dot{\omega}(\sigma, a)$ from (ii) to obtain a relationship, $\sigma(a)$, for σ in terms of a. Normally no powers of σ greater than unity should be retained in the solution.

(iv) Calculate time from the expression

$$t(a) = \int_{a(0)}^{a} \{1/a\sigma(a)\}\, \mathrm{d}a.$$

From this one can calculate directly $a(t)$, $\sigma(t)$, $\theta(t)$, $\omega(t)$ and thus $x(t)$.

Examples of the application of these methods are given later where they are compared with solutions obtained by the approaches discussed in the next two sections.

9.3 The exponential describing function

An exponential describing function was first defined and used in reference 10 for a SVNL with an input $x(t) = Ee^{-t/\tau}$. Obviously in this form the applicability of the DF is basically limited to systems where the transient input to the nonlinearity is nonoscillatory. Here we present a general theory for a transient input of the form $x(t) = Ee^{-t/\tau}\sin(\omega t + \phi)$ which in principle is also applicable to frequency dependent and DVNLs.

Substituting $\theta = \omega t$ and $\alpha = 1/\omega\tau$ in the above expression for $x(t)$ we obtain a general representation for the input in the form

$$x(\theta) = \mathrm{e}^{-\alpha\theta}(a\cos\theta + b\sin\theta), \tag{9.8}$$

where $a = E\sin\phi$ and $b = E\cos\phi$. By analogy with the DF presentation of section 3.3.1 we choose to represent a SVNL by a linear model $K_1 x(\theta)$ and a DVNL or frequency dependent nonlinearity by $K_1 x(\theta) + \omega K_2 x'(\theta)$ where $x'(\theta)$ denotes $\mathrm{d}x(\theta)/\mathrm{d}\theta$ which is given by

$$x'(\theta) = \mathrm{e}^{-\alpha\theta}\{-a(\sin\theta + \alpha\cos\theta) + b(\cos\theta - \alpha\sin\theta)\}. \tag{9.9}$$

It should be noted that unlike the DF situation if we choose to represent a SVNL by the second model, that is $K_1 x(\theta) + \omega K_2 x'(\theta)$, a finite, not zero, value may be obtained for K_2.

Considering the two gain model the error between the nonlinearity output and the model output is given by

$$e(\theta) = y(\theta) - K_1 x(\theta) - \omega K_2 x'(\theta). \tag{9.10}$$

As we wish to minimise some function of $e(\theta)$ over a range of θ from zero to infinity an obvious choice, since the mean squared error is zero over the infinite range, is the integral squared error. Thus if

$$I = \int_0^\infty e^2(\theta)\,\mathrm{d}\theta = \int_0^\infty \{y(\theta) - K_1 x(\theta) - \omega K_2 x'(\theta)\}^2\,\mathrm{d}\theta,$$

then differentiating separately with respect to the gains and equating $\mathrm{d}I/\mathrm{d}K_1$ and $\mathrm{d}I/\mathrm{d}K_2$ to zero gives

$$\int_0^\infty y(\theta)x(\theta)\,\mathrm{d}\theta = K_1 \int_0^\infty x^2(\theta)\,\mathrm{d}\theta + \omega K_2 \int_0^\infty x(\theta)x'(\theta)\,\mathrm{d}\theta, \tag{9.11}$$

and

$$\int_0^\infty y(\theta)x'(\theta)\,\mathrm{d}\theta = K_1 \int_0^\infty x(\theta)x'(\theta)\,\mathrm{d}\theta + \omega K_2 \int_0^\infty \{x'(\theta)\}^2\,\mathrm{d}\theta. \tag{9.12}$$

If the expressions for $x(\theta)$ and $x'(\theta)$ are substituted then the right hand side integrals can be evaluated and formulae for K_1 and K_2 obtained in terms of the integrals $\int_0^\infty y(\theta)x(\theta)\,\mathrm{d}\theta$ and $\int_0^\infty y(\theta)x'(\theta)\,\mathrm{d}\theta$ which depend upon the specific nonlinearity used. For this general case the results are probably too involved to be used in a practical situation. If, however, $a = 0$, corresponding to an input signal $x(\theta) = b\mathrm{e}^{-\alpha\theta}\sin\theta$, since $\int_0^\infty x(\theta)x'(\theta)\,\mathrm{d}\theta = (1/2)\{x^2(\infty) - x^2(0)\}$ is then zero, the solutions are considerably simplified.

For the general case with a SVNL the solution for K_1, which will also be denoted by $N_E(a, b, \alpha)$, is

$$K_1 = N_E(a, b, \alpha) = \int_0^\infty y(\theta)x(\theta)\,\mathrm{d}\theta \Big/ \int_0^\infty x^2(\theta)\,\mathrm{d}\theta, \tag{9.13}$$

which on evaluating the denominator integral becomes

$$N_E(a, b, \alpha) = \frac{4\alpha(1 + \alpha^2)}{(b + \alpha a)^2 + a^2(1 + \alpha^2)} \int_0^\infty y(\theta)x(\theta)\,\mathrm{d}\theta. \tag{9.14}$$

For the common situation of the initial condition $x'(0) = 0$, $b = a\alpha$ so that for a transient oscillation with α small, b can be neglected relative to a, and $x(\theta)$ may be assumed equal to $x(\theta) = a\mathrm{e}^{-\alpha\theta}\cos\theta$. The expression for the describing function is then

$$N_E(a, \alpha) = \{4\alpha(1 + \alpha^2)/a^2(1 + 2\alpha^2)\} \int_0^\infty y(\theta)x(\theta)\,\mathrm{d}\theta. \tag{9.15}$$

The describing function for the nonoscillatory exponential input $x = ae^{-\alpha\theta}$ can be obtained from equation (9.11) with $K_2 = 0$ and is

$$K_1 = N_E(a) = (2\alpha/a^2)\int_0^\infty y(\theta)x(\theta)\,d\theta, \tag{9.16}$$

an expression which in fact is independent of α and is also given by

$$N_E(a) = \lim_{\alpha\to\infty} N_E(a, \alpha). \tag{9.17}$$

A short list of the exponential describing functions, EDFs, $N_E(a)$ is given in table 9.1.

TABLE 9.1

Exponential describing functions

Nonlinearity		$N_E(a)$
Name	No.—Appendix A1	
Relay with dead zone	3	$(2h/a^2)(a - \delta)$
Ideal saturation	7	$(m\delta/a^2)(2a - \delta)$
Gain changing nonlinearity	9	$(\delta/a^2)(m_1 - m_2)(2a - \delta) + m_2$
$y = x^m$	17	$2a^{m-1}/(m + 1)$; $m \neq -1$
$y = A_m \sin mx$	20	$(2A_m/ma^2)(1 - \cos ma)$

Example

Evaluate $N_E(a, \alpha)$ and $N_E(a)$ for an ideal relay.

Since the input is $x(\theta) = ae^{-\alpha\theta}\cos\theta$, it is zero for $\theta = \pi/2, 3\pi/2, 5\pi/2\ldots$. Thus

$$\begin{aligned}\int_0^\infty y(\theta)x(\theta)\,d\theta &= ha\left\{\int_0^{\pi/2} e^{-\alpha\theta}\cos\theta\,d\theta - \int_{\pi/2}^{3\pi/2} e^{-\alpha\theta}\cos\theta\,d\theta\right.\\ &\quad \left. + \int_{3\pi/2}^{5\pi/2} e^{-\alpha\theta}\cos\theta\,d\theta - \cdots\right\}\\ &= \{ah/(1+\alpha^2\}\{\alpha + e^{-\alpha\pi/2} - (-e^{-3\alpha\pi/2} - e^{-\alpha\pi/2})\\ &\quad + (e^{-5\alpha\pi/2} + e^{-3\alpha\pi/2}) - \cdots\}\\ &= \{ah/(1+\alpha^2)\}\{\alpha + \operatorname{cosech}(\alpha\pi/2)\}.\end{aligned}$$

Substituting this value in equation (9.15) gives

$$N_E(a, \alpha) = \{4h\alpha/a(1 + 2\alpha^2)\}\{\alpha + \operatorname{cosech}(\alpha\pi/2)\}.$$

Applying equation (9.17)

$$\begin{aligned}N_E(a) &= (4h/a)\lim_{\alpha\to\infty}[\{\alpha^2 + \alpha\operatorname{cosech}(\alpha\pi/2)\}/(1 + 2\alpha^2)]\\ &= (4h/a)(1/2) = 2h/a.\end{aligned}$$

9.4 The half period exponential describing function

In using a describing function to represent a quasilinear model for a nonlinearity with a transient input, there is no justification, unlike the situation for steady state inputs, for keeping a fixed describing function throughout the complete response. We can instead minimise some function of $e(\theta)$ in equation (9.10) over a finite range, say from 0 to θ_1, and use the resulting describing function to compute the transient in that range, then repeat the process from θ_1 to θ_2 and so on until the complete transient is obtained. Obviously if the ranges are chosen too short the computations may tend to become excessive.

When a minimum mean squared error criterion over the range θ_1 to θ_2 is chosen it is easy to show that the gains K_1 and K_2 will again be defined by equations (9.11) and (9.12) but with the integrals taken from θ_1 to θ_2 rather than from 0 to ∞. Examination of these equations suggests two conditions for keeping the computations within reasonable limits, namely taking:

(a) ranges for which the same describing function expression, but with different parameters, can be used; and

(b) $x(\theta)$ so that $\int_{\theta_1}^{\theta_2} x(\theta)x'(\theta)\,d\theta$ is zero; a condition which is desirable for DVNLs only.

Both these conditions are satisfied when the nonlinearity is symmetrical odd for $x(\theta) = ae^{-\alpha\theta}\sin\theta$ over ranges of π from 0 to π, π to 2π, etc. since $ae^{-\alpha(n\pi+\theta)}\sin(n\pi+\theta) = (ae^{-n\alpha\pi}\cos n\pi)e^{-\alpha\theta}\sin\theta$ which is a similar input but of amplitude $|ae^{-n\alpha\pi}\cos n\pi|$. If the nonlinearity is asymmetrical two describing functions will be required, one for the positive and the other for the negative half cycles. The conditions are also obviously satisfied for $x(\theta) = ae^{-\alpha\theta}\cos\theta$ over ranges $\pi/2$ to $3\pi/2$, $3\pi/2$ to $5\pi/2$, etc., since the waveform in these ranges only differs in amplitude from that of the previous case. Condition (a) only is satisfied for the input $x(\theta) = ae^{-\alpha\theta}\cos\theta$ over the ranges 0 to π, π to 2π, etc. making this input representation suitable for SVNLs only.

A detailed presentation of the half period describing function assuming an input $x(\theta) = ae^{-\alpha\theta}\sin\theta$ and using the above ranges is given in references 11, 12 and 13 for both SVNLs and DVNLs, where it is named a half period transient gain. For this case equations (9.11) and (9.12) become

$$K_1 = \int_0^\pi y(\theta)x(\theta)\,d\theta \Big/ \int_0^\pi x^2(\theta)\,d\theta, \tag{9.18}$$

$$\omega K_2 = \int_0^\pi y(\theta)x'(\theta)\,d\theta \Big/ \int_0^\pi \{x'(\theta)\}^2\,d\theta. \tag{9.19}$$

Substituting the expressions for $x(\theta)$ and $x'(\theta)$ and evaluating the denominator integrals gives the solutions

$$K_1 = 4\alpha(1+\alpha^2)I_s/a(1-e^{-2\alpha\pi}), \tag{9.20}$$

and

$$\omega K_2 = 4\alpha(I_c - \alpha I_s)/a(1-e^{-2\alpha\pi}), \tag{9.21}$$

where

$$I_s, I_c = \int_0^{\pi} y(\theta)\mathrm{e}^{-\alpha\theta} \frac{\sin\theta}{\cos\theta} \,\mathrm{d}\theta. \quad (9.22)$$

For a DVNL, such as that shown in figure 3.2, the numerator of equation (9.19) may be written

$$\int_0^{\pi} y(\theta)x'(\theta)\,\mathrm{d}\theta = \int_0^{x_m} n_1(x)\,\mathrm{d}x + \int_{x_m}^{\pi} n_2(x)\,\mathrm{d}x$$

$$= \int_0^{x_m} \{n_1(x) - n_2(x)\}\,\mathrm{d}x,$$

where x_m is the maximum value of $x(\theta)$. Thus ωK_2 is alternatively given by

$$\omega K_2 = \{4\alpha/a^2(1 - \mathrm{e}^{-2\alpha\pi})\} \int_0^{x_m} \{n_1(x) - n_2(x)\}\,\mathrm{d}x$$

$$= \{4\alpha/a^2(1 - \mathrm{e}^{-2\alpha\pi})\}\{\text{Area of portion of DVNL loop for } x > 0\}. \quad (9.23)$$

This shows that for a SVNL, K_2 is zero and the half period exponential describing function is thus

$$K_1 = N_{hp}(a, \alpha) = 4\alpha(1 + \alpha^2)I_s/a(1 - \mathrm{e}^{-2\alpha\pi}). \quad (9.24)$$

A simple check on the expressions for K_1 and ωK_2 in equations (9.20), (9.21) and (9.23) is obtained by taking the limit as α tends to zero, in which case as expected the normal DF expressions are obtained.

If for a SVNL an input $x(\theta) = a\mathrm{e}^{-\alpha\theta}\cos\theta$ is taken evaluation of equation (9.18) gives

$$K_1 = N_{hpc}(a, \alpha) = 4\alpha(1 + \alpha^2)I'_c/a(1 + 2\alpha^2)(1 - \mathrm{e}^{-2\alpha\pi}),$$

where I'_c denotes equation (9.22) with $y(\theta)$ found for the above cosine input. This expression again produces the DF result if the limit as α tends to zero is taken.

Table 9.2 gives a list of the integral I_s for some specific nonlinearities from which the corresponding half period exponential describing function can be found. In agreement with the DF it is easily shown for both the exponential and half period exponential describing function that the describing function for a parallel combination of nonlinear elements is the sum of their separate describing functions.

It should be emphasised that in the half period describing function method, as with the dynamic DF, not only is the nonlinearity input waveform assumed to be represented in a certain manner for describing function purposes but the solution obtained does have that representation for the waveform over each half period of the response. Also since at the end of a half period the representation for the response changes, the two representations which meet at a half period changeover point must have the same value for the response and its derivative at that point. As the changeover point for the representation $x(\theta) = a\mathrm{e}^{-\alpha\theta}\sin\theta$

TABLE 9.2

Integral I_s for certain nonlinearities

Nonlinearity		I_s
Name	No.—Appendix A1	
Relay with dead zone	3	$\{h/(1+\alpha^2)\}\{g_1 e^{-\alpha\theta_1} - g_2 e^{-\alpha\theta_2}\}$ where $g = \cos\theta + \alpha\sin\theta$ and θ_1, θ_2 are the solutions of $ae^{-\alpha\theta}\sin\theta = \delta$ for $0 \le \theta \le \pi$
Ideal saturation	7	$\{m\delta/(1+\alpha^2)\}\{g_1 e^{-\alpha\theta_1} - g_2 e^{-\alpha\theta_2}\} + \{am/4(1+\alpha^2)\}\{\alpha^{-1}(1 - e^{-2\alpha\theta_1} + e^{-2\alpha\theta_2} - e^{-2\alpha\pi}) + 2g_2 e^{-2\alpha\theta_2}\sin\theta_2 - 2g_1 e^{-2\alpha\theta_1}\sin\theta_1\}$
Polynomial $y = x^n$ $x > 0$, n odd	18	$\dfrac{a^n n!(1 - e^{-(n+1)\alpha\pi})}{\alpha[\{(n+1)^2\alpha^2 + (n+1)^2\}\ldots\{(n+1)^2\alpha^2 + 2^2\}}$
Polynomial $y = x^n$ $x > 0$, n even	19	$\dfrac{a^n(n+1)!(1 + e^{-(n+1)\alpha\pi})}{\{(n+1)^2\alpha^2 + (n+1)^2\}\ldots\{(n+1)^2\alpha^2 + 1^2\}}$

is at zero values of the response the matching of these end conditions for this description is easier than for $x(\theta) = ae^{-\alpha\theta}\cos\theta$.

9.5 Some specific applications

To illustrate the use of the concepts presented in the three previous sections two examples, one concerning the step response of a stable system and the other the build up of a limit cycle in an oscillatory system, are considered. They have been studied earlier in reference 9 using the dynamic DF for both systems and the EDF, $N_E(a)$, for the stable system only.

9.5.1 *Step response of a stable system*

The system to be considered is shown in figure 9.2 with $G(s) = K/s(s+1)$. We wish to determine its response from the initial conditions $c(0) = K$, $\dot{c}(0) = 0$.

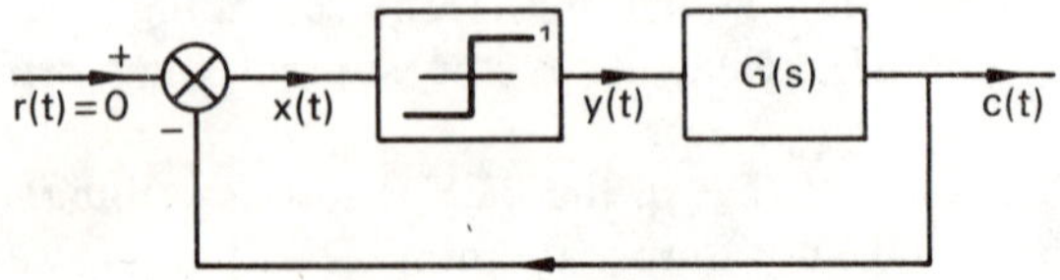

Figure 9.2 Relay control system

If the nonlinearity is replaced by a gain, N, then for the above conditions

$$C(s) = \frac{(s+1)K}{s^2 + s + KN}, \tag{9.25}$$

and in addition we have the relationship

$$s(s+1)X(s) = -KY(s). \tag{9.26}$$

(a) DF or quasi-static method

Here $N(a) = 4/a\pi$ and using this expression for N in equation (9.25) with $a = K$, gives

$$C(s)/K = (s+1)/(s^2 + s + 4/\pi).$$

Taking the inverse Laplace transform

$$c(t)/K = e^{-t/2}(\cos 1{\cdot}01t + 0{\cdot}495 \sin 1{\cdot}01t), \tag{9.27}$$

which is shown plotted, curve (a), in figure 9.3.

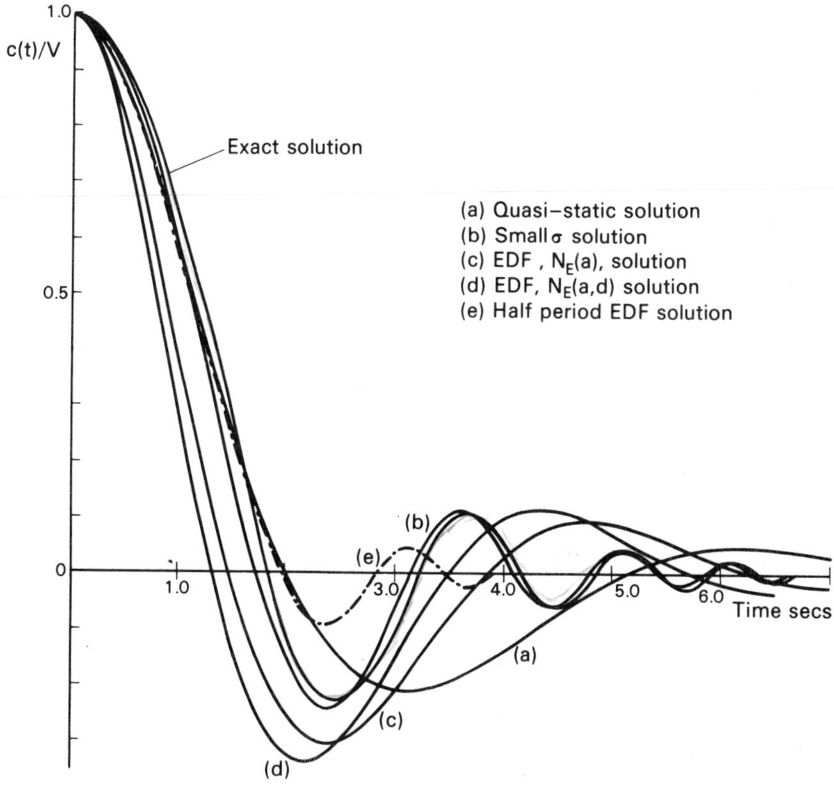

Figure 9.3 Step responses for system of section 9.5.1

(b) Small σ method

Using the replacement rules derived in section 9.2 in equation (9.26) gives

$$s^2 + \dot{s} + s = -KN.$$

Substituting the value of the DF, N, and $(\sigma + j\omega)$ for s yields the equations

$$\sigma^2 - \omega^2 + \dot{\sigma} + \sigma + 4K/\pi a = 0 \tag{9.28}$$

from the real terms and

$$2\sigma\omega + \dot{\omega} + \omega = 0 \tag{9.29}$$

from the imaginary terms. Proceeding as described in section 9.2, equation (9.28), on dropping $\dot{\sigma}$ terms gives

$$\omega = (\sigma^2 + \sigma + 4K/\pi a)^{1/2}, \tag{9.30}$$

which on differentiating with respect to time gives

$$\dot{\omega} = (2\sigma\dot{\sigma} + \dot{\sigma} - 4K\dot{a}/\pi a^2)/2\omega.$$

Again dropping terms in $\dot{\sigma}$ and remembering $\dot{a}/a = \sigma$ this expression reduces to

$$\dot{\omega} = -2K\sigma/\omega a\pi. \tag{9.31}$$

Using the above values of ω and $\dot{\omega}$ in equation (9.29) gives

$$\sigma a\pi(2\sigma^2 + 3\sigma + 1) + 6\sigma K + 4K = 0,$$

from which σ can be found as a function of a. Examination of this equation, however, shows that the constant value of $\sigma = -2/3$ is an approximate solution. This approximation can be seen in a different light if a is eliminated between equations (9.30) and (9.31) to give

$$\sigma^3 + \sigma^2 - 2\dot{\omega}\omega = \sigma\omega^2,$$

which on substituting for $\dot{\omega}$ in equation (9.29) gives

$$3\sigma/2 + 1 + (1/2)(\sigma/\omega)^2(1 + \sigma) = 0.$$

Here $\sigma = -2/3$ is a solution if (σ/ω) may be assumed small, which is the case for a lightly damped transient. Thus, using this value for σ the corresponding solution for ω from equation (9.30) is

$$\omega = \{(36K - 2a\pi)/9a\pi\}^{1/2}. \tag{9.32}$$

Since $-c(t)$ is the input to the nonlinearity the solution for $c(t)$ can be written

$$c(t) = \text{Re}\{a(t)e^{j\theta(t)}\} = a(t)\cos\theta(t). \tag{9.33}$$

Differentiating

$$\begin{aligned}\dot{c}(t) &= \dot{a}(t)\cos\theta(t) - a(t)\dot{\theta}(t)\sin\theta(t)\\ &= a(t)\{\sigma(t)\cos\theta(t) - \omega(t)\sin\theta(t)\}.\end{aligned}$$

Now at $t = 0$, the initial conditions are $c(0)/K = 1$, $\dot{c}(0) = 0$, so that

$$\{a(0)/K\}\cos\theta(0) = 1,$$

$$a(0)\{\sigma(0)\cos\theta(0) - \omega(0)\sin\theta(0)\} = 0.$$

Substituting $\sigma = -2/3$ and for $\omega(0)$ in terms of $a(0)$ from equation (9.32), the above equations give the solutions

$$a(0)/K = 1{\cdot}247, \qquad \theta(0) = -36{\cdot}68°.$$

Integrating

$$\dot{a}(t) = \sigma a(t)$$

gives

$$a(t) = a(0)e^{-2t/3},$$

which becomes, on substituting for $a(0)$,

$$a(t)/K = 1{\cdot}247e^{-2t/3}. \tag{9.34}$$

Using this expression for $a(t)$ in equation (9.32) gives

$$\omega(t) = \{(4/1{\cdot}247\pi)e^{2t/3} - (2/9)\}^{1/2}. \tag{9.35}$$

Now

$$\dot{\theta}(t) = \omega(t),$$

so that

$$\theta(t) = \theta(0) + \int_0^t \omega(\tau)\,d\tau,$$

which on substituting for $\theta(0)$, and $\omega(\tau)$ from equation (9.35), and integrating gives

$$\theta(t) = -0{\cdot}641 + 3\omega(t) - \sqrt{2}\tan^{-1}(3\omega(t)/\sqrt{2}) - 3\omega(0) + \sqrt{2}\tan^{-1}(3\omega(0)/\sqrt{2}) \quad \text{radians}, \tag{9.36}$$

where $\omega(t)$ is given by equation (9.35). Equations (9.33), (9.34) and (9.36) define the solution which is shown plotted, curve (b), in figure 9.3. The solution is seen to be particularly good despite the fact that the initial portion of the response is quite well damped. The relative changes in amplitude per quarter period, $\pi|\sigma|/2\omega$, and frequency per quarter period $\pi|\dot{\omega}|/2\omega^2$ are certainly not negligible, that is much less than unity, initially, as is preferable in using this DF approach.

(c) EDF using $N_E(a)$

For the relay characteristic, $N_E(a) = 2h/a$, which when substituted in equation (9.25) with $h = 1$ and $K = a$ gives

$$C(s)/K = (s + 1)/(s^2 + s + 2),$$

which has the solution

$$c(t)/K = e^{-t/2}(\cos 1{\cdot}32t + 0{\cdot}378 \sin 1{\cdot}32t). \tag{9.37}$$

The result is shown plotted, curve (c), in figure 9.3.

(d) EDF using $N_E(a, \alpha)$

To satisfy both the initial conditions of $c(0)/K = 1$ and $\dot{c}(0) = 0$ an input of the form given in equation (9.8) should be assumed and the describing function $N_E(a, b, \alpha)$ found from equation (9.14). Since this expression is rather more complicated and α is expected to be fairly small the simpler form $N_E(a, \alpha)$ given by

$$N_E(a, \alpha) = \{4h\alpha/a(1 + 2\alpha^2)\}\{\alpha + \operatorname{cosech}(\alpha\pi/2)\}$$

is assumed. With $s = \sigma + j\omega = (-\alpha + j)\omega$ and $N = N_E(a, \alpha)$ substituted in the characteristic equation the resulting equations from the real and imaginary parts are

$$\alpha^2\omega^2 - \omega^2 - \alpha\omega + KN_E(a, \alpha) = 0, \tag{9.38}$$

$$2\alpha\omega^2 - \omega = 0. \tag{9.39}$$

Substituting in equation (9.38) the value of ω from equation (9.39) and the expression for $N_E(a, \alpha)$, with a taken equal to K, gives

$$16\alpha^3\{\alpha + \operatorname{cosech}(\alpha\pi/2)\} = (1 + 2\alpha^2)(1 + \alpha^2),$$

which has the solution $\alpha = 0{\cdot}346$, so that $\omega = 1/2\alpha = 1{\cdot}45$. Using these results in equation (9.25) gives

$$C(s)/K = (s + 1)/\{(s + 1/2)^2 + (1{\cdot}45)^2\},$$

with the solution

$$c(t)/K = \mathrm{e}^{-t/2}\{\cos 1{\cdot}45t + 0{\cdot}346 \sin 1{\cdot}45t\}. \tag{9.40}$$

The solution is shown plotted, curve (d), in figure 9.3.

(e) The half period exponential describing function

Considering an input of the form $-x(\theta) = c(\theta) = ae^{-\alpha\theta} \sin\theta$ the half period exponential describing function for the relay, $N_{hp}(a, \alpha)$ is given by

$$N_{hp}(a, \alpha) = 4h\alpha/a(1 - \mathrm{e}^{-\alpha\pi}). \tag{9.41}$$

The above expression for $c(\theta)$ with $a = a_1$, $\alpha = \alpha_1$, denoted by $c_1(\theta)$, represents the output from time $t_1 = 0$ to π/ω_1 where $\theta = \omega_1 t_1$. Now

$$c_1'(\theta) = a_1\mathrm{e}^{-\alpha_1\theta}(-\alpha_1 \sin\theta + \cos\theta),$$

so that to satisfy the initial condition $c_1'(\theta) = 0$, $\tan\theta = 1/\alpha_1$ and the corresponding value of $c_1(\theta)$ is $a_1(1 + \alpha_1^2)^{-1/2}\mathrm{e}^{-\alpha_1 \tan^{-1}(1/\alpha_1)}$ which must equal the initial condition magnitude of K. This gives

$$a_1/K = (1 + \alpha_1^2)^{1/2}\mathrm{e}^{\alpha_1 \tan^{-1}(1/\alpha_1)}. \tag{9.42}$$

The output over the interval of t_1 from $(1/\omega)\tan^{-1}(1/\alpha_1)$, which corresponds to zero real time, t, to π/ω_1 thus corresponds to values of θ from $\tan^{-1} 1/\alpha_1$ to π not 0 to π for which the describing function is obtained. This means, of course, that apart from a time shift of $(1/\omega_1)\tan^{-1}(1/\alpha_1)$ the computed response will be identical with that for initial conditions of $c_1(\theta) = 0$, $c_1'(\theta) = a_1$. The remainder of the output, however, is described from each zero crossing by a complete half period of an expression, $c(\theta)$, of the above form.

From the true time t_c to $t_c + \pi/\omega_2$ the response will be $c_2(\theta)$ and α and a will have the values α_2 and a_2 respectively, where t_c the first changeover time is equal to $(1/\omega_1)\{\pi - \tan^{-1}(1/\alpha_1)\}$ secs. To provide a continuous response the derivatives of the time responses at t_c must be equal, that is

$$\omega_1 c_1'(\theta)|_{\theta=\pi} = \omega_2 c_2'(\theta)|_{\theta=0}. \tag{9.43}$$

To obtain the solution, equations (9.38) and (9.39) still apply with the relevant value of $N_{hp}(a, \alpha)$ substituted for $N_E(a, \alpha)$. Putting the value of $N_{hp}(a, \alpha)$ for the first portion of the response and $\omega = 1/2\alpha$ from equation (9.39) in equation (9.38) gives

$$(\alpha_1^2 + 1)/4\alpha_1^2 + 4\alpha_1 K/a_1(1 - e^{-\alpha_1 \pi}) = 0. \tag{9.44}$$

Solving equations (9.42) and (9.44) gives $\alpha_1 = 0{\cdot}489$, $a_1 = 1{\cdot}921K$, so that

$$c_1(\theta)/K = 1{\cdot}921\, e^{-0{\cdot}489\theta} \sin\theta,$$

or, in terms of time, since $\theta = \omega_1 t_1$ and $\alpha_1\omega_1 = 1/2$,

$$c_1(t_1)/K = 1{\cdot}921\, e^{-t_1/2} \sin 1{\cdot}023 t_1, \tag{9.45}$$

where $t = t_1 - (1/\omega_1)\tan(1/\alpha_1)$. For the nth half period equation (9.44) becomes

$$(\alpha_n^2 + 1)/4\alpha_n^2 + 4\alpha_n K/a_n(1 - e^{-\alpha_n \pi}) = 0, \tag{9.46}$$

and the general form of equation (9.43) gives

$$a_n = a_{n-1}\alpha_n\, e^{-\alpha_{n-1}\pi}/\alpha_{n-1}. \tag{9.47}$$

Successively solving these two equations gives $\alpha_2 = 0{\cdot}138$, $a_2 = 0{\cdot}117$; $\alpha_3 = 0{\cdot}094$, $a_3 = 0{\cdot}0516$ and so on. Since $\alpha_n\omega_n = 1/2$ for all n the corresponding solutions are

$$c_2(t_2)/K = -0{\cdot}117\, e^{-t_2/2} \sin 3{\cdot}62 t_2, \tag{9.48}$$

where $t_2 = t - t_c$,

$$c_3(t_3)/K = 0{\cdot}0516 e^{-t_3/2} \sin 5{\cdot}32 t_3, \tag{9.49}$$

where $t_3 = t - t_c - 2\alpha_2\pi$.

The response, curve (e), is shown plotted in figure 9.3. Perhaps the most general comment that can be made with regard to the results shown in figure 9.3 is that in the light of alternative approaches the use of describing function methods for the accurate evaluation of transient responses is rarely justifiable. One cannot generalise from the results of a single example but the small σ solution certainly comes out way ahead here although much better accuracy has been reported using the half period describing function in other examples.(13) The main problem with this latter method in the above example appears to be the fact that in obtaining a good estimate of the first portion of the response the slope at the first changeover point, which is a critical parameter, is appreciably in error and causes a poor estimate from thereon. It is certainly obvious from the exact response that no method which fails to account for the change in frequency of the transient will produce a good solution.

9.5.2 *Limit cycle start up*

The previous example shows that the small σ method yields the solution for $a(t)$ much quicker than the complete solution for $x(t)$. In some cases, for example variations in a system limit cycle, only the envelope $a(t)$ may be of interest. To illustrate this situation we consider the build up from zero initial conditions of the limit cycle in the system of figure 9.2 with $G(s) = K/s(s^2/\omega_0^2 + 2\zeta s/\omega_0 + 1)$.

(a) Small σ method

Applying the rules of section 9.2 to the specific configuration gives

$$s^3 + 3s\dot{s} + \ddot{s} + 2\zeta\omega_0(s^2 + \dot{s}) + s\omega_0^2 + 4hK\omega_0^2/a\pi = 0. \tag{9.50}$$

Substituting $s = \sigma + j\omega$, the equations from the real and imaginary terms are respectively

$$\sigma^3 - 3\sigma\omega^2 + 3\sigma\dot{\sigma} - 3\omega\dot{\omega} + \ddot{\sigma} + 2\zeta\omega_0(\sigma^2 - \omega^2 + \dot{\sigma}) + \sigma\omega_0^2 + 4hK\omega_0^2/a\pi = 0, \tag{9.51}$$

$$3\sigma^2\omega - \omega^3 + 3\sigma\dot{\omega} + 3\omega\dot{\sigma} + \ddot{\omega} + 2\zeta\omega_0(2\sigma\omega + \dot{\omega}) + \omega\omega_0^2 = 0. \tag{9.52}$$

The limit cycle solution is obtained by dropping all the derivatives and putting $\sigma = 0$ in these equations and is

$$\omega = \omega_0, \qquad a = 2hK/\zeta\omega_0\pi = a_0. \tag{9.53}$$

For the small σ solution dropping derivatives and powers of σ greater than two in equation (9.51) results in

$$-3\sigma\omega^2 + 2\zeta\omega_0(\sigma^2 - \omega^2) + \sigma\omega_0^2 + 4hK\omega_0^2/a\pi = 0,$$

which gives

$$\omega^2 = (4hK\omega_0^2/a\pi + \sigma\omega_0^2 + 2\zeta\omega_0\sigma^2)/(2\zeta\omega_0 + 3\sigma), \tag{9.54}$$

and dropping derivatives other than $\dot{\omega}$ in equation (9.52) gives

$$3\sigma^2\omega - \omega^3 + 3\sigma\dot{\omega} + 2\zeta\omega_0(2\sigma\omega + \dot{\omega}) + \omega\omega_0^2 = 0,$$

which may be written

$$3\sigma^2 - \omega^2 + \omega_0^2 + 4\zeta\sigma\omega_0 + (\dot{\omega}/\omega)(3\sigma + 2\zeta\omega_0) = 0. \tag{9.55}$$

Differentiating equation (9.54) to obtain $2\omega\dot{\omega}$, dividing by $2\omega^2$ and neglecting derivatives of σ gives

$$\dot{\omega}/\omega = (-2hK\sigma/\pi a)/\{4hK/\pi a + \sigma + (2\zeta/\omega_0)\sigma^2\}. \tag{9.56}$$

Substituting the expressions of equations (9.54) and (9.56) for ω^2 and $\dot{\omega}/\omega$ respectively in equation (9.55), rearranging the terms and neglecting powers of σ greater than two gives

$$2\sigma^2(1 + 6\zeta^2 + 8\zeta^2/\rho) + 2\zeta\omega_0\sigma(1 + 1/\rho + 6\zeta^2/\rho) + (4\zeta^2\omega_0^2/\rho)(1 - 1/\rho) = 0, \quad (9.57)$$

where $\rho = a/a_0$, a normalised amplitude parameter. Since $\sigma = \dot{a}/a = \dot{\rho}/\rho$, equation (9.57) is a nonlinear equation relating $\dot{\rho}$ and ρ the solution of which is shown plotted in figure 9.4 for $\omega_0 = 1$ and $\zeta = 5$. If one of the graphical methods described in section 2.4.3 to determine time on a phase plane trajectory is used then $\rho(t)$ can be obtained. The experimental and theoretical solutions are compared in figure 9.5(a) and are seen to be in close agreement. Since $|\sigma|/\omega$ and $|\dot{\omega}|/\omega^2$ are both small in this case, the result is not unexpected.

(b) IDF method

The rate of change of the limit cycle amplitude for a small perturbation about the limit cycle condition, that is $\dot{\rho}/\rho$ near $\rho = 1$ in figure 9.4 can be found using the IDF. That is we consider the effect on the system of a small perturbation at the limit cycle frequency of the form $be^{\sigma t} \sin \omega_0 t$. The IDF for this input is equal to that for a small sinusoid at the same frequency so that

$$N_{ib}(a) = (2h/a\pi)(1 - e^{-j2\phi}),$$

which at the limit cycle condition of $a = a_0 = 2hK/\zeta\omega_0\pi$ becomes

$$N_{ib}(a_0) = (\zeta\omega_0/K)(1 - e^{-j2\phi}).$$

To find the time constant of the perturbation, we determine the solutions for σ of the synchronous IDF characteristic equation

$$1 + N_{ib}(a_0)G(\sigma + j\omega_0). \quad (9.58)$$

The graphical solution is illustrated for a typical situation in figure 9.6. For any given nonlinearity and amplitude a_0 the locus $-1/N_{ib}(a_0)$ with ϕ a parameter

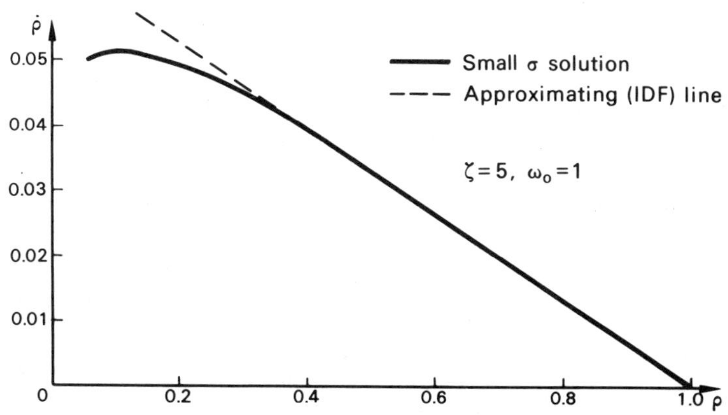

Figure 9.4 Phase trajectory for the amplitude of the transient oscillation

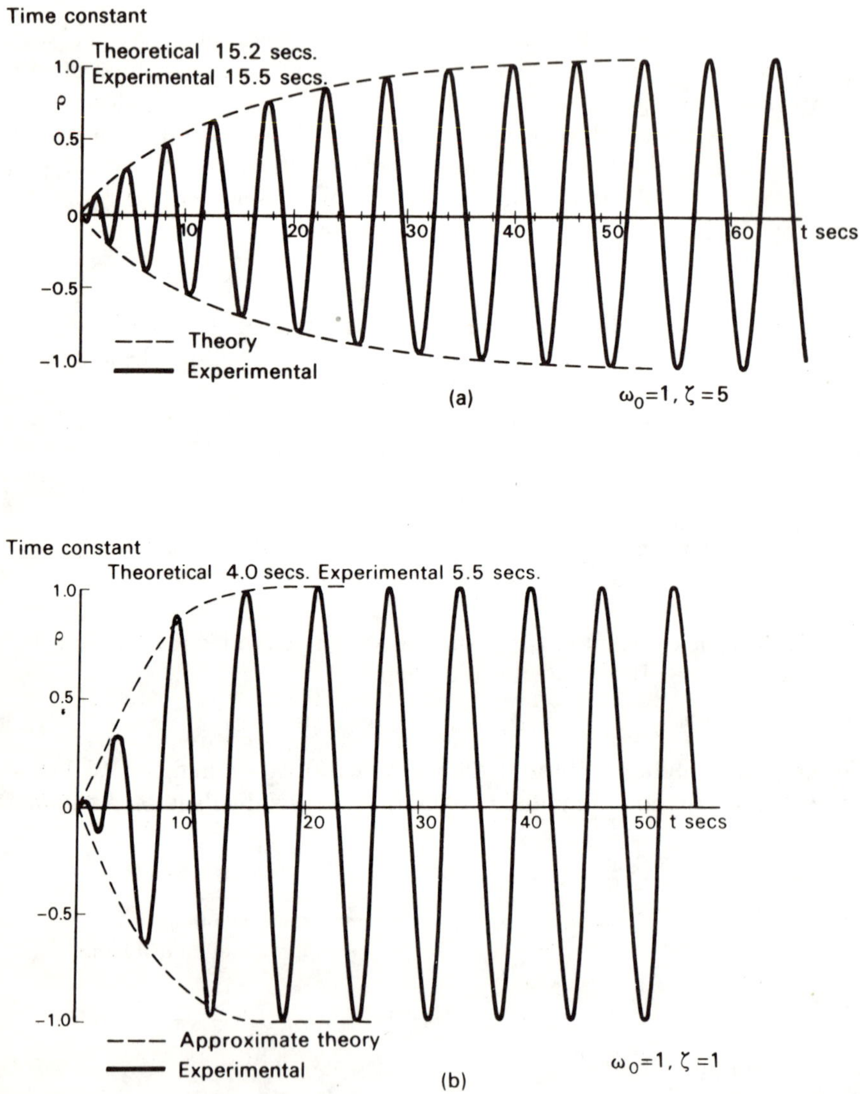

Figure 9.5 Growth of a limit cycle

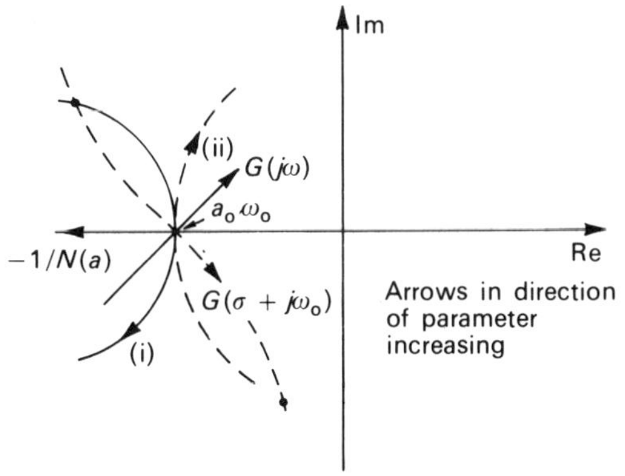

Figure 9.6 Illustrations of solutions for the synchronous IDF characteristic equation

is always a circle† passing through the solution point $G(j\omega_0) = -1/N(a_0)$ since $N_{ib}(a_0)|_{\phi=\pi/2} = N(a_0)$. For soft (or hard) spring nonlinearities the circles lie to the left (or right) of the solution point. The conformal transformation from the s to the $G(s)$ plane indicates, since $G(\sigma + j\omega_0)$ is initially orthogonal to $G(j\omega)$ at the solution point, at least one other solution for σ in addition to $\sigma = 0$. For high order transfer functions $G(s)$ more than one further solution for σ is in fact possible.(17)

For the particular example being considered an analytical solution can be obtained for σ by substituting in equation (9.58) for $N_{ib}(a_0)$ and $G(\sigma + j\omega_0)$. Separating the equation into its real and imaginary parts with $\omega = \omega_0$ gives

$$\sigma^3 - 3\sigma\omega_0^2 + 2\zeta\omega_0(\sigma^2 - \omega_0^2) + \sigma\omega_0^2 + \zeta\omega_0^3(1 - \cos 2\phi) = 0,$$

$$3\sigma^2\omega_0 - \omega_0^3 + 4\zeta\sigma\omega_0^2 + \omega_0^3 + \zeta\omega_0^3 \sin 2\phi = 0.$$

Eliminating ϕ and neglecting terms of order greater than σ^2 gives

$$4\sigma^2\omega_0^4 + 12\sigma^2\zeta^2\omega_0^4 + 4\sigma\zeta\omega_0^5 = 0,$$

i.e.

$$\sigma = 0 \quad \text{and} \quad -\zeta\omega_0/(1 + 3\zeta^2),$$

where the first solution corresponds to the damping of the limit cycle and the second, being the one required, to that of the limit cycle transient. The time constant, τ, of the transient is thus given by

$$\tau = (1 + 3\zeta^2)/\zeta\omega_0. \tag{9.59}$$

This result can also be obtained from equation (9.57).

† $-1/N_{ib}(a_0)$ for the ideal relay is a limiting case being a straight line through the solution point orthogonal to the real axis.

Near the limit cycle solution $\rho = 1$, $\dot{\rho} = 0$ the initial slope of the $\dot{\rho} - \rho$ curve of figure 9.4 is constant, showing that $\rho(t)$ varies exponentially, and is equal to

$$\left.\frac{d\dot{\rho}}{d\rho}\right|_{\dot{\rho}=0} = \left.\frac{d(\rho\sigma)}{d\rho}\right|_{\sigma=0} = [\sigma + \rho(d\sigma/d\rho)]\Big|_{\sigma=0} = \left.\frac{d\sigma}{d\rho}\right|_{\sigma=0}. \tag{9.60}$$

Differentiating equation (9.57) and putting $\rho = 1$ gives

$$2\zeta\omega_0(2 + 6\zeta^2)\left.\frac{d\sigma}{d\rho}\right|_{\sigma=0} = -4\zeta^2\omega_0^2,$$

that is

$$(d\sigma/d\rho)|_{\sigma=0} = -\zeta\omega_0/(1 + 3\zeta^2),$$

and

$$\tau = -(d\rho/d\sigma)|_{\sigma=0} = (1 + 3\zeta^2)/\zeta\omega_0.$$

Substituting the numerical values of $\omega_0 = 1$, $\zeta = 5$ the time constant τ is 15·2 secs. This value is a very good approximation to the time constant for the growth of the limit cycle of figure 9.5(a). For intermediate values of ζ, however, where the limit cycle steady state amplitude is reached in fewer oscillations, as in figure 9.5(b) where $\omega_0 = 1, \zeta = 1$, the time constant given by equation (9.59), since it neglects the nonlinearity in the $\dot{\rho} - \rho$ curve, predicts the limit cycle growth less accurately.

9.6 Limit cycling systems

In this section we examine further aspects of the dynamic behaviour of limit cycling systems including the closed loop behaviour of these systems with transient inputs. First, however, we consider the effect on a limit cycle of changes in a system parameter.

9.6.1 *Limit cycle dynamics*

Consider for example a closed loop limit cycling system with variable loop gain, K. For any given value of K we can evaluate the amplitude and frequency of the system limit cycle, however when K is changed from one value to another the amplitude and frequency do not change instantaneously. For small variations the $\dot{a} - a$ relationship, as in figure 9.4, may be assumed linear and the time constant, τ, governing changes in both amplitude and frequency is given, similar to equation (9.60) by

$$-\tau^{-1} = \left.\frac{d\dot{a}}{da}\right|_{\dot{a}=0} = a_0\left.\frac{d\sigma}{da}\right|_{\sigma=0}. \tag{9.61}$$

Thus, the transfer function relating a limit cycle amplitude change from a to $a + \Delta a$ at the nonlinearity input for a change in the loop gain K from K to $K + \Delta K$ is

$$\frac{\Delta a}{\Delta K} = \frac{(\Delta a/\Delta K)_{ss}}{1 + s\tau}, \tag{9.62}$$

where the gain $(\Delta a/\Delta K)_{ss}$ is the steady state change in a for a given change in K, and for small changes is the slope of the static limit cycle amplitude versus loop gain characteristic for the system at the loop gain K. The value of τ can be found using either the small σ solution approach and evaluating $(\mathrm{d}\sigma/\mathrm{d}a)|_{\sigma=0}$ from an equation similar to equation (9.57) or by using the IDF method. A direct approach to the evaluation of the parameters of equation (9.62), which we outline below, is given in reference 9.

When all derivatives of σ and those greater than one of ω are neglected, the equations from the real and imaginary terms of the differential equation obtained as outlined in section 9.2 may be written

$$R(\sigma, \omega, a, \dot{\omega}) = 0, \tag{9.63}$$

$$I(\sigma, \omega, a, \dot{\omega}) = 0. \tag{9.64}$$

Taking a differential of the equation of reals, solving it for $\mathrm{d}\omega$ and $\mathrm{d}\dot{\omega}$ with the derivatives of σ and $\dot{\omega}$ neglected and substituting the values in the differential of the equation of imaginaries gives $\mathrm{d}\sigma/\mathrm{d}a$. Using equation (9.61) then gives

$$\tau = \frac{\dfrac{1}{a_0}\dfrac{\partial I}{\partial \sigma} - \dfrac{\partial I}{\partial \dot{\omega}}\left(\dfrac{\partial R}{\partial \omega}\right)^{-1}\dfrac{\partial R}{\partial a} - \dfrac{\partial I}{\partial \omega}\left(\dfrac{\partial R}{\partial \omega}\right)^{-1}\left\{\dfrac{1}{a_0}\dfrac{\partial R}{\partial \sigma} - \dfrac{\partial R}{\partial \dot{\omega}}\left(\dfrac{\partial R}{\partial \omega}\right)^{-1}\dfrac{\partial R}{\partial a}\right\}}{\dfrac{\partial I}{\partial a} - \dfrac{\partial I}{\partial \omega}\left(\dfrac{\partial R}{\partial \omega}\right)^{-1}\dfrac{\partial R}{\partial a}}, \tag{9.65}$$

where the partial derivatives of the functions are evaluated at the limit cycle condition $\sigma = 0$, $\omega = \omega_0$, $a = a_0$ and $\dot{\omega} = 0$.

The steady state gain $(\Delta a/\Delta K)_{ss}$ of the transfer function can be obtained in a similar manner. Considering the general case where the system variable parameter is P, not necessarily the gain K, the steady state equations yielding the limit cycle may be written

$$R(a, \omega, P) = 0, \tag{9.66}$$

$$I(a, \omega, P) = 0. \tag{9.67}$$

Again taking differentials and eliminating $\mathrm{d}\omega$ to obtain the steady state gain of the amplitude transfer function $\mathrm{d}a/\mathrm{d}P$ gives

$$\frac{\mathrm{d}a}{\mathrm{d}P} = \frac{\dfrac{\partial I}{\partial \omega}\left(\dfrac{\partial R}{\partial \omega}\right)^{-1}\dfrac{\partial R}{\partial P} - \dfrac{\partial I}{\partial P}}{\dfrac{\partial I}{\partial a} - \dfrac{\partial I}{\partial \omega}\left(\dfrac{\partial R}{\partial \omega}\right)^{-1}\dfrac{\partial R}{\partial a}}. \tag{9.68}$$

If the frequency is of interest $\mathrm{d}\omega/\mathrm{d}P$ may be found in a similar manner.

9.6.2 *Limit cycle stability*

The stability of limit cycles has already been discussed using the DF in section 4.3 and the IDF in section 5.10.1. In both these methods a quasi-static approach is used, the effect of any changes in frequency when the limit cycle is perturbed being neglected. Usually these two methods give the same result for the stability of a limit cycle although there are some situations, see for examples problems 5.8 and 5.9, where this is not the case. This discrepancy occurs because the DF approach, or equivalently the Loeb criterion, ignores the location of system poles other than those specifying the oscillation.

In the previous section we have developed a procedure for finding the time constant, τ, associated with the perturbation of a limit cycle and obviously if this should turn out to be negative rather than positive, the limit cycle is unstable. Since this time constant takes into account the derivatives of frequency it will differ from τ_{qs} obtained by the quasi-static solution of section 4.2 and should provide a more accurate stability criterion.

In practice, however, it is difficult to find examples where the two criteria differ on the assessment of the stability of a limit cycle; although the magnitudes of the time constants τ and τ_{qs} predicted by the two methods may differ appreciably their signs are usually the same. The time constant, τ_{qs}, predicted by the quasi-static approach of section (4.2) is given from equations (4.5) and (4.6) by†

$$\tau_{qs} = \frac{1}{a_0}\frac{\Delta a}{\Delta \sigma} = \left. \frac{\left\{\left(\dfrac{\partial X}{\partial \omega}\right)^2 + \left(\dfrac{\partial Y}{\partial \omega}\right)^2\right\}}{a_0\left\{\dfrac{\partial X}{\partial a}\dfrac{\partial Y}{\partial \omega} - \dfrac{\partial X}{\partial \omega}\dfrac{\partial Y}{\partial a}\right\}}\right|_s . \tag{9.69}$$

For the example considered in the previous section substitution of the partial derivatives gives $\tau_{qs} = (1 + 4\zeta^2)/\zeta\omega_0$. This solution not only agrees with τ in sign but differs in magnitude from τ by less than 5 per cent for $\zeta < 1/4$. An example is given in reference 9 for which the two criteria do not agree for certain values of the system parameters, however it may also be shown that the IDF does not agree with the DF solution. Since the system of the example has very poor filtering and the nonlinearity is severe, changing from a positive to a negative slope, a DF analysis cannot be expected to produce good results. Certainly these inaccuracies do not allow one to conclude that the above stability criterion of $\tau > 0$ produces a better result than the IDF.‡

9.6.3 *Forced response of limit cycling systems*

In previous discussions of the input output behaviour of limit cycling control systems we have basically been concerned with the steady state properties in response to sinusoidal or random inputs. Here we wish to develop a quasilinear model of a limit cycling system suitable for evaluating the input output behaviour for transient inputs.

Consideration of the SBDF expressions $N(a, \gamma)$ and $N_\gamma(a, \gamma)$ for various nonlinearities indicates that for small values of the ratio γ/a they usually remain almost constant. For the particular case of the ideal relay both $N(a, \gamma)$ and $N_\gamma(a, \gamma)$ change by less than 6 per cent from their $\gamma = 0$ values of $N(a)$ and $N_{i\gamma}(a)$ for

† Note that for the notation used in section 4.2 $\Delta a/\Delta\sigma$ positive corresponds to a stable system.
‡ See also reference 17.

$\gamma/a < 1/3$. Therefore if γ is equal to $\gamma(t)$ a slowly varying function of time provided $|\gamma(t)|$ for all t is less than some fraction of a, which will depend upon the nonlinearity, the gain to $\gamma(t)$ may be taken as $N_{i\gamma}(a)$ and the limit cycle may be assumed constant. Figure 9.7 summarises this situation where system (a) is the original limit cycling system and system (b) its quasilinear input output equivalent for

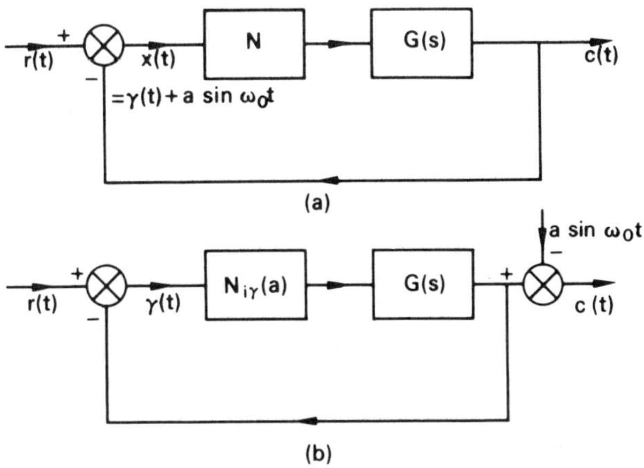

Figure 9.7 Nonlinear system and quasilinearised equivalent

transient inputs. The time variation of $\gamma(t)$, and therefore the validity of the model, will depend on the dynamics of the quasilinear system. Thus if $\gamma(t)$ is an exponential $\gamma e^{-t/\tau}$ it could be regarded as slowly varying if the decay lasts over several limit cycle periods. Reference 15 suggests a minimum of two periods which corresponds to

$$4\tau = 2(2\pi/\omega_0),$$

or

$$1/\tau = \omega_0/3. \tag{9.70}$$

This requirement will thus be satisfied for any $\gamma(t)$ provided the closed loop poles of the quasilinear system are in the region of the s plane bounded by the lines $\sigma = 0$ and $\sigma = -\omega_0/3$, as shown in figure 9.8. On the other hand if the poles are in the right half s plane the assumed limit cycle behaviour is not stable whilst if they are to the left of the line $\sigma = -\omega_0/3$ then the transients will be faster than our assumption allows.

The open loop transfer function for the linearised system is $N_{i\gamma}(a)G(s)$, which at the limit cycle frequency ω_0 has a value

$$N_{i\gamma}(a)G(j\omega_0) = -N_{i\gamma}(a)/N(a),$$

since the limit cycle is given by $N(a)G(j\omega_0) = -1$, where in general $N(a)$ may be complex. The above expression is seen to be a function of the nonlinearity and the limit cycle amplitude only.

A particular interesting situation arises with the ideal relay since for this characteristic $N_{i\gamma}(a)/N(a) = 1/2$, which is independent of a. The open loop gain of the linearised system is thus $-1/2$ at $\omega = \omega_0$ independent of the plant transfer

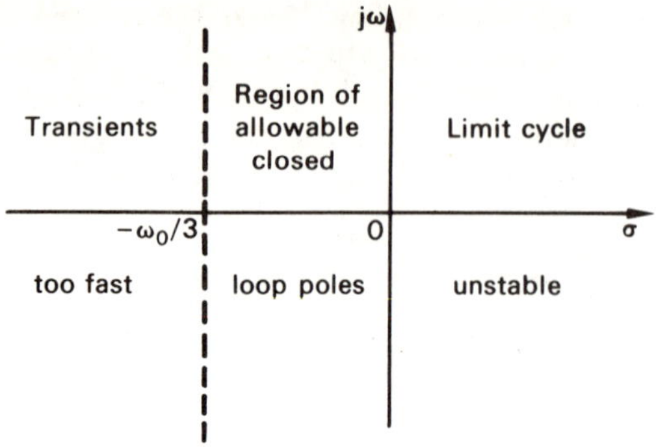

Figure 9.8 Pole locations in s plane

function $G(s)$. Changes in the parameters of the plant will thus cause a change in the limit cycle amplitude and frequency but the IDF of the relay changes to maintain the open loop gain of $-1/2$ at $\omega = \omega_0$. Further if only the gain of the plant changes the frequency of oscillation ω_0, to the accuracy of the DF approximation, will remain constant. This inherent adaptivity of a limit cycling relay control system has been used in the design of an autopilot(18) for a missile; a plant which has large parameter variations during flight. A detailed discussion of the design of limit cycling adaptive systems may be found in references 15 and 19.

To amplify the above comments we again consider the system of figure 9.2 with $G(s) = K\omega_0^2/s(s^2 + 2\zeta s\omega_0 + \omega_0^2)$. For the ideal relay

$$N_{i\gamma}(a) = 2h/a\pi,$$

and the open loop transfer function of the linearised system is

$$N_{i\gamma}(a)G(s) = 2hK\omega_0^2/a\pi s(s^2 + 2\zeta s_0 + \omega_0^2),$$

which on substituting for the limit cycle amplitude $a = 2hK/\pi\zeta\omega_0$ gives

$$N_{i\gamma}(a)G(s) = \zeta\omega_0^3/s(s^2 + 2\zeta s\omega_0 + \omega_0^2).$$

This expression is independent of the gain hK, showing that the system is adaptive with respect to these parameters. The closed loop transfer function is

$$C(s)/R(s) = \zeta\omega_0^3/(s^3 + 2\zeta s^2\omega_0 + \omega_0^2 s + \zeta\omega_0^3), \qquad (9.71)$$

with poles approximately located at

$$s \approx -\zeta\omega_0, -(\zeta\omega_0/2) \pm j\omega_0\{1 - (\zeta^2/4)\}^{1/2}$$

for small ζ. These all lie within the required s plane band provided $\zeta < 1/3$.

For this simple system a step input of magnitude greater than one-third that of the limit cycle violates the amplitude condition and equation (9.71) is not valid. It is however interesting to compare the step response of the nonlinear system for various values of hK with that of its quasilinear equivalent given by equation (9.71). Experimental results for this situation with $\zeta = 0{\cdot}18$ are shown in figure 9.9 and indicate that for all three values of hK, covering a three to one range, the response time is comparable with that given by equation (9.71). In contrast if the steady state open loop gain of the linear system described by equation (9.71) is doubled it becomes unstable and if reduced by a factor of two the response becomes much slower as shown by curve (e) of the figure.

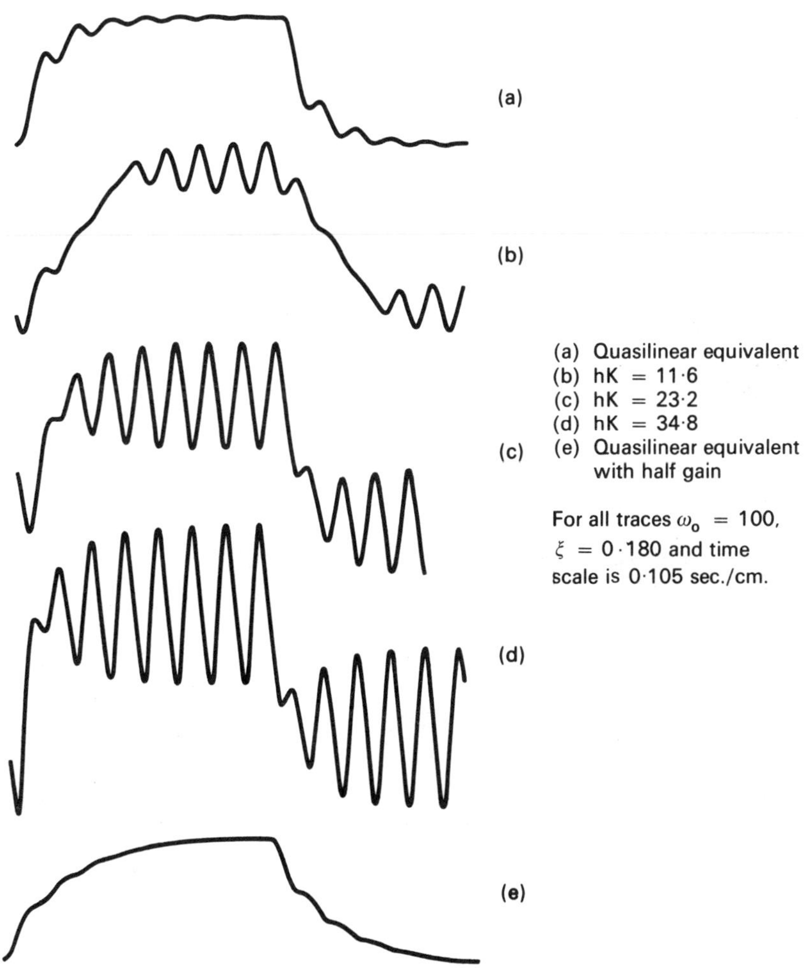

Figure 9.9 Step responses of nonlinear limit cycling system and quasilinearised equivalent

Because the linear transfer function contains an integrator the system has no steady state error for step inputs as seen in the figure. For a ramp input $r(t) = Rt$ the system output in the steady state will also be a ramp with a limit cycle superimposed. A bias of γ_0 at the relay output is integrated by the plant to produce a component $K\gamma_0 t$ in $c(t)$. Thus in the steady state

$$R = K\gamma_0,$$

$$\gamma_0 = \gamma N_\gamma(a, \gamma),$$

and

$$N(a, \gamma)G(j\omega) = -1.$$

Substituting for $N(a, \gamma)$, $N_\gamma(a, \gamma)$ and $G(j\omega)$ gives

$$\omega = \omega_0,$$

$$(4h/a^2\pi)(a^2 - \gamma^2)^{1/2} = 2\zeta\omega_0/K,$$

$$R/K = (2h/\pi)\sin^{-1}\gamma/a,$$

from which

$$a = \frac{2hK}{\pi\zeta\omega_0}\cos\frac{\pi R}{2hK} \quad \text{and} \quad \gamma = \frac{hK}{\pi\zeta\omega_0}\sin\frac{\pi R}{hK},$$

a solution which is obviously only possible for $R \leq hK$ since the maximum value of γ_0 is h. For a given system the steady state error γ increases and the limit cycle amplitude a decreases as R increases.

REFERENCES

1. Chen, K.: 'Quasi-linearization techniques for transient study of nonlinear feedback control systems', *A.I.E.E. Trans.*, 1955, 74, pt. II, pp. 354–365.
2. Chen, K.: 'Quasi-linearization design of nonlinear control systems', *Proc. J.A.C.C.*, Minneapolis, Minn., June, 1963, pp. 340–346.
3. Gelb, A. and Vander Velde, W. E.: *Multiple-input describing functions and nonlinear system design* (McGraw-Hill, New York, 1968, pp. 439–454).
4. Grensted, P. E. W.: 'Analysis of the transient response of nonlinear control systems', *Trans. A.S.M.E.*, 1958, 80, pp. 427–432.
5. Clauser, F. H.: 'The transient behaviour of non-linear systems', *I.R.E. Trans.*, 1960, pp. 446–458.
6. Popov, E. P.: 'Approximate study of transients in nonlinear automatic control systems by harmonic linearization', *Bull. Acad. Sci. U.S.S.R., Tech. Sci. Sec.*, 1956, no. 9, pp. 3–23; no 12, pp. 30–47.
7. Voronov, A. A.: 'An approximate determination of the self-oscillation startup process in certain nonlinear automatic control systems', *Automation and Remote Control*, 1958, 18, pp. 683–694.
8. Siljak, D. D.: 'Analysis and synthesis of feedback control systems in the parameter plane: III—Nonlinear systems', *I.E.E.E. Trans. Applications and Industry*, 1964, 83, pp. 466–473.
9. Gelb, A. and Vander Velde, W. E.: *Multiple-input describing functions and nonlinear system design* (McGraw-Hill, New York, 1968, Chapter 4).
10. Bickart, T. A.: 'The exponential describing function in the analysis of nonlinear systems', *I.E.E.E. Trans.*, 1966, AC-11, pp. 491–497.
11. Freeman, E. A. and Cox, C. S.: 'Dynamic analysis of nonlinear control systems', *Proc. I.E.E.*, 1967, 114, pp. 129–138.

12. Freeman, E. A.: 'Characterisation of nonlinearity for transient processes: its evaluation and application', *I.E.E.E. Trans.*, 1967, AC-12, pp. 491–501.
13. Freeman, E. A. and Cox, C. S.: 'Evaluation of transient-gain functions for nonlinear elements and their use in step-function analysis', *Proc. I.E.E.*, 1967, 114, pp. 1772–1780.
14. Dorrah, H. T. and Hamza, M. H.: 'Analysis of transient oscillations in nonlinear control systems', *Automatic Control Theory and Applications*, 1972, Vol. 1, No. 3, pp. 1–8.
15. Gelb, A. and Vander Velde, W. E.: 'On limit cycling control systems', *I.E.E.E. Trans.*, 1963, AC-8, pp. 142–157.
16. Gelb, A. and Vander Velde, W. E.: *Multiple input describing functions and nonlinear system design* (McGraw-Hill, New York, 1968, Chapter 6, pp. 317–340).
17. Choudhury, S. K. and Atherton, D. P.: Limit cycles in high order nonlinear systems, *Proc. I.E.E.*, 1974, 121, pp. 717–724.
18. Shuck, O. H.: 'Honeywell's history and philosophy in the adaptive control field', *Wright Air Develop. Center, Tech. Report*, January 1959, Wright Patterson Air Force Base, Ohio.
19. Gelb, A.: 'The analysis and design of limit cycling adaptive automatic control systems', *Sc.D. Thesis, M.I.T., Cambridge, Mass.*, 1961.

PROBLEMS

P.9.1 Show that the EDF, $N_E(a)$, is also given by

$$N_E(a) = (2/a^2)\int_0^a n(x)\,dx.$$

P.9.2 Following the procedures of section 9.3 determine the gains K_1 and K_2 for the EDF model $K_1x + K_2\dot{x}$ of a dynamic nonlinearity $n(x, \dot{x})$ with input $x = ae^{-t/T}$.

P.9.3 A system has a saturation nonlinearity, with unit gain in the linear regime, followed by a linear transfer function $G(s) = 4/s(s + 8)$ in the feedforward portion of the loop and a feedback transfer function of unity. Evaluate the responses to step inputs of various magnitudes using (a) the EDF, $N_E(a)$ and (b) an exact method.

P.9.4 Determine the step response of the system of figure 4.38 to inputs of 20 and 40 units by (a) the small σ method and (b) the half period EDF if $K_1 = 1$, $K_2 = \omega_0^2$, $T = \sqrt{2}/\omega_0$ and the nonlinearity is an ideal relay with unit magnitude output levels.

P.9.5 For the system of figure 9.2 with $G(s) = K(s + 2)/s(s + 1)$ find (a) the small σ solution and (b) the half period EDF solution for a step input. Compare the solutions with the exact solution.

P.9.6 If in the system of figure 9.2 $G(s) = K/(s - 3)(s^2 + 2s + 10)$, sketch the root locus of the quasilinear system for transient inputs. Is the system limit cycle stable?

P.9.7 It is required to stabilise the limit cycle in problem P.9.6 by introducing a compensator, $G_c(s)$, before the relay. Suggest a suitable compensator.

P.9.8 Obtain the transfer function for a limit cycle amplitude change of Δa corresponding to a loop gain change of ΔK for the system of figure 9.2 if $G(s) = K(s + 4)/s(s + 1)^2$.

P.9.9 A nonlinear system contains the nonlinearity $n(x) = x - \varepsilon x^3$, a transfer function $G(s) = K/s(s + 1)^2$ and a feedback transfer function of unity.

Show that a stable limit cycle only exists for

(a) $2 < K < 4$ by the IDF method,

(b) $2 < K < 7$ using Gelb's indirect method used in section 9.5.2(a), and

(c) $2 < K < 34/7$ using Gelb's direct method (equation (9.65)).

P.9.10 If the two gain model of equation (9.10) is chosen to minimise a weighted integral squared error of the form $I = \int_0^\infty e^2(\theta)\omega(\theta)\,d\theta$ obtain expressions for K_1 and K_2. This representation has been suggested[14] for a complete or half period transient describing function for an input $x(\theta) = ae^{-\alpha\theta}\sin\theta$ and $\omega(\theta) = e^{\alpha\theta}$. Show that in this case K_2 is not zero for a SVNL and evaluate K_1 and K_2 for the half period case for an ideal relay.

P.9.11 Show that if only one solution other than that for $\sigma = 0$ exists to the synchronous IDF equation (9.58) the condition for stability it produces is in agreement with the Loeb criterion.

Chapter 10

Multivariable Nonlinearities

10.1 Introduction

As its title suggests this chapter deals entirely with the nonlinearity problem, that is considerations of the response of nonlinearities whose output may be expressed as a function of several variables. Procedures for determining autocorrelation functions for the output as well as the equivalent gains of the nonlinearity are discussed.

The work commences with an examination of the transform method developed by Rice[1] for determining the autocorrelation function of the output of a nonlinear element with a sinusoidal signal plus Gaussian noise input. This leads to a proof of the modified nonlinearity approach for additive inputs, which has earlier been used for determining describing functions. Although the transform method can be extended for use in the more general case of the nonlinearity output being a function of many variables, not necessarily additive, the opportunity is taken to introduce an alternative approach based on orthogonal polynomials. The use of orthogonal polynomials for a single sinusoidal or Gaussian input has been discussed in the relevant earlier chapters. Several cases where the variables are related, which as seen earlier correspond to situations commonly required in the analysis and design of nonlinear feedback systems, are considered in some detail.

The general multivariable nonlinearity is then considered and its equivalent gains evaluated. The equivalent gain model is shown to be the optimum linear model in the mean squared sense when the 'inputs' belong to the class of separable random processes. These results show the properties and general nature of describing functions which have previously been derived in terms of specific input situations. The concepts of a modified nonlinearity and d.c. incremental gain are extended to the general multivariable case and their uses discussed. Finally some additional results are given which may be used for the evaluation of the describing functions of nonlinearities with two inputs.

10.2 The transform method

In previous chapters we have used the modified nonlinearity approach to evaluate describing functions and harmonic output components for a nonlinearity with two input signals. A proof of the validity of this approach for determining describing functions can be obtained by extending the procedure of section 8.2 to two input variables;[2] a problem which we defer until the case of n inputs is considered in section 10.5. The transform method,[1,3] or as it is occasionally called the characteristic function method, is important both historically and because of the alternative expressions it yields for determining the output terms from a nonlinearity. We will also use it to prove the validity of the modified nonlinearity approach for two unrelated input signals.

Consider a nonlinearity $n(x)$ with an input consisting of the addition of a sinusoidal signal and Gaussian noise. Defining the nonlinearity by

$$n(x) = \begin{cases} n_+(x) & x > 0 \\ n_-(x) & x < 0, \end{cases}$$

we may take the bilateral Laplace transform[4,5] (see appendix B) of the characteristic to yield

$$\begin{aligned} N(p) &= \int_{-\infty}^{\infty} n(x)\mathrm{e}^{-px}\,\mathrm{d}x \\ &= \int_{-\infty}^{0} n_-(x)\mathrm{e}^{-px}\,\mathrm{d}x + \int_{0}^{\infty} n_+(x)\mathrm{e}^{-px}\,\mathrm{d}x = N_-(p) + N_+(p) \qquad (10.1) \end{aligned}$$

and

$$n(x) = (1/2\pi j)\int_{C_-} N_-(p)e^{px}\,dp + (1/2\pi j)\int_{C_+} N_+(p)e^{px}\,dp, \qquad (10.2)$$

where the transform variable p rather than s has been used to avoid confusion, since the latter variable is used for the harmonic number of a sinusoidal signal.

If the input signal is denoted by z then the output autocorrelation function is

$$R_0(\tau) = \int_{-\infty}^{\infty}\int_{-\infty}^{\infty} n(z_1)n(z_2)p(z_1, z_2, \tau)\,dz_1\,dz_2,$$

where $p(z_1, z_2, \tau)$ is the joint probability density function of the input signal. Using equation (10.2) we obtain

$$\begin{aligned} R_0(\tau) = \int_{-\infty}^{\infty}\int_{-\infty}^{\infty} & (1/4\pi^2j^2)\left\{\int_{C_+} N_+(p_1)e^{p_1z_1}\,dp_1 + \int_{C_-} N_-(p_1)e^{p_1z_1}\,dp_1\right\} \\ & \times \left\{\int_{C_+} N_+(p_2)e^{p_2z_2}\,dp_2 + \int_{C_-} N_-(p_2)e^{p_2z_2}\,dp_2\right\} p(z_1, z_2, \tau)\,dz_1\,dz_2 \\ = (1/4\pi^2j^2)\Bigg\{ & \int_{C_+} N_+(p_1)\,dp_1 \int_{C_+} N_+(p_2)M(p_1, p_2, \tau)\,dp_2 \\ & + \int_{C_-} N_-(p_1)\,dp_1 \int_{C_+} N_+(p_2)M(p_1, p_2, \tau)\,dp_2 \\ & + \int_{C_+} N_+(p_1)\,dp_1 \int_{C_-} N_-(p_2)M(p_1, p_2, \tau)\,dp_2 \\ & + \int_{C_-} N_-(p_1)\,dp_1 \int_{C_-} N_-(p_2)M(p_1, p_2, \tau)\,dp_2\Bigg\}, \end{aligned}$$

where $M(p_1, p_2, \tau)$ is the joint characteristic function, with p_1 and p_2 replacing ju and jv, of z.

As the two input signals are uncorrelated

$$M(p_1, p_2, \tau) = M_x(p_1, p_2, \tau)M_y(p_1, p_2, \tau),$$

where from equations (7.106) and (7.119)

$$M_x(p_1, p_2, \tau) = \sum_{s=0}^{\infty} \varepsilon_s I_s(ap_1)I_s(ap_2)\cos s\omega\tau$$

for the sine wave, and

$$M_y(p_1, p_2, \tau) = e^{\sigma^2p_1^2/2}e^{\sigma^2p_2^2/2}\sum_{k=0}^{\infty}\{p_1p_2R_i(\tau)\}^k/k!$$

for the Gaussian signal with autocorrelation function $R_i(\tau)$. Considering a particular term in the expression for $R_0(\tau)$, we have

$$(1/4\pi^2 j^2)\int_{C_+} N_+(p_1)\,\mathrm{d}p_1 \int_{C_-} N_-(p_2) M(p_1, p_2, \tau)\,\mathrm{d}p_2$$

$$= (1/4\pi^2 j^2)\int_{C_+} N_+(p_1)\,\mathrm{d}p_1 \int_{C_-} N_-(p_2) e^{\sigma^2 p_1^2/2} e^{\sigma^2 p_2^2/2} \sum_{k=0}^{\infty} p_1^k p_2^k \{R_i^k(\tau)/k!\}$$

$$\times \sum_{s=0}^{\infty} \varepsilon_s I_s(ap_1) I_s(ap_2) \cos s\omega\tau\,\mathrm{d}p_2$$

$$= \sum_{s=0}^{\infty}\sum_{k=0}^{\infty} \varepsilon_s\{R_i^k(\tau)/k!\} \cos s\omega\tau \left\{(1/2\pi j)\int_{C_+} N_+(p_1) e^{\sigma^2 p_1^2/2} p_1^k I_s(ap_1)\,\mathrm{d}p_1\right\}$$

$$\times \left\{(1/2\pi j)\int_{C_-} N_-(p_2) e^{\sigma^2 p_2^2/2} p_2^k I_s(ap_2)\,\mathrm{d}p_2\right\}$$

$$= \sum_{s=0}^{\infty}\sum_{k=0}^{\infty} \varepsilon_s \alpha_{+sk}\alpha_{-sk}\{R_i^k(\tau)/k!\} \cos s\omega\tau,$$

where

$$\alpha_{\pm sk} = (1/2\pi j)\int_{C_\pm} N_\pm(p) p^k e^{\sigma^2 p^2/2} I_s(ap)\,\mathrm{d}p. \tag{10.3}$$

Using the similar expressions found for the other terms, the resulting expression for $R_0(\tau)$ becomes

$$R_0(\tau) = \sum_{s=0}^{\infty}\sum_{k=0}^{\infty} \{\varepsilon_s \alpha_{sk}^2 R_i^k(\tau) \cos s\omega\tau\}/k!, \tag{10.4}$$

where

$$\alpha_{sk} = \alpha_{+sk} + \alpha_{-sk}.$$

Alternatively we may write

$$R_0(\tau) = \alpha_{00}^2 + \sum_{s=1}^{\infty} \varepsilon_s \alpha_{s0}^2 \cos s\omega\tau + \sum_{k=1}^{\infty} \alpha_{0k}^2\{R_i^k(\tau)/k!\}$$

$$+ \sum_{s=1}^{\infty}\sum_{k=1}^{\infty} \varepsilon_s \alpha_{sk}^2 \{R_i^k(\tau)/k!\} \cos s\omega\tau. \tag{10.5}$$

In this form it is seen that the first term is the output bias variance, the second term contains sinusoidal terms only, the third noise terms only and the last cross-products of the sine and noise. The component gains of the sine plus Gaussian describing function, the RSDF, are respectively $(2\alpha_{10}/a)$ for the sine and α_{01} for the Gaussian signal.

Obviously for a symmetrical characteristic one need only evaluate α_{+sk} and then use the properties of the symmetry. Also α_{-sk} can, if desired, be evaluated from an 'equivalent' positive characteristic as described in section 5.4. When the integral of equation (10.3) has no singularities within the rectangular strip bounded by the lines of the C_+ or C_- contours then Cauchy's theorem can be used to give

$$\alpha_{\pm sk} = (1/2\pi) \int_{-\infty}^{\infty} N_{\pm}(j\omega) e^{-\sigma^2\omega^2/2} (j\omega)^k j^s J_s(a\omega) \, d\omega,$$

in which case

$$\alpha_{sk} = (1/2\pi) \int_{-\infty}^{\infty} N(j\omega) e^{-\sigma^2\omega^2/2} (j\omega)^k j^s J_s(a\omega) \, d\omega. \qquad (10.6)$$

For the particular case of a single sine input

$$\alpha_s = (a_s/2) = (1/2\pi) \int_{-\infty}^{\infty} N(j\omega) j^s J_s(a\omega) \, d\omega, \qquad (10.7)$$

and a single Gaussian input

$$\alpha_k = (1/2\pi) \int_{-\infty}^{\infty} N(j\omega)(j\omega)^k e^{-\sigma^2\omega^2/2} \, d\omega. \qquad (10.8)$$

It is further clear that the method can be applied to two incommensurate sinusoidal inputs to give

$$R_0(\tau) = \sum_{s=0}^{\infty} \sum_{k=0}^{\infty} \varepsilon_s \varepsilon_k \alpha_{sk}^2 \cos s\omega\tau \cos k\omega\tau, \qquad (10.9)$$

with

$$\alpha_{sk} = (1/2\pi) \int_{-\infty}^{\infty} N(j\omega) j^{s+k} J_s(a\omega) J_k(b\omega) \, d\omega, \qquad (10.10)$$

and can be extended to more inputs, say, two sinusoids plus Gaussian noise.

10.2.1 *Some theoretical results*

For many nonlinearities evaluation of α_{sk} from equation (10.6) requires the expansion of $N(j\omega)$ as a power series in ω and use of Hankel's exponential integral[6]

$$\int_0^{\infty} \omega^n e^{-\sigma^2\omega^2/2} J_s(a\omega) \, d\omega = \frac{\rho^s \Gamma\{(s+n+1)/2\}\, {}_1F_1\{(s+n+1)/2; s+1; -\rho^2\}}{2s!(\sigma/\sqrt{2})^{n+1}}, \qquad (10.11)$$

where ρ is the signal to noise ratio $(a/\sigma\sqrt{2})$, and ${}_1F_1$ denotes the confluent hypergeometric function[7] defined in equation (8.76).

Ideal saturation

For this characteristic

$$N_{+}(p) = (m/p^2)(1 - e^{-p\delta}),$$

and

$$N(p) = (m/p^2)(e^{p\delta} - e^{-p\delta}).$$

Since the pole at the origin of $N_{+}(p)$ is of order one and the integrand of equation (10.3) has a zero of order $s + k$ at the origin when $N_{+}(p)$ is not included, α_{sk} can be evaluated from equation (10.6) for $s + k \geq 1$.

Therefore, substituting for $N(j\omega)$ in equation (10.6) gives

$$\alpha_{sk} = m \int_{-\infty}^{\infty} \sum_{n=0}^{\infty} \frac{2(j\omega\delta)^{2n-1}}{(2n+1)!} \delta^2 (j\omega)^k j^s e^{-\sigma^2\omega^2/2} J_s(a\omega)\, d\omega,$$

which on taking account of symmetry and using equation (10.11) yields the solution

$$\alpha_{sk} = \frac{m\rho^s}{s!} \sum_{n=0}^{\infty} \frac{\delta^{2n+1}\Gamma\{(s+k+2n)/2\}\, {}_1F_1\{(s+k+2n)/2; s+1; -\rho^2\}}{(2n+1)!(\sigma/\sqrt{2})^{2n+k}} \quad \text{for } s+k \text{ odd,}$$

$$0 \qquad \text{for } s + k \text{ even.} \tag{10.12}$$

Harmonic nonlinearity

For $n(x) = \sin mx$,

$$N_{+}(p) = m/(p^2 + m^2),$$

and

$$N_{-}(p) = -m/(p^2 + m^2).$$

Both $N_{+}(p)$ and $N_{-}(p)$ have poles on the imaginary axis at $p = \pm jm$. From equation (10.3),

$$\alpha_{+sk} = (1/2\pi j) \int_{C_+} F(p)\, dp, \quad \text{and} \quad \alpha_{-sk} = (-1/2\pi j) \int_{C_-} F(p)\, dp,$$

where

$$F(p) = \{m/(p^2 + m^2)\} p^k e^{\sigma^2 p^2/2} I_s(ap).$$

Applying Cauchy's theorem around a closed rectangular contour with sides C_+ and C_-, and since the contributions at infinity are zero,

$$(1/2\pi j)\int_{C_+} F(p)\,\mathrm{d}p - (1/2\pi j)\int_{C_-} F(p)\,\mathrm{d}p = I,$$

where I is the sum of the residues of $F(p)$ at $p = \pm jm$. Thus

$$\alpha_{sk} = \alpha_{+sk} + \alpha_{-sk} = I,$$

and

$$\begin{aligned}\alpha_{sk} &= e^{-\sigma^2 m^2/2}\{(jm)^k I_s(jma) - (-jm)^k I_s(-jma)\}/2j \\ &= m^k J_s(ma)\mathrm{e}^{-\sigma^2 m^2/2}\{j^k(-j)^s(-1)^s - (-j)^k(-j)^s\}/2j \\ &= m^k J_s(ma)\mathrm{e}^{-\sigma^2 m^2/2}\{j^{k+s} - (-j)^{k+s}\}/2j,\end{aligned}$$

which may be written

$$\alpha_{sk} = m^k J_s(ma)\mathrm{e}^{-\sigma^2 m^2/2}\sin\{(s+k)\pi/2\}. \tag{10.13}$$

A procedure for evaluating α_{sk} based on the transform method for a sinusoidal plus Gaussian input to the general quantised nonlinearity is given in reference 8.

10.2.2 *Modified nonlinearity method*

When equation (10.6) is valid,

$$N(j\omega) = \int_{-\infty}^{\infty} n(z)\mathrm{e}^{-j\omega z}\,\mathrm{d}z,$$

so that equation (10.6) may be written

$$\alpha_{sk} = (1/2\pi)\int_{-\infty}^{\infty}\int_{-\infty}^{\infty} n(z)\mathrm{e}^{-j\omega z}\,\mathrm{d}z(j\omega)^k\mathrm{e}^{-\sigma^2\omega^2/2}j^s J_s(a\omega)\,\mathrm{d}\omega.$$

Changing the order of integration gives

$$\alpha_{sk} = \int_{-\infty}^{\infty} n(z)\,\mathrm{d}z\int_{-\infty}^{\infty}(1/2\pi)(j\omega)^k\mathrm{e}^{-\sigma^2\omega^2/2}j^s J_s(a\omega)\mathrm{e}^{-j\omega z}\,\mathrm{d}\omega,$$

that is

$$\alpha_{sk} = \int_{-\infty}^{\infty} n(z)f(z)\,\mathrm{d}z,$$

where

$$f(z) = (1/2\pi)\int_{-\infty}^{\infty}(j\omega)^k\mathrm{e}^{-\sigma^2\omega^2/2}j^s J_s(a\omega)\mathrm{e}^{-j\omega z}\,\mathrm{d}\omega.$$

Denoting the Fourier transform with reverse signs on the exponentials from the usual case by F_-, then

$$F_-\{f(z)\} = (j\omega)^k\mathrm{e}^{-\sigma^2\omega^2/2}j^s J_s(a\omega).$$

But[9]

$$F_-^{-1}\{(j\omega)^k e^{-\sigma^2\omega^2/2}\} = (1/2\pi)\int_{-\infty}^{\infty} (j\omega)^k e^{-\sigma^2\omega^2/2} e^{-j\omega y}\,dy = (1/\sigma^k)H_k(y/\sigma)q(y),$$

and

$$F_-^{-1}\{j^s J_s(a\omega)\} = (1/2\pi)\int_{-\infty}^{\infty} j^s J_s(a\omega)e^{-j\omega x}\,dx = T_s(x/a)r(x).$$

Therefore

$$F_-\{f(z)\} = F_-\{T_s(x/a)r(x)\}F_-\{(1/\sigma^k)H_k(y/\sigma)q(y)\}.$$

Using the convolution theorem for $f(z)$ and substituting back in the expression for α_{sk} gives

$$\alpha_{sk} = \int_{-\infty}^{\infty}\int_{-\infty}^{\infty} n(z)T_s\{(z - y)/a\}r(z - y)(1/\sigma^k)H_k(y/\sigma)q(y)\,dy\,dz,$$

and substituting $z = x + y$, the required result,

$$\alpha_{sk} = (1/\sigma^k)\int_{-\infty}^{\infty}\int_{-\infty}^{\infty} n(x + y)T_s(x/a)H_k(y/\sigma)r(x)q(y)\,dx\,dy, \tag{10.14}$$

is obtained. The solution is seen to be in agreement with those given previously for a single sinusoidal or Gaussian signal. Similarly the solution for two incommensurate sinusoidal inputs is

$$\alpha_{sk} = \int_{-b}^{b}\int_{-a}^{a} n(x + y)T_s(x/a)T_k(y/b)r(x)r(y)\,dx\,dy. \tag{10.15}$$

Here it should be noted that as the α_{sk}'s are coefficients in an autocorrelation function, equation (10.4), we can find the magnitude of the various output frequency components but not as yet their phases. This point is considered further in section 10.4.

The required formula for the extension of equation (10.14) to three or more uncorrelated inputs is obvious.

As discussed in section 5.8 the two stage modified nonlinearity approach for the computation of α_{sk} from equation (10.14) is

$$n(a, \gamma)_s = \int_{-a}^{a} n(x + \gamma)T_s(x/a)r(x)\,dx, \tag{10.16}$$

$$\alpha_{sk} = (1/\sigma^k)\int_{-\infty}^{\infty} n(a, y)_s H_k(y/\sigma)q(y)\,dy, \tag{10.17}$$

or

$$n(\sigma, \gamma)_k = \int_{-\infty}^{\infty} n(y + \gamma)H_k(y/\sigma)q(y)\,dy, \tag{10.18}$$

$$\alpha_{sk} = (1/\sigma^k)\int_{-a}^{a} n(\sigma, \gamma)_k T_s(x/a)r(x)\,dx. \tag{10.19}$$

Example

Consider the harmonic nonlinearity sin mx. From equation (5.16),

$$n(a, \gamma)_s = (-1)^s J_s(ma) \sin\{m\gamma - (s\pi/2)\}.$$

Therefore

$$\alpha_{sk} = (-1)^s J_s(ma)(1/\sigma^k) \int_{-\infty}^{\infty} \sin\{my - (s\pi/2)\} H_k(y/\sigma) q(y)\, dy,$$

which, using equation (8.26) with $m\gamma$ replaced by $-s\pi/2$, gives

$$\alpha_{sk} = (-1)^s J_s(ma)(1/\sigma^k)(-1)^k (m\sigma)^k e^{-\sigma^2 m^2/2} \sin\{-(s + k)\pi/2\}.$$

That is

$$\alpha_{sk} = m^k e^{-\sigma^2 m^2/2} J_s(ma) \sin\{(s + k)\pi/2\},$$

in agreement with equation (10.13).

10.3 Orthogonal polynomials

Previously we have restricted our consideration of multiple inputs to a nonlinearity to those situations where all the input signals are added on a single channel to give an output of the form $n(x + y)$. Many situations exist where a nonlinearity occurs as a function $n(x, y)$ of two or more variables, where for example x and y may both be input signals or x may be an input signal and y some internal nonlinearity parameter. Although the transform method can be extended to cover this situation by taking a bilateral Laplace transform in two variables[10] of $n(x, y)$, there are physical advantages to working in the real nonlinearity domain exemplified by the modified nonlinearity approach. The expansion of the nonlinearity output $n(x, y)$ in terms of orthogonal polynomials provides a suitable procedure for examining the nonlinearity response using the real domain.

Consider the nonlinear function $n(x, y)$ where the two variables x and y may be statistically dependent so that $p(x, y) \neq p(x)p(y)$. According to the Schmidt process of orthogonalisation,[11,12] which may be extended to cover any number of variables, if $g_0(x, y), g_1(x, y) \ldots$ form a set of functions integrable with their squares over a region R with no relation of linear dependence connecting any finite number of them, it is possible to form a normalised orthogonal sequence $G_0(x, y), G_1(x, y) \ldots$ in which $G_n(x, y)$ is a linear combination of $g_0(x, y), g_1(x, y) \ldots$. In particular if R is finite and if $p(x, y)$ is a non-negative integrable function having a positive integral over R, application of the orthogonalisation procedure to the linearly independent functions $p^{1/2}, p^{1/2}x, p^{1/2}y, p^{1/2}x^2, p^{1/2}xy \ldots$ taken in this order gives a sequence of polynomials $q_{mn}(x, y)$, $m = 0, 1, 2, \ldots, n = 0, 1, 2, \ldots,$ such that

$$\int_R \int q_{ij}(x, y) q_{mn}(x, y) p(x, y)\, dx\, dy = \delta_{im}\delta_{jn}, \tag{10.20}$$

where $q_{ij}(x, y)$ is a polynomial of degree j in y and i in x and y.

Also any arbitrary function $n(x, y)$ can be formally expanded in a series of the form

$$n(x, y) = \sum_{i=0}^{\infty} \sum_{j=0}^{i} c_{ij} q_{ij}(x, y), \tag{10.21}$$

where

$$c_{ij} = \int_R \int n(x, y) q_{ij}(x, y) p(x, y) \, dx \, dy. \tag{10.22}$$

Let $S_n(x, y)$ denote the partial sum of the series for $n(x, y)$ to terms of the nth degree. That is

$$S_n(x, y) = \sum_{i=0}^{n} \sum_{j=0}^{i} c_{ij} q_{ij}(x, y).$$

If $f_n(x, y)$ is any other polynomial of degree n in x and y it may be written in the form

$$f_n(x, y) = \sum_{i=0}^{n} \sum_{j=0}^{i} d_{ij} q_{ij}(x, y).$$

Then it can be shown that of all such polynomials f_n, the one that provides the least mean squared error on R with respect to $p(x, y)$ is $S_n(x, y)$.

10.3.1 *Equivalent gains*

Using the above theory of orthogonal polynomials we approximate the nonlinearity $n(x, y)$ by $S_1(x, y)$ where

$$S_1(x, y) = c_{00} q_{00}(x, y) + c_{10} q_{10}(x, y) + c_{11} q_{11}(x, y).$$

The q's are obtained from equation (10.20) with $p(x, y)$ the joint probability density of x and y, and the c's from equation (10.22).

Assuming, as required by the aforementioned theory,

$$q_{00} = \alpha_0,$$

$$q_{10} = \alpha_1 + \alpha_2 x,$$

$$q_{11} = \alpha_3 + \alpha_4 x + \alpha_5 y,$$

repeated application of equation (10.20) with different q's gives

$$\int_{-\infty}^{\infty} \int_{-\infty}^{\infty} \alpha_0^2 p(x, y) \, dx \, dy = 1$$

so that $\alpha_0 = 1$,

$$\int_{-\infty}^{\infty} \int_{-\infty}^{\infty} (\alpha_1 + \alpha_2 x) p(x, y) \, dx \, dy = 0,$$

thus $\alpha_1 = 0$ if $E(x) = 0$,

$$\int_{-\infty}^{\infty}\int_{-\infty}^{\infty} \alpha_2^2 x^2 p(x, y)\,dx\,dy = 1,$$

therefore $\alpha_2 = 1/R_{xx}^{1/2}(0)$,

$$\int_{-\infty}^{\infty}\int_{-\infty}^{\infty} (\alpha_3 + \alpha_4 x + \alpha_5 y)p(x, y)\,dx\,dy = 0,$$

thus $\alpha_3 = 0$ if $E(x) = E(y) = 0$,

$$\int_{-\infty}^{\infty}\int_{-\infty}^{\infty} \alpha_2 x(\alpha_4 x + \alpha_5 y)p(x, y)\,dx\,dy = 0$$

which, on substituting for α_2 and integrating, gives $\alpha_4 = -\alpha_5 R_{xy}(0)/R_{xx}(0)$. Finally

$$\int_{-\infty}^{\infty}\int_{-\infty}^{\infty} (\alpha_4 x + \alpha_5 y)^2 p(x, y)\,dx\,dy = 1.$$

which gives

$$\alpha_4^2 R_{xx}(0) + \alpha_5^2 R_{yy}(0) + 2\alpha_4\alpha_5 R_{xy}(0) = 1.$$

Solving for α_4 and α_5 gives

$$\alpha_4 = -R_{xy}(0)/R_{xx}^{1/2}(0)\{R_{xx}(0)R_{yy}(0) - R_{xy}^2(0)\}^{1/2},$$

and

$$\alpha_5 = R_{xx}^{1/2}(0)/\{R_{xx}(0)R_{yy}(0) - R_{xy}^2(0)\}^{1/2}.$$

Hence

$$S_1(x, y) = c_{00} + c_{10}\alpha_2 x + c_{11}(\alpha_4 x + \alpha_5 y),$$

where the values of α_2, α_4 and α_5 are as above. From equation (10.22)

$$c_{00} = \int_{-\infty}^{\infty}\int_{-\infty}^{\infty} n(x, y)p(x, y)\,dx\,dy = 0$$

and for a symmetrical odd characteristic

$$c_{10} = \int_{-\infty}^{\infty}\int_{-\infty}^{\infty} n(x, y)\alpha_2 x p(x, y)\,dx\,dy = \alpha_2 I_x,$$

where

$$I_x = \int_{-\infty}^{\infty}\int_{-\infty}^{\infty} x n(x, y)p(x, y)\,dx\,dy,$$

and

$$c_{11} = \int_{-\infty}^{\infty}\int_{-\infty}^{\infty} n(x, y)(\alpha_4 x + \alpha_5 y)p(x, y)\,dx\,dy = \alpha_4 I_x + \alpha_5 I_y,$$

where

$$I_y = \int_{-\infty}^{\infty}\int_{-\infty}^{\infty} yn(x, y)p(x, y)\,dx\,dy.$$

Therefore $S_1(x, y) = x\{\alpha_2^2 I_x + \alpha_4^2 I_x + \alpha_5\alpha_4 I_y\} + y\{\alpha_5^2 I_y + \alpha_4\alpha_5 I_x\}$, which means the equivalent gains K_x and K_y to the signals x and y respectively are

$$K_x = (\alpha_2^2 + \alpha_4^2)I_x + \alpha_4\alpha_5 I_y,$$

and

$$K_y = \alpha_5^2 I_y + \alpha_4\alpha_5 I_x.$$

Substituting for the coefficients the gains are

$$K_x = \{R_{yy}(0)I_x - R_{xy}(0)I_y\}/\{R_{xx}(0)R_{yy}(0) - R_{xy}^2(0)\}, \tag{10.23}$$

and

$$K_y = \{R_{xx}(0)I_y - R_{xy}(0)I_x\}/\{R_{xx}(0)R_{yy}(0) - R_{xy}^2(0)\}. \tag{10.24}$$

Obviously the approximation $S_n(x, y)$ can be evaluated for values of n greater than one to yield the terms neglected in a quasilinear analysis. The procedure becomes rather unwieldy however unless a particular case with a given $p(x, y)$ is considered.

If the covariance between x and y is zero, that is $R_{xy}(0) = 0$, equations (10.23) and (10.24) become

$$K_x = I_x/R_{xx}(0) = c_{10}/R_{xx}^{1/2}(0), \tag{10.25}$$

and

$$K_y = I_y/R_{yy}(0) = c_{11}/R_{yy}^{1/2}(0). \tag{10.26}$$

10.3.2 *Sine and Gaussian signals*

When the nonlinearity output $n(x, y)$ is a function of a sinusoidal signal x and Gaussian signal y then

$$R_{xy}(\tau) = 0,$$

and

$$p(x, y) = r(x)q(y).$$

Since $p(x, y)$ is separable it is known that[11]

$$q_{mn}(x, y) = t_{m-n}(x)u_n(y),$$

where $t_i(x)$ and $u_i(y)$ are orthogonal polynomials in one variable defined respectively by

$$\int_{-\infty}^{\infty} t_i(x)t_j(x)r(x)\,dx = \delta_{ij}, \tag{10.27}$$

and

$$\int_{-\infty}^{\infty} u_i(y)u_j(y)q(y)\,dy = \delta_{ij}. \tag{10.28}$$

Assuming $t_0(x) = \alpha_0$, $t_1(x) = \alpha_1 + \alpha_2 x$, $t_2(x) = \alpha_3 + \alpha_4 x + \alpha_5 x^2 \ldots$ the polynomials can be found. However from sections 3.2.2 and 8.3.1 the results are known to be

$$t_i(x) = \sqrt{\varepsilon_i}\,T_i(x/a), \tag{10.29}$$

and

$$u_i(y) = (1/i!)^{1/2} H_i(y/\sigma). \tag{10.30}$$

Thus the orthogonal polynomials $q_{mn}(x, y)$ in the expansion are given by

$$q_{mn}(x, y) = \sqrt{\varepsilon_{m-n}}\,T_{m-n}(x/a)(1/n!)^{1/2} H_n(y/\sigma), \tag{10.31}$$

so that

$$n(x, y) = \sum_{m=0}^{\infty} \sum_{n=0}^{m} c_{mn}\sqrt{\varepsilon_{m-n}}\,T_{m-n}(x/a)(1/n!)^{1/2} H_n(y/\sigma). \tag{10.32}$$

The output autocorrelation function $R_0(\tau)$ is given by

$$R_0(\tau) = \int_{-\infty}^{\infty}\int_{-\infty}^{\infty}\int_{-\infty}^{\infty}\int_{-\infty}^{\infty} n(x_1, y_1)n(x_2, y_2)r(x_1, x_2, \tau) \times q(y_1, y_2, \tau)\,dx_1\,dx_2\,dy_1\,dy_2.$$

Substituting for $n(x, y)$ from equation (10.32), $r(x_1, x_2, \tau)$ from equation (7.107), $q(y_1, y_2, \tau)$ from equation (7.117) and integrating, making use of the orthogonal properties of the polynomials, gives

$$R_0(\tau) = \sum_{m=0}^{\infty} \sum_{n=0}^{m} c_{mn}^2 R_i^n(\tau) \cos\{(m-n)\omega\tau\}, \tag{10.33}$$

where

$$c_{mn} = (\varepsilon_{m-n}/n!)^{1/2} \int_{-\infty}^{\infty}\int_{-a}^{a} n(x, y)T_{m-n}(x/a)H_n(y/\sigma)r(x)q(y)\,dx\,dy. \tag{10.34}$$

Thus

$$c_{s+k,k} = (\varepsilon_s/k!)^{1/2}\alpha_{sk}, \tag{10.35}$$

and if this value of $c_{s+k,k}$ is substituted in equation (10.33) it will be seen to be identical with equation (10.4) obtained for the less general case of $n(x, y)$ equal to $n(x + y)$. Substituting in equations (10.25) and (10.26) the equivalent gains are

$$K_x = (2/a^2) \int_{-\infty}^{\infty}\int_{-a}^{a} x n(x, y)r(x)q(y)\,dx\,dy, \tag{10.36}$$

and

$$K_y = (1/\sigma^2)\int_{-\infty}^{\infty}\int_{-a}^{a} yn(x, y)r(x)q(y)\,\mathrm{d}x\,\mathrm{d}y. \tag{10.37}$$

10.3.3 *Signal and its derivative*

A particular case of some interest occurs when one signal is the derivative of the other,[13] for example if $y = \dot{x}$. Then by definition

$$R_{xy}(0) = R_{x\dot{x}}(0) = \lim_{T\to\infty}(1/2T)\int_{-T}^{T} x(t)\dot{x}(t)\,\mathrm{d}t,$$

which on integrating by parts gives

$$2R_{x\dot{x}}(0) = \lim_{T\to\infty}(1/2T)[x^2(t)]_{-T}^{T} = 0.$$

Thus provided $p(x, \dot{x})$ is known the equivalent gains for this case can be evaluated from equations (10.25) and (10.26). The particular cases of the signal either Gaussian or a sinusoid are considered below.

Gaussian signal

For the Gaussian signal,

$$q(y_1, y_2, \tau) = \frac{1}{2\pi\sigma_1\sigma_2\{1-\rho^2(\tau)\}^{1/2}}$$
$$\times \exp\left[\frac{-1}{2\{1-\rho^2(\tau)\}}\{(y_1/\sigma_1)^2 - (2\rho(\tau)y_1y_2/\sigma_1\sigma_2) + (y_2/\sigma_2)^2\}\right],$$

so that

$$q(y, \dot{y}) = q(y, \dot{y}, 0) = (1/2\pi\sigma_y\sigma_{\dot{y}})\exp[(-1/2)\{(y/\sigma_y)^2 + (\dot{y}/\sigma_{\dot{y}})^2\}]$$
$$= q(y)q(\dot{y}). \tag{10.38}$$

Thus, as in the previous case, the orthogonal polynomial q_{mn} can be written in terms of the orthogonal polynomials for the single variable cases of y and $\dot{y}$ respectively. Hence

$$q_{mn}(y, \dot{y}) = H_{m-n}(y/\sigma_y)H_n(\dot{y}/\sigma_{\dot{y}})/\{(m-n)!n!\}^{1/2}, \tag{10.39}$$

and the expansion for the nonlinearity output is

$$n(y, \dot{y}) = \sum_{m=0}^{\infty}\sum_{n=0}^{m} c_{mn}H_{m-n}(y/\sigma_y)H_n(\dot{y}/\sigma_{\dot{y}})/\{(m-n)!n!\}^{1/2}, \tag{10.40}$$

where

$$c_{mn} = \{1/(m-n)!n!\}^{1/2}\int_{-\infty}^{\infty}\int_{-\infty}^{\infty} n(y, \dot{y})H_{m-n}(y/\sigma_y)H_n(\dot{y}/\sigma_{\dot{y}})q(y)q(\dot{y})\,\mathrm{d}y\,\mathrm{d}\dot{y}. \tag{10.41}$$

The equivalent gains are

$$K_y = (1/\sigma_y^2) \int_{-\infty}^{\infty} \int_{-\infty}^{\infty} y n(y, \dot{y}) q(y) q(\dot{y}) \, dy \, d\dot{y}, \tag{10.42}$$

and

$$K_{\dot{y}} = (1/\sigma_{\dot{y}}^2) \int_{-\infty}^{\infty} \int_{-\infty}^{\infty} \dot{y} n(y, \dot{y}) q(y) q(\dot{y}) \, dy \, d\dot{y}. \tag{10.43}$$

With $y_1 = y(t)$ and $y_2 = y(t + \tau)$ the output autocorrelation function is

$$R_0(\tau) = \int \int_{-\infty}^{\infty} \int \int n(y_1, \dot{y}_1) n(y_2, \dot{y}_2) q(y_1, \dot{y}_1, y_2, \dot{y}_2) \, dy_1 \, d\dot{y}_1 \, dy_2 \, d\dot{y}_2,$$

where[14]

$$q(y_1, \dot{y}_1, y_2, \dot{y}_2) = \frac{1}{(2\pi)^4 \det \mathbf{R}} \exp\{-\mathbf{w}\mathbf{R}^{-1}\mathbf{w}^T/2\},$$

$$\mathbf{w} = (y_1, \dot{y}_1, y_2, \dot{y}_2),$$

and $\mathbf{R}$ is the correlation matrix given by

$$\mathbf{R} = \begin{pmatrix} \sigma_y^2 & 0 & R_{yy}(\tau) & R_{y\dot{y}}(\tau) \\ 0 & \sigma_{\dot{y}}^2 & R_{\dot{y}y}(\tau) & R_{\dot{y}\dot{y}}(\tau) \\ R_{yy}(\tau) & R_{\dot{y}y}(\tau) & \sigma_y^2 & 0 \\ R_{y\dot{y}}(\tau) & R_{\dot{y}\dot{y}}(\tau) & 0 & \sigma_{\dot{y}}^2 \end{pmatrix}.$$

Using a series expansion for $q(y_1, \dot{y}_1, y_2, \dot{y}_2)$ and substituting for n from equation (10.40), it can be shown that the output autocorrelation function is given by

$$R_0(\tau) = \sum_{k=0}^{\infty} \sum_{\substack{p=0 \\ p+q+r+s=k}}^{k} \sum_{q=0}^{k} \sum_{r=0}^{k} \sum_{s=0}^{k}$$

$$\{c_{k,r+s} c_{k,q+s} N_{pqrs} R_{yy}^p(\tau) R_{y\dot{y}}^q(\tau) R_{\dot{y}y}^r(\tau) R_{\dot{y}\dot{y}}^s(\tau)\}/\sigma_y^{2p+q+r} \sigma_{\dot{y}}^{2s+q+r}, \tag{10.44}$$

where

$$N_{pqrs} = \{(p+q)!(r+s)!(p+r)!(q+s)!\}^{1/2}/p!q!r!s!. \tag{10.45}$$

A similar result may also be found using an extension of Price's theorem.[15] It should be noted that equation (10.44) contains crosscorrelation functions unlike the incorrect solution of reference 16 where the invalid assumption

$$q(y_1, \dot{y}_1, y_2, \dot{y}_2) = q(y_1, \dot{y}_1) q(y_2, \dot{y}_2),$$

which is true only for $\tau = 0$, was made.

Sinusoidal signal

When $x = a\cos\omega t$, $(\dot{x}/\omega) = -a\sin\omega t$ and the relationship between x and $\dot{x}/\omega$ is the circle $x^2 + (\dot{x}/\omega)^2 = a^2$. Clearly the joint density $r(x, \dot{x}/\omega)$ is finite on the circle and zero elsewhere, so that it may be written

$$r(x, \dot{x}/\omega) = (v_m/Lv)\delta[\{x^2 + (\dot{x}/\omega)^2\}^{1/2} - a],$$

where

$$L = \text{the length of the circle,}$$
$$v_m = \text{average speed of movement along the circle,}$$
$$v = \text{speed at a point } x, \dot{x}/\omega,$$

and δ denotes the Dirac delta function. Since $v = v_m$ and $L = 2\pi a$,

$$r(x, \dot{x}/\omega) = (1/2\pi a)\delta[\{x^2 + (\dot{x}/\omega)^2\}^{1/2} - a]. \tag{10.46}$$

Now

$$\int_{-a}^{a} r(x, \dot{x}/\omega)\, d(\dot{x}/\omega) = (1/2\pi a)\int_{-a}^{a} \delta[\{x^2 + (\dot{x}/\omega)^2\}^{1/2} - a]\, d(\dot{x}/\omega),$$

which on substituting $r^2 = x^2 + (\dot{x}/\omega)^2$ gives

$$\int_{-a}^{a} r(x, \dot{x}/\omega)\, d(\dot{x}/\omega) = \int_{r_1}^{r_2} (1/2\pi a)\delta(\pm r - a) r(r^2 - x^2)^{-1/2}\, dr$$
$$= \frac{1}{2\pi a}\,\frac{2a}{(a^2 - x^2)^{1/2}} = \frac{1}{\pi(a^2 - x^2)^{1/2}},$$

as required. Similarly one can show that

$$\int_{-a}^{a} r(x, \dot{x}/\omega)\, dx = \frac{1}{\pi\{a^2 - (\dot{x}/\omega)^2\}^{1/2}}.$$

Then using the above expression for $r(x, \dot{x}/\omega)$ the equivalent gains can be found from

$$K_x = (2/a^2)\int_{-a}^{a}\int_{-a}^{a} x n(x, \dot{x}) r(x, \dot{x}/\omega)\, dx\, d(\dot{x}/\omega), \tag{10.47}$$

and

$$K_{\dot{x}} = (2/a^2\omega^2)\int_{-a}^{a}\int_{-a}^{a} \dot{x} n(x, \dot{x}) r(x, \dot{x}/\omega)\, dx\, d(\dot{x}/\omega). \tag{10.48}$$

To determine the orthogonal expansion for the nonlinearity we first note that since

$$\dot{x}^2 = \omega^2(a^2 - x^2),$$

the powers of $\dot{x}$, namely $\dot{x}^m$, for $m > 2$ can be written in the form

$$\dot{x}^m = \sum_{i=0}^{p} (c_{i1}x^i\dot{x} + c_{i0}x^i),$$

where $p = m$ or $(m - 1)$ whichever is even. As a consequence of this linear dependence only the polynomials $q_{m0}(x, \dot{x})$ and $q_{m1}(x, \dot{x})$ as defined earlier in the section are necessary to generate a complete orthonormal set. Using the orthogonal conditions the polynomials may be shown to be

$$q_{m0}(x, \dot{x}) = (\varepsilon_m)^{1/2} T_m(x/a), \tag{10.49}$$

$$q_{m1}(x, \dot{x}) = (\varepsilon_m)^{1/2} T'_m(x/a)(\dot{x}/a\omega m) = (\varepsilon_m)^{1/2} D_m(x/a), \tag{10.50}$$

where

$$T'_m(x/a) = \frac{\mathrm{d}}{\mathrm{d}u}\{T_m(u)\}\bigg|_{u=x/a}. \tag{10.51}$$

Thus

$$n(x, \dot{x}) = \sum_{m=0}^{\infty} c_{m0}(\varepsilon_m)^{1/2} T_m(x/a) + c_{m1}(\varepsilon_m)^{1/2} T'_m(x/a)(\dot{x}/a\omega m), \tag{10.52}$$

with

$$c_{mn} = \int_{-a}^{a}\int_{-a}^{a} n(x, \dot{x}) q_{mn}(x, \dot{x}) r(x, \dot{x}/\omega)\,\mathrm{d}x\,\mathrm{d}(\dot{x}/\omega). \tag{10.53}$$

The output autocorrelation function may be shown, in a similar manner to that used for the Gaussian input, to be

$$R_0(\tau) = \sum_{m=0}^{\infty} (c_{m0}^2 + c_{m1}^2) \cos m\omega\tau,$$

a result which is obvious from harmonic considerations.

From the results given above using orthogonal polynomials in two variables it is easy to see how the method can be applied to nonlinear functions of more than two variables, particularly when several of them are uncorrelated.

10.4 Related sinusoidal signals

In this section we examine in more detail the output of the nonlinearity $n(x, y)$ when x and y are the commensurate sinusoidal signals $x = a\cos\theta_a$, $y = b\cos(\theta_b + \phi)$ with $\theta_a = \omega_a t$, $\theta_b = \omega_b t$ and $\theta_a/\theta_b = m/n$, a ratio of integers. It was seen earlier that since various forms of synchronisation can take place in feedback systems, the evaluation of quasilinear models for such situations is important. The model will have four linear gains K_x, $K_{\dot{x}}$, K_y and $K_{\dot{y}}$ as the DFs for both signals are complex. The DFs and their component linear gains are in general functions of the four parameters a, b, ϕ and m/n. Since there is no simple solution to this problem, we discuss various possible alternative approaches below.

10.4.1 *Direct expansion*

For simple functions, such as the cubic with two additive inputs considered in section 5.7, direct expansion of the output and collection of the desired frequency terms yields the solution.

10.4.2 *Numerical computation*

This is a 'sledge hammer' approach which can always be used but requires a digital computer. For specific values of the parameters a, b, ϕ and m/n the waveform $n(x, y)$ can be evaluated at a sufficient number of equidistant points in the period of the lowest common factor frequency. Fourier analysis of the point representation of $n(x, y)$ will then yield the required DFs.

10.4.3 *Output coefficients* α_{sk}

For two unrelated sinusoids the procedure of section 10.3.2 may be used to show that for the nonlinearity $n(x, y)$ the output autocorrelation function is given by

$$R_0(\tau) = \sum_{s=0}^{\infty} \sum_{k=0}^{\infty} \alpha_{sk}^2 \varepsilon_s \varepsilon_k \cos s\omega_a \tau \cos k\omega_b \tau, \tag{10.9}$$

with

$$\alpha_{sk} = \int_{-b}^{b} \int_{-a}^{a} n(x, y) T_s(x/a) T_k(y/b) r(x) r(y) \, dx \, dy, \tag{10.54}$$

where it can be shown that $\alpha_{sk} = 0$ for $s + k$ even with a symmetrical odd $n(x, y)$. For the particular case of the output being $n(x + y)$ the expression

$$\alpha_{sk} = (1/2\pi) \int_{-\infty}^{\infty} N(j\omega) j^{s+k} J_s(a\omega) J_s(b\omega) \, d\omega \tag{10.10}$$

may also be used. Inspection of equation (10.9) shows that the nonlinearity output may be written

$$n(x, y) = \sum_{s=0}^{\infty} \sum_{k=0}^{\infty} \alpha_{sk} \varepsilon_s \varepsilon_k \cos s(\theta_a + \psi_{as}) \cos k(\theta_b + \psi_{bk}), \tag{10.55}$$

where the angles ψ_{as} and ψ_{bk} are unknown because the process of correlation destroys all phase information.

Consider now the nonlinearity output expressed as a power series that is

$$n(x, y) = \sum_{i=0}^{\infty} \sum_{j=0}^{\infty} c_{ij} x^i y^j,$$

which for the given x and y becomes

$$n(x, y) = \sum_{i=0}^{\infty} \sum_{j=0}^{\infty} c_{ij} \{a \cos \theta_a\}^i \{b \cos(\theta_b + \phi)\}^j.$$

Using trigonometric identities this can be written in the general form

$$n(x, y) = \sum_{s=0}^{\infty} \sum_{k=0}^{\infty} a_{sk} \cos s\theta_a \cos k(\theta_b + \phi), \tag{10.56}$$

which is valid for any ω_a and ω_b. Comparison of equations (10.55) and (10.56) shows that

$$a_{sk} = \varepsilon_s \varepsilon_k \alpha_{sk} \qquad \psi_{as} = 0 \quad \text{and} \quad \psi_{bk} = \phi.$$

Thus, we conclude that the nonlinearity output is

$$n(x, y) = \sum_{s=0}^{\infty} \sum_{k=0}^{\infty} \varepsilon_s \varepsilon_k \alpha_{sk} \cos s\theta_a \cos k(\theta_b + \phi), \tag{10.57}$$

with α_{sk} given by equation (10.54) whether ω_a and ω_b are commensurate or not. If the substitutions $x = a\cos\theta_a$, $y = b\cos\theta_b'$, that is $\theta_b' = \theta_b + \phi$, are made in equation (10.54) we have the alternative expression

$$\alpha_{sk} = (\varepsilon_s \varepsilon_k/\pi^2) \int_0^{\pi} \int_0^{\pi} n(a\cos\theta_a, b\cos\theta_b') \cos s\theta_a \cos k\theta_b' \, d\theta_a \, d\theta_b'. \tag{10.58}$$

When the frequencies are incommensurate each term in equation (10.57) yields a unique frequency, whereas if the frequencies are commensurate a theoretically infinite number of terms yield the same frequency. Thus to determine the equivalent gains for the commensurate case all the terms which contribute to the output at the frequencies $m\omega$ and $n\omega$ are required. The skth term, α_{sk}, yields frequencies $(s\omega_a \pm k\omega_b) = (sm \pm kn)\omega$, and these will be at the frequency $m\omega$ if

$$(sm \pm kn)\omega = \pm m\omega,$$

which since s and k can only take positive values becomes

$$k = (s \pm 1)m/n, \qquad s > 0.$$

With this value of k, the skth term is

$$a_{sk} \cos sm\theta \cos k(n\theta + \phi) = (a_{sk}/2)\{\cos k\phi[\cos m\theta + \cos(2s \pm 1)m\theta] - \sin k\phi[\sin(\mp m\theta) + \sin(2s \pm 1)m\theta]\}.$$

For $s, k \neq 0$ this term has a component of frequency $m\omega$ and amplitude $2\alpha_{sk}\cos k\phi$ in phase with x, and a component of frequency $m\omega$ and amplitude $2\alpha_{sk}\sin k\phi$ in antiphase or phase with $\dot{x}/m\omega$. The total nonlinearity output at frequency $m\omega$ is then obtained by summing all in phase and quadrature terms of the above type. Thus

$$K_x = \frac{1}{a}\{\text{output at frequency } m\omega \text{ in phase with } x\}$$

$$= \frac{2}{a} \sum_{s=0}^{\infty} \sum_{k=I_1(s)} \alpha_{sk} \cos k\phi, \tag{10.59}$$

and

$$K_{\dot{x}/m\omega} = \frac{1}{a}\{\text{output at frequency } m\omega \text{ in phase with } (\dot{x}/m\omega)\}$$

$$= \frac{2}{a}\sum_{s=0}^{\infty}\sum_{k=I_1(s)} \mp \alpha_{sk} \sin k\phi, \tag{10.60}$$

where

$$I_1(s) = \text{positive integers of } (s \pm 1)m/n.$$

The DF, $N_a(a|b)$, is then given by

$$N_a(a|b) = K_x + jK_{\dot{x}/m\omega} = K_x + jm\omega K_{\dot{x}}. \tag{10.61}$$

Following a similar procedure for the $k$sth term with $k = (s \pm 1)n/m$ all the components which contribute to the output at frequency $n\omega$ can be found yielding the results

$$K_y = \frac{2}{b}\sum_{s=0}^{\infty}\sum_{k=I_2(s)} \alpha_{ks} \cos(s \pm 1)\phi, \tag{10.62}$$

$$K_{\dot{y}/n\omega} = \frac{2}{b}\sum_{s=0}^{\infty}\sum_{k=I_2(s)} \mp \alpha_{ks} \sin(s \pm 1)\phi, \tag{10.63}$$

where $I_2(s)$ = positive integers of $(s \pm 1)n/m$. Also the DF, $N_b(a|b)$, is given by

$$N_b(a|b) = K_y + jK_{\dot{y}/n\omega} = K_y + jn\omega K_{\dot{y}}. \tag{10.64}$$

Because of the rapid fall off in the magnitude of the higher order coefficients, α_{sk}, only the first few terms in the above series for the equivalent gains are usually significant. This method is therefore often quite feasible in practice.

10.4.4 *Double Fourier series*

The double Fourier series expansion is given in references 17 and 18. With a nonlinearity input of $a \sin \theta_a + b \sin(\theta_b + \phi)$ the output is sought in the form

$$n\{a \sin \theta_a + b \sin(\theta_b + \phi)\} = \sum_{s=0}^{\infty} \sum_{\substack{k=-\infty (s\neq 0)\\ k=0(s=0)}}^{\infty} P_{sk} \sin(s\theta_a + k\theta_b') + Q_{sk}\cos(s\theta_a + k\theta_b'),$$

where

$$\theta_b' = \theta_b + \phi.$$

The Fourier coefficients P_{sk} and Q_{sk} are found in the normal manner by multiplying both sides of the above equation by the multiplier of the coefficient which is to be found, and integrating the result throughout the square bounded by $\theta_a = \pm\pi, \theta_b' = \pm\pi$. This gives

$$P_{sk} = (1/2\pi^2)\int_{-\pi}^{\pi}\int_{-\pi}^{\pi} n\{a \sin \theta_a + b \sin \theta_b'\} \sin(s\theta_a + k\theta_b')\, d\theta_a\, d\theta_b',$$

and

$$Q_{sk} = (1/2\pi^2) \int_{-\pi}^{\pi} \int_{-\pi}^{\pi} n\{a \sin \theta_a + b \sin \theta'_b\} \cos(s\theta_a + k\theta'_b)\, d\theta_a\, d\theta'_b.$$

For a static single valued nonlinearity it can be shown that $P_{sk} = 0$ for $s + k$ even and $Q_{sk} = 0$ for $s + k$ odd. If, in addition, the nonlinearity is odd it can be proved that all Q_{sk} are zero. The DFs for unrelated inputs are then seen to be P_{10}/a and P_{01}/b respectively.

The above expansion is slightly more complicated than that given in the previous section using α_{sk}, particularly when for related inputs one has to pick out output terms of the same frequency. The reason for this is the choice of sinusoidal input terms, since unlike the cosine function $(\sin \theta)^n$ cannot be expressed in terms of $\sum_{m=0}^{n} a_m \sin m\theta$. If instead $x = a \cos \theta_a$, $y = b \cos \theta'_b$ and an output expansion for $n(x, y)$ is sought in the form of equation (10.56), the result of carrying out the procedure to determine α_{sk} is, of course, equation (10.58). The form of the solution given in the previous section is therefore to be preferred.

10.4.5 *Orthogonal polynomials and Fourier series*

In section 10.3 integral expressions giving the gains K_x and $K_{\dot{x}}$ for the particular case $m = n$ and $\phi = \pi/2$ of the general situation have been obtained. Unfortunately the form of $r(x, \dot{x}/\omega)$ is such that the integrals involved are not always easy to evaluate. Situations when $m = n$, $\phi \neq \pi/2$ can also be handled by this procedure since the input, y, can be expressed in terms of the input, x, as

$$y = (xb/a) \cos \phi + (\dot{x}b/a\omega) \sin \phi,$$

and the nonlinearity output again modelled as $K_x x + K_{\dot{x}} \dot{x}$. For cases when $m \neq n$ (or $m = n$, $\phi = \pi/2$) since $R_{xy}(0) = 0$ the four (or two) linear gains are given by

$$K_x = (2/a^2) \int_{-b}^{b} \int_{-a}^{a} x n(x, y) r(x, y)\, dx\, dy, \tag{10.65}$$

$$K_{\dot{x}} = (2/m^2 a^2 \omega^2) \int_{-b}^{b} \int_{-a}^{a} \dot{x} n(x, y) r(x, y)\, dx\, dy, \tag{10.66}$$

$$K_y = (2/b^2) \int_{-b}^{b} \int_{-a}^{a} y n(x, y) r(x, y)\, dx\, dy, \tag{10.67}$$

and

$$K_{\dot{y}} = (2/n^2 b^2 \omega^2) \int_{-b}^{b} \int_{-a}^{a} \dot{y} n(x, y) r(x, y)\, dx\, dy. \tag{10.68}$$

Since the point x, y traces out a Lissajous figure in the $x - y$ plane, $r(x, y)$ is zero at all points in the plane except those on the trajectory. The trajectory for general values of m and n is extremely complex as is the resulting expression for $r(x, y)$.

Using the variable $z = \cos\theta$, where $\theta_a = m\theta$ and $\theta_b = n\theta$, x and y are given by

$$x = a\cos m\theta = aT_m(z),$$
$$y = b\cos(n\theta + \phi) = b\{T_n(z)\cos\phi + (1 - z^2)^{1/2}\{T_n'(z)/n\}\operatorname{sgn}\dot{z}\sin\phi\},$$

where

$$T_n'(z) = \frac{\mathrm{d}}{\mathrm{d}z}\{T_n(z)\} = nU_{n-1}(z).$$

Developing an expression for $r(x, y)$ and using equation (10.65) with the substitution for x and y in terms of z, it can be shown that[19]

$$K_x = (1/a)\int_{-1}^{1} T_m(z)[n\{aT_m(z), y_+(z)\} + n\{aT_m(z), y_-(z)\}]r(z)\,\mathrm{d}z, \tag{10.69}$$

where

$$y_+(z) = y(z), \quad \text{for } \dot{z} > 0,$$
$$y_-(z) = y(z), \quad \text{for } \dot{z} < 0,$$

and

$$r(z) = 1/\pi(1 - z^2)^{1/2}.$$

Alternatively performing a Fourier analysis of the output expressed as a function of θ, one can obtain the equivalent expression

$$K_x = (1/a\pi)\int_{-\pi}^{\pi} n\{a\cos m\theta, b\cos(n\theta + \phi)\}\cos m\theta\,\mathrm{d}\theta. \tag{10.70}$$

Using the notation

$$n_+(z) = n\{aT_m(z), y_+(z)\},$$

and

$$n_-(z) = n\{aT_m(z), y_-(z)\},$$

the expressions for the other gains are

$$K_{\dot{x}/m\omega} = (1/a\pi)\int_{-1}^{1}\{n_+(z) - n_-(z)\}T_m'(z)\,\mathrm{d}z, \tag{10.71}$$

$$K_y = (1/b^2)\int_{-1}^{1}\{y_+(z)n_+(z) + y_-(z)n_-(z)\}r(z)\,\mathrm{d}z, \tag{10.72}$$

and

$$K_{\dot{y}/n\omega} = (1/b^2\pi)\int_{-1}^{1}\{y_+'(z)n_+(z) - y_-'(z)n_-(z)\}\,\mathrm{d}z, \tag{10.73}$$

where the prime denotes a derivative with respect to z. The expressions similar to equation (10.70) and equivalent to equations (10.71) to (10.73) are obvious.

Although the above equations provide single integral expressions for the gains, closed form solutions for linear segmented characteristics are prohibitive because of the need to break down the range of integration into various ranges, with limits for example dependent on $n_+(z)$ having a certain value.

Example

To illustrate some of the methods outlined above consider the rather trivial example

$$n(x, y) = x + y + \mu(x^2 + y^2)$$

with

$$x = a\cos\theta,\ y = b\cos(2\theta + \phi).$$

Using the output coefficient method, from equation (10.54) we have

$$\alpha_{00} = \mu(a^2 + b^2)/2 \qquad \alpha_{11} = \mu ab/2$$

$$\alpha_{10} = a/2 \qquad \alpha_{20} = \mu a^2/4$$

$$\alpha_{01} = b/2 \qquad \alpha_{02} = \mu b^2/4$$

with all other $\alpha_{ij} = 0$. Applying equation (10.59), with $m = 1, n = 2$

$$K_x = (2/a)\{\alpha_{11}\cos\phi + \alpha_{10}\} = 1 + \mu b\cos\phi.$$

Similarly

$$K_{\dot{x}/\omega} = (2/a)\{\alpha_{11}\sin\phi\} = \mu b\sin\phi$$

$$K_y = (2/b)\{\alpha_{20}\cos\phi + \alpha_{01}\} = 1 + (\mu a^2/2b)\cos\phi$$

$$K_{\dot{y}/2\omega} = (2/b)\{\alpha_{20}\sin -\phi\} = -(\mu a^2/2b)\sin\phi.$$

Alternatively in terms of $z = \cos\theta$

$$x = az$$

$$y_\pm = b\{(2z^2 - 1)\cos\phi \pm 2z(1 - z^2)^{1/2}\sin\phi\}$$

and substituting in equation (10.69)

$$K_x = (1/a)\int_{-1}^{1} z\{x + y_+ + \mu(x + y_+)^2 + x + y_- + \mu(x + y_-)^2\}r(z)\,dz$$

but

$$\{x + y_+ + \mu(x + y_+)^2 + x + y_- + \mu(x + y_-)^2\}$$
$$= 2x^2\mu + 2x + (y_+ + y_-)(1 + 2x\mu) + \mu(y_+^2 + y_-^2)$$
$$= 2\mu a^2z^2 + 2az + (1 + 2a\mu z)2b(2z^2 - 1)\cos\phi + 2b^2(2z^2 - 1)^2\mu$$
$$\times \cos^2\phi + 8\mu b^2z^2(1 - z^2)\sin^2\phi,$$

which on substituting in the above formula and integrating gives

$$K_x = (1/a)[2a(1/2) + 4ab\mu\{2(1/2)(3/4) - (1/2)\}\cos\phi] = 1 + \mu b\cos\phi.$$

Also, from equation (10.70),

$$K_x = (1/a\pi)\int_{-\pi}^{\pi} [a\cos\theta + b\cos(2\theta+\phi) + \mu\{a^2\cos^2\theta + 2ab\cos\theta\cos(2\theta+\phi)$$
$$+ b^2\cos^2(2\theta+\phi)\}]\cos\theta\,d\theta$$
$$= (1/a\pi)\{a\pi + \mu 2ab(\pi/2)\cos\phi\} = 1 + \mu b\cos\phi.$$

Similarly the results for $K_{\dot{x}/\omega}$, K_y and $K_{\dot{y}/2\omega}$ may also be found to agree with the values given above.

10.5 The equivalent gains of the multivariable nonlinearity

A multivariable nonlinearity, or as it is often called, a multidimensional nonlinearity is shown in figure 10.1. The nonlinearity has m inputs and r outputs,

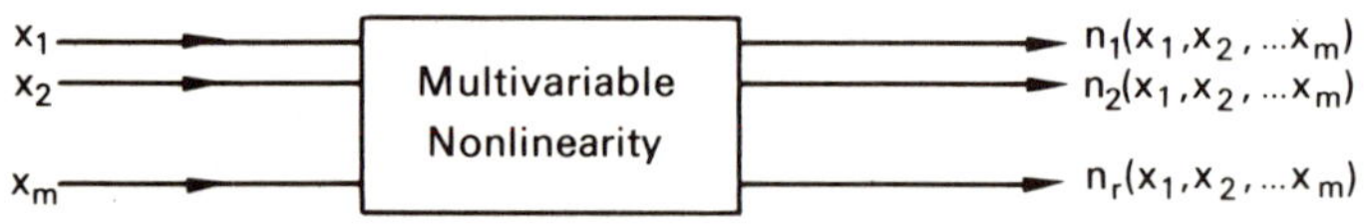

Figure 10.1 The multivariable nonlinearity

with each of the r outputs a function of the m inputs. As the number of outputs only adds to the complexity of the quasilinear model, without causing any other additional problems we consider the single output $n(x_1, x_2, \ldots, x_m)$ only. In the succeeding sections, following reference 15, we determine the equivalent gain model and examine its significance when the inputs are completely separable. It is possible[15] to derive the autocorrelation function of the output $n(x_1, x_2, \ldots, x_m)$ for various combinations of sinusoidal and Gaussian inputs x_i using techniques similar to those considered in the previous sections, however since the results are rather cumbersome and in many cases obvious from earlier considerations the interested reader is referred to the reference cited. Also, of course, autocorrelation functions are not normally required in a quasilinear analysis; hopefully however they give some indication of the distortion term magnitudes and therefore of the accuracy of the analysis.

Let the equivalent gains of the quasilinear model shown in figure 10.2 be $K_{x_1}, K_{x_2}, \ldots, K_{x_m}$ for the m inputs $x_1, x_2, \ldots, x_m$. Then the error e is given by

$$e = n(x_1, x_2, \ldots, x_m) - \sum_{j=1}^{m} K_{x_j} x_j,$$

and its mean squared value is

$$E(e^2) = E\left[\left\{n(x_1, x_2, \ldots, x_m) - \sum_{j=1}^{m} K_{x_j} x_j\right\}^2\right],$$

where E denotes the time or ensemble average since the inputs are assumed to be ergodic random processes. $E(e^2)$ is a minimum when $\{\partial E(e^2)/\partial K_{x_i}\} = 0$ for

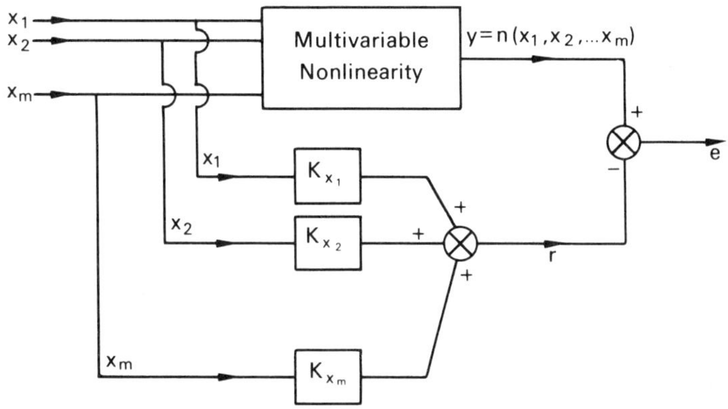

Figure 10.2 The equivalent gain model of a multivariable nonlinearity

$i = 1, 2, \ldots, m$, that is

$$\partial E(e^2)/\partial K_{x_i} = E\left[2\left\{n(x_1, x_2, \ldots, x_m) - \sum_{j=1}^{m} K_{x_j}x_j\right\}(-x_i)\right], \tag{10.74}$$

which may be written as

$$E\{n(x_1, x_2, \ldots, x_m)x_i\} = \sum_{j=1}^{m} K_{x_j}E(x_jx_i),$$

that is

$$E(x_in) = \sum_{j=1}^{m} K_{x_j}R_{x_ix_j}(0) \quad \text{for } i = 1, 2, \ldots, m.$$

Using matrix notation this becomes

$$\mathbf{R}_{xn} = \mathbf{RK},$$

where

$$\mathbf{R}_{xn} = \{E(x_1n), E(x_2n), \ldots, E(x_mn)\}^T,$$

$$\mathbf{R} = \begin{pmatrix} R_{x_1x_1}(0) & R_{x_1x_2}(0) & \ldots & R_{x_1x_m}(0) \\ R_{x_2x_1}(0) & R_{x_2x_2}(0) & \ldots & R_{x_2x_m}(0) \\ \vdots & \vdots & \vdots & \vdots \\ R_{x_mx_1}(0) & R_{x_mx_2}(0) & \ldots & R_{x_mx_m}(0) \end{pmatrix},$$

and $\mathbf{K} = (K_{x_1}, K_{x_2}, \ldots, K_{x_m})^T$, with the solution $\mathbf{K} = \mathbf{R}^{-1}\mathbf{R}_{xn}$. Alternatively, we have

$$K_{x_i} = \sum_{j=1}^{m} (\Delta_{ij}/\Delta)E(x_jn), \tag{10.75}$$

where

$$\Delta = \det \mathbf{R},$$

and Δ_{ij} is the cofactor of $R_{x_ix_j}(0)$ in $\mathbf{R}$. It should be noted that equation (10.74) implies

$$R_{x_ie}(0) = E\left[x_i\left\{n - \sum_{j=1}^{m} K_{x_j}x_j\right\}\right] = 0.$$

That is, the m inputs $x_1, x_2, \ldots, x_m$ are all uncorrelated with the error e for zero time shift, when the equivalent gains are chosen according to equation (10.75). The above result can also be obtained using orthogonal polynomials in m dimensions.[10] Equation (10.75) for the equivalent gains was first given by Kazakov.[20] The result, however, has been known for a long time in statistics[21] where K_{x_i} is the regression coefficient of x_i on $n(x_1, x_2, \ldots, x_m)$.

10.6 Completely separable random processes

Using the notation $x_i'(t) = x_i(t + \tau)$, the process x_1, as defined by Nuttall[22] (see section 8.3), is said to be separable if

$$\int_{-\infty}^{\infty} x_1 p(x_1, x_1')\,dx_1 = p(x_1')x_1' R_{x_1x_1}(\tau)/R_{x_1x_1}(0). \tag{10.76}$$

This concept of separability of a process can be generalised to include m processes.[15] Consider the m processes $x_1, x_2, \ldots, x_m$ with joint probability density $p(x_1, x_2, \ldots, x_m)$, then the process x_k is defined to be separable with respect to the m processes if

$$\int_{-\infty}^{\infty} x_k p(x_k, x_1', x_2', \ldots, x_m')\,dx_k = p(x_1', x_2', \ldots, x_m') \sum_{i=1}^{m}\sum_{j=1}^{m} \alpha_{ij} x_j' R_{x_kx_i}(\tau),$$

or

$$\int_{-\infty}^{\infty} x_k p_{m+1}\,dx_k = p_m \sum_{i=1}^{m}\sum_{j=1}^{m} \alpha_{ij} x_j' R_{x_kx_i}(\tau), \tag{10.77}$$

where α_{ij} are constants defined below. If in addition equation (10.77) is true for all k then the processes $x_1, x_2, \ldots, x_m$ are defined to be completely separable.

It should be noted that this definition of separability differs from that of Nuttall. Nuttall extended equation (10.76) for the single process to cover the case of two processes by defining the process x_1 to be separable with respect to the process x_2 if

$$\int_{-\infty}^{\infty} x_1 p(x_1, x_2')\,dx_1 = p(x_2')x_2' R_{x_1x_2}(\tau)/R_{x_1x_2}(0). \tag{10.78}$$

Here the process x_1 is said to be separable with respect to both x_1 and x_2 if

$$\int_{-\infty}^{\infty} x_1 p(x_1, x_1', x_2')\,dx_1 = p(x_1', x_2') \sum_{i=1}^{2}\sum_{j=1}^{2} \alpha_{ij} x_j' R_{x_1x_i}(\tau). \tag{10.79}$$

Finally if x_2 is also separable with respect to both x_1 and x_2 the processes x_1 and x_2 are completely separable.

Determination of α_{ij}

Consider $R_{x_k x_n}(\tau) = E(x_k x_n')$, the crosscorrelation function between the kth and nth signals which is given by

$$R_{x_k x_n}(\tau) = \int_{-\infty\,m+1}^{\infty} \int_{-\infty}^{\infty} x_k x_n' p_{m+1}\, \mathrm{d}x_k\, \mathrm{d}x_1 \ldots \mathrm{d}x_m'.$$

If x_k is separable with respect to the m processes then the above equation can be rewritten using equation (10.77) as

$$R_{x_k x_n}(\tau) = \int_{-\infty\,m}^{\infty} \int_{-\infty}^{\infty} x_n' p_m \left\{ \sum_{i=1}^{m} \sum_{j=1}^{m} \alpha_{ij} x_j' R_{x_k x_i}(\tau) \right\} \mathrm{d}x_1'\, \mathrm{d}x_2' \ldots \mathrm{d}x_m',$$

which on evaluating the integrals gives

$$R_{x_k x_n}(\tau) = \sum_{i=1}^{m} \sum_{j=1}^{m} \alpha_{ij} R_{x_j x_n}(0) R_{x_k x_i}(\tau).$$

Since this equation is valid for all τ,

$$\sum_{j=1}^{m} \alpha_{ij} R_{x_j x_n}(0) = \delta_{in}$$

for all $i, j, n = 1, 2, \ldots, m$. For a particular i this equation is

$$\alpha_{i1} R_{x_1 x_n}(0) + \alpha_{i2} R_{x_2 x_n}(0) + \cdots + \alpha_{im} R_{x_m x_n}(0) = \delta_{in}.$$

If n is now allowed to take the values $1, 2, \ldots, m$, the above equation produces a set of m simultaneous equations with m unknowns α_{ij}, $j = 1, 2, \ldots, m$. The solution of the set is

$$\alpha_{ij} = \Delta_{ij}/\Delta \quad \text{for } j = 1, 2, \ldots, m,$$

where Δ_{ij} and Δ are as previously defined. It is evident that the above is true for every $i = 1, 2, \ldots, m$. Hence

$$\alpha_{ij} = \Delta_{ij}/\Delta \quad \text{for all } i, j = 1, 2, \ldots, m. \tag{10.80}$$
$$= (\mathbf{R}^{-1})_{i,j} \quad \text{since } \mathbf{R} \text{ is symmetric.}$$

Thus using this value of α_{ij} in equation (10.77), the process x_k is defined to be separable with respect to all the m processes $x_1, x_2, \ldots, x_m$ if

$$\int_{-\infty}^{\infty} x_k p_{m+1}\, \mathrm{d}x_k = p_m \sum_{i=1}^{m} \sum_{j=1}^{m} (\Delta_{ij}/\Delta) x_j' R_{x_k x_i}(\tau). \tag{10.81}$$

Since the characteristic function is the Fourier transform of the probability density function, it can be shown that the equivalent equation (10.81) for x_k separable with respect to the m processes is

$$\left.\frac{\partial M_{m+1}}{\partial \omega_k}\right|_{\omega_k=0} = \sum_{i=1}^{m} \sum_{j=1}^{m} (\Delta_{ij}/\Delta) \frac{\partial M_m}{\partial \omega_j'} R_{x_k x_i}(\tau), \tag{10.82}$$

where the characteristic function

$$M_{m+1} = \int_{-\infty\, m+1}^{\infty} \cdots \int_{-\infty}^{\infty} p_{m+1} \exp\left\{ j\omega_k x_k + j \sum_{i=1}^{m} \omega_i' x_i' \right\} \mathrm{d}x_k \, \mathrm{d}x_1' \, \mathrm{d}x_2' \ldots \mathrm{d}x_n'.$$

Using the above definition of equations (10.81) or (10.82) it can be shown that two examples of completely separable processes are m jointly Gaussian processes and the two processes $x, \dot{x}$ where $x = a \cos \omega t$. In addition, if the processes $x_1, x_2, \ldots, x_k$ and $x_{k+1}, x_{k+2}, \ldots, x_{k+m}$ are each completely separable and if the random variable x_i for $i \leq k$ is statistically independent of x_j for $k < j < k + m$ then the processes $x_1, x_2, \ldots, x_{k+m}$ are completely separable.

10.6.1 *Input error crosscorrelation*

The crosscorrelation between x_k and e in figure 10.2 is

$$R_{x_k e}(\tau) = \int_{-\infty\, m+1}^{\infty} \cdots \int_{-\infty}^{\infty} x_k \left\{ n(x_1', x_2', \ldots, x_m') - \sum_{i=1}^{m} K_{x_i} x_i' \right\} p_{m+1} \, \mathrm{d}x_k \, \mathrm{d}x_1' \ldots \mathrm{d}x_n'.$$

If x_k is separable with respect to $x_1, x_2, \ldots, x_m$ this reduces to

$$\begin{aligned} R_{x_k e}(\tau) &= \int_{-\infty\, m}^{\infty} \cdots \int_{-\infty}^{\infty} n' p_m \left\{ \sum_{i=1}^{m} \sum_{j=1}^{m} (\Delta_{ij}/\Delta) x_j' R_{x_k x_i}(\tau) \right\} \mathrm{d}x_1' \ldots \mathrm{d}x_m' - \sum_{i=1}^{m} K_{x_i} R_{x_k x_i}(\tau) \\ &= \sum_{i=1}^{m} \left\{ \sum_{j=1}^{m} (\Delta_{ij}/\Delta) E(x_j n) - K_{x_i} \right\} R_{x_k x_i}(\tau) \\ &= 0 \end{aligned}$$

for all τ since

$$K_{x_i} = \sum_{j=1}^{m} (\Delta_{ij}/\Delta) E(x_j n) \quad \text{by choice.}$$

Alternatively if it is required that $R_{x_k e}(\tau) = 0$ for all τ, then

$$R_{x_k e}(\tau) = E\left\{ x_k \left(n' - \sum_{i=1}^{m} K_{x_i} x_i' \right) \right\} = 0,$$

that is

$$E(x_k n') - \sum_{i=1}^{m} K_{x_i} R_{x_k x_i}(\tau) = 0.$$

Substituting for K_{x_i} from equation (10.75) gives

$$\int_{-\infty\, m+1}^{\infty} \cdots \int_{-\infty}^{\infty} x_k n' p_{m+1} \, \mathrm{d}x_k \, \mathrm{d}x_1' \ldots \mathrm{d}x_m' - \sum_{i=1}^{m} \sum_{j=1}^{m} (\Delta_{ij}/\Delta) E(x_j' n') R_{x_k x_i}(\tau) = 0,$$

that is

$$\int_{-\infty\, m}^{\infty} \cdots \int_{-\infty}^{\infty} n' \, \mathrm{d}x_1' \ldots \mathrm{d}x_m' \left\{ \int_{-\infty}^{\infty} x_k p_{m+1} \, \mathrm{d}x_k - p_m \sum_{i=1}^{m} \sum_{j=1}^{m} (\Delta_{ij}/\Delta) x_j' R_{x_k x_i}(\tau) \right\} = 0.$$

Since this must be satisfied for all n and τ,

$$\int_{-\infty}^{\infty} x_k p_{m+1}\, dx_k = p_m \sum_{i=1}^{m} \sum_{j=1}^{m} (\Delta_{ij}/\Delta) x_j' R_{x_k x_i}(\tau),$$

or x_k is separable with respect to all the m processes. Thus with the equivalent gains chosen according to equation (10.75) a necessary and sufficient condition for $R_{x_k e}(\tau)$ to be zero for all τ is that x_k be separable with respect to all the m processes.

As a consequence of the above result, one can readily show with respect to figure 10.2 that

$$R_{yy}(\tau) = R_{rr}(\tau) + R_{ee}(\tau),$$

and

$$R_{re}(\tau) = 0 \quad \text{for all } \tau,$$

whenever x_k is separable with respect to $x_1, x_2, \ldots, x_m$ for all k.

10.6.2 *The optimum linear system*

A necessary and sufficient condition for $\mathbf{h}(t)$ to be the impulse response of the optimum realisable linear model of the multivariable nonlinearity in the Wiener sense[23] is

$$\mathbf{R}_{xy}(\tau) - \int_{-\infty}^{\infty} \mathbf{R}_{xx}(\tau - t)\mathbf{h}^T(t)\, dt = 0 \quad \text{for } \tau > 0, \tag{10.83}$$

where

$$\mathbf{R}_{xx}(\tau) = \begin{pmatrix} R_{x_1x_1}(\tau) & R_{x_1x_2}(\tau) & \cdots & R_{x_1x_m}(\tau) \\ R_{x_2x_1}(\tau) & R_{x_2x_2}(\tau) & \cdots & R_{x_2x_m}(\tau) \\ \vdots & \vdots & \vdots & \vdots \\ R_{x_mx_1}(\tau) & R_{x_mx_2}(\tau) & \cdots & R_{x_mx_m}(\tau) \end{pmatrix},$$

and

$$\mathbf{R}_{xy}(\tau) = \{R_{x_1y}(\tau), R_{x_2y}(\tau), \ldots, R_{x_my}(\tau)\}^T.$$

Considering the left hand side of equation (10.83) with the substitution

$$\mathbf{h}(t) = \{K_{x_1}\, \delta(t), K_{x_2}\, \delta(t), \ldots, K_{x_n}\, \delta(t)\},$$

one has

$$\begin{aligned}
&\mathbf{R}_{xy}(\tau) - \int_{-\infty}^{\infty} \mathbf{R}_{xx}(\tau - t)\mathbf{h}^T(t)\, dt \\
&= \begin{pmatrix} R_{x_1y}(\tau) - K_{x_1}R_{x_1x_1}(\tau) - K_{x_2}R_{x_1x_2}(\tau) & \cdots & -K_{x_m}R_{x_1x_m}(\tau) \\ R_{x_2y}(\tau) - K_{x_1}R_{x_2x_1}(\tau) - K_{x_2}R_{x_2x_2}(\tau) & \cdots & -K_{x_m}R_{x_2x_m}(\tau) \\ \vdots & \vdots & \vdots \\ R_{x_my}(\tau) - K_{x_1}R_{x_mx_1}(\tau) - K_{x_2}R_{x_mx_2}(\tau) & \cdots & -K_{x_m}R_{x_mx_m}(\tau) \end{pmatrix} \\
&= \{R_{x_1e}(\tau), R_{x_2e}(\tau), \ldots, R_{x_me}(\tau)\}^T = 0,
\end{aligned}$$

if the inputs are completely separable. Thus, the equivalent gain model is the best linear representation of the multivariable nonlinearity if and only if the inputs are completely separable.

10.7 The modified nonlinearity

The concept of a modified nonlinearity can be readily extended to the case of a multivariable nonlinearity. Consider such a nonlinearity $n(x_1, x_2, \ldots, x_p, x_{p+1}, \ldots, x_q)$ with $p + q$ inputs. If the p inputs $x_1, x_2, \ldots, x_p$ are statistically independent of the remaining q inputs $x_{p+1}, x_{p+2}, \ldots, x_{p+q}$ then

$$R_{x_ix_j}(\tau) = 0 \quad \text{for all } \tau, \quad \text{for i} \leq p, p + q \geq j > p,$$

so that

$$\mathbf{R} = \left(\begin{array}{cccc|ccc} R_{x_1x_1}(0) & \cdots & R_{x_1x_p}(0) & & & & \\ R_{x_2x_1}(0) & \cdots & R_{x_2x_p}(0) & & & 0 & \\ \vdots & & \vdots & & & & \\ R_{x_px_1}(0) & \cdots & R_{x_px_p}(0) & & & & \\ \hline & & & & R_{x_{p+1}x_{p+1}}(0) & \cdots & R_{x_{p+1}x_{p+q}}(0) \\ & 0 & & & \vdots & \vdots & \vdots \\ & & & & R_{x_{p+q}x_{p+1}}(0) & \cdots & R_{x_{p+q}x_{p+q}}(0) \end{array}\right)$$

$$= \left(\begin{array}{c|c} \mathbf{R}_1 & 0 \\ \hline 0 & \mathbf{R}_2 \end{array}\right).$$

The $p + q$ equivalent gains are given by

$$Kx_i = \sum_{j=1}^{p+q} (\Delta_{ij}/\Delta)E(x_jn).$$

It is easy to check for the above $\mathbf{R}$ that

$$\Delta_{ij} = 0 \quad \text{for } i > p, \quad j \leq p, \quad \text{and for } i \leq p, \quad j > p,$$

as a consequence of which it follows that

$$K_{x_i} = \sum_{j=1}^{p} (\Delta^1_{ij}/\Delta^1)E(x_jn) \qquad \text{for } i \leq p$$

$$\sum_{j=1}^{q} (\Delta^2_{i-p,j}/\Delta^2)E(x_{p+j}n) \quad \text{for } i > \text{p},$$

where Δ^1_{ij} and Δ^2_{ij} are the cofactors of the $i - j$th elements of $\mathbf{R}_1$ and $\mathbf{R}_2$ respectively,

$$\Delta^1 = \det \mathbf{R}_1 \quad \text{and} \quad \Delta^2 = \det \mathbf{R}_2.$$

However

$$E(x_jn) = \int_{-\infty_{p+q}}^{\infty} \cdots \int_{-\infty}^{\infty} x_jn(x_1, \ldots, x_{p+q})p_pp_q\, dx_1 \ldots dx_{p+q},$$

so that for $j \leq p$,

$$E(x_j n) = \left\{\int_{-\infty_p}^{\infty} \cdots \int_{-\infty}^{\infty} x_j p_p \, dx_1 \ldots dx_p\right\}\left\{\int_{-\infty_q}^{\infty} \cdots \int_{-\infty}^{\infty} n p_q \, dx_{p+1} \ldots dx_{p+q}\right\}.$$

Therefore defining

$$n_{mq}(x_1, x_2, \ldots, x_p) = \int_{-\infty_q}^{\infty} \cdots \int_{-\infty}^{\infty} n p_q \, dx_{p+1} \ldots dx_{p+q}$$

as the modified nonlinearity, one obtains

$$K_{x_i} = \sum_{j=1}^{p} (\Delta_{ij}^1/\Delta^1) E(x_j n_{mq}) \quad \text{for } i \leq p.$$

By a similar reasoning,

$$K_{x_i} = \sum_{j=1}^{q} (\Delta_{i-p,j}^2/\Delta^2) E(x_{p+j} n_{mp}) \quad \text{for } i > p,$$

where

$$n_{mp}(x_{p+1}, x_{p+2}, \ldots, x_{p+q}) = \int_{-\infty_p}^{\infty} \cdots \int_{-\infty}^{\infty} n p_p \, dx_1 \, dx_2 \ldots dx_p$$

is the nonlinearity as modified by the p inputs $x_1, x_2, \ldots, x_p$. It will be appreciated that the above modified nonlinearities are entirely consistent with the earlier formulation where two additive inputs only were considered and the introduction of the bias γ assisted in the physical interpretation. A consequence of the above theorem is that it provides a very simple conceptual method for evaluating multiple input describing functions for any number of dependent or independent inputs. Although in some cases the integrations involved are difficult, approximate solutions can always be obtained. Further the formulation is such that not all the variables need necessarily be input signals to the nonlinearity, they can for example be internal parameters such as time varying coefficients.

10.7.1 *D.c. incremental gain*

It is possible to extend the d.c. incremental gain result of section 8.4.1 for a Gaussian input to the multivariable case. By definition, the d.c. incremental gain $g(\gamma_i)$ is given by

$$\begin{aligned} g(\gamma_i) &= \frac{\partial}{\partial \gamma_i} \int_{-\infty_m}^{\infty} \cdots \int_{-\infty}^{\infty} n(y_1, y_2, \ldots, y_i + \gamma_i, \ldots, y_m) q(y_1 \ldots y_m) \, dy_1 \ldots dy_m \\ &= E(\partial n/\partial \gamma_i) = E(\partial n/\partial y_i). \end{aligned}$$

That is

$$\begin{aligned} g(\gamma_i) &= \int_{-\infty_m}^{\infty} \cdots \int_{-\infty}^{\infty} \frac{\partial n}{\partial y_i}(y_1 \ldots y_m) q(y_1 \ldots y_m) \, dy_1 \ldots dy_m \\ &= -\int_{-\infty_m}^{\infty} \cdots \int_{-\infty}^{\infty} n(y_1 \ldots y_m) \frac{\partial q}{\partial y_i}(y_1 \ldots y_m) \, dy_1 \ldots dy_m. \end{aligned}$$

From equation (7.115)

$$q(y_1 \dots y_m) = \{(2\pi)^m \Delta\}^{-1/2} \quad \exp\left\{-\sum_{i,j=1}^{m} \Delta_{ij} y_i y_j / 2\Delta\right\}.$$

Therefore

$$\frac{\partial q}{\partial y_i} = q(y_1 \dots y_m) \frac{\partial}{\partial y_i}\left\{-\sum_{i,j=1}^{m} \Delta_{ij} y_i y_j / 2\Delta\right\}$$

$$= q(y_1 \dots y_m) \sum_{j=1}^{m} -\Delta_{ij} y_j / \Delta \quad \text{since } \Delta_{ij} = \Delta_{ji}.$$

Substituting in the equation for $g(\gamma_i)$ gives

$$g(\gamma_i) = \int_{-\infty m}^{\infty} \int_{-\infty}^{\infty} n(y_1 \dots y_m) q(y_1 \dots y_m) \sum_{j=1}^{m} (\Delta_{ij} y_j / \Delta)\, \mathrm{d}y_1 \dots \mathrm{d}y_m$$

$$= \sum_{j=1}^{m} (\Delta_{ij}/\Delta) E(n y_j) = K_{y_i}.$$

That is

$$K_{y_i} = E(\partial n / \partial y_i) = g(\gamma_i). \tag{10.84}$$

This result is of considerable utility. Although in the above theory all the inputs have been assumed to be jointly normally distributed it is also possible to show that if p inputs are jointly normally distributed and statistically independent of the other $(m - p)$ inputs the equivalent gains to the normally distributed inputs can be found from $K_{y_i} = E(\partial n/\partial y_i)$, $i \leq p$. The expression of equation (10.84) is often easier to evaluate than equation (10.75), whilst experimental measurement of $g(\gamma_i)$ is relatively straightforward. If the output of a multivariable nonlinearity is given by $n(\sum_{i=1}^{p} y_i, x_{p+1}, \dots, x_m)$, the equivalent gains K_{y_i} for $i \leq p$ are all the same and equal to $E(\partial n/\partial y_i)$.

10.7.2 *The gain to any small unrelated input*

In section 5.2 it was stated without proof that the gain to any small additional unrelated input to a nonlinearity was independent of the form of the input. Physically for the case of a small input in the presence of another signal this gain was interpreted as the slope of the modified nonlinearity at the origin. Here we prove the general case for the nonlinearity whose output is $n(y_1, y_2, \dots, y_m)$. If a small unrelated input, δ, is added to y_i then the gain to δ is given by

$$K_\delta = (1/\mu_2) \int_{-\infty m+1}^{\infty} \int_{-\infty}^{\infty} \delta n(y_1, y_2, \dots, y_i + \delta, \dots, y_m) p(\delta) q(y_1, y_2, \dots, y_m)$$
$$\mathrm{d}\delta\, \mathrm{d}y_1 \dots \mathrm{d}y_m,$$

where μ_2 is the mean squared value of δ.

Expanding the nonlinear function in a Taylor series and neglecting terms of order δ^2 gives

$$K_\delta = (1/\mu_2)\int_{-\infty\, m+1}^{\infty}\cdots\int_{-\infty}^{\infty}\delta\left\{n+\delta\frac{\partial n}{\partial y_i}\right\}p(\delta)q(y_1\ldots y_m)\,d\delta\,dy_1\ldots dy_m,$$

where $n = n(y_1, y_2, \ldots, y_m)$.

Integrating with respect to $p(\delta)$ gives

$$K_\delta = \int_{-\infty}^{\infty}\cdots\int_{-\infty}^{\infty}\frac{\partial n}{\partial y_i}q(y_1,\ldots,y_m)\,dy_1\ldots dy_m, \tag{10.85}$$

if δ has zero mean. Thus

$$K_\delta = g(0) = E(\partial n/\partial y_i), \tag{10.86}$$

so that the gain to the small input δ is equal to the gain $\lim_{\gamma_i\to 0} g(\gamma_i)$ for a small bias input γ_i.

10.8 Some additional two input results

The purpose of this section is to provide additional results on two input nonlinearity characterisations, some of which are used by examples in later chapters. First, approximate methods for the evaluation of two input DFs, which are particularly useful when one input is much smaller than the other, are considered. Next formulae are derived for IDFs in the presence of two other inputs. It was seen in section 5.11.2 how such DFs are required to evaluate the stability of combined mode periodic solutions. Finally the orthogonal expansion for a nonlinearity with two jointly Gaussian correlated inputs is considered. The result is particularly useful for estimating the distortion in nonlinear multiple input systems where such correlated signals occur.

10.8.1 *Series expansions for the evaluation of two input DFs*

When the input to a nonlinearity is the sum of two signals it is possible to obtain series expansions for the DFs. Consider the two inputs to be x and y, then

$$K_x = \{1/\mu_2(a)\}\int_{-\infty}^{\infty} xn(x+y)p(x)q(y)\,dx\,dy, \tag{10.87}$$

and

$$K_y = \{1/\mu_2(b)\}\int_{-\infty}^{\infty} yn(x+y)p(x)q(y)\,dx\,dy, \tag{10.88}$$

where the signals x and y have amplitude probability densities $p(x)$ and $q(y)$ with second moments $\mu_2(a)$ and $\mu_2(b)$ respectively. If x is assumed to be the larger signal and $n(x+y)$ is expanded in the Taylor series,

$$n(x+y) = \sum_{k=0}^{\infty} (y^k/k!)(\mathrm{d}^k n(x)/\mathrm{d}x^k). \tag{10.89}$$

On substituting in equations (10.87) and (10.88) we obtain

$$K_x = \{1/\mu_2(a)\} \sum_{k=0}^{\infty} (\mu_k(b)/k!)A_k(a), \tag{10.90}$$

and

$$K_y = \{1/\mu_2(b)\} \sum_{k=0}^{\infty} (\mu_{k+1}(b)/k!)B_k(a), \tag{10.91}$$

where $\mu_k(b)$ is the kth moment of y,

$$A_k(a) = \int_{-\infty}^{\infty} x(\mathrm{d}^k n(x)/\mathrm{d}x^k)p(x)\,\mathrm{d}x \tag{10.92}$$

and

$$B_k(a) = \int_{-\infty}^{\infty} (\mathrm{d}^k n(x)/\mathrm{d}x^k)p(x)\,\mathrm{d}x. \tag{10.93}$$

Since

$$\mathrm{d}^k n(x)/\mathrm{d}x^k = \lim_{\gamma\to 0}(\partial^k n(x+\gamma)/\partial\gamma^k),$$

the expressions for $A_k(a)$ and $B_k(a)$ may also be written as

$$\begin{aligned} A_k(a) &= \lim_{\gamma\to 0}\int_{-\infty}^{\infty} x(\partial^k n(x+\gamma)/\partial\gamma^k)p(x)\,\mathrm{d}x \\ &= \lim_{\gamma\to 0}\{a(\partial^k n(a,\gamma)_1/\partial\gamma^k), \end{aligned} \tag{10.94}$$

where $n(a,\gamma)_1$ is the first modified nonlinearity defined by

$$n(a,\gamma)_1 = \int_{-\infty}^{\infty} n(x+\gamma)(x/a)p(x)\,\mathrm{d}x, \tag{10.95}$$

and

$$B_k(a) = \lim_{\gamma\to 0}\{\partial^k n(a,\gamma)/\partial\gamma^k\} \tag{10.96}$$

where the modified nonlinearity $n(a,\gamma)$ is given by

$$n(a,\gamma) = \int_{-\infty}^{\infty} n(x+\gamma)p(x)\,\mathrm{d}x. \tag{10.97}$$

Thus for any given inputs the series of equations (10.90) and (10.91) can be evaluated using the values of $A_k(a)$ and $B_k(a)$ obtained from either equations (10.92) and (10.93) or (10.94) and (10.96). For the usual situation where $q(y)$ is even and $n(x)$ odd, equations (10.90) and (10.91) contain only $A_k(a)$ for k even and $B_k(a)$ for k odd respectively. Also the first terms in the series are seen to be

$$A_0(a) = \mu_2(a)N(a), \tag{10.98}$$

and

$$B_1(a) = N_{i\gamma}(a). \tag{10.99}$$

For specific forms of $p(x)$ it may be possible to evaluate $A_k(a)$ and $B_k(a)$ more rapidly from differential recursion relationships. The two important cases of x, a sinusoid or Gaussian signal, are considered below.

(i) x sinusoidal

Here $p(x) = 1/\{\pi(a^2 - x^2)^{1/2}\}$, and the transform domain equivalents of equations (10.92) and (10.93) are

$$A_k(a) = (ja/2\pi)\int_{-\infty}^{\infty} (j\omega)^k N(j\omega)J_1(a\omega)\,\mathrm{d}\omega,$$

and

$$B_k(a) = (1/2\pi)\int_{-\infty}^{\infty} (j\omega)^k N(j\omega)J_0(a\omega)\,\mathrm{d}\omega.$$

Differentiation of these integrals with respect to a leads to the recursion relationships

$$A_{k+2}(a) = (\mathrm{d}^2 A_k(a)/\mathrm{d}a^2) - (1/a)(\mathrm{d}A_k(a)/\mathrm{d}a), \tag{10.100}$$

and

$$B_{k+2}(a) = (\mathrm{d}^2 B_k(a)/\mathrm{d}a^2) + (1/a)(\mathrm{d}B_k(a)/\mathrm{d}a). \tag{10.101}$$

(ii) x Gaussian

As $p(x) = (1/\sigma\sqrt{2\pi})\mathrm{e}^{-x^2/2\sigma^2}$ the transform domain relationships for A_k and B_k are

$$A_k(\sigma) = (\sigma^2/2\pi)\int_{-\infty}^{\infty} (j\omega)^{k+1} N(j\omega)\mathrm{e}^{-\sigma^2\omega^2/2}\,\mathrm{d}\omega = \sigma^2\alpha_{k+1},$$

and

$$B_k(\sigma) = (1/2\pi)\int_{-\infty}^{\infty} (j\omega)^k N(j\omega)\mathrm{e}^{-\sigma^2\omega^2/2}\,\mathrm{d}\omega = \alpha_k.$$

It is easily shown that

$$B_{k+2}(\sigma) = (1/\sigma)(\mathrm{d}B_k(\sigma)/\mathrm{d}\sigma), \tag{10.102}$$

and

$$A_k(\sigma) = \sigma^2 B_{k+1}(\sigma). \tag{10.103}$$

Similar expansions and recursion relationships can be derived to obtain all the output coefficients $\alpha_{s_1 s_2}$ and α_{sk}.[24]

10.8.2 *The IDF*

It is probably obvious to the reader how many of the results for the IDF given in section 5.9.1 can be extended to multiple input situations. Indeed in section 5.11.2 the expression for the ISSDF was given in terms of the SSDF without proof but simply by analogy with the single sine input case. Here some results are briefly derived for the general two input nonlinearity in a manner such that their extension to other cases, or more than two inputs, is readily apparent.

Let a bias γ be added to the signal x, then by definition

$$N_{i\gamma_x} = \lim_{\gamma \to 0} \int\int (\partial n(x+\gamma, y)/\partial\gamma) p(x) q(y)\, \mathrm{d}x\, \mathrm{d}y,$$

that is

$$N_{i\gamma_x} = \int\int (\partial n(x, y)/\partial x) p(x) q(y)\, \mathrm{d}x\, \mathrm{d}y. \tag{10.104}$$

Consider now various cases:

(i) x and y sinusoidal

In this case

$$N_{i\gamma_a} = N_{i\gamma_a}(a, b) = \int_{-b}^{b} \int_{-a}^{a} (\partial n(x, y)/\partial x) r(x) r(y)\, \mathrm{d}x\, \mathrm{d}y.$$

Now by definition

$$N_a(a, b) = (2/a^2) \int_{-b}^{b} \int_{-a}^{a} x n(x, y) r(x) r(y)\, \mathrm{d}x\, \mathrm{d}y,$$

which on performing an integration by parts with respect to x, as done in section 5.2, yields

$$(a^2/2) N_a(a, b) = \int_{-b}^{b} \int_{-a}^{a} \{(a^2 - x^2)^{1/2}/\pi\} (\partial n(x, y)/\partial x) r(y)\, \mathrm{d}x\, \mathrm{d}y.$$

Differentiating both sides with respect to a gives

$$a N_a(a, b) + (a^2/2)(\partial N_a(a, b)/\partial a) = a \int_{-b}^{b} \int_{-a}^{a} (\partial n(x, y)/\partial x) r(x) r(y)\, \mathrm{d}x\, \mathrm{d}y,$$

so that

$$N_{i\gamma_a}(a, b) = N_a(a, b) + (a/2)(\partial N_a(a, b)/\partial a), \tag{10.105}$$

the required result.

Similarly if the bias γ is added to the signal of amplitude b one obtains

$$N_{i\gamma_b}(a, b) = N_b(a, b) + (b/2)(\partial N_b(a, b)/\partial b). \tag{10.106}$$

For the special case when the inputs x and y are summed so that $n(x, y)$ is equal to $n(x + y)$ it is obvious that

$$N_{i\gamma_a}(a, b) = N_{i\gamma_b}(a, b). \tag{10.107}$$

When the incremental input is related to one of the other sinusoidal inputs it is easily shown that the required IDF is found by replacing the DF in the corresponding result of section 5.9 by the appropriate dual input DF.

(ii) x sinusoidal, y Gaussian

In this case, proceeding as above, it follows that

$$N_{i\gamma_a}(a, \sigma) = N_a(a, \sigma) + (a/2)(\partial N_a(a, \sigma)/\partial a). \tag{10.108}$$

(iii) x Gaussian, y sinusoidal

From the result of section 10.7.1 it follows directly that

$$N_{i\gamma_\sigma}(a, \sigma) = N_\sigma(a, \sigma). \tag{10.109}$$

10.8.3 *Orthogonal expansion of $n(x, y)$ for correlated Gaussian inputs*

The equivalent gain expressions for a nonlinearity with correlated Gaussian inputs have been derived earlier and are given in equations (10.23) and (10.24). The purpose of this section is to derive the complete orthogonal expansion for the nonlinearity output $n(x, y)$. From equation (10.21) the expansion is

$$n(x, y) = \sum_{i=0}^{\infty} \sum_{j=0}^{i} c_{ij} q_{ij}(x, y), \tag{10.21}$$

where

$$c_{ij} = \int_{-\infty}^{\infty} \int_{-\infty}^{\infty} n(x, y) q_{ij}(x, y) q(x, y)\, dx\, dy, \tag{10.22}$$

$$q(x, y) = \frac{1}{2\pi\sigma_x\sigma_y(1 - \rho^2)^{1/2}} \exp[-\{(x/\sigma_x)^2 - (2\rho xy/\sigma_x\sigma_y) + (y/\sigma_y)^2\}/2(1 - \rho^2)] \tag{7.116}$$

for two correlated Gaussian signals with $\rho = R_{xy}(0)/\sigma_x\sigma_y$, and the polynomials $q_{ij}(x, y)$ are required.

Consider a transformation of x and y to two new Gaussian signals u and v which are to be uncorrelated for a zero time shift. Let

$$u = x, \tag{10.110}$$

$$v = y - \alpha x, \tag{10.111}$$

then

$$R_{uv}(0) = R_{xy}(0) - \alpha R_{xx}(0),$$

which is zero when

$$\alpha = R_{xy}(0)/R_{xx}(0) = \rho\sigma_y/\sigma_x. \tag{10.112}$$

Also

$$\sigma_u = \sigma_x,$$

and

$$\sigma_v = \sigma_y(1 - \rho^2)^{1/2}.$$

Now equation (10.39) is valid for any two uncorrelated Gaussian inputs, so that

$$q_{ij}(u, v) = H_{i-j}(u/\sigma_u)H_j(v/\sigma_v)/\{(i-j)!j!\}^{1/2},$$

which on substituting for u and v in terms of x and y from equations (10.110) and (10.111) gives

$$q_{ij}(x, y) = H_{i-j}(x/\sigma_x)H_j\{(y - \alpha x)/\sigma_y(1 - \rho^2)^{1/2}\}/\{(i-j)!j!\}^{1/2} \tag{10.113}$$

as the required result.

REFERENCES

1. Rice, S. O.: 'Mathematical analysis of random noise', *Bell System Tech. J.*, 1944, 23, p. 282.
2. Somerville, M. J. and Atherton, D. P.: 'Multi-gain representation for a single valued nonlinearity with several inputs, and the evaluation of their equivalent gains by a cursor method', *Proc. I.E.E.*, 1958, 105C, pp. 537–549.
3. Atherton, D. P.: 'The evaluation of the response of single-valued nonlinearities to several inputs', *Proc. I.E.E.*, 1962, 109C, pp. 146–157.
4. Van der Pol, B. and Bremmer, H.: *Operational calculus based on the two-sided Laplace integral*, 2nd ed. (Cambridge University Press, Cambridge, 1955).
5. McCausland, I.: *Introduction to optimal control, Appendix B* (Wiley, New York, 1969).
6. Watson, G. N.: *A treatise on the theory of Bessel functions* (Cambridge U.P., 1944, p. 393).
7. Erdelyi, A. et al: *Higher transcendental functions* (McGraw-Hill, New York, 1953, Vol. 1).
8. Atherton, D. P.: 'Rapid evaluation of the auto-correlation function of the output of single-valued nonlinearities in response to sinusoidal and Gaussian signals', *Proc. I.E.E.*, 1962, 109C, pp. 656–664.
9. Erdelyi, A., et al.: *Table of integral transforms* (McGraw-Hill, New York, 1954, Vol. 1).
10. Balasubramanian, R.: 'Analysis of systems containing a multidimensional nonlinearity', *Ph.D. Thesis, University of New Brunswick*, 1968.
11. Jackson, D.: 'Formal properties of orthogonal polynomials in two variables', *Duke Mathematical Journal*, 1936, 2, p. 423.
12. Brown, J. L.: 'Application of the theory of orthogonal polynomials in two variables to a multigain equivalent linearization problem', *Proc. I.E.E.*, 1960, 108C, p. 115.
13. Balasubramanian, R. and Atherton, D. P.: 'Quasi-linear analysis of feedback systems containing a nonlinearity with output dependent upon both the input and its derivative', *Paper No. 6, Session B2, Proc. A.I.C.A. Congress*, Lausanne, August 1967.
14. Davenport, W. B. and Root, W. L.: *An introduction to the theory of random signals and noise* (McGraw-Hill, New York, 1958, Chapter 8, p. 151).
15. Balasubramanian, R. and Atherton, D. P.: 'Response of multidimensional nonlinearities to inputs which are separable processes', *Proc. I.E.E.*, 1968, 115, pp. 581–590.
16. Sawaragi, Y., et al.: 'Statistical analyses of control systems containing a nonlinear element with hysteresis characteristic', *Technical report of Engineering Research Institute, Kyoto University*, report 97, 1962.

17. Bennett, W. R.: 'New results in the calculation of modulation products', *Bell System Tech. J.*, 1933, 12, pp. 228–243.
18. Kalb, R. M. and Bennett, W. R.: 'Ferromagnetic distortion of a two-frequency wave', *Bell System Tech. J.*, 1935, 14, pp. 322–359.
19. Feiguth, W.: 'Multi-input quasi-linearization', *Ph.D. Thesis, University of New Brunswick*, 1968.
20. Kazakov, I. E.: 'Statistical analysis of systems with multidimensional nonlinearities', *Automation and Remote Control*, 1965, 26, pp. 458–464.
21. Cramer, H.: *Mathematical methods of statistics* (Princeton U.P., 1954, Chapter 23).
22. Nuttall, A. H.: 'Theory and application of the separable class of random processes', *Technical report of Research Laboratory of Electronics*, M.I.T., report 343, 1958.
23. Kavanagh, R. J.: 'The determination of linear multivariable models for complex processes', *Trans. Engineering Inst. Canada*, 1961, 5, pp. 140–147.
24. Atherton, D. P.: 'On the output autocorrelation function of a single-valued nonlinearity with signal plus noise as input', *I.E.E.E. Trans.*, 1974, IT-20, pp. 263–265.

BIBLIOGRAPHY

Atherton, D. P.: 'Series expansion for two input describing functions', *Electronics Letters*, 1973, 9, pp. 266–267.

Blachman, N. M.: 'The uncorrelated output components of a nonlinearity', *I.E.E.E. Trans.*, 1968, IT-14, pp. 250–255.

Blachman, N. M.: 'The signal *X* signal, noise *X* noise and signal *X* noise output of a nonlinearity', *I.E.E.E. Trans.*, 1968, IT-14, pp. 21–27.

Brown, J. L., Jr.: 'A generalized form of Price's Theorem and its converse', *I.E.E.E. Trans.*, 1967, IT-13, pp. 27–30.

Deutsch, R.: *Nonlinear transformations of random processes* (Prentice-Hall, Englewood Cliffs, N.J., 1962).

Jones, J. J.: 'Hard-limiting of two signals in random noise', *I.E.E.E. Trans.*, 1963, IT-9, pp. 34–42.

Kazakov, I. E.: 'Generalization of the method of statistical linearization to multidimensional systems', *Automation and Remote Control*, 1965, 26, pp. 1201–1206.

Kazakov, I. E.: 'Statistical analysis of nonlinear systems with random parameters', *Proc. Third I.F.A.C. Congress*, London, June, 1966, Paper 8.F.

Kazakov, I. E.: 'Statistical analysis of systems with ambiguous nonlinearities', *Engineering Cybernetics*, 1966, 3, pp. 384–391.

Kirlin, R. L.: 'Hard-limited intermodulation with low input signal-to-noise ratio', *I.E.E.E. Trans.*, 1967, COM-15, pp. 653–654.

Mahalanabis, A. K. and Nath, A. K.: 'A multiple-input quasilinearization method with applications', *Int. J. Control*, 1966, 4, pp. 73–86.

Mohan, M. A. and Krishna, G.: 'On some aspects of statistical linearization of non-linear elements', *Int. J. of Control*, 1966, 3, pp. 541–551.

Nath, A. K. and Mahalanabis, A. K.: 'Method of statistical linearization', *Proc. I.E.E.*, 1966, 113, pp. 2081–2086.

Obakov, V. K.: 'Description of Nonlinear Dependencies in Automatic Control Systems', *Automation and Remote Control*, 1974, 35, pp. 329–332.

Pawula, R. F.: 'A modified version of Price's Theorem', *I.E.E.E. Trans.*, 1967, IT-13, pp. 285–288.

Pervozvanskii, A. A.: *Random processes in non-linear control systems* (Academic Press, New York, 1965).

Price, R.: 'A useful theorem for non-linear devices having Gaussian inputs', *I.R.E. Trans.*, IT-4, 1958, pp. 69–72.

Rauch, S. and Fukada, M.: 'Output signal-to-noise ratio of an error function limiter', *Int. J. Control*, 1967, 5, pp. 47–54.

Sea, R. G. and Vacroux, A. G.: 'Steady state analysis of non-linear systems and multiple input describing functions', *Fourth I.F.A.C. Congress*, Warsaw, June 1969, Session 41, pp. 61–78.

Sevy, J. L.: 'The effect of multiple CW and FM signals passed through a hard-limiter or TWT', *I.E.E.E. Trans.*, 1966, COM-14, pp. 568–578.

Shaft, P. D.: 'Limiting of several signals and its effect on communications system performance', *I.E.E.E. Trans.*, 1965, COM-13, pp. 504–512.

Shutterly, H. B.: 'General results in the mathematical theory of random signals and noise in non-linear devices', *I.E.E.E. Trans.*, IT-9, 1963, pp. 74–84.

West, J. C.: 'The equivalent gain matrix of a multivariable non-linearity', *U.K.A.C. Conference*, Leicester, England, April, 1968.

PROBLEMS

P.10.1 Show using the transform method that for the ideal relay

$$\alpha_{sk} = h\rho^s{}_1F_1\{(s+k)/2;\, s+1;\, -\rho^2\}/s!(\sigma/\sqrt{2})^k\Gamma\{1-(s+k)/2\}.$$

P.10.2 A nonlinear phase advance compensator known as an ideal driven limiter is defined by

$$n(x, \dot{x}) = \begin{cases} x \operatorname{sgn}(x + T\dot{x}) & \text{for } |x| < |x + T\dot{x}| \\ x + T\dot{x} & \text{for } |x| > |x + T\dot{x}|. \end{cases}$$

Show that for a sine input its equivalent gains are $K_x = 2\tan^{-1}(2\pi/\omega T)$ and $K_{\dot{x}} = T\tan^{-1}(2\pi/\omega T)$. What is the phase lead of the compensator? How does it vary with the amplitude of the input sine wave?

P.10.3 Show that for a Gaussian input the ideal phase advance driven limiter of the previous question has equivalent gains of $K_x = 2\tan^{-1}(2\sigma_x\pi/\sigma_{\dot{x}}T)$ and $K_{\dot{x}} = T\tan^{-1}(2\sigma_x\pi/\sigma_{\dot{x}}T)$.

P.10.4 A nonlinear network consists of a cascade combination of two cubic nonlinearities with an ideal phase lead of transfer function $(1 + sT)$ between them. Express the network output as a function of the input and its first derivative and show that for a sinusoidal input its equivalent gains are

$$K_x = 63a^8(1 + 3\omega^2T^2)/128 \quad \text{and} \quad K_{\dot{x}} = 9a^8T(7 + 9\omega^2T^2)/128.$$

P.10.5 Evaluate the equivalent gain model of a network consisting of the nonlinearity $\sin mx$, an ideal phase advance $K(1 + sT)$ and an ideal relay $h \operatorname{sgn} x$. Does the model depend upon K and if so why?

P.10.6 Use equations (10.90) and (10.91) to obtain $N_a(a, \sigma)$ and $N_\sigma(a, \sigma)$ for an ideal relay. Why must the expansion be used in terms of $A_k(\sigma)$ and $B_k(\sigma)$?

P.10.7 Obtain an approximate solution for $N_a(a, \sigma)$ and $N_\sigma(a, \sigma)$ for an ideal relay for large signal to noise ratios ρ by using equations (10.90) and (10.91) expanded in terms of $A_k(a)$ and $B_k(a)$. Compare the solution with the exact result. What are the advantages of this approximate approach?

P.10.8 Evaluate $N_a(a, b)$ and $N_b(a, b)$ from equations (10.90) and (10.91) for
(a) an ideal relay
(b) $n(x) = \sin mx$
(c) $n(x) = x^3$.

P.10.9 Obtain series expansions for the describing function of a sine wave in the presence of an unrelated larger amplitude uniformly distributed signal for the same characteristics as the previous question.

P.10.10 Find $N_{i\gamma_a}(a, b)$ and $N_{i\gamma_b}(a, b)$ for $n(x, y) = x + x^2y^3$.

P.10.11 Evaluate $N_{i\gamma}(a, b)$ for two summed sinusoidal inputs to (a) an ideal relay and (b) a cubic.

P.10.12 A more general definition for a driven limiter (see P.10.2) is

$$n(x_1, x_2) = x_2 \qquad |x_2| < |x_1|$$
$$|x_1| \operatorname{sgn} x_2 \quad \text{if } |x_2| > |x_1|$$

with $X_2(s) = G_2(s)X_1(s)$. Show that for a sinusoidal input its linear gain model $K_{x_1}x_1 + K_{x_2}x_2$ has $K_{x_1} = (1/\pi)[\tan^{-1}\{(1 + R)/I\} - \tan^{-1} \times \{(1 - R)/I\}]$ and $K_{x_2} = (1/\pi)[\tan^{-1}\{(1 + R)/I\} + \tan^{-1}\{(1 - R)/I\}]$ where $G_2(j\omega) = R + jI$.

P.10.13 Show that an alternative linear gain model for the nonlinearity of P.10.12 is $K'_{x_1}x_1 + K'_{\dot{x}_1}\dot{x}_1$ where $K'_{x_1} = (1/\pi)[(1 + R)\tan^{-1}\{(1 + R)/I\} -(1 - R)\tan^{-1}\{(1 - R)/I\}]$ and $K'_{\dot{x}_1} = (I/\omega\pi)[\tan^{-1}\{(1 + R)/I\} + \tan^{-1}\{(1 - R)/I\}]$.

Chapter 11

Applications of Multiple Input Describing Functions

11.1 Introduction

The previous chapter investigated in considerable detail the evaluation of the output and the quasilinear model of a nonlinear function of several signals. Multiple input describing functions for unrelated inputs were seen to be logical extensions of the single input case particularly when making use of the modified nonlinearity concept. On the other hand describing functions for related inputs, which regularly occur in feedback systems, were more difficult to evaluate analytically. The purpose of this and the following chapters is to make use of the results of the previous chapter to study more complex nonlinear systems. In the following sections various diverse examples and phenomena are discussed whereas the latter chapters are basically restricted to a specific topic.

In section 11.2 we return to the topic of jump resonance, examining both possible design approaches for its elimination and synthesis procedures for its use as a frequency detection mechanism. The next three sections deal with limit cycles which require two input DFs for their evaluation and the applicability of DF methods to the analysis of systems containing nonlinear functions of system parameters as well as signals is illustrated. As a final application of two input DFs the evaluation of the equivalent gains and the analysis of the performance of a system with a nonlinear phase advance compensator is discussed.

11.2 Further aspects of jump resonance

The purpose of this section is to examine additional aspects of the jump resonance phenomenon. Since this is done using the IDF approach a brief review of the geometric properties of IDF loci is appropriate. From equation (5.59) we know for a single valued nonlinearity that

$$N_{ib}(a) = N(a) + (a/2)(\mathrm{d}N(a)/\mathrm{d}a)(1 + \mathrm{e}^{-j2\phi}), \tag{5.59}$$

where the incremental input b has the same frequency as a and a phase difference of ϕ. Letting $K_1 = N(a)$ and $K_2 = N(a) + a(\mathrm{d}N(a)/\mathrm{d}a)$ as done in section 4.7.5 equation (5.59) becomes

$$N_{ib}(a) = (K_1 + K_2)/2 + \{(K_2 - K_1)/2\}\mathrm{e}^{-j2\phi}. \tag{11.1}$$

As ϕ varies one can show that the locus of $-1/N_{ib}(a)$ for a fixed value of a is a circle in the complex plane with centre on the real axis at the point $\{-(K_1 + K_2)/2K_1K_2, 0\}$ and radius $(|K_1 - K_2|)/2K_1K_2$ in agreement with equation (4.44) found by Hatanaka's[1] method. Since equations (5.12) and (5.13) yield

$$K_1 = (2/a^2)\int_{-a}^{a} (a^2 - x^2)n'(x)r(x)\,\mathrm{d}x, \tag{5.12}$$

and

$$K_2 = (2/a^2)\int_{-a}^{a} x^2 n'(x)r(x)\,\mathrm{d}x, \tag{11.2}$$

the following qualitative properties of the $-1/N_{ib}(a)$ loci are apparent:

(i) For $n'(x)$ positive for all x, K_1 and K_2 are positive so that the loci lie to the left of the origin.

(ii) For a hard spring nonlinearity, since $n'(x)$ is a monotonically increasing function of x, $K_1 < K_2$ for a given value of a and both are monotonically increasing functions of a. The loci tend towards the origin as a increases.

(iii) For a soft spring nonlinearity, since $n'(x)$ is a monotonically decreasing function of x, $K_1 > K_2$ for a given value of a and both are monotonically decreasing functions of a. The loci tend away from the origin as a increases.

It is also possible to show[2] from equation (5.58) that the $-1/N_{ib}(a)$ loci for a DVNL are again circular. The closed form expressions for the centres, which are not in this case on the negative real axis, and the radii are complicated and not readily usable. The best approach for a given a is probably to take three different values of ϕ, say 0, $\pi/4$ and $\pi/2$, to find three points on $-1/N_{ib}(a)$ which are then sufficient to define the circle.

11.2.1 *Removal of jump resonance*

A sufficient condition, within the limits of DF accuracy, to ensure that jump resonance will not occur in the nonlinear system of figure 11.1 is that the open loop frequency response locus $G'(j\omega)$, where $G'(j\omega) = G_c(j\omega)G(j\omega)$, should not

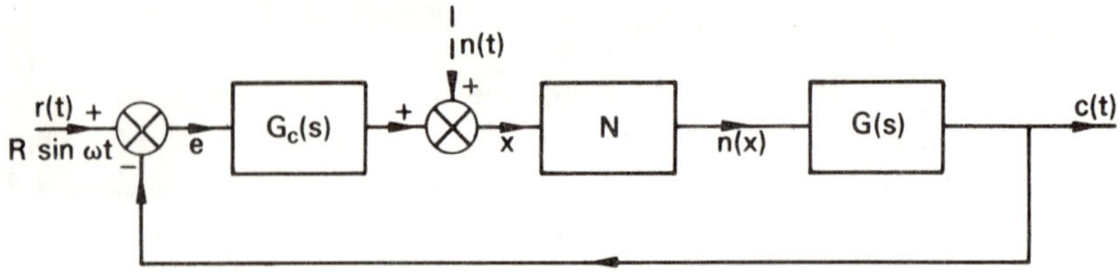

Figure 11.1 Nonlinear feedback system

intersect the IDF of the nonlinearity, N. If compensation is required to avoid an intersection, the transfer function of the linear compensator, $G_c(s)$, or the IDF loci may be changed. The IDF loci can be altered by a technique such as that of nonlinear model feedback described in section 4.5 or by the introduction of a high frequency modifying signal. It can be shown for the latter technique that as the amplitude of the modifying signal increases the IDF loci tend to move towards the origin for a hard spring nonlinearity and conversely away from the origin for a soft spring nonlinearity. Complete absence of jump resonance is an unnecessarily severe requirement in many cases, since a system is unlikely to be subjected to the entire spectrum of input amplitudes and frequencies. A more reasonable objective, which we now consider, is the removal of jump resonance over a limited range of input amplitudes.[2]

The design requirement may possibly be stated as follows: specify either a stabilising signal or linear compensator, $G_c(s)$, such that the system of figure 11.1 has no jump resonance at any frequency for the range of system amplitudes $0 \leq R \leq R_{max}$. As the IDF loci are plotted in terms of the nonlinearity input amplitude a, it is necessary to obtain a relationship between the amplitudes R and a. From the figure, with $G_c(s) = 1$

$$y^2 + (x + 1/K_1)^2 = R^2/K_1^2 a^2, \tag{11.3}$$

where $G(j\omega) = x + jy$. This is the equation of a circle, to be called the input circle, with centre at $-1/K_1$ and radius $R/K_1 a$. If the system is to have no jumps for $R \leq R_{max}$, then for an arbitrary a, say a_1, there must be no intersections of $G(j\omega)$ and the IDF circle for $a = a_1$ anywhere inside the input circle for $a = a_1$, and $R = R_{max}$. A series of input circles for different values of a will define a forbidden region in the $G(j\omega)$ plane which must be avoided by the plot of $G(j\omega)$. A typical input circle and the portion of the IDF circle with which intersection by $G(j\omega)$ is forbidden are illustrated in figure 11.2, where, for simplicity, only a single set of circles is shown.

The step-by-step procedure for using signal stabilisation to remove jumps for a given range of input amplitudes is as follows:

(a) assume an amplitude probability density and magnitude for the stabilising signal, $n(t)$;
(b) compute the modified nonlinearity and plot its IDF loci;
(c) for each IDF circle, draw the corresponding input circle;

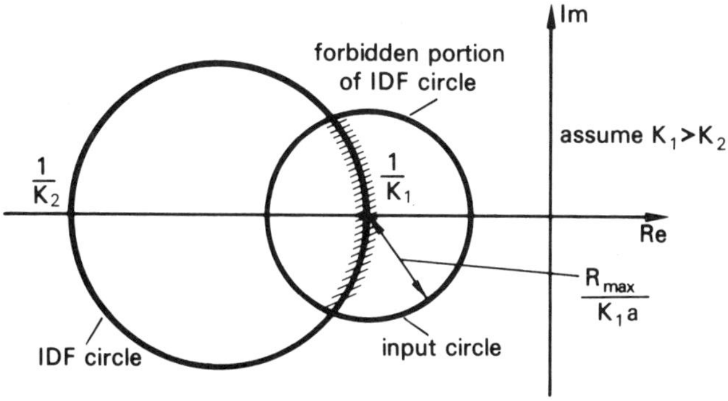

Figure 11.2 Input circle defining forbidden region for $G(j)$

(d) if $G(j\omega)$ intersects the forbidden region defined by the input circles, increase the assumed magnitude of $n(t)$ and repeat steps (b), (c) and (d).

As for the case of the complete removal of the jump phenomenon, there is no guarantee that this procedure will ultimately yield the desired solution. The effects of $n(t)$ on the movement of the IDF loci, mentioned earlier, should be continually borne in mind.

Application of the method of linear compensation is somewhat complicated by the fact that the procedure depends on the intended location of the compensator, $G_c(j\omega)$, in the system loop. If the compensator is located in the feedback path or between the nonlinearity and $G(j\omega)$ the design is easiest, simply requiring $G_c(j\omega)$ to be chosen so that $G'(j\omega) = G_c(j\omega)G(j\omega)$ does not intersect the forbidden region.

The more usual case, however, is for the compensator to be located immediately preceding the nonlinearity in the low power error signal path as shown in figure 11.1. Equation (11.3) is now not valid because the error signal, e, no longer constitutes the nonlinearity input. The difficulty can be resolved by shifting the compensator through the summing junction to give the system shown in figure 11.3 which is equivalent to the original system of figure 11.1. The block in the

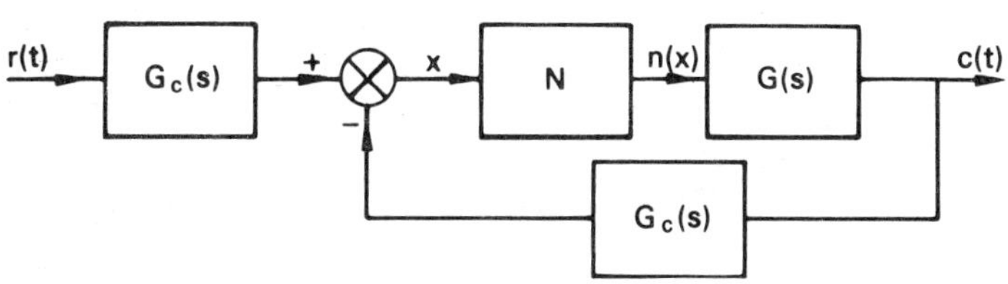

Figure 11.3 System equivalent to figure 11.1 with $n(t) = 0$

feedback path combines with $G(j\omega)$ to give the equivalent open loop linear part of the system, $G'(j\omega) = G_c(j\omega)G(j\omega)$, as discussed above, and the input signal amplitude at the summing junction is $|G_c(j\omega)|R$. The radii of the input circles are thus $|G_c(j\omega)|R_{max}/K_1a$. The compensator $G_c(j\omega)$ must now be chosen such that for every ω, $G'(j\omega)$ does not intersect the forbidden region defined by the input circles. The difference in this case, then, is that the forbidden region is ω-dependent.

The situation is not as complicated as it may appear, because $G(j\omega)$ will be in the critical region for a very limited range of frequencies. Over this frequency range $|G_c(j\omega)|$, and hence the size of the forbidden region, will not change by much. At most a few sets of input circles need be drawn, and usually the single most pessimistic case will suffice.

Example

The closed loop system of figure 11.1 with $n(t) = 0$, the nonlinearity, N, ideal saturation of unit slope and limit levels $\pm 1{\cdot}5$ and $G(s) = 40(s + 5)(s + 50)/s^3$ is to be compensated, if necessary, such that no jumps can occur for sinusoidal inputs of any frequency, and of peak amplitude less than 4·5 volts.

Figure 11.4 shows the IDF loci for the saturation nonlinearity and the $G(j\omega)$ locus for the linear plant. Superimposed on the plots is the forbidden region defined by the input circles for $R_{max} = 4{\cdot}5$, found as discussed above. Since $G(j\omega)$ passes through the forbidden region, some form of compensation is clearly required.

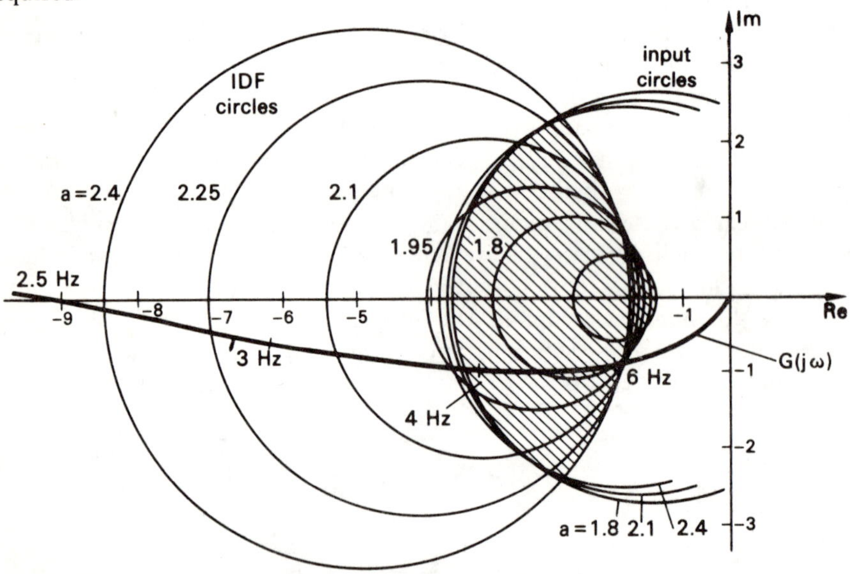

Figure 11.4 Uncompensated frequency response and forbidden region for example of section 11.2.1

Suppose passive phase advance compensation is chosen then figure 11.4 shows that approximately 30° of phase shift are required in the vicinity of 4 Hz. This is provided by the compensator

$$G_c(j\omega) = \frac{j\omega + 4\pi}{j\omega + 12\pi}. \tag{11.4}$$

Figure 11.5 shows the resulting open loop frequency response $G'(j\omega) = G_c(j\omega)G(j\omega)$ and the boundaries of the forbidden regions. Since $G'(j\omega)$ does not enter the forbidden regions the design should be satisfactory as was confirmed by analog simulation.[2]

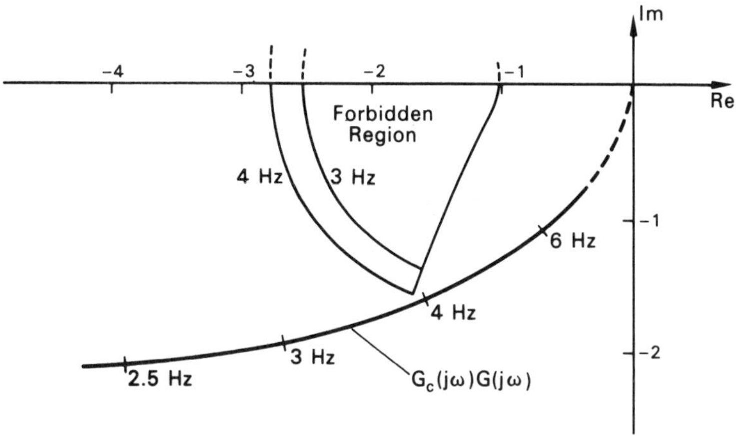

Figure 11.5 Compensated frequency response and forbidden regions for example of section 11.2.1

11.2.2 *Isolated jump resonance*

For a nonlinear feedback system with a prefilter such as that shown in figure 11.6 it is possible to obtain different forms of jump phenomena behaviour. The jump behaviour of the system can be predicted by superimposing the frequency response of the input filter on the jump boundaries plotted in an $R - \omega$ plane.

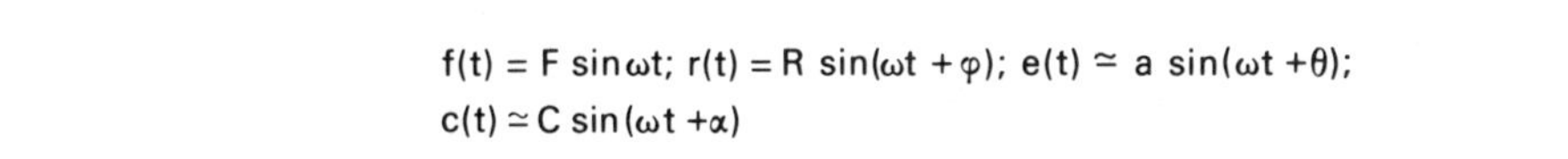

$f(t) = F \sin\omega t$; $r(t) = R\sin(\omega t + \varphi)$; $e(t) \simeq a \sin(\omega t + \theta)$;
$c(t) \simeq C \sin(\omega t + \alpha)$

Figure 11.6 A nonlinear feedback system with prefilter

From equations (4.44) for the IDF loci and (11.3) expressions for x and y can be found as functions of R and a. They are

$$x = -\{1 + \{K_2/(K_1 - K_2)\}(R^2/a^2)\}/K_1, \tag{11.5}$$

$$y = \pm R\{1 - \{K_2/(K_1 - K_2)\}^2(R^2/a^2)\}^{1/2}/aK_1. \tag{11.6}$$

For fixed values of R and an ideal saturation nonlinearity the equations yield the curves, on which a is a parameter, shown in figure 11.7 on the $x - jy$ Nyquist plane. The jump points in the $R - \omega$ plane are then found from intersections with the $G(j\omega)$ locus as shown in figure 11.7. For example, at $\omega = \omega_1$ the two

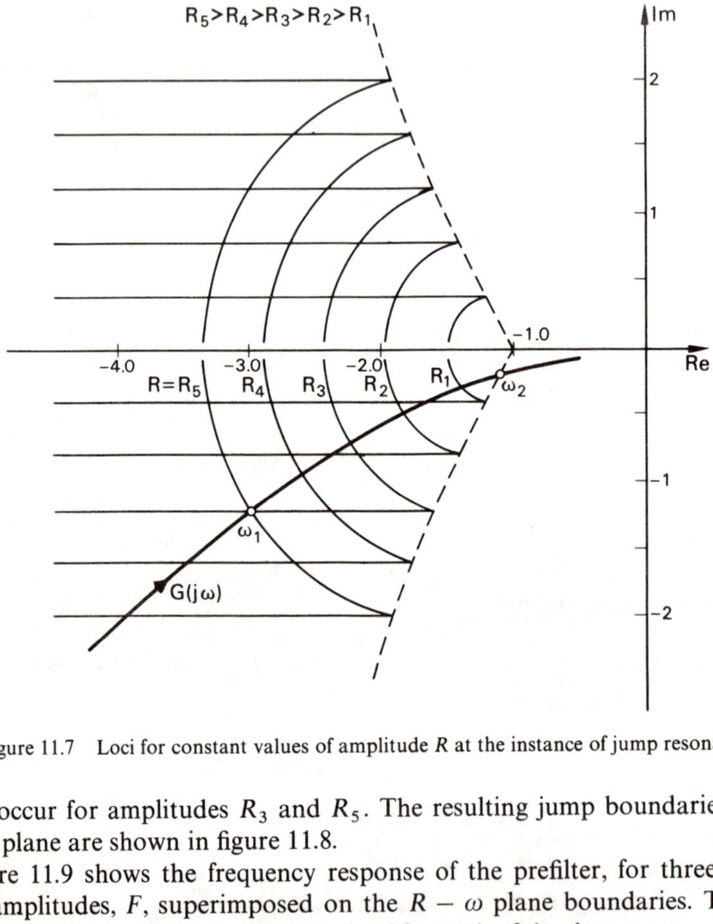

Figure 11.7 Loci for constant values of amplitude R at the instance of jump resonance

jumps occur for amplitudes R_3 and R_5. The resulting jump boundaries of the $R - \omega$ plane are shown in figure 11.8.

Figure 11.9 shows the frequency response of the prefilter, for three chosen input amplitudes, F, superimposed on the $R - \omega$ plane boundaries. The type of jump resonance curves that are obtained for each of the three cases are shown in figure 11.10 where the system output amplitude C is plotted versus input frequency. The shapes of the response curves in figure 11.10 have been chosen rather arbitrarily, the significant feature being the jump frequencies.

Examination of the curves for the cases A and B reveals a situation which is not met in conventional jump resonance. It is observed that once the system has jumped to its upper mode, it can revert to the lower mode by either a decrease or an increase in frequency. For the case C the filter input amplitude is insufficient to cause the operating point to cross the jump up locus and so no jumps occur. Suppose, however, that for the case C, with an input frequency of ω_5, the amplitude of the filter input is temporarily raised so that the operating point moves from

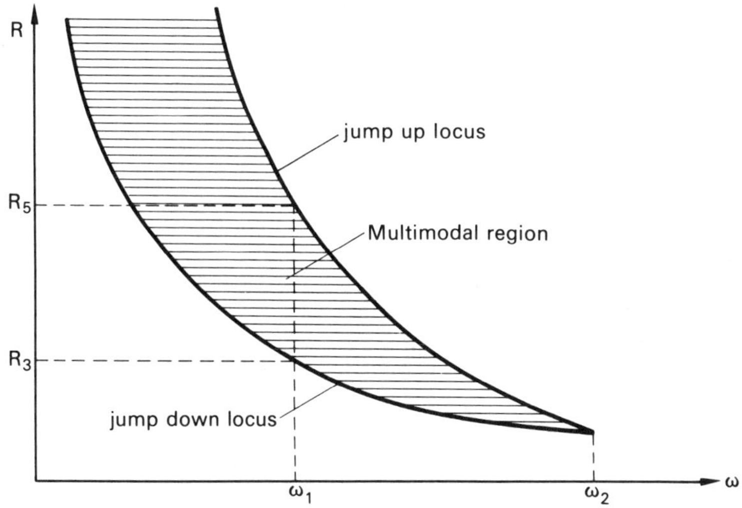

Figure 11.8 Multimodal region in $R - \omega$ plane

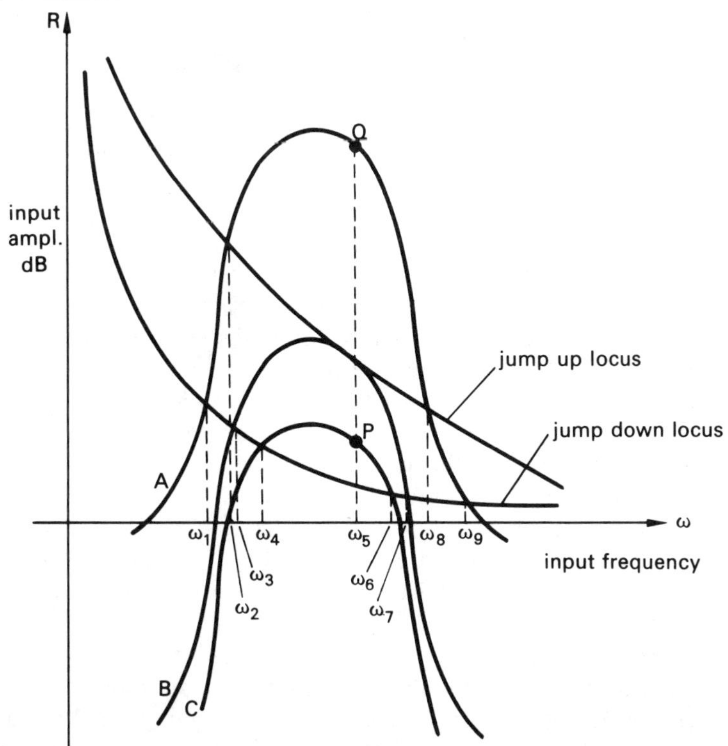

Figure 11.9 Operating point loci with input filter

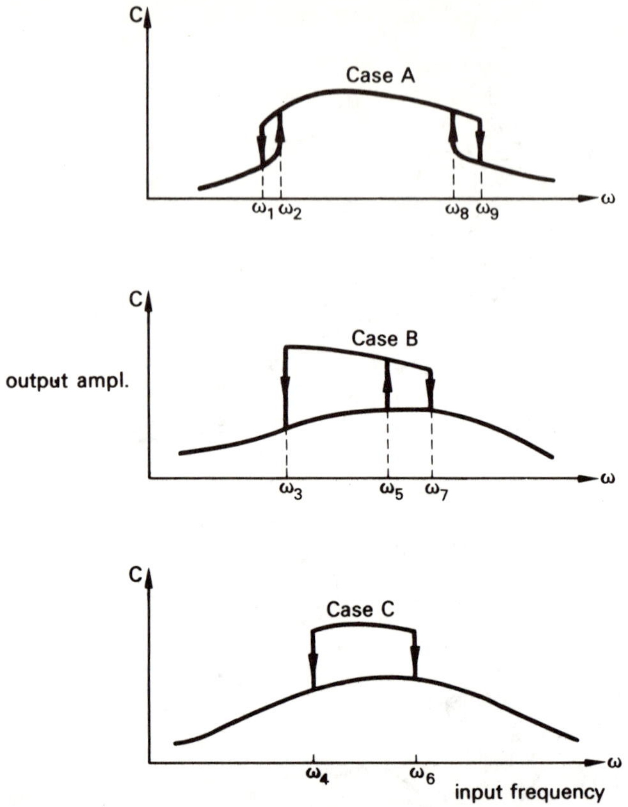

Figure 11.10 System response with input filter

P to Q and back again. In consequence, the system will jump to its upper mode and remain there even after the input amplitude is reduced to its original value. It may now be observed that the system will again be able to revert to its lower mode by either an increase in frequency to ω_6 or a decrease to ω_4. This particular type of behaviour is possible only if the filter is excited by an input whose frequency is in the range ω_4 to ω_6 and whose amplitude is at least enough to cause the operating point to move inside the multi-modal region.

The phenomenon in which a system is in a non-self-excitable mode such that jumps down occur for either an increase or decrease in frequency has been termed the 'isolated jump resonance phenomenon'. Frequency responses of a similar nature are also known to exist in the response of quasilinear second order differential difference equations.[(3)]

Since jump resonance in a nonlinear system occurs at very precisely determined frequencies, it has been possible to utilise the isolated jump resonance behaviour of the systems of the type shown in figure 11.6 for detecting small deviations[(4)] in the frequency of the input signal $f(t)$. Such frequency monitoring devices may find a range of applications in instrumentation and control.

11.3 Oscillations in the non-autonomous Van der Pol equation

In this section the effects of an external signal on the oscillations of a system described by the Van der Pol equation are considered. A general form of the equation is

$$\ddot{x} - (\alpha - 3\gamma x^2)\dot{x} + x = f(t), \tag{11.7}$$

where $f(t)$ is the input. The nonlinearity may be taken as

$$n(x, \dot{x}) = -(\alpha - 3\gamma x^2)\dot{x} + x, \tag{11.8}$$

so that the corresponding block diagram is as shown in figure 11.11(a). Alternatively the equation may be written

$$\ddot{x} - \alpha\dot{x} + x + \gamma\frac{\mathrm{d}}{\mathrm{d}t}(x^3) = f(t), \tag{11.9}$$

which can be represented by the block diagram of figure 11.11(b) with the single cubic nonlinearity γx^3, the plant, $G(s)$, given by

$$G(s) = s/(s^2 - \alpha s + 1), \tag{11.10}$$

and the input filter, $H(s)$, by

$$H(s) = 1/(s^2 - \alpha s + 1). \tag{11.11}$$

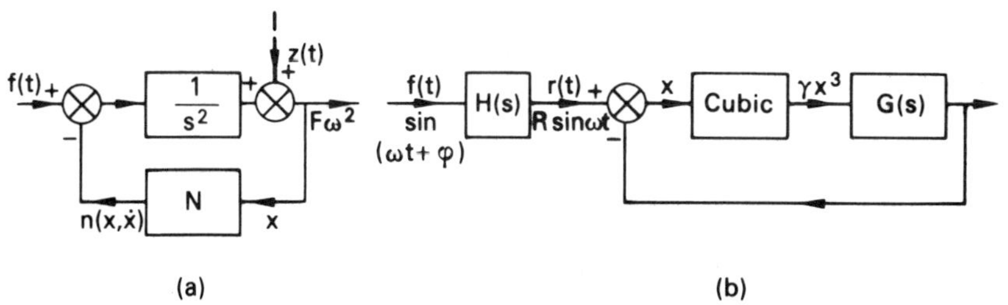

Figure 11.11 Equivalent block diagrams for Van der Pol equation

Using this representation the results of reference 5 are given for a harmonic input. The effect of an external random input is then examined using the former representation where the nonlinearity (equation (11.8)) is a function of two variables.

11.3.1 *Effect of harmonic input*

For the block diagram of figure 11.11(b) with no input signal the autonomous oscillation occurs at the frequency at which $G(j\omega)$ has 180° phase shift. Now

$$G(j\omega) = j\omega/(1 - \omega^2 - j\alpha\omega),$$

which gives $\omega = 1$ as the oscillation frequency and a corresponding gain of $1/\alpha$.

The amplitude, b, of the self oscillation is thus given by

$$(3b^2\gamma/4) = \alpha, \quad \text{i.e.} \quad b^2 = 4\alpha/3\gamma. \tag{11.12}$$

Figure 11.12 shows a sketch of the frequency response locus $G(j\omega)$ and $C(a) = -1/N(a)$ for the cubic nonlinearity. From the diagram it is seen that a linear system with an open loop frequency response $KG(j\omega)$ will be stable for $K > \alpha$.

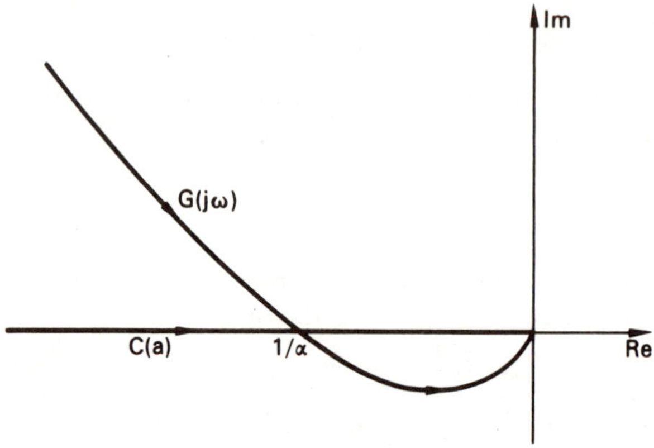

Figure 11.12 Sketch of $G(j\omega)$ and $C(a)$ for figure 11.11(b)

As the effect of an additional input to a cubic is to increase the gain to the existing signal, the harmonic input $f(t)$ will eventually quench the loop oscillation, which requires a gain of α through the cubic. The DF description for the sinusoidal component of the input $f(t)$ with magnitude a at the nonlinearity will be a combination of the single input DF and the SSDF. If the loop oscillation has a magnitude b at the nonlinearity input then

$$N_b(a, b) = (3b^2 + 6a^2)\gamma/4,$$

and

$$N_a(a, b) = (3a^2 + 6b^2)\gamma/4.$$

For the loop oscillation to be maintained $N_b(a, b) = \alpha$ giving

$$3b^2 + 6a^2 = 4\alpha/\gamma, \tag{11.13}$$

and

$$N_a(a, b) = 2\alpha - (9/4)\gamma a^2. \tag{11.14}$$

From equation (11.13) the loop oscillation is quenched, that is b becomes zero, when $a^2 = 2\alpha/3\gamma$. The DF for a is therefore

$$N_a(a, b) \quad \text{for } a^2 \leq 2\alpha/3\gamma,$$

and

$$N(a) \quad \text{for } a^2 \geq 2\alpha/3\gamma,$$

which has the value $\alpha/2$ at the quenching amplitude of $a^2 = 2\alpha/3\gamma$. To investigate the stability of the combined mode for $a^2 < 2\alpha/3\gamma$ and the synchronised forced oscillation for $a^2 > 2\alpha/3\gamma$ the appropriate IDFs must be evaluated. For a perturbation at an unrelated frequency we have

$$N_{i\gamma}(a) = 3a^2\gamma/2, \tag{11.15}$$

and

$$\begin{aligned} N_{i\gamma}(a, b) &= (3a^2 + 6b^2)(\gamma/4) + (a/2)(3a\gamma/2) \\ &= 2\alpha - (3a^2\gamma/2), \end{aligned} \tag{11.16}$$

after substitution for b^2 from equation (11.13). Both modes are therefore stable for a perturbation at an unrelated frequency since for $a^2 < 2\alpha/3\gamma$, $N_{i\gamma}(a, b)$ is greater than α and for $a^2 > 2\alpha/3\gamma$, $N_{i\gamma}(a)$ is greater than α.

To test the stability of the solutions for related frequency perturbations the corresponding IDF loci must be evaluated and drawn at the appropriate frequencies on the polar locus $G(j\omega)$. The only remaining frequency of significance, however, is usually that at the forcing frequency; the stability of which can also be evaluated from the sign of dR/da. If dR/da is less than zero then the solution is unstable. Since dR/dF for a given ω is positive the stability can also be found from the sign of dF/da at the solution amplitude. Substituting the appropriate DF for a it is easily shown that

$$(F^2\omega^2/a^2) = (1 - \omega^2)^2/\omega^2 + \alpha^2\{1 - a^2/(4\alpha/3\gamma)\}^2 \tag{11.17}$$

is the expression relating the amplitudes a and F during the synchronised mode and

$$(F^2\omega^2/a^2) = (1 - \omega^2)^2/\omega^2 + \alpha^2\{1 - 3a^2/(4\alpha/3\gamma)\}^2 \tag{11.18}$$

is the equivalent relationship during the combined mode operation. Both equations (11.17) and (11.18) are cubics in a^2 for given values F and ω of the harmonic input, so that three real roots for a^2 may exist indicating multimodal operation.

To illustrate the type of behaviour possible in more detail the specific case of a variable amplitude harmonic input at a frequency $\omega = 1{\cdot}25$ for the equation with the parameters $\alpha = 1$ and $\gamma = 4/9$ is considered using the universal chart approach. For these parameters the above results give the autonomous self oscillation amplitude as $\sqrt{3}$ and the quenching amplitude of the forcing signal at the input to the nonlinearity as $\sqrt{1{\cdot}5}$. Figure 11.13 shows the DF, N, for the forcing signal, with the critical point C where it changes from the SSDF to the DF, the frequencies and phase angles for points on the linear locus and the universal chart curves, corresponding to a phase angle of 156°, for various amplitudes R. From the figure it is seen that for small R only the combined mode solution exists. As R is increased three combined mode solutions, as shown for $\log R = \bar{1}{\cdot}61$, can exist only two of which are of course stable whilst further increase in R produces a single combined mode and then a jump to a synchronised solution.

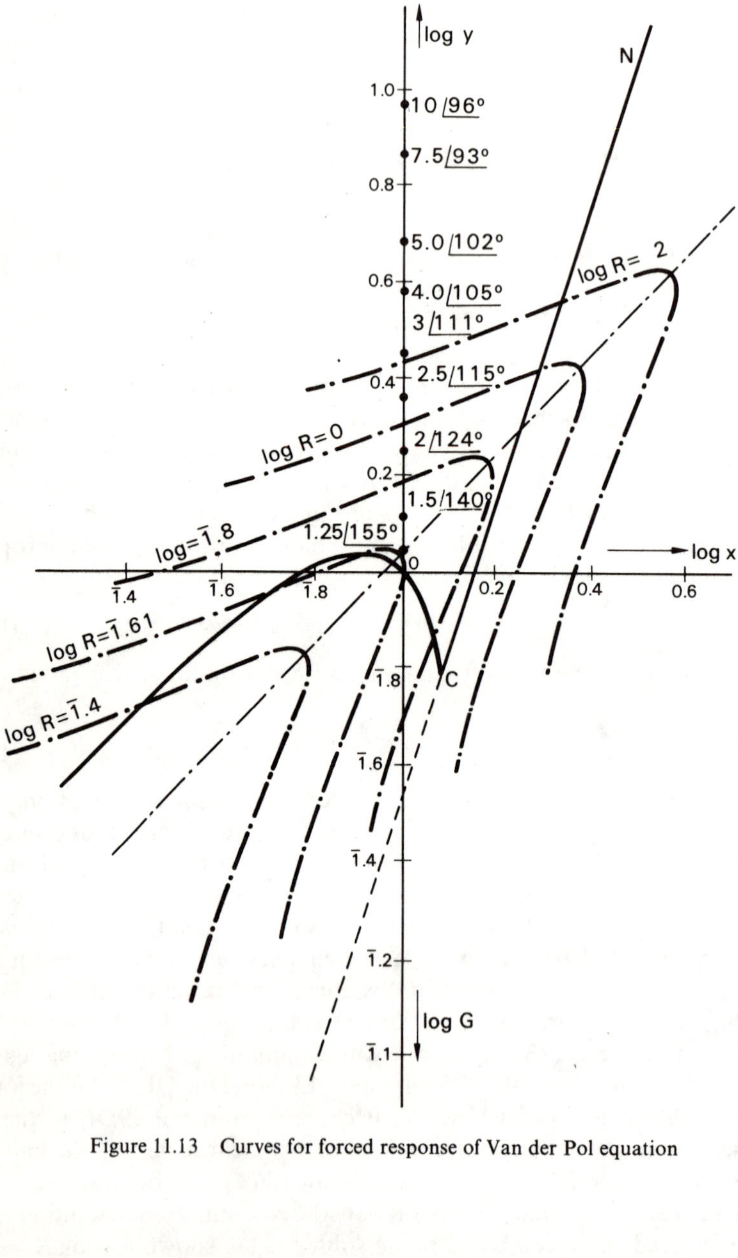

Figure 11.13 Curves for forced response of Van der Pol equation

For the combined mode the corresponding value of the oscillation amplitude b can be found from equation (11.13). On decreasing R from a large value the transition to the combined mode takes place at a slightly lower amplitude than the reverse transition mentioned above for increasing R. The complete behaviour is summarised in figure 11.14 where the dotted lines denote the third theoretical solution in the multimodal regions.

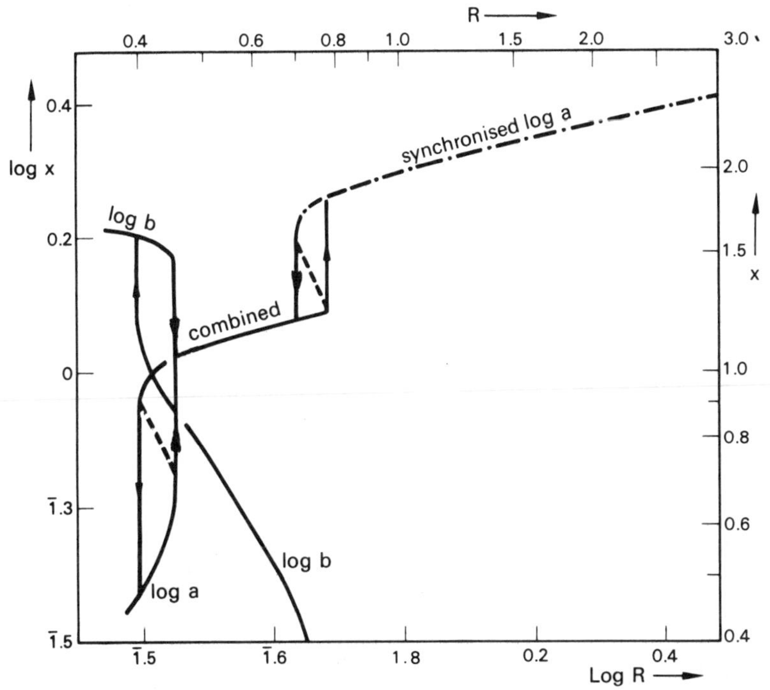

Figure 11.14 Forced response of Van der Pol equation at the frequency, $\omega = 1{\cdot}25$ rads/sec

11.3.2 *Random input*

Here we consider the effect of adding a high frequency Gaussian signal $z(t)$ to the loop[6] with $f(t) = 0$ as shown in figure 11.11(a). As the signal is at a high frequency its component at the input to $n(x, \dot{x})$ is essentially $z(t)$. Denoting the oscillation sinusoid at the input to the nonlinearity $n(x, \dot{x})$ by $x = a \cos \omega t$, the expression from section 10.7 for the nonlinearity modified by the Gaussian signal is

$$n_m(x, \dot{x}) = \int_{-\infty}^{\infty} \int_{-\infty}^{\infty} n(x + z, \dot{x} + \dot{z}) q(z) q(\dot{z}) \, dz \, d\dot{z},$$

where $q(z)$ and $q(\dot{z})$ are Gaussian distributions for the signals $z(t)$ and $\dot{z}(t)$ which have zero mean and mean squared values of σ_z^2 and $\sigma_{\dot{z}}^2$ respectively.

For the particular parameters $\alpha = 1, \gamma = 1/3$ the above expression becomes

$$n_m(x, \dot{x}) = \int_{-\infty}^{\infty}\int_{-\infty}^{\infty} [-\{1 - (x + z)^2\}(\dot{x} + \dot{z}) + x + z]q(z)q(\dot{z})\,dz\,d\dot{z}$$

$$= -(1 - x^2 - \sigma_z^2)\dot{x} + x. \tag{11.19}$$

The equivalent gains to the sinusoid can now be evaluated using equations (10.47) and (10.48) or by direct expansion of the above expression with $x = a\cos\omega t$. The results are

$$K_x = 1, \tag{11.20}$$

and

$$K_{\dot{x}} = -\{1 - (a^2/4) - \sigma_z^2\}. \tag{11.21}$$

Since from figure 11.11(a) the required 180° phase shift for the loop oscillation is produced by the double integrator, the requirement that $K_{\dot{x}} = 0$ gives the effect of σ_z on the oscillation amplitude as

$$a = 2(1 - \sigma_z^2)^{1/2}. \tag{11.22}$$

The oscillation frequency is given by $K_x/\omega^2 = 1$, i.e. $\omega = 1$. Figure 11.15 shows good agreement between the theoretical results and those found by analog

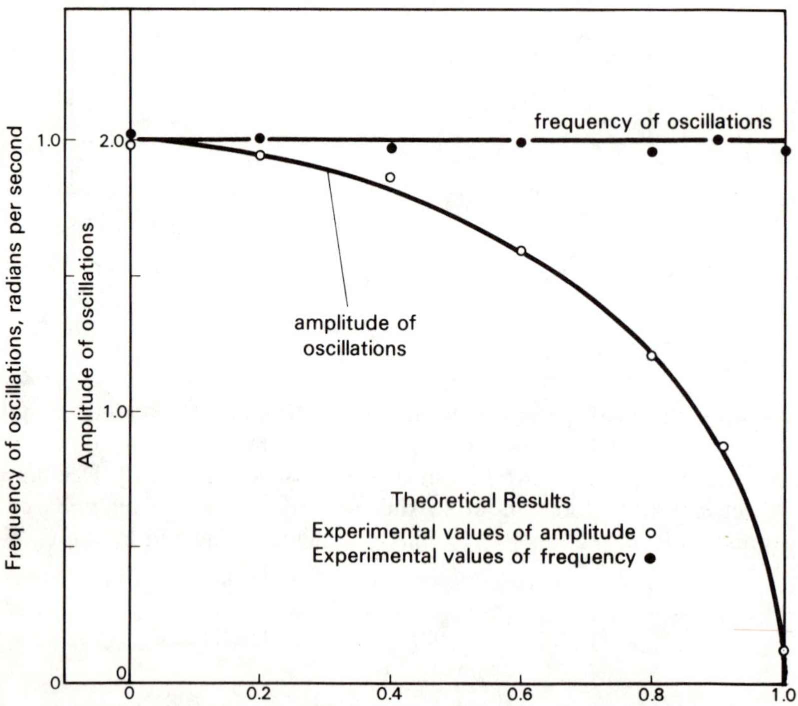

Figure 11.15 Amplitude and frequency of oscillations for the system of section 11.3.2

simulation. This is to be expected since the distortion factor, d, of section 4.8.1 evaluated for the feedback signal with the nonlinearity $n_m(x, \dot{x})$ is always less than 10 per cent. The oscillation may be shown to be stable using the Loeb criterion.

11.4 Oscillations in a time varying system

Many systems have parameters which vary with time either in a periodic[7] or random manner. Systems with periodically varying parameters have been studied for many years[8,9,10,11] but the work on systems with randomly varying parameters is much more recent.[12,13,14,15,16,17] It is not appropriate here to go into the details of these specialised methods but simply to draw attention to the possible application of multiple input describing functions to these problems. In the previous chapter the groundwork for such an approach was laid by assuming the nonlinearity to be any function of several variables, which can be system parameters, as well as input signals. There are, of course, limitations to the applicability of describing functions, perhaps the most obvious being the handling of multiplicative terms with unrelated inputs. For example, in the study of systems with randomly varying parameters the occurrence of terms such as $\alpha(t)u(t)$ is common and the DF model of $K_u = E\{\alpha(t)\}$, $K_\alpha = E\{u(t)\}$ may be useless, certainly so in relation to the alternative approaches.[12]

Although it may not have been apparent to the reader considerable use has already been made of DFs for systems with periodically varying parameters. If the simple nonlinear system of figure 11.16 has a periodic solution $x^*(t)$ then the equation of the first variation[11] about the periodic solution is

$$P(D)\gamma(t) + Q(D)n'\{x^*(t)\}\gamma(t) = 0, \tag{11.23}$$

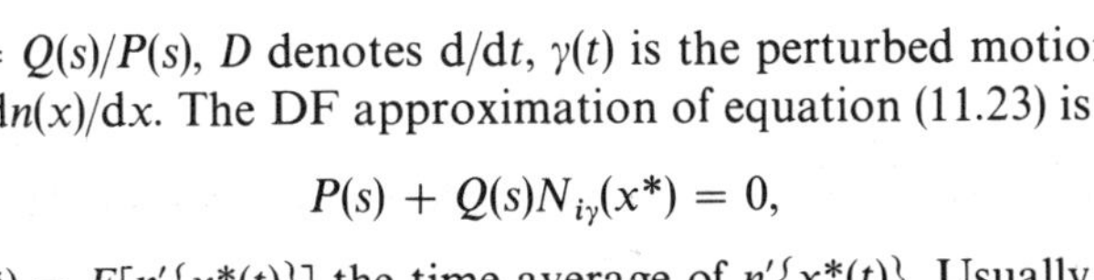

Figure 11.16 Nonlinear system with limit cycle

where $G(s) = Q(s)/P(s)$, D denotes d/dt, $\gamma(t)$ is the perturbed motion about $x^*(t)$ and $n'(x) = \mathrm{d}n(x)/\mathrm{d}x$. The DF approximation of equation (11.23) is

$$P(s) + Q(s)N_{i\gamma}(x^*) = 0, \tag{11.24}$$

where $N_{i\gamma}(x^*) = E[n'\{x^*(t)\}]$ the time average of $n'\{x^*(t)\}$. Usually since $x^*(t)$ is not known exactly $N_{i\gamma}(x^*)$ cannot be calculated, except in special cases such as for relay systems as discussed in section 6.9.3. If however the oscillation $x^*(t)$ is approximated by the DF solution $a \cos \omega t$, $N_{i\gamma}(x^*)$ becomes $N_{i\gamma}(a)$ with

$$N_{i\gamma}(a) = E\{n'(a \cos \omega t)\} = \int_{-a}^{a} n'(x)r(x)\,\mathrm{d}x,$$

in agreement with equation (5.11) for the IDF, and equation (11.24) becomes

$$P(s) + Q(s)N_{iy}(a) = 0,$$

in agreement with equation (5.60). The IDF can thus be regarded as the average of the periodically time varying parameter in the approximate variational equation obtained assuming a DF solution for the periodic mode.

Some results obtained using DFs to determine oscillations in systems with periodically varying parameters can be found in references 18 and 19. An example is now considered which illustrates the application of DF methods to a system with either periodically or random time varying parameters.

Figure 11.17 shows a feedback system with two time varying parameters, an ideal relay and a transfer function $G(s)$.[20] The gain $K(t)$ is equal to $K \cos \omega_1 t$ and the gain $K_2(t)$ to $K_0 + y(t)$, where in the first case considered $y(t)$ is assumed to be a high frequency Gaussian variation of mean squared value σ_y^2 and in the second case a periodic variation $b \sin \omega_n t$ with $\omega_n \gg \omega_1$ and $j\omega_n \neq k\omega_1$ where j and k are integers. In this latter case the system could represent an autonomous parametric amplifier with a heavily saturating gain.

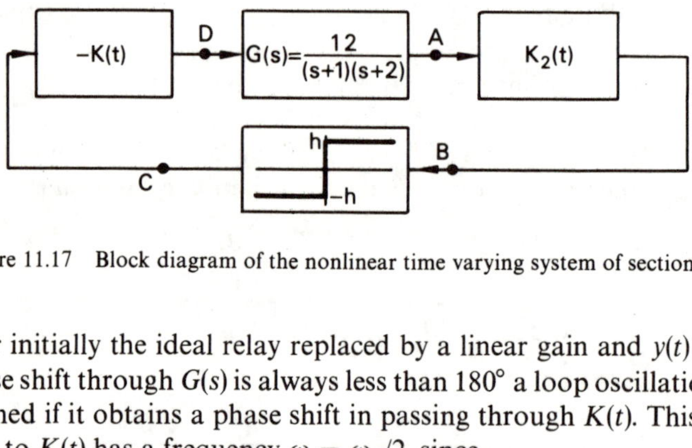

Figure 11.17 Block diagram of the nonlinear time varying system of section 11.4

Consider initially the ideal relay replaced by a linear gain and $y(t)$ to be zero. As the phase shift through $G(s)$ is always less than 180° a loop oscillation can only be maintained if it obtains a phase shift in passing through $K(t)$. This will occur if the input to $K(t)$ has a frequency $\omega = \omega_1/2$, since

$$\begin{aligned} K(t)\cos(\omega t + \phi) &= K \cos 2\omega t \cos(\omega t + \phi) \\ &= (K/2)\{\cos(\omega t - \phi) + \cos(3\omega t - \phi)\}, \end{aligned}$$

which shows a phase change of 2ϕ for the fundamental frequency ω. Similar arguments apply for the actual system of figure 11.17 with the loop oscillation frequency equal to $\omega_1/2$ and independent of $y(t)$.

Assuming the signal at A to be $x = a\cos(\omega t + \phi)$ then that at the output of the ideal relay is $n\{x(K_0 + y)\}$ where

$$\begin{aligned} n\{x(K_0 + y)\} &= h \quad \text{for } x(K_0 + y) > 0 \\ &\quad -h \quad \text{for } x(K_0 + y) < 0. \end{aligned} \tag{11.25}$$

For the signal x transmission from A to C, the nonlinearity output, is equivalent to passage through the modified nonlinearity

$$n_m(x) = \int_{-\infty}^{\infty} n\{x(K_0 + y)\}q(y)\,dy,$$

which on substituting the values of equation (11.25) gives

$$n_m(x) = h_1 \operatorname{sgn} x, \tag{11.26}$$

an ideal relay with

$$h_1 = h \operatorname{erf}(K_0/\sigma\sqrt{2}). \tag{11.27}$$

The signal at C thus contains all the odd harmonics of the sinusoid x together with random and random sinusoid cross product terms. Of the sinusoidal terms only those at frequencies ω and 3ω will produce components of frequency ω at D, the output of $K(t)$, and because of the low pass nature of $G(s)$ the other signal components which are mainly at higher frequencies may be neglected as a first approximation.

The appropriate signal at D is thus

$$(4h_1/\pi)\{\cos(\omega t + \phi) - (1/3)\cos(3\omega t + 3\phi)\}K(t), \tag{11.28}$$

which has a component at the frequency ω of

$$(4h_1K/\pi)\{1/2)\cos(\omega t - \phi) - (1/6)\cos(\omega t + 3\phi)\}$$
$$= (4h_1K/3\pi)\{\cos(\omega t + \phi)\cos 2\phi + 2\sin(\omega t + \phi)\sin 2\phi\}.$$

Relative to the signal $a\cos(\omega t + \phi)$ at A the signal at D thus has an amplitude of $(4h_1K/3\pi)(1 + 3\sin^2 2\phi)^{1/2}$ and a phase lag of $\tan^{-1}(2\tan 2\phi)$.

Assuming $\omega_1 = \pi$, then at the frequency $\omega = \pi/2$, $G(j\omega)$ has a phase lag of $\tan^{-1}(\pi/2) + \tan^{-1}(\pi/4) = 95{\cdot}6°$ and a gain of 2·54. For the oscillation to be maintained

$$\tan^{-1}(2\tan 2\phi) = 180 - 95{\cdot}6°,$$
$$\text{i.e. } \phi = 39{\cdot}4°,$$

and

$$a = \{(4 \times 2{\cdot}54)/3\pi\}hK \operatorname{erf}(K_0/\sigma\sqrt{2})\{1 + 3\sin^2 2\phi\}^{1/2},$$
$$\text{i.e. } a = 2{\cdot}12hK \operatorname{erf}(K_0/\sigma\sqrt{2}). \tag{11.29}$$

If the component of frequency 3ω in expression (11.28) is ignored the corresponding values for ϕ and a are

$$\phi = 42{\cdot}2°,$$

and

$$a = 1{\cdot}62hK \operatorname{erf}(K_0/\sigma\sqrt{2}),$$

which is in error by about 25 per cent. Figure 11.18(a) shows the theoretical solution for a from equation (11.29) with $h = 1{\cdot}2$ and experimental results obtained by analog simulation.

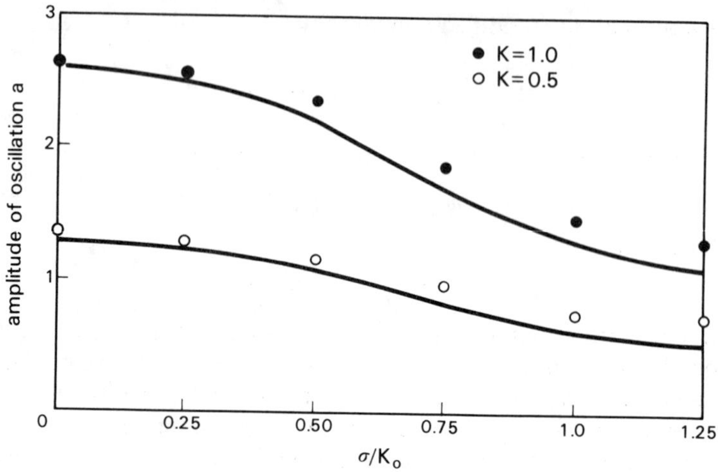

(a)

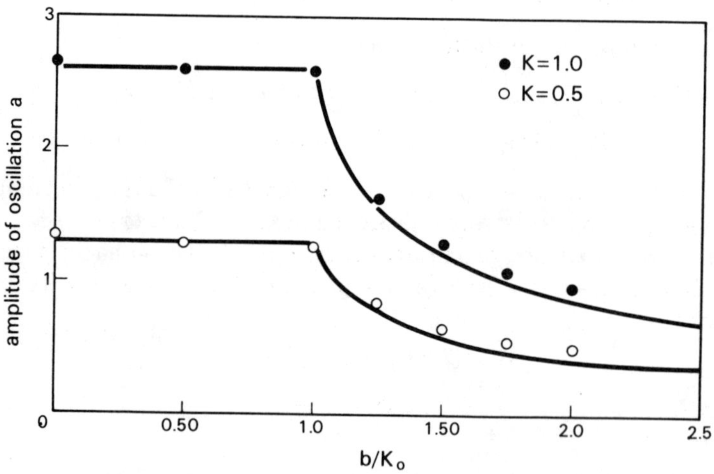

(b)

Figure 11.18 Variation of the amplitude of oscillation with $y(t)$ for the system of figure 11.17

When $y(t) = b \sin \omega_n t$ the only change in the above analysis is in equation (11.26) for the modified nonlinearity which becomes

$$n_m(x) = h_2 \operatorname{sgn} x, \tag{11.30}$$

where

$$h_2 = \begin{cases} h & \text{for } b < K_0 \\ (2h/\pi)\sin^{-1}(K_0/b) & \text{for } b > K_0. \end{cases} \tag{11.31}$$

This gives the result

$$a = \begin{cases} 2{\cdot}12hK & \text{for } b < K_0 \\ 2{\cdot}12hK(2/\pi)\sin^{-1}(K_0/b) & \text{for } b > K_0. \end{cases} \tag{11.32}$$

Figure 11.18(b) again shows a comparison of the theoretical solution, equation (11.32), and simulation results.

11.5 A system with two nonlinearities

The problem of cascaded nonlinearities separated by linear networks was discussed in section 4.6.1. It was pointed out then that the DF approach should be used with care, since an assumption that the inputs to all the nonlinearities involved are sinusoidal may be erroneous. To illustrate this point we consider the feedback system[6] of figure 11.19 which has ideal phase advance stabilisation and saturation of the error and power amplifier, by no means an uncommon occurrence. This system is seen to be stable when both nonlinearities are represented by their DFs since the loop phase shift is always less than 180°.

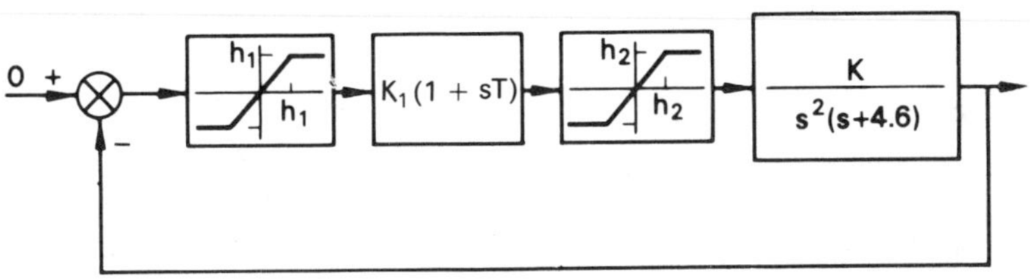

Figure 11.19 Block diagram of the closed loop system of section 11.5

Consider the nonlinearity $n(x, \dot{x})$ to be from the input to the first saturation to the output of the second then

$$n(x,\dot{x}) = \begin{cases} K_1(x + T\dot{x}) & \text{if } |x| < h_1 \text{ and } |K_1(x+T\dot{x})| < h_2 \\ h_2 \operatorname{sgn}\{K_1(x+T\dot{x})\} & \text{if } |x| < h_1 \text{ and } |K_1(x+T\dot{x})| > h_2 \\ K_1 h_1 \operatorname{sgn} x & \text{if } |x| > h_1 \text{ and } K_1 h_1 < h_2 \\ h_2 \operatorname{sgn} x & \text{if } |x| > h_1 \text{ and } K_1 h_1 > h_2. \end{cases}$$

The equivalent gains K_x and $K_{\dot{x}}$ can be found from equations (10.47) and (10.48) using equation (10.46) and the above expression for $n(x, \dot{x})$. Due to the linear segmented nature of the saturation characteristics, specifically the first, the actual evaluation is quite involved requiring the double integrals of equations (10.47) and (10.48) to be split up to cover all the different regions in the $x - \dot{x}$ plane for which $n(x, \dot{x})$ has different values. Figure 11.20 gives a comparison of the theoretical and experimental results for the DF, $M(a)e^{-j\psi(a)}$, where $M(a) = (K_x^2 + K_{\dot{x}}^2\omega^2)^{1/2}$ and $\psi(a) = \tan^{-1}(K_{\dot{x}}\omega/K_x)$. It is also interesting to note the considerable effect of the gain K_1 on $M(a)$ and $\psi(a)$ shown in figure 11.21 for

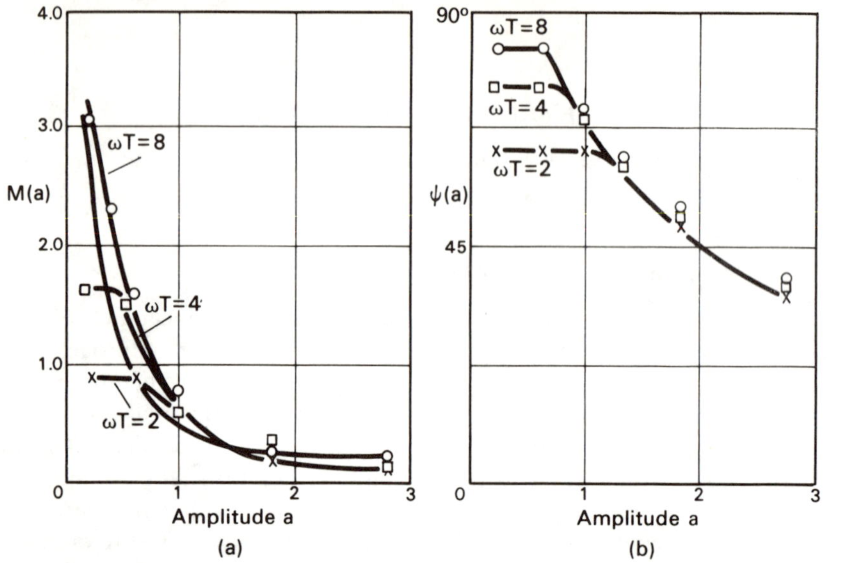

Figure 11.20 Describing function for $n(x, \dot{x})$ of section 11.5 with $K_1 = 0{\cdot}4$, $h_1 = 0{\cdot}6$ and $h_2 = 0{\cdot}8$

a fixed input amplitude and frequency. Figure 11.22 shows the solution for the oscillation frequency in the K_x, $K_{\dot{x}}$ parameter plane for $K_1 = 4{\cdot}0$ and $K = 5{\cdot}0$. As the DF for the nonlinearity is a function of both a and ω it is shown plotted with a a parameter for two values of ω. The oscillation, perhaps surprisingly, is seen to be unstable. Figure 11.23 shows a comparison of the theoretical and simulation results for the amplitude and frequency of oscillation as a function of the plant gain K. The experimental solutions were obtained by simulating the system in reverse time to get stable limit cycles.

11.6 Nonlinear phase advance compensation

In section 4.5 the attention of the reader was drawn to the possibility of using nonlinear compensators and indeed a specific application was discussed. The purpose of this section is to investigate a class of nonlinear phase advance compensators. Linear compensators, which are minimum phase networks, have the inherent disadvantage that their gain and phase characteristics are related. For example, if phase lead is introduced then so is an increased gain, whereas the ideal situation is a network with independent gain and phase characteristics.

Levinson[21] gives several examples of nonlinear circuits exhibiting gain reduction with frequency without phase change and two specific networks, namely the span filter[22] and driven limiter,[22] have been studied in some detail for control system applications. Figure 11.24 shows block diagrams of the span filter and driven limiter, both of which are split-path nonlinear circuits, the configuration description which led to the acronym span. Usually $G_1(s)$ will be a minimum phase network so that without loss of generality it will be assumed

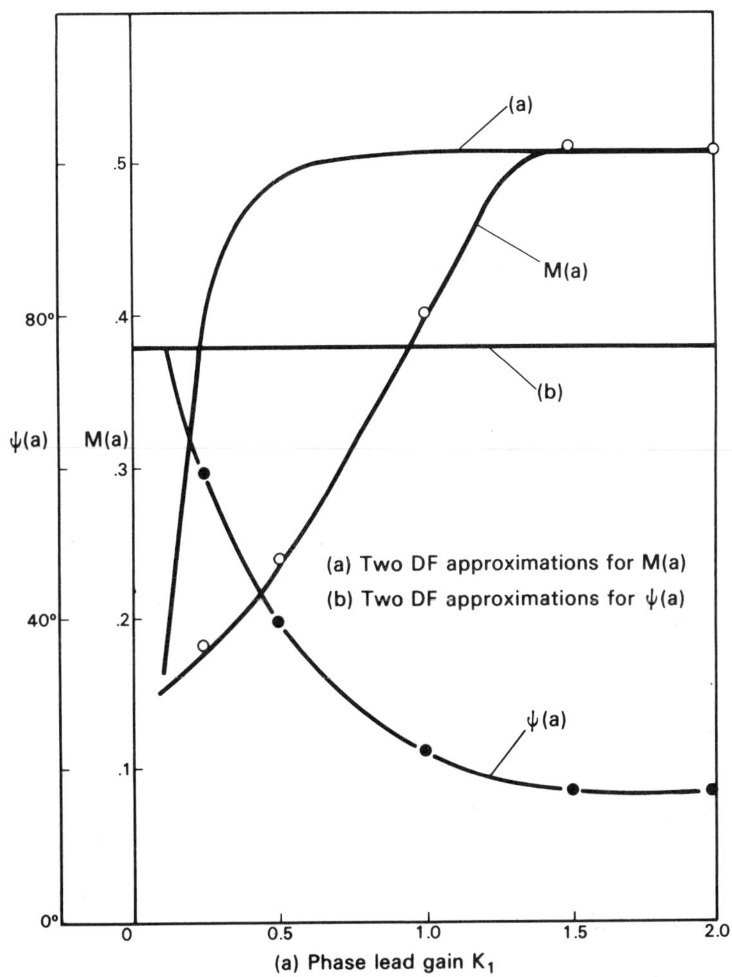

Figure 11.21 Dependence of DF for $n(x, \dot{x})$ of section 11.5 on phase lead gain K_1 for $\omega = 0{\cdot}4$, $a = 2$, $h_1 = 0{\cdot}6$ and $h_2 = 0{\cdot}8$

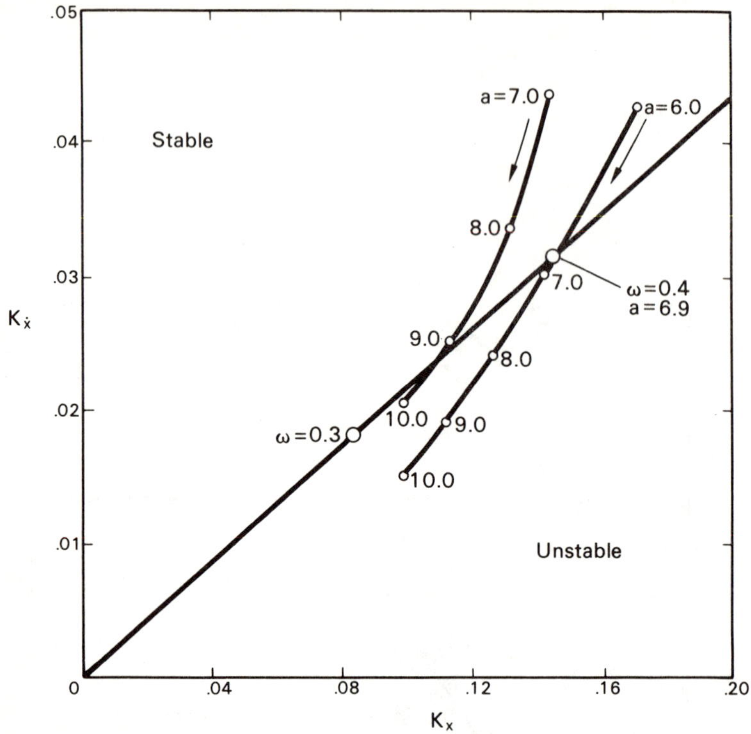

Figure 11.22 Parameter plane diagram for finding the frequency and amplitude of oscillation of the system of figure 11.17 with $K_1 = 4{\cdot}0$, $K = 5{\cdot}0$, $T = 1$, $h_1 = 0{\cdot}6$ and $h_2 = 0{\cdot}8$

to be a unit gain as it can always be removed from the configuration, combined with the system transfer function and $G_2(s)$ replaced by $G_2(s)/G_1(s)$.

In both references 22 and 23 the DF representations for the networks with $G_2(s)$ a phase lead were obtained by drawing the network waveforms and evaluating the fundamental of the signal obtained at the output by Fourier analysis. Both networks were shown to have DFs which were independent of the input amplitude and to produce phase lead without the corresponding increase in gain. Using the methods of the previous chapter the equivalent linear gain models of these networks can be evaluated analytically for various types of input. Assuming $G_2(s) = 1 + sT$, an ideal phase advance, it can be seen from figure 11.24 for the span filter that it is defined by

$$\begin{aligned} n(x, \dot{x}) &= |x| \quad &&\text{for } x + T\dot{x} > 0 \\ & -|x| \quad &&\text{for } x + T\dot{x} < 0, \end{aligned} \tag{11.33}$$

and similarly the driven limiter is defined by

$$\begin{aligned} n(x, \dot{x}) &= x\,\mathrm{sgn}(x + T\dot{x}) \quad &&\text{for } |x| < |x + T\dot{x}| \\ & x + T\dot{x} \quad &&\text{for } |x| > |x + T\dot{x}|. \end{aligned} \tag{11.34}$$

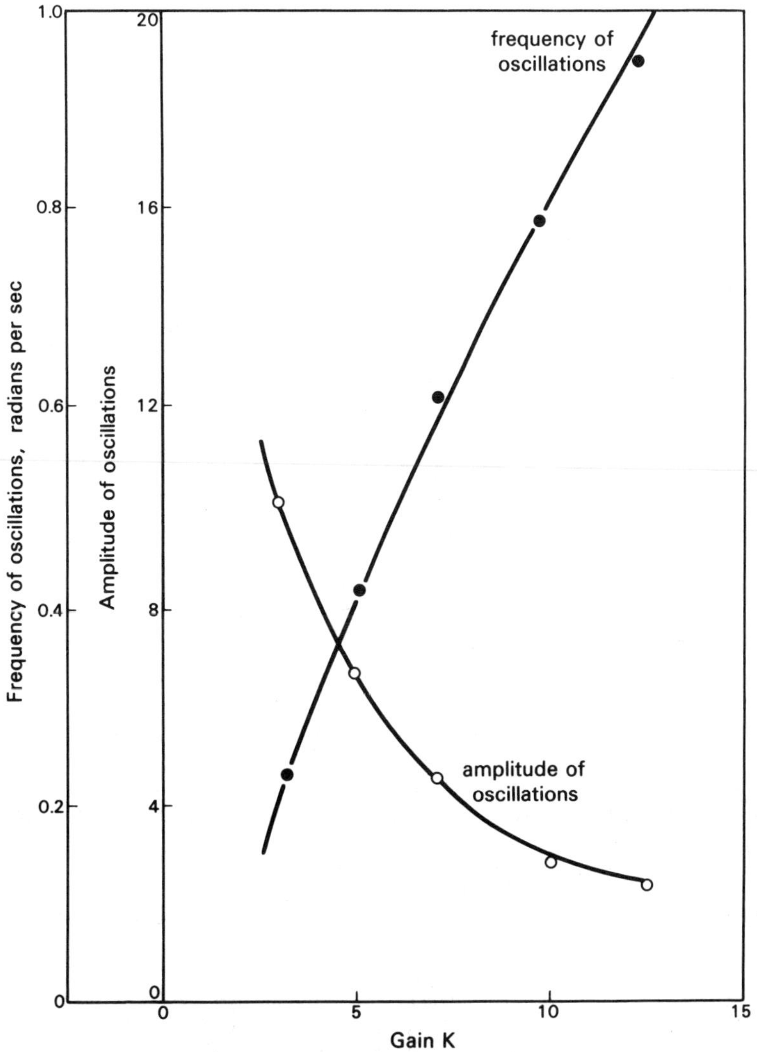

Figure 11.23 Amplitude and frequency of oscillations of the system of figure 11.17 with $K_1 = 4{\cdot}0$, $T = 1$, $h_1 = 0{\cdot}6$ and $h_2 = 0{\cdot}8$

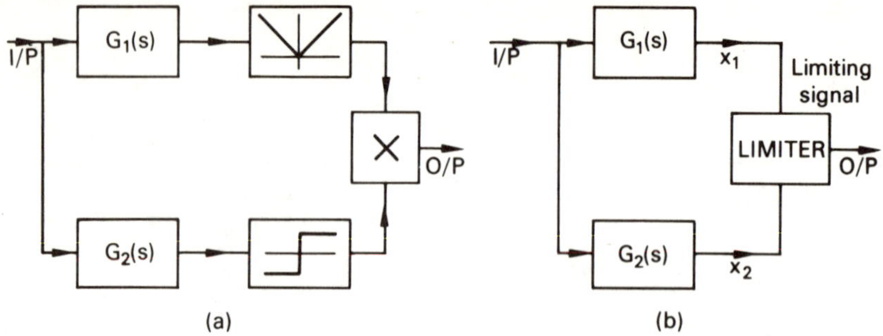

Figure 11.24 Block diagrams of (a) span filter and (b) driven limiter

The derivations of the equivalent gains for the driven limiter are given as problems in the previous chapter so that here we consider only the span filter. Figure 11.25 shows the values of $n(x, \dot{x})$ for the span filter in the various regions of the $x - \dot{x}$

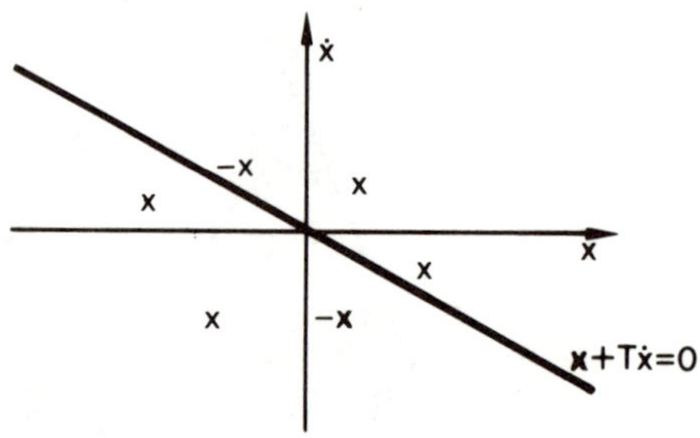

Figure 11.25 Values of span filter in $x - \dot{x}$ plane

plane. From the symmetry it can be seen that for a sine input using equation (10.47)

$$K_x = (4/a^2)\left\{\int_{-a}^{a}\int_{0}^{a} x^2 r(x, \dot{x}/\omega)\, d(\dot{x}/\omega)\, dx - 2\int_{-a}^{0}\int_{0}^{-x/\omega T} x^2 r(x, \dot{x}/\omega)\, d(\dot{x}/\omega)\, dx\right\}$$

$$= (4/a^2)\left\{(a^2/8) + 2\int_{0}^{a}\int_{0}^{x/\omega T} x^2 r(x, \dot{x}/\omega)\, d(\dot{x}/\omega)\, dx\right\}.$$

Since K_x will be continuous in T and when $T = 0$, $K_x = 1$ as the network is a linear gain, differentiating the above expression with respect to T gives

$$dK_x/dT = (8/a^2)\int_{0}^{a} (-x^3/\omega T^2) r(x, x/\omega T)\, dx,$$

which on substituting for $r(x, \dot{x}/\omega)$ gives

$$dK_x/dT = (-4/a^3\pi\omega T^2)\int_0^a x^3\delta\{(x^2 + x^2/\omega^2T^2)^{1/2} - a\}\,dx.$$

Writing $\omega T = \alpha$ and substituting $r^2 = x^2 + x^2/\alpha^2$ this becomes

$$dK_x/dT = (-4/a^3\pi\omega T^2)\int_0^{r_1} \{\alpha^2/(1 + \alpha^2)\}^2 r^3\delta(\pm r - a)\,dr$$

$$= -4\omega\alpha^2/\pi(1 + \alpha^2)^2,$$

that is

$$dK_x/d\alpha = -4\alpha^2/\pi(1 + \alpha^2)^2.$$

Integrating and evaluating the integration constant from $K_x = 1$ when $\alpha = 0$ gives

$$K_x = (2/\pi)\{\tan^{-1}(1/\alpha) + \alpha/(1 + \alpha^2)\}. \qquad (11.35)$$

Similarly using equation (10.48) it can be shown that

$$K_{\dot{x}} = (2T/\pi)\{\alpha/(1 + \alpha^2)\}. \qquad (11.36)$$

Using equations (10.42) and (10.43) the corresponding gains K_y and $K_{\dot{y}}$ for a Gaussian input, y, are found to be identical with equations (11.35) and (11.36) but with $\alpha = \sigma_{\dot{y}}T/\sigma_y$. Thus the equivalent gains to both sine and Gaussian inputs are independent of their magnitudes.

The DF model for the span filter with ideal phase advance is thus

$$(2/\pi)\{\tan^{-1}(1/\alpha) + \alpha/(1 + \alpha^2) + j\alpha^2/(1 + \alpha^2)\}.$$

The polar frequency response of this expression is shown in figure 11.26 together with that of the driven limiter with $G_2(s) = 1 + sT$ and of the ideal phase advance

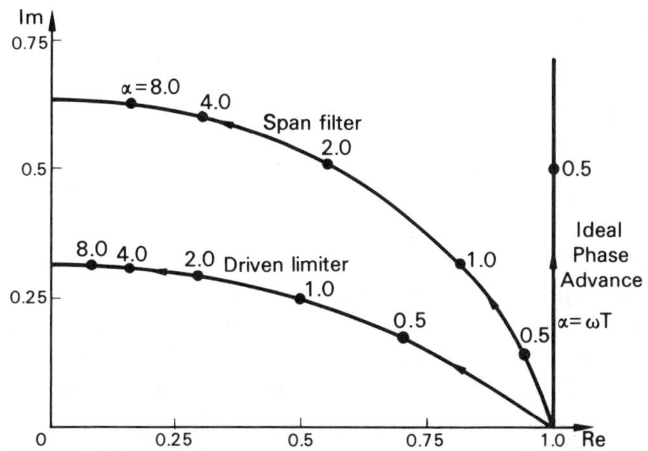

Figure 11.26 Frequency response loci of compensators

$1 + sT$. The corresponding gains for the driven limiter are $K_x = (2/\pi)\tan^{-1}(1/\alpha)$ and $K_{\dot{x}} = (T/\pi)\tan^{-1}(1/\alpha)$. Although the above results have been derived using an ideal phase advance transfer function, it is only slightly more difficult to obtain them for the non-ideal case using the techniques of the previous chapter for evaluating equivalent gains with correlated inputs (see question P.10.12). The conclusion that the networks give phase lead without the corresponding increased gain of the linear transfer function, as seen in figure 11.26, remains true.

It can be shown[20] that the two properties: (a) the equivalent gains to both sine and Gaussian inputs are independent of the magnitude of the inputs and (b) the equivalent gain expressions for sinusoidal inputs can be obtained by substituting $\omega = \sigma_{\dot{y}}/\sigma_y$ in the equivalent gain expressions for a Gaussian input, are satisfied by the general class of nonlinearities defined by

$$\begin{aligned} n(x, \dot{x}) &= a_i x + b_i \dot{x} \quad \text{if } x > 0;\ m_i x < \dot{x} < m_{i-1} x \\ &\ a_i x + b_i \dot{x} \quad \text{if } x < 0;\ m_i x > \dot{x} > m_{i-1} x \\ &\ i = 1, 2, \ldots, n. \end{aligned} \tag{11.37}$$

We note that these nonlinearities have odd symmetry in x and $\dot{x}$ and are linear in x and $\dot{x}$ in the various regions of the $x - \dot{x}$ plane which are subdivided by straight lines through the origin. The corresponding equivalent gains for a sine input are

$$\begin{aligned} K_x = \sum_{i=1}^{n} (a_i/\pi)&\left\{\frac{\omega m_{i-1}}{\omega^2 + m_{i-1}^2} - \frac{\omega m_i}{\omega^2 + m_i^2} + \tan^{-1}(\omega/m_i) - \tan^{-1}(\omega/m_{i-1})\right\} \\ &+ (b_i/\pi)\left\{\frac{\omega m_i^2}{\omega^2 + m_i^2} - \frac{\omega m_{i-1}^2}{\omega^2 + m_{i-1}^2}\right\}, \end{aligned} \tag{11.38}$$

and

$$\begin{aligned} K_{\dot{x}} = \sum_{i=1}^{n} (a_i/\pi)&\left\{\frac{\omega}{\omega^2 + m_i^2} - \frac{\omega}{\omega^2 + m_{i-1}^2}\right\} + (b_i/\pi)\left\{\tan^{-1}(\omega/m_i)\right. \\ &\left. - \tan^{-1}(\omega/m_{i-1}) + \frac{\omega m_i}{\omega^2 + m_i^2} - \frac{\omega m_{i-1}}{\omega^2 + m_{i-1}^2}\right\}. \end{aligned} \tag{11.39}$$

We now consider briefly two applications of the span filter. Compensation of a linear system with $G(s) = 5(1 - 0{\cdot}1s)/s(1 + 0{\cdot}05s)$ and unity feedback is discussed in reference 22. Figure 11.27 shows Bode diagrams for $G(s)$ before and after compensation and figure 11.28 shows corresponding step responses. Although the uncompensated system is stable its step response has an appreciable overshoot and relatively long settling time. There is not much that can be done to improve the response with linear compensation, a lag network increases the step response overshoot and the behaviour with a linear lead of transfer function $(1 + 0{\cdot}05s)/(1 + 0{\cdot}005s)$, as shown in the figures, is little better. The gain margin remains at 6 dB and the initial opposite side overshoot is increased due to the improved high frequency gain. The addition of the span filter with $G_2(s) = (1 + 0{\cdot}1s)/(1 + 0{\cdot}01s)$

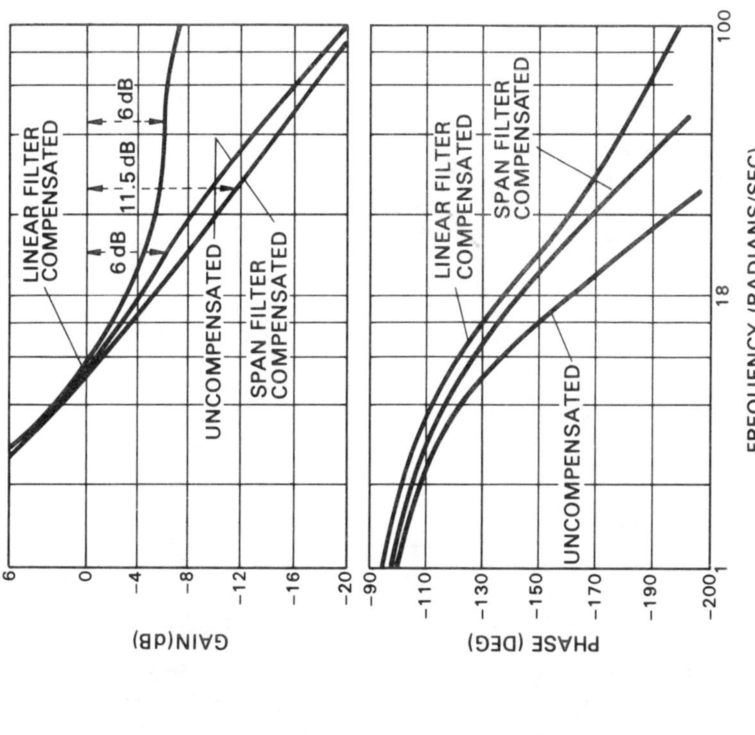

Figure 11.27 Bode diagrams for systems with nonminimum phase $G(s)$

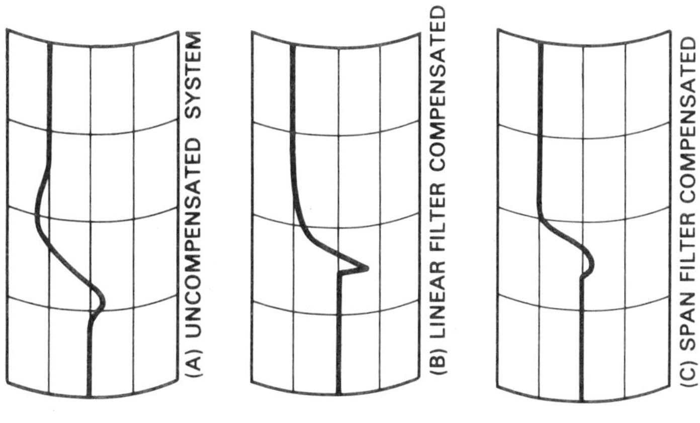

Figure 11.28 Step responses of uncompensated and compensated system

however, increases the gain margin to 11·5 dB and considerably improves the step response which is amplitude independent.

The behaviour of nonlinear phase advance compensators in systems with Gaussian inputs was studied in references 20 and 24. The results of the latter paper compared mean squared errors for a system with unity feedback and a feedforward transfer function consisting of a compensator and $G(s) = 10/s(s + 1)$. White noise was applied at the system input through a transfer function of $1/(s + b)^2$. The compensator was chosen to be either a phase lead with transfer function $(1 + sT)/(1 + \alpha sT)$ and $\alpha = 0{\cdot}4$ or span filter with $G_2(s)$ the same transfer function. The theoretical mean squared errors were calculated for both the linear system and the system with span filter using its equivalent gain model assuming $\alpha = 0$. T was chosen equal to 0·2 for the phase lead, which corresponds to a phase margin of 50°, and three different values for the span filter. The first value of T being 0·2, the second 0·5 which gives a phase margin of 50° for the system with the span filter, and finally the optimum value of 0·8 which gives a minimum mean squared error with the span filter when $b = 1{\cdot}0$. Figure 11.29 shows the per unit mean squared errors for these cases as a function of the input

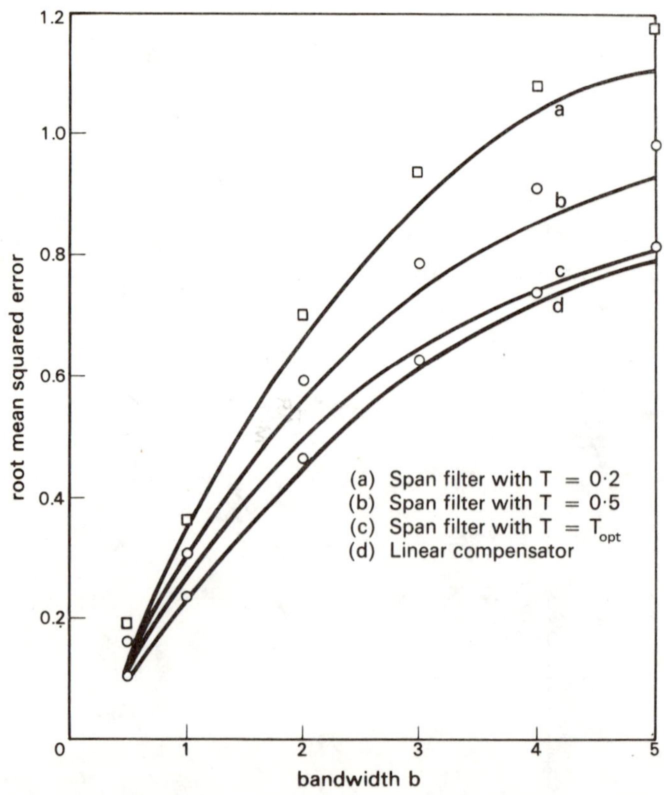

Figure 11.29 Variation of root mean squared error with input bandwidth for the system of section 11.6

bandwidth, *b*. The poor showing of the span filter is presumably due to the fact that the increased loop gain of the linear phase advance is desirable for reducing the mean squared error. Thus even though the nonlinear phase advance compensators discussed above are essentially amplitude independent these results suggest that it is dangerous to draw conclusions regarding their performance with any input from that found to a specific input.

REFERENCES

1. Hatanaka, H.: 'The frequency responses and jump resonance phenomena of nonlinear feedback control systems', *Trans. A.S.M.E., J. Basic Eng.*, 1963, D-85, pp. 236–242.
2. Fieguth, W.: 'Multi-input quasilinearization', *Ph.D. Thesis, University of New Brunswick*, 1968, pp. 174–178.
3. Anderson, R. A. and Soudack, A. C.: 'On the resonance properties of quasilinear second order differential-difference equations', *Int. J. Control*, 1970, 11, pp. 33–48.
4. Lamba, S. S. and Kavanagh, R. J.: 'Phenomenon of isolated jump resonance and its applications', *Proc. I.E.E.*, 1971, 118, pp. 1047–1050.
5. Rajagopalan, P. K. and Singh, Y. P.: 'Analysis of harmonic and almost periodic oscillations in forced self-oscillating systems', *Paper 41.6, I.F.A.C. Congress*, Warsaw, June 1969.
6. Balasubramanian, R. and Atherton, D. P.: 'Quasilinear analysis of feedback systems containing a nonlinearity with output dependent upon both the input and its derivative', *Paper No. 6, Session B2, Proc. A.I.C.A. Congress*, Lausanne, 1967.
7. Tucker, D. G.: *Circuits with periodically-varying parameters* (Macdonald, London, 1964).
8. Mathieu, E.: *J. de Lionville*, 1868, 13, p. 137.
9. Hill, G. W.: *Amer. J. Math.*, 1877, 1, pp. 129 and 245.
10. Rayleigh, Lord: *Theory of sound* (Macmillan and Co., Ltd., 1893, 1, 2nd Ed.).
11. Coddington, E. A. and Levinson, N.: *Theory of ordinary differential equations* (McGraw-Hill, 1955, Section 13.2).
12. Caughey, T. K. and Dienes, J. K.: 'The behaviour of linear systems with random parametric excitation', *J. of Math. and Phys.*, 1962, 41, p. 300.
13. Gray, A. H., Jr.: 'Behaviour of linear systems with random parametric excitation', *J. Acoust. Soc. Amer.*, 1965, 37, pp. 235–239.
14. Ariaratnam, S. T. and Graefe, P. W. U.: 'Linear systems with stochastic coefficients', *Int. J. of Control*, 1965, 1, pp. 239–249.
15. Bogdanoff, J. L. and Kozin, F.: 'Moments of the output of linear random systems', *J. Acoust. Soc. Amer.*, 1962, 34, pp. 1063–1066.
16. Kozin, F.: 'Stability of stochastic systems', *Proc. I.F.A.C. Conference*, London, 1966, pp. 3A.1–3A.8.
17. Leffen, D. J. S.: 'Measurements of random systems', *M.Sc. Thesis, University of New Brunswick*, 1969.
18. Leonhard, A.: 'Extension of the describing function method to the investigation of parametric oscillations', *Proc. I.F.A.C.*, Basel, Switzerland, Butterworth, London, 1963, pp. 21–28.
19. Gelb, A. and Vander Velde, W. E.: *Multiple-input describing functions and nonlinear system design* (McGraw-Hill, 1968, pp. 291–295).
20. Balasubramanian, R.: 'Analysis of systems containing a multidimensional nonlinearity', *Ph.D. Thesis, University of New Brunswick*, 1968.
21. Levinson, E.: 'Gain-phase relations of nonlinear circuits', *I.R.E. Nat. Conv. Rec.*, part 4, 1958, pp. 141–152.
22. Foster, W. C., Gieseking, D. L. and Waymayer, W. K.: 'A nonlinear filter for independent gain and phase (with applications)', *Trans. A.S.M.E., J. Basic Eng.*, 1966, 88, pp. 457–462.
23. Bailey, A. R.: 'Stabilization of control systems by the use of driven limiters', *Proc. I.E.E.*, 1966, 113, pp. 169–174.
24. Balasubramanian, R. and Atherton, D. P.: 'Performance of feedback systems with nonlinear phase advance compensators to Gaussian inputs', *Electronics Letters*, 1967, 3, pp. 563–565.

BIBLIOGRAPHY

Balasubramanian, R. and Atherton, D. P.: 'Prediction of jump resonance in systems containing certain multidimensional nonlinearities', *Proc. I.E.E.*, 1968, 115, pp. 1369–1372.

Bonenn, Z.: 'Relative stability of oscillations in non-linear control systems', *Proc. I.F.A.C.*, Basel, Switzerland, 1963, pp. 214/1–4.

Freeman, E. A. and Devine, M. L.: 'Calculation and advance of jump resonance in feedback amplifier circuits', *Proc. I.E.E.*, 1968, 115, pp. 229–235.

Freeman, E. A. and Barney, G. C.: 'Divergent oscillations and their excitation in control systems with two saturation type non-linear elements', *Proc. I.E.E.*, 1963, 110, pp. 1096–1106.

Giridharagopal, K.: 'Synthesis of a class of multimodal systems', *M.Sc.E. Thesis, University of New Brunswick*, 1971.

Johnson, R. A. and Leach, B. W.: 'Stability of oscillations in low-order non-linear systems', *I.E.E.E. Trans.*, 1972, AC-17, pp. 672–675.

Lamba, S. S.: 'Partial avoidance of jump resonance in feedback systems', *International J. of Control*, 1968, 8, pp. 537–542.

Lamba, S. S. and Kavanagh, R. J.: 'Jump resonance criteria for systems containing double-valued and frequency-dependent non-linearities', *Proc. I.E.E.*, 1969, 116, pp. 1225–1228.

Lamba, S. S. and Kavanagh, R. J.: 'Synthesis of certain nonlinear systems for multimodal operation', *Proc. I.E.E.*, 1973, 120, pp. 919–920.

Lamba, S. S. and Avtar Singh: 'Synthesis of hysteresis free non-linear sharp cut-off filters', *Electronics Letters*, 1972, 8, pp. 488–489.

Levinson, E.: 'Non-linear feedback control systems', *Part 3–5 Electro Technology*, 72, (Sept. 1962), pp. 136–146 (Oct. 1962), pp. 131–140 (Nov. 1962), pp. 71–82.

Nath, N. G., Bhattacharyya, A. K. and Choudrey, A. K.: 'Transient response improvement of servo systems by discrete control of a parameter', *Journal of Electronics and Control*, 1964, 4, pp. 449–464.

Ochiai, S. and Oldenburger, R.: 'The second describing function and removal of jump resonance by signal stabilization', *J.A.C.C.*, University of Washington, 1966, pp. 569–578.

Riemenschneider, A. L. and Mastascusa, E. J.: 'A correlation between the Fukuma–Matsubara and circle criteria for the determination of jump resonance', *I.E.E.E. Trans.*, 1970, AC-15, pp. 507–508.

Sawaragi, Y. and Sugai, N.: 'The stability of problems associated with steady state responses of a non-linear control system to a random signal', *Engineering Research Institute*, Kyoto University, 1960, 10, pp. 89–98.

Smetanin, N. A.: 'Harmonic linearization of a component with variable structure', *Auto. and Remote Control*, 1965, 26, pp. 998–1001.

Summers, C. M.: 'An unstable non-linear circuit', *A.I.E.E.*, 1940, pp. 273–276.

West, J. C. and Jayawant, B. V.: 'A new linear oscillating motor', *Proc. I.E.E.*, 1962, 109A, pp. 292–300.

Zhiltson, K. K.: 'Harmonic linearization of systems with logical control rule', *Auto. and Remote Control*, 1966, 27, pp. 286–289.

PROBLEMS

P.11.1 Show that for any odd symmetrical single valued nonlinearity, $n(x)$, confined within the sector for $x > 0$ defined by the lines k_1x and k_2x that $k_1 < K_1 < k_2$ and $k_1 < K_2 < k_2$ for $K_2 > K_1$. Hence show that all the IDF loci, $-1/N_{ib}(a)$, for the nonlinearity lie within a circle centred on the negative real axis and which intersects it at $-1/k_1$ and $-1/k_2$. (Note the similarity with the Zames circle in absolute stability theory.)

P.11.2 For the system shown in figure 11.1 with $G_c(s) = 1$, $n(x) = x^3$, $G(s) = 1/s(1 + 0{\cdot}02s)(1 + 0{\cdot}01s)(1 + 0{\cdot}005s)$ and $n(t)$ a Gaussian signal of mean squared value σ, obtain a curve showing the maximum input amplitude, R_{max}, for which there are no resonant jumps as a function of σ. Take values of σ from 0 to 4. Does the noise signal make the system more or less stable?

P.11.3 Figure P.11.1 shows the block diagram of a feedback system which has been designed to achieve an optimum step response by switching the output derivative feedback according to the law

$$y = K_1\dot{x} \quad \text{if } x + 0{\cdot}1\dot{x} > 0$$
$$K_2\dot{x} \quad \text{if } x + 0{\cdot}1\dot{x} < 0.$$

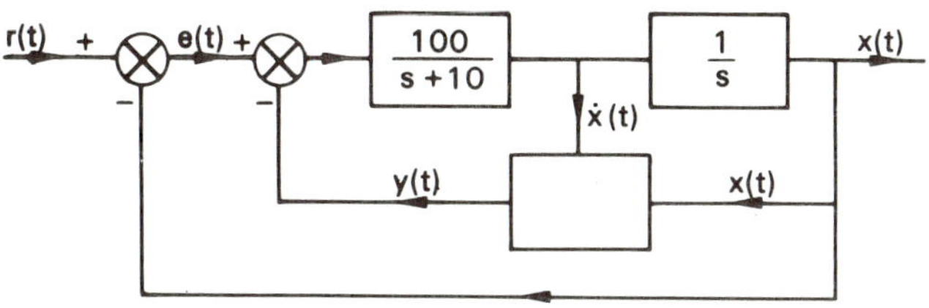

Figure P.11.1

Show that for a random input $r(t)$, the system behaves essentially as a linear system with $y = \{(k_1 + k_2)/2\}\dot{x}$. What is the mean squared error for white noise input?

P.11.4 Evaluate the maximum value of the gain K for which the system shown in figure P.11.2 will be stable. Does the system exhibit a stable limit cycle for values of K greater than this critical value?

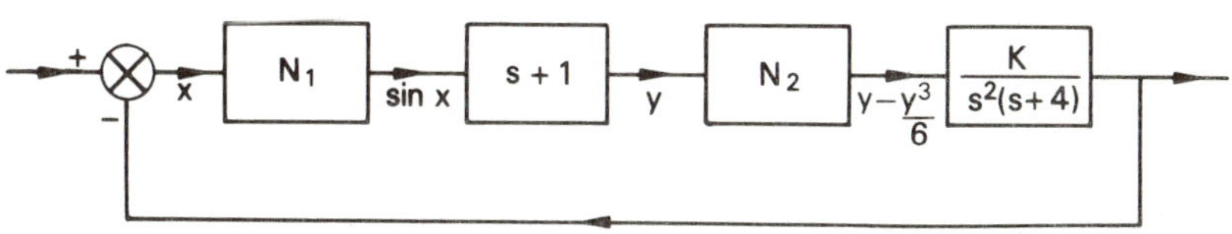

Figure P.11.2

P.11.5 In the system of figure 11.1, $G_c(s) = 1$, $n(t) = 0$, $n(x)$ is ideal saturation of unit gain and $G(s) = 1/s(s + 0{\cdot}1)$. With a sinusoidal input $R \sin \omega t$ jump down occurs in the frequency response on decreasing ω when $\omega = \omega_1$. If the same experiment is repeated first with a triangular wave of peak T and then a square wave of peak S with jump down occurring in both cases at the frequency ω_1 how are T and S related to R?

Chapter 12

Nonlinear Multivariable Systems

12.1 Introduction

The material presented to this point has been concerned with problems in nonlinear feedback systems containing plants, $G(s)$, with a single input and output. In many control situations however the plant has several input and output signals. The magnitude and frequency of the output voltage from an alternator may be controlled by the input speed and field current; an example of a plant with two input and two output variables. On the other hand chemical processes often have a large number of input and output variables. In the general case each input will affect all the outputs and the plant in the linear time invariant case will be described by a matrix of transfer functions. The analysis of multivariable systems can be accomplished as a relatively straightforward extension of single variable theory using matrix transfer functions. The satisfactory engineering design of linear multivariable systems on the other hand is much more difficult and considerable effort is currently being devoted to this topic. Although elegant analytical solutions can be obtained using optimal control theory the industrial realisations of such schemes often prove both uneconomical and unsatisfactory in performance.

The purpose of this chapter is primarily to investigate describing function analysis methods for nonlinear two variable systems. Since describing function methods rely on the frequency domain description of the linear elements of a multivariable system, it is appropriate to briefly review the basic concepts of frequency domain methods. This is done in the next section, after which we examine the stability of nonlinear systems, including exact methods for the determination of oscillations in relay systems using an extension of Tsypkin's method and finally the determination of mean squared errors for systems with random inputs. The reader should regard this work as a brief introduction to an area which as yet has received little attention and is fraught with difficulties. Because of their uniqueness it has been seen how difficult it is to generalise results for single variable nonlinear systems; it can therefore be appreciated how much more difficult the task will be for more complex nonlinear structures.

12.2 Transfer function analysis

Consider the general multivariable feedback system shown in figure 12.1. $\mathbf{G}(s)$ is an $m \times n$ plant transfer function matrix, $\mathbf{K}(s)$ is an $n \times m$ matrix of controller transfer functions and $\mathbf{H}(s)$ an $m \times m$ matrix of feedback transfer functions.

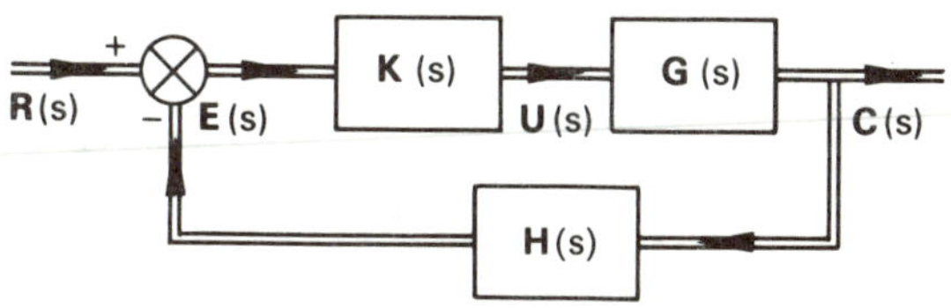

Figure 12.1 Linear multivariable system

The input, output, and error vector transforms $\mathbf{R}(s)$, $\mathbf{C}(s)$ and $\mathbf{E}(s)$ are m vectors and the plant input $\mathbf{U}(s)$ is an n vector. The signal vectors and transfer function matrices are related by expressions similar to those for single variable systems, for example

$$\mathbf{C}(s) = \mathbf{G}(s)\mathbf{U}(s), \tag{12.1}$$

or, in terms of the individual components,

$$\begin{pmatrix} C_1(s) \\ C_2(s) \\ \vdots \\ C_m(s) \end{pmatrix} = \begin{pmatrix} G_{11} & G_{12} & \cdots & G_{1n} \\ G_{21} & G_{22} & \cdots & G_{2n} \\ \vdots & \vdots & \vdots & \vdots \\ G_{m1} & G_{m2} & \cdots & G_{mn} \end{pmatrix} \begin{pmatrix} U_1(s) \\ U_2(s) \\ \vdots \\ U_n(s) \end{pmatrix}. \tag{12.2}$$

If the closed loop of figure 12.1 is broken at the error point and a signal $\mathbf{E}(s)$ injected, the signal fed back to the break point with $\mathbf{R}(s) = 0$ is

$$-\mathbf{H}(s)\mathbf{G}(s)\mathbf{K}(s)\mathbf{E}(s),$$

so that minus the ratio of this fedback signal to that injected is

$$\mathbf{T}_e(s) = \mathbf{H}(s)\mathbf{G}(s)\mathbf{K}(s). \tag{12.3}$$

The matrix $\mathbf{T}_e(s)$ is called the return-ratio matrix for the error vertex.[1] Since matrix multiplication is not commutative it should be noted that if the loop is broken at other points, using a similar notation

$$\mathbf{T}_u(s) = \mathbf{K}(s)\mathbf{H}(s)\mathbf{G}(s), \tag{12.4}$$

and

$$\mathbf{T}_c(s) = \mathbf{G}(s)\mathbf{K}(s)\mathbf{H}(s). \tag{12.5}$$

Now

$$\mathbf{E}(s) = \mathbf{R}(s) - \mathbf{H}(s)\mathbf{C}(s), \tag{12.6}$$

and

$$\mathbf{C}(s) = \mathbf{G}(s)\mathbf{K}(s)\mathbf{E}(s), \tag{12.7}$$

so that premultiplying equation (12.7) by $\mathbf{K}^{-1}(s)\mathbf{G}^{-1}(s)$ and eliminating $\mathbf{E}(s)$ between the equations gives

$$\mathbf{R}(s) = \{\mathbf{H}(s) + \mathbf{K}^{-1}(s)\mathbf{G}^{-1}(s)\}\mathbf{C}(s),$$

which may be written

$$\mathbf{C}(s) = \{\mathbf{I}_m + \mathbf{G}(s)\mathbf{K}(s)\mathbf{H}(s)\}^{-1}\mathbf{G}(s)\mathbf{K}(s)\mathbf{R}(s).$$

Thus if the closed loop transfer function matrix relating $\mathbf{C}(s)$ and $\mathbf{R}(s)$ is denoted by $\mathbf{W}(s)$, then

$$\mathbf{W}(s) = \mathbf{F}_c^{-1}(s)\mathbf{G}(s)\mathbf{K}(s), \tag{12.8}$$

where

$$\mathbf{F}_c(s) = \mathbf{I}_m + \mathbf{G}(s)\mathbf{K}(s)\mathbf{H}(s) = \mathbf{I}_m + \mathbf{T}_c(s), \tag{12.9}$$

and is defined as the system return-difference matrix[1] measured at the output of the plant. The corresponding return difference matrices for other positions in the loop are

$$\mathbf{F}_e(s) = \mathbf{I}_m + \mathbf{H}(s)\mathbf{G}(s)\mathbf{K}(s) = \mathbf{I}_m + \mathbf{T}_e(s), \tag{12.10}$$

and

$$\mathbf{F}_u(s) = \mathbf{I}_n + \mathbf{K}(s)\mathbf{H}(s)\mathbf{G}(s) = \mathbf{I}_n + \mathbf{T}_u(s). \tag{12.11}$$

Similarly it is easily shown that alternative expressions for the closed loop transfer function are

$$\mathbf{W}(s) = \mathbf{G}(s)\mathbf{K}(s)\mathbf{F}_e^{-1}(s), \tag{12.12}$$

and

$$\mathbf{W}(s) = \mathbf{G}(s)\mathbf{F}_u^{-1}(s)\mathbf{K}(s). \tag{12.13}$$

A simple application of standard formulae[2] for partitioned determinants gives the important result, to be used in later stability considerations, that

$$\det \mathbf{F}_c(s) = \det \mathbf{F}_u(s) = \det \mathbf{F}_e(s). \tag{12.14}$$

If, as is often the case, $\mathbf{H}(s) = \mathbf{I}_m$, then

$$\mathbf{T}_c(s) = \mathbf{T}_e(s) = \mathbf{T}_0(s), \quad \text{say} \tag{12.15}$$

$$\mathbf{F}_c(s) = \mathbf{F}_e(s) = \mathbf{F}(s) = \mathbf{I}_m + \mathbf{T}_0(s), \tag{12.16}$$

and

$$\mathbf{W}(s) = \mathbf{F}^{-1}(s)\mathbf{Q}(s) = \mathbf{Q}(s)\mathbf{F}^{-1}(s), \tag{12.17}$$

where

$$\mathbf{Q}(s) = \mathbf{G}(s)\mathbf{K}(s) \tag{12.18}$$

is the feedforward transfer function matrix.

Ideally in multivariable systems it would be desirable to have each system input vary only one system output. For example, variations in $r_1(t)$ would affect $c_1(t)$ only, those of $r_2(t)$ affect $c_2(t)$ only and so on. This will, of course, be the situation when the closed loop transfer function $\mathbf{W}(s)$ is diagonal; the system is then said to have non-interaction. Indeed much of the early work in multivariable system design was directed towards the achievement of non-interaction and resulted in several satisfactory practical applications(3,4,5) although the procedure is not a satisfactory general method.(6) Certainly from an engineering viewpoint the achievement of complete non-interaction is an ideal comparable with requiring a single variable system to follow a step input exactly. In the majority of situations steady state non-interaction is a necessity but small off diagonal transients, that is transients produced in all $c_m(t)$ except $c_i(t)$ when $r_i(t)$ is varied, may be tolerated. Designs based on concepts such as diagonal dominance(7) are a natural extension of the non-interaction approach.

12.2.1 *Stability*

Since $\mathbf{F}_c^{-1}(s) = \text{adj}\,\mathbf{F}_c(s)/\det \mathbf{F}_c(s)$ then from equation (12.8) if $\mathbf{K}(s)$ and $\mathbf{G}(s)$ have no right half plane poles the closed loop transfer function, $\mathbf{W}(s)$, will have no right half plane poles provided $\det \mathbf{F}_c(s)$ has no zeros with positive real parts. Thus the stability of the multivariable system depends upon the properties of the return-difference determinant, $\det \mathbf{F}(s)$, which as a consequence of equation (12.14) may be found at any point in the loop. From equation (12.17)

$$\det \mathbf{F}(s) = \det \mathbf{Q}(s)/\det \mathbf{W}(s), \tag{12.19}$$

whilst it may also be shown that(8,9,10)

$$\det \mathbf{F}(s) = \frac{\text{C.L.C.P.}}{\text{O.L.C.P.}}, \tag{12.20}$$

where C.L.C.P. and O.L.C.P. are respectively the closed loop and open loop characteristic polynomials of the signal flow graph of the multivariable system.

Let $w = \det \mathbf{F}(s)$ map s taken around the usual Nyquist contour, N, in the s plane into the locus $\det \mathbf{F}(j\omega)$ in the complex w (Nyquist) plane. Then if $\det \mathbf{F}(j\omega)$ encircles the origin n_f times clockwise, and if z_0 and z_c are respectively the number of right half plane zeros of the open loop and closed loop characteristic polynomials it follows from application of the principle of the argument[11] that

$$n_f = z_c - z_0,$$

where clockwise encirclements are counted as positive. Since z_c must be zero for absolute stability the encirclement relationship gives us the necessary and sufficient condition for closed loop stability

$$n_f = -z_0. \tag{12.21}$$

This is a multivariable form of the Nyquist stability criterion and in particular if the system is open loop stable $z_0 = 0$ and the system is stable if det $\mathbf{F}(j\omega)$ does not encircle the origin.

An alternative form of the criterion, which in general is much more useful for design purposes, may be obtained in terms of the eigenvalues of $\mathbf{T}(s)$, where $\mathbf{T}(s)$ is any return ratio matrix. Let the eigenvalues of $\mathbf{F}(s)$, which will be functions of s, be $\mu_i(s)$ and those of $\mathbf{T}(s)$ be $\lambda_i(s)$, then with

$$\mathbf{F}(s) = \mathbf{I}_m + \mathbf{T}(s),$$

it is easily shown that

$$\mu_i(s) = 1 + \lambda_i(s) \quad \text{for all } i = 1, 2, \ldots, m. \tag{12.22}$$

Now

$$\det \mathbf{F}(s) = \prod_{i=1}^{m} \mu_i(s),$$

so that if each eigenvalue $\mu_i(s)$ maps the Nyquist contour N into the locus $\mu_i(j\omega)$ which encircles the origin n_{fi} times in a clockwise direction

$$n_f = \sum_{i=1}^{m} n_{fi}. \tag{12.23}$$

Also as a consequence of equation (12.22) under the same mapping each eigenvalue $\lambda_i(s)$ will give a locus $\lambda_i(j\omega)$ which encircles the $(-1, 0)$ point n_{ti} times in a clockwise direction with

$$n_{ti} = n_{fi}. \tag{12.24}$$

Thus from equations (12.21), (12.23) and (12.24) the necessary and sufficient condition for closed loop stability is

$$\sum_{i=1}^{m} n_{ti} = -z_0. \tag{12.25}$$

The eigenvalue loci $\lambda_i(j\omega)$ appear to have first been introduced and used for stability and compensation studies for multivariable systems by Bohn and his associates.[12,13,14,15,16] The approach then received little attention for several years[5,18] due to the strong emphasis on time domain methods but recent work by MacFarlane[1,9,18,19] appears to have aroused new interest in the procedure as a design tool. In this recent work the eigenvalues $\lambda_i(j\omega)$ are called characteristic loci and the corresponding eigenvectors, which are also functions of frequency, are called characteristic signal directions.

References 5 and 9 provide more detailed reviews of linear multivariable theory and also contain excellent bibliographies.

Example[17]

Determine the conditions for stability of a two variable system with

$$\mathbf{G} = \begin{pmatrix} G_{11} & G_{12} \\ G_{21} & G_{22} \end{pmatrix}, \qquad \mathbf{H} = \begin{pmatrix} e^{-sT} & 0 \\ 0 & e^{-sT} \end{pmatrix}, \qquad \mathbf{K} = \begin{pmatrix} 1 & 0 \\ 0 & 1 \end{pmatrix}$$

if $G_{11} = G_{22} = K/(1 + 10s)^2$, and $G_{21} = -G_{12} = Ka/(1 + 10s)^2$.

Since **K** and **H** are diagonal the open loop return ratio matrix **T** is the same for each vertex in the loop and given by

$$\mathbf{T}(s) = \{Ke^{-sT}/(1 + 10s)^2\} \begin{pmatrix} 1 & -a \\ a & 1 \end{pmatrix}.$$

The eigenvalues, λ, are given by

$$\{Ke^{-sT}/(1 + 10s)^2\} \begin{vmatrix} 1 - \lambda & -a \\ a & 1 - \lambda \end{vmatrix} = 0,$$

that is

$$\lambda = \frac{(1 \pm ja)Ke^{-sT}}{(1 + 10s)^2}.$$

To see how the cross coupling factor a affects stability it is convenient to plot the three loci $g(\omega) = Ke^{-j\omega T}/(1 + 10j\omega)^2$, $\lambda_1(a) = -1/(1 + ja)$ and $\lambda_2(a) = -1/(1 - ja)$, as shown in figure 12.2, rather than the characteristic loci $\lambda(j\omega) =$

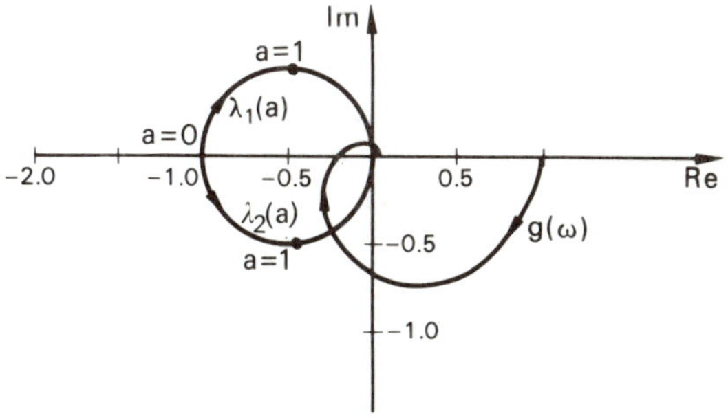

Figure 12.2 Loci for example of section 12.2.1

$(1 \pm ja)Ke^{-j\omega T}/(1 + 10j\omega)^2$ for various values of a. Since the loci $\lambda_1(a)$ and $\lambda_2(a)$ form a circle of radius 0·5 centred at the point $(-0{\cdot}5, 0)$ the system is stable for a given a if at the frequency ω_1 which is the solution of

$$180\omega_1 T/\pi + 2\tan^{-1} 10\omega_1 = \pi - \tan^{-1} a,$$

$$K/(1 + 100\omega_1^2) < (1 + a^2)^{-1/2}.$$

In particular the uncoupled system, $a = 0$, is stable for $K < 20{\cdot}7$ and the coupled system with $a = 1$ for $K < 3{\cdot}23$.

12.3 Stability of nonlinear multivariable systems

The use of DFs for the investigation of the stability of nonlinear multivariable systems is discussed below. Similar procedures may also be used to obtain the closed loop frequency response of multivariable systems, where in some cases graphical solution methods,[20] similar to those used for the single variable case, may be used to solve the nonlinear algebraic equations of the analysis.

The general nonlinear multivariable system may be considered to have the configuration shown in figure 12.3. Although both the nonlinearity and the plant

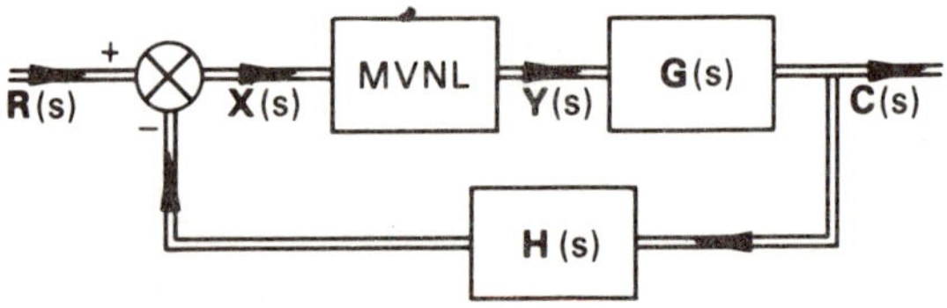

Figure 12.3 Nonlinear multivariable system

are assumed to have n inputs and outputs here, squareness is not necessary. The multivariable nonlinearity is of the form considered in section 10.5 where each output is a function of all the inputs, that is

$$y_i = n_i(x_1, x_2, \ldots, x_n), \qquad i = 1 \ldots n. \tag{12.26}$$

The quasilinear model of the nonlinearity is thus an n square matrix of n input DFs. A less general situation is when each output, y_i, depends on the sum of separate nonlinear functions of each input, that is

$$y_i = \sum_{k=1}^{n} n_{ik}(x_k), \qquad i = 1, 2 \ldots n, \tag{12.27}$$

in which case the quasilinear model is an n square matrix of single input DFs. The simplest case occurs when no cross coupling exists in the nonlinearity so that

$$y_i = n_{ii}(x_i) = n_i(x_i), \qquad i = 1, 2 \ldots n, \tag{12.28}$$

and the quasilinear model is a diagonal matrix of single input DFs. To investigate oscillations in a nonlinear multivariable system the usual describing function assumption, that the nonlinearity inputs are sinusoidal, must be made. As this also implies a single frequency oscillation, the assumption is both more difficult to justify and more significant than in the single variable case. However, since, apart from possible pole cancellations the individual transfer functions of a multivariable plant have the same poles,[21] dominant diagonal elements resonant at different frequencies are uncommon. Correspondingly the possibility of multiple frequency or harmonically related oscillations is low, and the assumption of a single frequency oscillation may be justified in many cases. On the other hand, it is an easy matter to synthesise coupled single variable systems for which the assumption would be invalid.

12.3.1 *The condition for a limit cycle*

The stability of a multivariable system, as for single variable systems, is easily found if it can be proved that no limit cycle exists. It only remains to apply the previously described Nyquist criterion to the system with the nonlinear elements replaced by any allowable gain.

Consider the system of figure 12.3 to have a MVNL of the form given by equation (12.27). Then for an oscillation to exist,

$$\mathbf{H}(j\omega)\mathbf{G}(j\omega)\mathbf{N}(\mathbf{a})\mathbf{x} = -\mathbf{x},$$

that is

$$\{\mathbf{I} + \mathbf{H}(j\omega)\mathbf{G}(j\omega)\mathbf{N}(\mathbf{a})\}\mathbf{x} = 0, \tag{12.29}$$

where $\mathbf{x}$ is a vector whose components, x_i, are complex representations of sinusoids, that is $x_i = a_i e^{j\theta_i}$ and $\mathbf{N}(\mathbf{a})$ is the DF matrix of the MVNL. For a limit cycle to exist equation (12.29) must have a nontrivial solution. Thus, assuming $\mathbf{H}(s) = \mathbf{I}$ without loss of generality, we require

$$\det\{\mathbf{I} + \mathbf{G}(j\omega)\mathbf{N}(\mathbf{a})\} = 0,$$

which may be written

$$\det\{\mathbf{G}(j\omega) + \mathbf{N}^{-1}(\mathbf{a})\} = 0. \tag{12.30}$$

If it can be shown that this equation cannot be satisfied, then no limit cycle will exist. On the other hand, if this equation can be satisfied then one can only conclude that a limit cycle may exist, as the equation may be satisfied by inconsistent $\mathbf{a}$ and ω. This latter aspect is delayed until we consider in more detail the significance of equation (12.30).

12.3.2 *Mees' criterion*

A method suggested by Mees,[22] which is in principle applicable to a system of any order with the MVNL given by equation (12.27), is first considered. Since equation (12.30) can only have a solution if $\mathbf{G}(j\omega) + \mathbf{N}^{-1}(\mathbf{a})$ has one or more zero eigenvalues, Mees made use of an eigenvalue theorem due to Gershgorin.[23] This states that for an $n \times n$ matrix $\mathbf{A}$ with elements a_{ik}, all the eigenvalues λ of $\mathbf{A}$ lie in the union of the discs

$$|\lambda - a_{kk}| \leq \sum_{i \neq k}^{n} |a_{ik}|, \qquad k = 1, 2 \ldots n. \tag{12.31}$$

Since $\mathbf{A}$ and $\mathbf{A}^T$ have the same eigenvalues, this means that the right hand side can also be written in terms of the column elements, then all the eigenvalues must lie in the union of the discs

$$|\lambda - a_{kk}| \leq \sum_{i \neq k}^{n} |a_{ki}|, \qquad k = 1, 2 \ldots n. \tag{12.32}$$

Therefore a sufficient condition for there to be no solution of equation (12.30) is that, for each ω, either

$$|G_{kk}(j\omega) + \hat{N}_{kk}(\mathbf{a})| > \sum_{i \neq k}^{n} |\hat{N}_{ik}(\mathbf{a}) + G_{ik}(\omega)|, \qquad k = 1, 2 \ldots n, \tag{12.33}$$

or

$$|G_{kk}(j\omega) + \hat{N}_{kk}(\mathbf{a})| > \sum_{i \neq k}^{n} |\hat{N}_{ki}(\mathbf{a}) + G_{ki}(\omega)|, \qquad k = 1, 2 \ldots n, \tag{12.34}$$

where $\hat{N}_{ik}(\mathbf{a})$ denotes the ikth term of the inverse DF matrix $\mathbf{N}^{-1}(\mathbf{a})$, and will in general be a function of all the input amplitudes. When the DF matrix is diagonal, however, $\hat{N}_{ik}(\mathbf{a}) = 0$ for $i \neq k$ and $\hat{N}_{kk}(\mathbf{a}) = -1/N_k(a_k)$ so that for a nonlinearity of the form of equation (12.28) equations (12.33) and (12.34) become

$$|G_{kk}(j\omega) + 1/N_k(a_k)| > \sum_{i \neq k}^{n} |G_{ik}(j\omega)|, \qquad k = 1, 2 \ldots n, \tag{12.35}$$

and

$$|G_{kk}(j\omega) + 1/N_k(a_k)| > \sum_{i \neq k}^{n} |G_{ki}(j\omega)|, \qquad k = 1, 2 \ldots n. \tag{12.36}$$

Since no solution to equation (12.30) exists if either of the above conditions is satisfied for each ω, one may investigate one condition for one value of ω and the other for a different value of ω. Often, in practice, for a given matrix one equation, say equation (12.35), is an obvious choice for use and the stability can be easily investigated graphically on a Nyquist diagram by plotting the DF locus $-1/N_k(a_k)$ and $G_{kk}(j\omega)$ surrounded by a band of halfwidth $\sum_{i \neq k}^{n} |G_{ik}(j\omega)|$. If for each $k = 1, 2 \ldots n$, the DF locus $-1/N_k(a_k)$ does not enter the band surrounding $G_{kk}(j\omega)$ then no limit cycle is possible. A possible disadvantage of this method is that if any $G_{ik}(\omega)$ with $i \neq k$ contains an integrator the width of the band tends to infinity as $\omega \to 0$. Other procedures, however, may rule out the necessity of considering frequencies less than a certain value.[24]

Returning to the less general case of a non-diagonal DF matrix a condition dependent on one amplitude component of **a** is obtained by applying the Gershgorin theorem to

$$\det\{\mathbf{G}^{-1}(j\omega) + \mathbf{N}(\mathbf{a})\} = 0. \tag{12.37}$$

Since the elements of the kth column of $\mathbf{N}(\mathbf{a})$ depend upon the input amplitude, a_k, only, the single condition

$$|N_{kk}(a_k) + \hat{G}_{kk}(j\omega)| > \sum_{i \neq k}^{n} |N_{ik}(a_k) + \hat{G}_{ik}(j\omega)|, \qquad k = 1, 2 \ldots n, \tag{12.38}$$

where $\hat{G}_{ik}(\omega)$ is the ikth element of $\mathbf{G}^{-1}$, is obtained. Equation (12.38) is obviously satisfied if

$$|N_{kk}(a_k) + \hat{G}_{kk}(j\omega)| > \sum_{i \neq k}^{n} |N_{ik}(a_k)| + \sum_{i \neq k}^{n} |\hat{G}_{ik}(j\omega)|. \qquad (12.39)$$

This result can be checked for each k and all ω by plotting on the inverse Nyquist diagram $\hat{G}_{kk}(j\omega)$ surrounded by a band of halfwidth $\sum_{i \neq k}^{n} |\hat{G}_{ik}(j\omega)|$ and the DF locus $N_{kk}(a_k)$ surrounded by a band of halfwidth $\sum_{i \neq k}^{n} |N_{ik}(a_k)|$. When the two bands, for each k, do not intersect then no limit cycle is possible. If the bands intersect, for some values of k it may be possible to adjust them and show that no limit cycle can in fact exist.(25) Consider the two square matrix

$$\begin{pmatrix} \hat{G}_{11}(j\omega) & \alpha\hat{G}_{12}(j\omega) \\ \hat{G}_{21}(j\omega)/\alpha & \hat{G}_{22}(j\omega) \end{pmatrix} + \begin{pmatrix} N_{11}(a_1) & \alpha N_{12}(a_2) \\ N_{21}(a_1)/\alpha & N_{22}(a_2) \end{pmatrix}.$$

Since it has the same eigenvalues as the matrix of equation (12.37), various values of α may be considered to see if the bands can be made nonintersecting.

Another approach(25) which can be used for a system with a non-diagonal DF matrix is to reformulate the problem as that of a system with a diagonal matrix but of the order-squared, a procedure which is always possible. The earlier results of equations (12.35) and (12.36) or those of reference 24 may then be used.

12.3.3 *Eigenvalue loci methods*

As the stability of a linear multivariable system may be found directly from its eigenvalue loci, one might hope that sets of eigenvalue and DF loci could be used to evaluate the stability of a nonlinear system. Unfortunately no simple general criterion of this form exists, although a procedure of this type can be used for two variable systems with a diagonal DF matrix.(26) Expanding equation (12.30) for this particular case gives

$$\{G_{11}(j\omega) + 1/N_1(a_1)\}\{G_{22}(j\omega) + 1/N_2(a_2)\} - G_{12}(j\omega)G_{21}(j\omega) = 0.$$

If two complex numbers, μ and K_0, are found such that

$$G_{22}(j\omega) + 1/N_2(a_2) = (G_{22}(j\omega) - \mu)/K_0,$$

and

$$G_{22}(j\omega) + 1/N_2(a_2) = (G_{22}(j\omega) - \mu)/K_0,$$

then the above equation becomes

$$(G_{11}(j\omega) - \mu)(G_{22}(j\omega) - \mu) - G_{21}(j\omega)G_{12}(j\omega) = 0,$$

so that $\mu = \lambda$, an eigenvalue of $\mathbf{G}(j\omega)$. Thus, an oscillation can only exist if it is possible, for some ω, a_1 and a_2, to find a K_0 which satisfies

$$\{\lambda(j\omega) - G_{11}(j\omega)\}K_0 = -1/N_1(a_1) - G_{11}(j\omega), \qquad (12.40)$$

$$\{\lambda(j\omega) - G_{22}(j\omega)\}/K_0 = -1/N_2(a_2) - G_{22}(j\omega). \qquad (12.41)$$

When the eigenvalue loci, $G_{11}(j\omega)$, $G_{22}(j\omega)$, $-1/N_1(a_1)$ and $-1/N_2(a_2)$ are plotted on a Nyquist diagram it becomes a relatively easy matter to investigate whether such a K_0 can exist. Typical situations are shown in figure 12.4 where for simplicity only one eigenvalue locus is shown, although because one locus is often dominant it may be all that is necessary in practice. Also although the condition must be investigated for all frequencies only the case for one specific frequency is shown. In each example no limit cycle is possible at the frequency considered. This is so because in figure 12.4

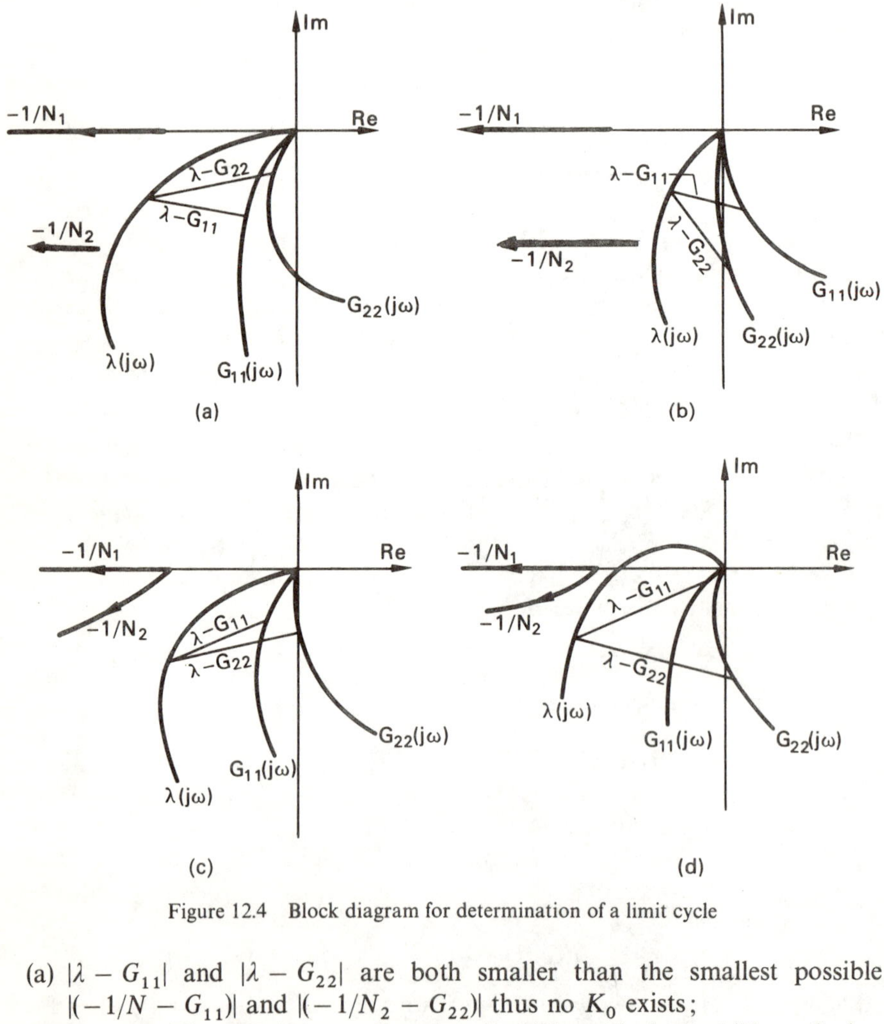

Figure 12.4 Block diagram for determination of a limit cycle

(a) $|\lambda - G_{11}|$ and $|\lambda - G_{22}|$ are both smaller than the smallest possible $|(-1/N - G_{11})|$ and $|(-1/N_2 - G_{22})|$ thus no K_0 exists;
(b) $|(\lambda - G_{11})(\lambda - G_{22})|$ is smaller than $|-1/N_1 - G_{11}|_{min}|-1/N_2 - G_{22}|_{min}$ so no K_0 exists;
(c) $\underline{/K_0}$ and $\underline{/1/K_0}$ are required to be in the same quadrant for a solution for K_0, clearly impossible;

(d) the required angle of K_0 to satisfy equation (12.40) will make it impossible to satisfy equation (12.41).

When at some frequency a K_0 can be found to satisfy equations (12.40) and (12.41) this does not assure the presence of an oscillation, since as pointed out in the previous section the values of a_1, a_2 and ω needed may not be consistent. Substituting equations (12.40) and (12.41) in equation (12.29), written in terms of the nonlinearity output vector $\mathbf{y}$, where $\mathbf{y} = \mathbf{N(a)x}$, for the particular two variable case being considered, gives

$$\begin{pmatrix} G_{11}(j\omega) - \lambda & G_{12}(j\omega) \\ G_{21}(j\omega) & G_{22}(j\omega) - \lambda \end{pmatrix} \begin{pmatrix} y_1 K_0 \\ y_2 \end{pmatrix} = 0.$$

Thus if $\{1, z(j\omega)\}^T$ is the eigenvector of $\mathbf{G}(j\omega)$ corresponding to $\lambda(j\omega)$ satisfaction of this equation requires

$$\frac{a_1|N_1(a_1)|\,|K_0|}{a_2|N_2(a_2)|} = \frac{1}{|z(j\omega)|}. \tag{12.42}$$

A limit cycle is therefore possible when equations (12.40), (12.41) and (12.42) are satisfied. This can be tested by finding for each ω the a_1, a_2 and K_0 which satisfy equations (12.40) and (12.41), then using these values in the left hand side of equation (12.42) to obtain a gain versus frequency curve which may be compared with the similar known right hand side function $|1/z(j\omega)|$, an intersection of the two curves gives the limit cycle frequency. If, when a limit cycle exists, the phase relationship of the two fundamental sinusoids at the nonlinearity outputs is required this can be found from the phase relationship equivalent to equation (12.42), that is

$$\frac{\underline{/y_1}\ \underline{/K_0}}{\underline{/y_2}} = \frac{1}{\underline{/z(j\omega)}}. \tag{12.43}$$

More significant results can be found in a few special cases which we now consider.

(a) Eigenvector of G of the form $(1, e^{j\theta})^T$, N_1 and N_2 identical

One assumes with reasonable confidence that since $|z(j\omega)| = 1$ for all ω that $a_1 = a_2$, and hence $N_1(a_1) = N_2(a_2)$, during any oscillation. Thus $K_0 = 1$ satisfies equation (12.42) and from equation (12.40) any intersection of $\lambda(j\omega)$ and $-1/N_1(a_1)$ determines the limit cycle. Forms of $\mathbf{G}$ which satisfy this eigenvector condition are

$$\begin{pmatrix} G_1 & G_2 \\ G_2 & G_1 \end{pmatrix}, \begin{pmatrix} G_1 & -G_2 \\ G_2 & G_1 \end{pmatrix}, \begin{pmatrix} G_1 & G_2 \\ G_1 & G_2 \end{pmatrix}, \begin{pmatrix} G_1 & G_2 \\ -G_1 & -G_2 \end{pmatrix}.$$

The first of these cases, the symmetrical system with eigenvalues $\lambda(j\omega) = G_1(j\omega) \pm G_2(j\omega)$, does occur in practical situations and has been studied by Lindgren.[27]

(b) $G_{11} = G_{22}$, N_1 and N_2 single valued

In this case it can be shown from equations (12.40) and (12.41) that no limit cycle is possible unless $\lambda(j\omega) - G_{11}(j\omega)$ cuts the $-1/N_1(a_1)$ or $-1/N_2(a_2)$ loci. For a **G** matrix which has two stable uncoupled nonlinear systems when $G_{12} = G_{21} = 0$; this condition reduces to, no limit cycle is possible unless $\lambda(j\omega)$ cuts $-1/N_1(a_1)$ or $-1/N_2(a_2)$ equivalent to the single loop criterion.

12.3.4 *Limit cycle computation*

In principle, since we can evaluate DFs for the MVNL, the problem of evaluating a limit cycle for the system of figure 12.3 becomes one of solving a set of nonlinear algebraic equations. However, to try to solve such a problem blindly is not advisable even when we know a solution, i.e. limit cycle, exists. The methods presented in the previous section can be used to obtain good estimates for the possible limit cycle parameters of a two variable system, but even for such a system with simple plant transfer functions two limit cycle solutions usually exist. If no solution is found to the set of nonlinear algebraic equations it may be because the search algorithm is inadequate or that no limit cycle exists. Often a realistic approach, particularly if the system is expected to be stable, is to adjust the system parameters until one or more limit cycles are found and then increment the parameters to find the stability boundary.

In order to illustrate this approach and to restrict the complexity the general two variable form of figure 12.3 is considered. The block diagram can be redrawn as shown in figure 12.5 with the MVNL inputs assumed to be $a \cos \omega t$

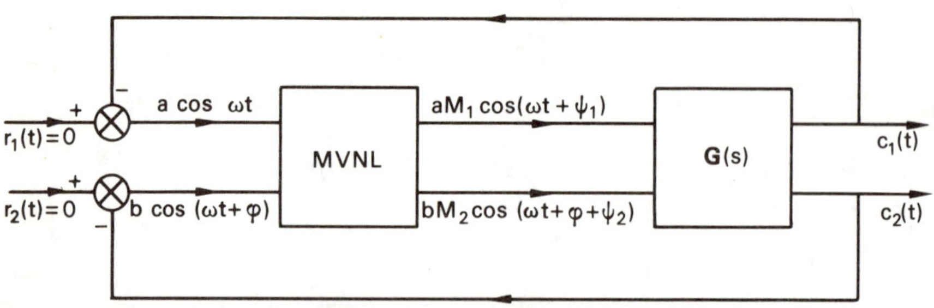

Figure 12.5 Nonlinear coupled system

and $b \cos(\omega t + \phi)$. For a limit cycle to exist the following equations must be satisfied

$$g_{11}M_1 a \cos(\omega t + \psi_1 + \theta_{11}) + g_{12}M_2 b \cos(\omega t + \phi + \psi_2 + \theta_{12}) = -a \cos \omega t, \quad (12.44)$$

$$g_{21}M_1 a \cos(\omega t + \psi_1 + \theta_{21}) + g_{22}M_2 b \cos(\omega t + \phi + \psi_2 + \theta_{22}) = -b \cos(\omega t + \phi), \quad (12.45)$$

where $G_{11}(j\omega) = g_{11}(\omega)e^{j\theta_{11}(\omega)}$, etc. and M_1, M_2, ψ_1 and ψ_2, which are functions of a, b and ϕ, can be evaluated for a given MVNL as described in section 10.4. The set of four nonlinear algebraic equations, that is (12.44), (12.45) and two complex DF relationships can in principle be solved for the parameters of any possible limit cycle.

12.4 Relay systems

Periodic modes in multivariable relay systems of any order may be investigated using extensions of the Tsypkin approach presented in chapter 6. Graphical methods can be used to obtain solutions for two variable systems containing relays without dead zone, but other situations normally require a digital computer to solve the set of nonlinear algebraic equations given by the theory.

The basic assumption of the method is that the output waveforms from all the relays have the same fundamental frequency and are of the form shown in figure 6.2. It is, of course, quite easy to get other forms of periodic behaviour in a multivariable relay system such as that shown in figure 6.6(b). For example, a two variable system in which $\mathbf{G}(s)$ is almost diagonal, and possesses single loop oscillations when $G_{12}(s) = G_{21}(s) = 0$ at frequencies near to f and $3f$, is not likely to satisfy the above assumption. Indeed if the oscillation waveforms were as shown in figure 6.2 but differed only in fundamental frequency by a factor of three the method to be discussed could be used with minor alterations. However there is a strong probability that one relay output might have a more complex waveform. such as that shown in figure 6.16(b). This situation again would not be impossible, though certainly more tedious to analyse, however the major difficulty at present is the lack of satisfactory criteria both for finding accurately the possible modes of oscillation which may exist and then for determining the stability of any solutions found for such modes. Symmetrical oscillations only will be considered in the following section although the extension of the method to asymmetrical situations, as found in section 6.7 for single variable systems, is straightforward.

12.4.1 *Self oscillations*

It was shown in section 6.6 that for the autonomous multivariable relay system of figure 6.6(b) any periodic mode of the assumed form, that is with the relay outputs of the same frequency and waveform as in figure 6.2, is given by the solution of

$$\sum_{k=1}^{n} h_k\{A_{G_{jk}}(\alpha_k - \alpha_j, \omega) + A_{G_{jk}}(-\beta_k - \alpha_j, \omega)\} \qquad \begin{array}{l} \text{R.P.} < 0 \\ \text{I.P.} = -\pi(\delta_j + \Delta_j)/2, \end{array} \tag{12.46}$$

and

$$\sum_{k=1}^{n} h_k\{A_{G_{jk}}(\alpha_k + \beta_j, \omega) + A_{G_{jk}}(-\beta_k + \beta_j, \omega)\} \qquad \begin{array}{l} \text{R.P.} < 0 \\ \text{I.P.} = \pi(\delta_j - \Delta_j)/2, \end{array} \tag{12.47}$$

for all the relays $j = 1$ to n. Thus, in the general case we have $2n$ nonlinear algebraic equations in the $2n + 1$ unknowns, α_j, β_j and ω which can be solved if one of α_j or β_j is chosen as the time reference. A solution procedure for the equations on a digital computer is relatively easy to implement using a Newton–Raphson method since expressions for $\partial\{A_G(\theta, \omega)\}/\partial\theta$ and $\partial\{A_G(\theta, \omega)\}/\partial\omega$ can be found in closed form for the common transfer functions $G(s)$. That such expressions are possible can be seen from the derivative relationships given in table 6.2 between the sine and cosine series.

If a relay has no dead zone then only one equation is used for the relay, so that in the special case of all the relays having no dead zone the n equations are

$$\sum_{k=1}^{n} 2h_k A_{G_{jk}}(\alpha_k - \alpha_j, \omega) \qquad \begin{array}{l}\text{R.P.} < 0,\\ \text{I.P.} = -\pi\Delta_j/2,\end{array} \tag{12.48}$$

since $\alpha_j = -\beta_j$. A rather interesting aspect of this situation is that these same n equations could also be obtained from n single variable systems containing relay j and a transfer function $F_j(s)$ for all $j = 1$ to n. Now from equation (6.38) $A_G(\theta, \omega)$ is the A locus for a transfer function $G(s)e^{-s\theta/\omega}$, so that if $F_j(s)$ is chosen such that

$$F_j(s) = \sum_{k=1}^{n} (h_k/h_j) G_{jk}(s) e^{-s(\alpha_k - \alpha_j)/\omega} \tag{12.49}$$

then the n equations given for the n single variable systems will be identical to equation (12.48).

It is possible that more than one frequency, ω, satisfies equations (12.46) and (12.47) for a particular system, which mode then exists in practice requires an as yet unknown criterion for predicting the stability of the periodic solutions.

Example[28]

A two variable relay control system has two identical on-off relays with hysteresis, having $h = \Delta = 1$ and a symmetrical plant transfer matrix

$$\mathbf{G}(s) = \begin{pmatrix} G_{11}(s) & G_{12}(s) \\ G_{21}(s) & G_{22}(s) \end{pmatrix}$$

with $G_{11}(s) = G_{22}(s) = 10/(1 + sT)^2$ and $G_{12}(s) = G_{21}(s) = 10a/(1 + sT)^2$.

For this case equation (12.48) applies with $n = 2$ and gives

$$A_G(0, \omega) + aA_G(\theta, \omega) \qquad \text{R.P.} < 0, \quad \text{I.P.} = -\pi/40,$$

and

$$A_G(0, \omega) + aA_G(-\theta, \omega) \qquad \text{R.P.} < 0, \quad \text{I.P.} = -\pi/40,$$

where the transfer function $G(s) = 1/(1 + sT)^2$ and $\theta = \alpha_2 - \alpha_1$. Now since the A loci are odd and of period 2π in θ (section 6.5), it is seen that the above equations are satisfied for $\theta = 0$ or π. The two possible solutions, which correspond to the relays switching either in phase or antiphase, are given by

$$1 \pm aA_G(0, \omega) \qquad \text{R.P.} < 0,$$
$$\text{I.P.} = -\pi/40,$$

where the positive sign applies when $\theta = 0$ and the negative when $\theta = \pi$.
Substituting from table 6.3 for $\text{Im}\{A_G(0, \omega)\}$ gives

$$-\tanh(\pi/2\lambda) + \pi/2\lambda \cosh^2(\pi/2\lambda) = -1/10(1 \pm a),$$

which may be written

$$\sinh 2u - 2u = \cosh^2 u/5(1 \pm a),$$

where $u = \pi/2\lambda$ and $\lambda = \omega T$. Also substituting for $\text{Re}\{A_G(0, \omega)\}$ the inequality which must be satisfied reduces to

$$-(1 \pm a)u^2/2 \cosh^2 u < 0,$$

which is true for all positive frequencies provided $a > -1$ for $\theta = 0$ and $a < 1$ for $\theta = \pi$.

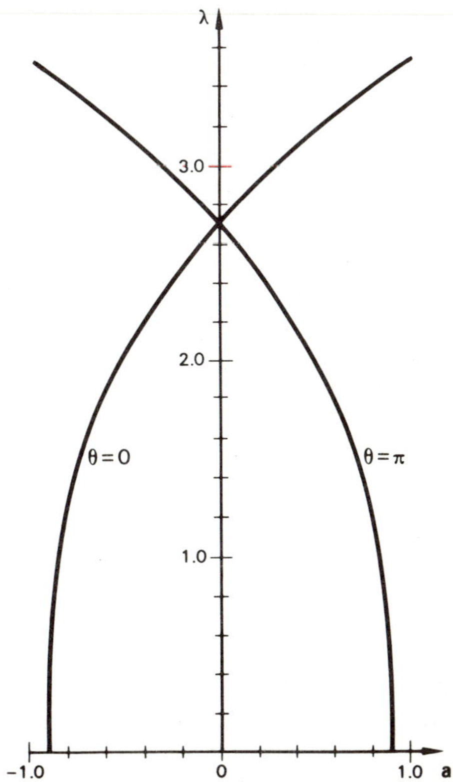

Figure 12.6 Oscillation frequencies for example of section 12.4.1

Figure 12.6 shows the variations of the predicted oscillation frequency, λ, with the cross coupling factor a for $|a| < 1$. Analog simulation results showed very good agreement with the high frequency value, the mode which was nearly always reached independent of the initial conditions. The exceptions were instances where the low frequency mode was occasionally obtained for values of a near zero.

Application of Tsypkin's stability criterion to the two equivalent single variable systems predicts both modes stable for the above range of a. In a similar problem discussed in reference 29 a heuristic but not very convincing argument was presented to predict which of the two oscillations would be stable.

12.4.2 *Forced oscillations*

If in figure 6.6(b) the input signals $r_j(t)$ are given by $r_j(t) = R_j \sin(\omega_f t + \phi_j)$ it is relatively easy to show that the equations giving the solution, corresponding to equations (12.46) and (12.47) are

$$(\pi/2)R_j(\pi + \phi_j + \alpha_j) + \sum_{k=1}^{n} h_k\{A_{G_{jk}}(\alpha_k - \alpha_j, \omega_f) + A_{G_{jk}}(-\beta_k - \alpha_j, \omega_f)\} \quad (12.50)$$

$$\text{R.P.} < 0, \qquad \text{I.P.} = -\pi(\delta_j + \Delta_j)/2,$$

and

$$(\pi/2)R_j(\pi + \phi_j - \beta_j) + \sum_{k=1}^{n} h_k\{A_{G_{jk}}(\alpha_k + \beta_j, \omega_f) + A_{G_{jk}}(-\beta_k + \beta_j, \omega_f)\} \quad (12.51)$$

$$\text{R.P.} < 0, \qquad \text{I.P.} = \pi(\delta_j - \Delta_j)/2$$

for all $j = 1$ to n and where the notation is the same as that used in section 6.9. When there are only two equations, as for example in a two variable system containing relays without dead zone, a graphical solution procedure similar to that described in figure 6.12 can be used. Reference 29 gives an example of this type using the solution equations in Tsypkin notation.

12.5 Systems with random inputs

The basic system to be considered is shown in figure 12.3 where the inputs $r(t)$ are Gaussian random signals. Again only an analysis problem, that of evaluating the mean squared errors, is considered. It is assumed that the system is stable and also that no oscillations are excited by the random inputs. The possibility of such an occurrence can be investigated by studying the stability of the system using the RSDF, or in many cases the IRDF will prove sufficient, for the multivariable nonlinearity. Thus with the inputs to the multivariable nonlinearity assumed to be jointly Gaussian, its equivalent gains may be evaluated as described in section 10.5, and the resulting linearised system is as shown in figure 12.7, with the nonlinearity replaced by the equivalent gain matrix **K**.

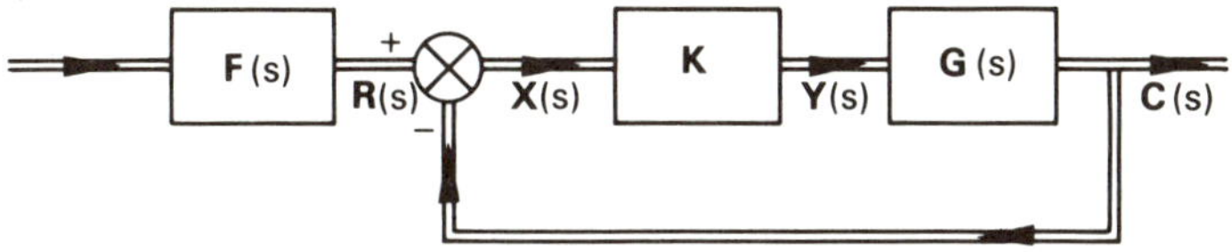

Figure 12.7 Quasilinear multivariable system

Now, from the figure

$$\mathbf{X}(s) = \mathbf{R}(s) - \mathbf{C}(s),$$
$$\mathbf{C}(s) = \mathbf{G}(s)\mathbf{K}\mathbf{X}(s),$$

so that

$$\mathbf{X}(s) = \{\mathbf{I} + \mathbf{G}(s)\mathbf{K}\}^{-1}\mathbf{R}(s), \tag{12.52}$$

where I is the indentity matrix. The power spectral density of $\mathbf{x}$ is given by

$$\mathbf{S}_{xx}(s) = \{\mathbf{I} + \mathbf{G}(-s)\mathbf{K}\}^{-1}\mathbf{S}_{rr}(s)\{(\mathbf{I} + \mathbf{G}(s)\mathbf{K}^{-1}\}^{T}, \tag{12.53}$$

where

$$\mathbf{S}_{xx}(s) = \begin{pmatrix} S_{x_1x_1}(s) & S_{x_1x_2}(s) & \dots & S_{x_1x_n}(s) \\ \vdots & \vdots & \vdots & \vdots \\ S_{x_nx_1}(s) & S_{x_nx_2}(s) & \dots & S_{x_nx_n}(s) \end{pmatrix},$$

and

$$\mathbf{S}_{rr}(s) = \begin{pmatrix} S_{r_1r_1}(s) & S_{r_1r_2}(s) & \dots & S_{r_1r_n}(s) \\ \vdots & \vdots & \vdots & \vdots \\ S_{r_nr_1}(s) & S_{r_nr_2}(s) & \dots & S_{r_nr_n}(s) \end{pmatrix},$$

which is the matrix of input power spectral densities.

If the random input signals $r_i(t)$ are obtained by passing Gaussian white noise signals through an input filter $\mathbf{F}(s)$ and if the power spectrum of the white noise is $\mathbf{S}_{kk}$, $\mathbf{S}_{rr}(s)$ is given by

$$\mathbf{S}_{rr}(s) = \mathbf{F}(-s)\mathbf{S}_{kk}\{\mathbf{F}(s)\}^{T}. \tag{12.54}$$

The error variance matrix is evaluated from the power spectral density matrix of equation (12.53) by

$$\mathbf{R}_{xx}(0) = (1/2\pi j)\int_{-j\infty}^{j\infty} \mathbf{S}_{xx}(s)\,\mathrm{d}s. \tag{12.55}$$

Consider now the application of the above procedure to a two variable system with input filter

$$\mathbf{F}(s) = \begin{pmatrix} F_1(s) & 0 \\ 0 & F_2(s) \end{pmatrix},$$

and linear plant

$$\mathbf{G}(s) = \begin{pmatrix} G_{11}(s) & G_{12}(s) \\ G_{21}(s) & G_{22}(s) \end{pmatrix}.$$

Then

$$\{\mathbf{I} + \mathbf{G}(s)\mathbf{K}\}^{-1} = \frac{1}{\Omega(s)} \times \begin{pmatrix} 1 + G_{21}(s)K_{12} + G_{22}(s)K_{22} & -\{G_{11}(s)K_{12} + G_{12}(s)K_{22}\} \\ -\{G_{21}(s)K_{11} + G_{22}(s)K_{21}\} & 1 + G_{11}(s)K_{11} + G_{12}(s)K_{21} \end{pmatrix}, \tag{12.56}$$

where

$$\Omega(s) = \{1 + G_{11}(s)K_{11} + G_{12}(s)K_{21}\}\{1 + G_{21}(s)K_{12} + G_{22}(s)K_{22}\} - \{G_{11}(s)K_{12} + G_{12}(s)K_{22}\}\{G_{21}(s)K_{11} + G_{22}(s)K_{21}\}.$$

Substituting equation (12.56) in equation (12.53), the power spectral densities $S_{x_ix_j}(s)$, $i, j = 1, 2$, are given by the expressions

$$S_{x_1x_1}(s) = \frac{1}{|\Omega(s)|^2}[S_{r_1r_1}(s)\{|1 + G_{21}(s)K_{12} + G_{22}(s)K_{22}|\}^2 - 2\,\mathrm{Re}\{(1 + G_{21}(-s)K_{12} + G_{22}(-s)K_{22})S_{r_1r_2}(s)(G_{11}(s)K_{12} + G_{12}(s)K_{22})\} + S_{r_2r_2}(s)\{|G_{11}(s)K_{12} + G_{12}(s)K_{22}|\}^2], \tag{12.57}$$

$$S_{x_1x_2}(s) = \frac{1}{|\Omega(s)|^2}[\{1 + G_{21}(-s)K_{12} + G_{22}(s)K_{22}\}\{S_{r_1r_2}(s)(1 + G_{11}(s)K_{11} + G_{12}(s)K_{21}) - S_{r_1r_1}(s)(G_{21}(s)K_{11} + G_{22}(s)K_{21})\} + \{G_{11}(-s)K_{12} + G_{12}(-s)K_{22}\}\{S_{r_2r_1}(s)(G_{21}(s)K_{11} + G_{22}(s)K_{21}) - S_{r_2r_2}(s)(1 + G_{11}(s)K_{11} + G_{12}(s)K_{21})\}], \tag{12.58}$$

$$S_{x_2x_1}(s) = S_{x_1x_2}(-s), \tag{12.59}$$

$$S_{x_2x_2}(s) = \frac{1}{|\Omega(s)|^2}[S_{r_1r_1}(s)\{|G_{21}(s)K_{11} + G_{22}(s)K_{21}|^2\} - 2\,\mathrm{Re}\{(G_{21}(-s)K_{11} + G_{22}(-s)K_{21})S_{r_1r_2}(s)(1 + G_{11}(s)K_{11} + G_{12}(s)K_{21})\} + S_{r_2r_2}(s)\{|1 + G_{11}(s)K_{11} + G_{12}(s)K_{21}|\}^2]. \tag{12.60}$$

If $\mathbf{S}_{kk}$ is given by

$$\mathbf{S}_{kk} = \begin{pmatrix} 1 & 1 \\ 1 & 1 \end{pmatrix} S_k,$$

where S_k is a constant, then, using equation (12.54),

$$\begin{aligned} S_{r_1r_1}(s) &= |F_1(s)|^2 S_k, \\ S_{r_1r_2}(s) &= F_1(-s)F_2(s)S_k, \\ S_{r_2r_1}(s) &= F_2(-s)F_1(s)S_k, \\ S_{r_2r_2}(s) &= |F_2(s)|^2 S_k. \end{aligned} \tag{12.61}$$

Using the integral of equation (12.55) on equations (12.57) to (12.60) with the above expressions for $\mathbf{S}_{rr}(s)$ substituted gives the variance matrix $\mathbf{R}_{xx}(0)$. Thus the evaluation of $\mathbf{R}_{xx}(0)$ involves the solution of seven nonlinear algebraic equations which may be written in implicit form as

$$\begin{aligned} S_{x_1x_1}(0) &= f_1(K_{11}, K_{12}, K_{21}, K_{22}), \\ S_{x_1x_2}(0) &= f_2(K_{11}, K_{12}, K_{21}, K_{22}), \\ S_{x_2x_2}(0) &= f_3(K_{11}, K_{12}, K_{21}, K_{22}), \\ K_{11} &= f_4\{S_{x_1x_1}(0), S_{x_1x_2}(0), S_{x_2x_2}(0)\}, \\ K_{12} &= f_5\{S_{x_1x_1}(0), S_{x_1x_2}(0), S_{x_2x_2}(0)\}, \\ K_{21} &= f_6\{S_{x_1x_1}(0), S_{x_1x_2}(0), S_{x_2x_2}(0)\}, \\ K_{22} &= f_7\{S_{x_1x_1}(0), S_{x_1x_2}(0), S_{x_2x_2}(0)\}. \end{aligned} \tag{12.62}$$

The first three equations are obtained from a linear analysis of the quasilinearised system and the last four from the equivalent gain analysis of the MVNL.

The analysis outlined above has been applied to two variable systems and the results compared with those obtained by simulation in reference 30. Two multivariable nonlinearities were used, the first consisting of four single valued nonlinearities as shown in figure 12.8 and the second given analytically by

$$y_1 = 0{\cdot}4x_1^2x_2 + 0{\cdot}6x_1x_2^2,$$

and

$$y_2 = 0{\cdot}5x_1^2x_2 + 0{\cdot}7x_1x_2^2. \tag{12.63}$$

Only plants $\mathbf{G}(s)$ consisting of first order transfer functions were used, however quite good agreement between calculated and measured mean squared values was obtained. Experimental measurements were made of the probability density functions and power spectral densities of the inputs to the nonlinearity and it was found that the quasilinear analysis solutions were more accurate the nearer these signals were to being jointly Gaussian.

In an endeavour to estimate the accuracy of the analysis several distortion factors were evaluated, using extensions of the ideas presented in section 4.8.1 for sinusoids, for this random multivariable case. Assuming that the input signals to the nonlinearity are jointly Gaussian with properties given by the quasilinear

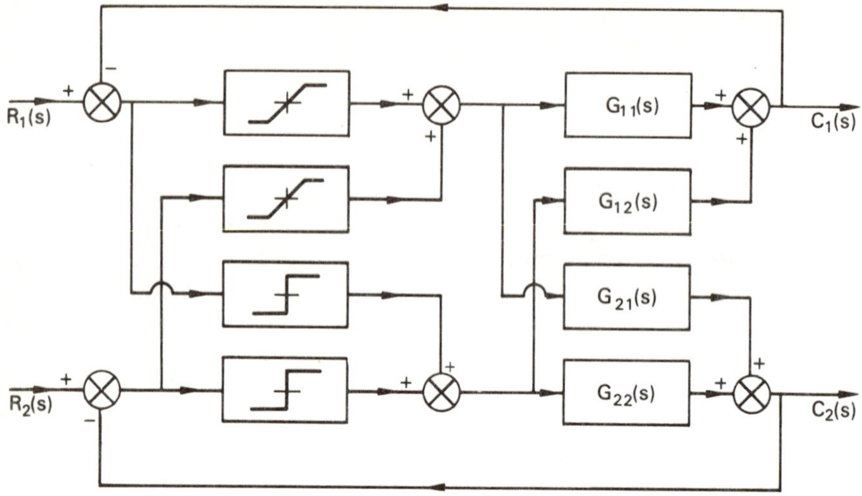

Figure 12.8 A system considered in reference 30

solution, the contribution $[\sigma_{c_i}^2]_3$ $(i = 1, 2)$ to the mean squared value of the system outputs due to the distortion terms in the nonlinearity output expansion with coefficients α_{sk} with $s + k = 3$ was calculated. The distortion measure

$$\delta_{c_i} = [\sigma_{c_i}^2]_3/[\sigma_{c_i}^2]_1, \tag{12.64}$$

where $[\sigma_{c_i}^2]_1$ is the mean squared value of c_i in the quasilinear analysis, was used. An additional distortion measure based on an amplitude probability density calculation for c_i and defined by

$$\varepsilon_{c_i} = \frac{\text{fourth moment of } c_i \text{ due to distortion terms}}{\text{fourth moment of } c_i \text{ found by quasilinearisation}}, \tag{12.65}$$

was also used. The numerator of this equation can be found using the orthogonal expansion for the nonlinearity outputs y_1 and y_2 in terms of Hermite polynomials of the correlated inputs x_1 and x_2 (see section (10.8.3)), again assumed to be the jointly Gaussian signals of the quasilinear solution. Using the expansions for y_1 and y_2 the fourth moments of c_1 and c_2 can be obtained from a Fourier transformation of the convolution integral expressions relating the inputs y_i and outputs c_i of $\mathbf{G}(s)$. Table 12.1 lists some of the results obtained in reference 30 for a system with the multivariable nonlinearity defined by equation (12.63). Both distortion criteria, which are of course zero for the linear case, are seen to give a good indication of the accuracy of the quasilinear analysis. Since the distortion terms are uncorrelated with the system input signals similar distortion expressions, which may be more appropriate, can easily be evaluated for the nonlinearity inputs.

TABLE 12.1
Sample of results obtained in reference 30

Inputs to the system	$\mathbf{F}(s) = \begin{bmatrix} \frac{40}{s+20} & 0 \\ 0 & \frac{900}{(s+20)^2} \end{bmatrix}$		Theory	Exp.
		σ_{r_1}	0·346	0·35
		σ_{r_2}	0·276	0·28
Linear plants $\mathbf{G}(s)$	(a) $\begin{bmatrix} \frac{7}{s+5} & \frac{8}{s+6} \\ \frac{8}{s+6} & \frac{7}{s+5} \end{bmatrix}$		(b) $\begin{bmatrix} \frac{70}{s+50} & \frac{80}{s+60} \\ \frac{80}{s+60} & \frac{70}{s+50} \end{bmatrix}$	
	Theory	Exp.	Theory	Exp.
σ_{x_1}	0·3267	0·32	0·2955	0·34
σ_{x_2}	0·2277	0·23	0·2026	0·24
σ_{y_1}	0·0414	0·047	0·0266	0·02
σ_{y_2}	0·05	0·055	0·0321	0·031
σ_{c_1}	0·0753	0·07	0·0807	0·072
σ_{c_2}	0·0752	0·07	0·0809	0·072
$E\{x_1x_1\}$	0·0485	0·0475	0·0334	0·035
Theoretical distortion coefficients				
$[\sigma^2_{c_1}]_1$	$5{\cdot}6654 \times 10^{-3}$		$0{\cdot}5661 \times 10^{-2}$	
$[\sigma^2_{c_1}]_3$	$0{\cdot}7069 \times 10^{-3}$		$0{\cdot}1340 \times 10^{-2}$	
δ_{c_1}	12·48%		20·59%	
$[\sigma^2_{c_2}]_1$	$6{\cdot}5061 \times 10^{-3}$		$0{\cdot}6551 \times 10^{-2}$	
$[\sigma^2_{c_2}]_3$	$0{\cdot}7060 \times 10^{-3}$		$0{\cdot}1344 \times 10^{-2}$	
δ_{c_2}	12·47%		20·51%	
ε_{c_1}	2·26%		14·46%	
ε_{c_2}	2·21%		15·49%	

REFERENCES

1. MacFarlane, A. G. J.: 'The return-difference and return-ratio matrices and their use in the analysis and design of multivariable feedback control systems', *Proc. I.E.E.*, 1970, 117, pp. 2037–2049.
2. Gantmacher, F. R.: *Theory of Matrices, Vol. 1* (Chelsea, New York, 1959).
3. Voznesenski, I. N.: 'Controlling machines with large numbers of controlled parameters', *Automat. Telemech.*, Moscow, 1938, 4–5.
4. Boksenbom, A. S. and Hood, R.: 'General algebraic method applied to control analysis of complex engine types', *N.A.C.A. Rep.* 980, April 1949.
5. Kavanagh, R. J.: 'The multi-variable problem', *Progress in Control Engineering* (Macmillan, R. H. et al., eds., 3, Heywood, London, 1966, pp. 93–129).
6. Rosenbrock, H. H.: 'On the design of linear multivariable control systems', *Proc. 3rd Congress I.F.A.C.*, London, Inst. of Mech. Engrs, London, 1966, Paper 1A.

7. Rosenbrock, H. H.: 'Design of multivariable control systems using the inverse Nyquist array', *Proc. I.E.E.*, 1969, 116, pp. 1929–1936.
8. MacFarlane, A. G. J.: 'Notes on the vector frequency response approach to the analysis and design of multivariable feedback systems', *Series of lectures, University of New Brunswick*, September 1972.
9. MacFarlane, A. G. J.: 'A survey of some recent results in linear multivariable feedback theory', *Automatica*, 1972, 8, pp. 455–492.
10. Popov, V. M.: 'Hyperstability and optimality of automatic systems with several control functions', *Revue Roumaine de Sciences Techniques—Electrotechnics and Energetics*, 1964, 9, Bucharest.
11. Ku, Y. H.: *Analysis and control of linear systems* (International Textbook Company, Scranton, Penn., Chapter 5).
12. Bohn, E. V.: 'Stabilization of linear multivariable feedback control systems', *I.R.E. Trans.*, 1960, AC-5, pp. 321–327.
13. Bohn, E. V.: 'Design and synthesis methods for a class of multivariable feedback control systems based on single variable methods', *A.I.E.E. Trans.*, 1962, 81, pt. II, pp. 109–115.
14. Bohn, E. V. and Kasvand, T.: 'Use of matrix transformations and system eigenvalues in the design of linear multivariable control systems', *Proc. I.E.E.*, 1963, 110, pp. 989–996.
15. Baird, C. R.: 'Design studies of a class of multivariable feedback control systems', *M.A.Sc. Thesis, University of British Columbia*, September 1962.
16. Kasvand, T.: 'Application of eigenvalue techniques in the analysis, compensation and synthesis of linear multivariable control systems', *Ph.D. Thesis, University of British Columbia*, September 1962.
17. Chung, W-K.: 'Multivariable control systems involving time delays', *M.Sc. Thesis, University of New Brunswick*, 1964.
18. Belletrutti, J. and MacFarlane, A. G. J.: 'Characteristic loci techniques in multivariable-control system design', *Proc. I.E.E.*, 1971, 118, 1291–1297.
19. MacFarlane, A. G. J.: 'Use of characteristic transfer functions in the design of multivariable control systems', *Proc. 2nd I.F.A.C. Conference on Multivariable Systems Theory*, Paper 1.3.4, Dusseldorf, 1971.
20. Nikiforuk, P. N. and Wintonyk, B. L. M.: 'Frequency response analysis of two-dimensional non-linear symmetrical and non-symmetrical control systems', *Int. J. Control*, 1968, 7, pp. 49–62.
21. Brockett, R. W.: 'Poles, zeros and feedback: State space interpretation', *I.E.E.E. Trans.*, 1965, AC-10, pp. 129–135.
22. Mees, A.: 'Describing functions, circle criteria and multiple-loop feedback systems', *Proc. I.E.E.*, 1973, 120, pp. 126–130.
23. Minc, H. and Marcus, M.: *A survey of matrix theory and matrix inequalities* (Prindle, Weber & Schmidt, Inc., Boston, Mass., 1964).
24. Ramani, N. and Atherton, D. P.: 'Stability of nonlinear multivariable systems', *Proceedings of the 3rd I.F.A.C. Symposium on Multivariable Technological Systems*, Paper S-10, Sept. 1974, Manchester, England.
25. Ramani, N., Atherton, D. P. and Mees, A.: 'Discussion of reference 22', *Proc. I.E.E.*, 1973, 120, p. 814.
26. Ramani, N. and Atherton, D. P.: 'Frequency response methods for nonlinear multivariable systems', *Paper No. 9.2, Canadian Conference on Automatic Control*, September 24–25, 1973, University of New Brunswick, Fredericton.
27. Lindgren, A. G.: 'Limit cycles in symmetric multivariable systems', *I.E.E.E. Trans.*, 1964, AC-9, pp. 119–120.
28. Atherton, D. P.: 'Conditions for periodicity in control systems containing several relays', *Proc. I.F.A.C.*, 1966, Paper 28E, London.
29. Nugent, S. T. and Kavanagh, R. J.: 'Self and forced oscillations in multivariable relay control systems', *Automatica*, 1969, 5, pp. 519–527.
30. Wan, Y. S.: 'Validity and accuracy of statistical linearization in the analysis of nonlinear multivariable control systems with random inputs', *M.Sc. Thesis, University of New Brunswick*, 1971.

BIBLIOGRAPHY

Chung, J. K-C. and Atherton, D. P.: 'Analytical design of multivariable relay control systems for non-interaction in the mean', *I.F.A.C. Symp. on Multivariable Control Systems*, Dusseldorf, 1968.

Cook, P. A.: 'Modified multivariable circle theorems', in *Recent Mathematical Developments in Control Theory* (ed. Bell, D. J.), Academic Press, London, pp. 367–372, 1973.

Cook, P. A.: 'Nyquist-plot methods of investigating multivariable feedback systems', *Proceedings I.F.A.C. Symposium on Multivariable Technological Systems*, Paper S-28, Manchester, September 1974.

Dewey, A. G.: 'Frequency domain stability criteria for nonlinear multivariable systems', *Int. J. Control*, 1967, 5, pp. 77–84.

Freeman, E. A.: 'Some control system stability and optimality results obtained via functional analysis', in *Recent Mathematical Developments in Control Theory* (ed. Bell, D. J.), Academic Press, London, pp. 45–68, 1973.

Fuller, A. T.: 'Analysis of nonlinear stochastic systems by means of the Fokker–Planck equation', *Int. J. Control*, 1969, 9, pp. 603–655.

Gelb, A.: 'Limit cycles in symmetric multiple nonlinearity systems', *I.E.E.E. Trans.*, 1963, AC-8, pp. 177–178.

Kazakov, I. E.: 'Statistical analysis of systems with multidimensional nonlinearities', *Automation and Remote Control*, 1965, 26, pp. 458–464.

Lighthill, J. and Mees, A. I.: 'Stability of nonlinear feedback systems', in *Recent Mathematical Developments in Control Theory* (ed. Bell, D. J.), Academic Press, London, pp. 1–19, 1973.

Lindgren, A. G. and Pinkos, R. F.: 'Stability of symmetric nonlinear multivariable systems', *J. Franklin Inst.*, 1966, 282, pp. 92–101.

Lindorf, D. P.: 'Control of nonlinear multivariable systems', *I.E.E.E. Trans.*, 1967, AC-12, pp. 506–515.

MacFarlane, A. G. J.: 'Relationships between recent developments in linear control theory and classical design techniques', *Proceedings I.F.A.C. Symposium on Multivariable Technological Systems*, Survey Paper, Manchester, September 1974.

Newman, D. B.: 'The analysis of cross-coupling effects on the stability of two dimensional, orthogonal, feedback control systems', *I.E.E.E. Trans.*, 1960, AC-5, pp. 314–320.

Rae, W. G. and MacLellan, G. D. S.: 'Stability of two classes of non-linear multivariable control systems', *Proc. I.F.A.C.*, 1966, Paper 12C, London.

Sandberg, I. W.: 'Some results on the theory of physical systems governed by nonlinear functional equations', *Bell Syst. Tech. J.*, 1965, 44, pp. 871–898.

Viswanadham, N., Gururaja, H. V. and Deekshatulu, B. L.: 'Design of nonlinear multivariable control systems', *Int. J. Control*, 1966, 4, pp. 153–166.

Viswanadham, N. and Deekshatulu, B. L.: 'Stability analysis of nonlinear multivariable systems', *Int. J. Control*, 1966, 5, pp. 369–375.

Morgan, B. S.: 'Multivariable systems', *I.E.E.E. Int. Conv. Record*, 1965, part 6, pp. 87–95.

PROBLEMS

P.12.1 Show for a symmetrical nonlinear system of the form discussed in section 12.3.3 that possible limit cycles have either in phase or antiphase sinusoids at the NL inputs if the NLs are identical.

P.12.2 Solve the example done in section 12.4 by the eigenvalue method.

P.12.3 Determine by the eigenvalue method the possible limit cycles in a two variable system with no cross coupling in the MVNL and the diagonal elements ideal relay characteristics if $G(s)$ is symmetric with $G_{11} = K/(s+1)^3$ and $G_{12} = \alpha K/(s+1)^3$.

P.12.4 Obtain exact solutions for problem P.12.3.

P.12.5 Prove the result given for the special case (b) in section 12.3.3.

P.12.6 Apply Mees' method to show that the example of section 12.4 can have a limit cycle.

P.12.7 Show for a system with $n_{11} = n_{22} = x^3$, $n_{12} = n_{21} = 0$, $G_{11} = G_{22} = K/(s+1)^2$, $G_{12} = \alpha K/(s+1)^2$, $G_{21} = \beta K/(s+1)^2$ and α, β, K are constants greater than zero that
(a) Mees' method indicates the possibility of a limit cycle.
(b) No limit cycle is possible according to the eigenvalue method.

P.12.8 What is the maximum possible K for stability for a system with $n_{11} = n_{22} = n(x)$ an ideal saturation characteristic, $n_{12} = n_{21} = 0$, $G_{11} = G_{22} = K/(s+1)^2$, $G_{12} = K/4(s+1)^2$ and $G_{21} = K/8(s+1)^2$ by
(a) Mees' method.
(b) The modification of Mees' method.
(c) The eigenvalue method.

P.12.9 Repeat problem P.12.8 with $G_{11} = G_{22} = K/(s+1)^3$, $G_{12} = K/4(s+1)^3$ and $G_{21} = K/8(s+1)^3$.

P.12.10 Compute possible limit cycles for problem P.12.9 with a value of K equal to 10.

P.12.11 For a nonlinear system with $n_{11} = n_{22} =$ relay with dead zone having $h = \delta = 1$, $n_{12} = n_{21} = 0$ and a linear plant as in problem P.12.9 compute the value of K for stability by Mees' method and the exact method of section 12.4.

Chapter 13

Double Valued Nonlinearities

13.1 Introduction

In earlier chapters DVNLs with a single deterministic input or a deterministic input plus bias were considered. Due to the added complications present with DVNLs the material presented for random and multiple inputs was restricted almost entirely to SVNLs. Here we are concerned solely with the quasilinear modelling of DVNLs; uses of such a model are obvious from earlier considerations. Two approaches to the problem have been taken. One procedure is to look for a configuration of SVNLs and linear elements with input–output behaviour identical to the DVNL and then perform a quasilinear analysis of that configuration. Unfortunately the resulting block diagram is usually a nonlinear feedback system with poor filtering characteristics for which quasilinear analysis is not too accurate. Figure 13.1 shows the block diagrams for an on-off relay with hysteresis and an inertia controlled backlash characteristic. The latter, which is of course multi rather than double valued, has been studied(1) using this procedure.

Alternatively a means of specifying the output from the actual double valued characteristic can be sought and then a suitable linearisation determined from it.

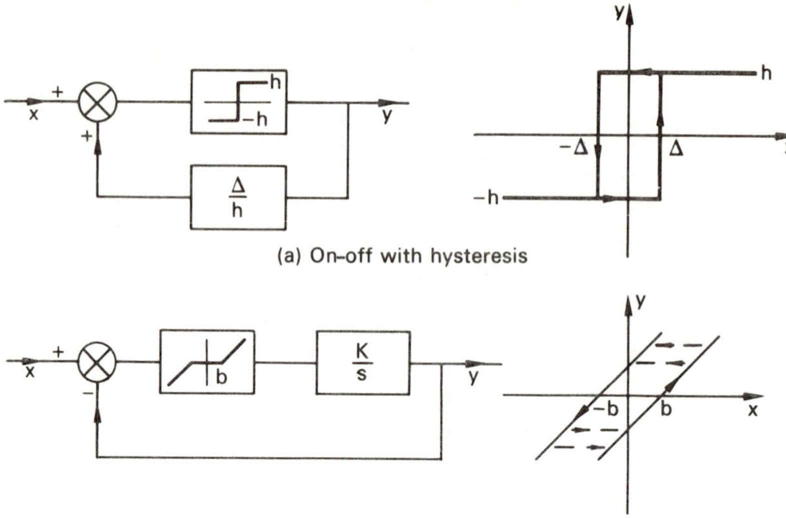

(b) Inertia controlled backlash

Figure 13.1 Block diagrams for nonlinearities

Several authors[2,3,4] have incorrectly assumed that the output of a relay with hysteresis can be expressed as a function $n(x, \dot{x})$ despite warnings to the contrary.[5,6] Theoretical results which are largely qualitative or approximate, but give equivalent gains in good agreement with experimental data, are presented in the following sections for a Gaussian input to the on-off relay with hysteresis and dead zone. The problem is shown to be extremely complex since the exact equivalent gain is found to depend upon the spectrum of the input signal. Use of the modified nonlinearity method for two sine or a sine plus Gaussian input is considered. Since analytical solutions are impossible and graphical results, say from experimental measurements, almost prohibitive to present due to the number of parameters which affect the DFs, the physical behaviour is discussed and satisfactory results obtained for certain limiting cases. Many of the problems encountered in feedback systems do not unfortunately fall into these limiting case categories. However, if all this chapter does is to indicate to the reader the true complexity of the problem and stop him from following erroneous ideas it will have served its purpose.

13.2 Relay with dead zone and hysteresis

The characteristic of a relay with hysteresis and dead zone is shown in figure 13.2.

True relay

The behaviour of the true relay is such that its state within a hysteresis region is determined only by the direction from which the input signal entered the

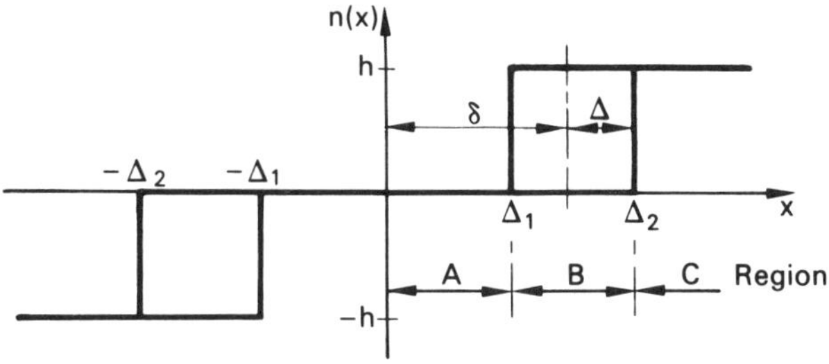

Figure 13.2 Relay with hysteresis and dead zone

region. That is for $x > 0$

$$n(x) = \begin{cases} 0 & \text{for } x < \Delta_1, \\ h & \text{for } x > \Delta_2, \\ 0 & \text{for } \Delta_1 < x < \Delta_2, \text{ if the presence of } x \text{ in this region was preceded by the presence of } x \text{ in the region } x < \Delta_1, \\ h & \text{for } \Delta_1 < x < \Delta_2, \text{ if the presence of } x \text{ in this region was preceded by the presence of } x \text{ in the region } x > \Delta_2. \end{cases}$$

$n(x, \dot{x})$ model

A widely used[2,3,4] mathematical description of the relay, which will be called the $n(x, \dot{x})$ model, is

$$n(x, \dot{x}) = (h/2)\{\text{sgn}(x - \delta - \Delta\,\text{sgn}\,\dot{x}) + \text{sgn}(x + \delta - \Delta\,\text{sgn}\,\dot{x})\}. \tag{13.1}$$

This equation implies that the state of the relay within the hysteresis region depends only on the direction of motion within the region. The $n(x, \dot{x})$ model can be simulated by combinations of SVNLs with no feedback paths as shown in figure 13.3. The outputs of the true relay and the $n(x, \dot{x})$ model in response

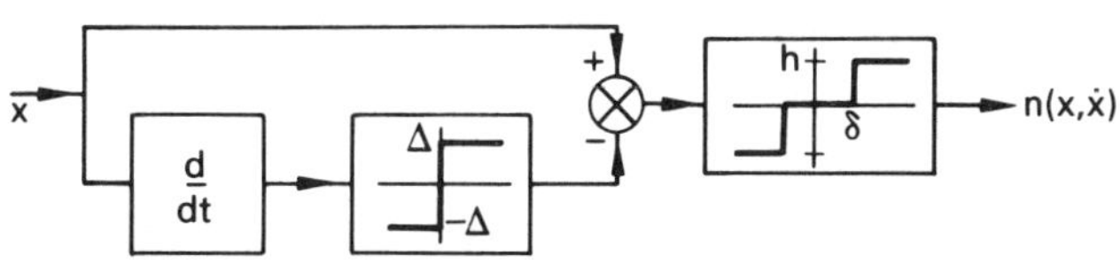

Figure 13.3 $n(x, \dot{x})$ model of relay in figure 13.2

to a typical random input are shown in figure 13.4. It is seen that the two nonlinearities behave differently whenever maxima and minima are encountered within the hysteresis region. The $n(x, \dot{x})$ model is therefore correct if x is a sinusoid and the true equivalent gains K_x and $K_{\dot{x}/w}$ can be found using equations (10.69) and (10.71) respectively.

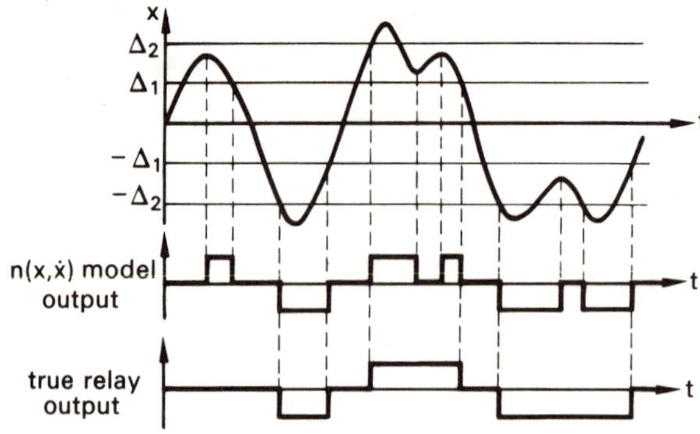

Figure 13.4 Output of the true relay and its $n(x, \dot{x})$ model

13.3 The $n(x, \dot{x})$ model

Although we have seen that the $n(x, \dot{x})$ model is an incorrect representation for a general input, its equivalent gain model is useful for comparative purposes.

13.3.1 *Method of Sawaragi*

Sawaragi[2] assumed the $n(x, \dot{x})$ model to be an exact description of a relay with hysteresis, and used the equivalent gains obtained in a theoretical study of such a nonlinearity in a feedback system with a Gaussian input. The equivalent gains were evaluated from equations (10.42) and (10.43), that is

$$K_y = (1/\sigma_y^2)\int_{-\infty}^{\infty}\int_{-\infty}^{\infty} yn(y, \dot{y})q(y)q(\dot{y})\,\mathrm{d}y\,\mathrm{d}\dot{y}, \tag{13.2}$$

and

$$K_{\dot{y}} = (1/\sigma_{\dot{y}}^2)\int_{-\infty}^{\infty}\int_{-\infty}^{\infty} \dot{y}n(y, \dot{y})q(y)q(\dot{y})\,\mathrm{d}y\,\mathrm{d}\dot{y}. \tag{13.3}$$

For the $n(x, \dot{x})$ model, $n(y, \dot{y})$ has the constant values in the various regions of the $y - \dot{y}$ plane, shown in figure 13.5. Using these values in equations (13.2)

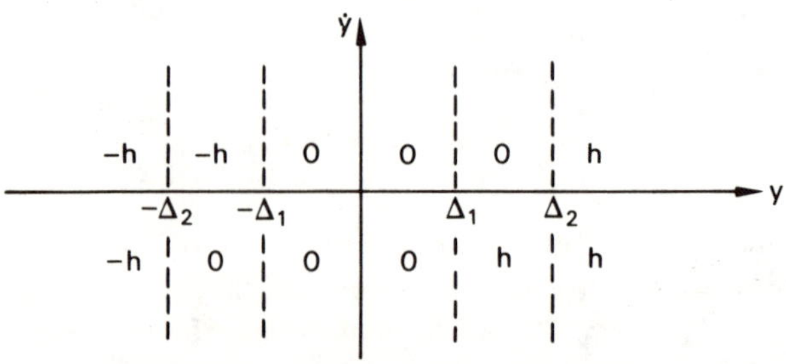

Figure 13.5 The values of $n(y, \dot{y})$ in the $y, \dot{y}$ plane

and (13.3) one obtains

$$K_y = (h/\sigma_y\sqrt{2\pi})\{e^{-\Delta_2^2/2\sigma_y^2} + e^{-\Delta_1^2/2\sigma_y^2}\} = h\{q(\Delta_2) + q(\Delta_1)\}, \tag{13.4}$$

and

$$K_{\dot{y}} = (h/\sigma_{\dot{y}}\sqrt{2\pi})\{\mathrm{erf}(\Delta_2/\sigma_y\sqrt{2}) - \mathrm{erf}(\Delta_1/\sigma_y\sqrt{2})\}, \tag{13.5}$$

in agreement with Sawaragi for the case of no dead zone, i.e. $\Delta_1 = -\Delta_2$.

13.3.2 *Method of Mahalanabis and Nath*

These authors(3,7) have evaluated equivalent gains for Gaussian and sine plus Gaussian inputs to several nonlinearities using the narrow band model of a Gaussian process. When a Gaussian signal is passed through a narrow band filter the output appears to be a sinusoidal wave with slowly varying envelope (amplitude) and phase. Thus a sample function of the narrow band random process is expressible in the form

$$y(t) = R(t)\cos\{\omega_c t + \delta(t)\},$$

where ω_c is the mean frequency of the spectral band, $R(t)$ the envelope and $\delta(t)$ the phase. Formally, such a representation need not be restricted to narrow band waves; the concept of an envelope has significance, however, only when the variation of $R(t)$ and $\delta(t)$ are slow compared to those of cos $w_c t$. It can be shown that(9) the envelope $R(t)$ and phase $\delta(t)$ have Rayleigh and uniform distributions respectively.

Thus

$$p(R) = (R/\sigma_y^2)e^{-R^2/2\sigma_y^2} \quad \text{for } R > 0, \tag{13.6}$$

where σ_y^2 is the mean squared value of the signal, y.

Considering an odd symmetrical nonlinearity with such a narrow band process as input, the in phase and quadrature outputs for an input $R\cos\theta$ are

$$a_1 = (2/\pi)\int_0^\pi n(R\cos\theta)\cos\theta\,d\theta, \tag{13.7}$$

$$b_1 = (2/\pi)\int_0^\pi n(R\cos\theta)\sin\theta\,d\theta. \tag{13.8}$$

The crosscorrelations Re R_{io} between the input and output and Im R_{io} between a signal lagging the input by 90° and the output can then be obtained by averaging with respect to the slowly varying amplitude R to give

$$\mathrm{Re}\,R_{io} = (1/2)\int_0^\infty Ra_1(R)p(R)\,dR, \tag{13.9}$$

$$\mathrm{Im}\,R_{io} = (1/2)\int_0^\infty Rb_1(R)p(R)\,dR. \tag{13.10}$$

Thus, the equivalent gains Re $K(y)$ and Im $K(y)$ are given by

$$\text{Re } K(y) = \{\text{Re } R_{i0}/\overline{y^2(t)}\} = \{1/\pi\sigma_y^2\} \int_0^\infty \int_0^\pi n(R\cos\theta)R\cos\theta\, p(R)\, \mathrm{d}\theta\, \mathrm{d}R, \quad (13.11)$$

and

$$\text{Im } K(y) = \{\text{Im } R_{i0}/\overline{y^2(t)}\} = \{1/\pi\sigma_y^2\} \int_0^\infty \int_0^\pi n(R\cos\theta)R\sin\theta\, p(R)\, \mathrm{d}\theta\, \mathrm{d}R. \quad (13.12)$$

When the input is a sinusoidal signal plus Gaussian noise, it can be shown that[9]

$$p(R) = (R/\sigma_y^2)\exp\{-(R^2 + a^2)/2\sigma_y^2\}I_0(aR/\sigma_y^2), \qquad R > 0, \qquad (13.13)$$

so that Re R_{i0} and Im R_{i0} can again be evaluated from equations (13.9) and (13.10) with (13.7) and (13.8) substituted. The individual equivalent gains for the sinusoid x and noise y can then be obtained by recognising that Re R_{i0} must be of the form

$$\text{Re } R_{i0} = \overline{x^2}\,\text{Re } K(x) + \overline{y^2}\,\text{Re } K(y),$$

with a similar result for Im R_{i0}. Therefore for the $n(x, \dot{x})$ model with a single Gaussian input, y, equations (13.11) and (13.12) can be evaluated to give†

$$\text{Re } K(y) = (h/\sigma_y\sqrt{2\pi})\{\mathrm{e}^{-\Delta_2^2/2\sigma_y^2} + \mathrm{e}^{-\Delta_1^2/2\sigma_y^2}\}, \qquad (13.14)$$

$$\text{Im } K(y) = (h/\sigma_y\sqrt{2\pi})\{\operatorname{erf}(\Delta_2/\sigma_y\sqrt{2}) - \operatorname{erf}(\Delta_1/\sigma_y\sqrt{2})\}. \qquad (13.15)$$

Comparing equations (13.14) and (13.15) with equations (13.4) and (13.5), it is seen that

$$\text{Re } K(y) = K_y,$$

$$\text{Im } K(y) = -\sigma_{\dot{y}}K_{\dot{y}}/\sigma_y.$$

These results are to be expected since equations (13.14) and (13.15) were derived for narrow band noise represented by a random sinusoid. For a sinusoidal signal x, $\dot{x}$ and x phase lagged by 90° are identical except for a minus sign and a scaling factor relating their amplitudes.

13.3.3 *Effect of input spectrum*

For a nonlinearity $n(y, \dot{y})$ the equivalent gains K_y and $K_{\dot{y}}$ may depend upon the input spectrum, since for a given σ_y, $\sigma_{\dot{y}}$ depends upon the spectrum of y. It is seen from equations (13.14) and (13.15) however that for the $n(y, \dot{y})$ model of the relay Re $K(y)$ and Im $K(y)$ depend only upon σ_y and are therefore independent of the input spectrum. This result is now shown to be true for any $n(y, \dot{y})$ whose dependence upon $\dot{y}$ is strictly a function of sgn $\dot{y}$.

---

† The solutions in reference 3 are evaluated incorrectly.

From equation (13.2)

$$\operatorname{Re} K(y) = K_y = (1/\sigma_y^2)\int_{-\infty}^{\infty} yq(y)\,\mathrm{d}y \int_{-\infty}^{\infty} n(y, \operatorname{sgn}\dot{y})q(\dot{y})\,\mathrm{d}\dot{y}$$

$$= (1/\sigma_y^2)\int_{-\infty}^{\infty} yq(y)\left\{\int_{-\infty}^{0} n(y,-1)q(\dot{y})\,\mathrm{d}\dot{y} + \int_{0}^{\infty} n(y,1)q(\dot{y})\,\mathrm{d}\dot{y}\right\}\mathrm{d}y$$

$$= (1/\sigma_y^2)\int_{-\infty}^{\infty} (y/2)\{n(y,-1) + n(y,1)\}q(y)\,\mathrm{d}y,$$

which is clearly independent of the spectrum of y.

Also from equation (13.3)

$$\operatorname{Im} K(y) = -\sigma_{\dot{y}}K_{\dot{y}}/\sigma_y = (-1/\sigma_y\sigma_{\dot{y}})\int_{-\infty}^{\infty} q(y)\left\{\int_{-\infty}^{\infty} \dot{y}n(y,\operatorname{sgn}\dot{y})q(\dot{y})\,\mathrm{d}\dot{y}\right\}\mathrm{d}y$$

$$= (-1/\sigma_y\sigma_{\dot{y}})\int_{-\infty}^{\infty} q(y)\left\{\int_{0}^{\infty} [n(y,1) - n(y,-1)]\dot{y}q(\dot{y})\,\mathrm{d}\dot{y}\right\}\mathrm{d}y$$

$$= (\overline{|\dot{y}|}/2\sigma_y\sigma_{\dot{y}})\int_{-\infty}^{\infty} \{n(y,-1) - n(y,1)\}q(y)\,\mathrm{d}y.$$

But for a Gaussian signal the form factor $F = \sigma_{\dot{y}}/\overline{|\dot{y}|} = (\pi/2)^{1/2}$ so that $\operatorname{Im} K(y)$ is also independent of the spectrum of y.

13.4 True relay

From figure 13.4 it is obvious that for the true relay the equivalent gains will depend upon the possibility of maxima and minima within the hysteresis range and therefore upon the input spectrum.

13.4.1 *Narrow band Gaussian input*

The method described in section 13.3.2, since it employs what may be called the cosine model of noise, is valid only for a very narrow band Gaussian input. It does however give correct solutions for the RDFs of single valued nonlinearities since the result for a narrow band Gaussian input is valid for any Gaussian input. The method is a general approach which can be used for a nonlinearity with a narrow band Gaussian input. The characteristic, in fact, need not be known analytically since the integrations with respect to θ in equations (13.7) and (13.8) are the standard fundamental output expressions for a cosine input and can be evaluated graphically as described in chapter 3.

For the true relay

$$\begin{aligned} a_1 &= RN_p(R) \\ &= 0 \qquad\qquad R < \Delta_2, \\ &= (2h/\pi R)\{(R^2 - \Delta_1^2)^{1/2} + (R^2 - \Delta_2^2)^{1/2}\} \quad R > \Delta_2, \end{aligned}$$

therefore

$$\operatorname{Re} K(y) = (h/\pi\sigma_y^2)\int_{\Delta_2}^{\infty} \{(R^2 - \Delta_1^2)^{1/2} + (R^2 - \Delta_2^2)^{1/2}\} (R/\sigma_y^2)e^{-R^2/2\sigma_y^2}\,dR,$$

which on integration gives

$$\operatorname{Re} K(y) = (2/\pi)^{1/2}(h/\sigma_y)\{e^{-\Delta_2^2/2\sigma_y^2}[(1/2) + (u/\pi)] + e^{-\Delta_1^2/2\sigma_y^2}[(1/2) - (1/2)\operatorname{erf} u]\}, \tag{13.16}$$

where

$$u = \{(\Delta_2^2 - \Delta_1^2)/2\sigma_y^2\}^{1/2}.$$

Also

$$\begin{aligned} b_1 &= -RN_q(R) \\ &= 0 \qquad\qquad\qquad\quad R < \Delta_2, \\ &= 2h(\Delta_2 - \Delta_1)/\pi R \quad R > \Delta_2, \end{aligned}$$

giving

$$\operatorname{Im} K(y) = (h/\pi\sigma_y^2)\int_{\Delta_2}^{\infty} (\Delta_2 - \Delta_1)(R/\sigma_y^2)e^{-R^2/2\sigma_y^2}\,dR,$$

that is

$$\operatorname{Im} K(y) = (h/\pi\sigma_y^2)(\Delta_2 - \Delta_1)e^{-\Delta_2^2/2\sigma_y^2}. \tag{13.17}$$

Equations (13.16) and (13.17) thus give exact solutions for Re $K(y)$ and Im $K(y)$ for the true relay with a narrow band Gaussian input.(10,11)

13.4.2 *General Gaussian input*(10,11)

By definition

$$K_y = R_{yz}(0)/\sigma_y^2 = (1/\sigma_y^2)\int_{-\infty}^{\infty} yn(y)q(y)\,dy.$$

Consider the behaviour for y positive. When it is in the range $\Delta_1 < y < \Delta_2$, it must have one of the four possible types of path illustrated in figure 13.6.

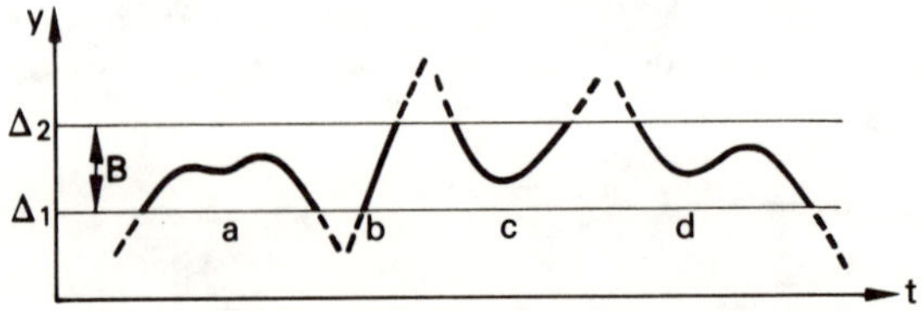

Figure 13.6 Possible signal paths in the region *B*

When $y < \Delta_1$ or is on a path of type a or b in the figure, $n(y) = 0$. Otherwise $n(y) = h$.

Therefore, for the complete relay

$$K_y = (2h/\sigma_y^2)\left\{\int_{\Delta_1}^{\Delta_2} yq_c(y)\,\mathrm{d}y + \int_{\Delta_1}^{\Delta_2} yq_d(y)\,\mathrm{d}y + \int_{\Delta_2}^{\infty} yq(y)\,\mathrm{d}y\right\},$$

that is

$$K_y = (2h/\sigma_y^2)\left\{\int_{\Delta_1}^{\Delta_2} y[q_c(y) + q_d(y)]\,\mathrm{d}y\right\} + (h/\sigma_y\sqrt{2\pi})\mathrm{e}^{-\Delta_2^2/2\sigma_y^2}, \qquad (13.18)$$

where $q_c(y)$ and $q_d(y)$ are the probability density functions for those values of y for which the signal is on paths c and d in figure 13.6, respectively. Equation (13.18) is an exact solution for K_y, but is not useful as it stands because $q_c(y)$ and $q_d(y)$ are unknown functions. Before considering possible approximations for equation (13.18), the imaginary part of the gain is studied.

Again, by definition

$$K_{\dot{y}} = (R_{\dot{y}z}(0)/\sigma_{\dot{y}}^2) = \lim_{T\to\infty} (1/T\sigma_{\dot{y}}^2)\int_0^T \dot{y}n(y)\,\mathrm{d}t. \qquad (13.19)$$

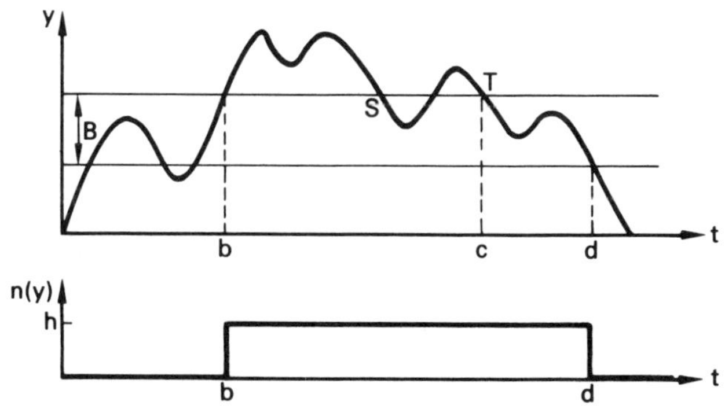

Figure 13.7 Signal for computing $R_{\dot{y}z}(0)$

With reference to figure 13.7

$$\int_b^c \dot{y}n(y)\,\mathrm{d}t = h\int_{\Delta_2}^{\Delta_2} \mathrm{d}y = 0,$$

and

$$\int_c^d \dot{y}n(y)\,\mathrm{d}t = h\int_{\Delta_2}^{\Delta_1} \mathrm{d}y = -h(\Delta_2 - \Delta_1).$$

Therefore contributions to $R_{\dot{y}z}(0)$ are made only when y passes through the hysteresis region from Δ_2 to Δ_1, and $R_{\dot{y}z}(0)$ can be expressed simply in terms of the average number of such crossings per second. That is

$$R_{\dot{y}z}(0) = -2mfh(\Delta_2 - \Delta_1), \tag{13.20}$$

where m is the average number of crossings per second of level Δ_2 with negative slope, and f is the fraction of such crossings which pass directly through the hysteresis region.

Rice[8] gives the following expression for m;

$$m = (e^{-\Delta_1^2/2\sigma_y^2}/2) \times (\text{average number of zeros of } y \text{ per second})$$
$$= (\sigma_{\dot{y}}/2\pi\sigma_y)e^{-\Delta_2^2/2\sigma_y^2}.$$

Substituting equation (13.20) with the above value of m in equation (13.19) gives

$$K_{\dot{y}} = -f(h/\pi\sigma_y\sigma_{\dot{y}})(\Delta_2 - \Delta_1)e^{-\Delta_2^2/2\sigma_y^2},$$

and converting to Im $K(y)$ by the ratio $-\sigma_{\dot{y}}/\sigma_y$ gives

$$\text{Im } K(y) = f(h/\pi\sigma_y^2)(\Delta_2 - \Delta_1)e^{-\Delta_2^2/2\sigma_y^2}, \tag{13.21}$$

an exact solution for Im $K(y)$. Again however the factor f is an unknown so that the usefulness of the expression is limited. For a narrow band signal since the cosine model of noise with $R(t)$ and $\theta(t)$ slowly varying allows no minima for $y > 0$, all crossings of Δ_2 with negative slope pass directly through the hysteresis making $f = 1$. It can be seen that equation (13.21) with $f = 1$ is identical with equation (13.17) for a narrow band input. Thus for other input bandwidths, since $f < 1$ the value of Im $K(y)$ for the narrow band input provides an upper bound.

13.4.3 *Approximate solution for* Re $K(y)$

In general, as will be seen later, Re $K(y)$ is greater and therefore more important than Im $K(y)$. It would therefore seem desirable to obtain an approximate solution for equation (13.18) without having to perform exhaustive measurements to find $q_c(y)$ and $q_d(y)$. A possible approach is suggested if we examine the method proposed independently by Pervozvanskii[5] and Hollingworth[12] for finding Re $K(y)$. In the hysteresis region B the relay output $n(y, \dot{y})$ may be h or 0, depending on the direction from which the signal entered the region. Let λ be the probability that, when y is in the region B, $n(y, \dot{y}) = h$. Then one can immediately write down

$$K_y = \lambda\,(\text{RDF for relay with dead zone, } \Delta_1)$$
$$+ (1 - \lambda)\,(\text{RDF for relay with dead zone, } \Delta_2),$$

that is

$$K_y = (2/\pi)^{1/2}(h/\sigma_y)\{\lambda e^{-\Delta_1^2/2\sigma_y^2} + (1 - \lambda)e^{-\Delta_2^2/2\sigma_y^2}\}. \tag{13.22}$$

Hollingworth and Pervozvanskii both made the intuitively appealing assumption that λ, the probability that the signal entered the hysteresis region B from the right, was equal to the ratio of the time of dwell of the signal to the right of B

to the total time of dwell outside the hysteresis region. That is, with reference to figure 13.8

$$\lambda = \frac{\text{Area I}}{\text{Area I} + \text{Area II}},$$

giving

$$\lambda = \left\{\int_{\Delta_2}^{\infty} q(y)\,\mathrm{d}y\right\} \Big/ \left\{\int_{-\infty}^{\Delta_1} q(y)\,\mathrm{d}y + \int_{\Delta_2}^{\infty} q(y)\,\mathrm{d}y\right\}. \tag{13.23}$$

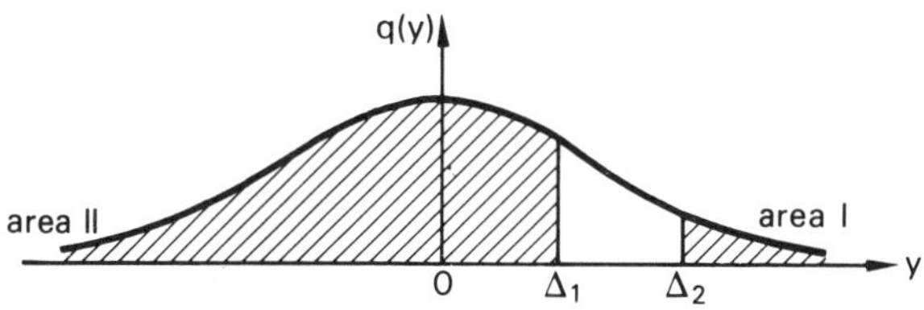

Figure 13.8 Probability areas for Pervozvanskii's method

Comparing equations (13.18) and (13.22) however they are seen to be identical if

$$\lambda q(y) = q_c(y) + q_d(y), \tag{13.24}$$

an expression which cannot in fact be satisfied by a constant value of λ. For example in the case of narrow band Gaussian noise, $q_c(y) = 0$ since there are no minima for $y > 0$, and $q_d(y)$ is given by

$$q_d(y) = (1/2\pi) \int_{\Delta_2}^{\infty} [1/R \cos\{\sin^{-1}(y/R)\}] p(R)\,\mathrm{d}R,$$

which is seen to be a monotonically increasing function of y. However, $q(y)$ is a monotonically decreasing function of y so clearly no constant value of λ is possible for this case. The simplicity of an approximation such as equation (13.24) is certainly appealing and one is led to ask how good a value of λ is given by equation (13.23), and whether it is the best possible choice. When Δ_1 and Δ_2 are approximately equal and large, equation (13.23) gives a small value of λ, whereas from physical considerations one would expect a value close to 0·5, whenever $\Delta_1 \to \Delta_2$; a fact which has been confirmed by experiment using a low pass Gaussian input.[10] More generally the experimental results showed that equation (13.23) was badly in error unless Δ_1 was near zero and Δ_2 not too large.

An alternative value for λ, obtained considering a simple triangular model for Gaussian noise, is given by Fieguth.[10] Figure 13.9 shows a triangular waveform, y, where all the slopes have equal magnitudes and the peaks alternate in sign. It can be shown that if the peaks are distributed according to the law

$$p(P) = (P^2/\sigma_y\sqrt{2\pi})\mathrm{e}^{-P^2/2\sigma_y^2},$$

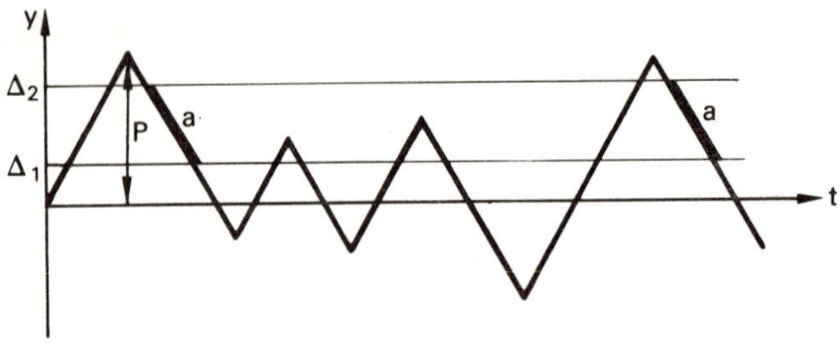

Figure 13.9 A triangular model of Gaussian noise

the probability density function of the signal is Gaussian with variance σ_y^2. The signal therefore constitutes a first order approximation to a Gaussian signal. The ratio of the time that the signal is on segments a in figure 13.9 to the total time the signal spends in region B, λ_T, can be shown to be given by

$$\lambda_T = \frac{(1/2)\text{Area I}}{\text{Area II}} = \frac{(1/2)(\Delta_2 - \Delta_1)q(\Delta_2)}{\int_{\Delta_1}^{\Delta_2} q(y)\,\mathrm{d}y}, \tag{13.25}$$

where the areas are shown in figure 13.10. Equation (13.25) is seen to yield the correct value of 0·5 for λ_T as $\Delta_1 \to \Delta_2$ and was found to agree quite closely with experimental values of λ for a low pass Gaussian input provided $\Delta_1 > 0$. An approximate solution for Re $K(y)$ is therefore

$$\text{Re}\, K(y) \simeq (2/\pi)^{1/2}(h/\sigma_y)\{\lambda_T \mathrm{e}^{-\Delta_1^2/2\sigma_y^2} + (1 - \lambda_T)\mathrm{e}^{-\Delta_2^2/2\sigma_y^2}\}, \tag{13.26}$$

where

$$\lambda_T = (\Delta_2 - \Delta_1)\mathrm{e}^{-\Delta_2^2/2\sigma_y^2}/[\sigma_y\sqrt{2\pi}\{\text{erf}(\Delta_2/\sigma_y\sqrt{2}) - \text{erf}(\Delta_1/\sigma_y\sqrt{2})\}]. \tag{13.27}$$

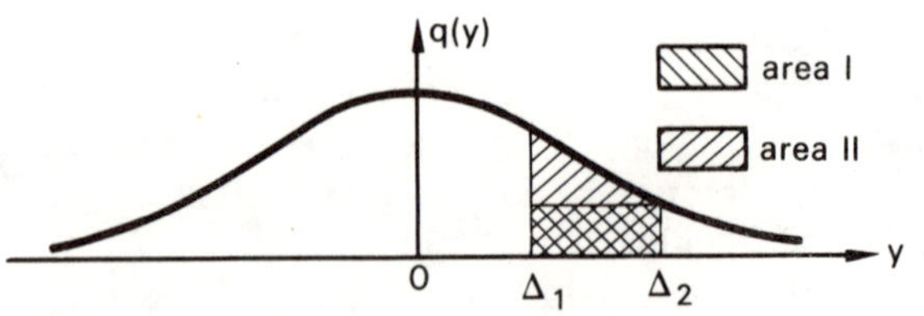

Figure 13.10 Improved probability areas

13.4.4 *Effect of input spectrum*

In Rice's[8] harmonic model of Gaussian random signals, bandpass processes with the same ratio ω_2/ω_1 and the same spectral shape, but different absolute values of ω_2 and ω_1, are simple time scaled versions of one another. A change in spectrum is therefore taken to mean a change in the ratio ω_2/ω_1. Thus all low pass processes, for which $\omega_1 \to 0$, are identical in the spectral sense. The statistical properties such as $q_c(y)$, $q_d(y)$ and f which affect the equivalent gains of the true relay will change with changes in input spectrum. Now

$$\mathrm{Re}\, K(y) = (2h/\sigma_y^2) \int_{\Delta_1}^{\Delta_2} y\{q_c(y) + q_d(y)\}\, \mathrm{d}y + (h/\sigma_y\sqrt{2\pi})\mathrm{e}^{-\Delta_2^2/2\sigma_y^2}, \tag{13.18}$$

so that the first term only will change with variations in the input spectrum, the second term being the RDF of a relay with dead zone Δ_2 will remain constant. Since $q_c(y)$ is zero for a narrow band signal it might be expected that the first term will increase as the input bandwidth increases. Also

$$\mathrm{Im}\, K(y) = \{fh(\Delta_2 - \Delta_1)\mathrm{e}^{-\Delta_2^2/2\sigma_y^2}\}/\pi\sigma_y^2, \tag{13.21}$$

where the fraction f is unity for a narrow band signal and decreases with increasing input bandwidth. It is expected therefore that Im $K(y)$ will decrease appreciably with increasing input bandwidth.

13.5 Comparison of theoretical and experimental results

Some authors[13,14] who have put forward equivalent gain models for double or multivalued nonlinearities and have performed any experimental work have usually done so with the nonlinearity included in a closed loop feedback system. Since a good feedback control system should not be too sensitive to internal loop gain changes, results given for say the mean squared error, do not alone provide a satisfactory assessment of the model. Although in an actual feedback system the nonlinearity input will not be truly Gaussian, it is usually preferable to have a model of known accuracy for a Gaussian input and to estimate the error caused by the distorted waveform in some other manner. Experimental results[10] derived for the true relay and the $n(x, \dot{x})$ model with a Gaussian input are compared with the theoretical solutions in figure 13.11. Except for the special case of Re $K(y)$ for the relay with hysteresis only, it is seen that the gains of the $n(x, \dot{x})$ model are very different from those of the true relay. That is, the $n(x, \dot{x})$ model is a poor approximation to the actual relay over a considerable range of input amplitudes. The input spectrum has no effect on the gains of the $n(x, \dot{x})$ model. For the true relay however Re $K(y)$ increases and Im $K(y)$ decreases from their theoretical values for a narrow band input as the bandwidth is increased. For a low pass input equation (13.26) gives very good agreement with experimental results for $\Delta_1 > 0$. The limitation on Δ_1 exists because the constant λ_T is only a suitable approximation to $\{q_c(y) + q_d(y)\}/q(y)$ for a low pass Gaussian signal and relay with $\Delta_1 > 0$.

Im $K(y)$ is seen to be much smaller for the actual relay than for the $n(x, \dot{x})$ model, particularly for wide band inputs. Using the experimental data for such an input it is easily shown that Im $K(y)$ makes up less than 3 per cent of $[\{\mathrm{Re}\, K(y)\}^2 + \{\mathrm{Im}\, K(y)\}^2]^{1/2}$ for all but very low values of σ_y. At these values of σ_y the distortion components in the relay output are considerably larger than the correlated components, so that the validity of a quasilinear analysis for a feedback system may be open to doubt.

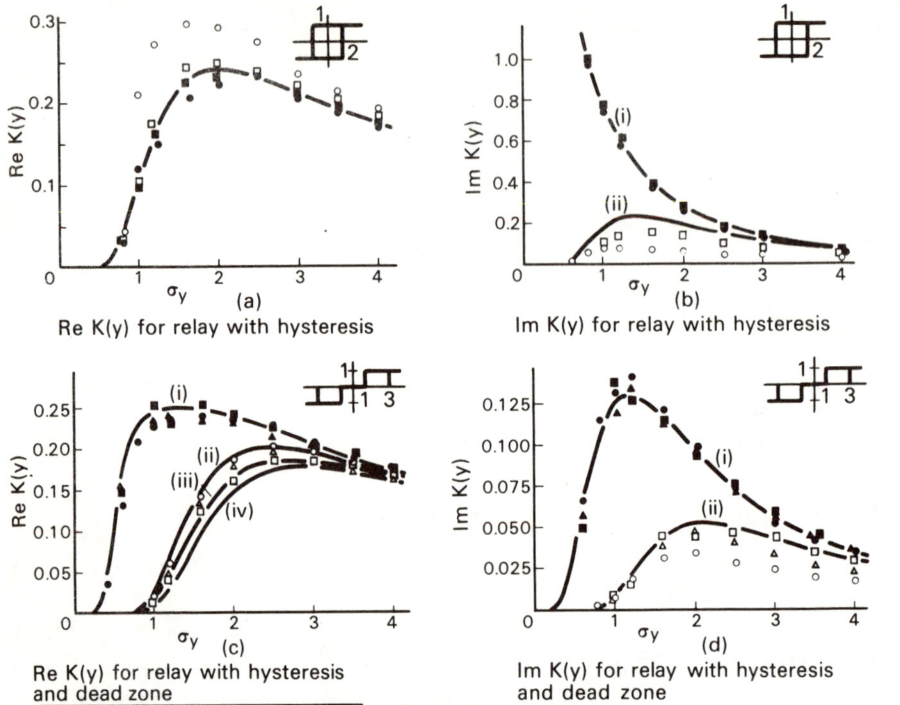

	$\omega_2 = 100$, $\omega_1 = 1$	$\omega_2 = 100$, $\omega_1 = 10$	Narrowband at ω_0 B.W. $= \omega_0/15$
True relay	○	△	□
$f(x, \dot{x})$ model	●	▲	■

Theoretical curves
(a) Eqns. (13.14), (13.16), (13.26) or (13.22) with (13.23)
(b) and (d) (i) Eqn. (13.15) and (ii) Eqn. (13.17)
(c) (i) Eqn. (13.14); (ii) Eqn. (13.26); (iii) Eqn. (13.16); (iv) Eqn. (13.22)

Figure 13.11 Comparison of experimental and theoretical solutions for Re $K(y)$ and Im $K(y)$

In summary Re $K(y)$ and Im $K(y)$ can easily be evaluated from equations (13.16) and (13.17) for a narrow band Gaussian input and provide a lower and upper bound respectively on the gains for other input bandwidths. For low pass inputs experimental results suggest the factor f will be within the range 0·3 to 0·7 and for $\Delta_1 > 0$ equation (13.26) provides a good approximation for Re $K(y)$.

13.6 Two sine inputs

In chapter 5 it was seen that the modified nonlinearity approach is a conceptually easy method for determining the response of a nonlinearity to the sum of two sinusoidal signals. The method for determining a component of the SSDF for sinusoidal inputs x and y of amplitudes a and b respectively into a SVNL may be summarised in the equations

$$n(a, \gamma) = \int_{-a}^{a} n(x + \gamma) r(x)\, \mathrm{d}x, \tag{13.28}$$

$$N_b(a, b) = (2/b^2) \int_{-b}^{b} n(a, y) y r(y)\, dy, \tag{13.29}$$

or alternatively

$$N_b(b, \gamma) = (2/b^2) \int_{-b}^{b} n(y + \gamma) y r(y)\, dy, \tag{13.30}$$

$$N_b(a, b) = \int_{-a}^{a} N_b(b, x) r(x)\, dx. \tag{13.31}$$

Using equations (13.28) and (13.29) the bias output $n(a, \gamma)$ in response to an input of x plus bias γ is found and then the DF of the modified nonlinearity $n(a, \gamma)$ with input y is evaluated to give $N_b(a, b)$. By the alternative procedure of equations (13.30) and (13.31) the DF $N_b(b, \gamma)$ for the input y in the presence of the bias is first found and then its value averaged for the input bias varying as x. Which of the above two procedures is chosen is usually influenced by the particular problem being considered. If the procedures just described verbally, rather than using equations (13.28) to (13.31) since they are not written to allow a complex DF, are carried out for a DVNL the solutions obtained by the two methods are not identical.[15]

Before doing calculations for an on-off relay with hysteresis let us examine the physical behaviour in order to explain the above result. A typical input with $\omega_a \gg \omega_b$ and the corresponding relay output are shown in figure 13.12. It is seen

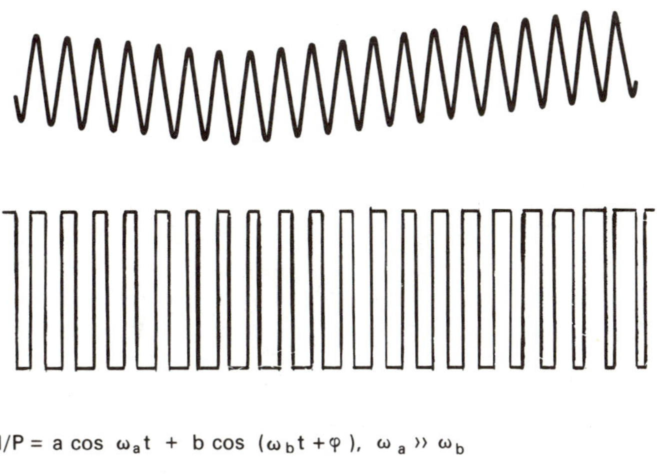

Figure 13.12 Two sinusoid input and output waveforms from relay with hysteresis

that the switching of the relay is controlled almost entirely by the high frequency signal, x. In effect the relay output is essentially $n(x + y, \dot{x})$ where $n(x, \dot{x})$ is given by equation (13.1) with $\delta = 0$. In equations (13.28) to (13.31) the relay switching is accounted for by x in equation (13.28) and y in equation (13.30).

Thus for $\omega_a \gg \omega_b$ it is expected that equations (13.28) and (13.29) will give the correct value of $N_b(a, b)$, which is indeed confirmed experimentally. It may therefore be concluded that the modified nonlinearity approach for finding the SSDF of a DVNL is usually valid,† when one input frequency is several times greater than the other, provided the first modification is carried out with respect to the higher frequency signal.

Consider now the actual solutions for the on-off relay with hysteresis. Since $\omega_a \gg \omega_b$ the first modification must be with respect to a. To determine the gain to b from section 5.3.4 we have, for $a < \Delta$,

$$n(a, \gamma) = \text{Relay with hysteresis } (\Delta - a),$$

and, for $a > \Delta$,

$$n(a, \gamma) = \begin{cases} (h/\pi)\{\sin^{-1}[(\Delta + \gamma)/a] + \sin^{-1}[(\gamma - \Delta)/a]\} & \text{for } -(a - \Delta) < \gamma < (a - \Delta), \\ h & \text{for } \gamma > (a - \Delta), \\ -h & \text{for } \gamma < -(a - \Delta). \end{cases} \tag{13.32}$$

Thus if $a < \Delta$, the gain to $bN_b(a, b)$ is given by

$$N_b(a, b) = \begin{cases} 0 & \text{for } b < \Delta - a, \\ (4h/\pi b^2)[\{b^2 - (\Delta - a)^2\}^{1/2} - j(\Delta - a)] & \text{for } b > \Delta - a. \end{cases}$$

If $a > \Delta$, using the odd symmetry of $n(a, \gamma)$

$$N_b(a, b) = (4h/b^2\pi) \int_0^b [\sin^{-1}\{(\Delta + y)/a\} + \sin^{-1}\{(y - \Delta)/a\}] y r(y)\, dy \quad \text{for } b < (a - \Delta),$$

and

$$N_b(a, b) = (4h/b^2\pi) \int_0^{a-\Delta} [\sin^{-1}\{(\Delta + y)/a\} + \sin^{-1}\{(y - \Delta)/a\}] y r(y)\, dy + (4/b^2) \int_{a-\Delta}^{b} h y r(y)\, dy \quad \text{for } b > (a - \Delta).$$

Since it is not possible to get simple closed form solutions to the above expressions for $N_b(a, b)$ the integrals were evaluated using cursors. Figure 13.13 gives theoretical and experimental results for $N_b(a, b)$, plotted in terms of gain and phase for two values of (b/Δ) as a function of a. The experimental measurements of the gain and phase shift, using a frequency ratio near to ten, are seen to be in good agreement with the theory.

† If the amplitude of x is very small the above argument does not apply.

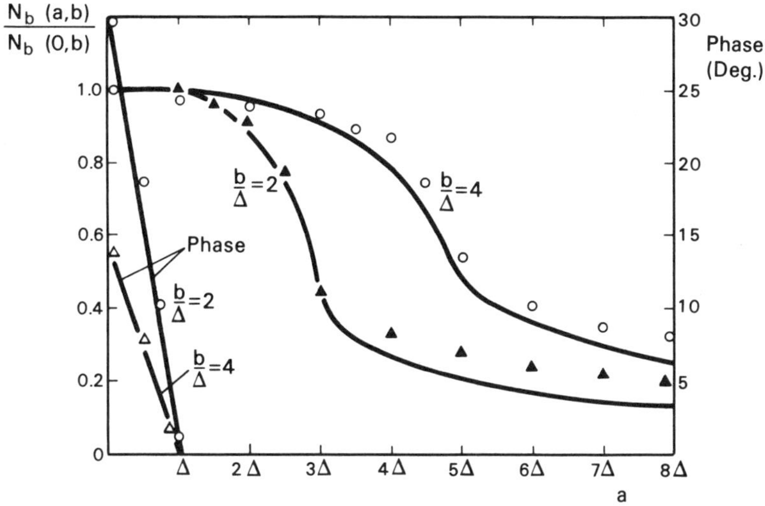

Figure 13.13 DF $N_b(a, b)$ for relay with hysteresis

To determine the gain to a we have, considering positive values of γ only since $N_a(a, \gamma)$ has even symmetry

$$N_a(a,\gamma) = 0 \qquad \text{for } \gamma < \Delta - a,$$
$$(2h/\pi a^2)[\{a^2 - (\Delta + \gamma)^2\}^{1/2} + \{a^2 - (\Delta - \gamma)^2\}^{1/2} - j2\Delta) \qquad \text{for } \Delta - a < \gamma < a - \Delta,$$
$$0 \qquad \text{for } \gamma > a - \Delta. \tag{13.33}$$

The DF for a in the presence of b is now found from $N_a(a, \gamma)$ by averaging with respect to γ taken as a sinusoid of amplitude b. Thus for $a > \Delta$

$$N_a(a,b) = (4h/\pi a^2)\int_0^b [\{a^2 - (\Delta + y)^2\}^{1/2} + \{a^2 - (\Delta - y)^2\}^{1/2} - j2\Delta]r(y)\,dy \qquad \text{for } b < a - \Delta,$$

$$= (4h/\pi a^2)\int_0^{a-\Delta} [\{a^2 - (\Delta + y)^2\}^{1/2} + \{a^2 - (\Delta - y)^2\}^{1/2} - j2\Delta]r(y)\,dy \qquad \text{for } b > a - \Delta.$$

Figure 13.14 shows the experimental and theoretical results for $N_a(a, b)$. It should be noted that averaging the modulus and phase of the DF[15] $N_a(a, \gamma)$ does not give quite the same result as the above procedure of averaging the complex DF. The former procedure can be shown to be incorrect.[10]

In reference 15 experimental results were also obtained with ω_a fixed and varying ω_b in the range $\omega_a/10$ to $10\omega_a$ from which it was found that the above theoretical results were quite accurate for frequency ratios greater than about

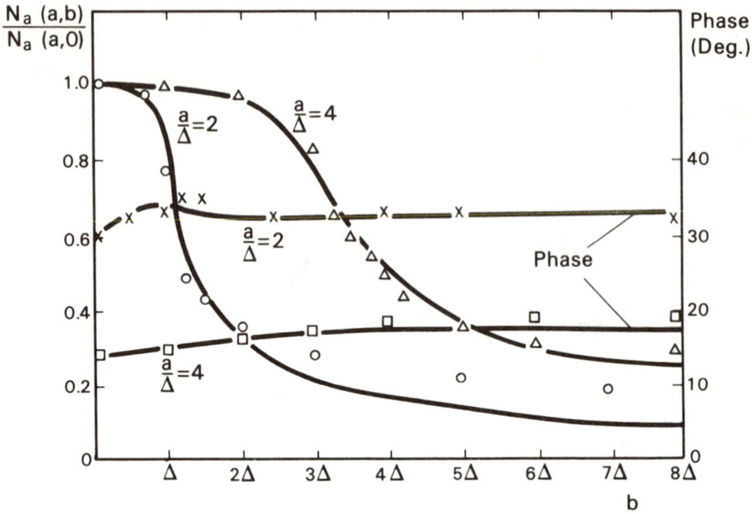

Figure 13.14 DF $N_a(a, b)$ for relay with hysteresis

five. The situation for lower frequency ratios is particularly complex. Since quasilinear analysis of feedback systems involving such frequency ratios may not be particularly accurate due to the presence of large distortion terms, a solution to the problem is probably more of academic than practical interest.

13.7 Sine plus Gaussian input

This problem is of considerable practical importance since it is applicable to the analysis of self oscillating systems with random inputs. Theoretical results can be derived for the $n(x, \dot{x})$ model, although those given in reference 7 are incorrect, but since the expressions are cumbersome and the model invalid they are of no practical use.

Consider the on-off relay with hysteresis to have a sine input of amplitude a and frequency ω plus a band pass Gaussian signal of r.m.s. value σ with lower and upper cut off frequencies Ω_1 and Ω_2 with Ω a representative mid band frequency. From previous considerations one can conclude that the equivalent gains will depend upon h, Δ, σ, a, ω/Ω and Ω_1/Ω_2. For a given relay this is a four dimensional function which it is impossible to calculate analytically and exceedingly awkward to present graphically after experimental measurement. No such complete measurement has been attempted. Instead a number of limiting cases are discussed below.

$\Omega \ll \omega$

In this case the Gaussian signal can be treated as a slowly varying bias so that the first modification should be performed with respect to the sinusoid giving $n(a, \gamma)$ and $N_a(a, \gamma)$ of equations (13.28) and (13.30) respectively. The gain to the random signal is then given by

$$N_\sigma(a,\sigma) = (1/\sigma^2)\int_0^{a-\Delta} (h/\pi)[\sin^{-1}\{(\Delta + y)/a\} + \sin^{-1}\{(y-\Delta)/a\}]$$

$$\times\; yq(y)\,\mathrm{d}y + (1/\sigma^2)\int_{a-\Delta}^{\infty} hyq(y)\,\mathrm{d}y, \quad \text{for } a > \Delta,$$

and to the sinusoid by

$$N_a(a,\sigma) = (4h/\pi a^2)\int_0^{a-\Delta} [\{a^2 - (\Delta + y)^2\}^{1/2} + \{a^2 - (\Delta - y)^2\}^{1/2} - j2\Delta]q(y)\,\mathrm{d}y$$

$$\text{for } a > \Delta.$$

From the form of the above expression it will be seen that the modulus of $N_a(a, \sigma)$ decreases monotonically and the phase increases slightly as the magnitude of the r.m.s. value of the noise increases. Experimental and theoretical results for this situation show good agreement.

$\Omega \gg \omega$

In this case a further difficulty exists, that of finding noise modified DVNLs. Previous results suggest that thc high frequency noise will reduce the hysteresis width seen by the sinusoid. As an approximation one might assume the hysteresis width to be reduced by an amount proportional to σ, with the constant of proportionality k dependent upon the frequency ratio Ω/ω. In this case when $k\sigma > \Delta$ the modified nonlinearity is single valued. The magnitude of the DF for the sinusoid would be expected to be somewhat greater than that for an ideal relay modified by the same noise signal. This is because the noise will begin to reduce the gain of the ideal relay for very low values of σ whereas it will not begin to have much effect on the magnitude of the gain of the relay with hysteresis until σ exceeds Δ/k.

Analog computer results[10] for the DF of a relay with hysteresis to a sinusoid in the presence of narrow band Gaussian noise are shown in figure 13.15. The results largely confirm the general picture presented above. The straight lines on the phase curve show estimates for k to fit the experimental data. Additional experiments carried out using wider bandwidth noise suggested that its effective frequency was near to its upper cut off frequency.

Figure 13.16 shows a self oscillating system with an additional Gaussian noise input of r.m.s. value σ and figure 13.17 measured results for the amplitude of the oscillation as a function of σ. In confirmation of the results for the relay with hysteresis given in figure 13.15 it is seen that the oscillation amplitude depends on the frequency spectrum of the noise being smaller for a higher frequency noise input of given r.m.s. value σ.

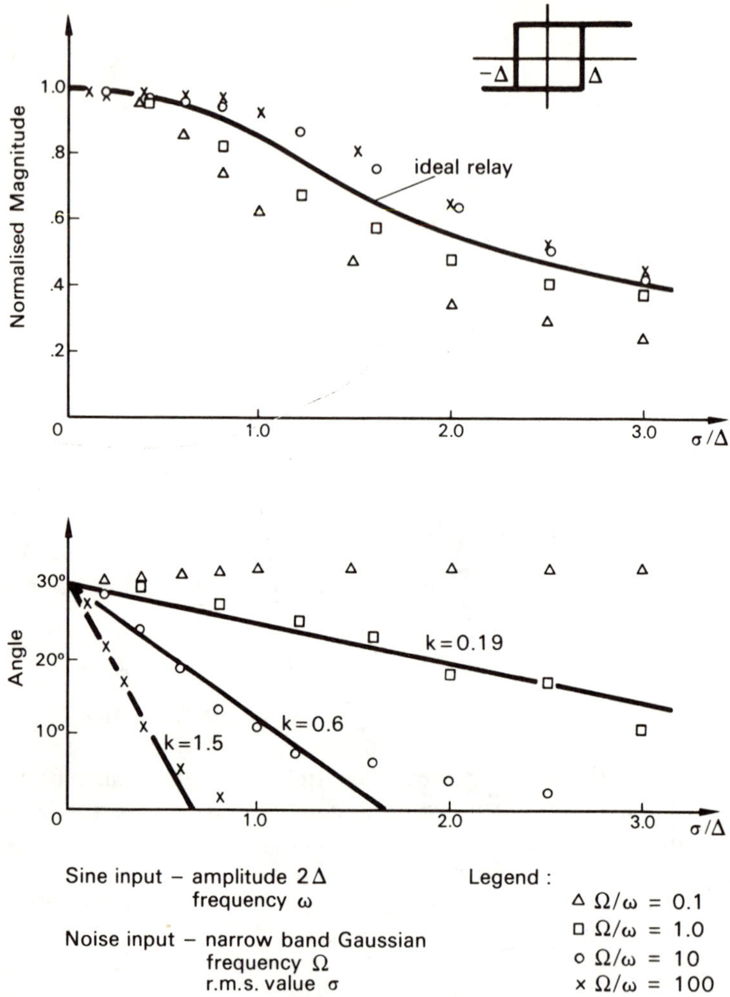

Figure 13.15 DF for sinusoidal input to relay with hysteresis in presence of noise

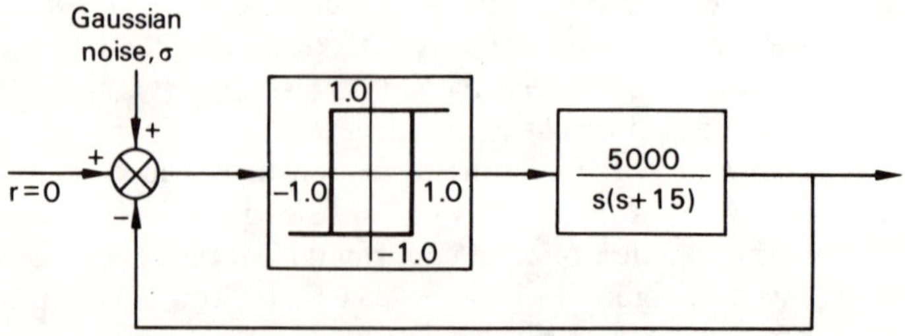

Figure 13.16 Self oscillating system

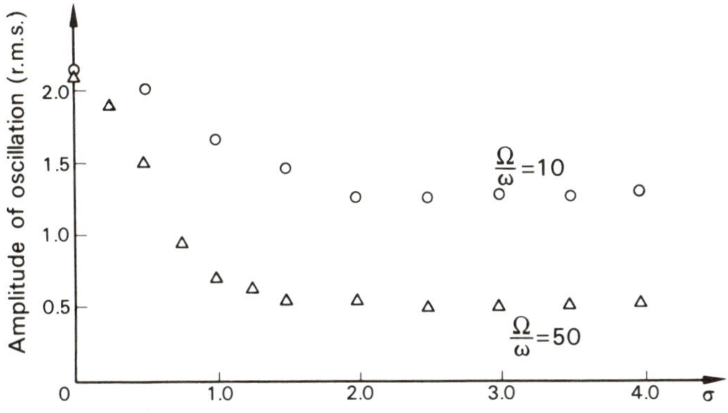

Figure 13.17 Dependence of oscillation amplitude on σ in system of figure 13.16.

REFERENCES

1. Leland, H. R.: 'Input–output cross-correlation functions for some memory-type nonlinear systems with Gaussian inputs', *A.I.E.E. Trans.*, 1960, 79, pp. 219–223.
2. Sawaragi, Y., Sunahara, Y., Nakamizo, T. and Hirakawa, K.: 'Statistical analysis of control systems containing a nonlinear element with hysteresis characteristic', *Tech. Rep. Engng. Res. Inst.*, Kyoto University, 1962, 12, pp. 87–105.
3. Mahalanabis, A. K. and Nath, A. K.: 'Method of statistical linearization', *Proc. I.E.E.*, 1966, 113, pp. 2081–2086.
4. Kazakov, I. Y.: 'Statistical analysis of systems with ambiguous nonlinearities', *Engng. Cybernetics*, 1966, 3, pp. 384–391.
5. Pervozvanskii, A. A.: *Random processes in nonlinear control systems* (Academic Press, 1965, p. 22).
6. Atherton, D. P.: 'The use of complex equivalent gain for a memory-type nonlinearity for the analysis of feedback control systems with random inputs', *I.E.E.E. Trans.*, 1965, AC-10, pp. 223–224.
7. Mahalanabis, A. K. and Nath, A. K.: 'A multi-input quasilinearization method with applications', *Int. J. Control*, 1966, 4, pp. 73–86.
8. Rice, S. O.: 'Mathematical analysis of random noise', *Bell Syst. Tech. J.*, 1944, 23, pp. 282–332, and 1945, 24, pp. 46–156. (Reprinted in Wax, N.: *Selected papers on noise and stochastic processes* (Dover, New York, 1954, pp. 133–294).)
9. Davenport, W. B. and Root, W. L.: *An introduction to the theory of random signals and noise* (McGraw-Hill, New York, 1958, Chapter 8).
10. Fieguth, W.: 'Multi-input quasilinearization', *Ph.D. Thesis, University of New Brunswick*, 1968.
11. Fieguth, W. and Atherton, D. P.: 'Double-valued relays with random inputs', *Proc. I.E.E.*, 1968, 115, pp. 355–361.
12. Hollingworth, R. G.: 'An equivalent gain for relay type nonlinearities with hysteresis', *M.Sc. Thesis, University of New Brunswick*, 1965.
13. Mahalanabis, A. K. and Nath, A. K.: 'Stability of hysteretic system with random signal inputs', *Int. J. Control*, 1965, 2, pp. 135–148.
14. Sen, A. K.: 'The use of complex equivalent gain of a memory-type nonlinearity for the analysis of feedback control systems with random inputs', *Trans. I.E.E.E.*, 1964, AC-9, pp. 245–248.
15. Atherton, D. P. and Turnbull, G. F.: 'The response of nonlinear characteristics to several inputs and the use of the modified nonlinearity concept in control systems', *Proc. I.E.E.*, 1964, 111, pp. 157–164.

Chapter 14

Nonlinear Carrier Systems

14.1 Introduction

There are many ways in which information can be transmitted, particularly by electrical means. In previous chapters we have been concerned with d.c. systems where the signals transmitted are directly proportional to the variables concerned. For reasons of economy, reliability, performance specifications, the particular situation and so on a design may involve other forms of signal transmission within the system. Some of the more common techniques used are those of amplitude, frequency, pulse height and pulse width modulation. The expanding use of digital techniques in control systems has led to the considerable theory of sampled data systems or equivalent systems employing pulse height modulation. Here we are concerned with systems, usually simply referred to as a.c. systems, where at some point amplitude modulated signals are present, and specifically with the effects of any nonlinearity in the amplitude modulated signal channel.

A.c. systems exist for several reasons. Fundamentally, they are used because of the large number of available transducers, which either generate or are driven by suppressed carrier amplitude modulated signals. Two examples of such devices are synchros and two phase servomotors. The latter is extremely useful at low power levels because of its simplicity, ruggedness and low shaft friction due to the absence of any rotor contacts. Another advantage is that a.c. amplifiers, which unlike d.c. amplifiers have no drift problems, can be used. Further if orthogonal modulation-demodulation techniques are employed the same amplifier may be used to amplify two essentially independent control signals for different output stages. Carrier frequencies for modulation and demodulation of 60 or 400 Hz. are usual and although periodic carrier signals other than sinusoids are used, the theory developed on the assumption of sinusoidal carrier signals can often be applied to the more general case.

A brief review of modulation theory and linear a.c. theory is given and the equivalent d.c. transfer function of an a.c. channel containing a modulator, linear network and demodulator is found.[1] Analysis of an a.c. system with nonlinearity in the d.c. or low frequency path thus becomes a conventional nonlinear d.c. problem after replacement of the a.c. channel by its d.c. equivalent.

Consideration of a nonlinear a.c. channel consisting of a modulator, nonlinearity and demodulator is also shown to be reducible to an equivalent nonlinear d.c. channel. When the signal input to the a.c. channel, that is the modulating signal, is unrelated to the carrier signals this equivalent nonlinear d.c. channel contains a modified nonlinearity which is a function of the original nonlinearity and the form of modulation and demodulation employed. Analysis of the a.c. system when the modulating signal is slowly varying relative to the carrier signals may then be done using conventional describing function methods for the modified nonlinearity. It is further shown how the above method can be used to predict carrier locked oscillations and low frequency output distortion which may occur with a high frequency system input.

14.2 Review of linear a.c. theory

Figure 14.1 shows the conventional form of a linear a.c. channel. In many situations the modulation is inherent in the detector as for example when synchros

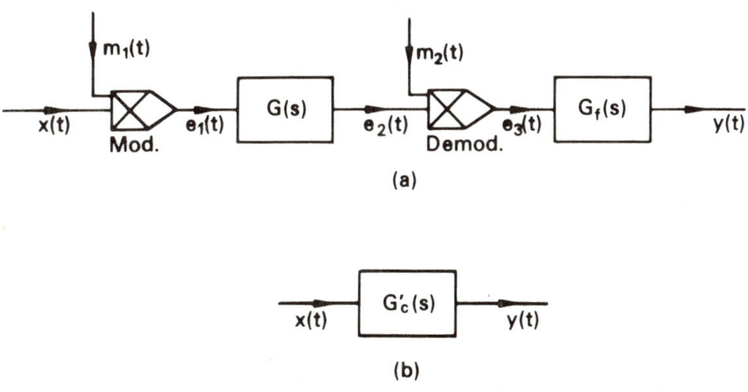

Figure 14.1 Linear a.c. channel and its d.c. equivalent

are used and the demodulation included in the output stage as is the case with a two phase servomotor. The filter $G_f(s)$ is included after the demodulator to remove any frequency components in $e_3(t)$ produced by the modulation demodulation process at frequencies much higher than those in the input signal $x(t)$. The equivalent d.c. transfer function $G_c'(s) = Y(s)/X(s)$ derived below assumes the existence of such filtering. It is normally assumed that the equivalent circuit of the two phase servomotor is an ideal demodulator, that is, one producing frequency components of the order of those in $x(t)$ only, followed by the conventional d.c. transfer function $K/s(1 + sT)$. The transfer function $G_c'(s)$ is then an exact representation before the servomotor. If however the modulated signal contains harmonics of the carrier, as is the case with nonsinusoidal modulation, the representation is only approximate since the higher frequencies can affect the servomotor performance.[(2)]

Modulation is performed using either a nonlinear element or a linear time varying network. In the former, the input to the nonlinear element with a finite α_{11} is $x(t) + m_1(t)$, and when $m_1(t) = M_1 \cos(\omega_c t + \phi_1)$, $x(t) = a \cos \omega_m t$ the output of the modulator contains the desired frequency components $\omega_c \pm \omega_m$ as well as many others, including in many instances the carrier frequency ω_c. For the linear time varying network the carrier $m_1(t)$ acts as a switching signal and the output of the modulator is the product $m_1(t)x(t)$. A general form of modulation is obtained when $m_1(t)$ in figure 14.1 is represented by the Fourier series $m_1(t) = \sum_{p=0}^{\infty} M_{1p} \cos(p\omega_c t + \phi_{1p})$, $\phi_{10} = 0$. This modulation is carrier suppressed (balanced modulation) and if $M_{10} = 0$ so that $m_1(t)$ has a zero average value, it is then full wave rather than half wave modulation. In practice $m_1(t)$ is usually either sinusoidal and equal to $M_1 \cos(\omega_c t + \phi_1)$ or a one to one mark to space square wave of period $2\pi/\omega_c$ possibly with a non-zero average value. Demodulation, although it is the reverse of modulation, is usually accomplished using a linear time variant modulator to give the output $e_3(t) = e_2(t)m_2(t)$.

Before deriving the equivalent d.c. transfer function $G_c'(s)$, it is advantageous to consider the frequency content of the signals at various positions in figure 14.1. Assuming that $x(t)$ has a spectrum from 0 to ω_m, where $\omega_m \ll \omega_c$, and balanced modulation is used with $m_1(t) = \sum_{p=1}^{\infty} M_{1p} \cos(p\omega_c t + \phi_{1p})$ then $e_1(t)$ contains frequencies in the spectral bands $p\omega_c - \omega_m$ to $p\omega_c + \omega_m$ for all p from 1 to ∞. The desired spectral components are those in the band $\omega_c - \omega_m$ to $\omega_c + \omega_m$ and filters are often introduced to reduce the amount of signal in the 'harmonic' bands. No 'harmonic' bands exist with a sinusoidal carrier $m_1(t) = M_1 \cos(\omega_c t + \phi_1)$. Taking all terms into account, if $m_2(t) = \sum_{q=0}^{\infty} \cos(q\omega_c t + \phi_{2q})$ the demodulator produces outputs in the frequency bands $(p \pm q)\omega_c - \omega_m$ to $(p \pm q)\omega_c + \omega_m$ for $p = 1$ to ∞ and $q = 1$ to ∞. In particular it is seen that the output contains signals in the desired band from 0 to ω_m for $p = q$ and all the other frequencies being greater than $\omega_c - \omega_m$ can be removed by $G_f(s)$. As the desired output in the band 0 to ω_m is for $p = q = 1$ only, it is seen that on those few occasions where $m_1(t)$ and $m_2(t)$ contain the same harmonic of ω_c it is necessary to provide filtering before the demodulator, often one which removes all frequencies except those in the band $\omega_c - \omega_m$ to $\omega_c + \omega_m$.

14.2.1 *Equivalent d.c. transfer function*

Referring to figure 14.1 and assuming sinusoidal modulation and demodulation

$$e_1(t) = x(t)M_1 \cos(\omega_c t + \phi_1),$$

so that

$$E_1(s) = (M_1/2)\mathscr{L}\{x(t)e^{j(\omega_c t + \phi_1)} + x(t)e^{-j(\omega_c t + \phi_1)}\}$$
$$= (M_1/2)\{X(s - j\omega_c)e^{j\phi_1} + X(s + j\omega_c)e^{-j\phi_1}\}, \quad (14.1)$$

similarly

$$e_3(t) = e_2(t)M_2 \cos(\omega_c t + \phi_2),$$

and

$$E_3(s) = (M_2/2)\{E_2(s - j\omega_c)e^{j\phi_2} + E_2(s + j\omega_c)e^{-j\phi_2}\}. \quad (14.2)$$

Thus using

$$E_2(s) = G(s)E_1(s) \quad (14.3)$$

in equation (14.2) and substituting for $E_1(s)$ from equation (14.1) gives

$$E_3(s) = (M_1 M_2/4)\{G(s + j\omega_c)X(s + 2j\omega_c)e^{-j(\phi_1 + \phi_2)} + G(s + j\omega_c)$$
$$\times X(s)e^{-j(\phi_2 - \phi_1)} + G(s - j\omega_c)X(s)e^{j(\phi_2 - \phi_1)} + G(s - j\omega_c)$$
$$\times X(s - 2j\omega_c)e^{j(\phi_2 + \phi_1)}\}.$$

We know however from the previous section that in this case, since $p = q = 1$, only, output spectra just exist in the bands from 0 to ω_m and $2\omega_c - \omega_m$ to $2\omega_c + \omega_m$. Obviously, therefore, the terms contributing to spectra in the band 0 to ω_m are those involving $X(s)$ in the above expression and the others will be filtered out by $G_f(s)$. Hence

$$G_c'(s) = \frac{Y(s)}{X(s)} = (M_1 M_2/4)\{G(s + j\omega_c)e^{-j(\phi_2 - \phi_1)} + G(s - j\omega_c)e^{j(\phi_2 - \phi_1)}\}, \quad (14.4)$$

which may also be written

$$G_c'(s) = (M_1 M_2/2)\,\mathrm{Re}_i\{G(s + i\omega_c)e^{-i(\phi_2 - \phi_1)}\}, \quad (14.5)$$

where Re_i denotes taking the real part with respect to i. The change of variable to i has been used so that s can be replaced by $j\omega$ to obtain the frequency response without confusion. Alternatively the equivalent transfer function $G_c(s)$ of the frequency dependent network $G(s)$ only is given by

$$G_c(s) = \mathrm{Re}_i\{G(s + i\omega_c)e^{-i(\phi_2 - \phi_1)}\}, \quad (14.6)$$

with

$$G_c'(s) = M_1 M_2 G_c(s)/2. \quad (14.7)$$

In particular it should be noted that the transfer function $G_c'(s)$ will be affected by changes in M_1, M_2, ϕ_1, ϕ_2, and ω_c.

It will be appreciated from the above discussion that the same frequency response can be obtained from the a.c. channel as from a d.c. transfer function $G_c(s)$, if a network $G(s)$ can be found with a frequency response about ω_c identical with that of $G_c(s)$ about zero, that is, it must have an even amplitude function and an odd phase function about ω_c. Unfortunately no such networks exist, although the characteristics are approximated for small variations about ω_c by networks having symmetry with respect to frequency plotted on a logarithmic scale, that is geometric rather than arithmetic symmetry. Several RC networks such as the twin or bridged T have such a property. Others can be found using the band pass transformation[3] illustrated in figure 14.2. For these symmetrical networks it can be shown that[1]

$$G_c(s) \approx G(s + i\omega_c)|_{\omega_c \gg |s|} \cos(\phi_2 - \phi_1).\dagger \tag{14.8}$$

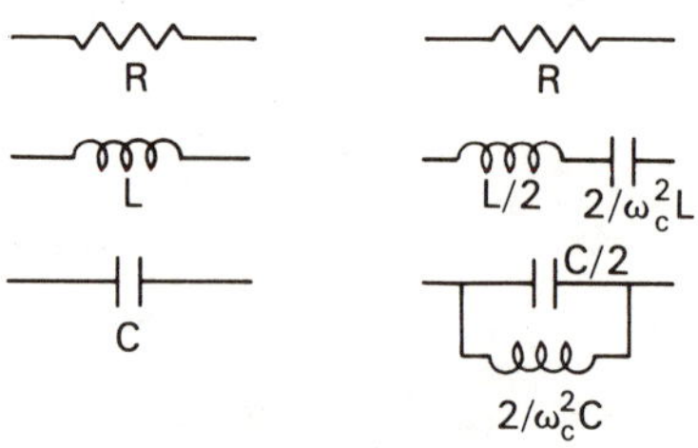

Figure 14.2 Network element changes for low pass to band pass transformation

When the above property is satisfied by a network having a constant gain K and a linear phase variation of slope m about ω_c, it can be shown that

$$G_c(s) = K \cos(\phi_2 - \phi_1 - \phi_c)e^{sm}, \tag{14.9}$$

where ϕ_c is the phase shift at ω_c.

Linear compensation of a.c. systems can be accomplished by adding networks in the a.c. channel or in some cases compensation is achieved using a demodulator and a conventional d.c. compensator followed by a modulator. Disadvantages of the latter approach are the effects of the zero drift and harmonic generation introduced by such circuits. A.c. compensation suffers from the disadvantages that (a) the required transfer function can often only be realised using RLC circuits with high Q coils, (b) RC structures can only be used for simple lead compensation and then at the expense of zero frequency gain, (c) the networks produce a quadrature output even with no quadrature input and (d) the network performance as seen above is dependent upon carrier stability. With reference to point (c), by taking $\phi_1 = \phi_2 = 0$ and $x(t) = a \cos \omega_m t$ in figure 14.1, we obtain

$$e_2(t) = (M_1 a/2)\{[G_+ \cos(\omega_m + \phi_+)t + G_- \cos(-\omega_m + \phi_-)t] \cos \omega_c t$$
$$- [G_+ \sin(\omega_m + \phi_+)t + G_- \sin(-\omega_m + \phi_-)t] \sin \omega_c t\},$$

† Using $\omega_c \gg |s|$ will result in a real $G_c(s)$, that is independent of i.

where

$$G_+ e^{j\phi_+} = G\{j(\omega_c + \omega_m)\},$$

and

$$G_- e^{j\phi_-} = G\{j(\omega_c - \omega_m)\}.$$

Unless the network has arithmetic symmetry, that is $G_+ = G_-$ and $\phi_+ = -\phi_-$, a requirement for the ideal compensator, the quadrature term involving $\sin \omega_c t$ will be non-zero. Although the term is rejected by the demodulator it should preferably be minimised to avoid amplifier saturation effects.

To avoid the disadvantages of the above methods synchronous networks or as they are often called demodulating compensators are finding increasing use. The latter name is somewhat misleading, however, as both the input and output signals are around the carrier frequency. These networks(4,5) contain passive components together with switches operating at the carrier frequency and for further information the reader is referred to the references cited.

Example

For the network $G(s)$ shown in figure 14.3(a) determine the equivalent d.c. transfer function $G_c(s)$

$$G(s) = \frac{R}{R + 2s/(s^2C + \omega_c^2 C)} = \frac{RC(s^2 + \omega_c^2)}{RC(s^2 + \omega_c^2) + 2s},$$

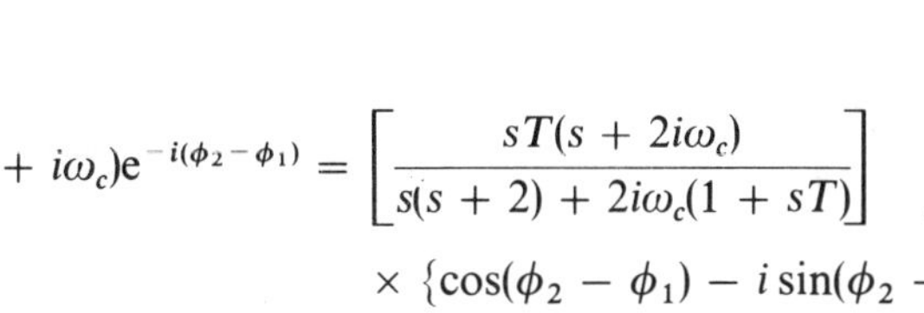

Figure 14.3 Band pass network and its low pass equivalent

and

$$G(s + i\omega_c)e^{-i(\phi_2 - \phi_1)} = \left[\frac{sT(s + 2i\omega_c)}{s(s + 2) + 2i\omega_c(1 + sT)}\right] \times \{\cos(\phi_2 - \phi_1) - i\sin(\phi_2 - \phi_1)\},$$

where

$$T = RC.$$

After further manipulation this gives

$$\begin{aligned} G_c(s) &= \mathrm{Re}_i\{G(s + i\omega_c)e^{-i(\phi_2 - \phi_1)}\} \\ &= [\{s^2(s + 2) + 4\omega_c^2(1 + sT)\}sT\cos(\phi_2 - \phi_1) + \{(s + 2) - sT(1 + sT)\} \\ &\quad \times 2s\omega_c \sin(\phi_2 - \phi_1)]\{s^2(s + 2)^2 + 4\omega_c^2(1 + sT)^2\}^{-1}. \end{aligned}$$

Making the usual assumption that the signal frequencies are much less than ω_c gives

$$G_c(s) \approx \{4\omega_c^2(1 + sT)sT\cos(\phi_2 - \phi_1) - 2s^2\omega_c T(1 + sT)\sin(\phi_2 - \phi_1)\}$$
$$\times \{4\omega_c^2(1 + sT)^2\}^{-1}$$
$$\approx \{sT/(1 + sT)\}\cos(\phi_2 - \phi_1).$$

A result which can be written down by inspection as the network is seen to be a band pass transformation of the network of figure 14.3(b).

Alternatively recognising that $G(s)$ has approximately symmetrical properties about ω_c one can use equation (14.8) to give

$$G_c(s) \approx \cos(\phi_2 - \phi_1)\left[\frac{sT(s + 2i\omega_c)}{s(s + 2) + 2i\omega_c(1 + sT)}\right]\Bigg|_{\omega_c \gg |s|}$$
$$= \frac{sT\cos(\phi_2 - \phi_1)}{(1 + sT)}.$$

14.3 The nonlinear a.c. channel

Nonlinear behaviour can occur in the a.c. channel of figure 14.1 due to nonlinearity in the modulator, demodulator or a.c. amplifier. In general the nonlinear a.c. channel will also be frequency dependent, however, when the nonlinear and frequency dependent networks can be separated, as shown in figure 14.4, it is usually possible to move the frequency dependent networks into the d.c. portion of the system where they are represented by equivalent d.c. transfer

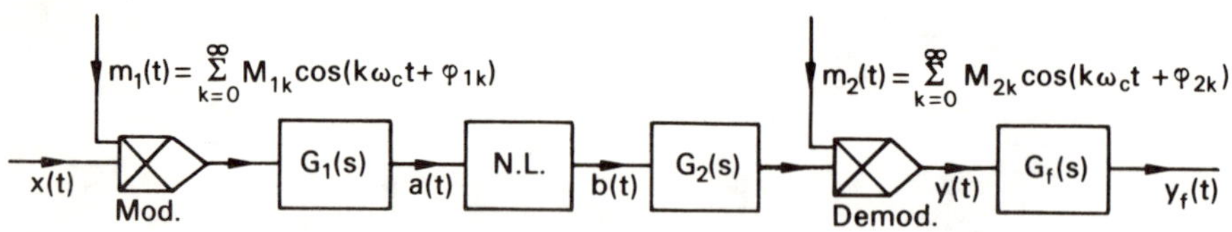

Figure 14.4 General nonlinear a.c. carrier process

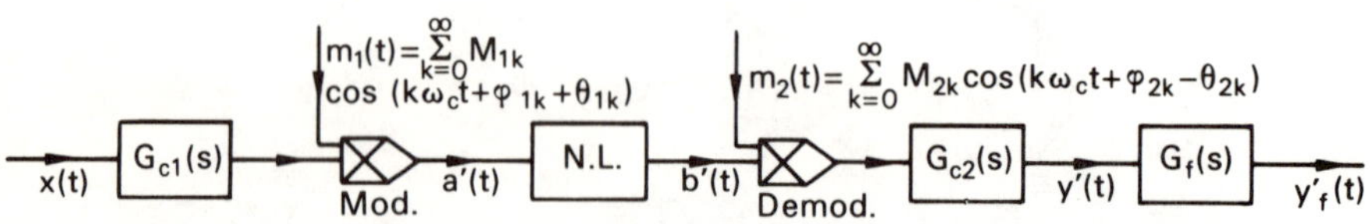

Figure 14.5 Nonlinear a.c. carrier process equivalent to that of figure 14.4

functions, as shown in figure 14.5. By obtaining expressions for the signals $a(t)$, $a'(t)$, $y(t)$ and $y'(t)$ in figures 14.4 and 14.5 when $x(t) = a\cos\omega_m t$ with $\omega_m < \omega_c/2$; setting $a(t) = a'(t)$, $y(t) = y'(t)$ it can be shown[6] that shifting of the linear elements will be valid if:

(a) Either $M_{1k} = M_{2k} = 0$ for all $k \neq 1$, that is sinusoidal modulation and demodulation must be used, or $G_1(j\omega)$ and $G_2(j\omega)$ must be essentially zero outside the range of frequencies $\omega_c/2 < \omega < 3\omega_c/2$.

(b) Over the range of frequencies $\omega_c - (\omega_m)_{\max} < \omega < \omega_c + (\omega_m)_{\max}$ $G_1(j\omega)$ and $G_2(j\omega)$ have either constant gain and linear phase characteristics or even symmetric gain and odd symmetric phase characteristics about $\omega = \omega_c$.

The above conditions governing the shifting of the linear networks out of the a.c. path are not excessively restrictive since as seen earlier condition (b) is normally satisfied by an a.c. type network. In this case the transformed system of figure 14.5 still contains the modulation–demodulation process so that the equivalent d.c. transfer functions of the networks do not contain the phases angles of the carrier signals. Thus for a network with even gain symmetry about $\omega = \omega_c$

$$G_{ci}(s) = G_i(s + i\omega_c)|_{\omega_c \gg |s|} \quad \text{for } i = 1, 2, \tag{14.10}$$

which should be compared with equation (14.8). The phase angles of the carrier signals will remain unchanged, that is $\theta_{ik} = 0$. For a network with constant gain and linear phase, the equivalent of equation (14.9) is

$$G_{ci}(s) = Ke^{sm}, \tag{14.11}$$

and the phase shift, ϕ_{ci}, of the network at the frequency ω_c will be included in the carrier, that is $\theta_{ik} = \phi_{ci}$.

We now turn our attention to the determination of the properties of the remaining nonlinear portion of the a.c. channel of figure 14.5 which is redrawn in figure 14.6. Assuming initially that the nonlinearity is single valued, it is seen

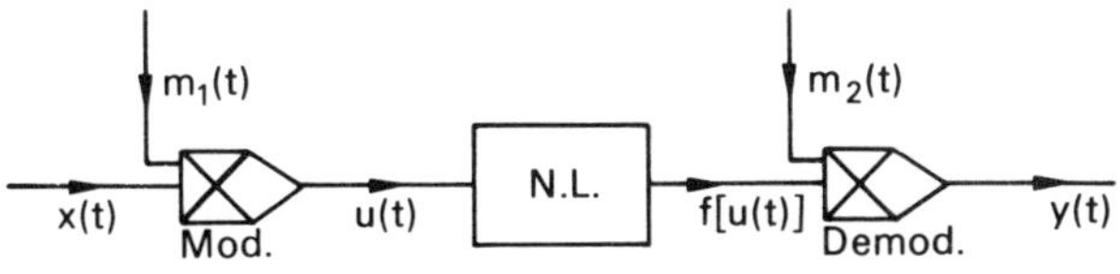

Figure 14.6 Basic nonlinear a.c. process

that the output y of the channel is a nonlinear function of the three signals x, m_1 and m_2, that is

$$y = g(x, m_1, m_2) = m_2 f(u) = m_2 f(xm_1). \tag{14.12}$$

Thus the DF $N_c(x)$ for the input signal x is given by

$$N_c(x) = E[xy]/E[x^2] = E[xm_2 f(xm_1)]/E[x^2]. \tag{14.13}$$

Writing this as an ensemble average

$$N_c(x) = \sigma_x^{-2} \int_{-\infty}^{\infty}\int_{-\infty}^{\infty}\int_{-\infty}^{\infty} xm_2 f(xm_1) p(x, m_1, m_2)\, dx\, dm_1\, dm_2, \tag{14.14}$$

and assuming x is unrelated to the carrier signals, equation (14.14) may be written

$$N_c(x) = \sigma_x^{-2} \int_{-\infty}^{\infty} x n_c(x) p(x) \, dx, \tag{14.15}$$

where

$$n_c(x) = \int_{-\infty}^{\infty} \int_{-\infty}^{\infty} m_2 f(xm_1) p(m_1, m_2) \, dm_1 \, dm_2, \tag{14.16}$$

and σ_x^2 is the mean squared value of x. $n_c(x)$ is a modified nonlinearity equivalent to the entire nonlinear process of figure 14.6 and it therefore allows the channel to be replaced in the remaining d.c. system by a single nonlinear element.

14.3.1 *The modified nonlinearity*

Before considering specific examples let us examine further the form of the modified nonlinearity. Writing

$$f(u) = f_e(u) + f_o(u),$$

where $f_e(-u) = f_e(u)$ is an even characteristic and $f_o(-u) = -f_o(u)$ is an odd characteristic, equation (14.16) becomes

$$n_c(x) = \int_{-\infty}^{\infty} \int_{0}^{\infty} m_2 [f_e(xm_1)\{p(m_1, m_2) + p(-m_1, m_2)\}$$
$$+ f_o(xm_1)\{p(m_1, m_2) - p(-m_1, m_2)\}] \, dm_1 \, dm_2.$$

When the carrier signals m_1 and m_2 are periodic signals with the same time period, the point (m_1, m_2) traces out a closed trajectory in the $m_1 - m_2$ plane. If the trajectory is symmetric through the origin, which will be the case when the Fourier series for m_1 and m_2 contain no even harmonics, then

$$\int_{-\infty}^{\infty} m_2 p(m_1, m_2) \, dm_2 = \int_{-\infty}^{\infty} -m_2 p(-m_1, m_2) \, dm_2,$$

and equation (14.16) reduces to

$$n_c(x) = \int_{-\infty}^{\infty} \int_{-\infty}^{\infty} m_2 f_o(xm_1) p(m_1, m_2) \, dm_1 \, dm_2.$$

Since the above conditions are satisfied in most practical situations, we see that $n_c(x)$ has odd symmetry independent of the form of $f(u)$. Only odd symmetrical characteristics $f(u)$, which will be the odd part of any asymmetrical characteristic in the channel, are thus considered further.

Although it is possible to formulate expressions for the joint probability density, $p(m_1, m_2)$, of the carrier signals (see section 10.4) and to use these in equation (14.16) to evaluate $n_c(x)$, it is usually simpler to use the time average equivalent of equation (14.16), namely

$$n_c(x) = \overline{m_2 f(xm_1)}, \tag{14.17}$$

where x is considered fixed during the averaging process. Obviously $n_c(x)$ will be dependent upon the carrier signals m_1 and m_2 so that the various cases must be considered separately. To avoid notation difficulties we consider positive values of x only in equation (14.17) and use the symmetry property $n_c(-x) = -n_c(x)$ to complete the nonlinear characteristic for all x.

(a) Sine-wave modulator and demodulator

Here

$$m_1(t) = M_1 \cos(\omega_c t + \phi_1) \quad \text{and} \quad m_2(t) = M_2 \cos(\omega_c t + \phi_2).$$

Expanding the signal $f(u)$ where $u = xM_1 \cos(\omega_c t + \phi_1)$ in a Fourier series gives

$$f(u) = \sum_{s=1(2)}^{\infty} a_s(U) \cos\{s(\omega_c t + \phi_1)\} \tag{14.18}$$

where $U = xM_1$. The time average of $m_2 f(u)$ is thus

$$\begin{aligned} n_c(x) &= \{M_2 a_1(xM_1) \cos(\phi_2 - \phi_1)\}/2 \\ &= K_1 a_1(xM_1) \quad \text{for } x > 0, \end{aligned} \tag{14.19}$$

where $a_1(xM_1)$ is the fundamental output from the characteristic $f(u)$ with an input sinusoid of peak magnitude xM_1 and $K_1 = (M_2/2)\cos(\phi_2 - \phi_1)$. Alternatively the result may be written in terms of the DF of $f(u)$ and is

$$n_c(x) = K_1 x M_1 N(xM_1). \tag{14.20}$$

(b) Sine wave modulator and square wave demodulator

In this case we take

$$m_1(t) = M_1 \cos(\omega_c t + \phi_1) \quad \text{and} \quad m_2(t) = M_2 \operatorname{sgn}\{\cos(\omega_c t + \phi_2)\},$$

and again equation (14.18) is valid for $f(u)$. Writing m_2 in a Fourier series, that is

$$m_2 = (4M_2/\pi) \sum_{s=1(2)}^{\infty} (1/s) \cos(s\omega_c t + \phi_2),$$

and taking the time average of $m_2 f(u)$ gives

$$n_c(x) = (2M_2/\pi) \sum_{s=1(2)}^{\infty} \{a_s(xM_1)/s\} \cos\{s(\phi_2 - \phi_1)\}, \tag{14.21}$$

where $a_s(U)$ is given by equation (3.16), that is

$$a_s(U) = 4 \int_0^U f(u) T_s(u/U) r(u)\, du. \tag{14.22}$$

(c) Square wave modulator and demodulator

For this situation

$$m_1(t) = M_1 \operatorname{sgn}\{\cos(\omega_c t + \phi_1)\} \quad \text{and} \quad m_2(t) = M_2 \operatorname{sgn}\{\cos(\omega_c t + \phi_2)\}$$

so that

$$f(u) = f(xM_1)\,\mathrm{sgn}\{\cos(\omega_c t + \phi_1)\}, \tag{14.23}$$

and the time average of $m_2 f(u)$ gives

$$n_c(x) = \{(\pi + 2\phi_1 - 2\phi_2)/\pi\}M_2 f(xM_1). \tag{14.24}$$

The modified nonlinearity, $n_c(x)$, is thus simply a scaled version of the original nonlinear characteristic, $f(u)$, in this case and its form is easily found.

(d) Square wave modulator and sine wave demodulator

In this case

$$m_1(t) = M_1\,\mathrm{sgn}\{\cos(\omega_c t + \phi_1)\} \quad \text{and} \quad m_2(t) = M_2 \cos(\omega_c t + \phi_2),$$

so that $f(u)$ is again as in equation (14.22) and the time average of $m_2 f(u)$ is

$$n_c(x) = (2/\pi)M_2 f(xM_1)\cos(\phi_2 - \phi_1). \tag{14.25}$$

Again, therefore, the modified nonlinearity is simply a scaled version of the original characteristic.

Thus once the modified nonlinearity has been evaluated, describing functions for any forms of the input x can be found either analytically or by making use of some of the approximate methods discussed in chapter 3. For analytically defined characteristics the modified nonlinearity for sine wave modulation can be found from DFs and harmonic coefficients listed in appendix A1.

14.3.2 *A particular example*

To illustrate the derivation of the various describing functions for the nonlinear a.c. channel the case of a cubic nonlinearity, that is $f(u) = u^3$, in a channel having sine wave modulation and square wave demodulation with $\phi_1 = \phi_2$ is considered. For this case, with $u = U\cos\omega_c t$

$$f\{u(t)\} = (U^3/4)\{3\cos\omega_c t + \cos 3\omega_c t\},$$

and equation (14.21) becomes

$$n_c(x) = (2M_2/\pi)(3/4 + 1/12)x^3 M_1^3 = 5M_2 M_1^3 x^3/3\pi.$$

Thus the modified nonlinearity is also a cubic and describing functions for different forms of input x can be found in the usual manner. For x a sinusoid

$$N_c(a) = 5M_2 M_1^3 a^2/4\pi,$$

and the IDF, $N_{cib}(a)$ is given by

$$N_{cib}(a) = (2 + e^{-j2\phi})5M_2 M_1^3 a^2/4\pi.$$

For x a Gaussian signal

$$N_c(\sigma) = 5M_2 M_1^3 \sigma^2/\pi,$$

and so on for x any combination of input signals.

14.3.3 *Double valued nonlinearities*

To this point we have assumed the nonlinearity in the a.c. channel to be single valued, however, the same basic approach can be used for a double valued nonlinearity. The procedure is illustrated by considering a particular case, that of an on-off characteristic with hysteresis in an a.c. sine modulated sine demodulated channel. If the signal $x(t)$ is slowly varying with respect to the carrier signal $m_1(t)$, it may be assumed that $m_1(t)$ does all the switching and $f(u)$ may be expanded as the Fourier series

$$f(u) = \sum_{s=1(2)}^{\infty} a_s(U)\cos\{s(\omega_c t + \phi_1)\} + b_s(U)\sin\{s(\omega_c t + \phi_1)\},$$

where $U = xM_1$. Taking the time average of $m_2 f(u)$ then gives for the solution to equation (14.17)

$$n_c(x) = (M_1M_2/2)\{xN_p(xM_1)\cos\phi - xN_q(xM_1)\sin\phi\}, \qquad (14.26)$$

where $\phi = \phi_2 - \phi_1$ and $N_p(U)$ and $N_q(U)$ are respectively the real and imaginary parts of the DF, $N(U)$, of $f(u)$ for an input $u = U\cos\omega t$.

Now for the specific characteristic

$$N_p(xM_1) = (4h/xM_1\pi)\{1 - (\Delta^2/x^2M_1^2)\}^{1/2},$$

and

$$N_q(xM_1) = 4h\Delta/xM_1\pi.$$

Substituting these values in equation (14.26) and the resulting expression for $n_c(x)$ in the equation

$$N_c(a) = (4/a^2)\int_0^a xn_c(x)r(x)\,\mathrm{d}x$$

for the DF, gives

$$N_c(a) = \frac{8hM_2}{a^2\pi^2}\left\{\int_b^a \frac{(x^2 - b^2)^{1/2}}{(a^2 - x^2)^{1/2}}\cos\phi\,\mathrm{d}x - \int_b^a \frac{b\sin\phi}{(a^2 - x^2)^{1/2}}\,\mathrm{d}x\right\},$$

where $b = \Delta/M_1$. Evaluation of the integrals yields the solution

$$N_c(a) = (8hM_2/a\pi^2)[\{E'(b/a) - (b/a)^2K'(b/a)\}\cos\phi - \{(\pi/2) - \sin^{-1}(b/a)\}(b/a)\sin\phi], \qquad (14.27)$$

where $E'(m)$ and $K'(m)$ are the standard elliptic integrals defined by

$$E'(m) = E(m_1) = E(1 - m) = \int_0^{\pi/2} (1 - m_1\sin^2\theta)^{1/2}\,\mathrm{d}\theta, \qquad (14.28)$$

and

$$K'(m) = K(m_1) = K(1 - m) = \int_0^{\pi/2} (1 - m_1\sin^2\theta)^{-1/2}\,\mathrm{d}\theta. \qquad (14.29)$$

Figure 14.7 shows experimental measurements for the DF, $N_c(a)$, with $x = a\cos\omega_m t$, $h = \Delta = M_1 = M_2 = 1$ and $\phi = 0$ for various frequency ratios ω_c/ω_m. It is seen that the magnitude of $N_c(a)$ for all the frequency ratios agrees closely with the theoretical curve of equation (14.27) although until ω_c/ω_m exceeds ten the DF possesses significant phase shift. Presumably for lower frequency ratios the assumption of $x(t)$ varying slowly with respect to $m_1(t)$ is poor.

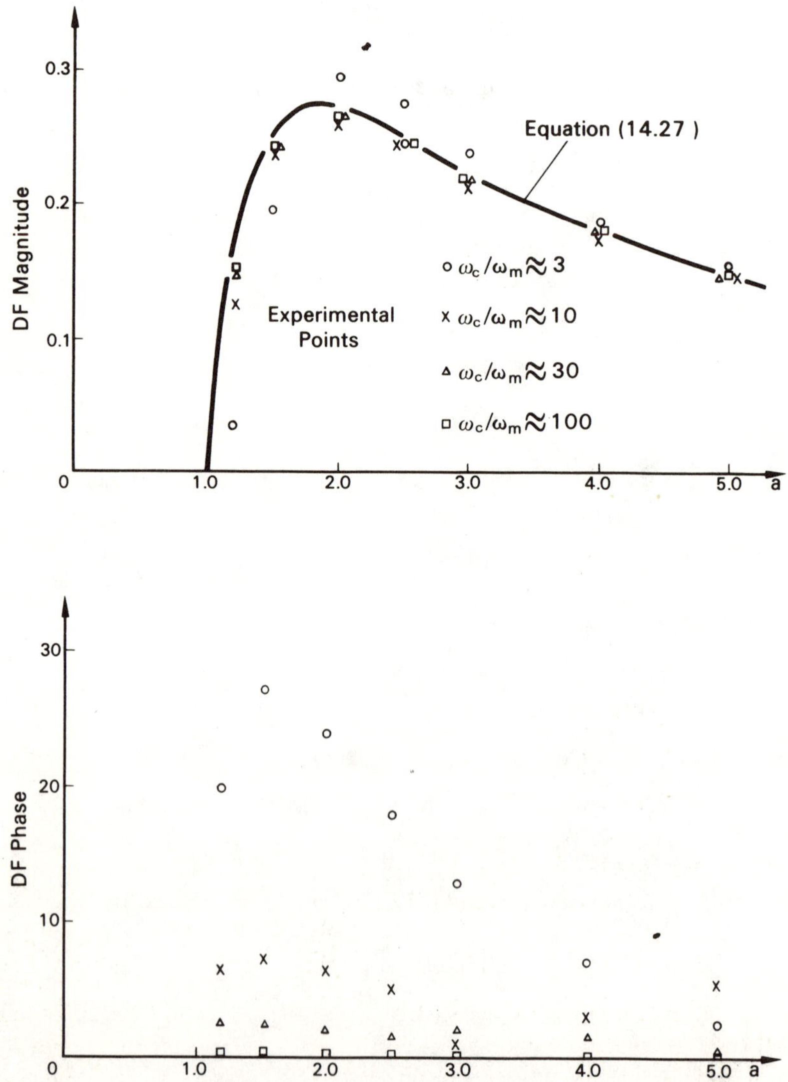

Figure 14.7 DF for a.c. channel containing on-off with hysteresis nonlinearity

14.4 Behaviour of some closed loop systems

In deriving the modified nonlinearity, $n_c(x)$, equivalent to the nonlinear a.c. channel it was assumed that the modulating input $x(t)$ was independent of and varied slowly relative to the carrier signals. When these assumptions are valid the behaviour of an a.c. system can be confidently predicted using normal describing function methods for the equivalent d.c. system. Some specific examples are discussed in the next section illustrating this fact. Two specific situations of interest where the above assumptions are not true are discussed in later sections. At the same time it is shown how the theory can be modified to predict the unique behaviour found in these two cases.

14.4.1 *Applications of the modified nonlinearity*

Using the modified nonlinearity for the a.c. channel the properties of some a.c. systems are compared with those of similar conventional nonlinear systems. The similar systems are obtained by replacing the modulator and demodulator by the appropriate gain factors which would apply for a linear a.c. channel and keeping the nonlinearity in the loop. This procedure is illustrated in figure 14.8.

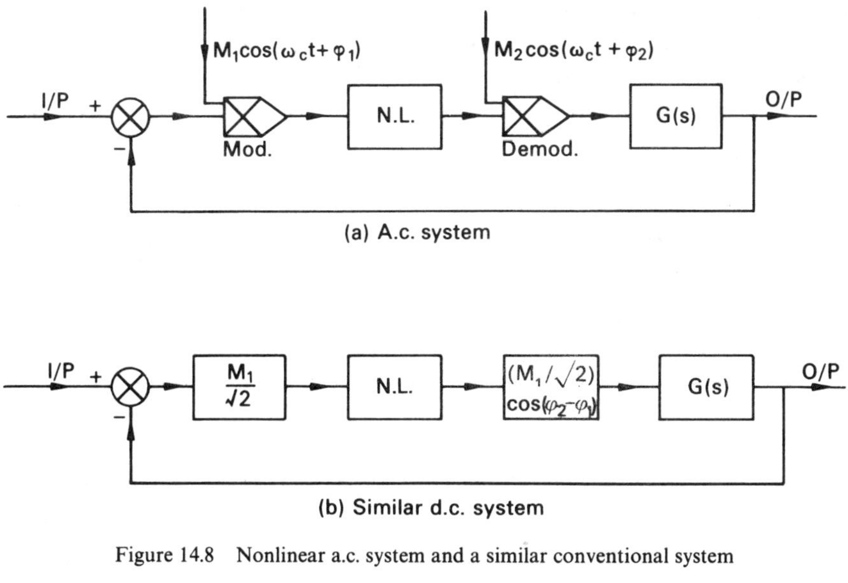

Figure 14.8 Nonlinear a.c. system and a similar conventional system

Consider the system of figure 14.8(a) with the nonlinearity an on-off relay with hysteresis and the linear transfer function $G(s) = K/s(s + a)$. In agreement with the DF theory the simulated system, unlike the similar d.c. system of figure 14.8(b), was found to be stable with no limit cycle for all values of K.

Figure 14.9 shows the block diagram of a nonlinear a.c. position control system.[6] The nonlinearity included in the system was the asymmetrical

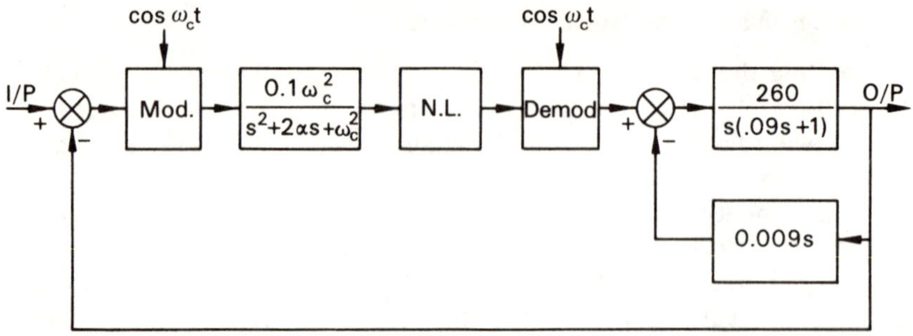

Figure 14.9 Block diagram of nonlinear a.c. position control system

characteristic, $n(x)$, where

$$n(x) = \begin{cases} 7{\cdot}5 & x > 1{\cdot}5, \\ 5x & -0{\cdot}5 < x < 1{\cdot}5, \\ -2{\cdot}5 & x < -0{\cdot}5, \end{cases}$$

and the parameter α was 50. The theoretical results obtained previously show that the modified nonlinearity for this a.c. channel is symmetrical, thus, provided the modulating signal changes slowly with respect to the carrier frequency the system input–output behaviour should have odd symmetry. Even under transient conditions this is normally the case as seen from figure 14.10 which shows the

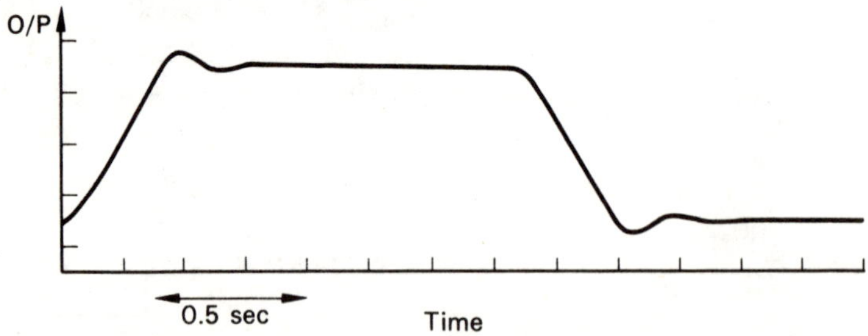

Figure 14.10 Step response of system of figure 14.9

response of the system to a low frequency square wave input. For comparison purposes the same response is shown for a similar conventional system in figure 14.11, where the asymmetry is clearly evident.

14.4.2 *Carrier locked self oscillations*

Experimental measurements[6,7] on a.c. systems show that when limit cycles exist they are often locked, at a relatively high sub-multiple frequency, to the carrier frequency. To explain this phenomenon we again consider the case of

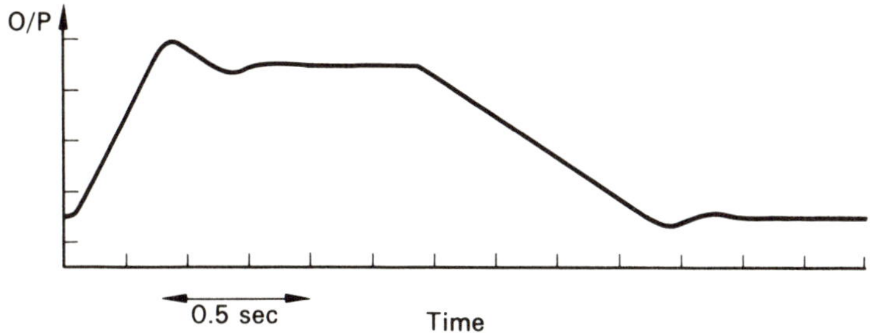

Figure 14.11 Step response of similar d.c. system to figure 14.9

sinusoidal modulation with $m_1(t) = M_1 \cos \omega_c t$ and a sinusoidal input $x(t) = a\cos(\omega_m t + \theta)$. The signal at the input to the nonlinearity in figure 14.6 then contains two sinusoids of different phase and with frequencies $\omega_c \pm \omega_m$. If $\omega_m = \omega_c/n$ the ratio of these frequencies is $(n + 1)/(n - 1)$ and as seen in section 10.4 coefficients for the nonlinearity other than α_{10} and α_{01} will produce componentes at the nonlinearity output at these two frequencies. The coefficients α_{sk}, however, which produce these output frequencies will have values of s and k greater than n and their contribution to the output will be small. Hence the phase change for the two frequencies in passing through a single valued nonlinearity will normally not exceed, say three degrees,[6] so that analysis using the DF method presented, which ignores the locking, is quite satisfactory for engineering purposes.

14.4.3 *An input with high frequency components*

Although the assumption that the frequency content of $x(t)$ will be much less than ω_c is valid for input signal components which the system is required to follow, noise disturbance signals may produce higher frequency components in $x(t)$. It would therefore seem appropriate to see if such noise will be suitably filtered by the system. We have seen in section 14.2 for the linear channel with sinusoidal modulation and demodulation that if $x(t)$ contains a signal with a bandwidth from 0 to ω_m the output of the a.c. channel contains not only components in the desired frequency band from 0 to ω_m but also components in the band from $2\omega_c - \omega_m$ to $2\omega_c + \omega_m$. Hence the need for the filter $G_f(s)$. Clearly therefore if $x(t)$ contains noise near to the frequency $2\omega_c$ it will be transmitted in the a.c. channel around the frequency ω_c and demodulated to give two channel output bands, one near $2\omega_c$ and one near zero. This latter low frequency component will not be filtered by $G_f(s)$ and will produce undesirable low frequency output variations.

In nonlinear a.c. systems the above effect extends to lower input frequencies due to the distortion terms generated by the nonlinearity. Assuming $x(t) = a\cos(\omega_m t + \theta)$ and $m_1(t) = M_1 \cos \omega_c t$ the nonlinearity input $u(t)$ in figure 14.6 is given by

$$u(t) = aM_1[\cos\{(\omega_c + \omega_m)t + \theta\} + \cos\{(\omega_c - \omega_m)t - \theta\}],$$

and the nonlinearity output by

$$f(u) = \sum_{s=0}^{\infty} \sum_{k=0}^{\infty} \varepsilon_s \varepsilon_k \alpha_{sk} \cos\{s(\omega_c + \omega_m)t + s\theta\} \cos\{k(\omega_c - \omega_m)t - k\theta\}. \quad (14.30)$$

The nonlinearity output thus has frequency components near ω_c if

$$s(\omega_c + \omega_m) \pm k(\omega_c - \omega_m) \approx \pm\omega_c,$$

that is

$$\omega_m = \left| \frac{s \pm k - 1}{-s \pm k} \right| \omega_c + \varepsilon, \quad (14.31)$$

where ε is the deviation from that input frequency which gives a nonlinearity output of exactly ω_c. This equation thus specifies which input frequencies ω_m, lead to demodulator output components near zero, specifically at the frequency $(s \mp k)\varepsilon$. Thus, for example, if the coefficient α_{11} for the nonlinearity is finite, substitution of $s = k = 1$ in equation (14.31) shows that input frequencies near to $\omega_c/2$ produce a near zero frequency output from the nonlinear channel.

Figure 14.12 shows experimental results for a high frequency noise input to the position control system of figure 14.9. The input noise used had a narrow band spectrum near to $(2/3)\omega_c$, the system parameter α was 100 and the nonlinearity was a symmetrical saturation characteristic with input limit levels of ± 2 and a linear gain of 4. From equation (14.31), $\omega_m = (2/3)\omega_c$ when $s = 2$ and $k = 1$ so that the low frequency noise output is due to the coefficient α_{21} of the saturation nonlinearity. When the noise input is sufficient to cause strong saturation the results show that the low frequency output variations become quite large.

REFERENCES

1. Ivey, K. A.: *A-c carrier control systems* (Wiley, New York, 1964).
2. Wilson D. R.: 'Harmonic stabilization of a.c. servos', *Electronics Letters*, 1967, 3, pp. 256–257.
3. Cassell, W. L.: *Linear electric circuits* (Wiley, New York, 1964, pp. 520–522).
4. Wilson, D. R.: 'An exact modulation model for a demodulating compensating network', *Paper 47C, Proc. 3rd I.F.A.C. Conference*, London, June, 1966.
5. Knox, A. and Murphy, G. J.: 'A method for the design of phase lead demodulating compensators for use in carrier control systems', *Trans. I.E.E.E.*, 1964, AC-8, pp. 333–338.
6. Fieguth, W.: 'Multi-input quasilinearization', *Ph.D. Thesis, University of New Brunswick*, 1968.
7. Fieguth, W. and Atherton, D. P.: 'Analysis of nonlinear a.c. control systems', *Paper 34.5, Proc. 4th I.F.A.C. Conference*, Warsaw, June, 1969.

BIBLIOGRAPHY

Barskii, A. G.: 'Periodic modes in two-channel automatic control systems containing a nonlinear element in the single-channel part of the circuit', *Automation and Remote Control*, Vol. 27, No. 9, pp. 1548–1555, September, 1966.

Krasovskii, A. A.: 'On the theory of two-channel follow up systems with a relay in the A-C circuit', *Automation and Remote Control*, Vol. 21, No. 9, pp. 911–919, September, 1960.

Weiss, G.: 'A network theory for carrier suppressed modulated systems', *I.R.E. Trans. of Automatic Control*, Vol. AC-6, pp. 54–66.

Weiss, G.: 'Synchronous networks', *I.E.E.E. Trans.*, 1962, AC-7, pp. 45–54.

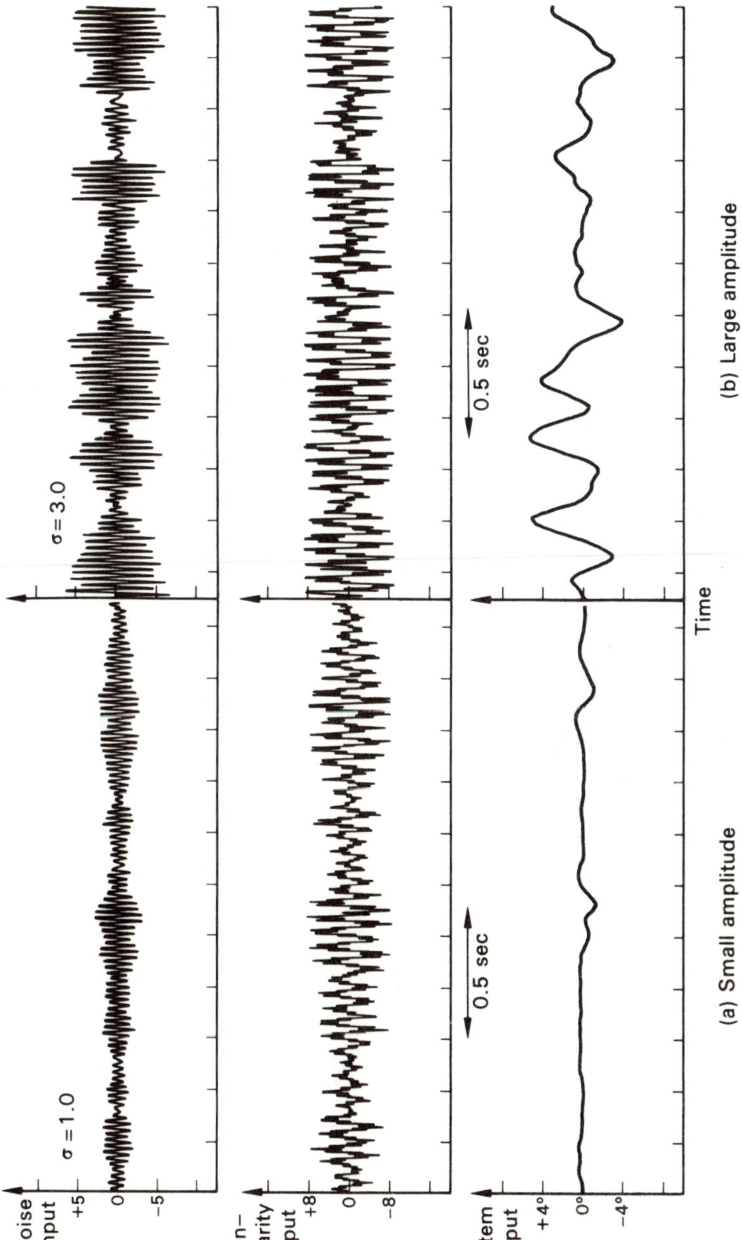

Figure 14.12 Response of system of figure 14.9 to a noise input

APPENDIX A1

TABLE OF DFs AND HARMONIC COEFFICIENTS FOR SYMMETRICAL ODD NONLINEARITIES

For an input $x = a\cos\theta$

$$a_s = (2/\pi)\int_0^{\pi} y(\theta)\cos s\theta\, d\theta = 4\int_0^{a} n_p(x)T_s(x/a)r(x)\, dx$$

$$b_s = (2/\pi)\int_0^{\pi} y(\theta)\sin s\theta\, d\theta = 4\int_0^{a} n_q(x)D_s(x/a)r(x)\, dx.$$

The DF is given by

$$N(a) = N_p(a) + jN_q(a) = (a_1 - jb_1)/a = (4/a\pi)\int_0^{\pi/2} y(\theta)e^{-j\theta}\, d\theta$$

or alternatively

$$N_p(a) = (4/a\pi)\int_0^{\pi/2} n_p(a\cos\theta)\cos\theta\, d\theta = (4/a^2)\int_0^{a} xn_p(x)r(x)\, dx$$

$$N_q(a) = -(4/a\pi)\int_0^{\pi/2} n_q(a\cos\theta)\sin\theta\, d\theta = -(4/a^2\pi)\int_0^{a} n_q(x)\, dx.$$

In this table use is made of the saturation functions

$$f_1(\rho) = \begin{cases} 1 & \rho > 1 \\ (1/\pi)(2\alpha + \sin 2\alpha) & |\rho| < 1 \\ -1 & \rho < -1 \end{cases}$$

$$f_s(\rho) = \begin{cases} 0 & |\rho| > 1 \\ (-1)^{(s-1)/2}\dfrac{2}{s\pi}\left[\dfrac{\sin(s+1)\alpha}{s+1} + \dfrac{\sin(s-1)\alpha}{s-1}\right] & |\rho| < 1 \end{cases}$$

where $\alpha = \sin^{-1}\rho$, $\rho = \delta/a$.

Since all the nonlinearities are assumed to have odd symmetry that is $n(-x) = -n(x)$, the expressions used in the following table define the characteristics for $x > 0$ only.

TABLE OF DFs AND HARMONIC COEFFICIENTS FOR SYMMETRICAL ODD NONLINEARITIES

(a) *Single valued nonlinearities*

Nonlinearity	Comments	$N_p(a) = a_1/a$	a_s for s odd
1. General quantiser 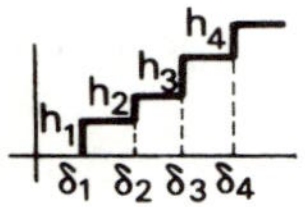	$a < \delta_1$	$N_p = 0$	$a_s = 0$
	$\delta_{M+1} > a > \delta_M$ See section 3.5.3	$N_p = (4/a^2\pi) \sum_{m=1}^{M} h_m(a^2 - \delta_m^2)^{1/2}$	$a_s = (4/s\pi) \sum_{m=1}^{M} h_m D_s(\delta_m/a)$
2. Uniform quantiser $h_1 = h_2 = \cdots h$ $\delta_m = (2m - 1)\delta/2$	$a < \delta$	$N_p = 0$	$a_s = 0$
	$(2M + 1)\delta > a > (2M - 1)\delta$ $n = (2m - 1)/2$	$N_p = (4h/a^2\pi) \sum_{m=1}^{M} (a^2 - n^2\delta^2)^{1/2}$	$a_s = (4h/s\pi) \sum_{m=1}^{M} D_s(n\delta/a)$
3. Relay with dead zone 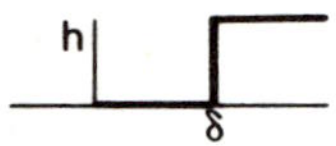	$a < \delta$	$N_p = 0$	$a_s = 0$
	$a > \delta$	$N_p = 4h(a^2 - \delta^2)^{1/2}/a^2\pi$	$a_s = 4hD_s(\delta/a)/s\pi$
4. Ideal relay 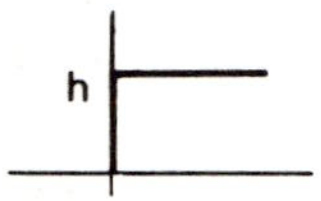		$N_p = 4h/a\pi$	$a_s = (-1)^{(s-1)/2}4h/s\pi$
5. Preload 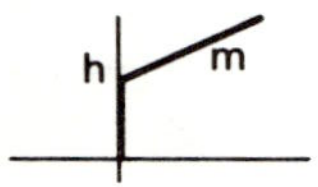		$N_p = (4h/a\pi) + m$	$a_s = (-1)^{(s-1)/2}4h/s\pi$

(a) *Single valued nonlinearities*—continued

Nonlinearity	Comments	$N_p(a) = a_1/a$	a_s for s odd
6. General piecewise linear 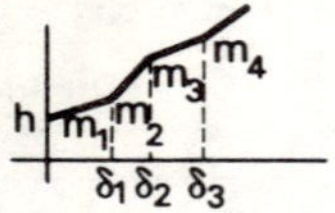	$a < \delta_1$ $\delta_{M+1} > a > \delta_M$	$N_p = (4h/a\pi) + m_1$ $N_p = (4h/a\pi) + m_{M+1}$ $+ \sum_{j=1}^{M} (m_j - m_{j+1}) f_1(\delta_j/a)$	$a_s = [(-1)^{(s-1)/2} 4h/s\pi]$ $a_s = [(-1)^{(s-1)/2} 4h/s\pi]$ $+ a \sum_{j=1}^{M} (m_j - m_{j+1}) f_s(\delta_j/a)$
7. Ideal saturation 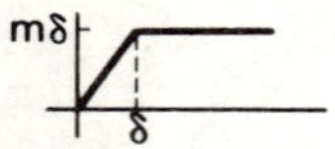	See section 3.4.1	$N_p = m f_1(\delta/a)$	$a_s = maf_s(\delta/a)$
8. Dead zone 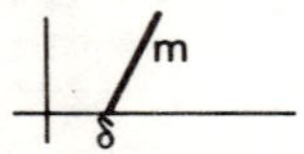	See section 3.6.2	$N_p = m\{1 - f_1(\delta/a)\}$	$a_s = -maf_s(\delta/a)$
9. Gain changing nonlinearity 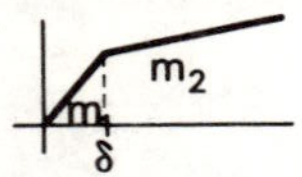		$N_p = (m_1 - m_2) f_1(\delta/a) + m_2$	$a_s = (m_1 - m_2) af_s(\delta/a)$
10. Saturation with dead zone 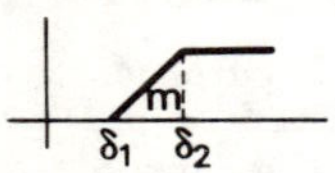		$N_p = m\{f_1(\delta_2/a) - f_1(\delta_1/a)\}$	$a_s = ma\{f_s(\delta_2/a) - f_s(\delta_1/a)\}$
11. 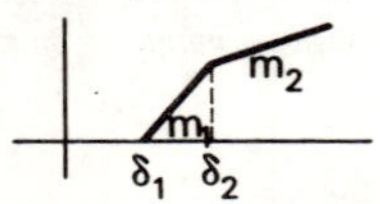		$N_p = -m_1 f_1(\delta_1/a) + (m_1 - m_2) f_1(\delta_2/a)$ $+ m_2$	$a_s = -m_1 af_s(\delta_1/a) + (m_1 - m_2) af_s(\delta_2/a)$

12. 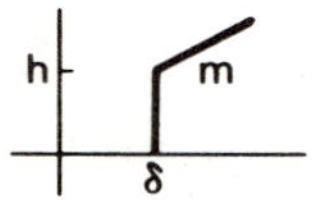	$a < \delta$	$N_p = 0$	$a_s = 0$
	$a > \delta$	$N_p = 4h(a^2 - \delta^2)^{1/2}/a^2\pi + m - mf_1(\delta/a)$	$a_s = -maf_s(\delta/a) + 4hD_s(\delta/a)/s\pi$
13. 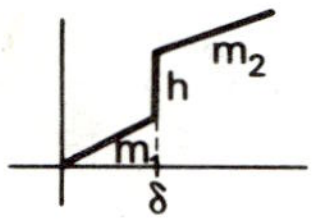	$a < \delta$	$N_p = m_1$	$a_s = 0$
	$a > \delta$	$N_p = (m_1 - m_2)f_1(\delta/a) + m_2 + 4h(a^2 - \delta^2)^{1/2}/a^2\pi$	$a_s = (m_1 - m_2)af_s(\delta/a) + 4hD_s(\delta/a)/s\pi$
14. 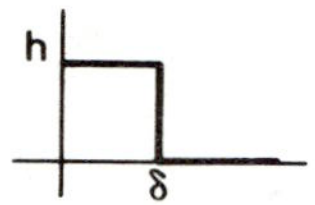	$a < \delta$	$N_p = 4h/a\pi$	$a_s = (-1)^{(s-1)/2}4h/s\pi$
	$a > \delta$	$N_p = 4h\{a - (a^2 - \delta^2)^{1/2}\}/a^2\pi$	$a_s = \{(-1)^{(s-1)/2} - D_s(\delta/a)\}4h/s\pi$
15. Limited field of view 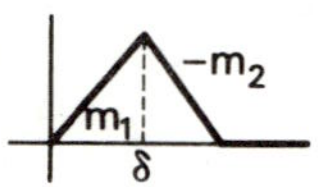	See section 3.6.2	$N_p = (m_1 + m_2)f_1(\delta/a) - m_2 f_1\{(m_1 + m_2)\delta/m_2 a\}$	$a_s = (m_1 + m_2)af_s(\delta/a) - m_2 af_s\{(m_1 + m_2)\delta/m_2 a\}$
16. 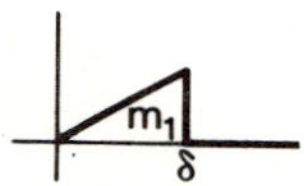	$a < \delta$	$N_p = m_1$	$a_s = 0$
	$a > \delta$	$N_p = m_1 f_1(\delta/a) - 4m_1\delta(a^2 - \delta^2)^{1/2}/a^2\pi$	$a_s = m_1 af_s(\delta/a) - 4m_1\delta D_s(\delta/a)/s\pi$

(a) *Single valued nonlinearities*—continued

Nonlinearity	Comments	$N_p(a) = a_1/a$	a_s for s odd
17. $y = x^m$	$m > -2$ Γ is the gamma function See section 3.5.2	$N_p = \dfrac{\Gamma(m+1)a^{m-1}}{2^{m-1}\Gamma[(3+m)/2]\Gamma[(1+m)/2]} = \dfrac{2}{\sqrt{\pi}} \dfrac{\Gamma[(m+2)/2]a^{m-1}}{\Gamma[(m+3)/2]}$	$a_s = \dfrac{a^m\Gamma(m+1)}{2^{m-1}\Gamma[(2+m+s)/2]\Gamma[(2+m-s)/2]}$
18. $y = x^n$	n odd integer, μ_n is the nth moment of the amplitude distribution function of a sinusoid See section 3.5.1	$N_p = \dfrac{n(n-2)(n-4)\dots 3}{(n+1)(n-1)\dots 4} a^{n-1} = \dfrac{2}{a^2}\mu_{n+1}$	Use 17
19. $y = x^n$	n even integer	$N_p = \dfrac{4}{\pi} \dfrac{n(n-2)\dots 2}{(n+1)(n-1)\dots 3} a^{n-1}$	Use 17
20. Harmonic nonlinearity $y = A_m \sin mx$	$J_s(ma)$ is the Bessel function of order s See section 3.5.2	$N_p = 2A_mJ_1(ma)/a$	$a_s = (-1)^{(s-1)/2}2A_mJ_s(ma)$
21. $y = A_m \sinh mx$	$I_s(ma)$ is the modified Bessel function of order s	$N_p = 2A_mI_1(ma)/a$	$a_s = 2A_mI_s(ma)$
22. Exponential saturation $y = 1 - e^{-cx}$	$I_1(ca)$ is the modified Bessel function of order 1. $S_1(ca)$ is the modified Struve function of order 1	$N_p = (2/a)\{I_1(ca) - S_1(ca)\}$	
23. $y = \dfrac{cx}{(1 + c^2x^2)}$	$K(k)$ and $E(k)$ are complete elliptic integrals $u = ca/(1 + c^2a^2)^{1/2}$	$N_p = (4/\pi)\{(-u/c^2a^3)K(u) + (1/au)E(u)\}$	

(b) *Double valued nonlinearities—DVNLs*

Nonlinearity	Comments	$N_p(a) = a_1/a$	$N_q(a) = -b_1/a$	a_s for s odd	b_s for s odd
24. On-off relay with hysteresis 	$a < \Delta$ $a > \Delta$ See section 3.4.3	$N_p = 0$ $N_p = 4h(a^2 - \Delta^2)^{1/2}/a^2\pi$	$N_q = 0$ $N_q = -4h\Delta/a^2\pi$	$a_a = 0$ $a_s = \{4hD_s(\Delta/a)\}/s\pi$	$b_s = 0$ $b_s = \{4hT_s(\Delta/a)\}/s\pi$
25. Friction-controlled backlash 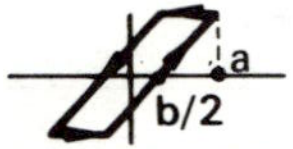	$a > b/2$ See section 3.6.2	$N_p = \{1 + f_1[(a - b)/a]\}/2$	$N_q = b(b - 2a)/a^2\pi$	$a_s = af_s\{(a - b)/a\}/2$	$b_s = (2/s\pi)\left\{a - (a - b)T_s\left(\frac{a - b}{a}\right)\right\}$ $-(2/a\pi)\int_{a-b}^{a} xU_{s-1}(x/a)\,dx$
26. 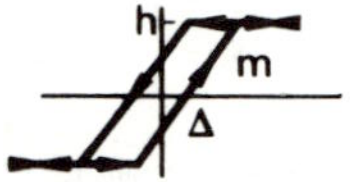	$a > \Delta + (h/m)$	$N_p = \frac{m}{2}\left\{f_1\left(\frac{h - m\Delta}{ma}\right) + f_1\left(\frac{h + m\Delta}{ma}\right)\right\}$	$N_q = -4h\Delta/a^2\pi$	$a_s = \frac{ma}{2}\left\{f_s\left(\frac{h - m\Delta}{ma}\right) + f_s\left(\frac{h + m\Delta}{ma}\right)\right\}$	$b_s = (2m/s\pi)[2\Delta T_s(c_1/a)$ $+ c_2\{T_s(c_2/a) - T_s(c_1/a)\}]$ $-(2m/a\pi)\int_{c_1}^{c_2} xU_{s-1}(x/a)\,dx$; $c_1 = (h - m\Delta)/m$, $c_2 = (h + m\Delta)/m$
27. 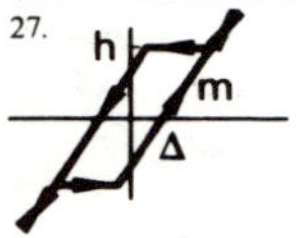	$a > \Delta + (h/m)$	$N_p = \frac{m}{2}\left\{2 - f_1\left(\frac{h + m\Delta}{ma}\right) + f_1\left(\frac{h - m\Delta}{ma}\right)\right\}$	$N_q = -4h\Delta/a^2\pi$	$a_s = (ma/2)\left\{f_s\left(\frac{h - m\Delta}{ma}\right) - f_s\left(\frac{h + m\Delta}{ma}\right)\right\}$	b_s – as for 26
28. 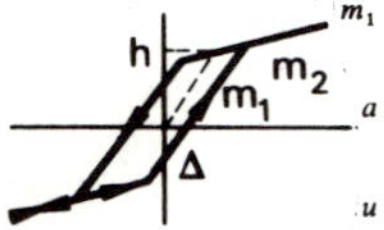	$m_1 > m_2$ $a > \frac{h}{m_1} + \frac{\Delta m_1}{(m_1 - m_2)}$ $u = \frac{\Delta m_1}{(m_1 - m_2)}$	$N_p = \left(\frac{m_1 - m_2}{2}\right)\left\{f_1\left(\frac{h + m_1 u}{m_1 a}\right)\right.$ $\left. + f_1\left(\frac{h - m_1 u}{m_1 a}\right)\right\} + m_2$	$N_q = -4h\Delta/a^2\pi$	$a_s = \left(\frac{m_1 - m_2}{2}\right)\left\{f_s\left(\frac{h + m_1 u}{m_1 a}\right)\right.$ $\left. + f_s\left(\frac{h - m_1 u}{m_1 a}\right)\right\}$	$b_s = (4/s\pi)[m_1\Delta T_s(c_1/a) + c_2\left(\frac{m_1 - m_2}{2}\right)$ $\times\left\{T_s(c_2/a) - T_s(c_1/a)\right\}$ $-\frac{2(m_1 - m_2)}{a\pi}\int_{c_1}^{c_2} xU_{s-1}(x/a)\,dx$ $c_2 = (h/m_1) + u$, $c_1 = (h/m_1) - u$.
29. 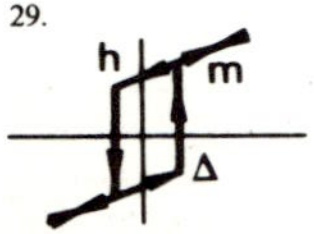	$a > \Delta$	$N_p = m + 4h(a^2 - \Delta^2)^{1/2}/a^2\pi$	$N_q = -4h\Delta/a^2\pi$	$a_s = \{4hD_s(\Delta/a)\}/s\pi$	$b_s = \{4hT_s(\Delta/a)\}/s\pi$

(b) *Double valued nonlinearities—DVNLs*—continued

Nonlinearity	Comments	$N_p(a) = a_1/a$	$N_q(a) = -b_1/a$	a_s for s odd	b_s for s odd
30. On-off relay with dead zone and hysteresis	$a < \delta + \Delta$	$N_p = 0$	$N_q = 0$		
	$a > \delta + \Delta$	$N_p = \frac{2h}{a^2\pi}[\{a^2 - (\delta + \Delta)^2\}^{1/2} + \{a^2 - (\delta - \Delta)^2\}^{1/2}]$	$N_q = -4h\Delta/a^2\pi$	$a_s = (2h/s\pi)\left\{D_s\left(\frac{\delta - \Delta}{a}\right) + D_s\left(\frac{\delta + \Delta}{a}\right)\right\}$	$b_s = (2h/s\pi)\left\{T_s\left(\frac{\delta + \Delta}{a}\right) - T_s\left(\frac{\delta - \Delta}{a}\right)\right\}$
31. Negative hysteresis	$a < \delta - \Delta$	$N_p = 0$	$N_q = 0$		
	$\delta - \Delta < a < \delta + \Delta$	$N_p = \frac{4h}{a^2\pi}\{a^2 - (\delta - \Delta)^2\}^{1/2}$	$N_q = 0$		
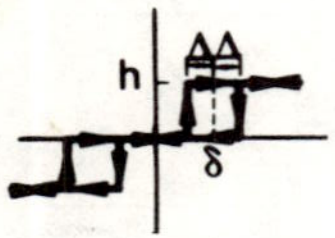	$a > \delta + \Delta$	$N_p = \frac{2h}{a^2\pi}[\{a^2 - (\delta - \Delta)^2\}^{1/2} + \{a^2 - (\delta + \Delta)^2\}^{1/2}]$	$N_q = 4h\Delta/a^2\pi$	$a_s = (2h/s\pi)\left\{D_s\left(\frac{\delta - \Delta}{a}\right) + D_s\left(\frac{\delta + \Delta}{a}\right)\right\}$	$b_s = (2h/s\pi)\left\{T_s\left(\frac{\delta - \Delta}{a}\right) - T_s\left(\frac{\delta + \Delta}{a}\right)\right\}$

Nonlinearity	Comments	$N_p(a) = a_1/a$	$N_q(a) = -b_1/a$
32. 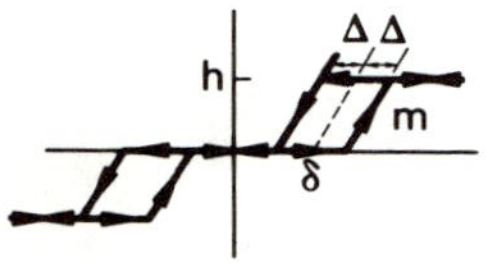	$a > \delta + \Delta + h/m$	$N_p = \frac{m}{2}\left[-f_1\left(\frac{\delta+\Delta}{a}\right) + f_1\left(\frac{h + m\delta + m\Delta}{ma}\right) + f_1\left(\frac{h + m\delta - m\Delta}{ma}\right) - f_1\left(\frac{\delta - \Delta}{a}\right)\right]$	$N_q = -4h\Delta/a^2\pi$
33. 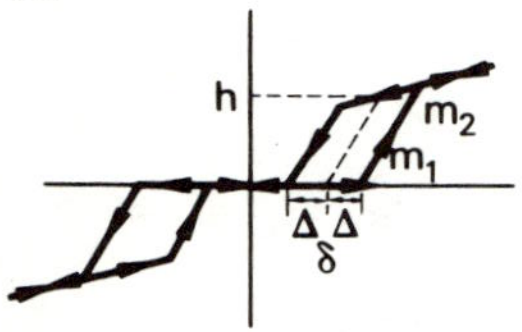	$m_1 > m_2$ $a > \delta + \frac{h}{m_1} + \frac{\Delta m_1}{m_1 - m_2}$ $u = \frac{\Delta m_1}{(m_1 - m_2)}$	$N_p = m_2 - \frac{m_1}{2}\left\{f_1\left(\frac{\delta+\Delta}{a}\right) + f_1\left(\frac{\delta-\Delta}{a}\right)\right\} + \left(\frac{m_1 - m_2}{2}\right)\left\{f_1\left(\frac{h + m_1\delta + m_1 u}{m_1 a}\right) + f_1\left(\frac{h + m_1\delta - m_1 u}{m_1 a}\right)\right\}$	$N_q = -4h\Delta/a^2\pi$

APPENDIX A2

TABLE OF SBDFs

For an input $z = \gamma + x$, where $x = a\cos\theta$

$$N_\gamma(a,\gamma) = (1/\gamma)\int_{-a}^{a} n(x+\gamma)r(x)\,dx = (1/\pi)\int_0^\pi y(\theta)\,d\theta$$

$$N_a(a,\gamma) = N_p(a,\gamma) + jN_q(a,\gamma) = (a_1 - jb_1)/a$$

where

$$a_1 = (2/a)\int_{-a}^{a} xn_p(x+\gamma)r(x)\,dx = (2/\pi)\int_0^\pi y(\theta)\cos\theta\,d\theta$$

$$b_1 = (2/a\pi)\int_{-a}^{a} n_q(x+\gamma)\,dx = (2/\pi)\int_0^\pi y(\theta)\sin\theta\,d\theta.$$

The functions

$$g(\rho) = \begin{cases} -1/2 & \rho < -1 \\ (1/\pi)\sin^{-1}\rho & |\rho| \le 1 \\ 1/2 & \rho > 1 \end{cases}$$

$$D_1(\rho) = \begin{cases} (1-\rho^2)^{1/2} & |\rho| \le 1 \quad \text{(see section 5.3.3)} \\ 0 & |\rho| > 1 \end{cases}$$

and the saturation functions

$$f_0(\rho) = \begin{cases} (2/\pi)\{\rho\sin^{-1}\rho + (1-\rho^2)^{1/2}\} & |\rho| \le 1 \\ |\rho| & |\rho| > 1 \end{cases}$$

$$f_1(\rho) = \begin{cases} -1 & \rho < -1 \\ (2/\pi)\{\sin^{-1}\rho + \rho(1-\rho^2)^{1/2}\} & |\rho| \le 1 \\ 1 & \rho > 1 \end{cases}$$

are used in the table.

TABLE OF SBDFs

(a) *SVNLs*

Nonlinearity	Comments	$N_y(a, \gamma)$ and $N_a(a, \gamma)$
1. General quantiser 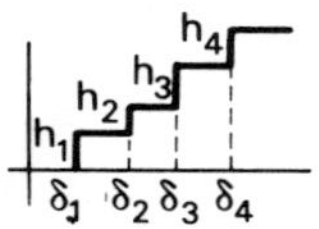	$\delta_{M+1} > a + \lvert\gamma\rvert > \delta_M$	$N_\gamma = (1/\gamma) \sum_{m=1}^{M} h_m \left\{ g\left(\frac{\delta_m + \gamma}{a}\right) - g\left(\frac{\delta_m - \gamma}{a}\right) \right\}$ $N_p = (2/a\pi) \sum_{m=1}^{M} h_m \left\{ D_1\left(\frac{\delta_m + \gamma}{a}\right) + D_1\left(\frac{\delta_m - \gamma}{a}\right) \right\}$
2. Uniform quantiser $h_1 = h_2 = \ldots h$ $\delta_m = (2m - 1)\delta/2$	$(2M + 1)\delta > a > (2M - 1)\delta$ $n = (2m - 1)/2$	$N_\gamma = (h/\gamma) \sum_{m=1}^{M} \left\{ g\left(\frac{n\delta + \gamma}{a}\right) - g\left(\frac{n\delta - \gamma}{a}\right) \right\}$ $N_p = (2h/a\pi) \sum_{m=1}^{M} \left\{ D_1\left(\frac{n\delta + \gamma}{a}\right) + D_1\left(\frac{n\delta - \gamma}{a}\right) \right\}$
3. Relay with dead zone 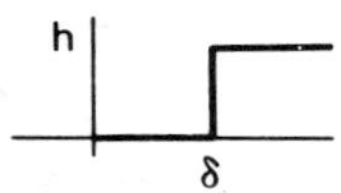		$N_\gamma = (h/\gamma) \left\{ g\left(\frac{\delta + \gamma}{a}\right) - g\left(\frac{\delta - \gamma}{a}\right) \right\}$ $N_p = (2h/a\pi) \left\{ D_1\left(\frac{\delta + \gamma}{a}\right) + D_1\left(\frac{\delta - \gamma}{a}\right) \right\}$
4. Ideal relay 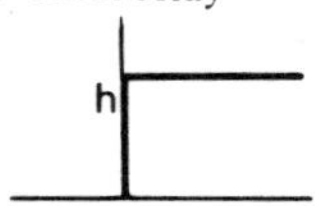		$N_\gamma = (2h/\gamma)g(\gamma/a)$ $N_p = (4h/a\pi)D_1(\gamma/a)$
5. Preload 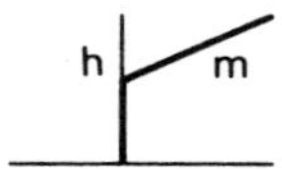		$N_\gamma = m + (2h/\gamma)g(\gamma/a)$ $N_p = m + (4h/a\pi)D_1(\gamma/a)$

(a) *SVNLs*—continued

Nonlinearity	Comments	$N_y(a,\gamma)$ and $N_a(a,\gamma)$
6. General piecewise linear 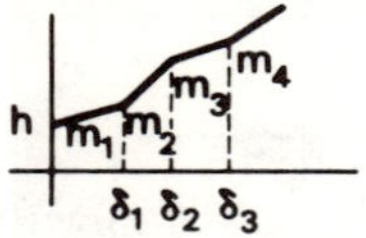	$\delta_{M+1} > a + \lvert\gamma\rvert > \delta_M$	$N_\gamma = (2h/\gamma)g(\gamma/a) + m_{M+1} + (a/2\gamma)\sum_{j=1}^{M}(m_j - m_{j+1})\left\{f_0\left(\frac{\delta_j+\gamma}{a}\right) - f_0\left(\frac{\delta_j-\gamma}{a}\right)\right\}$ $N_p = (4h/a\pi)D_1(\gamma/a) + m_{M+1} + (1/2)\sum_{j=1}^{M}(m_j - m_{j+1})\left\{f_1\left(\frac{\delta_j+\gamma}{a}\right) + f_1\left(\frac{\delta_j-\gamma}{a}\right)\right\}$
7. Ideal saturation 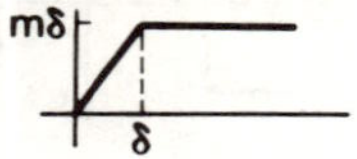	See section 5.3.3	$N_\gamma = (ma/2\gamma)\left\{f_0\left(\frac{\delta+\gamma}{a}\right) - f_0\left(\frac{\delta-\gamma}{a}\right)\right\}$ $N_p = (m/2)\left\{f_1\left(\frac{\delta+\gamma}{a}\right) + f_1\left(\frac{\delta-\gamma}{a}\right)\right\}$
8. Dead zone 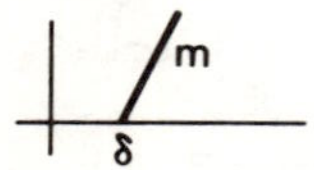		$N_\gamma = m - (ma/2\gamma)\left\{f_0\left(\frac{\delta+\gamma}{a}\right) - f_0\left(\frac{\delta-\gamma}{a}\right)\right\}$ $N_p = m - (m/2)\left\{f_1\left(\frac{\delta+\gamma}{a}\right) + f_1\left(\frac{\delta-\gamma}{a}\right)\right\}$
9. Gain changing nonlinearity 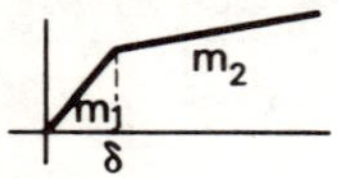		$N_\gamma = \{(m_1 - m_2)a/2\gamma\}\left\{f_0\left(\frac{\delta+\gamma}{a}\right) - f_0\left(\frac{\delta-\gamma}{a}\right)\right\} + m_2$ $N_p = \{(m_1 - m_2)/2\}\left\{f_1\left(\frac{\delta+\gamma}{a}\right) + f_1\left(\frac{\delta-\gamma}{a}\right)\right\} + m_2$
10. Saturation with dead zone 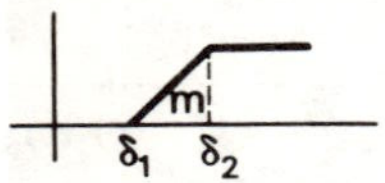		$N_\gamma = (ma/2\gamma)\left\{f_0\left(\frac{\delta_2+\gamma}{a}\right) - f_0\left(\frac{\delta_2-\gamma}{a}\right) - f_0\left(\frac{\delta_1+\gamma}{a}\right) + f_0\left(\frac{\delta_1-\gamma}{a}\right)\right\}$ $N_p = (m/2)\left\{f_1\left(\frac{\delta_2+\gamma}{a}\right) + f_1\left(\frac{\delta_2-\gamma}{a}\right) - f_1\left(\frac{\delta_1+\gamma}{a}\right) - f_1\left(\frac{\delta_1-\gamma}{a}\right)\right\}$

11.

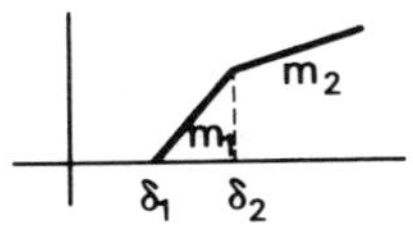

$$N_\gamma = (a/2\gamma)\left[-m_1\left\{f_0\left(\frac{\delta_1+\gamma}{a}\right) - f_0\left(\frac{\delta_1-\gamma}{a}\right)\right\} + (m_1 - m_2)\left\{f_0\left(\frac{\delta_2+\gamma}{a}\right) - f_0\left(\frac{\delta_2-\gamma}{a}\right)\right\}\right] + m_2$$

$$N_p = (1/2)\left[-m_1\left\{f_1\left(\frac{\delta_1+\gamma}{a}\right) + f_1\left(\frac{\delta_1-\gamma}{a}\right)\right\} + (m_1 - m_2)\left\{f_1\left(\frac{\delta_2+\gamma}{a}\right) + f_1\left(\frac{\delta_2-\gamma}{a}\right)\right\}\right] + m_2$$

12.

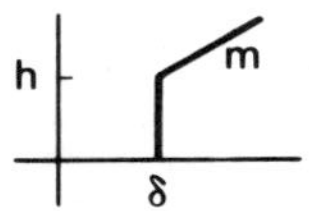

$$N_\gamma = (h/\gamma)\left\{g\left(\frac{\delta+\gamma}{a}\right) - g\left(\frac{\delta-\gamma}{a}\right)\right\} + m - (ma/2\gamma)\left\{f_0\left(\frac{\delta+\gamma}{a}\right) - f_0\left(\frac{\delta-\gamma}{a}\right)\right\}$$

$$N_p = (2h/a\pi)\left\{D_1\left(\frac{\delta+\gamma}{a}\right) + D_1\left(\frac{\delta-\gamma}{a}\right)\right\} + m - (m/2)\left\{f_1\left(\frac{\delta+\gamma}{a}\right) + f_1\left(\frac{\delta-\gamma}{a}\right)\right\}$$

13.

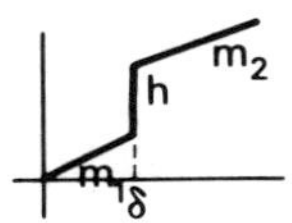

$$N_\gamma = (h/\gamma)\left\{g\left(\frac{\delta+\gamma}{a}\right) - g\left(\frac{\delta-\gamma}{a}\right)\right\} + m_2 + \{m_1 - m_2)a/2\gamma\}\left\{f_0\left(\frac{\delta+\gamma}{a}\right) - f_0\left(\frac{\delta-\gamma}{a}\right)\right\}$$

$$N_p = (2h/a\pi)\left\{D_1\left(\frac{\delta+\gamma}{a}\right) + D_1\left(\frac{\delta-\gamma}{a}\right)\right\} + m_2 + \{(m_1 - m_2)/2\}\left\{f_1\left(\frac{\delta+\gamma}{a}\right) + f_1\left(\frac{\delta-\gamma}{a}\right)\right\}$$

14.

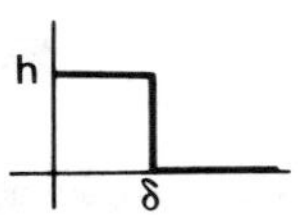

$$N_\gamma = (h/\gamma)\left\{2g(\gamma/a) - g\left(\frac{\delta+\gamma}{a}\right) + g\left(\frac{\delta-\gamma}{a}\right)\right\}$$

$$N_p = (2h/a\pi)\left\{2D_1(\gamma/a) - D_1\left(\frac{\delta+\gamma}{a}\right) - D_1\left(\frac{\delta-\gamma}{a}\right)\right\}$$

(a) *SVNLs*—continued

Nonlinearity	Comments	$N_\gamma(a,\gamma)$ and $N_a(a,\gamma)$
15. Limited field of view 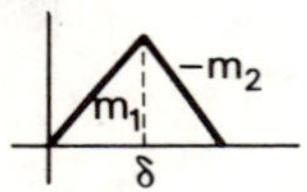	$\alpha = (m_1 + m_2)/m_2$	$N_\gamma = \{(m_1+m_2)a/2\gamma\}\left\{f_0\left(\frac{\delta+\gamma}{a}\right) - f_0\left(\frac{\delta-\gamma}{a}\right)\right\} - (m_2 a/2\gamma)\left\{f_0\left(\frac{\alpha\delta+\gamma}{a}\right) - f_0\left(\frac{\alpha\delta-\gamma}{a}\right)\right\}$ $N_p = \{(m_1+m_2)/2\}\left\{f_1\left(\frac{\delta+\gamma}{a}\right) + f_1\left(\frac{\delta-\gamma}{a}\right)\right\} - (m_2/2)\left\{f_1\left(\frac{\alpha\delta+\gamma}{a}\right) + f_1\left(\frac{\alpha\delta-\gamma}{a}\right)\right\}$
16. 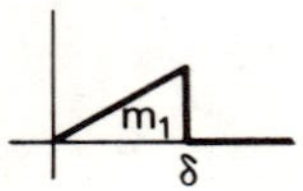		$N_\gamma = (m_1 a/2\gamma)\left\{f_0\left(\frac{\delta+\gamma}{a}\right) - f_0\left(\frac{\delta-\gamma}{a}\right)\right\} - (m_1\delta/\gamma)\left\{g\left(\frac{\delta+\gamma}{a}\right) - g\left(\frac{\delta-\gamma}{a}\right)\right\}$ $N_p = (m_1/2)\left\{f_1\left(\frac{\delta+\gamma}{a}\right) + f_1\left(\frac{\delta-\gamma}{a}\right)\right\} - (2m_1\delta/a\pi)\left\{D_1\left(\frac{\delta+\gamma}{a}\right) + D_1\left(\frac{\delta-\gamma}{a}\right)\right\}$
18, 19. $y = x^n$	See section 5.3.1 n integer	$N_\gamma = \sum_{k=0}^{n} {}_nC_k \mu_k \gamma^{n-k-1}$ $N_p = (2/a^2)\sum_{k=0}^{n} {}_nC_k \mu_{k+1}\gamma^{n-k}$
20. Harmonic nonlinearity $y = A_m \sin mx$	See section 5.3.2	$N_\gamma = (1/\gamma)A_m J_0(ma)\sin m\gamma$ $N_p = (2/a)A_m J_1(ma)\cos m\gamma$
21. $y = A_m \sinh mx$		$N_\gamma = (1/\gamma)A_m I_0(ma)\sinh m\gamma$ $N_p = (2/a)A_m I_1(ma)\cosh m\gamma$
34. Asymmetrical gain 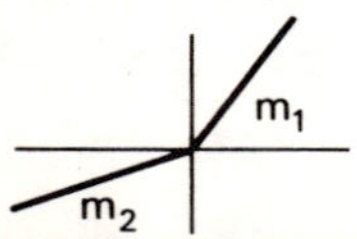		$N_\gamma = \{(m_1+m_2)/2\} + \{(m_1-m_2)a/2\gamma\}f_0(\gamma/a)$ $N_p = \{(m_1+m_2)/2\} + \{(m_1-m_2)/2\}f_1(\gamma/a)$

35. Absolute value 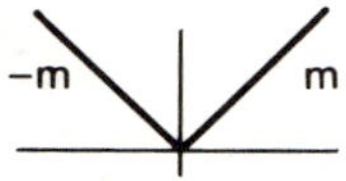		$N_{\gamma} = (ma/\gamma) f_0(\gamma/a)$ $N_p = mf_1(\gamma/a)$
36. Ideal relay 		$N_{\gamma} = (h/\gamma)\left\{1 - g\left(\frac{\delta + \gamma}{a}\right) - g\left(\frac{\delta - \gamma}{a}\right)\right\}$ $N_p = (2h/a\pi)\left\{D_1\left(\frac{\delta - \gamma}{a}\right) - D_1\left(\frac{\delta + \gamma}{a}\right)\right\}$
37. Biased ideal relay 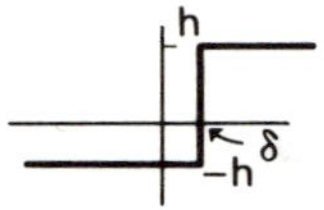		$N_{\gamma} = (2h/\gamma) g\left(\frac{\gamma - \delta}{a}\right)$ $N_p = (4h/a\pi) D_1\left(\frac{\delta - \gamma}{a}\right)$

(b) *DVNLs*

24. On-off relay with hysteresis 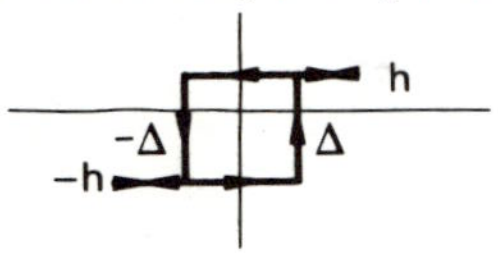	See section 5.3.4 $a - \lvert\gamma\rvert > \Delta$	$N_{\gamma} = (h/\gamma)\left\{g\left(\frac{\Delta + \gamma}{a}\right) - g\left(\frac{\Delta - \gamma}{a}\right)\right\}$ $N_p = (2h/a\pi)\left\{D_1\left(\frac{\Delta + \gamma}{a}\right) + D_1\left(\frac{\Delta - \gamma}{a}\right)\right\}$ $N_q = -4h\Delta/a^2\pi$
25. Friction controlled backlash 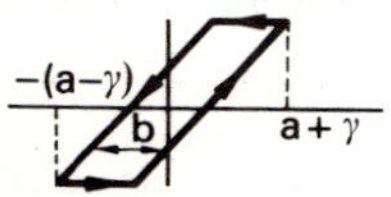	$a - \lvert\gamma\rvert > b/2$	$N_{\gamma} = 1$ $N_p = (1/2)[1 + f_1\{(a - b)/a\}]$ $N_q = b(b - 2a)/a^2\pi$

(b) *DVNLs*—continued

Nonlinearity	Comments	$N_\gamma(a, \gamma)$ and $N_a(a, \gamma)$
28. 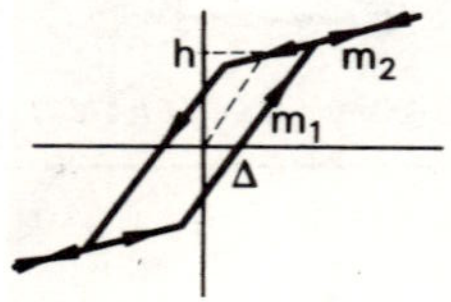	$m_1 > m_2$ $a - \lvert\gamma\rvert > \frac{h}{m_1} + \frac{m_1\Delta}{m_1 - m_2}$ $u = \frac{m_1\Delta}{m_1 - m_2}$	$N_\gamma = \{(m_1 - m_2)a/2\gamma\}\left\{f_0\left(\frac{h + m_1u + m_1\gamma}{m_1a}\right) - f_0\left(\frac{h + m_1u - m_1\gamma}{m_1a}\right) + f_0\left(\frac{h - m_1u + m_1\gamma}{m_1a}\right) - f_0\left(\frac{h - m_1u - m_1\gamma}{m_1a}\right)\right\}$ $N_p = \{(m_1 - m_2)/4\}\left\{f_1\left(\frac{h + m_1u + m_1\gamma}{m_1a}\right) + f_1\left(\frac{h + m_1u - m_1\gamma}{m_1a}\right) + f_1\left(\frac{h - m_1u + m_1\gamma}{m_1a}\right) + f_1\left(\frac{h - m_1u - m_1\gamma}{m_1a}\right)\right\}$ $N_q = -4h\Delta/a^2\pi$
30. On-off relay with dead zone and hysteresis 	$a - \lvert\gamma\rvert > \delta + \Delta$	$N_\gamma = (h/2\gamma)\left\{g\left(\frac{\delta - \Delta + \gamma}{a}\right) - g\left(\frac{\delta - \Delta - \gamma}{a}\right) + g\left(\frac{\delta + \Delta + \gamma}{a}\right) - g\left(\frac{\delta + \Delta - \gamma}{a}\right)\right\}$ $N_p = (h/a\pi)\left\{D_1\left(\frac{\delta - \Delta + \gamma}{a}\right) + D_1\left(\frac{\delta - \Delta - \gamma}{a}\right) + D_1\left(\frac{\delta + \Delta + \gamma}{a}\right) + D_1\left(\frac{\delta + \Delta - \gamma}{a}\right)\right\}$ $N_q = -4h\Delta/a^2\pi$

APPENDIX A3

TABLE OF SSDFs

For two incommensurate sinusoidal input signals the SSDFs can be found (see section 5.8) from the following expressions

$$N_a(a, b) = (2/a^2) \int_{-b}^{b} \int_{-a}^{a} x n(x + y) r(x) r(y) \, dx \, dy$$

or

$$N_a(a, b) = (j/a\pi) \int_{-\infty}^{\infty} N(j\omega) J_1(a\omega) J_0(b\omega) \, d\omega$$

and

$$N_b(a, b) = [N_a(a, b)]_{\substack{a \to b \\ b \to a}}$$

TABLE OF SSDFs

Nonlinearity	Comments	$N_a(a, b)$
3. Relay with dead zone 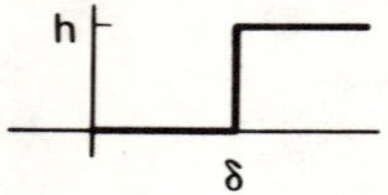	See figure A3.1	$(2h/\pi a)\int_{-\infty}^{\infty}\{(\cos\delta\omega)/\omega\}J_0(b\omega)J_1(a\omega)\,d\omega$
4. Ideal relay 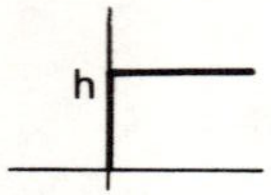	$E(k)$ and $B(k)$ are elliptic integrals See section 5.8.3 See figure A3.2	$(8h/b\pi^2)B(a/b)$ for $a < b$ $(8h/a\pi^2)E(b/a)$ for $a > b$
7. Ideal saturation 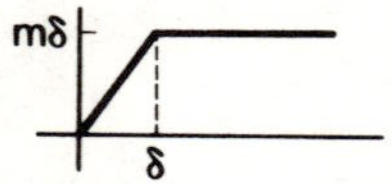		$(2m/\pi a)\int_{-\infty}^{\infty}\{(\sin\delta\omega)/\omega^2\}J_0(b\omega)J_1(a\omega)\,d\omega$
18, 19. $y = x^n$	See section 5.8.1	$(2/a^2)\sum_{k=0}^{n} {}_nC_k\mu_k(b)\mu_{n-k+1}(a)$
20. Harmonic nonlinearity $y = A_m \sin mx$	See section 5.8.2	$(2A_m/a)J_0(mb)J_1(ma)$
21. $y = A_m \sinh mx$		$(2A_m/a)I_0(mb)I_1(ma)$

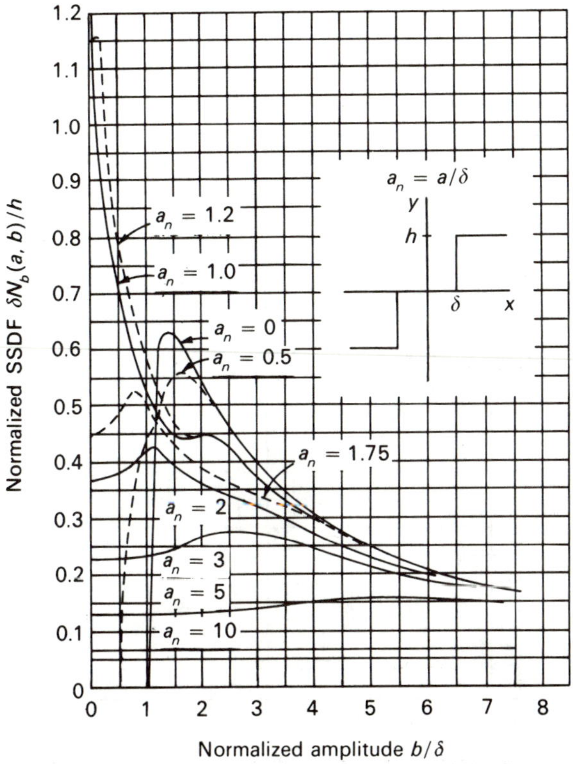

Permission of Gelb and Vander Velde (Ref. 9 of Chap. 5)

Figure A3.1 Normalized relay with dead zone SSDF

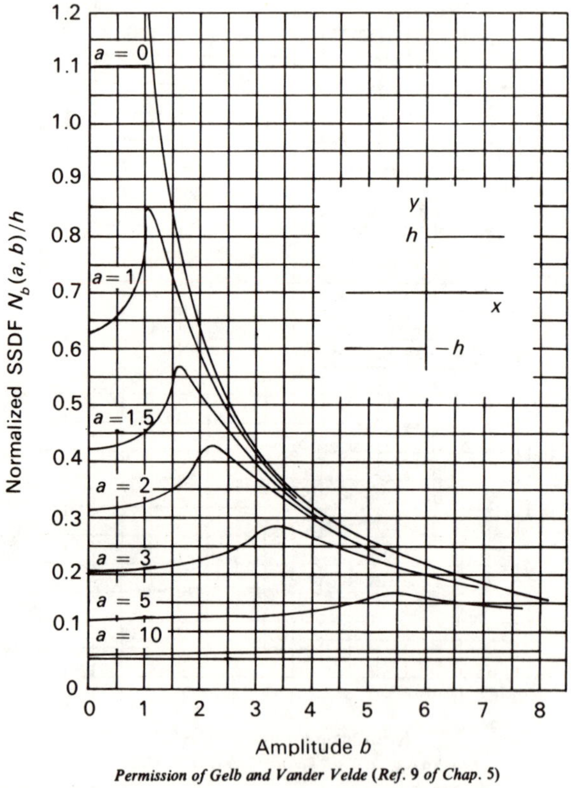

Permission of Gelb and Vander Velde (Ref. 9 of Chap. 5)

Figure A3.2 Normalized ideal relay SSDF

APPENDIX A4

TABLE OF COEFFICIENTS OF SYMMETRICAL ODD NONLINEARITIES FOR IDF DETERMINATION

The IDF, $N_{ib}(a)$, is given by

$$N_{ib}(a) = \{c'_0 + (c'_s/2)e^{j(\phi'_s - 2\phi)}\}_{s=2n/m}$$

for the input $a\cos m\theta + b\cos(n\theta + \phi)$ with $b \to 0$. For a SVNL

$$c'_s = a'_s = 2\varepsilon_s \int_0^a (\mathrm{d}n(x)/\mathrm{d}x)T_s(x/a)r(x)\,\mathrm{d}x$$

and for a DVNL

$$a'_s = 2\varepsilon_s \int_0^a (\mathrm{d}n_p(x)/\mathrm{d}x)T_s(x/a)r(x)\,\mathrm{d}x$$

$$b'_s = 4\int_0^a (\mathrm{d}n_q(x)/\mathrm{d}x)D_s(x/a)r(x)\,\mathrm{d}x$$

$$c'_s = \{(a'_s)^2 + (b'_s)^2\}^{1/2}, \qquad \phi'_s = -\tan^{-1}(b'_s/a'_s).$$

Also used are $T_s(x)$, $D_s(x)$, $r(x)$ and ε_s (see chapter 3) and the Kronecker delta

$$\delta_{jk} = \begin{cases} 1 & j = k \\ 0 & j \neq k. \end{cases}$$

TABLE OF COEFFICIENTS OF SYMMETRICAL ODD NONLINEARITIES FOR IDF DETERMINATION

(a) *Single valued nonlinearities*

Nonlinearity	Comments	a_s' s even
1. General quantiser 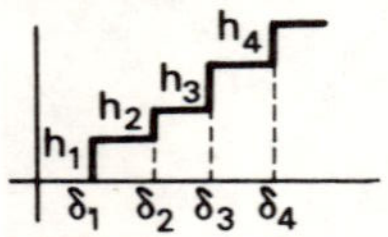	$\delta_{M+1} > a > \delta_M$	$2\varepsilon_s \sum_{m=1}^{M} h_m T_s(\delta_m/a) r(\delta_m)$
2. Uniform quantiser $h_1 = h_2 = \dots h$ $\delta_m = (2m - 1)\delta/2$	$\delta_{M+1} > a > \delta_M$	$2\varepsilon_s h \sum_{m=1}^{M} T_s(\delta_m/a) r(\delta_m)$
3. Relay with dead zone 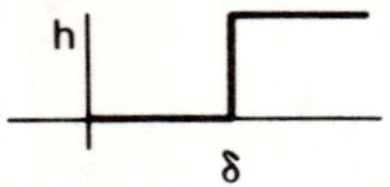		$2h\varepsilon_s T_s(\delta/a) r(\delta)$
4. Ideal relay 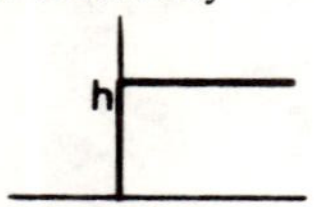		$2h\varepsilon_s T_s(0)/a\pi$
5. Preload 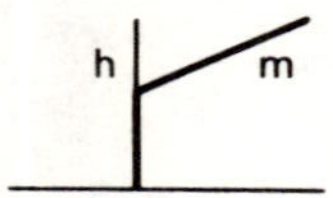		$(2h\varepsilon_s T_s(0)/a\pi) + m\delta_{s0}$

6. General piecewise linear 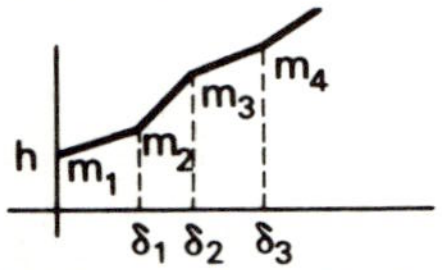	$\delta_{M+1} > a > \delta_M$	$(2h\varepsilon_s T_s(0)/a\pi) + m_1\delta_{s0} + (2\varepsilon_s/s\pi) \sum_{j=1}^{M} (m_{j+1} - m_j)D_s(\delta_j/a)$
7. Ideal saturation 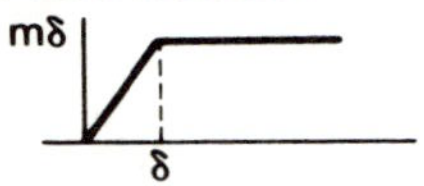	$a > \delta$	$m\delta_{s0} - (2m\varepsilon_s/s\pi)D_s(\delta/a)$
8. Dead zone 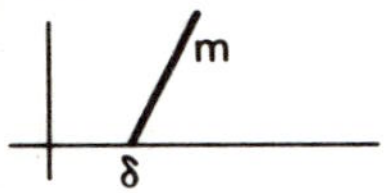	$a > \delta$	$(2m\varepsilon_s/s\pi)D_s(\delta/a)$
9. Gain changing nonlinearity 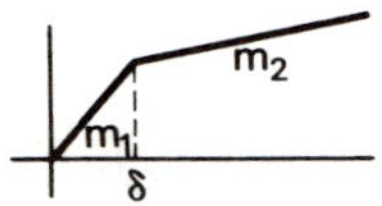	$a > \delta$	$m_1\delta_{s0} + (2\varepsilon_s/s\pi)(m_2 - m_1)D_s(\delta/a)$
10. Saturation with dead zone 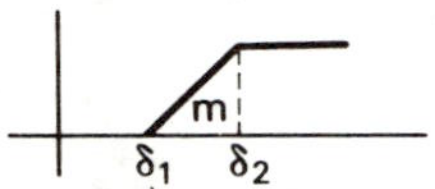	$a > \delta_2$	$(2\varepsilon_s m/s\pi)\{D_s(\delta_1/a) - D_s(\delta_2/a)\}$
11. 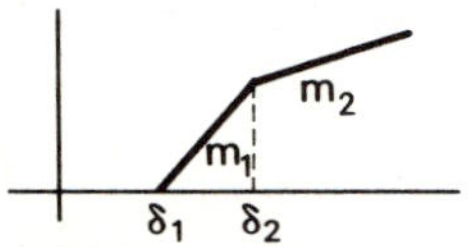	$a > \delta_2$	$(2\varepsilon_s/s\pi)\{m_1 D_s(\delta_1/a) + (m_2 - m_1)D_s(\delta_2/a)\}$

(a) *Single valued nonlinearities*—continued

Nonlinearity	Comments	a'_s s even
12. 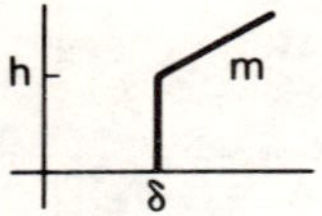	$a > \delta$	$2h\varepsilon_s T_s(\delta/a)r(\delta) + (2m\varepsilon_s/s\pi)D_s(\delta/a)$
13. 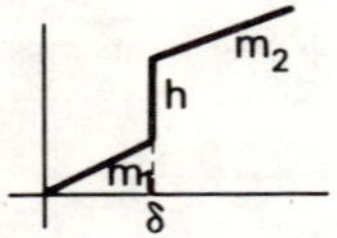	$a > \delta$	$2h\varepsilon_s T_s(\delta/a)r(\delta) + m_1\delta_{s0} + (2\varepsilon_s/s\pi)(m_2 - m_1)D_s(\delta/a)$
14. 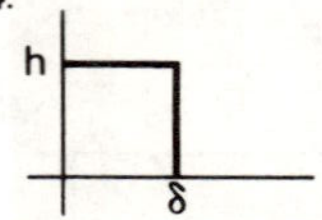	$a > \delta$	$2\varepsilon_s h\{(1/a\pi) - T_s(\delta/a)r(\delta)\}$
15. Limited field value 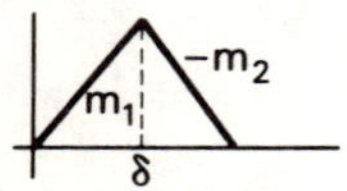	$a > \delta(m_1 + m_2)/m_2$	$m_1\delta_{s0} + (2\varepsilon_s/s\pi)[m_2 D_s\{\delta(m_1 + m_2)/m_2 a\} - (m_1 + m_2)D_s(\delta/a)]$
16. 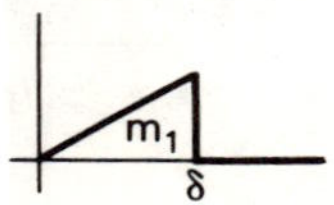	$a > \delta$	$m_1\delta_{s0} - (2m_1\varepsilon_s/s\pi)D_s(\delta/a) - 2m_1\delta\varepsilon_s T_s(\delta/a)r(\delta)$

17. $y = x^m$	$m > -1$	$\{\varepsilon_s m a^{m-1}\Gamma(m)\}/2^{m-1}\Gamma\left(\frac{1+m+s}{2}\right)\Gamma\left(\frac{1+m-s}{2}\right)$
20. Harmonic nonlinearity $y = A_m \sin mx$		$mA_m\varepsilon_s(-1)^{s/2}J_s(ma)$
21. $y = A_m \sinh mx$		$mA_m\varepsilon_s I_s(ma)$

(b) *Double valued nonlinearities*

Nonlinearity	Comments	a_s' (s even)	b_s' (s even)
24. On-off relay with hysteresis 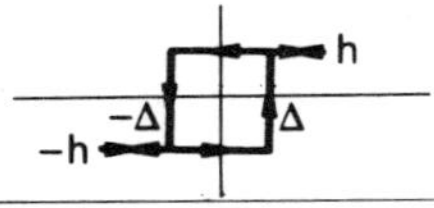	$a > \Delta$	$2h\varepsilon_s T_s(\Delta/a)r(\Delta)$	$4hD_s(\Delta/a)r(\Delta)$
25. Friction-controlled backlash 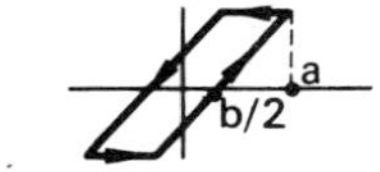	$a > b/2$	$\delta_{s0} - (2\varepsilon_s/s\pi)D_s\{(a-b)/a\}$	$(-2/s\pi)[1 - T_s\{(a-b)/a\}]$
28. 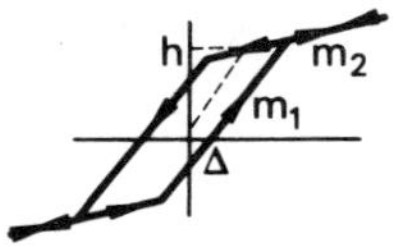	$a > u + (h/m_1)$ $u = \Delta m_1/(m_1 - m_2)$	$m_1\delta_{s0} - \frac{(m_1-m_2)\varepsilon_s}{s\pi}\left\{D_s\left(\frac{h+m_1u}{m_1a}\right) + D_s\left(\frac{h-m_1u}{m_1a}\right)\right\}$	$(4/s\pi)\left[m_2 + (-1)^{s/2}m_1 + \left(\frac{m_2-m_1}{2}\right) \times \left\{T_s\left(\frac{h-m_1u}{m_1a}\right) + T_s\left(\frac{h+m_1u}{m_1a}\right)\right\}\right]$
30. On-off relay with dead zone and hysteresis 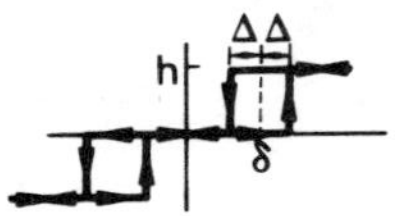	$a > \delta + \Delta$	$h\varepsilon_s[T_s\{(\delta+\Delta)/a\}r(\delta+\Delta) + T_s\{(\delta-\Delta)/a\}r(\delta-\Delta)]$	$2h[D_s\{(\delta-\Delta)/a\}r(\delta-\Delta) - D_s\{(\delta+\Delta)/a\}r(\delta+\Delta)]$

APPENDIX A5

TABLE OF RDFs FOR SYMMETRICAL ODD NONLINEARITIES

For a Gaussian input y

$$\alpha_1 = N(\sigma) = (2/\sigma^2) \int_0^\infty yn(y)q(y)\,\mathrm{d}y,$$

where

$$q(y) = (1/\sigma\sqrt{2\pi})\mathrm{e}^{-y^2/2\sigma^2},$$

or in terms of a Laplace transform

$$N(\sigma) = \frac{2}{\sigma^3\sqrt{2\pi}}[\mathscr{L}\{n(2t)^{1/2}\}]_{s=1/\sigma^2}.$$

In addition

$$\operatorname{erf} y = (2/\sqrt{\pi}) \int_0^y \mathrm{e}^{-u^2}\,\mathrm{d}u.$$

TABLE OF RDFs

Nonlinearity	Comments	$N(\sigma)$
1. General quantiser 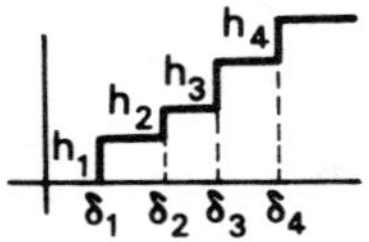	See section 8.6.1	$2 \sum_{m=1}^{M} h_m q(\delta_m)$
2. Uniform quantiser $h_1 = h_2 = \ldots h$ $\delta_m = (2m - 1)\delta/2$		$2h \sum_{m=1}^{M} q\{(2m - 1)\delta/2\}$
3. Relay with dead zone 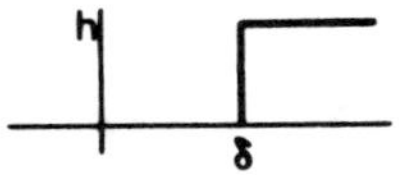	See section 8.5.4	$2hq(\delta)$
4. Ideal relay 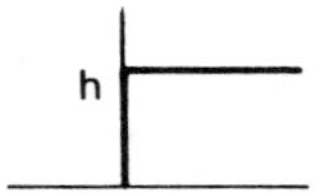		$(2/\pi)^{1/2}(h/\sigma)$
5. Preload 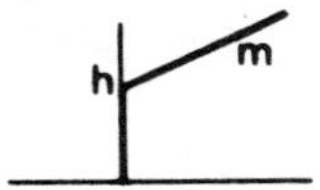		$m + (2/\pi)^{1/2}(h/\sigma)$

TABLE OF RDFs—continued

Nonlinearity	Comments	$N(\sigma)$
6. General piecewise linear 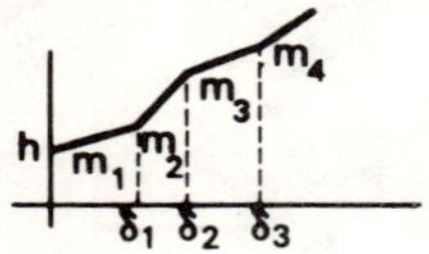		$(2/\pi)^{1/2}(h/\sigma) + m_{m+1} + \sum_{j=1}^{M} (m_j - m_{j+1}) \operatorname{erf}(\delta_j/\sigma\sqrt{2})$
7. Ideal saturation 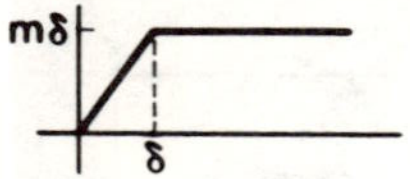	See section 8.5.3	$m \operatorname{erf}(\delta/\sigma\sqrt{2})$
8. Dead zone 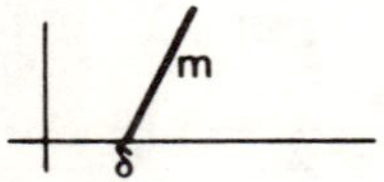		$m\{1 - \operatorname{erf}(\delta/\sigma\sqrt{2})\}$
9. Gain changing nonlinearity 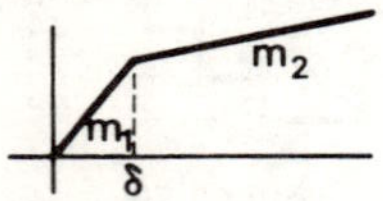		$(m_1 - m_2) \operatorname{erf}(\delta/\sigma\sqrt{2}) + m_2$
10. Saturation with dead zone 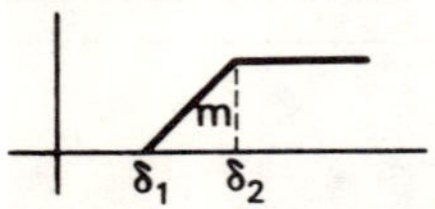		$m\{\operatorname{erf}(\delta_2/\sigma\sqrt{2}) - \operatorname{erf}(\delta_1/\sigma\sqrt{2})\}$

11.	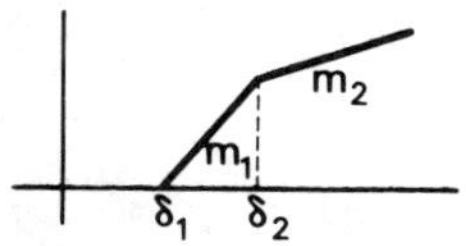	$m_2 - m_1 \operatorname{erf}(\delta_1/\sigma\sqrt{2}) + (m_1 - m_2) \operatorname{erf}(\delta_2/\sigma\sqrt{2})$
12.	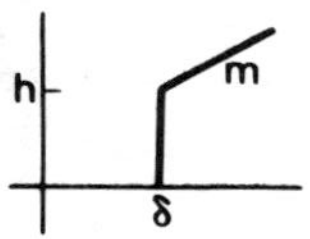	$2hq(\delta) + m - m \operatorname{erf}(\delta/\sigma\sqrt{2})$
13.	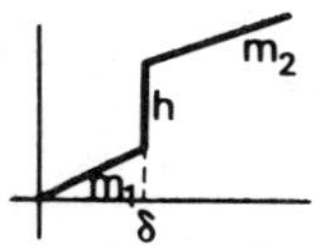	$2hq(\delta) + m_2 + (m_1 + m_2) \operatorname{erf}(\delta/\sigma\sqrt{2})$
14.	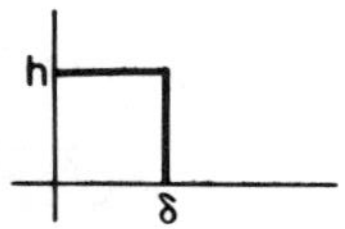	$2h\{q(0) - q(\delta)\}$
15. Limited field of view	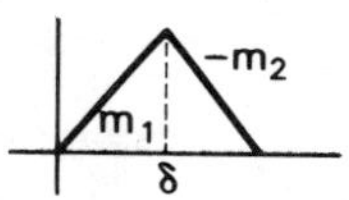	$(m_1 + m_2) \operatorname{erf}(\delta/\sigma\sqrt{2}) - m_2 \operatorname{erf}\{(m_1 + m_2)\delta/m_2\sigma\sqrt{2}\}$

TABLE OF RDFs—continued

Nonlinearity	Comments	$N(\sigma)$
16. 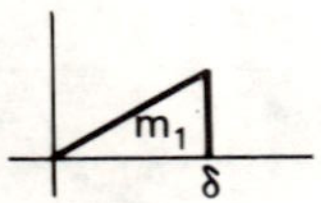		$m_1 \operatorname{erf}(\delta/\sigma\sqrt{2}) - 2m_1 q(\delta)$
17. $y = x^m$	Γ is the gamma function	$(2/\pi)^{1/2} 2^{m/2} \Gamma\{(2 + m)/2\} \sigma^{m-1}$
18. $y = x^n$	See section 8.5.1 n odd integer	$n(n - 2)(n - 4) \ldots 1\sigma^{n-1}$
19. $y = x^n$	n even integer	$(2/\pi)^{1/2} n(n - 2)(n - 4) \ldots 2\sigma^{n-1}$
20. Harmonic nonlinearity $y = A_m \sin mx$	See section 8.5.2	$mA_m e^{-m^2\sigma^2/2}$
21. $y = A_m \sinh mx$		$mA_m e^{m^2\sigma^2/2}$
22. Exponential saturation $y = 1 - e^{-cx}$		$ce^{c^2\sigma^2/2}\{1 - \operatorname{erf}(c\sigma/\sqrt{2})\}$

APPENDIX A6

TABLE OF RBDFs

For a Gaussian input y

$$\alpha_0 = \gamma N_\gamma(\sigma, \gamma) = \int_{-\infty}^{\infty} n(y + \gamma)q(y)\,\mathrm{d}y$$

and

$$\alpha_1 = N_\sigma(\sigma, \gamma) = (1/\sigma^2)\int_{-\infty}^{\infty} yn(y + \gamma)q(y)\,\mathrm{d}y$$

or

$$N_\sigma(\sigma, \gamma) = \partial\alpha_0/\partial\gamma = N_\gamma + \gamma(\partial N_\gamma/\partial\gamma),$$

where

$$q(y) = (1/\sigma\sqrt{2\pi})\mathrm{e}^{-y^2/2\sigma^2}.$$

Also used are the functions

$$\operatorname{erf} y = (2/\sqrt{\pi})\int_0^y \mathrm{e}^{-u^2}\,\mathrm{d}u,$$

$$F(y) = q(y) + (y/2\sigma)\operatorname{erf}(y/\sigma\sqrt{2}),$$

$$G(\alpha, \gamma) = \operatorname{erf}\left(\frac{\delta + \gamma}{\sigma\sqrt{2}}\right) + \operatorname{erf}\left(\frac{\delta - \gamma}{\sigma\sqrt{2}}\right)$$

and

$$H(\delta, \gamma) = F(\delta + \gamma) - F(\delta - \gamma).$$

TABLE OF RBDFs

Nonlinearity	Comments	N_γ and N_σ
1. General quantiser 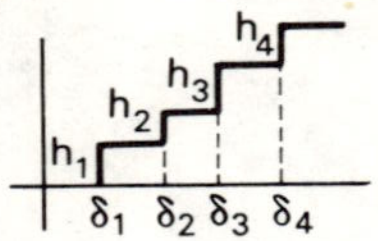		$N_\gamma = (1/2\gamma) \sum_{m=1}^{M} h_m \left\{\mathrm{erf}\left(\frac{\delta_m + \gamma}{\sigma\sqrt{2}}\right) - \mathrm{erf}\left(\frac{\delta_m - \gamma}{\sigma\sqrt{2}}\right)\right\}$ $N_\sigma = \sum_{m=1}^{M} h_m\{q(\delta_m + \gamma) + q(\delta_m - \gamma)\}$
2. Uniform quantiser $h_1 = h_2 = \ldots h$ $\delta_m = (2m - 1)\delta/2$		$N_\gamma = (h/2\gamma) \sum_{m=1}^{M} \left[\mathrm{erf}\left\{\frac{(2m-1)\delta + 2\gamma}{2\sqrt{2}\sigma}\right\} - \mathrm{erf}\left\{\frac{(2m-1)\delta - 2\gamma}{2\sqrt{2}\sigma}\right\}\right]$ $N_\sigma = h \sum_{m=1}^{M} \left[q\left\{\frac{(2m-1)\delta}{2} + \gamma\right\} + q\left\{\frac{(2m-1)\delta}{2} - \gamma\right\}\right]$
3. Relay with dead zone 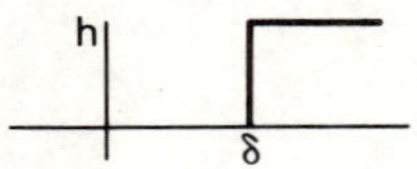	See section 8.5.4	$N_\gamma = (h/2\gamma)\left\{\mathrm{erf}\left(\frac{\delta + \gamma}{\sigma\sqrt{2}}\right) - \mathrm{erf}\left(\frac{\delta - \gamma}{\sigma\sqrt{2}}\right)\right\}$ $N_\sigma = h\{q(\delta + \gamma) + q(\delta - \gamma)\}$
4. Ideal relay 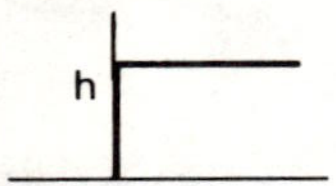		$N_\gamma = (h/\gamma)\,\mathrm{erf}(\gamma/\sigma\sqrt{2})$ $N_\sigma = 2hq(\gamma)$
5. Preload 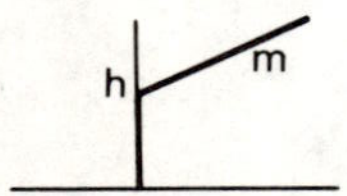		$N_\gamma = m + (h/\gamma)\,\mathrm{erf}(\gamma/\sigma\sqrt{2})$ $N_\sigma = m + 2hq(\gamma)$

6. General piecewise linear

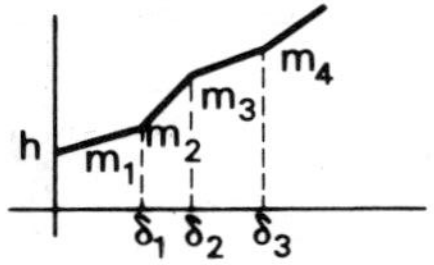

$$N_\gamma = (h/\gamma)\,\mathrm{erf}(\gamma/\sigma\sqrt{2}) + m_{M+1} + \sum_{j=1}^{M} \{(m_j - m_{j+1})\sigma/\gamma\} H(\delta_j, \gamma)$$

$$N_\sigma = 2hq(\gamma) + m_{M+1} + \sum_{j=1}^{M} \{(m_j - m_{j+1})/2\} G(\delta_j, \gamma)$$

7. Ideal saturation

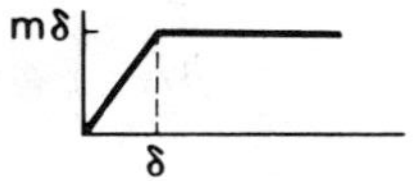

See section 8.5.3

$$N_\gamma = (m\sigma/\gamma)\{F(\delta + \gamma) - F(\delta - \gamma)\} = (m\sigma/\gamma)H(\delta, \gamma)$$

$$N_\sigma = (m/2)\left\{\mathrm{erf}\left(\frac{\delta + \gamma}{\sigma\sqrt{2}}\right) + \mathrm{erf}\left(\frac{\delta - \gamma}{\sigma\sqrt{2}}\right)\right\} = (m/2)G(\delta, \gamma)$$

8. Dead zone

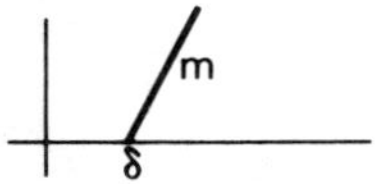

$$N_\gamma = m - (m\sigma/\gamma)H(\delta, \gamma)$$

$$N_\sigma = m - (m/2)G(\delta, \gamma)$$

9. Gain changing nonlinearity

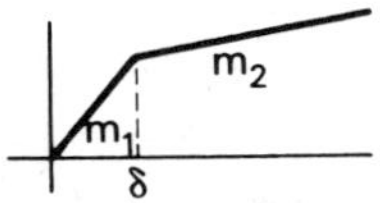

$$N_\gamma = m_2 + \{(m_1 - m_2)\sigma/\gamma\} H(\delta, \gamma)$$

$$N_\sigma = m_2 + \{(m_1 - m_2)/2\} G(\delta, \gamma)$$

10. Saturation with dead zone

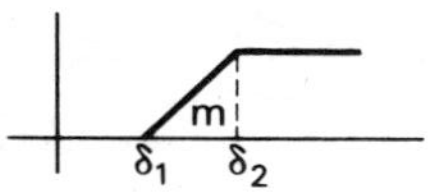

$$N_\gamma = (m\sigma/\gamma)\{H(\delta_2, \gamma) - H(\delta_1, \gamma)\}$$

$$N_\sigma = (m/2)\{G(\delta_2, \gamma) - G(\delta_1, \gamma)\}$$

TABLE OF RBDFs

	Nonlinearity	Comments	N_γ and N_σ
11.	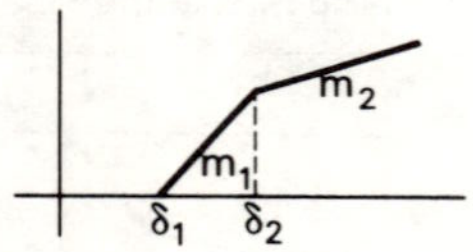		$N_\gamma = m_2 - (m_1\sigma/\gamma)H(\delta_1, \gamma) + \{(m_1 - m_2)\sigma/\gamma\}H(\delta_2, \gamma)$ $N_\sigma = m_2 - (m_1/2)G(\delta_1, \gamma) + \{(m_1 - m_2)/2\}G(\delta_2, \gamma)$
12.	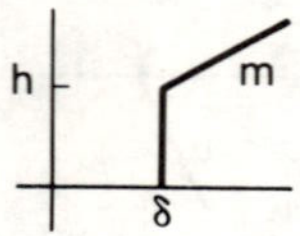		$N_\gamma = (h/2\gamma)\left\{\mathrm{erf}\left(\frac{\delta+\gamma}{\sigma\sqrt{2}}\right) - \mathrm{erf}\left(\frac{\delta-\gamma}{\sigma\sqrt{2}}\right)\right\} + m - (m\sigma/\gamma)H(\delta, \gamma)$ $N_\sigma = h\{q(\delta+\gamma) - q(\delta-\gamma)\} + m - (m/2)G(\delta, \gamma)$
13.	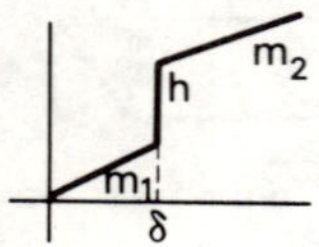		$N_\gamma = (h/2\gamma)\left\{\mathrm{erf}\left(\frac{\delta+\gamma}{\sigma\sqrt{2}}\right) - \mathrm{erf}\left(\frac{\delta-\gamma}{\sigma\sqrt{2}}\right)\right\} + m_2 + \{(m_1 - m_2)\sigma/\gamma)\}H(\delta, \gamma)$ $N_\sigma = h\{q(\delta+\gamma) - q(\delta-\gamma)\} + m_2 + \{(m_1 - m_2)/2\}G(\delta, \gamma)$
14.	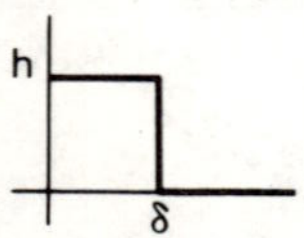		$N_\gamma = (h/2\gamma)\left\{2\,\mathrm{erf}\left(\frac{\gamma}{\sigma\sqrt{2}}\right) - \mathrm{erf}\left(\frac{\delta+\gamma}{\sigma\sqrt{2}}\right) + \mathrm{erf}\left(\frac{\delta-\gamma}{\sigma\sqrt{2}}\right)\right\}$ $N_\sigma = h\{2q(\gamma) - q(\delta+\gamma) - q(\delta-\gamma)\}$

15. Limited field of view 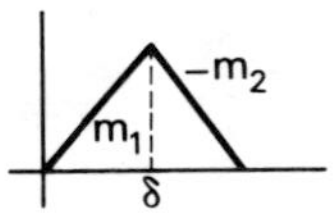		$N_\gamma = \{(m_1 + m_2)\sigma/\gamma\}H(\delta, \gamma) - (m_2\sigma/\gamma)H\{(m_1 + m_2)\delta/m_2, \gamma\}$ $N_\sigma = \{(m_1 + m_2)/2\}G(\delta, \gamma) - (m_2/2)G\{(m_1 + m_2)\delta/m_2, \gamma\}$
16. 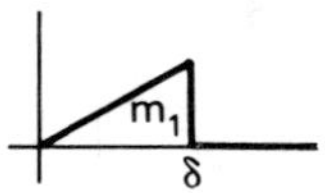		$N_\gamma = (m_1\sigma/\gamma)H(\delta, \gamma) - (m_1\delta/2\gamma)\left\{\mathrm{erf}\left(\frac{\delta + \gamma}{\sigma\sqrt{2}}\right) - \mathrm{erf}\left(\frac{\delta - \gamma}{\sigma\sqrt{2}}\right)\right\}$ $N_\sigma = (m_1/2)G(\delta, \gamma) - m_1\delta\{q(\delta + \gamma) - q(\delta - \gamma)\}$
18, 19. $y = x^n$	See section 8.5.1 n integer	$N_\gamma = \sum_{j=0}^{n} {}_nC_j\mu_{n-j}\gamma^{j-1}$ $N_\sigma = \sum_{j=0}^{n} {}_nC_j\mu_{n-j-1}\gamma^j$
20. Harmonic nonlinearity $y = A_m \sin mx$	See section 8.5.2	$N_\gamma = (A_m/\gamma)e^{-\sigma^2m^2/2} \sin m\gamma$ $N_\sigma = A_m m e^{-\sigma^2m^2/2} \cos m\gamma$
21. $y = A_m \sinh mx$		$N_\gamma = (A_m/\gamma)e^{\sigma^2m^2/2} \sinh m\gamma$ $N_\sigma = A_m m e^{\sigma^2m^2/2} \cosh m\gamma$
34. Asymmetrical gain 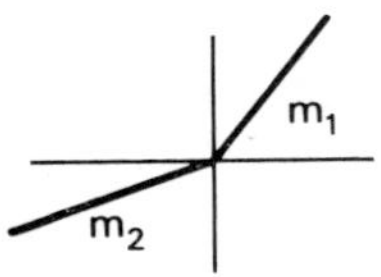		$N_\gamma = \{(m_1 + m_2)/2\} + \{(m_1 - m_2)/2\}\{\mathrm{erf}(\gamma/\sigma\sqrt{2}) + (2\sigma^2/\gamma)q(\gamma)\}$ $N_\sigma = \{(m_1 + m_2)/2\} + \{(m_1 - m_2)/2\}\mathrm{erf}(\gamma/\sigma\sqrt{2})$

TABLE OF RBDFs

Nonlinearity	Comments	N_γ and N_σ
35. Absolute value 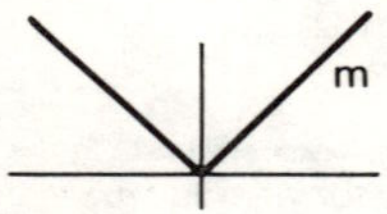		$N_\gamma = m\{\text{erf}(\gamma/\sigma\sqrt{2}) + (2\sigma^2/\gamma)q(\gamma)\}$ $N_\sigma = m\,\text{erf}(\gamma/\sigma\sqrt{2})$
36. Ideal relay 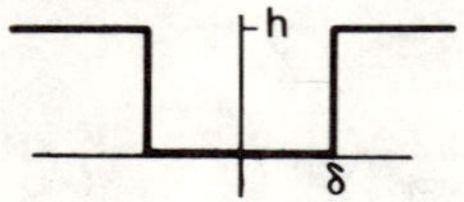		$N_\gamma = (h/\gamma)\{1 - G(\delta, \gamma)\}$ $N_\sigma = h\{q(\delta - \gamma) - q(\delta + \gamma)\}$
37. 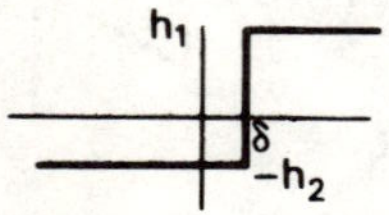		$N_\gamma = (1/2\gamma)\left\{h_1 - h_2 - (h_1 + h_2)\,\text{erf}\left(\dfrac{\delta - \gamma}{\sigma\sqrt{2}}\right)\right\}$ $N_\sigma = (h_1 + h_2)q(\delta - \gamma)$

APPENDIX A7

TABLE OF RSDFs

The components of the RSDF, $N_a(a, \sigma)$ and $N_\sigma(a, \sigma)$ can be evaluated (see section 10.2) from

$$N_a(a, \sigma) = 2\alpha_{10}/a$$

$$N_\sigma(a, \sigma) = \alpha_{01},$$

where

$$\alpha_{sk} = (1/2\pi)\int_{-\infty}^{\infty} N(j\omega)e^{-\sigma^2\omega^2/2}(j\omega)^k j^s J_s(a\omega)\,d\omega$$

or

$$\alpha_{sk} = (1/\sigma^k)\int_{-\infty}^{\infty}\int_{-a}^{a} n(x+y)T_s(x/a)H_k(y/\sigma)r(x)q(y)\,dx\,dy$$

(see also references 3 and 8 of chapter 10). $\rho = a/\sigma\sqrt{2}$ is the signal to noise ratio.

TABLE OF RSDFs

Nonlinearity	Comments	$N_a(a,\sigma)$ and $N_\sigma(a,\sigma)$
1. General quantiser 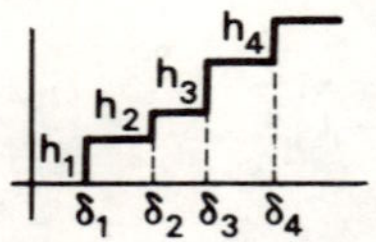		$N_a = (\sqrt{2}/\sigma)\sum_{m=1}^{M} h_m \sum_{n=0}^{\infty} \frac{(\sqrt{2}\delta_m/\sigma)^{2n}{}_1F_1\{n+1/2;2;-\rho^2\}}{(2n)!\Gamma\{(1-2n)/2\}}$ $N_\sigma = (\sqrt{2}/\sigma)\sum_{m=1}^{M} h_m \sum_{n=0}^{\infty} \frac{(\sqrt{2}\delta_m/\sigma)^{2n}{}_1F_1\{n+1/2;1;-\rho^2\}}{(2n)!\Gamma\{(1-2n)/2\}}$
3. Relay with dead zone 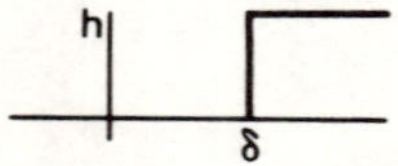		$N_a = (\sqrt{2}h/\sigma)\sum_{n=0}^{\infty} \frac{(\sqrt{2}\delta/\sigma)^{2n}{}_1F_1\{n+1/2;2;-\rho^2\}}{(2n)!\Gamma\{(1-2n)/2\}}$ $N_\sigma = (\sqrt{2}h/\sigma)\sum_{n=0}^{\infty} \frac{(\sqrt{2}\delta/\sigma)^{2n}{}_1F_1\{n+1/2;1;-\rho^2\}}{(2n)!\Gamma\{(1-2n)/2\}}$
4. Ideal relay 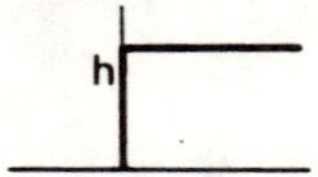	See section 8.13.1	$N_a = (2/\pi)^{1/2}(h/\sigma){}_1F_1(1/2;2;-\rho^2)$ $N_\sigma = (2/\pi)^{1/2}(h/\sigma){}_1F_1(1/2;1;-\rho^2)$
5. Ideal saturation 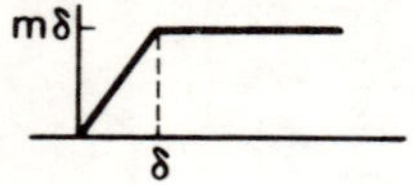	See section 10.2.1	$N_a = m\sum_{n=0}^{\infty}\left(\frac{\sqrt{2}\delta}{\sigma}\right)^{2n+1}\frac{\Gamma(n+1/2){}_1F_1(n+1/2;2;-\rho^2)}{(2n+1)!}$ $N_\sigma = m\sum_{n=0}^{\infty}\left(\frac{\sqrt{2}\delta}{\sigma}\right)^{2n+1}\frac{\Gamma(n+1/2){}_1F_1(n+1/2;1;-\rho^2)}{(2n+1)!}$
18. $y = x^n$		$N_a = (2/a^2)\sum_{k=0}^{n} {}_nC_k\mu_k(\sigma)\mu_{n-k+1}(a)$ $N_\sigma = (1/\sigma^2)\sum_{k=0}^{n} {}_nC_k\mu_k(a)\mu_{n-k+1}(\sigma)$

20. Harmonic nonlinearity $y = A_m \sin mx$	See section 10.2.2	$N_a = (2A_m/a)e^{-\sigma^2 m^2/2} J_1(ma)$ $N_\sigma = mA_m e^{-\sigma^2 m^2/2} J_0(ma)$
21. $y = A_m \sinh mx$		$N_a = (2A_m/a)e^{\sigma^2 m^2/2} I_1(ma)$ $N_\sigma = mA_m e^{\sigma^2 m^2/2} I_0(ma)$

APPENDIX B

THE BILATERAL LAPLACE TRANSFORM

A brief resumé of the bilateral Laplace transform is given below; for a more detailed discussion of the topic the interested reader is referred to reference 1. The Laplace transform familiarly used in control is the single sided transformation defined by

$$F(s) \triangleq \int_0^\infty f(t)e^{-st}\,dt \tag{B.1}$$

and the corresponding inverse transformation is given by the relationship

$$f(t) = (1/2\pi j)\int_{c-j\infty}^{c+j\infty} F(s)e^{st}\,ds. \tag{B.2}$$

The transformation equation (B.1) is only defined over those values of s for which the integral converges. Thus if we can find real positive constants M and α such that

$$|f(t)| < Me^{-\alpha t} \quad \text{for all } t > 0, \tag{B.3}$$

the transformation will exist for $\operatorname{Re} s > -\alpha$ and the value of c in equation (B.2) must be taken greater than $-\alpha$. This requirement may alternatively be stated as, the line for the integral of equation (B.2) should lie to the right of the poles of $F(s)$. Evaluation of the integral is then accomplished using Cauchy's theorem over a closed contour which encloses the poles of $F(s)$ and consists of the line and an infinite semicircle in the left half s plane.

The bilateral Laplace transform is defined in a similar manner to equation (B.1) by

$$F(s) \triangleq \int_{-\infty}^\infty f(t)e^{-st}\,dt. \tag{B.4}$$

If

$$\begin{aligned} f_+(t) &= f(t) \quad \text{for } t > 0 \\ &\quad\; 0 \quad\;\; \text{for } t < 0 \end{aligned}$$

and

$$\begin{aligned} f_-(t) &= 0 \quad\;\; \text{for } t > 0 \\ &\quad f(t) \quad \text{for } t < 0, \end{aligned}$$

then equation (B.4) may be written

$$\begin{aligned} F(s) &= \int_{-\infty}^0 f_-(t)e^{-st}\,dt + \int_0^\infty f_+(t)e^{-st}\,dt \\ &= F_-(s) + F_+(s) \end{aligned} \tag{B.5}$$

and the inverse transformation is given by

$$f(t) = (1/2\pi j)\int_{c_2 - j\infty}^{c_2 + j\infty} F_-(s)e^{-st}\,ds + (1/2\pi j)\int_{c_1 - j\infty}^{c_1 + j\infty} F_+(s)e^{-st}\,ds. \qquad (B.6)$$

If

$$|f(t)| < M_1 e^{-\alpha_1 t} \quad \text{for } t > 0$$

and

$$|f(t)| < M_2 e^{\alpha_2 t} \quad \text{for } t < 0,$$

then the integrals for $F_+(s)$ and $F_-(s)$ converge for Re $s > -\alpha_1$ and Re $s < \alpha_2$ respectively. Therefore for the line integrals of equation (B.6) the conditions $c_1 > -\alpha_1$ and $c_2 < \alpha_2$ must apply and the inversion integral may be alternately written as

$$f(t) = (1/2\pi j)\int_{c - j\infty}^{c + j\infty} F(s)e^{-st}\,ds, \qquad (B.7)$$

where $-\alpha_1 < c < \alpha_2$ and the rectangular strip between the lines $-\alpha_1 + j\omega$ and $\alpha_2 + j\omega$ as ω goes from $-\infty$ to ∞ is the region of definition of the transformation. Normally α_1 and α_2 will be positive constants, that is $F_+(s)$ and $F_-(s)$ will have their poles in the left half and right half s plane respectively, so that c may be taken equal to zero for the evaluation of equation (B.7). The theory of analytic continuation,[2] however, allows results to be obtained for the limiting case of $F_+(s)$ and $F_-(s)$ having poles on the imaginary axis of the s plane. Equation (B.6) may be evaluated from the individual solutions of the two line integrals obtained using Cauchy's theorem over the lines and around infinite semicircles taken in a direction so that the complete path encloses the poles of the integrand.

Since the inversion integral is rarely used in practice, the time functions corresponding to the more common forms of $F_+(s)$ being easily recognisable, it is appropriate to examine the bilateral Laplace transform of the function $f_-(t) = e^{bt}$. Since

$$F_-(s) = \int_{-\infty}^{0} f_-(t)e^{-st}\,dt$$

substituting $u = -t$ gives

$$F_-(s) = \int_{0}^{\infty} f_-(-u)e^{-(-s)u}\,du.$$

Thus, rather than evaluating the integral the above expression shows that the transform may be obtained by the following steps:

1. Reflect $f_-(t)$ about the vertical axis {i.e. form $f_-(-u)$}.
2. Obtain the normal single sided Laplace transform of $f_-(-u)$; a function which is zero for $u < 0$.
3. Replace s by $-s$ in the result obtained in 2.

For the case of $f_-(t) = e^{bt}$ this procedure gives $f_-(-u) = e^{-bu}$ which has a Laplace transform of $1/(s+b)$ defined for $\text{Re} - s > -b$ so that the result is $F_-(s) = 1/(b-s)$ defined for $\text{Re} - s > -b$, that is $\text{Re}\, s < b$. Therefore from a short table of transform pairs $f_-(t)$, $F_-(s)$ the inversion integral of equation (B.7) can be evaluated when the function $F(s)$ is factored into $F_+(s)$ which has poles in the left half plane only and $F_-(s)$ with poles in the right half plane only.

Particular care needs to be exercised in dealing with transforms of functions having discontinuities at the origin. The impulse, $\delta(t)$, and unit step function, $u(t)$, are defined as shown in figure B.1. Denoting the double sided Laplace transformation by $\mathscr{L}_D$,

$$\begin{aligned}\mathscr{L}_D\{\delta(t)\} &= \Delta_+(s) + \Delta_-(s)\\ &= 1/2 + 1/2 = 1\end{aligned}$$

and

$$\begin{aligned}\mathscr{L}_D\{u(t)\} &= U_+(s) + U_-(s)\\ &= 1/s + 0 = 1/s.\end{aligned}$$

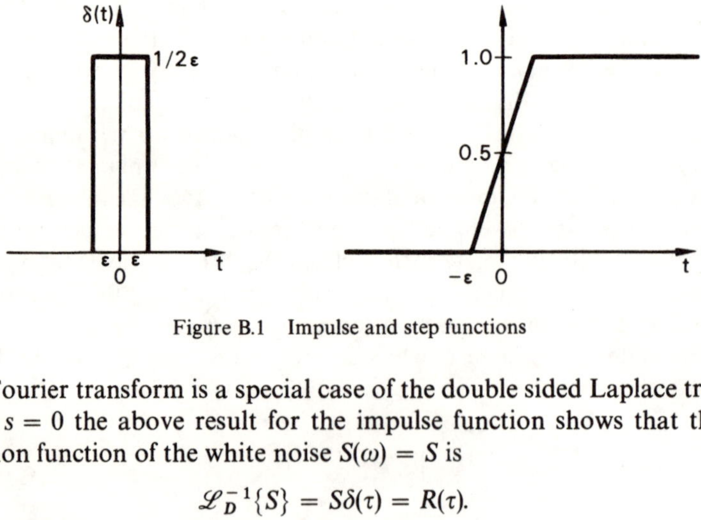

Figure B.1 Impulse and step functions

As the Fourier transform is a special case of the double sided Laplace transform with $\text{Re}\, s = 0$ the above result for the impulse function shows that the autocorrelation function of the white noise $S(\omega) = S$ is

$$\mathscr{L}_D^{-1}\{S\} = S\delta(\tau) = R(\tau).$$

REFERENCES

1. Van der Pol, B. and Bremmer, H.: *Operational calculus based on the two-sided Laplace transform*, 2nd edition (Cambridge University Press, Cambridge, 1955).
2. Churchill, R. V.: *Complex variables and applications* (McGraw-Hill, New York, 1960, Chapter 12).

APPENDIX C

ABSOLUTE STABILITY OF NONLINEAR SYSTEMS

C.1 Introduction

The purpose of this appendix is to provide a brief review of the major results available on the absolute stability of nonlinear systems. No proofs of the theorems are given and for more detailed information the reader is referred to the specialised texts in this area listed in the bibliography. Although many of the methods are applicable to the more general class of nonlinear time varying systems these comments are restricted to the time invariant case. Many of the results because of their generality are quite conservative; however, it is often useful to compare the solutions, especially the frequency domain criteria of Popov[1] and Zames,[2] with those obtained for particular cases by DF or other methods. We begin with a brief review of the method of Lyapunov.[3]

C.2 Lyapunov's method

The direct or second method of Lyapunov is a time domain method based on a nonlinear state space description of the system. It is extremely important from a mathematical viewpoint since it has provided a base from which other results have been obtained. The problem is to determine the stability of an equilibrium state (singular point), which will be assumed to be the origin of the state space,† of the autonomous system

$$\dot{\mathbf{x}} = \mathbf{F}(\mathbf{x}). \tag{C.1}$$

C.2.1 *Stability definitions*

Definition 1

The equilibrium state $\mathbf{0}$, or the equilibrium solution $\mathbf{x}(t) = \mathbf{0}$, is called stable if for any given positive ε, there exists a positive δ such that

$$\|\mathbf{x}(t_0)\| < \delta$$

implies

$$\|\mathbf{x}(t)\| < \varepsilon$$

for all $t > t_0$, where $\| \ \|$ denotes the Euclidean norm.

It should be noted that a system exhibiting a limit cycle or solutions for which $\|\mathbf{x}(t)\|$ increases temporarily will be considered stable by this definition, when ε and δ are appropriately chosen, but it rules out solutions which grow without bound. Typical situations are illustrated in figure C.1. For a nonlinear system stability may only exist within some domain $\|\mathbf{x}\| < R$ of the state space.

† A transformation of coordinates can be used to move any equilibrium state to the origin.

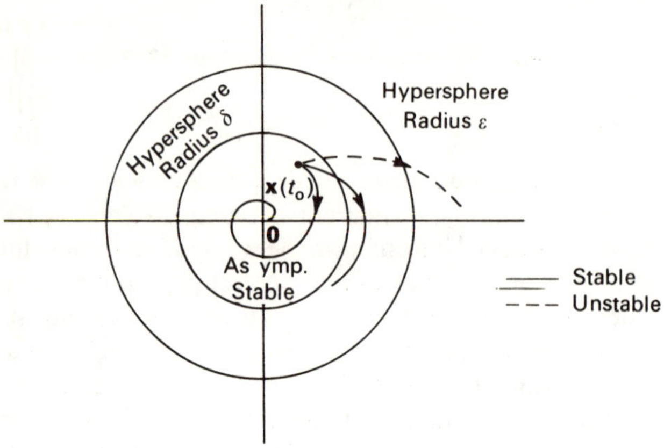

Figure C.1

Definition 2

The equilibrium state $\mathbf{0}$ is called asymptotically stable, if it is stable and $\mathbf{x}(t) \to \mathbf{0}$ as $t \to \infty$.

Definition 3

The equilibrium state $\mathbf{0}$ is called asymptotically stable in the large, or globally asymptotically stable if it is asymptotically stable and the domain R covers the whole state space.

C.2.2 *Positive definite functions*

The notion of positive definite functions is important for the understanding of the Lyapunov theorems of the next section. Let $V(\mathbf{x})$ be a time invariant scalar function of the vector $\mathbf{x}$ and S be a closed bounded region in the $\mathbf{x}$ space containing the origin.

Definition 4

The function $V(\mathbf{x})$ is positive definite (or semi-definite) in S if, for all $\mathbf{x}$ in S

(i) $V(\mathbf{x})$ has continuous partial derivatives with respect to the components of the vector $\mathbf{x}$

(ii) $V(\mathbf{0}) = 0$

(iii) $V(\mathbf{x}) > 0$ for $\mathbf{x} \neq 0$ (or $V(\mathbf{x}) \geq 0$).

If the inequality signs in (iii) are reversed then the conditions for a negative definite (or semi-definite) function are obtained.

Consider the function

$$V(\mathbf{x}) = x_1^2 + x_2^2$$

which is zero only when $x_1 = x_2 = 0$. It is therefore positive definite in a two-dimensional space but only positive semi-definite in spaces of more dimensions.

On the other hand

$$V(\mathbf{x}) = (x_1 + x_2)^2$$

is only positive semi-definite in two space, since it is zero everywhere along the line $x_1 + x_2 = 0$.

The quadratic form scalar function

$$Q(\mathbf{x}) = \mathbf{x}^T\mathbf{Q}\mathbf{x} \tag{C.2}$$

where

$$\mathbf{Q} = \begin{pmatrix} q_{11} & q_{12} & \cdots & q_{1n} \\ q_{21} & q_{22} & \cdots & q_{2n} \\ \vdots & \vdots & \cdots & \\ q_{n1} & q_{n2} & \cdots & q_{nn} \end{pmatrix}$$

is often used. It may also be written

$$Q(\mathbf{x}) = \sum_{i=1}^{n} \sum_{j=1}^{n} q_{ij} x_i x_j,$$

from which it can be seen that the same $Q(\mathbf{x})$ can be realised by all choices of $\mathbf{Q}$ for which $q_{ij} + q_{ji}$ are the same. $\mathbf{Q}$ is normally therefore chosen to be symmetric, that is $q_{ij} = q_{ji}$, in which case the sign definiteness of $Q(\mathbf{x})$ can be obtained from the coefficients of $\mathbf{Q}$.

Sylvester's Theorem

A necessary and sufficient condition for the quadratic form of equation (C.2), with $\mathbf{Q}$ symmetric, to be positive definite, is that each of the following determinants, from the leading principle minors of $\mathbf{Q}$, be positive:

$$\det(q_{11});\ \det\begin{pmatrix} q_{11} & q_{12} \\ q_{21} & q_{22} \end{pmatrix};\ \det\begin{pmatrix} q_{11} & q_{12} & q_{13} \\ q_{21} & q_{22} & q_{23} \\ q_{31} & q_{32} & q_{33} \end{pmatrix};\ldots;\ \det(\mathbf{Q}).$$

$Q(\mathbf{x})$ is positive semi-definite if the determinants of all the principle minors, not just the leading ones, are non-negative. $\mathbf{Q}$ is negative definite if $-\mathbf{Q}$ is positive definite.

C.2.3 *Lyapunov's theorems*

Theorem 1

The null solution, or the equilibrium state at the origin of the system of equation (C.1), is stable if in some neighbourhood of the origin there exists a positive definite function $V(\mathbf{x})$ such that its derivative $\dot{V}(\mathbf{x})$ with respect to the solutions of equation (C.1) is negative semi-definite in that region.

Theorem 2

The null solution of equation (C.1) is asymptotically stable if in some neighbourhood of the origin there exists a positive definite function $V(\mathbf{x})$ such that its derivative $\dot{V}(\mathbf{x})$ with respect to the solutions of equation (C.1) is negative definite in that region.

The following theorem, due to LaSalle,[4] enables the conditions on the derivative $\dot{V}(\mathbf{x})$ in the asymptotic stability theorem 2 to be relaxed.

Theorem 3

The null solution of equation (C.1) is asymptotically stable if, in some neighbourhood R of the origin, there exists a positive definite function $V(\mathbf{x})$ such that its derivative $\dot{V}(\mathbf{x})$ is negative semi-definite in R and is non-zero in R along any solution of equation (C.1) other than the null solution.

If the conditions of theorem 3 are satisfied for all $\mathbf{x}$ then the equilibrium state of equation (C.1) is asymptotically stable in the large. A function $V(\mathbf{x})$ which satisfies the conditions of theorem 1 is known as a Lyapunov function. It is important to note that the above theorems provide only sufficient conditions for stability. If, for a particular application, we are unable to find a Lyapunov function it does not prove instability. It simply means that we have failed to prove stability. For a stable system a Lyapunov function exists but finding it is the problem. Also the choice of a particular Lyapunov function may result in a very conservative condition for stability.

C.2.3.1 *Linear system applications*

Lyapunov's method is easily applied to find the stability of a linear system. Consider a linear autonomous system

$$\dot{\mathbf{x}} = \mathbf{A}\mathbf{x}$$

and choose $V(\mathbf{x}) = \mathbf{x}^T\mathbf{Q}\mathbf{x}$.

Differentiating and substituting for $\dot{\mathbf{x}}$ and $\dot{\mathbf{x}}^T$ gives

$$\begin{aligned}\dot{V}(\mathbf{x}) &= \mathbf{x}^T\{\mathbf{A}^T\mathbf{Q} + \mathbf{Q}\mathbf{A}\}\mathbf{x}\\ &= \mathbf{x}^T\mathbf{P}\mathbf{x}\end{aligned}$$

where

$$\mathbf{P} = \mathbf{A}^T\mathbf{Q} + \mathbf{Q}\mathbf{A}. \tag{C.3}$$

Asymptotic stability of the linear system will thus be assured if with $\mathbf{P}$ a chosen negative definite symmetric matrix and $\mathbf{A}$ the given system matrix, a symmetric positive definite matrix $\mathbf{Q}$ can be found. Since the matrix equation is equivalent to a set of $n(n+1)/2$ linear equations in $n(n+1)/2$ unknowns for an nth order system, given $\mathbf{P}$ one can in principle solve for $\mathbf{Q}$. Two theorems on the matrix equation (C.3), known as the Lyapunov equation, are particularly important.

Theorem 4

If $\lambda_1, \lambda_2 \ldots \lambda_n$ are the eigenvalues of the matrix **A**, then equation (C.3) has a unique solution **Q** if and only if $\lambda_i + \lambda_j \neq 0$ for all $i, j = 1, 2 \ldots n$.

Theorem 5

With **P** a negative definite symmetric matrix, the solution **Q** of equation (C.3)
(a) is positive definite if **A** has only eigenvalues with negative real parts,
(b) is negative definite if **A** has only eigenvalues with positive real parts, and
(c) is indefinite if **A** has some eigenvalues with both positive and negative real parts.

A Lyapunov function $V(\mathbf{x}) = \mathbf{x}^T\mathbf{Q}\mathbf{x}$ for a linear system can thus be found from the solution of equation (C.3) with **P** taken equal to $-\mathbf{I}$. For example consider the second order system with

$$\mathbf{A} = \begin{pmatrix} 0 & 1 \\ -2 & -3 \end{pmatrix}.$$

Let

$$\mathbf{Q} = \begin{pmatrix} a & b \\ b & c \end{pmatrix},$$

then substituting in equation (C.3) with $\mathbf{P} = -\mathbf{I}$ gives

$$\begin{pmatrix} 0 & -2 \\ 1 & -3 \end{pmatrix}\begin{pmatrix} a & b \\ b & c \end{pmatrix} + \begin{pmatrix} a & b \\ b & c \end{pmatrix}\begin{pmatrix} 0 & 1 \\ -2 & -3 \end{pmatrix} = \begin{pmatrix} -1 & 0 \\ 0 & -1 \end{pmatrix},$$

which has the solution $a = 5/4$, $b = c = 1/4$. Thus the Lyapunov function is

$$V(\mathbf{x}) = (1/4)\{5x_1^2 + 2x_1x_2 + x_2^2\}. \qquad \text{(C.4)}$$

On the other hand if we had simply assumed any positive definite quadratic form for $V(\mathbf{x})$ it may not have been possible to prove stability. For example, consider

$$V(\mathbf{x}) = x_1^2 + x_2^2, \qquad \text{(C.5)}$$

then

$$\begin{aligned} \dot{V}(\mathbf{x}) &= 2x_1\dot{x}_1 + 2x_2\dot{x}_2 \\ &= 2x_1x_2 + 2x_2(-2x_1 - 3x_2) = -(2x_1x_2 + 6x_2^2), \qquad \text{(C.6)} \end{aligned}$$

which is not negative definite and no conclusion can be drawn about stability.

If the elliptical contours for constant values of V in expression (C.4) are drawn on a phase portrait for the system with the above A matrix, it will be found that all trajectories cut a contour only once. This then shows the physical interpretation of a Lyapunov function; since it has a $\dot{V}(\mathbf{x})$ which is negative all trajectories move towards the singular point through successively decreasing values of V. On the other hand if the circles of expression (C.5) for constant values of V are drawn a trajectory may cut more than once.

C.2.3.2 *Nonlinear system applications*

The difficulty of applying Lyapunov's method to nonlinear systems, as has no doubt been envisaged by the reader, is the problem of finding a satisfactory V function. This is a problem which continues to exercise the ingenuity of mathematicians and only a brief review of two approaches will be given here.

Variable gradient method[5]

Consider the autonomous nonlinear system

$$\dot{\mathbf{x}} = \mathbf{f}(\mathbf{x}).$$

Then for a function $V(\mathbf{x})$

$$\begin{aligned}\dot{V}(\mathbf{x}) &= \frac{\partial V}{\partial x_1}\frac{dx_1}{dt} + \frac{\partial V}{\partial x_2}\frac{dx_2}{dt} + \cdots + \frac{\partial V}{\partial x_n}\frac{dx_n}{dt} \\ &= \text{grad } V(\mathbf{x}) \,.\, \dot{\mathbf{x}} \\ &= \mathbf{g}(\mathbf{x}) \,.\, \mathbf{f}(\mathbf{x}),\end{aligned} \qquad \text{(C.7)}$$

where . denotes a vector dot product and $\mathbf{g}(\mathbf{x}) = \text{grad } V(\mathbf{x})$. Use is then made of the following.

Lemma

A necessary and sufficient condition for a continuous vector function $\mathbf{g}(\mathbf{x})$ to be the gradient of a scalar function is that the matrix

$$\mathbf{M} = \begin{pmatrix} \partial g_1/\partial x_1 & \partial g_2/\partial x_1 & \cdots & \partial g_n/\partial x_1 \\ \partial g_1/\partial x_2 & \partial g_2/\partial x_2 & \cdots & \partial g_n/\partial x_2 \\ \vdots & \vdots & & \vdots \\ \partial g_1/\partial x_n & \partial g_2/\partial g_n & \cdots & \partial g_n/\partial x_n \end{pmatrix} \qquad \text{(C.8)}$$

be symmetric where $g_1, g_2 \ldots g_n$ denote the components of the vector function $\mathbf{g}(\mathbf{x})$.

The function $V(\mathbf{x})$ can be computed from

$$V(\mathbf{x}) = \int_{\mathbf{0}}^{\mathbf{x}} \mathbf{g}(\mathbf{x})\, d\mathbf{x}$$

along any path joining the origin of the state space to the point $\mathbf{x}$. Usually this is done along the axes, that is

$$\begin{aligned}V(\mathbf{x}) = &\int_0^{x_1} g_1(x_1, 0 \ldots 0)\, dx_1 + \int_0^{x_2} g_2(x_1, x_2, 0 \ldots 0)\, dx_2 \\ &+ \cdots + \int_0^{x_n} g_n(x_1, x_2, \ldots x_n)\, dx_n.\end{aligned} \qquad \text{(C.9)}$$

The method then consists of selecting a vector function $\mathbf{g(x)}$ with some adjustable parameters such that the matrix $\mathbf{M}$ of equation (C.8) is symmetric and $V(\mathbf{x})$ is a Lyapunov function; that is $\dot{V}(\mathbf{x})$ of equation (C.7) is negative definite and $V(\mathbf{x})$ of equation (C.9) positive definite.

Example

Consider a nonlinear second order system given by

$$\dot{x}_1 = x_2$$

$$\dot{x}_2 = -\alpha x_2 - x_1 - x_1^3$$

and take

$$\mathbf{g}(\mathbf{x}) = \begin{pmatrix} ax_1 + bx_2 \\ cx_1 + dx_2 \end{pmatrix},$$

where a, b, c and d may be functions of x_1 and x_2. The matrix $\mathbf{M}$ is symmetric for all x_1 and x_2 if

$$\partial g_1/\partial x_2 = \partial g_2/\partial x_1,$$

that is

$$x_1 \frac{\partial a}{\partial x_2} + b + x_2 \frac{\partial b}{\partial x_2} = c + x_1 \frac{\partial c}{\partial x_1} + x_2 \frac{\partial d}{\partial x_1}. \tag{C.10}$$

Using equation (C.7)

$$\begin{aligned}\dot{V}(\mathbf{x}) &= (ax_1 + bx_2)x_2 + (cx_1 + dx_2)(-\alpha x_2 - x_1 - x_1^3) \\ &= x_1x_2(a - \alpha c - d - dx_1^2) + (b - \alpha d)x_2^2 - cx_1^2 - cx_1^4\end{aligned}$$

which is clearly negative definite if b, c and d are constants with

$$\alpha d > b, \qquad c > 0 \quad \text{and} \quad a = \alpha c + d + dx_1^2.$$

Using these parameters in equation (C.10) gives $b = c$, so that we require

$$\alpha d > b = c > 0.$$

From equation (C.9)

$$V(\mathbf{x}) = \int_0^{x_1} ax_1 \, \mathrm{d}x_1 + \int_0^{x_2} (cx_1 + dx_2) \, \mathrm{d}x_2$$

and on substituting for a and integrating

$$V(\mathbf{x}) = (\alpha cx_1^2/2) + (dx_1^2/2) + (dx_1^4/4) + cx_1x_2 + dx_2^2/2.$$

This is clearly positive definite if $\alpha c + d > 0$ and $(\alpha c + d)d > c^2$ which will be satisfied by the previous conditions when α and d are positive. Thus $V(\mathbf{x})$ is a Lyapunov function for this choice of the parameters and the system is stable for $\alpha > 0$.

Lur'e approach[6]

Lur'e considered the stability of a feedback system with a single nonlinearity defined by

$$\dot{\mathbf{x}} = \mathbf{A}\mathbf{x} + \mathbf{b}u$$
$$u = n(\sigma) \tag{C.11}$$
$$\sigma = \mathbf{c}^T\mathbf{x}$$

using a V function of the type quadratic form plus integral of the nonlinearity, that is

$$V(\mathbf{x}, \sigma) = \mathbf{x}^T\mathbf{Q}\mathbf{x} + \beta \int_0^\sigma n(\sigma)\, d\sigma \tag{C.12}$$

where β is a positive constant.

He was able to obtain necessary conditions for $V(\mathbf{x}, \sigma)$ to be a Lyapunov function with (a) the nonlinearity confined to the sector $[0, K]$ and all the eigenvalues of **A** in the left half plane and (b) the nonlinearity confined to the sector $[\varepsilon, K]$ and all the eigenvalues of **A** in the left half plane or on the imaginary axis. Application of this method to specific examples is by no means as simple as might be inferred from the above and the interested reader is referred to the appropriate literature.[6,7]

C.3 Stability of motion

In the previous sections we have only been concerned with the stability of an equilibrium point, however, Lyapunov's approach may also be used to investigate the stability of a fixed motion of a free or forced system. In this case stability is defined in terms of the deviation from the fixed motion when the system is perturbed. Consider the system

$$\dot{\mathbf{x}}(t) = \mathbf{f}(\mathbf{x}, \mathbf{u}, t) \tag{C.13}$$

to have a specific motion $\mathbf{x}_f(t)$ and other motions $\mathbf{x}(t)$ which start near to $\mathbf{x}_f(t)$ at $t = t_0$ and are subject to the same forcing functions $\mathbf{u}(t)$. The stability problem consists of investigating the behaviour of the deviation vector $\Delta\mathbf{x}(t) = \mathbf{x}(t) - \mathbf{x}_f(t)$ as time proceeds. Now from equation (C.13)

$$\Delta\dot{\mathbf{x}}(t) = \mathbf{f}(\mathbf{x}_f + \Delta\mathbf{x}, \mathbf{u}, t) - \mathbf{f}(\mathbf{x}_f, \mathbf{u}, t).$$

Since $\mathbf{f}(\mathbf{x}_f, \mathbf{u}, t)$ is a fixed function of time the right hand side can be written as $\mathbf{g}(\Delta\mathbf{x}, t)$ with $\mathbf{g}(\mathbf{0}, t) = \mathbf{0}$.

Thus the problem becomes one of investigating the stability of the equilibrium point ${}_{f_1}(\mathbf{0}, t) = \mathbf{0}$ of the autonomous time varying system

$$\dot{\mathbf{x}} = {}_{f_1}(\mathbf{x}, t). \tag{C.14}$$

For this case similar Lyapunov theorems exist to those for the time independent situation and application of Lyapunov's method proceeds in a similar manner but with a V function, $V(\mathbf{x}, t)$, that is now also a function of time.

C.4 Frequency domain criteria

Although Lyapunov's method is a useful procedure for determining the stability of a nonlinear system, a major disadvantage is the need to obtain a suitable V function. Absolute stability criteria, which can be used in a simple manner with frequency response data, as derived by Popov and others are therefore usually more useful from an engineering viewpoint. Indeed, since it can be shown that the Popov criterion is the best result that can be obtained using a Lur'e type Lyapunov function, there are few occasions where the practising engineer will require the Lyapunov approach. The Popov criterion, which can be proved directly by functional analysis or using Lyapunov theory,[7] is valid provided that for the system shown in figure C.2:

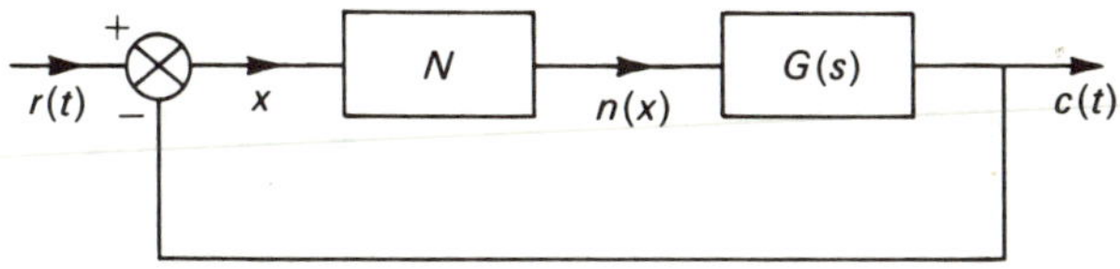

Figure C.2

(i) $G(s)$ is linear, time invariant, causal, and finite dimensional with all poles in the left half plane,

(ii) the input $r(t)$ and any disturbances entering the system are bounded, uniformly continuous, and square integrable over the time interval $[0, \infty)$,

(iii) the nonlinearity does not explicitly depend on time, must be memoryless and a piecewise continuous function of x in the sector $[0, K]$ with bounded slope $\mathrm{d}n(x)/\mathrm{d}x$, that is

$$0 < yx < Kx^2.$$

When $G(s)$ has poles on the imaginary axis the sector is changed to $[\varepsilon, K]$ where $\varepsilon > 0$.

From a control applications point of view it is important to note that $r(t)$ cannot be a bias signal.

Theorem 6

For any initial state a sufficient condition for the output of the system of figure 2 to be bounded and tend to zero as time tends to infinity is that a real number q and an arbitrary small $\delta > 0$ can be found such that for all $\omega \geq 0$

$$\mathrm{Re}\{(1 + j\omega q)G(j\omega)\} + \frac{1}{K} \geq \delta > 0. \tag{C.15}$$

The criterion has a very simple graphical interpretation. Consider the Popov locus $G^*(j\omega) = X^*(j\omega) + jY^*(j\omega)$ where $X^*(j\omega) = \text{Re}\{G(j\omega)\}$ and $Y^*(j\omega) = \omega\,\text{Im}\{G(j\omega)\}$ then equation (C.15) becomes

$$X^*(j\omega) - qY^*(j\omega) + \frac{1}{K} \geq \delta > 0.$$

This condition will be satisfied if the $G^*(j\omega)$ locus lies strictly to the right of the Popov line

$$X^*(j\omega) - qY^*(j\omega) + \frac{1}{K} = 0 \tag{C.16}$$

which passes through the point $(-1/K, 0)$ and has a slope $1/q$. Figure C.3 shows the two possibilities where (a) a q can be found and (b) a q cannot be found. It is important to note that since the criterion only gives a sufficient condition the system portrayed in (b) may be stable.

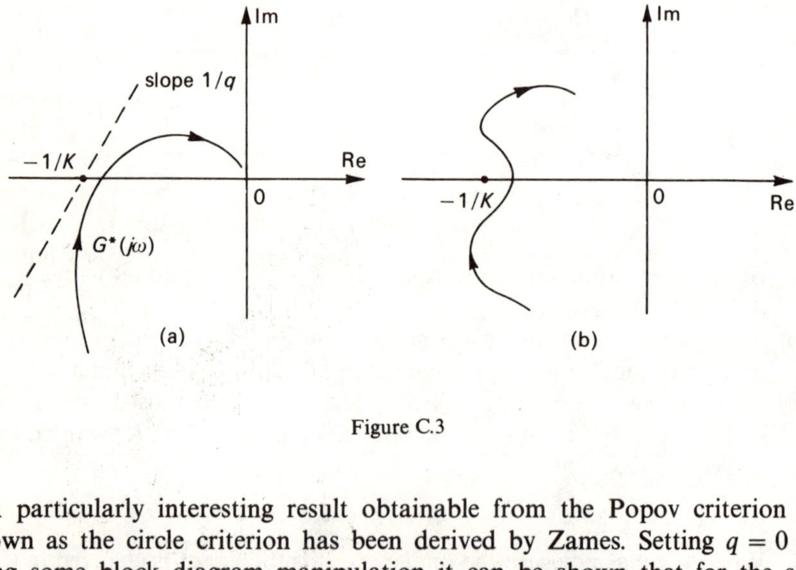

Figure C.3

A particularly interesting result obtainable from the Popov criterion and known as the circle criterion has been derived by Zames. Setting $q = 0$ and using some block diagram manipulation it can be shown that for the same assumptions as the Popov criterion, except that $G(s)$ may have poles on the imaginary axis and the nonlinearity is confined to the sector $[\alpha, \beta]$, the following theorem applies.

Theorem 7

For any initial state a sufficient condition for the output of the system of figure C.2 to be bounded and tend to zero as time tends to infinity is that the frequency response locus $G(j\omega)$ should not touch or encircle the disc with centre at the point $\{-(\alpha + \beta)/2\alpha\beta, 0\}$ and radius $(\beta - \alpha)/2\alpha\beta$.

This critical disc, which is shown in figure C.4, cuts the negative real axis at $-1/\alpha$ and $-1/\beta$.

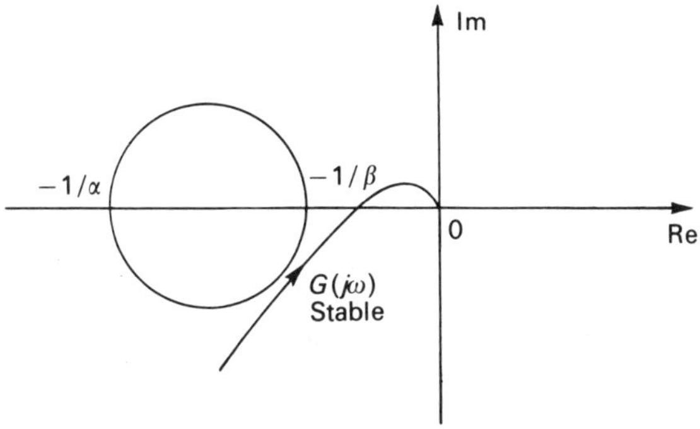

Figure C.4

A few comments on the relationship between the Popov criterion, Aizermann conjecture[8] and DF stability criterion are useful. For a nonlinearity confined to the sector $[0, K]$ Aizermann's conjecture is satisfied when the point $-1/K$ is not enclosed by the $G(j\omega)$ locus. Since $\text{Re}\{G(j\omega)\} = \text{Re}\{G^*(j\omega)\}$ a system satisfying Aizermann's conjecture will have the point $-1/K$ to the left of the intersection of $G^*(j\omega)$ with the negative real axis. Thus Aizermann's hypothesis is a necessary, but not sufficient, condition for Popov's criterion to be satisfied. In figure C.3 both (a) and (b) satisfy the Aizermann conjecture but only (a) satisfies the Popov criterion. Similarly the Aizermann hypothesis is a necessary, but not sufficient, condition for the circle criterion to be satisfied, since the $-1/\beta$ point is equivalent to $-1/K$. For a nonlinearity confined to a sector $[\alpha, \beta]$ then the DF is bounded by $\alpha \leq N(a) \leq \beta$ for all a. Whether $N(a)$ actually attains the limits α and β, however, depends upon the nonlinearity. Thus using the DF a system may be indicated to be stable:

(a) if it does not satisfy the Aizermann conjecture (and therefore Popov criterion) and
(b) if the $G(j\omega)$ locus enters the critical disc possibly cutting the negative real axis between $-1/\alpha$ and $-1/\beta$.

C.5 Multivariable systems

Jury and Lee,[9] in 1964, were the first to obtain an absolute stability criterion for the nonlinear multivariable problem. It was discussed, with examples, by Dewey[10] and an alternative proof was given by McGee and MacLellan.[11] Consider the system of figure C.5 where $\mathbf{r}$, $\mathbf{x}$, $\mathbf{y}$ and $\mathbf{c}$ are m-vectors, $\mathbf{N}$ a set of memoryless nonlinear elements, whose outputs are given by $y_i = n_i(x_i)$ and $\mathbf{G}(s)$ a matrix of linear transfer functions. Provided:

(i) the elements, $G_{ij}(s)$, of $\mathbf{G}(s)$ are linear, time invariant, causal, finite dimensional and have only left half plane poles,

(ii) the input $\mathbf{r}(t)$ is bounded, uniformly continuous and square integrable over the time interval $[0, \infty)$, and

(iii) the elements of $\mathbf{N}$ satisfy $0 \leq y_i x_i < k_i x_i^2$ for $i = 1, 2 \ldots m$, that is $n_i(x_i)$ is confined to the sector $[0, k_i]$, and $n_i(x_i)$ is a time invariant memoryless piecewise continuous function of x_i,

then the stability theorem can be stated as follows.

Theorem 8

If there exists a real $\mathbf{Q} = \text{diag}\{q_i\}$ such that for all $\omega \geq 0$, $\mathbf{P}(\omega)$ is positive definite then for any initial state the output of the system is bounded and tends to zero as time goes to infinity. Further for $\mathbf{r}(t) = \mathbf{0}$, the solution $\mathbf{c}(t) = \mathbf{x}(t) = \mathbf{y}(t) = \mathbf{0}$ is asymptotically stable in the large. Here

$$\mathbf{P}(\omega) = \mathbf{A}(j\omega) + \mathbf{A}^*(j\omega)$$

$$\mathbf{A}(j\omega) = (1 + j\omega\mathbf{Q})\mathbf{G}(j\omega) + \mathbf{K}^{-1}$$

where $\mathbf{K} = \text{diag}\{k_i\}$ and * denotes the conjugate transpose.

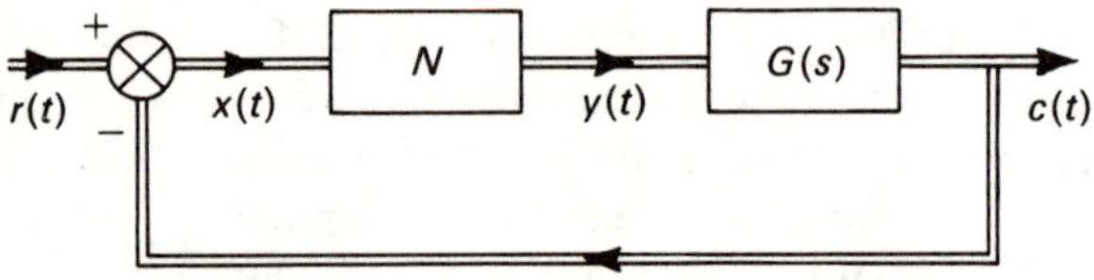

Figure C.5

Obviously for the single variable case the result becomes identical with that of Popov, however, except for a few special cases[12] no graphical interpretation of the above criterion is possible.

A graphical interpretation, on the other hand, is possible for the multivariable equivalent of the circle theorem derived by Rosenbrock[13] but the results appear in general to be more conservative. Again under the same conditions as theorem 8, with the exceptions that the nonlinearities are confined to the sector $[\alpha_i, \beta_i]$ with $\alpha_i > 0$ and $G_{ij}(s)$ may have poles on the imaginary axis, Rosenbrock proved the following criterion.

Theorem 9

The stability of the system of figure C.5 is assured if the area defined by $G_{ii}(j\omega) + d_i(\omega)e^{j\phi}$ ($\phi = 0$ to 2π) does not touch or encircle the critical disc with centre at the point $\{-(1/2\alpha_i\beta_i)(\alpha_i + \beta_i), 0\}$ and radius $(1/2\alpha_i\beta_i)\theta_i(\beta_i - \alpha_i)$ for all $i = 1, 2 \ldots m$. $d_i(\omega) = \sum_{k \neq i}^{m} |G_{ik}(j\omega)|$ and the θ_i are chosen to satisfy $\sum_{i=1}^{m} \theta_i^{-2} \leq 1$.

REFERENCES

1. Popov, V. M.: 'Absolute stability of nonlinear control systems of automatic control', *Automation and Remote Control*, 1962, 22, pp. 857–875.
2. Zames, G.: 'On the input-output stability of time-varying nonlinear feedback systems', *I.E.E. Trans. on Automatic Control*, 1966, AC-11, pp. 228–238 and 465–476.

3. Liapunov, M. A.: 'Problème general de la stabilité du mouvement', *Annals of Mathematical Studies*, 1949, 17 (Princeton University Press, Princeton, N.J.).
4. LaSalle, J. P. and Lefshetz, S.: *Stability by Liapunov's direct method* (Academic Press, New York, 1961).
5. Schultz, D. G. and Gibson, J. E.: 'The variable gradient method for generating Liapunov functions', *Trans. I.E.E.*, 1962, 81 (II), pp. 203–210.
6. Lur'e, Q. I.: *Einige nichtlineare problem aus der theorie der automatischen regelung* (Akademie Verlag, Berlin, 1957).
7. Aizermann, M. A. and Gantmacher, F. R.: *Absolute stability of regulator systems* (Holden-Day, San Francisco, 1964).
8. Aizermann, M. A.: 'On a problem concerning the stability in the large of dynamic systems', *Usp. Mat. Nauk.*, 1949, 4, pp. 187–188.
9. Jury, E. I and Lee, B. W.: 'The absolute stability of systems with many nonlinearities', *Automation and Remote Control*, 1965, 26, pp. 943–961.
10. Dewey, A. G.: 'Frequency domain stability criteria for nonlinear multivariable systems', *Int. J. Control*, 1967, 5, pp. 77–84.
11. McGee, R. W. and MacLellan, G. D. S.: 'Stability of nonlinear multivariable control systems', *Proc. I.E.E.*, 1968, 115, pp. 590–591.
12. Lindgren, A. G. and Pinkos, R. F.: 'Stability of symmetric nonlinear multivariable systems', *J. of Franklin Inst.*, 1966, 282, pp. 92–101.
13. Rosenbrock, H. H.: 'Multivariable circle theorems', *presented at the IMA conference on 'Recent Developments in Control*', Bath, 1972.

BIBLIOGRAPHY

Brockett, R. W. and Lee, H. B.: 'Frequency-domain stability criteria for time-varying and nonlinear systems, *Proc. I.E.E.*, 1967, 55, pp. 604–619.

Brockett, R. W. and Willems, J. L.: 'Frequency domain stability criteria', *I.E.E. Trans. on Automatic Control*, 1965, AC-10, pp. 255–261 and 407–413.

Cook, P. A.: 'Modified multivariable circle theorems', in *Recent Mathematical Developments in Control Theory* (ed. Bell, D. J.), Academic Press, London, pp. 367–372, 1973.

Freeman, E. A.: 'Some control system stability and optimality results obtained via functional analysis', in *Recent Mathematical Developments in Control Theory* (ed. Bell, D. J.), Academic Press, London, pp. 45–68, 1973.

Hahn, W., *Theory and application of Liapunov's direct method* (Prentice-Hall, Englewood Cliffs, N.J., 1963).

Hahn, W.: *Stability of motion* (Springer-Verlag, Berlin, 1967).

Kalman, R. E. and Bertram, J. E.: 'Control system analysis and design via the second method of Liapunov', *Trans. A.S.M.E., J. Basic. Eng.*, 1960, 82, p. 371–392.

LaSalle, J. P.: 'Some extensions of Liapunov's second method', *I.R.E. Trans. Circuit Theory*, 1960, CT-7, pp. 520–527.

Lefshetz, S.: *Stability of nonlinear control systems* (Academic Press, New York, 1965).

Letov, A. M.: *Stability in nonlinear control systems* (Princeton University Press, Princeton, N.J., 1961).

Lighthill, J. and Mees, A. I.: 'Stability of nonlinear feedback systems', in *Recent Mathematical Developments in Control Theory* (ed. Bell, D. J.), Academic Press, London, pp. 1–19, 1973.

Parks, P. C.: 'Stability analysis for linear and non-linear systems using Liapunov's second method', in R. H. MacMillan (ed.), *Progress in Control Engineering*, vol. 2 (Academic Press, New York, 1964, pp. 29–64).

Rosenbrock, H. H. and Storey, C.: *Mathematics of dynamical systems* (Nelson, London, 1970).

Sandberg, I. W.: A frequency domain condition for the stability of feedback systems containing a single time-varying nonlinear element', *Bell System Technical Journal*, 1964, 43, pp. 1901–1908.

Stern, T. E.: *Theory of nonlinear networks and systems* (Addison-Wesley, Reading, Mass., 1965).

ANSWERS TO SELECTED PROBLEMS

CHAPTER 2

P.2.2 $x \simeq 4{\cdot}3$, unstable.
P.2.6 (a) $u_0\omega_0$, (b) $4{\cdot}6u_0\omega_0$.
P.2.7 $\{\zeta + (\zeta^2 - 1)^{1/2}\}/\{\zeta - (\zeta^2 - 1)^{1/2}\}$.
P.2.9 $-3{\cdot}86°$.
P.2.10 6π rads/sec.
P.2.11 Amplifier gain 1225 mA/V. Limit voltage 4·3.
P.2.13 (c) (i) $6h$, (ii) $2h$, time = 1·67 secs.
P.2.16 $2\delta/hT$.
P.2.18 $m_1 = 1$, $m_2 = 1/3$, break point (4, 4). 32 units, 8 secs., 8 secs.; 24 units, 6·99 secs., 6·93 secs.; 8 units, 4 secs., 4 secs.
P.2.19 Overshoot +5, time 2 9/16 sec. and undershoot −1, time 3 13/16 sec. Overshoot +3.32, time 1·83 sec. and undershoot −1, time > 2·84 sec.
P.2.20 5 secs., 12·5 secs.

CHAPTER 3

P.3.2 $a_1 = 7{\cdot}72/\pi$, $b_1 = 0{\cdot}667/\pi$; $a_1 = 7{\cdot}60/\pi$, $b_1 = 0{\cdot}40/\pi$.
P.3.3 $n(x) = (h/2)\{\text{sgn}\,(x + \delta - \Delta\,\text{sgn}\,\dot{x}) + \text{sgn}\,(x - \delta - \Delta\,\text{sgn}\,\dot{x})\}$.
P.3.2 $n(x) = 1{\cdot}05x - 0{\cdot}15x^3$.
P.3.5 (a) $n_p(x) = 0{\cdot}620x - 0{\cdot}0156x^3$.
$n_q(x) = (1{\cdot}533 - 0{\cdot}1826x^2 + 0{\cdot}00623x^4)\{1 - (x^2/16)\}^{1/2}$.
(b) $n_p(x) = -0.05x + 0.00423x^3 - 0{\cdot}0015x^5$.
$n_q(x) = (1{\cdot}48 + 0{\cdot}0625x^2)\{1 - (x^2/16)\}^{1/2}$.
P.3.7 $K = 3h/2a$.
P.3.8 $(4h^3/a\pi) + 3h^2 + (8ha/\pi) + (3a^2/4)$.

P.3.11 $k_{n+1} + \sum_{j=1}^{n} (k_j - k_{j+1})f_1(\alpha_j/a)$.

P.3.18 $a_1 = 6{\cdot}32$, $b_1 = 3{\cdot}06$.
P.3.20 $a_1 = (4/a\pi)\{(a^2 - 1)^{1/2} + 0{\cdot}5(a^2 - 4)^{1/2} + (a^2/4)\sin^{-1} 2/a\}$, $b_1 = 8/a\pi$.
P.3.22 $a_1 = 0$, $b_1 = -2ha/\pi$.

CHAPTER 4

P.4.1 $\pi^2\delta/4hT$; $\omega = \pi/2T$, $a = 0{\cdot}81hKT - 0{\cdot}62\delta^2/hKT$.
P.4.2 3·6, 1·17 rads/sec., $u_0 = 2{\cdot}90$, 1·26 rads/sec.
P.4.3 Limit cycle for $T > 0{\cdot}75\{\pi\Delta/hK\}^{2/3}$, 1·08 at 12·37 rads/sec.

P.4.4 (a) $\{(\Delta + 1)/(\Delta - 1)\}d$, $\omega_0(1 - \zeta^2)^{1/2}$ where $\Delta = e^{-\zeta\pi/(1-\zeta^2)^{1/2}}$.
(b) $2d/\zeta\pi$, ω_0; unstable $2{\cdot}60d$, $0{\cdot}97\omega_0$, $2{\cdot}55d$, ω_0.
P.4.5 0·1 sec., 1·3 rad/sec., 5·95, stable.
P.4.8 1·21 secs.; 2·67, 0·575 rads/sec.
P.4.10 $u^2 = (1/a\pi)\{(a - b)\cos\alpha + a\alpha + a\pi/2\}$, $u = 4b(2a - b)/a^2\pi$ where $u = \omega/\omega_0$ and $\alpha = \sin^{-1}\{(a - b)/a\}$.
P.4.11 $3\pi/4a\omega_0^2$.
P.4.12 2 rads/sec., $2{\cdot}86\delta$. Nyquist perturbation—unstable, Loeb—stable. Actually unstable—see section 5.11.1
P.4.14 $x_2 = 2{\cdot}35$, ω indeterminate.
P.4.15 Stable for $a < 1{\cdot}33$. Limit cycle at 0·577 rads/sec. is unstable.
P.4.16 Unstable limit cycle of 0·825 rads/sec. and amplitude $\{1{\cdot}33(\mu - 1{\cdot}12)/\mu\}^{1/2}$.

CHAPTER 5

P.5.2 Oscillation ceases when $p = 4$, $\gamma = 1{\cdot}86\Delta$ and $\omega = 3{\cdot}05$.
P.5.10 Solutions for ω when $\omega T_d + \tan^{-1}\omega = (2n + 1)\pi$, n all positive integers.
P.5.11 Solution for a from $2\alpha + \sin 2\alpha = 2\pi d/h$, where $\alpha = \sin^{-1}(d/a)$. Oscillation at 1 rad/sec. ceases when $d > 0{\cdot}5h$.
P.5.16 66·8°C, 0·374 rads/sec.

CHAPTER 6

P.6.2 Peak amplitude = 2·94; $\omega = 1{\cdot}27$ rad/sec.
P.6.3 Peak amplitude = 0·167; $\omega = 15{\cdot}0$ rad/sec.
P.6.4 $0 < k < 3{\cdot}01\omega_0\,\delta/h$, for $\zeta = 1$.
P.6.12 $\omega = 1{\cdot}967$ (exact); 2·0 (DF).
P.6.13 $\omega = 0{\cdot}980$ (exact); 1·0 (DF).

CHAPTER 7

P.7.11 3·286.
P.7.12 ·00032.
P.7.13 $2/3(2 + s)$.
P.7.15 $R_x = 2aA^2/B^2$; $R_y = 2A^2(a + T)/B^2$.
P.7.17 2·732; 0·618.
P.7.18 $0{\cdot}64(s - 4)/(4s + 1)$.

CHAPTER 8

P.8.4 $p(z) = 0{\cdot}1\{1 + \delta(z \pm 4)\}$, $-4 \le z \le 4$.
P.8.5 0·167.
P.8.14 $\{n!/(2n + 1)!\sqrt{2}\}H_{2n+1}(x/\sqrt{2})$; $(\pi/2\alpha^2)^{1/2}\,e^{\alpha^2x^2/2}\,\mathrm{erf}\,(\alpha x/\sqrt{2})$.

P.8.16 (i) (a) $(h^2/\pi)(\pi - 2\phi - 2\omega\tau)$, (b) $(hb/\pi)\cos(\phi + \omega\tau)$; (ii) as (i) (a) in both cases.

P.8.18 $N(a) = (3a^2/4) + 3[h^2\{1 - \text{erf}(h/\sigma\sqrt{2})\} + \sigma^2 \text{erf}(h/\sigma\sqrt{2}) - (2/\pi)^{1/2}\sigma h\, e^{-h^2/2\sigma^2}]$

$IDF = N(a) + (a/2)\{dN(a)/da\}$.

P.8.19 $3(\sigma^2 + h^2) + (2/\sigma^2\pi)^{1/2}(h^3 + 3h\sigma^2)$.

P.8.23 0·493.

P.8.27 3·9.

CHAPTER 9

P.9.2 x and $\dot{x}$ are linearly dependent, therefore only one $K = (2/a^2)\int_0^a n(x, \dot{x})\,dx$, is required.

P.9.10 $K_1 = 2h(e^{-\pi\alpha} - 1)/\{a\pi(1 + \alpha^2)\}$, $\omega K_2 = \alpha K_1$.

CHAPTER 12

P.12.3 $\omega = 1{\cdot}732$ rad/sec.; $a = K(1 + \alpha)/\pi$ (in phase) or $K(1 - \alpha)/\pi$ (antiphase).

P.12.8 (a) 65·4.
(b) 122·0.
(c) ∞.

P.12.9 (a) 5·15.
(b) 5·75.
(c) 6·80.

P.12.10 $\omega = 1{\cdot}732$ rad/sec.; $a = 1{\cdot}89$ and $1{\cdot}68$ (in phase) or $1{\cdot}14$ and $0{\cdot}68$ (antiphase).

Index